MIND OVER MAGMA

MIND OVER MAGMA

THE STORY OF IGNEOUS PETROLOGY

Davis A. Young

PRINCETON UNIVERSITY PRESS
Princeton and Oxford

Copyright © 2003 by Princeton University Press
Published by Princeton University Press, 41 William Street, Princeton,
New Jersey 08540
In the United Kingdom: Princeton University Press, 3 Market Place,
Woodstock, Oxfordshire OX20 1SY

All Rights Reserved

Library of Congress Cataloging-in-Publication Data

Young, Davis A.

Mind over magma : the story of igneous petrology / Davis A. Young.

p. cm.

Includes bibliographical references and index.

ISBN 0-691-10279-1

1. Rocks, Igneous. I. Title.

QE461 .Y66 2003

552'.1—dc21   2002030790

British Library Cataloging-in-Publication Data is available

This book has been composed in Berkeley and Albertus

Printed on acid-free paper. ∞

www.pupress.princeton.edu

Printed in the United States of America

1 3 5 7 9 10 8 6 4 2

Dedicated to the memory of

RICHARD HENRY JAHNS

(1915–1983)

Superb geologist, teacher, administrator

Above all, he made petrology fun!

CONTENTS

List of Figures	xi
List of Tables	xiii
Preface	xv

THE FOUNDATIONAL ERA

1.	Breath-Pipes and Ignivomous Mountains—Early Concepts of Volcanism	3
2.	Basaltes Prismatiques—Lava, Columnar Basalt, and Ancient Volcanoes	16
3.	Fire or Water?—The Debate over the Origin of Basalt	34
4.	The "Insolently Triumphant Dogma"—The Collapse of the Concept of Aqueous Basalt	50
5.	Subterraneous Lava—The Recognition of Intrusive Granite	62

THE PRIMITIVE ERA

6.	Wet or Dry?—The Origin of Granite	81
7.	Classes and Orders—Petrography and Classification in the Early Nineteenth Century	104
8.	Basalt and Trachyte—Early Theories of Diversity	125

THE MICROSCOPE ERA

9.	Minute Objects, Great Conclusions—The Rise of Microscopic Petrography	143
10.	Basalt or Melaphyre?—Igneous Rocks in Time	167
11.	Provinces and Plugs—Igneous Rocks in Space	182

CONTENTS

12. Spaltung und Kerne—The Emergence of the Theory of Differentiation — 199
13. Physical Chemistry and Petrology—The Mechanism of Differentiation — 215
14. The Language of Petrology—Nomenclature and Classification — 231
15. Meldometers and Thermocouples—Early Experimental Petrology — 264

THE EXPERIMENTAL ERA

16. Clearing the Mists—The Theory of Crystallization–Differentiation — 283
17. An Unsurpassed Natural Laboratory—Differentiated Sills and Layered Intrusions — 313
18. Cone Sheets and Cauldrons—The Mechanics of Igneous Intrusion — 333
19. Magma or Emanations?—The Beginnings of the Granite Controversy — 350
20. Pontiffs and Soaks—The Resolution of the Granite Controversy — 368
21. Modes and Norms—Classification in Crisis — 389

THE GEOCHEMICAL ERA

22. Structures and Spectra—The Geochemical Revolution in Petrology — 411
23. The Sea Below, the Heavens Above—The Extension of Petrology — 431
24. Delta and Epsilon—Stable and Radiogenic Isotopes in Igneous Petrogenesis — 448
25. Mathematical Modeling—Trace-Element Studies in Igneous Petrogenesis — 472
26. Bombs and Buffers—Experimental Petrology after Bowen — 498
27. Classification Salvaged?—IUGS to the Rescue — 527

THE FLUID DYNAMICAL ERA

28. Rayleigh and Reynolds—The Fluid Dynamics of Magma Chambers — 549

29.	Paradigm Lost, Paradigm Regained?—Cumulate Theory under Fire	578
30.	Past and Future—Some Concluding Remarks	602

Bibliography	615
Index of Names	675
Index of Subjects	681

FIGURES

1. Extinct volcanic cones in the Massif Central, France — 18
2. Portrait of Abraham G. Werner — 27
3. Portrait of Robert Jameson — 29
4. Portrait of James Hutton — 31
5. Junction of granite and schist at Glen Tilt locality studied by Hutton — 66
6. Portrait of Theodor Scheerer — 83
7. Portrait of Robert Bunsen — 102
8. Thin sections made by Nicol and Nicol's polarizing calcite prism — 146–47
9. Portrait of Ferdinand Zirkel — 152
10. Portrait of Harry Rosenbusch — 155
11. Portrait of Alfred Lacroix — 160
12. Portrait of Alfred Harker — 186
13. Portrait of Reginald A. Daly — 196
14. Molecular variation diagram used by Iddings for Yellowstone rocks — 211
15. Igneous rock classifications of Michel-Lévy and Rosenbusch — 234–35
16. Rose diagrams for plotting chemical composition of various rocks by Hobbs — 236
17. Authors of the American quantitative system: Cross, Iddings, Pirsson, and Washington — 241
18. Portrait of Arthur Holmes — 258
19. Johannsen's igneous rock classification — 260
20. The original Geophysical Laboratory — 274
21. Portrait of Norman L. Bowen — 288
22. Phase diagram of plagioclase feldspars determined by Bowen — 289
23. Phase diagram of the liquidus surface in the ternary system albite–anorthite–diopside determined by Bowen — 292
24. Reaction series of Bowen — 298
25. Portrait of J. Frank Schairer — 303
26. Phase diagram of the liquidus surface in "petrogeny's residua system" determined by Bowen — 308

FIGURES

27.	Portrait of Laurence R. Wager	320
28.	Schematic cross section of the Skaergaard Intrusion	321
29.	AFM diagram of Skaergaard liquids by Wager and Deer	324
30.	Anderson's diagram for stress trajectories associated with cone sheets and ring dikes	342
31.	Portrait of N. L. Bowen and O. Frank Tuttle	383
32.	Effect of water-vapor pressure on isobaric minimum points in the simple granite system determined by Tuttle and Bowen	384
33.	Contour diagram of granite compositions on simple granite system	385
34.	Igneous rock classifications of Niggli and Tröger	391
35.	Igneous rock classification of Nockolds	393
36.	Igneous rock classifications of Williams and Moorhouse	394
37.	Igneous rock classifications of Kupletsky and Ginsburg	395
38.	Niggli molecular number diagrams	399
39.	Igneous rock compositions on Zavaritsky diagrams	403
40.	Variation diagrams of Peacock	405
41.	Major-element variation in the Southern California batholith by Nockolds and Allen	427
42.	Variation of trace-element concentration in solidifying melt	481
43.	Rare earth element distributions in the Southern California batholith	487
44.	Trace-element distribution in ophiolites and possible parental magmas	491
45.	Spider diagram of Wood	493
46.	Trace-element distributions in granitic rocks from different tectonic environments	494–95
47.	Basalt tetrahedron of Yoder and Tilley	501
48.	Experimental oxygen buffer apparatus developed by Eugster	504
49.	Effect of addition of anorthite to granite system	512
50.	Granodiorite-water T–X section of Robertson and Wyllie	519
51.	Model rock–water diagram of Robertson and Wyllie	520
52.	Mineralogical classification diagram for plutonic rocks by IUGS	533
53.	Mineralogical classification diagram for volcanic rocks by IUGS	535
54.	TAS classification diagram for volcanic rocks by IUGS	537
55.	Synopsis of classifications of granitoid rocks	545
56.	Capture of settling phenocrysts in sheet-like magma body	568
57.	Portrait of Harry Hammond Hess	579
58.	Schematic diagram of magmatic processes in Skaergaard Intrusion	601

TABLES

1. Partial rock classification of von Leonhard   110
2. Partial rock classification of Naumann   113
3. Partial rock classification of Senft   115
4. Partial rock classification of von Cotta   117
5. Partial rock classification of Roth   119
6. Partial igneous rock classification of Scheerer   120
7. Partial rock classification of Haüy   121
8. Partial rock classification of Brongniart   122
9. Partial rock classification of Cordier   123

PREFACE

The budding and tentative early growth of geology as a scientific discipline were accompanied by relatively limited endeavors to trace its early history. The most important early work on the history of geology consisted primarily of the historical summaries incorporated in the first volume of *Principles of Geology* (Lyell, 1830) and in the third volume of *History of the Inductive Sciences* (Whewell, 1837), and of lengthier treatises such as *The History of Geology and Paleontology* (von Zittel, 1899), *The Founders of Geology* (Geikie, 1905), and, more recently, *The Birth and Development of the Geological Sciences* (Adams, 1938). In contrast, the luxuriant growth and full flowering of the many branches of geology in the latter twentieth century have been followed by an accelerating fecundity of the discipline of history of geology. In just the last three decades, interest in various aspects of the history of geology has blossomed dramatically. Although recent years have witnessed the publication of still more overviews of the history of geology such as *It Began with a Stone* (Faul and Faul, 1983), *A History of Geology* (Gohau, 1990), and *History of Geology* (Ellenberger, 1996, 1999) and of the monumental bibliographic encyclopedia on the history of geology by Sarjeant (1980, 1987, 1996), the maturation of historical reflection upon the science of geology has been signaled by the proliferation of very detailed historical investigations of specific aspects of geology.

Among the detailed studies of major epochs in the development of geology have been such volumes as *When Geologists Were Historians, 1665–1750* (Rappaport, 1997), *The Making of Geology* (Porter, 1977), *From Mineralogy to Geology* (Laudan, 1987), and *Geology in the Nineteenth Century* (Greene, 1982). Hallam (1983) in *Great Geological Controversies* and Oldroyd (1996) in *Thinking About the Earth* have recounted several key episodes in the history of geology. The growing field of history of geology has been enriched by a wealth of studies of nineteenth century British stratigraphy, geomorphology, and paleontology, including *Earth in Decay* (Davies, 1969); *The Great Chain of History* (Rupke, 1983), an examination of William Buckland and the Oxford school of geology; *Controversy in Victorian Geology* (Secord, 1986), an investigation of the dispute between Adam Sedgwick and Roderick Murchison over the Cambrian–Silurian boundary; *The Great Devonian Controversy* (Rudwick, 1985), a thorough analysis of the establishment of the Devonian System; *The Highlands Controversy* (Oldroyd, 1990), a discussion of the deciphering of structure and stratigraphy in the Scottish Highlands and the establishment of the existence of the Moine thrust; *Archetypes and Ancestors* (Desmond, 1982) and *Richard Owen* (Rupke, 1994), books about Richard Owen and paleontology in Victorian England; and *James Hutton and the History of Geology* (Dean, 1992).

A large number of biographical studies of geologists have also emerged for English-speaking audiences, including works on Georges Cuvier (Outram, 1984), James Dwight Dana (Prendergast, 1978), J. William Dawson (Sheets-Pyenson, 1996), G. K. Gilbert (Pyne, 1980), Clarence King (Wilkins, 1988), Andrew C. Lawson (Vaughan, 1970), Joseph Le Conte (Stephens, 1982), Charles Lyell (Wilson, 1972, 1998), Gideon Mantell (Dean, 1999), Roderick Murchison (Stafford, 1989), Raphael Pumpelly (Champlin, 1994), and Charles Walcott (Yochelson, 1998).

Historical studies and reminiscences about plate tectonics (Takeuchi et al., 1967; Wood, 1985; Menard, 1986), the dinosaur extinction controversy (Glen, 1994), and lunar exploration (Wilhelms, 1993) have also appeared recently.

Other indicators of the increased interest in the history of geology are a volume on the history of the Geological Society of America (Eckel, 1982) and a volume of studies on two hundred years of geology in America (Schneer, 1979). Articles on the history of geology regularly occur in such journals as *British Journal for the History of Science*, *Isis*, and *Annals of Science*. Most telling, however, was the establishment in 1982 of *Earth Sciences History*, a journal devoted exclusively to the history of geology.

Accompanying these publication ventures has been the erection of organizational structures for promoting the history of geology such as the History of Earth Sciences Society (HESS), the International Society for the History of Geology (INHIGEO), and the history of geology division of the Geological Society of America.

Within the corpus of historical literature some attention has been given to the development of the subdiscipline of igneous petrology. In particular, the controversies of the late eighteenth and early nineteenth centuries involving the vulcanist, neptunist, and plutonist parties have been reviewed incessantly (Adams, 1938; Greene, 1982; Hallam, 1983; Laudan, 1987; Dean, 1992; den Tex, 1996). Occasionally, articles on the impact of microscopy on petrology have appeared (Hamilton, 1982; Geschwind, 1994). There are also valuable older sources of information pertaining to specific aspects of igneous petrology. For example, Iddings (1892b) and Teall (1901) provided historical overviews of the causes of igneous rock diversity. In conjunction with the radical igneous rock classification scheme proposed by Cross, Iddings, Pirsson, and Washington, Cross (1902) wrote a comprehensive review of nineteenth-century systematic petrography and classification schemes. An abundance of historical information is contained in the four volumes of *The Descriptive Petrography of the Igneous Rocks* by Johannsen (1931–1937). In recent years, studies of the institutional role of the Geophysical Laboratory in the advancement of igneous petrology have been published (Servos, 1983, 1990; Yochelson and Yoder, 1994; Good, 1994; Geschwind, 1995). I recently reexamined the development of ideas about the diversity of igneous rocks (Young, 1999a, b) and, in particular, have written at length about the development of the theory of crystallization-differentiation in the hands of Norman L. Bowen (Young, 1998). Biographical studies of geologists with strong interests in igneous petrology are also beginning to appear. In addition to my book on Bowen, these include studies of the lives of Laurence R. Wager (Hargreaves, 1991), Victor M. Goldschmidt (Mason, 1992), and Arthur Holmes (Lewis, 2000). A recent conference was also devoted to the history of petrology and related disciplines (Fritscher and Henderson, 1998).

Despite these indications of emerging interest in the history of igneous petrology, the history of geology still suffers from relative silence on that topic. There is, for example, no comprehensive history of igneous petrology in print. The only overviews of the history of igneous petrology of which I am aware are those of Loewinson-Lessing (1954) and Fischer (1961). *A Historical Survey of Petrology*, originally written in Russian in 1936 by petrologist F. Y. Loewinson-Lessing, was translated into English in 1954 by Serge Tomkeieff. Loewinson-Lessing's little book contains much valuable historical information, but it is fundamentally a compilation of names, ideas, and dates rather than a coherent narrative historical analysis. Although most of the 85-page book concerns igneous petrology, there are a few pages on sedimentary and metamorphic rocks. Considering the spectacular advances in igneous petrology since 1936, including, for example, the granite controversy, the plate tectonics revolution, the development of trace-element geochemistry, and the use of stable and radiogenic isotopes, it is clear that the work of Loewinson-Lessing needs to be brought up to date.

Of more recent vintage is the German work of Walther Fischer (1961) entitled *Gesteins- und Lagerstättenbildung im Wandel der wissenschaftlichen Anschauung*. Fischer's work is divided about equally between the history of petrography and that of ore deposits. The part on petrography, however, deals primarily with sedimentary rocks. After about 35 pages devoted to petrography in former times and to the vulcanist–neptunist controversy, Fischer wrote 50 pages on the history of the study of the eruptive rocks.

The brief histories of Loewinson-Lessing and Fischer may be supplemented by the summary articles of Harker (1896a), Pirsson (1918), Bascom (1927), and Knopf (1941). The recent "timetable of petrology" by Yoder (1993) gives a very useful summary of significant dates and events in the history of both igneous and metamorphic petrology but provides no narrative. The brevity of these accounts, most of them quite dated, underscores the need for a more extensive overview of the history of igneous petrology.

In addition to the general paucity of historical studies in the field of igneous petrology, igneous petrology textbooks of recent decades, such as those by Turner and Verhoogen (1960), Carmichael et al. (1974), Barker (1983), Hyndman (1985), Philpotts (1990), Raymond (1995), and Winter (2001), have largely ignored the historical development of the discipline. As an example, *Principles of Igneous and Metamorphic Petrology*, the textbook by Philpotts (1990), contains no reference pertinent to igneous rocks prior to a paper by Clough et al. (1909) on cauldron subsidence at Glen Coe, Scotland, except for that of J. Willard Gibbs (1875) on the phase rule. There are no references to such great igneous petrologists as Vogt, Brögger, Iddings, Loewinson-Lessing, Rosenbusch, Zirkel, Judd, Harker, Teall, or Doelter. In most recent English-language petrology texts, there are very few references to petrological literature prior to 1950. Students, among whom are the future petrologists, who use such texts are unintentionally given the impression that nothing significant happened in igneous petrology during the nineteenth century and the early part of the twentieth century. Such students will likely come away with little appreciation for the struggles of earlier generations of geologists to arrive at conclusions that may seem obvious today. A grasp of the history of igneous petrology should encourage future petrologists to be less dogmatic about their own ideas. In recognizing that some ideas that now strike us as ludicrous once seemed perfectly reasonable to our petrological predecessors, among them many outstanding and capable scientists, historically aware petrologists will realize that many contemporary ideas that seem so compelling may also ultimately provide a few chuckles for our petrological descendants. A knowledge of the history of petrology, therefore, should help petrologists take themselves less seriously and be more open to rival concepts. To achieve this end, it would be desirable if future petrology textbooks provided brief summaries of the history of igneous petrology and referred students to appropriate historical studies. But, first, there must be a history of igneous petrology to which one may refer.

This book, therefore, attempts to fill the need for an overview of the history of the development of igneous petrology. *Mind over Magma* chronicles the struggles and hard-won successes of the human intellect in its search for solutions to the problems posed by rocks of magmatic origin. The book not only reviews intellectual milestones but also seeks to point out along the way how petrological concepts and individual petrologists fit into a larger framework of political and institutional factors, advances in technology, and discoveries in other sciences, such as physical chemistry and fluid mechanics. I have partitioned the narrative into several sections corresponding to eras or phases of development that were characterized by some leading or dominant petrological methodology or approach. Thus I have written about the "microscope era," the "experimental era," the "fluid dynamical era," and so on. My division of the history of igneous petrology into eras by no means implies that microscopical petrography, for example, passed into obsolescence during the succeeding "experimental era." The "experimental era" was, however, characterized by the emergence and dominance of a new, powerful methodology that yielded valuable data and dramatic insights into the diversity of igneous rocks. For a time, phase-equilibrium experiments represented the cutting edge of igneous petrology. Similarly, the "fluid dynamical era" is not characterized by the obsolescence of phase-equilibrium studies or isotopic studies or even microscopic petrography but is marked by the dramatic development of groundbreaking and stimulating applications of an already established discipline to igneous petrology with the result that there has been a conceptual revolution in thinking about processes in magma chambers, particularly within large layered intrusions.

Although the narrative is broadly chronological, large segments of time are traversed more than once in different chapters that discuss distinct but roughly contemporaneous major developments. For example, the chapter on the classification of the igneous rocks in the mid-twentieth century discusses nomenclatural developments in the 1920s even though other important events of the 1920s such as x-ray crystallography, phase-equilibrium studies, and the mechanics of igneous intrusion were reviewed in previous chapters. In other words, the text often jumps back to a period of time that was previously examined from a different point of view than the one under immediate consideration.

Even before embarking on this project, I understood the impossibility of digesting the sheer mass of significant petrological information. As a result, some important events, petrologists, and concepts have undoubtedly been given insufficient attention or even omitted altogether. Such omissions have their dangers, of course, particularly in regard to recent history. With so many outstanding practitioners of igneous petrology actively contributing to the science today, readers will likely be distressed to discover that I have neglected to mention the important contributions of many highly competent, highly respected individuals. Perhaps those very individuals will also be distressed that neither they nor the insights that have been so important in their careers have been mentioned. Such neglect should be regarded not as a matter of malice but of practicality. If I included everything important, I would never complete the book, which, after all, is not a gigantic review article, but a summary of the history of the discipline. I must also plead ignorance; like every other petrologist, I do not have the time to read everything petrological. Choices must be made. In the end, then, I have emphasized events, ideas, and individuals that seemed to me to be particularly important.

Some readers will probably be annoyed that I gave little attention to the history of ideas about specific rock types such as carbonatite, komatiite, ophiolite, anorthosite, mangerite, lamprophyre, kimberlite, diorite, andesite, syenite, melteigite, or whatever their favorite igneous rock type might be. I have discussed ideas about granite and basalt at some length, but even here so much more can be said. I decided to discuss the origins of specific rock types only in regard to their impact on the development of igneous petrology as a whole.

Nor have I given much attention to the petrology of a vast array of interesting regions around the world. I have said quite a bit about some of the regions or igneous rock bodies that have played a particularly crucial role. The Skaergaard Intrusion is discussed in many places, but the Bushveld Complex fares poorly in comparison in my treatment. I have totally ignored many other important layered intrusions. I have discussed the Tertiary Hebridean volcanic centers but have ignored other important volcanic centers like those of Nigeria. I have commented briefly on the Sierra Nevada Batholith of California and the Central Batholith of Peru, but I have said nothing about the Coast Range Batholith of British Columbia or about the batholiths of Australia and Japan. I encourage experts to think about the development of ideas of these important regions of igneous rocks that I have ignored.

It will not be difficult to find deficiencies such as those noted above in the text. I have not attempted, however, to provide an exhaustive (or exhausting!) history. I leave it to other petrologists and historians of science to make corrections and additions as they see fit. Indeed, I urge and challenge them to do so. This book is only a foundation on which and a framework around which more detailed studies may be constructed. It is my hope that the book will serve as a stimulus to petrologists to write historically about specific episodes, individuals, or ideas in the history of petrology in much greater detail than has been done in this volume.

In light of the extreme complexity of igneous petrology, no individual can possibly master every aspect of the field. As a result, the input of many other petrologists has been required in order to correct mistakes, supply additional insights and information, and to clarify interpretations. It is, therefore, my great pleasure to acknowledge the very generous assistance of the following individuals for their valuable comments on individual chapters that touch on their areas of expertise: Bernard Bonin, Ian H. Campbell, Ian S. E. Carmichael, Dennis R. Dean, Donald J. DePaolo, Carl-Henry Geschwind, L. Peter Gromet, Anthony Hallam, Paul C. Hess, Herbert E. Huppert, Rebecca A. Lange, Rachel Laudan, Bruce D. Marsh, Brian Mason, Alexander McBirney, Stearns A. Morse, Nick Petford, Wallace S. Pitcher, Denis Shaw, Henning Sørensen, R. Stephen J. Sparks, Hugh P. Taylor, Jr., S. Ross Taylor, and Peter J. Wyllie. Hatten S. Yoder, Jr., was particulary helpful in earlier stages of the project when I was dealing specifically with Bowen. Despite the eagle eyes of these scholars, technical errors and grammatical infelicities unquestionably remain. I am also very grateful to the late Julian Goldsmith, Ikuo Kushiro, Ian McCallum, Michael O'Hara, Julian Pearce, the late Francis Pettijohn, and Kenzo Yagi for generously

PREFACE   xxi

providing information about themselves or about other petrologists with whose work they are or were familiar. Their assistance was extremely beneficial.

Donald McIntyre kindly provided me with his photograph of a locality in Glen Tilt observed by James Hutton, and Pierre Boivin generously sent me two photographs of Puy de Dôme and environs in the Massif Central of France. Other photographs, predominantly portraits of prominent petrologists, were made available by the Bergakademie of Freiberg, the Picture Library of the National Galleries of Scotland, the National Museums of Scotland, the University of Leipzig, the Archives of the University of Heidelberg, the University of Cambridge, the Geophysical Laboratory, the Royal Society of London, Harvard College Library, and Princeton University. Many thanks to them. The following publishers and institutions are also gratefully acknowledged for granting permission to reproduce illustrations: the Geological Society of America, Elsevier Science, Oxford University Press, Springer-Verlag, *American Journal of Science*, E. Schweizerbart'sche Verlagsbuchhandlung, Americal Geophysical Union, The University of Chicago Press, American Geological Institute, the Royal Society of Edinburgh, and the Danish Polar Center. Portions of chapters 8, 12, and 13 have been expanded from articles that were printed in the journal of the History of Earth Sciences Society, *Earth Sciences History*.

I am also delighted to acknowledge the assistance of numerous staff librarians at the University of Illinois, the Geophysical Laboratory, the University of Cambridge, and the University of Michigan. Archivists at Princeton University, Harvard University, and Yale University have been especially helpful. I am particularly grateful to Nancy Romero of the rare book room at the University of Illinois, Nanci Young of the Princeton University library archives, and Shaun Hardy, librarian at the Geophysical Laboratory for their generous assistance. Kathy Struck, the inter-library loan librarian at the Hekman Library of Calvin College has uncomplainingly fed me an interminable diet of photocopies of geological articles in ancient and obscure journals.

Special thanks are also due to Steve Laurie, a curator at the Sedgwick Museum of the University of Cambridge. Steve graciously took me on a personal tour to see the thousands of rock specimens collected by the Cambridge petrologists T. G. Bonney, J.J.H. Teall, A. Harker, and C. E. Tilley as well as the thin sections made from those rocks. He also kindly showed me several of the rocks that Charles Darwin collected during the voyage of H.M.S. *Beagle*, including specimens of lava from the Galapagos Islands. These specimens are now stored at the Bullard Laboratories of the University of Cambridge. I was struck by the small size of Darwin's samples, typically only about 5 centimeters on a side.

For translations of several of the French and German articles consulted in preparation of the manuscript I am greatly indebted to former Calvin College undergraduates Sarah Vanderhill, Joel Schickel, Chris Blauwkamp, and Chandra Pasma.

I am also greatly indebted to Virginia Dunn and Jill Harris for guiding the manuscript through the editorial process.

Finally, this work would not have been possible without the generous financial assistance provided by National Science Foundation grants SBR-9601203 and SES-9905627. Calvin College also provided me with reduced teaching loads on a number of occasions through Calvin Research Fellowships that aided immeasurably in making progress on the project. Funds from the Calvin Alumni Association Faculty Research Grant program made visits to the University of Illinois and the Geophysical Laboratory possible.

The field of igneous petrology has experienced a remarkable and fascinating history. I hope that my readers will enjoy the guided tour.

Grand Rapids, Michigan
January 2003

The Foundational Era

1 ❖ BREATH-PIPES AND IGNIVOMOUS MOUNTAINS

Early Concepts of Volcanism

The science of igneous petrology requires three things: igneous rocks, human beings, and an appropriate methodology for the acquisition of knowledge and understanding about the igneous rocks. Magmatic activity and igneous rocks have characterized the crust throughout the 4.5 billion years of Earth history. Despite the existence of a planet that throbs with volcanic activity and that is decorated with innumerable masses of granite and gabbro, rhyolite and basalt, anorthosite and carbonatite, a science of igneous petrology cannot exist in the absence of humans. Anatomically modern humans probably appeared on Earth more than 100,000 years ago. Some of them observed volcanic eruptions, and many of them used igneous rocks such as obsidian. That was certainly the case a few millennia ago when civilizations sprang up in the ancient Near East. Yet the science of igneous petrology finally emerged only when people began to devise systematic ways of observing and thinking about igneous rocks. This book tells the story of how the science that we now call igneous petrology came into being and how it subsequently evolved.

VOLCANOES IN THE WORLD OF THE ANCIENTS

Human beings have long been curious about the behavior and significance of volcanoes, the great "fire-breathing" mountains that instill both terror and awe when they violently spew noxious fumes and flaming rocks. Because some early civilizations were located in regions of vigorous tectonic activity, our distant predecessors had ample opportunity to observe and reflect on volcanic phenomena. Although the ancient peoples dwelling on the vast alluvial plains of the Nile, Tigris, Euphrates, and Indus Rivers had little acquaintance with volcanoes, the citizens of classical Greece and Rome as well as the inhabitants of ancient Israel, living near junctions of tectonic plates, were certainly familiar with volcanism. The interpretations of volcanic phenomena entertained by these civilizations were variously mythological, theological, or naively scientific.

The Old Testament contains a sufficient number of references to "burning sulfur" or "brimstone" (Genesis 19:24; Deuteronomy 29:23; Job 18:15; Psalms 11:6; Isaiah 30:33, 34:9; and Ezekiel 38:22) and to "smoking mountains" (Exodus 19:18, 20:18; Psalms 104:32, 144:5) to suggest that the ancient Israelites were acquainted with volcanic phenomena. Israel's interest in the phenomena of nature, however, was largely theological. The Old Testament writers did not indulge in scientific theorizing about volcanoes, and no other ancient literature indicates the extent to which Hebrew scholars may have speculated about such matters. The book of Genesis envisioned the destruction of the ancient cities of Sodom and Gomorrah by fire and burning sulfur as an event of divine retribution for their unparalleled wickedness. Although the description in the book of Exodus of the giving of the ten commandments at Mount Sinai by Yahweh suggests the possibility of accompanying volcanic activity, the writer of Exodus regarded Yahweh as the one who caused the mountains to smoke and the earth to tremble. Neither Genesis nor Exodus inquired into the nature of smoking mountains or the eruption of burning sulfur in terms of mechanistic explanations or natural causes. More recently, Neev and Emery (1995) suggested that Sodom and Gomorrah were destroyed by an earthquake and that the flaming materials might have been ignited hydrocarbons that had been released from reservoirs by the seismic activity.

In contrast to the monotheistic Hebrews, both the classical Greek and Roman civilizations were steeped in pagan polytheism and its attendant mythology. Volcanic phenomena were, therefore, commonly interpreted in mythological terms. For example, the Roman writer, Publius Vergilius Maro, better known as Virgil (70–19 B.C.), penned a graphic description of the eruption of Mount Etna, the great volcano of Sicily, in his epic *Aeneid*. Although describing terrifying crashes, black clouds smoking with pitch-black eddies, glowing ashes, balls of flame licking the stars, and the violent vomiting of molten rock boiling up from the lowest depths, Virgil (Fairclough, 1994) linked these occurrences to the story about the giant Enceladus (Typhon to the Greeks) being scathed by a thunderbolt of Jupiter and weighed down by the mass of Mount Etna. Every time that the giant shifted from one side to the other, Virgil said, the region moaned, trembled, and veiled the sky in smoke. Virgil also located the entrance to the underworld in the vicinity of Lago Averno (Lake Avernus) in the volcanic terrain known as Campi Phlegraei (the Phlegraean Fields) west of Naples, Italy. The lake occupies a volcanic crater.

The Greeks designated Hephaestus, and later Roman myths designated Vulcan, as the god of fire. Roman poets linked Vulcan's workshop with active volcanoes, allegedly the chimneys of Vulcan's forge. The island of Vulcano, one of the Lipari Islands north of Sicily, received its name because it was considered to be the location of Vulcan's forge. From the Roman god Vulcan and the island Vulcano the name of "volcano" has since been applied to smoking, cinder-erupting, lava-emitting mountains around the world.

Mythological conceptions of volcanoes were not confined to the classical civilizations. More recent volcano myths existed in Tonga, Samoa, Indonesia, Iceland, and among the Aztecs of Mexico. Myths about volcanoes persist in Japan and Hawaii. Many native Hawaiians still envision eruptions of Kilauea as expressions of the wrath of the goddess Pele, who is said to reside in the caldera of Kilauea. They believe that Pele is able to make islands and mountains, melt rocks, and destroy forests.

Despite the mythological overtones in much classical writing about the natural world, some authors did provide explanations for natural phenomena, including volcanoes, in terms of natural, material causes. Several Greek writers made important observations about geological phenomena such as erosion, sedimentation, fossilization, the interchange of land and sea, and earthquakes (Desmond, 1975). In addition, many of the Greek and Roman references to Mount Vesuvius and the Phlegraean Fields in Italy, Mount Etna in Sicily, the Lipari Islands, and Santorini near Crete contain descriptions of landforms and eruptions. In some cases, the writers attempted to explain volcanic behavior in terms of physical causes.

Prior to the eruption of Mount Vesuvius in A.D. 79, Mount Etna, known for its frequent emissions, drew the widest attention. The philosopher Empedocles of Agrigentum (492–432 B.C.) presumably died by falling into the crater of Mount Etna. Diodorus of Sicily (1st century B.C.) referred to fiery eruptions from Mount Etna that laid waste to regions along the sea (Oldfather, 1933). He wrote that such great torrents of lava were poured forth from Mount Etna that the people known as the Sicani had to move to the western part of Sicily. The geographer Strabo (63 B.C.–A.D. 21) discussed volcanoes in considerable detail. In his *Geography*, he noted that an earlier writer, Poseidonius, had commented that the fields of the Catanaeans were covered to great depth by volcanic ash during eruptions of Mount Etna. Strabo observed that volcanic ash ultimately proved to be beneficial because it rendered the land so fertile (Jones, 1931). When lava solidified, he found that the surface became stony to such a depth that quarrying was necessary to uncover the original land surface. He reported that the melted liquid that pours over the rim of Mount Etna's craters is a "black mud" that flows down the mountain and solidifies to millstone. Strabo provided an extensive description of Mount Etna, which evidently supported a significant summit-climbing trade. He observed that the top of the mountain experienced many changes. The fire concentrated first in one crater, then in another. The mountain sent forth lava on one occasion, flames and fiery smoke on another, and red-hot masses on still another. Strabo conjectured that Sicily was not a piece of land that had broken away from Italy but a landmass that might have been elevated out of the sea because of the volcanic eruptions of Mount Etna. He proposed a similar origin for the Lipari Islands and described other instances of land emerging from the sea by volcanic action.

Many references to Mount Vesuvius and the nearby Phlegraean Fields appear in classical literature written prior to the cataclysm of A.D. 79. Diodorus of Sicily referred to the arrival of Hercules at the Phlegraean plain, so named from the mountain that in old times had erupted in a huge fire as Mount Etna did in Sicily (Oldfather, 1933). Diodorus noted that the mountain was called Vesuvius in his day and that it showed many signs of the fire that raged in ancient times. The Roman architect Vitruvius (d. 25 B.C.) described volcanic products in the vicinity of Mount Vesuvius such as pozzolana and pumice. Vitruvius noted that "in ancient times the tides of heat, swelling and overflowing from under Mount Vesuvius, vomited forth fire from the mountain upon the neighbouring country" (Morgan, 1960, p. 47). As a result, he surmised that "sponge-stone" (Pompeian pumice) had been formed by burning of some other kind of rock. Strabo wrote that the summit of Mount Vesuvius was mostly flat and unfruitful, appeared ash-colored, and had pore-like cavities in soot-colored masses of rock that looked as though they had been eaten out by fire (Jones, 1931). He inferred that the district had once been on fire and that the fire was quenched after the fuel gave out.

Understandably, the cataclysmic eruption of Mount Vesuvius that buried the cities of Pompeii and Herculaneum in A.D. 79 attracted considerable attention. The eruption was described in graphic and dramatic detail by Gaius Plinius Caecilius Secundus, better known as Pliny the Younger (A.D. 61–114), who was only seventeen years old at the time. In a letter to Cornelius Tacitus, Pliny wrote about the death of his uncle, Gaius Plinius Secundus (Pliny the Elder) (A.D. 23–79), the author of the monumental *Natural History*. The younger Pliny wrote prophetically that his uncle died in a catastrophe that was so spectacular that his name would likely live forever (Radice, 1969). He reported that his uncle, the commander of the fleet at Misenum, ordered his ships across the Bay of Naples for a closer look at the developing eruption. Eventually, according to the nephew, the flames and smell of sulfur became so strong that most people took flight and roused Pliny the Elder to stand up. As he was leaning on two slaves, his nephew wrote, he suddenly collapsed, most likely because the dense fumes had choked his breathing. When daylight had returned two days later, the body of Pliny the Elder was found intact, uninjured, fully clothed, and looking asleep rather than dead. In a later letter to Tacitus, Pliny the Younger told of his own escape from the dense black cloud of falling ash. When the darkness finally dispersed, he wrote, the sun was shining much as it does during an eclipse. Pliny admitted to being terrified to see that everything had been buried deep in ashes like snowdrifts. Sigurdsson (1999) has written a detailed account of the famous eruption.

A remarkable aspect of Greco-Roman writing about volcanoes is that at least three rudimentary theories of volcanism were proposed: the exhalation theory originated by Aristotle, the fuel theory of Strabo, Vitruvius, and Seneca, and the organismal-breathing theory of Ovid. The most elaborate of these theories, expounded in *Meteorologica*, was developed by Aristotle (B.C. 384–322) in the context of his general theory of exhalations. He interpreted volcanic eruptions as the end result of the movements of subterranean exhalations (winds) that gave rise to earthquakes. Aristotle cited an eruption in the Lipari Islands as an example. He said that the wind that caused earthquakes on the islands broke out like a hurricane. The earth swelled and rose into a lump with a noise. The swelling finally exploded so that a large amount of wind, cinders, and ash broke forth, smothering the city of Lipara and extending as far as Italy. In this case, Aristotle said, the breaking up of the air into tiny particles caused them to catch fire within the earth (Lee, 1952). T. Lucretius Caro (Lucretius) (c. 99-c. 55 B.C.), a Roman poet, adopted an essentially Aristotelian explanation in his account of the nature of volcanoes in *De Rerum Natura*. Mount Etna, Lucretius suggested, is hollow underneath, and all its caves are filled with wind (Bailey, 1947). When agitated, the wind becomes hot, eventually heating all the rocks and earth that it touches. In time, the agitated wind heats the rocks so much that they emit swift flames. The wind, Lucretius wrote, drives itself through the mountain's jaws carrying heat, scattering ash and smoke with thick murky darkness, and hurling heavy rocks.

Strabo, too, linked volcanic flames with winds. He claimed that the flames at Mount Etna and on the volcanic island of Thermessa (Hiera) were stimulated along with winds. When the winds died away, so, too, did the flames. Strabo, however, also maintained that volcanic eruptions feed on some kind of fuel. He referred to the exhaustion of fuel at Mount Vesuvius and suggested that volcanic fires might be kindled by fuel just as the wind is fueled by evaporation from the sea. Vitruvius was much more explicit about this alternative view. In discussing "pozzolana" he noted that the soil in the vicinity of Mount Vesuvius is hot and full of hot springs. Such a condition would not exist, he claimed, "unless the mountains had beneath them huge fires of burning sulphur or alum or asphalt" (Morgan, 1960, p. 47). Aristotle's exhalation explanation was combined with the fuel theory of Vitruvius by Lucius Annaeus Seneca (c. 2 B.C.–A.D. 65). Seneca suggested in *Naturales Quaestiones* that when subterranean winds rush through underground cavities containing sulfur and other combustible materials, these flammable substances are ignited by friction (Corcoran, 1971).

A third view was suggested by Publius Ovidius Naso (Ovid) (43 B.C.–A.D. 17), the Roman author of the epic poem *Metamorphoses*. Suggesting that Earth is a great organism, he likened volcanic eruptions to the breathing of an animal. Ovid, however, combined the organic theory with Aristotle's exhalation theory and Vitruvius' fuel theory: He asserted that if the earth is like a living animal with many breathing-holes that can exhale flames, then it can close up those holes and open new ones when it shakes itself. Moreover, the friction of pent-up winds in caverns driving against the rocks could cause fire. After the winds had spent their force, however, the caverns would cool and the flames would be extinguished. Ovid also said that pitchy substances and yellow sulfur might serve as sustenance for the fires, and when these nourishing substances were exhausted then the fires would die out (Miller, 1977). Sigurdsson (1999) has discussed classical Greek and Roman ideas about volcanoes more thoroughly.

VOLCANISM IN THE MIDDLE AGES

The writers of ancient Greece and Rome provided descriptions of volcanoes and volcanic eruptions, suggested several causes for volcanic activity, and referred to emission products such as lava (ruax), scoria, pozzolana, and ash. In the early centuries of the Christian era, conditions that fostered scholarly thought began to deteriorate. Economic decline and political turmoil plagued the Roman Empire until it fell to barbarian invaders. Contact between the Latin West and Greek East diminished. The early Christian movement, typically popular among the poorer classes and faced with much persecution, had no time for leisurely reflection about the natural world. As Christianity continued to spread, however, it became the dominant cultural force in the western world. While Christianity encouraged literacy and education, particularly through the monastic movement, theological questions absorbed most of the intellectual energy of scholars. Moreover, the church was ambivalent about the knowledge acquired by the pagan Greeks. As historian of science David Lindberg (1992, p. 151) observed in his discussion of the development of science from classical times to the Middle Ages, if the ancient church is compared to the National Science Foundation, it will "prove to have failed abysmally as a supporter of science and natural philosophy." He pointed out, however, that in relation to its contemporaries, "the church was one of the major patrons—perhaps *the* major patron—of scientific learning." Although scientific thought in general and thought about volcanoes in particular stagnated during the first several centuries of the Christian era, some medieval Christian scholars retained an interest in geological matters. These individuals speculated about fossils and the effects of the biblical flood and compiled encyclopedic lists of stony objects (Adams, 1938). So far as we know, however, medieval Christian writers said little that was new about volcanoes, although Albertus Magnus (c. 1200–1280) attempted an experiment to determine whether volcanic action resulted from subterranean steam pressure (Koch, 1966).

Although the Latin West experienced some decline in scholarly activity, knowledge of classical Greek scholarship persisted in the eastern Mediterranean. During the Islamic conquest, much Greek learning was absorbed into that expanding culture, and original scientific activity eventually flourished in the Islamic world. To the extent that Arab writers of the Middle Ages like Ibn

Sina (Avicenna) devoted attention to geological features, they were preoccupied with erosion and sedimentation. Nevertheless, the theory of petrifying juices was developed to explain the formation of some rocks. Much classical literature, preserved by Islamic scholars, was reintroduced into western Christendom during the thirteenth century. In the latter Middle Ages, Christian scholars uncritically accepted much of the science of the classical writers, including the descriptions and theories of volcanoes of Aristotle and Strabo. They added virtually nothing new regarding the nature of volcanoes. Scientific interest in volcanic activity may also have been dormant throughout the Middle Ages because of the linkage of the horrors of hell to volcanic fires in much medieval theology.

VOLCANISM IN THE SIXTEENTH AND SEVENTEENTH CENTURIES

After the Renaissance, even such astute sixteenth-century observers as Bernard Palissy (1510–1590), Conrad Gesner (1516–1565), and Johann Kentmann (1518–1568) added little to the body of knowledge about volcanoes. Georg Bauer (1494–1555), also known as Agricola, wrote at length about rocks and minerals and speculated about the role of fluids in their formation, yet he barely mentioned volcanic activity. In *De Re Metallica* (Bandy and Bandy, 1955), Agricola referred to millstones that were formed from molten rock erupted at burning places. Rivers of such molten rock were noted to flow away as at Mount Etna.

The ascendancy of the mechanical philosophy in the seventeenth century was accompanied, however, by intense interest in the nature, behavior, origin, and history of the entire terrestrial globe, as indicated by the studies of Galileo Galilei (1564–1642) and Isaac Newton (1642–1727) on the place of Earth in the heavens and by the speculations of René Descartes (1596–1650) about the formation of Earth from mechanically interacting particles from vortices in space. Later in the seventeenth century, global histories based on mechanical principles were formulated by Thomas Burnet (1635–1715), William Whiston (1667–1752), and many others (Rappaport, 1997). Several of these global theories were accompanied by illustrations of Earth's interior. Among the scholars participating in the movement to understand and illustrate the globe as a whole was Athanasius Kircher.

Athansius Kircher (1602–1680) was a prolific German Roman Catholic priest of encyclopedic interests who spent most of his career at the Jesuit College of Rome (Reilly, 1974). A polymathic scholar who wrote 44 books, Kircher addressed the subject of volcanoes in a volume, *Mundus Subterraneus*, devoted to understanding Earth as a globe (Kircher, 1664). Kircher had visited both Mount Vesuvius and Mount Etna in 1638 while working on this massive study of underground phenomena. An extract of material on volcanoes from this book appeared five years later as *The Vulcano's or Burning and Fire-vomiting Mountains, Famous in the World: with their Remarkables* (Kircher, 1669). Kircher described Mount Vesuvius and Mount Etna, reviewed known volcanoes, and provided a general theory of volcanic action in terms of subterranean fire. Kircher (1669, p. 3) believed that there are two "Associates, and Agents of Nature," namely, fire and water, agents that "sweetly conspire together in mutual service, with an inviolable friendship and wedlock, for the good of the whole in their several and distinct private-lodgings." The various interactions of the two agents moving through an elaborate system of interconnected subterranean channels and vents were thought to lead to a host of terrestrial phenomena, such as "Minerals, Juyces, Marles, Glebes, and other soyls, with ebullitions, and bublings up of Fountains."

Apart from that point, Kircher's views on volcanoes largely represented an amalgam of the conceptions of Aristotle, Strabo, Vitruvius, and other classical writers. Referring to volcanoes as "ignivomous" or "fire-vomiting" mountains, Kircher (1669, p. 5) claimed that they are "the vent-holes, or breath-pipes of Nature, to give vent to the superfluous choaking fumes and smoaky vapours, which fly upwards." Upon emission of the fumes, the entrance into Earth is made possible for "friendly cherishing Air to revive and ventilate those suffocating flames" which would otherwise "continually shake the foundations of the Ground with intolerable commotions and Earthquakes." As an Aristotelian, Kircher considered volcanoes from the perspective of various causes. The formal cause of the eruptions, he said, is the fire. The material cause is sulfur, salt, niter, and bitumen in the dark recesses of Earth. The instrumental cause, Kircher (1669, pp. 58–59) claimed, is "the Cavernous nature of the place . . . oppressed with Sulphureous Smoak and Soot," and, lastly, the efficient cause consists of "Winds and Blasts," issuing bellows-like from the cavernous interior, stirring the dormant fires, and kindling the matter.

For the most part, studies of volcanoes in the seventeenth and eighteenth centuries, primarily by Italian writers, focused on the volcanoes themselves apart from any consideration of the global context. Mount Vesuvius had experienced some minor events during the Middle Ages, but a violent eruption in 1631 elicited descriptive books from Antonio Santorelli (1632) and Giulio Cesare Recupito (1635). Subsequent eruptions in the late seventeenth and eighteenth centuries resulted in additional studies by Paragallo (1705), Sorrentino (1734), Serao (1738), and della Torre (1755). Giovanni Maria della Torre described white polygonal garnet-like crystals in the lava (leucite) and observed that thick lava from an eruption in 1754 split into pieces when it was raised with a pole. The numerous eruptions of Mount Etna throughout the seventeenth century attracted the interest of Giovanni Alfonso Borelli (1670) and the Englishman Heneage Finch Winchilsea (1669). The eruption on Santorini between 1707 and 1711 made a strong impression on Abbé Anton-Lazzaro

Moro (1687–1764) of Venice. So did an account of the 1538 eruption of the cinder cone, Monte Nuovo, in the Phlegraean Fields. As a result, Moro (1740) attributed virtually all stratified rocks to the elevating effects of volcanic agency.

Despite a growing awareness of the worldwide distribution of volcanoes, the most detailed descriptions of the latter eighteenth century were confined predominantly to Mount Vesuvius and to Iceland. Much information about volcanoes was obtained by academics and government officials and recorded in scientific letters, scientific travelogs, and communications to *Philosophical Transactions of the Royal Society of London* (Rappaport, 1997). Several such reports were sent to *Philosophical Transactions* by Sir William Hamilton (1730–1803), the envoy of Great Britain to the court of Naples (Sleep, 1969; Carozzi, 1972). Hamilton (1772) eventually compiled these letters and reports into a book about Mount Vesuvius, Mount Etna, and other volcanoes.

Hamilton arrived in Naples on November 17, 1764. On June 10, 1766, he wrote to the President of the Royal Society of London that Mount Vesuvius had recently made a large eruption. He reported that nothing terribly interesting other than a lot of smoking had happened on the mountain for several months until late in March 1766, when Mount Vesuvius erupted ash and lava began to boil out of its crater. Hamilton passed a night on the mountain while the eruption proceeded. He approached the mouth of the volcano as closely as he dared and observed that the lava looked like a river of red hot liquid metal. As the eruption proceeded, he said, some English observers who accompanied him got too close and were injured by flying rocks. Hamilton described the development of crust on the cooling lava and the existence of lava tubes.

In a letter of February 3, 1767, he wrote that he had collected various salts, sulfur, lava, and cinders for later analysis. He was amazed to discover that none of the local chemists had ever bothered to analyze any of the eruption products of Mount Vesuvius. On December 29, 1767, Hamilton wrote to the Royal Society that a violent eruption had commenced on October 19, 1767, lasted seven days, and emitted about three times as much lava as had the ten-month-long 1766 eruption. He shipped an extensive collection of eruption products from Mount Vesuvius to the British Museum. Having painstakingly assembled the samples during the past three years, he expressed the hope that some of his learned countrymen might put the collection to use in making some important discoveries about volcanoes.

A couple of years later, Hamilton wrote on October 17, 1769, that after having spent almost five years examining volcanic phenomena all around Naples, including the nearby island of Ischia, he thought it would be a good idea to visit Mount Etna. He reported that on June 24, 1769, he set out with some companions from Catania, Sicily, to climb the mountain. He succeeded in climbing to the edge of the crater where he spent three hours. In a masterpiece of understatement he wrote that "the steep ascent, the keenness of the air, the vapours of the sulphur, and the violence of the wind, which obliged us several times to throw ourselves flat upon our faces to avoid being overturned by it, made this latter part of our expedition rather inconvenient and disagreeable" (Hamilton, 1772, p. 59). On this excursion, Hamilton also spent three days in the Lipari Islands, where he saw Vulcano and observed Stromboli in eruption. Upon his return to Naples, he had the opportunity to watch the excavations of Pompeii at a time when both human and horse skeletons were encountered. He ascertained that the skulls of the dead had been fractured by the falling rocks.

Hamilton speculated that "subterraneous fire" acted as the "great plough" used by nature to overturn the bowels of Earth, thereby exposing fresh fields on which to work. Because of his more than twenty ascents of Mount Vesuvius, his detailed observations of actual eruptions, his descriptions of volcanic products, his collection of specimens, and his letters published in *Philosophical Transactions of the Royal Society of London*, Hamilton became a recognized authority on volcanic eruptions.

Hamilton had plenty of company on top of Mount Vesuvius, including Ferber, Spallanzani, and Breislak. The Swede Johann Jacob Ferber (1743–1790), Professor of Natural History at Mietau, had been educated at Uppsala, Sweden, where Linnaeus, Cronstedt, and Wallerius had systematized the different kingdoms of nature. On his travels throughout Europe during 1771 and 1772, Ferber (1776) climbed to the summit of Mount Vesuvius twice. On one occasion he entered into its crater. Like Della Torre, he reported seeing the garnet-like crystals embedded in the lavas.

The Abbé Lazarro Spallanzani (1729–1799), Professor-Royal of Natural History in the University of Pavia and Superintendent of the Imperial Museum in that city, undertook a tour of Italy and Sicily in 1788 to improve the collections of the museum with the addition of some volcanic products (Dolman, 1975). During six months at Mount Vesuvius, Spallanzani (1798) tried to reach the summit but lava showers and sulfurous gases forced him to retreat so that he was denied the "pleasure" of getting near the edge of the crater. He was successful, however, in coming remarkably close to flowing lava and measuring its velocity. He reported that when he was only five feet from the lava, the "caloric" was sufficiently vehement to encourage him to back away. He measured the velocity of the lava currents by throwing cooled chunks into the flow and observing how long it took them to be carried a specified distance.

Spallanzani lacked sufficient time to measure the temperature of the lava, but he outlined a detailed scheme for how to make such measurements. He also issued a challenge and warning for potential students of volcanoes. "I do not deny," Spallanzani (1798, p. 19) said,

> but that these and other similar experiments are difficult, offensive, and, in some degree, even dangerous; but what experiment can be undertaken perfectly free from inconvenience, and all fear of danger, on mountains which vomit forth fire? I would

certainly advise the philosopher who wishes always to make his observations entirely at his ease, and without risk, never to visit volcanoes.

Spallanzani knew from firsthand experience what he was talking about, for on one occasion he burned out his shoes. That he repeatedly exposed himself to considerable danger in the interests of a scientific investigation of Mount Vesuvius is confirmed by the fact that many of his gas collection experiments occurred in the Grotta del Cane. He reported that if a dog were brought into this cavern and its nose held to the ground, it would soon begin to breathe with difficulty and die if not speedily removed into the open purer air. Spallanzani measured the chemical composition of volcanic gases of Mount Vesuvius jointly with Scipione Breislak (1750–1826), a priest and teacher at the Military Academy of Naples (Francani, 1970). They determined that the gas consisted of 10 percent "vital air, or oxygenous gas," 40 percent "fixed air, or carbonic acid," and 50 percent "phlogisticated air, or azotic gas" (Spallanzani, 1798, p. 91). Breislak later witnessed the 1794 eruption of Mount Vesuvius and observed the volcanic phenomena of the Phlegraean Fields and Pozzuoli.

One of the most respected geological writers of the era, Déodat Guy Silvain Tancréde Gratet de Dolomieu (1750–1801), was a military officer in the Knights of Malta with a flair for natural history (K. L. Taylor, 1971c). Dolomieu studied the lavas of Mount Etna and many of the recent volcanic fields of Italy and speculated on the nature of lava and the causes of volcanism in a series of articles and travel books.

Iceland, too, had begun to draw the attention of scientific parties. Of particular importance was a 1772 expedition under the supervision of Sir Joseph Banks (1743–1820), a prominent botanist who had participated in the first voyage of Captain James Cook between 1769 and 1771 (Cameron, 1952; Foote, 1970). Banks was accompanied by the natural historian Daniel Carl Solander (1733–1782) and Uno von Troil (1746–1803), later to become Archbishop of Uppsala. Von Troil (1780) published an account of the Icelandic expedition. "There was not one," said von Troil (1780, p. 5) of his enthusiastic and determined party, the first ever to attain the summit of Mount Hekla, "who did not wish to have his cloaths a little singed, only for the sake of seeing Heckla in a blaze." Although they were not rewarded by seeing an eruption, they described eruption products, hot springs, and geysers. Von Troil summarized the historical eruptions of Mount Hekla and surmised that its activity had been more intense than that of either Mount Vesuvius or Mount Etna.

Every late eighteenth-century European natural philosopher knew of the existence of the Mediterranean and Icelandic volcanoes, but reports of volcanoes had come to western Europe from other parts of the world during the past three centuries. European naturalists knew of eruptions in the Azores and the Canary Islands. By the end of the eighteenth century, the exploration of the New World had disclosed the existence of many volcanoes in the West

Indies, Mexico, and along the length of the Andes Mountains in South America, and volcanoes had also been observed by European voyagers in Japan, the Philippines, and the East Indies (Scrope, 1825). Toward the end of the eighteenth century, the Hawaiian volcanoes came to the attention of the western world thanks to the voyages of James Cook. Even if a natural philosopher had never seen an active volcano, he had read accounts of them and undoubtedly had seen many specimens of volcanic products in museum collections. Naturalists of the latter eighteenth century knew very well what lava, scoria, pumice, obsidian, and volcanic ash looked like.

Since the Renaissance, naturalists had also arrived at a moderate consensus on the cause of volcanism. Following the lead of Strabo, Vitruvius, and Seneca rather than Aristotle, writers generally conceived that volcanoes owed their origin to the combustion of some sort of fuel, perhaps coal beds, most likely triggered by the decomposition of pyrite not far below Earth's surface. Although he had closely linked volcanoes to subterranean fire in intimately interconnected passageways, Kircher (1669, p. 59) also regarded the matter of the subterraneous fire as "Sulphur, Bitumen, Pit-Coals, and also Allom, Salt, Nitre, Coaly Earth, and Calcanthum or Vitriol, and such kind of Metals."

Kircher's contemporary, Martin Lister (1638–1712), was one of the premier English natural historians of the seventeenth century (Carr, 1973). Lister (1683) issued a series of three papers in *Philosophical Transactions* in which he espoused pyrite as the cause of thunder and lightning, earthquakes, and volcanic eruptions. He found other potential causes of volcanism wanting. It could not have been the sun, he said, for Mount Hekla, located in an extremely cold climate, was just as active as Mount Etna. Human activity could not have caused eruptions because some volcanoes seem to have fired before the world was populated everywhere. Moreover, eruptions occur at the very tops of vast high mountains, places that are unfit for human habitation. Only pyrite, he said, provided so general and lasting a fuel that could cause lightning, thunder, earthquakes, and volcanoes.

Lister's proposal received experimental support at the hands of the French chemist Nicholas Lémery (1645–1715). Lémery demonstrated experimentally that sufficiently massive, pasty mixtures of water with iron and powdered sulfur, the constituents of pyrite, eventually burst into flame (Hannaway, 1973). Lémery argued that the heat and fire were generated by the violent friction of pointed acid particles of sulfur acting upon particles of iron. This experiment, he concluded, sufficiently demonstrated how fermentations, shocks, and conflagrations occur in the bowels of Earth at places like Mount Vesuvius and Mount Etna (Lémery, 1700). For years thereafter, writers asserted the importance of pyrite decomposition in volcanic action, assuming that Lémery's experiment provided vindication of the hypothesis that pyrite served to fuel volcanic fires.

One such writer was Richard Kirwan (1733–1812), a prominent Irish mineralogist, Inspector General of Mines for Ireland, and long-time President of the Royal Irish Academy (Scott, 1973b). Kirwan (1799) maintained that the heat of lavas originated from the decomposition of water by pyrite. Flame would arise, Kirwan proposed, where heated pyrite had been in contact with heated coal, black wad, or petrol. He envisioned that bituminous, sulfurated masses of stone and earth were gradually softened and liquefied in the process. Moreover, he noted, volcanic ejections never contained any undecomposed pyrite.

More ambivalent, Ferber (1776) asserted that many possible causes of subterraneous inflammations had been demonstrated by experimental philosophy, but he was uncertain whether these causes were ultimately traceable to the interaction of water and pyrite or to the fermentation of calcareous materials produced by acids or waters. He did note, however, that pyrite was reported to be thrown out of volcanoes in small quantities and that sublimated sulfur is very common.

By the end of the eighteenth century, numerous volcanoes had been discovered around the globe; exploration and description of Mediterranean and Icelandic volcanoes had become quite commonplace; some lava and volcanic gas had been chemically analyzed; individual crystals within lava had been noted; the velocity of lava currents had been measured; and a commonly held theory that attributed volcanism to the interaction of decomposing pyrite with carbonaceous matter had emerged. The infant science of volcanology had been born.

2 ✣ BASALTES PRISMATIQUES
Lava, Columnar Basalt, and Ancient Volcanoes

Throughout the eighteenth century, the vast majority of naturalists believed that water, in the form of the Noachian deluge, a diminishing ocean, or marine incursions of the land, was the dominant geological agent in the course of Earth history. In particular, the viewpoint prevailed during much of the century that mechanically accumulated strata were deposited in, and that many crystalline materials such as basalt and granite were precipitated from, an essentially global ocean. It certainly was evident to naturalists that fossiliferous sandstone, puddingstone (conglomerate), limestone, and clay had all been deposited from water. The role of heat as a significant geological agent, however, was generally restricted to the present and the very recent past. In the eyes of most naturalists, the construction of Earth's rocky layers and crystalline masses had little to do with volcanic eruptions. So long as volcanic eruptions were treated only as recent or contemporary sporadic accidents, the aspect of geology that we now designate as igneous petrology would play a very subsidiary role within the burgeoning field of inquiry into the nature of Earth.

THE DISCOVERY OF EXTINCT VOLCANOES

That state of affairs began to change in 1751 when two members of the Académie Royale des Sciences de Paris, Jean-Étienne Guettard (1715–1786) and his friend, Chrétien-Guillaume de Lamoignon de Malesherbes (1721–1786), were traveling in the Auvergne, a region in the Precambrian and Paleozoic crystalline *massif central* of south-central France. In that era, many important scientific advances were made by individuals who either were independently wealthy, were supported by the wealthy, or held administrative or diplomatic positions that made it possible for them to travel widely and meet a variety of influential people. Guettard, for example, had previously exercised his passion for various branches of natural history and medicine by serving in positions such as Curator of Réaumur's natural history collection, member of the Faculté de Médecine de Paris, Adjunct Botanist to the Académie des Sciences, and, most importantly, medical botanist for a French family of nobles who supported him financially and provided him with a suitable laboratory (Rappaport, 1972). One of Guettard's major achievements had been the compilation of geological information for a mineralogical map of France, a preliminary version of which he presented to the Académie in 1746. Malesherbes was the son of a Chancellor of France. He himself occupied a number of government posts in the *ancien régime* (Rappaport, 1974). His status afforded him the opportunity to indulge a passion for botany, breeding, and natural history. Although Malesherbes never became a major scientific figure, he traveled considerably and helped to gather data for Guettard's mineralogical map. In 1792, during the revolutionary turmoil in France, Malesherbes volunteered to serve as the defense counsel for King Louis XVI. The less-than-grateful Revolutionary Tribunal repaid Malesherbes in 1794 by separating him from his head.

Guettard and Malesherbes were impressed by the extent to which a black cellular rock was used as a building stone throughout the Auvergne for houses, fountains, and public edifices, such as the strikingly dark Cathédrale de Notre Dame de l'Assomption in Clermont-Ferrand (de Beer, 1962). Suspecting that this dark building stone is volcanic, they inquired about its source and were directed eight kilometers north of Clermont-Ferrand to the village of Volvic, the present-day site of mineral springs that support a thriving bottled water industry. Near Volvic, they discovered quarries that had been excavated into dark lava that had obviously flowed from the crater of a nearby volcanic cone. The presence of lava currents, associated scoria, and a summit crater on a conical hill confirmed for them a site of extinct volcanic activity. From the summit of Puy-de-Dôme, the 1465-meter trachytic dome 6 kilometers west of Clermont-Ferrand, they saw a chain of other vegetated cones, many with craters and lava flows, stretching both north and south (Figure 1). Guettard and Malesherbes already knew that Blaise Pascal had conducted measurements of air pressure at the summit with a primitive barometer in 1648. During their sojourn, they learned that Puy-de-Dôme had been climbed by other scientific men as well. No one, however, had breathed a word about its volcanic nature. After more extensive exploration, Guettard concluded that the region combined remnants of an ancient volcanic terrane that was no longer active. He reported the discoveries to the Académie des Sciences in 1752 and published a paper on the topic four years later (Guettard, 1756).

In 1763, Nicolas Desmarest (1725–1815), another government employee with a fascination for natural history, visited the Auvergne to gather industrial information for the Director of Commerce. Prior to his visit, Desmarest had published studies on earthquake propagation and marine erosion. He had not been especially interested in volcanoes and regarded them as accidental phenomena (K. L. Taylor, 1969, 1971b). In the Auvergne, however, he was struck by the wealth of volcanic features. He returned the following year with a cartographer and an artist to map the area. After presenting a preliminary report on his observations in 1765 before the Académie des Sciences, Desmar-

1. Extinct volcanic cones in the Massif Central of south-central France. Reproduced by permission of Pierre Boivin.

est spent a year touring volcanic districts, including the Euganean Hills near Padua in northeastern Italy. He returned to the Auvergne to complete his mapping project. The first published notice of his work appeared in 1768, and a major memoir was subsequently presented to the Académie in 1771 and published three years later (Desmarest, 1774). Like many of his predecessors and contemporaries, he imagined that the lavas had been heated by the accidental combustion of subterranean coal seams. Despite his remarkable discoveries, Desmarest continued to maintain that volcanic action had not played a major role in Earth history. He was fundamentally a neptunist who regarded water as the primary agent of geological change.

Intrigued by the claims of Guettard and Desmarest, French naturalists began to pour into the *massif central*. In 1766, Jacques Montet noted the presence of volcanic mountains south of the Auvergne near Montpellier, a region later to be visited by Guettard. Barthélemy Faujas de St. Fond (1741–1819) published a memoir on the volcanic mountains of Vivarais and Velay, southeast of the Auvergne (Faujas de St. Fond, 1778), and Jean-Louis Giraud Soulavie (1752–1813) also explored Vivarais and Velay during the 1770s and concluded that the region had been affected by past volcanic activity (Soulavie, 1780–1784).

The English, too, got into the act. After his own tours of the Auvergne in 1770 and 1772, evidently after having learned of Guettard's and Desmarest's conclusions, the Bishop of Derry, Frederick Augustus Hervey (1730–1807), became an enthusiast for the volcanic interpretation of the area (R.D., 1891). Hervey's subsequent communications with John Strange (1732–1799), British minister in Venice, stimulated the latter to visit the Auvergne in 1773 (T.S., 1898). Rudel (1962) and de Goër et al. (1994) have discussed the details of the volcanic geology of the Auvergne.

The *massif central* of France, however, was not the only place where extinct volcanism was being deciphered. Taking advantage of his post in Venice, Strange (1775) also described numerous volcanic features near Padua, Verona, and Vicenza in northeastern Italy. He was following the lead of Giovanni Arduino (1714–1795), the Italian mining engineer who worked out the geology of various copper, lead, and mercury mines and made a particularly valuable contribution to stratigraphy by developing a primitive geological timescale (Rodolico, 1970). Arduino had recognized the volcanic origin of the Euganean Hills near Padua as early as 1759. By 1773, J. J. Ferber had also agreed on the volcanic origin of these Italian landforms. Dolomieu reported the existence of ancient lavas in northern Italy and in Portugal.

Extinct volcanic terranes were also recognized in Germany. Rudolf Erich Raspe (1737–1794), widely known as the author of the Baron von Munchhausen tales, was a clerk in the Royal Library of Hannover. His early interests in geology were stimulated by the great 1755 Lisbon earthquake (Carozzi, 1975). Influenced by Desmarest's 1768 article on volcanism in the Auvergne and by his own visit to Italy in 1771, Raspe became a convert to the idea of extinct volcanism. He later recognized that Habichtswald, near Cassel, Germany, had been the site of past volcanic eruptions. The area, Raspe (1771, p. 29) believed, had once been a burning volcano that had "at different times, and by different craters, vomited sand, ashes, brimstones, flags, and lavas," all of which he believed he found in the area. He commented that he, Desmarest, and Ferber were all convinced of ancient volcanism. Raspe suggested that lava was caused by "strong heating fermentation" and maintained that the hypothesis of the subterranean fire lacked sufficient evidence. He speculated that the mountains in Saxony were also of volcanic origin.

By 1780, therefore, the existence of several terranes of extinct volcanism throughout Europe was widely accepted. Even Richard Kirwan (1784), an opponent of most igneous theories, conceded in his *Elements of Mineralogy* that ancient extinguished volcanoes had been discovered in the inland parts of most countries. For decades to come, a trek to the extinct cones of the French *massif central* became virtually mandatory for those who were geologically curious. The discovery of extinct volcanic regions in France, Germany, and Italy led to the realization that volcanism was not a strictly contemporary phenomenon.

THE PROPOSAL THAT BASALT REPRESENTS SOLIDIFIED LAVA

Despite being extinct, the volcanic fields of south-central France and northeastern Italy appeared to be geologically more recent than the Primitive, Secondary, and Tertiary rocks. For the most part, the volcanoes and their eruption products had neither been substantially eroded nor buried beneath later deposits. As a result, the notion of relatively recently extinct volcanic fields met with general acceptance. A few eighteenth-century investigators precipitated controversy with the proposal that the rock type known as "basalt" had at one time been lava. The controversy was heightened because they found some basalt to be interlayered with Primitive, Secondary, and Tertiary rocks. The radical implication, therefore, was that some lavas might be very old.

Basalt was accepted as a common rock widely distributed throughout Europe. This dense, generally nonvesicular, compact, dark, finely crystalline rock, believed to be composed mainly of the minerals feldspar, augite, and hornblende, had long been known to students of nature. The term "basalt" had its roots in classical Greece and Rome and had been used by more recent writers like Agricola and Kentmann. Johannsen (1937) and Harrell (1995) discussed the etymology of the term "basalt." In the latter eighteenth century, the term "basalt" was frequently used interchangeably with different names. In Great Britain, very dense, slightly more coarsely crystalline basalt that occurs in sheets and dikes was designated as "whinstone" or "whin." Scandinavian naturalists and quarrymen referred to the rock as "trapp." Germans often referred to basalt as "Wacken," "Grünstein" (greenstone), or "Trapp." Today, of course, basalt is defined as a fine-grained crystalline rock that is composed essentially of the minerals plagioclase and pyroxene with lesser amounts of olivine, iron oxides, and glass, and both vesicular and nonvesicular varieties of basalt are recognized.

Many occurrences of basalt, most notably that of the Giant's Causeway along the northern coast of County Antrim in Ireland, are distinguished by the presence of great polygonal columns oriented at high angles to the layers of basalt. Long a cause for wonder and speculation, the Giant's Causeway had baffled many a brilliant mind. One early writer stated that "the *Giants Causway* of *Ireland* may very well be esteemed one of the greatest *Wonders*, Nature, or the first Cause of all things, has produced" (Molyneux, 1698, p. 210). Several papers about the strange formation, typically by medical doctors or Anglican ministers, appeared in *Philosophical Transactions* during the seventeenth and eighteenth centuries (e.g., Foley, 1694). Some observers had suggested that the great polygonal columns of basalt are giant crystals or possibly parts of corals. Molyneux (1694, pp. 176–177) took the trouble to explain at length why the causeway could not be a human production:

> there is not the least sign of Morter, or any equivalent Cement, to joyn the Commissures or Sides of the Columns together; . . . there are no foot-steps of the strokes of Tools or Chissels, in the Surface of any part of the Stone . . . If, I say, one will but a little consider these Circumstances, I am sure he can't imagine that Men could have the least Design in putting all this useless Lumber in this most wonderful manner together in so Remote and Desolate a place.

On the other hand, Molyneux (1694, p. 177) argued, there were many other natural productions that might not be exactly like the basalt columns at Giant's Causeway but so closely resembled them in composition and shape that both must be "the Architecture of the Regular Hand of Nature." Molyneux also likened the columns to a variety of stones such as *Entrochus* and *Astroites*, now recognized as types of echinoderms characterized by pentagonal symmetry. Reflecting the contemporary tendency to blur distinctions between the organic and inorganic realms as Rudwick (1972) has pointed out, he saw a similarity to plants such as reeds and, particularly, bamboo.

The solution to the enigma of the Giant's Causeway appeared on the horizon during the 1763 travels of Nicolas Desmarest through the Auvergne. As he left the city of Clermont-Ferrand to see Puy-de-Dôme, he encountered an abundance of columnar basalt that he designated "basaltes prismatiques." Desmarest immediately recognized the great similarity between these prismatic rocks and the famous formation of the "Chaussée de Geans" in Ireland. He proposed the astonishing claim that columnar basalt is a volcanic rock on the grounds that the prismatic formations could be traced to and found in lava currents that had issued from the extinct volcanic cones of the Auvergne (Desmarest, 1774). Desmarest also found basalt sheets capping hills and buttes that are not obviously connected to any volcano. He proposed that prismatic basalt, lava, scoria, and other volcanic materials stemmed from different epochs of volcanic activity in the region. The underlying primitive granite on which the puys were constructed, he suggested, was the source rock that was melted to yield the lava. Not until the twentieth century was this suggestion ruled out completely by the melting experiments of Greig et al. (1931).

Desmarest's compatriot Guettard had seen examples of these prismatic basalts during his initial foray around Clermont-Ferrand, but he failed to make any connection to the volcanoes. Because he discovered no reason to regard basalt as the product of fusion, Guettard restricted the assignment of volcanic origin to obvious lava flows and scoria. As a neptunist, he had long regarded basalt as the product of aqueous precipitation. During visits to volcanic terranes in Italy in 1771 and 1772, however, Guettard began to have doubts about the origin of basalt. These doubts were intensified during still later studies undertaken near Montelimar in Dauphiné. By the time he published his *Mémoire de la Minéralogie au Dauphiné*, Guettard (1779) had converted to the volcanic origin of basalt.

Although originally accepting the neptunist origin of basalt, Raspe also came out in favor of the idea that basalt is a volcanic rock after he read Desmarest's original report (Carozzi, 1969; Sigurdsson, 1999). Near Cassel, in lower Hesse, Raspe found layers of massive basalt lying above layers of columnar basalt, slabs of which the locals used for paving stones they called "Wacken." Raspe understood that this columnar basalt resembled the occurrences at the Giant's Causeway. Aware that Desmarest had found basalt columns toward the terminal margins of lava flows in the Auvergne, Raspe wrote to Sir William Hamilton in 1769 to ask if he had found any prismatic lava flows among the deposits of Mount Etna or of Mount Vesuvius (Carozzi, 1969). Hamilton responded that although he regarded basalt as a type of lava, he had not seen columnar basalt at Mount Vesuvius, Mount Etna, or on the island of Ischia. Raspe also learned from Ferber that he had never seen any old or recent flows of prismatic lava. On the basis of their testimony about contemporary activity at Mount Vesuvius and Mount Etna, Raspe (1776) concluded that subaerial volcanism does not produce prismatic flows. In contrast, Raspe pointed out, the prismatic basalts that occur at the Giant's Causeway, on the island of Staffa off the west coast of Scotland, on Iceland, and in the Veronese area of Italy are all very close to the sea. Consequently, he argued that submarine eruptions must have been responsible for the production of columnar basalt and that massive basalt was the product of subaerial volcanic eruption. In Raspe's view, either the Auvergne volcanoes were anomalous or they had erupted under the sea.

Raspe's friend J. J. Ferber also came to accept the volcanic theory for at least some occurrences of basalt. In 1768, Ferber had visited the German basalts with Raspe. At first, he was skeptical of their volcanic origin. For the next few years until 1773, he traveled throughout Europe, interrupting his travels for study at the Mining Academy at Freiberg and brief service as Assessor in the Royal Department of Mines. His visits to Mount Vesuvius and other ancient volcanic terranes of Italy rendered him more favorable to the vulcanist viewpoint. As previously noted, Ferber concluded that the hills of Padua, Vicenza, and Verona were formerly parts of volcanoes. He also recognized the basaltic nature of the hills and observed for the first time the close association of columnar basalt and of "rude, unformed, lava masses" (Ferber, 1776, p. 62). Ferber shrunk from drawing the general conclusion that all basalt was derived from volcanoes, conceding the possibility that some of it might have been formed by crystallization from water. That was the original opinion that he held upon seeing the German basalts a few years earlier. Now he confessed that he wished he could reexamine them. Puzzled as to why volcanism did not always produce prismatic basalt and skeptical that all basalt is a true lava, Ferber cautiously conjectured that nature worked the same effects by means of both fire and water. John Strange (1775) advocated the view that the common occurrences of prismatic basalt and granitic-looking rock in the Euganean Hills had formed from fusion.

Staffa, a tiny island in Scotland's Inner Hebrides just west of the Isle of Mull, was visited on August 13, 1772, during the Iceland expedition of Sir Joseph Banks. The island was just beginning to attract attention as the site of Fingal's Cave and the locus of spectacular basalt columns. Banks had written a description of the columns in *A Tour in Scotland, 1769* (Pennant, 1771) and concluded that the prismatic basalt of Fingal's Cave is volcanic. A few years later, the account of Iceland and its volcanoes by von Troil (1780) incorporated another description of Staffa by Banks. Von Troil thought that the pillars of Staffa were even more impressive than those of the Giant's Causeway. Von Troil (1780, p. 266) also noted that most mineralogists had considered the basaltic pillars "a kind of chrystalization." When Desmarest claimed a volcanic origin for columnar basalt, von Troil observed, many natural historians at first thought the idea absurd on the grounds that there had never been any volcanoes where the basalt pillars are found. But von Troil was convinced that the pillars of Staffa are the remains of an ancient volcano. He ventured the further opinion that no one could entertain the least doubt that subterranean fire had formerly operated in those places where the basalt pillars stand at Stolpenstein, in Lusatia, in Bohemia, in Hesse, in Sicily, in numerous localities in Italy, in Iceland, and in the western islands of Scotland.

Von Troil wrote that he kept in contact with the respected Swedish chemist and mineralogist Torbern Bergman (1735–1784) about the matter of these basalt columns. Bergman, successor to J. G. Wallerius as Professor of Chemistry at the University of Uppsala, had distinguished himself in the field of mineralogy by developing new methods in blowpipe analysis and by carrying out many qualitative and quantitative chemical analyses of minerals and rocks (Smeaton, 1970). Although generally a neptunist in his geological views, Bergman responded to von Troil in a letter on June 12, 1776, that he rejected the idea that the columns are crystals precipitated from water. Although uncertain about the details of the process, Bergman claimed that there were undoubtedly some connections between the pillars and the effects of subterranean fire inasmuch as the pillars "are found in places where the signs of fire are yet visible" (von Troil, 1780, p. 381).

Barthélemy Faujas de St. Fond, who had accepted the volcanic origin of basalt during his earlier studies of the extinct volcanic terranes of Vivarais and Velay, started out as a lawyer (Challinor, 1971). Under the influence of Comte de Buffon, he was granted a position in 1778 as Assistant Naturalist at the Muséum d'Histoire Naturelle in Paris. In 1785, he became Royal Commissioner of Mines and, in 1793, Professor of Geology at the museum, a post he held for the remainder of his life. In 1784, Faujas de St. Fond had the opportunity to travel through Great Britain. Unfortunately, the publication of the account of his travels was delayed until 1799 owing to the turmoil of the French Revolution (Faujas de St. Fond, 1799). Faujas de St. Fond had read Sir Joseph Banks' report on Staffa and was eager to see it for himself. Upon his arrival in England, he met with Banks and also made the acquaintance of John Whitehurst (1713–1788), another proponent of volcanic basalt. Accompanied by three other naturalists, Faujas de St. Fond set off for northern England, various parts of Scotland, and the Isle of Staffa. In all these areas he reported seeing lava, and he also unhesitatingly referred to black prismatic basalt as lava. He was mightily impressed by Staffa. As the sun was going down, Faujas de St. Fond (1799, p. 34) lamented that he hated to leave "a place which presented scenes so striking and volcanic phenomena so remarkable." He described Fingal's Cave as a "superb monument of a grand subterranean conflagration." Of the Isle of Staffa generally Faujas de St. Fond (1799, p. 43) enthused that "I have seen many ancient volcanoes: I have described and made known some superb basaltic causeways and fine caverns in the midst of lavas; but I have never found anything which comes near this one."

John Whitehurst, his new friend, was a practical natural historian who excelled as a clockmaker (Challinor, 1976; Craven, 1996). He had gained some attention with his studies of the geology of Derbyshire that were incorporated into a book entitled *Inquiry into the Original State and Formation of the Earth* (Whitehurst, 1778). In that text he came out for the volcanic origin of basalt. A second edition concluded with a chapter on the strata of northern Ireland and focused especially on the phenomena associated with the Giant's Causeway (Whitehurst, 1786). To Whitehurst it was evident that the abundant sheets of basalt and whinstone were so many ancient lavas. The highly vesicular lavas, he reported, contain abundant "bladder holes," some of which are partly or completely filled with zeolite minerals. The whole mountain above the Giant's Causeway, he said, was formed by several successive eruptions. Locally he found that coal had been converted into charcoal adjacent to contacts with basalt. He perceived this phenomenon to provide evidence of the origin of basalt in a state of fusion. Whitehurst proposed the modern view that the basalt columns formed by contraction upon cooling of a fused mass of liquid in the interior of the mountain. He was also impressed by von Troil's observation that Iceland abounds with basalt columns of various shapes and sizes although it is composed principally of lava. Untroubled by the lack of a visible crater in the vicinity of the Giant's Causeway "nor the least vestige of an extinguished volcano . . . from whence such immense torrents could have flowed," Whitehurst (1786, p. 256) postulated that the volcano had once existed toward the north but since sunk completely into the ocean. He even linked his extinct volcano with the sunken continent of Atlantis.

Jean-André Deluc (1727–1817), a respected Swiss geologist who left Geneva in 1773 for England, where he was soon appointed as reader for Queen Charlotte, a position that provided sufficient income to allow for numerous natural history tours of Europe, came to regard basalt as volcanic as early as 1778 (Beckinsale, 1971). A tour of the Auvergne by Dominique Reynaud, Comte de Montlosier (1755–1838), a French politican and mineralogist, convinced him of the volcanic origin of basalt (Montlosier, 1802). Dolomieu observed prismatic basalts among the ancient lavas of Mount Etna and described sheets of basalt interbedded with limestones at Val di Noto in Sicily which he interpreted as submarine volcanics. Other visits to Lisbon, Isle de Ponce, the Auvergne, and elsewhere in Europe contributed to his acceptance of the volcanic theory of basalt. He attributed the prismatic form of basalt to sudden contraction from the cooling effects of water. Jean-Claude de Lamétherie (1743–1817) was for many years a scientific journalist on the staff of *Journal de Physique* and Adjunct Professor of Mineralogy and Geology at the Collège de France (K. L. Taylor, 1973). He was another essentially neptunian writer who accepted the volcanic origin of basalt by 1795.

In the latter eighteenth century, a growing host of individuals, primarily from the British Isles, Scandinavia, France, Germany, and Italy, devoted themselves to the pursuit of natural history thanks to the leisure afforded them by virtue of their privileged positions as medical doctors, ministers, lawyers, mining engineers, or government officials. They communicated the fruits of their studies primarily in books of scientific travels and in the journals of learned societies in Paris and London. By the 1780s, most of those who had attained any expertise in geology typically recognized signs of extinct volcanism throughout Europe and believed that basalt is a volcanic rock. Although there is a common perception among contemporary geologists that neptunists were adamantly opposed to the volcanic origin of basalt, that certainly was not the case for many of them in the 1770s and 1780s. Most of the authorities noted above who adopted the volcanic origin of basalt were fundamentally neptunists on most geological questions. Despite their acceptance of the igneous origin of basalt, by no means had they yet perceived the pervasive role that igneous activity has played throughout Earth history. Neptunist opinion about basalt, however, was soon to change dramatically.

CHAPTER 2

THE WERNERIAN REACTION

Writing on the basalts of Saxony in 1803, the same year in which he began service with the mining administration of France, Jean François d'Aubuisson de Voisins (1769–1841) expressed considerable doubt about the applicability of the volcanic theory to that particular region (Birembaut, 1970a). After six years in military service and some time spent as a mathematics teacher, d'Aubuisson de Voisins studied for a couple of years at the Bergakademie, a highly regarded mining school in Freiberg, Saxony, from which he emerged as an adherent of the aqueous theory of the origin of basalt. In his 1803 book on Saxon basalts, d'Aubuisson de Voisins wrote that evidences for volcanism had been overestimated. According to the English translation, d'Aubuisson de Voisins (1814, p. 154) suggested that the vulcanist hypothesis was popular "because it delights the imagination to suppose the existence of former volcanoes on the spot of our peaceful habitations." He alleged that the volcanic theory was in decline because, although he could look back on a time a few decades earlier when all the mineralogists of Europe considered basalt as of undoubted volcanic origin, most of them had since abandoned that view. People like Kirwan, Dolomieu, and H. B. de Saussure, he claimed, had retreated from earlier vulcanist positions because they had "latterly been staggered by the arguments of the Neptunists" (d'Aubuisson de Voisins, 1814, p. 5). After his visit to the Auvergne in 1798, de Saussure, for example, supposedly could no longer bring himself to admit the volcanic origin of basalt. Assuming the validity of d'Aubuisson de Voisins's observations, what engendered the loss of nerve in viewing basalt as a product of volcanism?

During the final decades of the eighteenth century, two competing schools of thought arose to issue a severe challenge to the volcanic theory of basalt: Wernerian neptunism and Huttonian plutonism. Huttonianism asserted that basalt had solidified from underground lava rather than from lava erupted onto the surface. In contrast, Wernerianism categorically rejected any igneous origin for basalt, insisted that the rock was an aqueous precipitate, and fomented a furious debate about the nature of basalt. These controversies have been thoroughly discussed and analyzed, among others, by Adams (1938), Greene (1982), Hallam (1983), Laudan (1987), Fritscher (1991), Dean (1992), den Tex (1996), and Sigurdsson (1999).

At the very beginning of his professional career, Abraham Gottlob Werner (1749–1817), a brilliant and influential Professor of Mining and Mineralogy at the Bergakademie, had accepted the volcanic theory and regarded it as established (Ospovat, 1971). After a visit to the basaltic mountain at Stolpen, Saxony, in 1776, however, Werner (Figure 2) could not see any sign of its volcanic origin. He began to teach that not all basalt could be volcanic.

In his *Short Classification and Description of the Various Rocks* published in 1786, Werner put forward a classification of rocks that was based on the earlier

2. Abraham G. Werner (1749–1817). Reproduced by permission of the Bergakademie, Freiberg.

stratigraphic work of Johann Lehmann and Torbern Bergman as well as his own studies (Ospovat, 1971). He placed rocks into one of four groups: Urgebirge (primitive mountains), Flötz (stratified), volcanic, and alluvial. Werner plainly acknowledged the existence of true volcanic rocks, but, like virtually all of his contemporaries, he considered volcanism to be an accidental and recent phenomenon, not a major agent in the formation of Earth. His category of volcanic rocks included true lavas, pumice, volcanic ash, and solidified ash, which he also called volcanic tuff or "trass." He claimed that true lavas have a "blistered" appearance. Sulfur, hot springs, and steam vents were said to be associated with true volcanic rocks. Under the category of volcanic rocks, Werner also discussed pseudo-volcanic rocks such as earth slag, porcelain jasper, and half burned clays. He commented that "the number of volcanic rocks so far presented here [is] very reduced, perhaps to the great displeasure of many mineralogists and geognosts who have a passion for fire" (Ospovat, 1971, p. 84). Conceding nothing to the volcanic enthusiasts, Werner did not even consider basalt to be a pseudo-volcanic rock. He noted that basalt, porphyry, and amygdaloid had been treated as lavas by Italian, French, German, and English mineralogists. Through his own investigations into these rocks in Saxony and through reading reports about these rocks from experts in other countries, however, he became convinced that they occur everywhere under the same conditions as other primitive and Flötz rocks. Basalt, porphyry, and amygdaloid, he insisted, show nothing that would hint at an origin through heat.

The question of the origin of basalt could not be divorced in Werner's mind from his theory of rock classification founded on neptunist principles. Basalt was listed as one of the primitive rocks along with granite, gneiss, mica slate, clay slate, porphyry slate, porphyry, amygdaloid, serpentine, primitive limestone, quartz, and topaz rock. These oldest rocks, found in the cores of mountains, were typically crystalline and allegedly showed all the characteristics of aqueous formation. Werner believed that such rocks had formed by precipitation from a gradually receding global ocean that was a potent aqueous solution rich in dissolved chemicals.

Any indecision that might have remained in Werner's thinking about basalt was removed during a visit to Scheibenberg, about 50 miles southwest of Freiberg, in the Erzgebirge in 1787. At this locality, an erosional remnant of columnar basalt caps a small hill and rests on layers of "wacken," clay, and sand. Werner was unable to detect any sharp boundary between the basalt and the underlying rocks. Instead he found only gradation between rock types. He saw no evidence of baking of the rocks beneath the basalt as might be expected if hot lava had flowed over them. Shortly after making this discovery, Werner issued some brief articles in support of the aqueous origin of basalt (Werner, 1788).

Werner realized that basalt could not always be considered as a Primitive rock because the basalt lay above Flötz strata at the Scheibenberg locality. He also became aware of many other instances of the association of basalt with younger stratified, mechanical deposits belonging to his Flötz category. As Werner revised his classification, he eventually added a category of "transition rocks" between the Primitive and Flötz, and he also included basalt and trap rocks among both the Primitive and the Flötz rocks (Flötz-trapp) where he still considered it as an aqueous precipitate.

Because of Werner's reputation as an outstanding teacher and scholar, he attracted large numbers of students from all parts of Europe to study with him at the Bergakademie. Imbued with devotion to their mentor and fired with enthusiasm for his comprehensive geological system, these students, many of whom belonged to the very front rank of geologists, such as d'Aubuisson de Voisins, Alexander von Humboldt, Leopold von Buch, and Robert Jameson,

3. "Robert Jameson (1774–1854). Mineralogist." Reproduced from a lithograph by Friedrich Emil Ernest Theodore Schenck after a drawing by W. Stewart by permission of the Scottish National Portrait Gallery.

often carried the banner for the aqueous origin of basalt more enthusiastically than the master himself.

Robert Jameson (1774–1854), Regius Professor of Natural History at the University of Edinburgh and Keeper of its Natural History Museum, was the very embodiment of Wernerianism (J. M. Eyles, 1973a). The interest of Jameson (Figure 3) in Werner's views began in 1793 through contact with some of the latter's students. He eventually decided to study in Freiberg between 1800 and 1802. His establishment of the Wernerian Natural History Society in Edinburgh in 1808 was symptomatic of the grip that Werner's views imposed on him. Chitnis (1970) has presented evidence that Jameson became such a rabid partisan in the controversy among Wernerians, vulcanists, and Huttonian plutonists that he used the natural history museum, not as a forum for debate, but as a weapon in the scientific struggle between 1804 and 1831 when he was finally ousted from his position. Eminent scholars like David Brewster complained that Jameson restricted access to the museum. Only collections that supported Wernerian concepts were displayed. Boxes of specimens that had been donated to the museum by Hutton and other opponents of Wernerianism were never opened. As a result of Jameson's excessive zeal, Werner's views gained a foothold in Great Britain and even in America.

The third volume of Jameson's *System of Mineralogy* was essentially an exposition of Werner's latest classification scheme (Jameson, 1808). In this system, basalt was included in a number of categories because of the increasing recognition that basaltic layers occur in several sequences. The classification included not only Primitive trap, where Werner had originally placed basaltic rocks, and the "newest Flötz-trap," which he had added in response to his analysis of the Scheibenberg occurrence, but also a new category of "transition-trap" that incorporated "transition-greenstone" and "transition-amygdaloid." Jameson added a group entitled "Flötz-trap formation" that he distinguished from Werner's "newest Flötz-trap" because it was found in conjunction with the oldest Flötz limestone. Much like his mentor, Jameson regarded basalt as a combined chemical precipitate and mechanical deposit. So confident was he of the nonvolcanic nature of basalt and trap that he asserted with supreme confidence in his *Mineralogical Travels through the Hebrides, Orkney, and Shetland Islands, and Mainland of Scotland* that volcanic rocks had never been discovered in Scotland (Jameson, 1813). In reaction to the vulcanist views of Faujas de St. Fond expressed in *A Journey through England and Scotland*, Jameson (1813, p. 5) charged its author with "rigid adherence to a theory which has no foundation in nature" and repeated his claim that "there is not in all Scotland the vestige of a volcano."

THE HUTTONIAN SCHOOL

The other school of thought to dispute the volcanic origin of basalt was initiated by James Hutton (1726–1797), a Scottish chemist, gentleman farmer, and natural philosopher. Dean (1992) has produced a thorough study of Hutton's life and views on geology. Hutton (Figure 4) and his leading protagonists, James Hall and John Playfair, sided with vulcanists in rejecting the Wernerian claim of the aqueous origin of basalt and insisting on its origin by fusion. The Huttonians, however, also sided with the Wernerian neptunists in rejecting the volcanic theory of basalt. Basalt was, for Hutton, the product of solidification of subterraneous lava.

The term "lava" is, of course, restricted by contemporary geologists to molten rock that appears on Earth's surface. In contrast, molten rock material, including suspended crystals, gas bubbles, entrained rock fragments, and so on, that is underground is always referred to as "magma." In the latter eighteenth and early nineteenth centuries, however, underground molten material was typi-

4. "James Hutton (1726–1797). Geologist." Reproduced from a painting by Sir Henry Raeburn by permission of the Scottish National Portrait Gallery.

cally referred to as "subterranean" or "subterraneous" lava. The Greek word μαγμα, signifying "paste," had been employed in pharmacy and chemistry in the eighteenth century in reference to a viscous paste or gelatin. Dolomieu evidently borrowed the term and used it in a geological context. He described "magmas of mother-waters, reduced to a pasty state by evaporation" (Dolomieu, 1794, p. 193). A couple of years later, he wrote of "a crystal of niter formed in the pasty magma of a mother-water" (Dolomieu, 1796, p. 375). These earliest recognized geological appearances of the term "magma" were employed in a strictly neptunist context, not at all in reference to igneous melts! Not until the 1830s and 1840s would the French geologists Fournet (1838) and Durocher (1845) introduce the term "magma" in the context of igneous rocks.

Hutton (1788) wrote a synopsis of a grand theory of Earth in which he emphasized the critical role that heat had played in the historical development of the planet. His theory was elaborated in detail in the much expanded *Theory of the Earth* (Hutton, 1795). Consistent with his emphasis on the role of heat, Hutton unquestionably accepted the idea of extinct volcanic terranes and probably departed farther than most of his contemporaries from the notion that volcanism is a recent, accidental phenomenon. He regarded volcanism as a "general operation" of the globe and recognized that naturalists had discovered "the most undoubted proofs of many ancient volcanos, which had not been before suspected." Such proofs indicated, contrary to common belief, that volcanoes are not simply a matter of accident, but rather are "general to the globe, so far as there is no place upon the earth that may not have an eruption of this kind" (Hutton, 1788, p. 274). The purpose of volcanoes, said Hutton (1788, p. 275), is not simply to have a burning mountain, or to frighten superstitious people into fits of piety and devotion, or to overwhelm cities with destruction. Rather "a volcano should be considered as a spiracle to the subterranean furnace, in order to prevent the unnecessary elevation of land, and fatal effects of earthquakes."

Hutton argued that if volcanic action is a general operation, great quantities of melted matter should occur among the strata of the earth even where there are no visible marks of a volcano. And we do find abundant examples of the products of melted matter, Hutton maintained, such as trap, amygdaloid, "Schwartzstein," and whinstone like that at Edinburgh's Salisbury Crags. Hutton drew a careful distinction between these "subterraneous lavas" and the proper lavas that had issued out of a volcano. Lava and whinstone have the same origin and are composed of the same material, Hutton (1788, p. 280) said, but formed under different circumstances:

> the one has been emitted to the atmosphere in its fluid state, the other only came to be exposed to the light in a long course of time, after it had congealed under the compression of an immense load of earth, and after certain operations, proper to the mineral regions, had been exercised upon the indurated mass.

For Hutton, the interpretation of specific field evidence pertinent to a solution of the narrow question of the origin of basalt was embedded within a much larger theoretical framework. His proposal of a plutonic origin for basalt was in accord with his attribution of virtually everything significant in the history of Earth to the action of subterraneous heat. He claimed that sediments deposited on the seabed were consolidated into rock by fusion of individual mineral grains. Flint nodules in limestone and agate nodules in basalt were attributed to the effects of fusion. Consolidated sedimentary rocks were said to be elevated from the seabed and tilted by the expansive power of subterranean heat. Moreover, Hutton (1795) had postulated what seemed to be an unending cycle of revolutions of erosion and heat-driven land elevation.

The implications of Hutton's theory for the history of the planet were alarming to those like Richard Kirwan who believed that neptunist views of Earth history were more compatible with prevailing biblically based conceptions of the origin of the world than was Hutton's quasi-eternalism. As a result, scholars who saw scientific flaws in other aspects of Hutton's larger theory or were put off by its philosophical implications were understandably reluctant to cast their vote for the origin of basalt as a subterranean melt. Even apart from the philosophical implications of Hutton's theory, the narrow question of the origin of basalt was sufficiently ambiguous on strictly observational grounds to guarantee vigorous discussion for several decades.

The igneous origin of basalt has been accepted since the 1820s (Dean, 1992). The igneous origin of basalt was widely accepted in the 1770s and 1780s. The intervening prominence of the Wernerian view that basalt has an aqueous origin serves as a reminder that the dismissal of a widely held idea by the scientific community provides no guarantee that the idea may not be resurrected. Such an example should serve as a caution to igneous petrologists against excessive dogmatism. A hypothesis that is ferociously held and defended may end up being discarded in favor of one that had previously been rejected.

3 ⁕ FIRE OR WATER?

The Debate over the Origin of Basalt

Basalt is a fine-grained igneous rock that is composed predominantly of plagioclase, pyroxene, and olivine and that contains approximately 50 weight percent of silica. Basaltic rocks occur in great abundance as lava flows and in sills and dikes on the ocean floors, on oceanic islands, in continental volcanic plateaus, and in association with strato-volcanoes and cinder cones. Basalt has also been discovered in vast quantities in the lunar maria. Given this contemporary understanding of the chemical composition, mineralogy, texture, field relations, and tectonic setting of basalt, it is difficult for contemporary petrologists to appreciate fully why a dispute over the origin of this rock should ever have consumed so much time and energy. The fact that we now regard basaltic rocks, on compelling grounds, as both volcanic and intrusive (in dikes and sills), but certainly not as aqueous precipitates, must not lead us to adopt Lyell's historiographic reconstruction of Wernerian neptunists as stubborn impediments to progress in geology (Ospovat, 1976). At the stage of advancement of mineralogy, chemistry, technology, and theory at the close of the eighteenth century, the solution to the question of the origin of basalt was anything but obvious. At root, the debate hinged on how closely "basalt," "whin," or "trapp" were thought to resemble lava. John Murray (d. 1820), a lecturer on natural philosophy, chemistry, and pharmacy at the University of Edinburgh and author of *A Comparative View of the Huttonian and Neptunian Systems of Geology* (Murray, 1802), claimed that mere superficial resemblance between whinstone and lava is a very weak proof of similar origin (B.B.W., 1894). As a result, proponents of volcanic origin were eager to demonstrate as many resemblances as possible. Wernerians and Huttonians, for differing reasons, stressed the differences (Murray, 1802). Participants in the debate disputed the significance of a host of features, including the topographic position of basalt, its associated products, the form of its bodies, the nature of its contacts, its mineralogy, its internal structures, its inclusions, its chemical composition, and experimental studies on basalt and lava. Positions that were diametrically opposed developed in regard to every one of these aspects of basaltic rocks.

TOPOGRAPHIC AND GEOLOGICAL SETTING

Opponents of volcanic basalt stressed the fact that outcrops of basalt, porphyry, trap, or whinstone typically occur far from any obviously volcanic terrane. Richard Kirwan (1799), without even bothering to offer argumentation against igneous basalt, simply assumed what was to be demonstrated by declaring that wherever trap or basalt columns form the body of a hill or mountain, then the hill or mountain should be deemed neptunian.

D'Aubuisson de Voisins (1814), one of Werner's former students, approached his study of Saxon basalts with a distinct Wernerian bias. He also explicitly confessed that he had not visited any volcanic terranes, and only later, after studying the geology of the Auvergne in person, would he come to appreciate the force of many of the arguments for volcanic basalt. Earlier in his career, d'Aubuisson de Voisins maintained that basalt lies only on the summits of hills above all types of rocks, such as granite, gneiss, mica slate, sandstone, and gravel, and that it is unrelated to them. Such a situation, he said, was a strike against the hypothesis of volcanic origin. If the volcanic hypothesis were true, he argued, then each basaltic summit would have to be either the remnant of a volcano or else the remnant of one great lava stream. Basaltic summits were unlikely to be remnants of volcanoes because he saw evidence for neither craters nor the "utter disorder" of blocks, pumice, cinders, scoria, and lava associated with volcanoes. If basalt were truly the remnant of a lava stream, he reasoned, the lava should have burst out at the foot of the mountain instead of the summit because of the density of lava. Besides, he pointed out, the extensive mine workings within the region had failed to penetrate anything resembling the interior of a volcano.

D'Aubuisson de Voisins further argued that, if all the basalt cappings represented a single eruption, then the lava torrent must have covered a chain of mountains more than 100 miles long and 40 to 50 miles wide. Instead, he regarded the isolated summits as remnants of a large, continuous sheet of sediment that had been precipitated from a solution that had once covered the entire region. He envisioned that the sheet had been extensively eroded to its present topography.

Theological concerns were brought to bear on the question of the origin of basalt by William Richardson (1740–1820), a graduate and fellow of Trinity College, Dublin, who later became the Rector of Moy and Clonfele in County Antrim, home of the Giant's Causeway (G.S.B., 1896). It was in County Antrim that Richardson's interest in the basalt question was piqued, and he became an active participant in the debate, contributing several papers on the topic in *Transactions of the Irish Academy of Science*, one of which was directed specifically against Desmarest's memoir on prismatic basalt. As is often the case in the history of science, views on a scientific question are not always shaped by strictly scientific considerations. Part of the reason for Richardson's hostility to the notion of a volcanic origin for basalt was that he, as a minister, was suspicious of Desmarest's intentions in making observations that would "enable him to combat the account, given by Moses, of the creation of the world, particularly in its date" (Richardson, 1806, p. 36). Fearing that the concept of volcanic basalt might pose unwelcome implications regarding the age of Earth, Richardson issued a host of challenges to that concept, one of which was based on topography. He emphasized that volcanic countries consist of isolated conical mountains that typically have summits, and he could claim from his own experience that such topography simply did not exist along the Antrim coast of Ireland.

Even the Huttonian plutonists agreed with their arch-opponents, the Wernerians, that the topographic setting of basalt did not support the volcanic theory. One of the leading plutonists was John Playfair (1748–1819), Professor of Mathematics and, later, Natural History, at the University of Edinburgh, editor of *Transactions of the Royal Society of Edinburgh*, close friend of James Hutton, and author of *Illustrations of the Huttonian Theory of the Earth*, a popularization of Hutton's theory (Challinor, 1975). Although whinstone is neither volcanic nor aqueous, it certainly is of igneous origin, Playfair (1802) wrote in his *Illustrations*, and one of the evidences he gave against the volcanic view is that the shape of whinstone hills does not resemble that of volcanoes.

On the vulcanist side, Spallanzani (1798) conceded that some basalts might be aqueous but insisted that many others were certainly the product of heat. After all, he noted, basalt is common in volcanic countries. Other supporters of volcanic basalt responded to Wernerian and Huttonian arguments that even if basalt occurred in an area that no longer had obvious volcanoes in it, the volcanic landforms had been eroded away to leave behind remnants of former lava currents in the form of discontinuous basalt sheets.

Plutonists, too, appealed to erosion. When called upon to explain how sheets of solidified subterranean lava got to the surface, Playfair (1802) postulated the extensive removal of rocks that had originally been superimposed on sheets of whinstone. The solution to the problem of the topographic position of basalt was, therefore, intimately related to the geomorphic problem of the efficacy and extent of erosion (Davies, 1969), another problem that had not yet been solved. Even though a Wernerian like d'Aubuisson de Voisins, a Huttonian like Playfair, and a vulcanist like Faujas de Saint Fond all called for extensive denudation in their discussion of the basalt problem, not everyone at the beginning of the nineteenth century was prepared to concede the reality that vast amounts of erosive work had occurred.

Christian Leopold von Buch (1774–1853), for example, was among those unwilling to make that concession (Nieuwenkamp, 1970). Von Buch had studied at the Bergakademie with Werner for three years beginning in 1790 and even lived in his home. As a devoted disciple of Werner, von Buch became a dedicated neptunist. In time, he gradually abandoned some of his teacher's theories but only with extreme reluctance. Owing to his independent wealth, von Buch, who had served as a mine inspector and conducted geognostic surveys for the Prussian civil service before 1798, carried out extensive field research throughout much of Europe. His research in the Phlegraean Fields near Mount Vesuvius undermined his Wernerian commitments when he realized that there was no evidence that the volcanic activity was the result of burning of coal deposits, a leading point in Werner's theory of volcanism. In 1802, von Buch visited the Auvergne and experienced further weakening of his Wernerian attachments. He was sufficiently persuaded by what he saw in the Auvergne to concede both the reality of ancient volcanic activity and the idea that some basalt might be volcanic in origin. Von Buch (1802), however, continued to insist that the basalts in Saxony were aqueous deposits on the grounds that there are no nearby volcanic mountains to produce such a tremendous outpouring of the alleged lavas. Because the basalts commonly form caps on the local hills, he wondered why, if the basalt had been a lava, the lava would flow over large areas only to stop abruptly at the edge of a bluff to form the precipitous cliffs now marked by huge prismatic columns. Von Buch was not yet conceptually prepared to envision a large-scale erosional process that wore away very extensive basalt lava sheets and carved out extensive valleys between the now separate basaltic hills. Like many of his contemporaries, von Buch failed to appreciate the vast amount of time required for such extensive erosion to occur.

PRODUCTS ASSOCIATED WITH BASALT AND LAVA

Discussants debated whether or not obvious volcanic eruption products occurred with basalt. Opponents of the volcanic theory like d'Aubuisson de Voisins (1814) were quick to point out the ubiquitous association of glass, scoria, and ash with known lava currents. He maintained a neptunian origin for the basalts of Saxony because he could find no hint of scoria, lava, or any of the other usual products of known volcanism in the area.

William Richardson (1806) belabored the point that Faujas de Saint Fond's description of the volcanoes of Vivarais never mentioned basalt in association with scoria, that Sir Joseph Banks made no mention of scoria in connection with the columnar basalt on the island of Staffa, that Dolomieu reported no scoria with basalt near Lisbon, and that Strange, Ferber, and de Saussure mentioned no instance of scoria in association with columnar basalt in the localities near Verona and Vicentin. Moreover, Richardson claimed that he had never seen any scoria associated with the basaltic rocks along the Antrim coast. For him, the absence of scoria was a clear indication of the lack of volcanism.

To the contrary, vulcanists asserted that such materials *are* occasionally found in association with basalt. Von Troil (1780) reported the presence of lava directly above the pillared basalt on Staffa, and Spallanzani (1798) pointed out the existence of columnar basalt in the crater of Vulcano. Nevertheless, Wernerians were correct in noting that scoria and glass had not been frequently discovered alongside basalt, and they remained unconvinced that those few instances where lava was known to occur in association with basalt offered any proof of the igneous origin of basalt.

THE FORM AND ORIENTATION OF BASALTIC ROCK BODIES

Richardson (1806) insisted that lava currents are irregular in nature and noted that they were often described as confusedly arranged heaps. In contrast, he remarked, basalt is characterized by a consummate regularity in which each stratum preserves its own place within a sequence of strata. Moreover, he said, lava currents vary in thickness and are narrower and deeper in the vicinity of the crater from which they issued, but basaltic strata are uniformly thick throughout their extent. Currents of lava, he maintained, are never parallel to one another, whereas basalt strata preserve a steady parallelism. In addition, lava currents were said to have a layer of "vegetable earth" sandwiched in between them, whereas basalt layers do not, but pass directly into one another.

Playfair (1802) argued for the intrusive nature of whinstone by claiming that the beds above and below whin layers are exactly alike. Why should this be so, he asked, on neptunian principles? In direct opposition to Richardson's later assertion of the uniform thickness of basalt, he maintained that neptunists were unable to account for the variations in thickness and the irregularities of basalt layers in terms of chemical precipitation from superincumbent water.

Of crucial relevance, too, especially for the plutonists, was the orientation of whinstone bodies. All participants in the debate realized that basaltic rocks did not always occur in conformable sheets. How was one to account for crosscutting dikes and veins of whinstone and trap? The Wernerians interpreted dikes as vein fillings of chemicals that had been precipitated from an aqueous fluid that had penetrated into crevices from above. Faujas de Saint Fond (1799) interpreted dikes in the Scottish Hebrides as the result of lava currents flowing downward into crevices in the ground, and Dolomieu (1788) had similarly interpreted dikes in the vicinity of Mount Vesuvius. Early in his career, James Hall (1800), one of Hutton's good friends, also accepted that interpretation. Later he adopted the Huttonian idea that the lava had been injected upward into fractures from below. For Playfair (1802, p. 267), "the veins of whinstone which intersect the strata, are the completest proofs of the [plutonist] theory here given of these rocks, and the most inconsistent, in all respects, with the hypothesis of their volcanic origin."

NATURE OF THE CONTACTS

The dispute also revolved around whether the boundaries between basalt and adjacent rocks are transitional or sharp and whether they display evidence of heating. Werner's investigation of the Scheibenberg basalt moved him toward full-fledged neptunism. There, he had claimed, one could see a gradual transition from underlying wacke and clay into overlying basalt. He could find no sharp, well-defined contact such as one would expect if a lava current had flowed onto the wacke and clay or if subterranean lava had been injected along bedding planes of these sedimentary rocks. Said Werner:

> finding there a perfect transition from pure sand to clayey sand, from that to a sandy clay, and from the sandy clay, through several gradations, to a greasy clay, to wacke, and finally to basalt.... I was, I say, irresistibly led to the following ideas: this basalt, this wacke, this clay, and this sand, are all one and the same formation; that they are all the effect of a precipitation by the wet way in one and the same submersion of this country. (Werner, 1791, p. 415; translated by the author from French version in Playfair, 1802, p. 281)

In a display of nonpartisanship unusual for the times, Spallanzani (1798) accepted Werner's interpretation of the contact relationships at Scheibenberg. Nevertheless, he rejected Werner's conclusion that all basalt was water-deposited, maintaining that many basalts may have derived their origin from water and that many others were certainly igneous.

Playfair (1802), however, was considerably less generous toward Werner. He countered that Werner's gradual contact was imaginary. He insinuated that Werner had not been sufficiently careful in examining the contact and intimated that he had failed to chip off a sufficient number of pieces of fresh rock with his hammer. If only he had done that, Playfair suggested, Werner could have seen that there really is a distinct, well-defined boundary between the underlying sedimentary materials and the basalt layer. Playfair (1802, pp. 284–285) conceded that there might be instances of gradational contacts, but that was only because of the heating effect of molten whinstone. "It is certain," he said,

> that the basis of whinstone, or the material out of which it is prepared by the action of subterraneous heat, is clay in some state or other, and probably in that of argillaceous schistus. It follows, of consequence, that argillaceous schistus may by heat be converted into whinstone. When, therefore, melted whinstone has been poured over a rock of such schistus, it may, by its heat, have converted a part of that rock into a stone similar to itself; and thus may now seem to be united, by an insensible gradation, with the stratum on which it is incumbent.

Wernerians also appealed to the absence of contact metamorphic effects as evidence that basalt is neither volcanic nor plutonic. Richardson (1806), for example, noted a complete lack of calcination of limestone adjacent to basalt in Antrim. Even though there are fragments of limestone embedded in the basalt, he argued, there is no sign of calcination in the fragments such as one finds in a known volcanic area like Mount Vesuvius.

Although admitting that he had not witnessed any volcanic eruptions, d'Aubuisson de Voisins pointed out that he had seen subterranean fires caused by the burning of thin beds of coal. He testified that he had seen coal fires "vitrify clays and earthy substances; melt the surface of masses of quartz: and roast ironstone, causing the reduced metal to flow out" (d'Aubuisson de Voisins, 1814, p. 20). If burning coal beds could produce such striking effects, he reasoned, it would be impossible for a thick stream of lava to flow over a bed of bituminous matter without radically altering it. And yet d'Aubuisson de Voisins maintained that wherever he saw basalt resting on other rocks he never perceived anything to indicate that those rocks had been covered by a stream of melted matter. Even where basalt lies immediately over beds of coal, he said that he observed no discernible effects of fire at the contact even though coal should necessarily be very sensitive to the action of intense heat. He concluded that the basaltic rocks he had observed had not resulted from subterranean fire.

Those who favored an igneous origin, however, were quick to point out the effects of baking at the contacts of whinstone. Playfair (1802), for example, observed that coal layers are altered to coke or charcoal next to whinstone contacts. Moreover, he suggested, disturbances of rocks adjacent to veins of whin were further proof of the original fluidity of the material in the veins.

John Pinkerton (1758–1826), a prolific author of poetry, Scottish history, and antiquaries in general (T.F.H., 1896), assessed the entire debate in a book entitled *Petralogy: a Treatise on Rocks*. Although the title of Pinkerton's book is spelled slightly differently, his is the first known usage of the term "petrology." It is ironic that the term was coined by a historian rather than a natural philosopher or geologist. Although Pinkerton's prolix volume dealt primarily with the classification of rocks, it was rarely considered very seriously by his contemporaries owing to his evident lack of geological knowledge. Nevertheless, he effectively summarized the dispute over contact phenomena. Coal, said Pinkerton,

> when in contact with what are called whin-dykes . . . is often observed to be decomposed; having lost its bitumen, and wearing the appearance of being charred. The Neptunists say, that the stone has absorbed the bitumen; while the Plutonists affirm that the melted stone, ejected from beneath, has caused the bitumen to evaporate. (Pinkerton, 1811, p. 251)

Not even contact phenomena lent themselves to unequivocal interpretation.

MINERAL CONTENT

Mineralogical observations provided one of the more hotly contested aspects of the dispute. Although hornblende is no longer regarded as an essential mineral in basalt, all parties at that time thought that basalt consisted of feldspar, augite, hornblende, and iron oxide. For the most part, the disagreements centered on the role of incidental minerals. Many participants in the debate believed that basalt is a dense, compact rock that commonly contains nodular clumps of zeolites, calcareous spar (calcite), quartz, and agate, whereas they thought that true lava is typically vesicular or scoriaceous and never contains calcareous spar or zeolites. James Hutton (1795), for example, said that he asked Sir William Hamilton if he had ever found veins and nodules of calcareous spar in the lava flows of Mount Vesuvius. Hamilton replied that he had not. For Wernerian neptunists and Huttonian plutonists alike, this fundamental mineralogical difference was sufficiently great to warrant the claim that the two lithologies could not have the same origin. Those who favored a volcanic origin, however, believed that basalt and lava did have the same origin and had an explanation why one rock contained zeolite, calcareous spar, or agate and the other did not.

Among the Wernerians, d'Aubuisson de Voisins (1814) noted that basalt contains crystals of hornblende, olivine, augite, mica, or feldspar. He argued that such crystals could not have crystallized from flowing lava because lava is too viscous for particles to form crystals. One would certainly not expect feldspar to crystallize from molten basalt, he said, because feldspar is known to be more fusible than basalt. Moreover, if a piece of feldspar had been picked up by flowing basaltic lava, it would have been melted. The presence of feldspar crystals in basalt seemed to him to be incompatible with the notion that basalt had once been molten like lava. Moreover, he pointed out that the lavas of Mount Vesuvius and Mount Etna lacked the balls of zeolite, calcareous spar, chalcedony, and quartz that were commonly found in basalt.

Richardson (1806) was convinced that basalt was never in a state of fusion on account of the mineral substances contained within it. Zeolite, he said, fuses in a moderate heat, calcareous spar calcines in moderate heat, and chalcedony loses its beauty in strong heat.

On the vulcanist side, Spallanzani (1798) reported that zeolites *are* frequently found in volcanic countries. He described beautiful quartzose bodies that were formed in lavas along the shores of Lipari. He concluded from his investigation in the Euganean Hills that the zeolite, calcite, and agate in basalt had formed when mineral-laden waters infiltrated the vesicles in lava and precipitated those substances. The zeolites did not derive their origin from heat after all, he asserted. They are, rather, "adventitious" to those localities. Nor were these minerals, he claimed, lying on the ground prior to volcanic eruptions and then later picked up by and incorporated into lava flows. Instead, Spallanzani maintained, zeolites were formed after the eruptions became extinct when their constituent parts were deposited by water in the cavities of the lava. He believed that the aqueous derivation of the zeolites was proved by their abundant water of crystallization.

Dolomieu agreed with Spallanzani that the zeolites and chalcedony were the result of aqueous deposition of those minerals within the vesicular cavities of already congealed lava flows. He, however, claimed that zeolite formation demanded submergence of the lava flows beneath the sea and that the chemicals requisite for zeolite crystallization were drawn from the seawater.

While the plutonists agreed with Spallanzani and Dolomieu that basalt is igneous, they were less excited about their neptunist appeals to an aqueous process to account for zeolitic production. In diametrical opposition to Spallanzani, James Hall (1800) stated that calcareous spar is never found in lava. He did concede a close resemblance between the vesicles in lava and the cavities left in whinstone after calcite nodules had been washed out by rain. Nevertheless, Hall labeled this close resemblance a "deception." Despite the resemblance, he said, the nodules of calcareous spar in whinstone were produced by a cause akin to what would today be designated liquid immiscibility, namely, the mutual repulsion of two fluids that were intermixed and not disposed to unite. Hall proposed that the calcareous compound, although molten, kept separate from the whinstone, "as oil separates from water through which it has been diffused, thus giving rise to the spherical form, which the nodules of calcareous spar generally exhibit with more or less regularity" (Hall, 1800, p. 62).

Playfair (1802) also disputed Spallanzani's claim that calcite crystals are the fillings of vesicles in lava. He argued that if the sites of zeolite and calcareous spar had originally been holes in lava, then these holes would have been unable to support the weight of the lava and should have closed up. By default, calcite and zeolite must be original crystallization products. He concluded that "the spar, then, may be considered as a proof, that the rocks in question are to be numbered with those unerupted lavas which have flowed deep in the bowels of the earth, and under a great compressing force" (Playfair, 1802, p. 262). Nor, Playfair argued, can the agate nodules in basalts have an aqueous origin. He failed to see how water could pass through solid basalt into holes to deposit anything, nor how water that might have originally been trapped inside a growing but hollow agate nodule could ultimately escape back out through the solid basalt.

The plutonists all argued that zeolite, calcareous spar, and agate solidified from fused melt along with all the other minerals in the once liquid basalt melt. Given that lava normally did not have calcareous spar, zeolite, or agate, they had to explain why such minerals developed in basalt but not in lava. Their consistent refrain was the difference in pressure under which the two igneous rocks formed. They argued, given that lava solidified on Earth's surface under atmospheric pressure, that calcareous spar would be unable to form. Upon cooling, the "fixed air," that is, carbon dioxide, component of calcareous spar would escape and be unable to combine with the lime or calcareous earth in the melt. The calcareous earth would, therefore, enter into other minerals in the cooling melt. In contrast, under the pressure existing at depth in the "mineral regions" the "fixed air" would be unable to escape and thus could combine with calcareous earth to form the calcareous spar. Hutton (1795) went so far as to say that a calcareous lava, once relieved from high pressure, would vigorously effervesce as a result of the explosive release of the fixed air. This process would result in the violent eruption of a volcano and the emission of much pumice and ash.

Wernerians were understandably skeptical of such arm waving in view of the complete lack of knowledge of what would actually happen underground. Thus John Murray (1802), who felt that lava might simply be melted basalt, responded to Playfair that if trap were fused in an open volcanic fire to become lava, then carbonic acid would be expelled from it and calcareous spar would be missing from the lava. Murray also doubted that the absence of compression would prevent agate formation in lava because, unlike calcareous spar, agate contains no volatile matter that could escape under low pressure conditions. Better, he thought, to maintain that agate owes its formation to the agency of water.

The mineralogical disputes were not confined to zeolite, calcareous spar, and agate. Inasmuch as many scholars had attributed volcanic action to the combustion of subterranean pyrite deposits, Playfair (1802) claimed that the presence of pyrite in basalt must be a mark of its origin by fusion. Murray (1802) claimed exactly the opposite. Because pyrite is more likely to have formed through an aqueous agency, he said, so is the whin that contains it. Werner claimed that, because hornblende occurred in basalt, it must be a product of aqueous precipitation rather than crystallization from melt. He reasoned that any hornblende found in lava must be a refractory mineral left over when the lava was melted out of a preexisting hornblende-bearing basalt layer.

INTERNAL STRUCTURES

As might be expected, the spectacular polygonal columns of Staffa, the Giant's Causeway, and other basalt occurrences continued to play an important role in the discussion about the origin of basaltic rocks. In a letter to von Troil written in 1776, Torbern Bergman suggested that basalt columns are analogous to the shrinkage cracks that develop during the thorough drying out of originally moist sediments (von Troil, 1780). He expected that similar shrinkage phenomena would form in basalt that was drying out after it had been precipitated from an aqueous solution.

After asserting that the neptunian origin of numerous columnar basalt occurrences had been demonstrated by several writers, Kirwan (1799) conceded that perhaps some columns had been found in association with lava. These columns, however, had probably been tossed out of a volcano in the same manner as other neptunian stones are sometimes ejected. In any case, Kirwan claimed that the columns supposedly found in lava are never erect.

Murray (1802), too, thought that columnar structure was at least very rare in lavas and was dubious about most of the occurrences for which columnar structure in lava had been claimed. For those rare instances where genuine columns might exist in lava, he suggested that the columnar structure in both basalt and lava was simply a function of the nearly identical chemical composition of the two rocks. Murray was also open to the shrinkage-crack idea, thinking that columnar structure might be due largely to the predominance of clay in the rocks inasmuch as many other argillaceous rocks, some of which had plainly never been melted, commonly contain shrinkage cracks.

Richardson (1806) insisted that nothing similar to the prisms of basalt was ever observed in genuine lava and that the restriction of these columns to basalt constituted a significant reason why basalt should not be considered as lava. He regarded cavities containing freshwater within the basalt columns as evidence for the neptunian origin of basalt. Such water must be coeval with the time of formation of the basalt and cannot be ascribed to an igneous origin, he insisted, because the water cannot have entered the solid rock by percolation at a later time.

Opponents of neptunism countered that columnar structure should be nearly as common in lava as it is in basalt given their similarity of composition. Desmarest, Dolomieu, and Playfair pointed to numerous occurrences where lavas erupted from known volcanoes do contain columnar structure. According to Spallanzani (1798), Dolomieu attributed the prisms to the sudden chilling and contraction of lava within the sea. With this opinion Spallanzani was in agreement, but he was not ready to insist that all prismatic lava required submarine development. On the one hand, he observed, some lavas do enter the sea without the development of prisms, and on the other hand, he wondered why sudden coagulation and contraction of lavas could not occur subaerially. Moreover, Spallanzani was not ready to concede that basalt was necessarily always volcanic in origin simply because some lavas contain articulated prisms. Maintaining that Swedish basalts are aqueous, he continued to insist that nature could achieve the same effect in two different ways.

INCLUSIONS AND FOSSILS

One of the more persuasive lines of evidence for the Wernerians was the abundance of organic remains that presumably had been found in basaltic rocks. According to Robert Jameson (1802b, p. 17), Werner's view of rock deposition made it evident "that we may expect to meet with vestiges of many of the organic and unorganic matter which at that time existed upon the crust of the earth, in the rocks of this formation." Jameson reported that Werner had discovered in "wacken" the horns of deer as well as remains of trees, including branches, leaves, and fruit. De Saussure was said to have discovered bones of quadrupeds in basalt. Jameson also said that he had seen a large number of occurrences of shells, *cornua ammonis* (ammonites), vegetable remains, and other organic materials in basalt. His listing was repeated by Murray (1802). D'Aubuisson de Voisins (1814) noted additional instances where petrifactions such as *chamites, cornua ammonis, gryphites*, shells, trees, bark, branches, and leaves allegedly occurred in a variety of basalts throughout Europe.

As a specific example, Richardson (1805) described organic remains at Portrush along the Antrim coast of northern Ireland. The paper, read before the Royal Society of Edinburgh in 1803, discussed "siliceous basalt," a rock said to be composed of strata of uniform thickness and constructed of large polygonal prisms. The beds of siliceous basalt were described as fine-grained and "remarkable for containing marine exuviae in great abundance, particularly impressions of *cornua ammonis*" (Richardson, 1805, p. 15). The flat shells, Richardson reported, are perpendicular to the axis of the prisms. These strata of siliceous basalt, he said, alternate with strata of coarse-grained basalt. Richardson was aware that several mineralogists denied that the shell-bearing stone is actually basalt and that others strenuously contended that it is. In the discussion of Richardson's paper, opponents commented that most of the alleged "siliceous basalt" did not resemble basalt. They asserted that the undisuputed, coarse-grained basalt contains no vestiges of animals. Moreover, they claimed, veins issue from the real basalt and penetrate the alleged siliceous basalt, but they said they had never seen veins proceeding from the siliceous basalt into the regular basalt. A year later, even d'Aubuisson de Voisins conceded that the shells at Portrush are not found in the basalt but in a rock resembling slate-clay. Despite these criticisms, in a later paper Richardson (1806) persisted in his rebuttal of the volcanic origin of basalt by maintaining that the basalt of Portrush along the Antrim coast was full of *pectinites, belemnites*, and *cornua ammonis* dispersed through the whole mass.

To the circulating neptunist claims, Playfair (1802) responded that it was unlikely that shells would be found in basalt. What likely happened, he suggested, is that shells might have been caught up into basalt where fossiliferous sediments at the contact with the basalt had been incorporated into the latter. In the case of pyritized ammonites along the Antrim coast, later reported by Richardson, Playfair suggested that the rock containing the shells is a "schistus" or "stratified stone" that served as the base onto which the basalt was injected. The shell-bearing rock then became extremely indurated by the intense heat of the melted mass of whinstone. In short, he intimated, the altered rocks adjacent to igneous basalt took on a basalt-like appearance and became difficult to distinguish from it. Murray (1802) conceded that Playfair's idea had merit in some cases but countered in a blatant appeal to authority that naturalists were not at liberty to suppose that Werner and de Saussure had committed so obvious a mistake in their observations. As a result, their observations were to be considered as decisive proof of the aqueous origin of the rocks.

Rock inclusions, too, entered into the argument. Playfair (1802, p. 72), for example, in arguing for the intrusive origin of basalt described "pieces of sandstone, or of the other contiguous strata, completely insulated, and having the appearance of fragments of rock, floating in a fluid sufficiently dense and ponderous to sustain their weight." Neptunists countered that inclusions of rock are only an illusion. If we could but have a three-dimensional exposure of these fragments, they contended, we would see that they are not completely detached and isolated at all. They only appear to be so in a two-dimensional cross section.

CHEMICAL COMPOSITION

In the late eighteenth century, chemical analyses of rocks were extremely difficult to obtain. Analytical determinations were tedious because of the chemical complexity of the rocks and because of the difficulty of dissolving them. Not until Robert Kennedy read a paper before the Royal Society of Edinburgh in late 1798 did the chemical composition of basalt and lava become a factor in the debate about their origin. Kennedy (1805) reported that he had undertaken the painstaking chemical analysis of several samples of lava and whinstone on which Sir James Hall had also performed melting experiments. These included a piece of basalt column from the island of Staffa and pieces of whinstone from Salisbury Crags and Calton Hill in Edinburgh. Kennedy also analyzed two samples of lava erupted from Mount Etna near Catania and at St. Venere, Piedmonte. Of particular importance were Kennedy's detection of a substantial amount of soda in all the rocks and the recognition of the close chemical similarity between the whinstones and the lavas. The range of values obtained for the five analyses were "silex" (silica), 46–51 weight percent; "argil" (alumina), 16–19 weight percent; "oxyd" of iron, 14.25–17 weight percent; lime, 3–10 weight percent; soda, 3.5–4 weight percent; and muriatic acid, about 1 weight percent. The major difference between whin and lava, Kennedy found, was in the amount of moisture and other volatile matter given off during initial heating of the sample. Otherwise, Kennedy concluded, whinstones and some lavas consist of the same component elements in nearly in the same proportions.

Kennedy's demonstration of the chemical similarity of basalt (whinstone) and lava was seized upon enthusiastically by igneous advocates as yet another evidence that basalt had an origin in fusion. But Wernerians retorted that such chemical similarity would be expected if lava originated by the melting of buried layers of basalt. Indeed, John Murray (1802) had previously argued that lava was simply remelted basalt. And d'Aubuisson de Voisins, too, had earlier suggested that the lavas of Mount Etna so closely resembled basalt because of melting of the basalt on which the volcano was situated.

Despite the virtual identity in chemical composition of lava and basalt, all other evidences were still said to point to aqueous beginnings by the Wernerians. Murray even turned Kennedy's discovery of soda in the rocks to neptunist advantage by pointing out that the presence of alkali in the whinstones could have contributed to the solubility of the constituents of whinstone in water.

EXPERIMENTAL EVIDENCE

For several decades, a number of experimenters, like Réaumur, Pott, and Keir, had been heating silicate materials in furnaces and examining the products. Heating experiments, therefore, showed promise for shedding some light on the problem of the origin of basalt. James Keir (1735–1820), for example, educated at the University of Edinburgh, was a pioneer industrial chemist who had developed the first successful commercial process for the synthesis of alkali (Scott, 1973a). Between 1771 and 1778, Keir managed a glass factory. During this period his investigations into the crystallization of glass led him to wonder if the great "crystals" of basalt that form the Giant's Causeway or the pillars of Staffa had been produced by the crystallization of a vitreous lava (Keir, 1776).

Despite Keir's musings, Wernerians objected to the igneous theory of basalt on the grounds that experiments showed that melted samples of basalt were converted into glass upon cooling rather than into a stony substance resembling the original basalt. They reasoned that the texture of basalt could not properly be attributed to cooling from melted material.

To investigate this matter further Sir James Hall (1761–1832) decided to carry out a series of experiments (V. A. Eyles, 1972). Hall, the son of a wealthy Scottish landowner, inherited a generous fortune and was thereby free to pursue his own interests. These interests included geology and chemistry, topics that came alive for Hall following his formal education during a three-year tour of Europe that began in 1783. The tour included meetings with eminent natural philosophers and visits to sites of volcanic activity in Italy, Sicily, and the Lipari Islands. Upon his return to Edinburgh, Hall became a member of the circle of scientific minds that included chemist Joseph Black, Hutton, and Playfair. Hall wondered what basalt would look like if it were cooled slowly after fusion. He knew of one report that accidental slow cooling of green bottle glass led to crystallization, but he decided to carry out his own experiments. Although Hall undertook his experiments in an effort to test and possibly confirm Hutton's views regarding subterranean lava, he received little encouragement from Hutton. Hall (1800) himself later pointed out that Hutton (1795, p. 251) had censured those who would "judge of the great operations of the mineral kingdom, from having kindled a fire, and looked into the bottom of a little crucible."

Undaunted by the skepticism of his friend, Hall began his experiments in 1790 by melting and cooling whinstone from a quarry near Edinburgh. Rapid cooling resulted in the formation of black glass. Hall then attempted a slow cooling experiment by rapidly removing a crucible with a quantity of molten whinstone from a reverberatory furnace to a large open fire surrounded by burning coals. The fire was maintained for several hours and then allowed to go out. He reported breaking open the crucible after it had cooled and finding that it contained a substance that differed in all respects from glass and completely resembled whinstone in texture. To forestall the objection that he had never destroyed the original whinstone texture during fusion, Hall first made a glass, then reheated and slowly cooled the glass. Following this procedure, Hall again succeeded in producing a crystalline texture.

Buoyed by his success, Hall further experimented with other samples of coarse-textured whinstone from Edinburgh Castle, Arthur's Seat, Salisbury Crags, and the water of Leith in Edinburgh, Duddingstone Loch, and the Island of Staffa. He obtained similar results. For comparison, Hall executed more experiments on six lava samples from Mount Etna, Mount Vesuvius, and Iceland. He achieved the same results as before and concluded that "the stony character of a lava is fully accounted for by slow cooling after the most perfect fusion . . . . Whin and lava . . . agree so exactly in all their properties which we have examined, as to lead to a belief of their absolute identity" (Hall, 1800, p. 60). As a plutonist, he was of the opinion that arguments against the idea of subterraneous fusion of whinstone had been completely refuted by his experiments.

Hall's close friend, Playfair (1802), of course, agreed, and predictably endorsed the results of Hall's experiments with enthusiasm. But neither Hall's experiments nor Playfair's endorsement were sufficient to convince the advocates of aqueous basalt. They replied that the experiments, like the identity in chemical composition, demonstrated that basalt could very likely have served as the source for the lava erupted at known volcanoes like Mount Vesuvius or the ancient volcanoes of the Auvergne. Either some inherent principle caused melting of the basalt, they suggested, or subterranean coal beds or pyrite deposits had ignited to melt the basalt and erupt the lava.

Murray (1802) did graciously concede that Hall's work had refuted the standard neptunist objection that if basalt were volcanic it would be a glass rather than a crystalline rock. He did not, however, consider that Hall's experiments provided proof of basalt's igneous origin. Because compounds derive their properties from their chemical composition, he reasoned, if their constituent parts could be united both by aqueous crystallization and by fusion, then a similar compound would be formed in both cases. Consequently, Murray claimed, the actual production of such a compound by one of these processes would not prove that it could not also be formed by the other process.

Further experiments on basalt were attempted by Gregory Watt (1777–1804), son of James Watt and graduate of Glasgow College (de Beer, 1957). Stimulated by Hall's work, Watt tried to duplicate Hall's results on a huge scale by melting several hundred pounds of basalt in an ironworks furnace (Watt, 1804). By cooling the melt at different rates, Watt was able to produce a series of products from glass to material containing spherical clumps of fibers to a more stony-textured substance. In the end, Watt concluded that the results could be compatible with either aqueous or igneous origin for basalt.

Different conceptions of the relative importance of heat ("fire") and water in Earth history as a whole; different conceptions of the location of volcanic heat, whether shallow or deep-seated (den Tex, 1996); different conceptions of the extent of erosion, in turn dependent on different conceptions of the length of Earth history; different views on the significance of experiments to geology; and the widely varying field experiences of the participants in the basalt controversy all contributed to the confusion and ambiguity in the interpretation of the origin of basaltic rocks. The fact that the investigation of such relatively fine-grained rocks was limited primarily to the unaided eye, the hand magnifier, and the hammer also hampered progress toward resolution of the issue. Only patient accumulation of more chemical and mineralogical information and, particularly, more careful field work and repeated exposure of more and more geologists to volcanic terranes would eventually tip the scales to a recognition of the igneous character of basalt.

4 ❖ THE "INSOLENTLY TRIUMPHANT DOGMA"

The Collapse of the Concept of Aqueous Basalt

The reasons for the surprising tenacity with which the concept of basalt as an aqueous precipitate was adopted by many geologists during the latter eighteenth and early nineteenth centuries are many. As Rachel Laudan (1987, p. 181) has pointed out with regard to basalt as interpreted in the late eighteenth century, "neither its place in the succession, nor its mineralogy, nor its stratigraphic relations, nor evidence from contemporary volcanoes decisively indicated its origin." As we have just seen, inherent ambiguity characterized the several lines of evidence available to participants in the basalt controversy.

The famous assertion of H. H. Read (1957, p. 3) that "the best geologist is he who has seen the most rocks" inevitably comes to mind in connection with the gradual resolution of the controversy. The literature written during the height of the furor contains numerous candid admissions that the writers were interpreting the geological phenomena in terranes that they had never seen. In some cases they were interpreting phenomena they had never observed in any terrane. A comment by the Reverend William Richardson in his lengthy article on the volcanic theory of basalt is representative. In a criticism of Desmarest's vulcanist view Richardson (1806, p. 55) wrote with words as sharp as a fresh splinter of basalt:

> As Mr. Desmarest has been pleased to give us so much information about our county of Antrim which he never saw: I think the least I can do is, to return the obligation, and give him some information, relative to his Auvergne, which I have never visited.

Richardson also confessed that he had never seen the basalts of Sicily, but that deficiency did not prevent him from disputing Dolomieu about their origin.

Desmarest and Richardson had company. Werner, of course, was long known to have had limited field experience outside Saxony. D'Aubuisson de Voisins (1814) confessed that he had never had the opportunity to witness an active volcano and expressed a desire some day to visit some volcanoes and their eruption products. He also wanted to see the basalts and extinct volcanoes of the Auvergne and the Vivarais. He conceded that he might then be better fitted to consider the entire basalt question and to judge properly what had been written about the subject.

Even antiquarian John Pinkerton, though not a geologist, recognized that firm geological convictions ought not to be uttered without the requisite field experience. He chided the French for being too much attached to the volcanic theory and the British for being too violently prejudiced against it. Above all, Pinkerton (1811, p. 274) mocked the "puerile ideas of those Wernerians who have never visited volcanic countries" and attributed the works of nature to "a few beds of coal!"

THE DEFECTION OF WERNER'S STUDENTS

Increased exposure to volcanic terranes played a critical role in the conversion or partial conversion of several of Werner's own Bergakademie students away from neptunist ideas about basalt. Several of them, unlike Werner and some of his most vehement advocates like Jameson, sooner or later spent time in the Auvergne, that "veritable graveyard of the Neptunian opinions" according to Adams (1938, p. 238), and came away persuaded of the existence of extinct volcanism and the volcanic nature of at least the basalt of the Auvergne. Among the earliest of these converts was Leopold von Buch, one of the premier geologists of his era.

In 1797 and 1798, von Buch traveled throughout Germany, visited extinct volcanic terranes in northeastern Italy, and spent several months around Mount Vesuvius. His commitment to neptunist thought at this time was transparent in the first volume of a work that described his travels (von Buch, 1802). He reported that evidence everywhere was directly opposed to the idea that basalt was erupted in molten condition or that basalt hills marked the site of volcanoes. At Mount Vesuvius, however, he looked in vain for extensive deposits of coal or pyrite that Wernerians maintained should have fueled the volcanic eruptions. As a result, some doubts about the neptunist view of basalt started to enter his mind. A trip to the Auvergne in 1802 sufficed to convince him of the volcanic nature of basalt at that location. In the Auvergne, he concluded again that volcanism was not fueled by the burning of coal because the volcanoes are situated on top of basement granite. Von Buch did persist in believing, however, that Saxon basalts were aqueous on the grounds that not enough erosion had occurred to decimate any volcanoes that would allegedly have produced the basalt. In 1805, von Buch and Alexander von Humboldt returned for a visit to Mount Vesuvius and witnessed an eruption. Stimulated by von Humboldt's subsequent studies of South American volcanoes, von Buch decided to investigate the geology of the Canary Islands. During his 1815 visit, he concluded that the islands had experienced considerable volcanic activity, and it was here that von Buch (1825) became fully persuaded of the volcanic nature of basalt.

Von Buch's friend Alexander von Humboldt (1769–1859) was arguably the most illustrious of all of Werner's students (Biermann, 1972). After his education at Frankfurt and Göttingen, von Humboldt studied at the Bergakademie in 1791 and 1792, and, like von Buch, lived for awhile with Werner. Even before entering the Bergakademie, von Humboldt (1790) had written a small volume challenging volcanic theories of basalt. After a few years as a mining director, he devoted his life to world travels and geographic studies. In 1799, he set sail with botanist Aimé Bonpland for Central and South America, where his geological studies focused predominantly on volcanoes and earthquakes. He returned to Europe in 1804 and, as noted, witnessed the 1805 eruption of Mount Vesuvius. The widespread distribution of volcanoes convinced von Humboldt of the importance of the action of heat and led him to reject the view held by Werner that volcanic phenomena were of small importance and developed only where localized coal deposits had ignited. He came to accept the igneous origin of basalt.

The year after his work on Saxon basalts was published, the wish of d'Aubuisson de Voisins to see the Auvergne was fulfilled. As a result of his visit, he came to regard the puys of that terrane as extinct volcanoes and the basalt of that area as lava. Although d'Aubuisson de Voisins conceded the volcanic origin of the Auvergne, he steadfastly maintained the neptunist origin of the basalts of Saxony. Pinkerton (1811), however, said that d'Aubuisson de Voisins told him that, after his visit to the Auvergne, he had already begun to have second thoughts even about Saxon basalts. The complete conversion of d'Aubuisson de Voisins (1819) became known to the public in his general geological text, *Traité de Géognosie*, in which he regarded the basalts of Saxony and the Auvergne to be identical.

Another disciple of Werner to depart from neptunist concepts was mineralogist Christian Samuel Weiss (1780–1856). Weiss graduated from the University of Leipzig in 1796 after studying medicine (Holser, 1976). He further pursued doctoral work in chemistry and physics at Leipzig and was admitted to the faculty in 1800. Before settling into his work, however, Weiss spent two years in Berlin learning mineral analysis. Here he became acquainted with von Buch who urged him to study with Werner. Weiss, therefore, spent a year in Freiberg, and toured Europe visiting other prominent mineralogists, including René Haüy. After 1803, Weiss taught chemistry, physics, and mineralogy at the University of Leipzig, but, thanks again to the influence of von Buch, he became Professor of Mineralogy and Curator of the Mineralogical Museum at the brand-new University of Berlin in 1810. He ultimately served as Rector of the university for five years. In Berlin Weiss taught several future illustrious mineralogists and developed such crystallographic concepts as axes, systems, zones, and Weiss parameters, a system of notation for specifying the orientation of crystal faces and planes. Weiss visited the Auvergne a short time after von Buch and was similarly persuaded of the volcanic character of the phenomena.

The defections were not confined to those who studied directly under Werner. Even his acquaintances, including the leading petrographer of the early nineteenth century, Karl Caesar von Leonhard (1779–1862), drifted away (Burke, 1973). Von Leonhard was educated at Marburg and later at Göttingen, where he studied under Johann Blumenbach, who stimulated his interest in minerals. Marriage put a halt to plans to study with Werner. Instead, von Leonhard entered the Hessian government service in 1800 and occupied a number of important posts, managing to pursue his interests in mineralogy in his spare time and maintaining correspondence with Werner, von Buch, and others. In 1807, he founded the Wernerian-oriented journal *Taschenbuch für die gesammte Mineralogie*, the forerunner of *Neues Jahrbuch für Mineralogie, Geologie, und Paleontologie*. In 1818, von Leonhard became Professor of Mineralogy and Geognosy at the University of Heidelberg and initiated the long-continued distinction of Heidelberg in the field of petrography, the study of rocks. Here he authored a host of technical and popular texts on geology, the most influential of which was *Charakteristik der Felsarten* (von Leonhard, 1823).

In 1803, von Leonhard became acquainted with basalt as a result of travels through Saxony and Thuringia. He published his neptunist views in a three-volume *Handbuch einer allgemeinen topographischen Mineralogie* (von Leonhard, 1805–1809). He later turned his attention specifically to basalts and volcanic rocks. His studies included visits to the Auvergne, Bohemia, and other volcanic districts. Thanks to his exposure to such volcanic centers, he swung over to the volcanic theory and announced his change of mind in *Zur Naturgeschichte der Vulkane* (von Leonhard, 1818). The conclusive demonstration of the volcanic origin of basalt in his *Die Basaltgebilde* (von Leonhard, 1832) served as a death knell for the neptunist idea of the origin of basalt throughout Europe.

THE DEFECTION IN THE BRITISH ISLES

A major contributor to the collapse of neptunism was George Poulett Scrope (1797–1876) (Anonymous, 1870; Rudwick, 1974; Page, 1975). Scrope, a graduate of the University of Cambridge, became fascinated by the continuous activity of Mount Vesuvius during a visit to Naples while an undergraduate. He returned for further study of the volcano in 1819, and between 1819 and 1823 he had opportunity for extended study of Mount Etna, the Lipari Isles, the Auvergne, the Euganean Hills, and the Eifel district. In October 1822, Scrope witnessed a catastrophic explosion that reduced the elevation of Mount Vesuvius by 600 feet and mightily impressed upon him the role of volatile material as a driving force in volcanic eruptions. He became an ardent vulcanist as a consequence of these field experiences and published a controversial volume on volcanoes in which he pointed out that modern volcanoes are far more numerous and widespread than was generally accepted (Scrope, 1825). This work was followed by a memoir on the extinct volcanoes of central France (Scrope, 1827). During the 1820s, Scrope was closely allied with Charles Lyell as an adherent of uniformity of causation in nature. After 1830, he became increasingly involved in the political arena and was elected to Parliament in 1833, where he served for the next 35 years. During this period his writings focused primarily on political economy. From time to time, however, he contributed geological papers and published revised editions of his earlier books on volcanoes.

In the preface to the second edition of his memoir on the volcanoes of central France, Scrope (1858, p. v) recounted that during the winters of 1817 through 1819 when he witnessed the volcanic phenomena of Mount Vesuvius, Mount Etna, and the Lipari Islands, "the doctrines of Werner were then so completely in the ascendant that it was considered little better than heresy to dispute any of them." In the light of defections on the part of several of Werner's own students and friends shortly after the beginning of the nineteenth century, Scrope's assertion seems somewhat exaggerated as far as Germany is concerned. The question may be raised, however, as to how the concept of the aqueous origin of basalt fared in the British Isles at that time.

Wernerianism certainly had ardent Scottish and Irish advocates like Richard Kirwan, John Murray, William Richardson, and Robert Jameson. Jameson's students at the University of Edinburgh, such as Ami Boué, Charles Darwin, Charles Daubeny, and Edward Forbes, were treated with generous doses of neptunism. It might seem, then, that Wernerianism was sufficiently alive and well in Britain by 1819 to provoke Scrope's comment. In 1819, however, Darwin and Forbes were only young children who had not yet sat at Jameson's feet, whereas Boué and Daubeny, who had already been his students, were on the verge of recanting their neptunist beliefs. Moreover, a host of British geologists other than Hutton, Hall, and Playfair had already rejected neptunism.

Although he conceded that Wernerianism was on the decline after 1819, Scrope's assertion about its "complete ascendancy" seems as overblown regarding Britain as it does regarding Germany. One wonders if Scrope (1827) exaggerated the strength of Wernerianism in order to inflate the importance of his own contribution to the discussion of the volcanoes of the Auvergne. Scrope plainly considered that his own effort had put the nails in the Wernerian coffin. He claimed that after the publication of that memoir,

> the Wernerian notion of the aqueous precipitation of "trap" has since that date never held up its head. And I had good grounds for believing that the publication of the first edition of this Memoir . . . did assist in no small degree in putting an extinguisher on the then fading and flickering, but shortly before insolently triumphant, dogma. (Scrope, 1858, pp. ix–x)

From the knowledge of igneous rocks that he had acquired in Italy, Scrope concluded that the dogma that denied a volcanic origin to basalt, clinkstone, and trachyte and affirmed them as precipitates from the ancient ocean was decidedly in error. In so writing, Scrope (1858, p. vi) noted that no less than Adam Sedgwick was in full agreement with him that the Wernerian propensity for "despising" the influence of volcanic forces was "a fatal bar to the progess of sound geological science" that needed to be removed. By no means, however, were Sedgwick and Scrope lone beacons in a presumed dark night of British neptunism. A host of British geologists, including adherents of Wernerian stratigraphic concepts, were constantly discovering evidence that many basalts represents former lava flows.

One such person was Sir George Steuart Mackenzie (1780–1848), who had studied mineralogy and geology with his friend, Robert Jameson (W.A.S.H., 1893). Mackenzie attracted the attention of the scientific community in 1800 by demonstrating the identity of diamond with carbon. In a series of experiments, Mackenzie successfully produced steel by combining iron with diamonds, purportedly taken from his mother's fine jewel collection! Mackenzie devoted his life to several scientific subjects, including geology, agriculture, and phrenology. In 1810, Mackenzie visited Iceland and the following year issued a book, *Travels in the Island of Iceland* (Mackenzie, 1811), describing the abundant volcanic features. In 1812, he visited the Faeroe Islands along with his friend, mineralogist Thomas Allan (1772–1833). This brief visit convinced him that the Faeroes, too, were the scene of abundant volcanic activity (Mackenzie, 1815). He was impressed by rocks that he called "amygdaloid," "tuffa," and "trap-tuff" that bore signs of origin by fusion. On his first visit to the Island of Naalsoe, Mackenzie observed the surface of a bed of amygdaloid that had been partially exposed by the removal of the bed above, thus displaying an exact replica of the lavas he had seen in Iceland. This discovery, he reported, forced an instantaneous conviction on his companions, none of whom had ever seen lava before, that heat must have caused the features they saw. The hand specimens that he brought back, he said, spoke a language that was "not to be misunderstood." These beds, he thought, had probably flowed in a molten condition on the bottom of the sea. He particularly regarded the successions of beds of trap or amygdaloid superimposed above one another as a fact that the theory of submarine volcanoes adequately explained.

In a paper accompanying Mackenzie's in the same issue of *Transactions of the Royal Society of Edinburgh*, Thomas Allan, namesake of the mineral allanite and possessor of probably the finest mineral collection in Scotland, described the mineralogy of the Faeroe Islands (Anonymous, 1885). He provided an account of some of the magnificent amygdalular crystal groups of stilbite, apophyllite, chabasie (chabazite), calcite, native copper, and mesotype (natrolite). Although claiming that he planned to describe his findings "without relation to theory" and modestly indicating that he did not want to embarrass himself by reducing observed phenomena to an existing theory, Allan concluded his paper in full agreement with Mackenzie that the facts encountered in the Faeroes removed any remaining doubts regarding the igneous formation of trap. "No production of a furnace can tell its tale in plainer language," Allan (1815, p. 265) argued, "nor any slag bear more distinct marks of the effects of heat."

Another skeptic regarding the Wernerian view of basalt was Robert Bakewell (1768–1843), a mineralogical surveyor who attained his reputation primarily as the author of the first modern textbook of geology (Hansen, 1970). *Introduction to Geology* (Bakewell, 1813), widely hailed for its clarity of presentation, went through five editions by 1838 as well as German and American editions. As early as 1813, Bakewell believed that Hutton's view of basalt had been fairly well established and grumbled that neither Werner nor his neptunist disciples bothered to visit active volcanoes.

During the 1810s, several individuals studied the geology along the northern coast of Ireland. Among them were the young geologists Buckland and Conybeare. During the summer of 1813, the Reverends William Buckland (1784–1856), then Reader in Mineralogy at Oxford, and William Conybeare (1787–1857) undertook a tour of the Antrim and Derry coastline along the north of Ireland, the region already well known as the site of Giant's Causeway and other notable occurrences of trap and basalt described by Richardson and others. Conybeare (1816) described the observations of their tour. The son of an Anglican minister, Conybeare was educated at Christ Church, Oxford, and then occupied a series of ecclesiastical posts, ultimately becoming Dean of Llandaff Cathedral in Cardiff, Wales (Rudwick, 1971). He published in the field of biblical and patristic theology, and yet, throughout his career, Conybeare was also active in geological studies and was a Fellow of both the Geological and Royal Societies of London. One of the leading diluvial catastrophists of the Oxford school, his interests lay primarily in the fields of stratigraphy and paleontology, and he gained considerable fame as co-author with William Phillips of *Outlines of the Geology of England and Wales* (Conybeare and Phillips, 1822).

Conybeare's early study in Ireland, however, exposed him to a very different kind of geology. After dutifully recording the bare phenomena pertaining to basalt, about halfway through the paper he inserted a nearly two-page footnote in which his interpretive convictions came pouring forth. Conybeare (1816, pp. 208–209) said that he had "studiously refrained from expressing the views which I have been led to form on the origin of basalt," but in the midst of his description of the Kenbaan cliffs, he could no longer resist the need to "express my full assent to the arguments of those who maintain the igneous origin of such formations." The formation, he said, possessed characters that were "directly opposed to those which all rocks undoubtedly of aqueous origin possess." He asserted that he thought that it was "impossible to conceive appearances more utterly irreconcilable with the hypothesis, that the basalt was deposited regularly above the chalk from a state of aqueous solution." In contrast, he said, if one imagined a priori the phenomena that would result from an eruption of a lava flow that covered the upper surface of the chalk while submerged beneath the sea, then these phenomena "would exactly accord with those which may actually be observed at Kenbaan." All along the coast he saw "negative evidence against the Neptunian hypothesis" as well as "positive in favor of the volcanists." Conybeare was especially impressed by the repeated evidences of severe alteration of Old Red Sandstone into hornstone, slate clay to flinty slate, coal to cinders, and chalk into granular marble. He suggested that the hypothesis that ascribed the formation of the Flötz trap rocks to ancient submarine volcanoes was more consistent and more satisfactory than any other. He also recognized that a flinty slate containing pyritized *cornua ammonis* had been identified as a variety of basalt by neptunists like Richardson. Conybeare concurred with Playfair that this "basalt" was only a rock that had been altered by its contact with hot basaltic lava.

The very highly respected Scottish geologist, John Macculloch (1773–1835), also entertained doubts about the Wernerian view of basalt (V. A. Eyles, 1973b). Macculloch, trained in medicine in Edinburgh, probably developed an interest in geology as a result of hearing the lectures of John Walker (1731–1803), Robert Jameson's predecessor as Professor of Natural History at the University of Edinburgh. After a stint as a military surgeon, Macculloch served as a chemist with the Board of Ordnance. In 1811, he completely abandoned his medical practice and performed tasks for the Board of Ordnance requiring geological knowledge. He also served as Geologist for the trigonometric survey of Scotland. In 1816, he became President of the Royal Society of London. Even after his employment with the Board of Ordnance was phased out after 1826, Macculloch continued independent geological mapping of Scotland. The first geological map of Scotland was published in 1836, shortly after his death.

Macculloch's survey work took him to the Hebrides. After extensive studies, he published an authoritative three-volume *Description of the Western Islands of Scotland* (Macculloch, 1819). As deeply eroded volcanic complexes, Skye, Mull, and the other Hebridean islands are, of course, full of basaltic rocks, providing Macculloch with an unparalleled opportunity to reflect on the problem of their origin. He asserted that he had found fossiliferous strata that had been altered by basalt but had never observed basalt that was fossiliferous and went so far as to conclude that basalt and lava are virtually identical without explicitly committing himself to the igneous view.

Charles Giles Bridle Daubeny (1795–1867), son of an Anglican clergyman, wanted to practice medicine (Thackray (1971). To prepare for that profession, he attended the University of Oxford where he took lectures in chemistry. At Oxford he met Buckland and Conybeare and developed an interest in geology. Between 1815 and 1818 he undertook further medical studies at the University of Edinburgh, where he also attended Robert Jameson's lectures. After completing medical studies in 1821, Daubeny practiced medicine until 1829. During this period, in 1822, he was also elected as Professor of Chemistry at Oxford. In 1834 the chair of botany at Oxford was added to his resumé, and in 1840 the chair of rural economy as well. Despite these professional commitments, Daubeny made numerous contributions to geology, particularly in the field of volcanology.

In the introductory remarks to his *A Description of Ancient and Extinct Volcanos*, Daubeny (1826) commented that when he was a student under Jameson in 1816, he was a convert to Jameson's neptunist position on the origin of basalt. Still, he confessed that despite his respect for Jameson and deference to his views, he "never rose from the enquiry without a conviction that something was yet wanting to compleat the chain of his proofs" (Daubeny, 1826, p. 4). It seemed to Daubeny that the most effective method of determining whether or not trap rocks were really of igneous origin would be to undertake a thorough comparison of trap rocks and basalts with products that were universally acknowledged to be volcanic. As a result, Daubeny set out to study firsthand as many of the world's known volcanoes as possible or else to glean information from those who had recently visited such localities. Although he was successful in visiting many volcanoes, including the obligatory pilgrimage to the Auvergne, Daubeny still had to rely upon detailed published descriptions of volcanoes in the Americas, Africa, and the Pacific.

In the decade between his studies with Jameson and publication of his book on volcanoes, Daubeny became a convert to the igneous origin of basalt. Intimations of that conversion appeared in a series of articles on the volcanoes of the Auvergne that he published in *Edinburgh Philosophical Journal* (Daubeny, 1820, 1821a). Daubeny had toured the French volcanic districts in the summer of 1819. Amid his descriptions of the various domes, craters, cinder deposits, and lavas, Daubeny took particular note of Puy Graveneire near Clermont-Ferrand, an extinct cone whose currents of scoriaceous lava he found to be intimately intermingled with compact, black basalt. Daubeny also noted that Puy Graveneire was not the only site at which vesicular lava was associated with basalts just like those in Scotland (Daubeny, 1820). In the sequel article, Daubeny (1821a) laid out his suspicions that the older trap rocks of the Auvergne would also prove to be of volcanic origin. By the time Daubeny (1826) issued his book on volcanoes, his conversion was essentially complete. He conceded the virtual identity of the mineralogy and chemical composition of lava and basalt and attributed the differences in texture to the application of higher pressure to the solidification of basalt, quite possibly as the result of submarine eruption.

Scrope's claim that, before 1819, "the doctrines of Werner were then so completely in the ascendant that it was considered little better than heresy to dispute any of them" was, therefore, not an accurate reflection of the actual state of affairs. Numerous leading British geologists did not hesitate to engage in the heresy of disputing Werner's doctrine of basalt. Moreover, the abandonment of Werner's views on basalt continued unabated in the British Isles after 1819. But what of France?

FRANCE

The vulcanist school had its beginnings in France. Desmarest, Faujas de Saint Fond, Montlosier, Montet, and others had all been adherents of the concept of extinct volcanism and the idea that basalt was virtually identical to lava. Although d'Aubuisson de Voisins was a Wernerian, he later converted to the igneous origin of basalt. In the years around 1819, other French scholars were also disputing the Wernerian conception of aqueous basalt.

One major French contribution to the understanding of the origin of basalt was made by Pierre-Louis-Antoine Cordier (1777–1861). Cordier, for whom the mineral cordierite was named, took his education at École des Mines, became a mining engineer, accompanied Dolomieu to Egypt and traveled throughout Europe before continuing his career as an Inspector of Mines (Burke, 1971). In 1819, Cordier became Professor of Geology at the Muséum d'Histoire Naturelle, and in 1832 he was President of the French Conseil des Mines. Cordier had plutonist leanings, attributing the action of volcanoes to Earth's internal heat.

In 1816, however, Cordier published the results of one of the more important studies on basaltic rocks. Recognizing that the controversy over basalt had gone on for decades without a satisfactory resolution, Cordier (1816) wanted to perform some definitive comparative mineralogical tests. He began by examining crushed powders of minerals from definite volcanic lava as a basis for comparison with the minerals in basalts. He distinguished pyroxene, feldspar, peridot (olivine), "titanium iron" (ilmenite), amphibole, mica, (presumably biotite), "amphigene" (leucite), and "oligist iron" (hematite) under the microscope on the basis of color, cleavage, and texture. After successfully recognizing the different mineral species in powdered form, he subjected them to blowpipe tests to determine their fusibilities in terms of the Wedgwood scale. He ranked the minerals in order of decreasing fusibility: amphibole, feldspar, pyroxene, titanium iron, oligist iron, mica, amphigene, and peridot. Similar tests were also performed on basalts and on lavas from extinct volcanoes. Cordier effectively demonstrated the near identity of mineralogy between lava and basalt. Despite concluding that basalt was of igneous origin, he decided not to propose a change in the name of the common volcanic mineral, pyroxene, meaning "stranger to fire," and he urged his readers to get used to the idea that volcanic "fires" are radically different from those generated by humans in furnaces.

The studies of Boué proceeded along geological rather than mineralogical lines. Amédée Boué (1794–1881), widely known as Ami (the diminutive form of Amédée) was a native of Hamburg, Germany (Birembaut, 1970b). After being orphaned at the age of eleven, he was raised by relatives in Geneva and then Paris. He began the study of medicine in 1814 at the University of Edinburgh, where his interest in geology was stimulated by Jameson. After receiving his medical degree in 1817, Boué continued medical studies in Paris, Berlin, and Vienna, but he devoted his time increasingly to geology. He traveled extensively throughout Europe and published descriptive books on the geology of both Scotland and Germany. In 1830, he moved to Paris, where he became a co-founder of the Societé Géologique de France. He served as its president in 1835.

In a book on the geology of Scotland (Boué, 1820a) and in one of his early geological articles, written upon his return from the Auvergne, Boué offered a comparison of volcanic rocks in France with similar rocks in Scotland. He pointed out the close resemblance of many of the basalts in Scotland and Ireland with the Auvergne rocks. Some Scottish basalts, he noted, could not be distinguished from some of the rocks near Clermont-Ferrand. Boué noted that several Scottish greenstone occurrences have no obvious counterparts in the Auvergne, yet he refrained from considering these greenstones as candidates for a Wernerian origin in spite of writing the article as a letter to his former teacher Jameson, then editor of *Edinburgh Philosophical Journal*. Quite the opposite. Boué (1820b, p. 331) observed that knowledge of volcanic rocks was still sufficiently imperfect that "we cannot maintain that all these . . . are of Neptunian origin."

Boué proposed arranging Scottish and Irish trap rocks into three categories: "evidently volcanic," "probably volcanic," and "doubtful rocks." In the first category he included the Calder basalt between Edinburgh and Glasgow and the Giant's Causeway. In the second category he included the columnar basalts of the Isle of Staffa, some basalts on the Isle of Mull, Arthur's Seat in Edinburgh, the trap rocks on the Isle of Arran in the Hebrides, and several other localities. Among rocks in the doubtful category he included the rock beneath Edinburgh Castle, the greenstone beneath Stirling Castle rock, Salisbury Crags in Edinburgh, and some other greenstones and amygdaloids. Despite his uncertainty about a number of occurrences, Boué's concession on the volcanic origins of several others is significant in view of his training as a neptunist.

Two years later, Boué (1822) published in the *Memoirs of the Wernerian Natural History Society*, of all places, an article on German geology with emphasis on the "igneous origin of trap." He now was more willing than ever to include numerous basalts and traps as igneous rocks. Some he regarded as volcanic and some, he thought, were even injected from below.

A host of European geologists were abandoning the Wernerian view of the origin of basalt throughout the early decades of the nineteenth century for a variety of reasons, several of which have been discussed by den Tex (1996), who also observed that most former neptunists had defected within a decade of Werner's death in 1817. Conybeare and Phillips (1822) noted that the geological evidence was overwhelmingly in favor of the igneous origin of basalt. In an article in *Edinburgh Philosophical Journal*, Boué (1825) summed up the situation by noting that the majority of geologists, especially the ones who had visited extinct volcanoes, believed in the igneous origin of Tertiary basalt. He further observed that many distinguished geologists considered it probable that the Secondary or Flötz trap rocks were also of igneous origin. In the same issue, Necker (1825) wrote that the long-running controversy between the neptunists and the vulcanists over the origin of basalt appeared to be drawing to a close. Even Robert Jameson could not hold out forever. He finally admitted that Hutton had correctly pointed out that chemical changes in some rocks had been brought about by plutonian agency (Jameson, 1833).

The battle had been long and taxing. By the early 1820s, however, basalt by whatever name, whether trap, whinstone, greenstone, or basalt, was firmly established as a rock of igneous origin. Scrope's 1827 work on the volcanoes of the Auvergne and von Leonhard's 1832 book on basalts only served as the culmination of a revolution in thinking about this once enigmatic rock. The geological community not only accepted the existence of extinct but relatively recent volcanic activity in places like the Auvergne, but also began to realize that igneous activity had occurred throughout the Tertiary ages and possibly even in the Secondary ages. They realized that volcanic activity not only was very widespread at present but had been in past geological ages as well inasmuch as Secondary or Tertiary basalt had been found all over Europe and elsewhere. Volcanism could no longer be regarded as an accidental, unimportant, highly localized phenomenon restricted to the present; it now had to be considered as a significant process throughout much of geological history. Moreover, the common lack of association of volcanic rocks with coal indicated that the driving force of volcanic action was not the combustion of local coal deposits just beneath a volcano (den Tex, 1996; Sigurdsson, 1999). Rather, a much deeper, more fundamental cause lay behind, and beneath, volcanic eruptions. By the 1820s, the foundation for a science of igneous petrology had been laid inasmuch as the existence of former volcanic rocks in the geological record had been satisfactorily demonstrated.

5 ⟡ SUBTERRANEOUS LAVA
The Recognition of Intrusive Granite

By the end of the 1820s, geologists had almost universally accepted the idea that the rock record, particularly Tertiary sequences, contains remnants of ancient lava flows, chiefly basaltic in character. They had also recognized that there is some degree of variability in the mineral constitution and chemical composition of recent and ancient lava flows and also of trap and whinstone formations. Because many of the ancient lava flows, for example, those in the Auvergne, are not composed of basalt in the strict sense, the term "trachyte," coined by mineralogist René Haüy around 1811, was increasingly used by geologists to refer to rocks somewhat lighter in color, less dense, richer in feldspar, and poorer in augite than basalt. Consequently, acceptance of the volcanic origin of basalt was simultaneously accompanied by acceptance of the volcanic origin of "trachyte."

Not only did geologists accept the notion that basalt and trachyte represent volcanic lava flows, but they also began to recognize that trap dikes and many sheets of basalt were formed from subterranean lava that had been injected into cross-cutting fissures or between already existing sedimentary layers while still underground. Experiences like those of George Mackenzie in the Faeroe Islands convinced geologists that at least some basaltic rocks were injected. Mackenzie was able to distinguish different kinds of basalt layers. Although some basaltic rocks might be attributed to submarine volcanic eruptions, he reasoned that such a theory could not be extended to the beds of trap that are interspersed between strata of sandstone because these beds commonly display indications of baking of the adjacent rock at both the upper and lower contacts. Such traps, Mackenzie (1815, p. 225) suggested, have been "thrust among the strata from below, and have probably been the cause of the strata being elevated above the sea, in which they were formed."

Earlier, when geologists widely believed that volcanic activity was a relatively recent and localized phenomenon, they did not expect to see evidence for what we now regard as widespread subsurface intrusive activity that had subsequently been exposed by erosion. In a sense, acceptance of basalt as a volcanic product made it easier to accept basaltic trap as an intrusive igneous product, because the recognition that basalt is a very widespread rock type that can be found even in some Tertiary rock sequences made it clear that volcanism was more than an accidental occurrence. As a result, one might now expect that, to get to the surface to be erupted, large quantities of lava must have migrated underground. It made sense, then, to think that some of that lava had never made it to the surface at all and had frozen underground. Moreover, the growing awareness of Earth's antiquity during the latter eighteenth and early nineteenth centuries made it much easier to accept the idea of extensive erosion that would expose some of these formerly underground bodies of once-molten material.

The discussion about the existence of intrusive, subterranean lava, however, was by no means confined to dikes and sheets of whinstone and trap. Plutonists had something much bigger in mind. As proponents of the idea of Earth's central heat, they conceived that the phenomena associated with subterreanean lava must be very widespread and included even such coarsely crystalline rocks as granite, composed primarily of quartz, feldspar, and some mica or hornblende, and syenite, composed mainly of feldspar and hornblende. Although geologists were finally ready in the 1820s to acknowledge the evidence for the intrusive origin of some basaltic rocks, the acceptance of the entire plutonist worldview was quite limited. Hutton, for example, wanted to make a sharp distinction between basalt and lava, but after his death geologists began arriving at the conviction that many basalt layers were once lava flows. Thus, although geologists might be willing to be plutonists on some points, they increasingly wanted to be vulcanists. By arguing for the intrusive origin of granite, plutonists managed to attract even greater hostility for their views. Acceptance of intrusive granite by some geologists would ultimately come, but again the plutonist victory was limited. As Dean (1992, pp. 188–189) pointed out in his review of the work of James Hutton, "once basaltic dykes and sills had been recognized as igneous, the acceptance of igneous granite soon followed, though never so completely."

HUTTON AND THE NOTION OF INTRUSIVE GRANITE

The work of Pierre Simon Pallas, Giovanni Arduino, Horace Benedict de Saussure, and other eighteenth-century investigators demonstrated that the cores of mountain ranges on which the stratified Secondary rocks rest consist largely of granite and similar coarsely crystalline rocks. On the principle of superposition, it was assumed that the granite bodies are older than the overlying Secondary strata. Granite, therefore, was generally regarded as Primitive. Because of the crystalline appearance of granite and the vague stratification in some granite masses, prevailing neptunist theories were readily applied to granite. In the Wernerian scheme, for example, granite, the dominant rock type in the Primitive foundation on which all other rocks were later deposited, was regarded as the product of precipitation of mineral crystals from the primeval ocean.

More detailed studies of Primitive terranes showed that masses of granite are not everywhere large and homogeneous. In 1776, de Saussure, for example, described granitic veins from the French Alps, and Werner described granitic veins in Saxony. Both men believed that such localized forms of granite originated when aqueous solutions infiltrated downward into sets of fractures within rocks below the surface. As the solutions cooled and drained, the dissolved mineral matter was said to be precipitated in the veins.

James Hutton, however, proposed a radical new notion that Primitive granite is the product of fusion. His views were set forth in a series of publications from 1785 to 1795 as well as in material that remained unpublished until well after his death (McIntyre and McKirdy, 1997). Dean (1992) has established the temporal sequence of the various Hutton writings and reconstructed the development of his views on the nature and origin of granite. On March 7, 1785, the chemist Joseph Black, Hutton's close friend, read before the Royal Society of Edinburgh the first part of a lengthy paper written by Hutton that outlined many of the views that would be set forth in detail a decade later in his famous magnum opus, *Theory of the Earth*. A month later, on April 4, Hutton completed the paper and presented an abstract of the entire contribution. The *Abstract of a Dissertation Concerning the System of the Earth, its Duration, and Stability* was published that same year (Hutton, 1785; White, 1973). Although the abstract indicated Hutton's plutonist emphasis on the fundamental roles of heat and fusion, he said nothing specific about granite. In the abstract, however, Hutton did note that his plutonic theory was confirmed by the existence of mineral veins containing matter foreign to the strata that they cut through in fissures. He maintained that this foreign matter was derived from the mineral region where the active power of "fire" and the expansive force of heat reside. Hutton did not, however, enlarge on the nature of veins in the abstract.

The complete paper, entitled "Theory of the Earth; or an investigation of laws observable in the composition, dissolution, and restoration of land upon the globe," was not published until three years later (Hutton, 1788; White, 1973). In this paper, Hutton advanced his first ideas concerning granite. He noted the common claim that granite makes up "a great part of the solid mass of this earth not properly comprehended among those bodies which have been thus proved to be consolidated by means of fusion" (Hutton, 1788, p. 254). This common opinion regarding granite he disputed. As evidence for his view, he cited a graphic granite occurring near Portsoy on the North Sea coast of Scotland, a few miles east of Inverness. He described the rock as having "the regular lineal appearance of types set in writing" (Hutton, 1788, p. 256). This rock of striking texture, of which Hutton included several excellent illustrations, was said to be continuous with the common granite of the vicinity. On the basis of its texture, Hutton claimed that the granite evidently had been in a fluid state. He said that the crystallization of the sparry substance, that is, the feldspar, had determined the regular structure of the quartz. Hutton (1788, p. 256) overconfidently insisted that "it is not possible to conceive any other way in which those two substances, quartz and feldspar, could be thus concreted, except by congelation from a fluid state, in which they had been mixed." From this conclusion regarding graphic granite, Hutton (1788, p. 257) moved to the much more sweeping conclusion that whether granite was considered as a stratum or not, "there is sufficient evidence of this body having been consolidated by means of fusion, and in no other manner."

Unfortunately, many of those who read Hutton's 1788 paper were probably unaware that he had made several critical field observations since having read the paper three years before. After making the observations, Hutton did write additional material detailing his findings and undoubtedly shared his thoughts with his closest friends, but publication was delayed. Early in his *Theory of the Earth*, Hutton (1795) indicated that he would be supplying details about granite "in the course of this work." Those details, however, never were forthcoming in the two volumes published in Hutton's lifetime. Nearly a century after Hutton's death, however, Canadian geologist Frank Dawson Adams came across a manuscript bound in book form in the library of the Geological Society of London. The manuscript consisted of six chapters and dealt primarily with the nature of granite and its contacts. Adams called the manuscript to the attention of Sir Archibald Geikie, then Director of the British Geological Survey. In 1899, Geikie edited the manuscript and published it as Volume 3 of *Theory of the Earth* (Geikie, 1899). Dean (1997) recently edited a reprint of Geikie's edition and provided details on the history of the Hutton manuscript. Hutton's own observations on the nature of granite were spelled out in chapters on "Observations made in a journey to the north alpine part of Scotland in the year 1785," on "Observations made in a journey to the south alpine parts of Scotland in the year 1786," and on "An examination of the mineral history of the island of Arran." The remaining three chapters dealt with Hutton's interpretation of observations on granite made by de Saussure, Dolomieu, and others.

When Hutton set forth the preliminary statement of his views in 1785, he had seen very little granite in the field. Later that summer, he set out into the Scottish Highlands in the company of his close friend, John Clerk of Eldin, with the express purpose of finding contacts between granite and the surrounding rocks. A competent artist, whose sketches in Hutton's work demonstrate considerable geological awareness, Clerk of Eldin showed Hutton exposures of granitic rock that he had previously encountered. In this way, Clerk of Eldin likely had considerable influence in shaping Hutton's thinking. Hutton and Clerk of Eldin gained access to the vast estate of the Duke of Atholl near Blair Atholl in Perthshire and explored the geology along the valley of the River Tilt. On the south side of Glen Tilt he found stratified "alpine schistus"

5. Locality in Glen Tilt near Blair Atholl, Scotland, described by James Hutton and drawn by John Clerk of Eldin in 1785. Junction of granite (bottom) with schist (above) showing truncation of layering in schist by granite and veins of granite in schist. Width of field about two feet. Photo by Donald B. McIntyre.

composed of quartzite and limestone. On the north side is red granite. The critical contact exposures occur in the river bed itself (McIntyre and Stephenson, 1997, for current understanding of the geology). Hutton understood that if the granite were truly primitive, that is, older than all other rocks, then "fragments of the granite might be found included in the schistus, but none of the schistus in the granite" (Geikie, 1899, p. 13). Instead, what he discovered in many places along the river was that the granite breaks and displaces the strata in every conceivable manner, that it includes fragments of the broken strata, and that it is interjected in every possible direction among the strata that appear (Figure 5). He concluded, therefore, that the granite must be posterior to the schist. Noting the appearances of fluidity of the granite and comparing them to similar appearances associated with basalt in other localities, Hutton reckoned that nature consolidates bodies by means of heat and fusion and moves large masses of fluid matter within the bowels of Earth.

A year later, in 1786, Hutton set out with John Clerk of Eldin to explore granite exposures on the Hebridean island of Arran. As it was too late in the year to explore the mountains, they contented themselves with looking for possible granite contacts on the mainland around Galloway. After observing a host of whinstone dikes, the pair discovered a great granite mass at Cairns Muir. At first, they confused the jointing of the granite with stratification, one of the characteristics claimed by many neptunists as evidence for its aqueous origin. Hutton concluded on closer inspection that the apparent stratification was produced by a process akin to the contraction that produced basalt columns. Better still, Hutton and Clerk of Eldin observed both the inclusion of schist fragments within the granite and veins of the granite penetrating the schist. "It was not possible," Hutton enthused, "to conceive an appearance more perfectly calculated to remove every doubt, or to command belief, with regard to an operation which perhaps never can be seen" (Geikie, 1899, p. 49). He failed, however, to notice one of the strongest indicators of the process he was so eager to demonstrate, namely, evidence that the schist had been metamorphosed by the heat of the invading granite. He and Clerk of Eldin discovered other exposures of granitic dikes and veins penetrating the schist throughout the region. Although he had not actually seen granite in a fluid state, Hutton claimed that he had observed every possible demonstration of the fact that granite had been forced to flow in a state of fusion among the broken and distorted strata.

In his chapter on the granitic rocks of Galloway, Hutton conceded that he did have some ambivalence about the origin of granite before he undertook his studies in Glen Tilt and Galloway. He acknowledged that he had been uncertain as to whether he should adopt a semi-neptunist position in which stratified granite should be viewed as matter that accumulated at the bottom of the sea and was later consolidated by fusion or whether he should adopt the view that granite acted as a mass of subterraneous lava that flowed in the same manner as whinstone or basalt. Without denying that some granites were stratified, Hutton appears to have abandoned any remnants of neptunist tendencies in his thought in the light of these studies.

A third critical exploration occurred in the late summer of 1787, when Hutton finally visited Arran in the company of John Clerk, Jr., who produced a number of geological sketches for Hutton during the expedition. Hutton described a large mass of granite at the northern end of the island that was girdled by a ring of "alpine schistus." Again, he attempted to find exposures of the junctions by hiking up several river courses into the interiors only to be disappointed because the contacts were totally covered by debris. Finally, along the North Sanox River, Hutton encountered abundant exposures of granite both cutting across the stratified schist and intruding along its bedding planes and enclosing fragments of the schist. He was so enthusiastic about his discoveries that he reported shipping a 600-pound specimen back to Edinburgh that showed granite traversing in all directions through the schist. Unfortunately, Archibald Geikie (1899, p. 226) recorded, a century later, that the "specimen is not now known to exist." These field observations reinforced Hutton's conviction that granite had been in a state of fusion caused by subterraneous heat.

Not until January 4, 1790, when Hutton read a very brief paper before the Royal Society of Edinburgh entitled "Observations on granite," were some of his conclusions about granite made public. The paper was published four years later (Hutton, 1794; White, 1973). In this paper, Hutton reported that he had made field excursions to Glen Tilt, Galloway, and Arran and commented that he had succeeded beyond his wildest expectations in looking into the natural history of granite. He formally admitted that he had not been perfectly decided about the origin of granite, noting that he had entertained the idea that granite might be an originally stratified body later consolidated by fusion. He commented on his good fortune in finding critical exposures of intrusive relationships in Glen Tilt. Had the junction been in the mountains, he said, it probably would have been covered with moss and heath, and he never would have seen it. Discovery of the nest of granitic veins that penetrated the alpine strata convinced Hutton that the granite had been in a fluid state, a conclusion that he had formerly drawn from the specimens of the Portsoy graphic granite. Significant advances in igneous petrology often came with the discovery of a critical exposure.

HUTTON'S SUPPORTERS AND OPPONENTS

After his visit to Galloway, Hutton described his findings to James Hall, who became another in Hutton's circle of close friends and protagonists. Having earlier been skeptical of Hutton's geological views, Hall went to Galloway in 1788 to see the junctions of granite with "alpine schistus" for himself. The results of his trip were read before the Royal Society of Edinburgh on January 4, 1790, the same day that Hutton read his paper on granite, and on March 1, 1790. A summary was published later (Hall, 1794). In an area 11 by 7 miles, Hall and T. Douglas, his field partner, found that wherever they observed the junction of the granite with the schist, veins of granite ranging from one-tenth of an inch to 50 yards in width appeared to run into the schist and pervade it in all directions. In confirmation of Hutton's conclusions, Hall (1794, pp. 8–9) asserted that these relationships "put it beyond all doubt, that the granite of these veins, and consequently of the great body itself . . . must have flowed in a soft or liquid state into its present position."

Hall was also acquainted with Hutton's argument for the origin of granite by fusion on the basis of the texture of the graphic granite at Portsoy. Opponents of igneous granite had objected that, because the feldspar had well-defined crystals with quartz molded onto those crystals, the granite could not have crystallized from a melt. They reasoned that because feldspar has a lower melting temperature than quartz, then quartz should crystallize before feldspar during cooling of a melt. Although the phenomenon of freezing-point depression had been observed in some aqueous solutions, and although the phenomenon of fluxing in ore processing was recognized, it was not yet fully appreciated by most of Hall's contemporaries that silicate melts are solutions in which the freezing temperatures of individual minerals may be drastically reduced when dissolved in the melt. Nor was the phenomenon of eutectic crystallization, displayed by quartz–feldspar mixtures, yet recognized. Therefore, it was not realized that the melting temperature of quartz can be drastically reduced in the presence of alkali feldspar and that feldspar can readily crystallize first from alkali-rich melted mixtures of quartz and alkali feldspar. Hall, however, did respond to the objection to igneous granite by suggesting that feldspar serves as a fluxing agent to quartz and that melted feldspar is a fluid in which quartz can dissolve somewhat like salt in water.

Hall also responded to the objection that, after granite is melted, it cools to a glass and does not crystallize. He noted a recent "accidental experiment" at the Leith glasshouse in which some common green glass had been allowed to cool gradually. The glass was reconstituted into a mass of crystals and a few cavities. Only upon remelting and rapid cooling did the material become glass. From this example, Hall asserted that if the glass produced by the melting and cooling of granite had been allowed to cool much more slowly, then it might have crystallized and produced a substance similar to the original granite.

The large host of writers who favored the aqueous origin of basalt at the end of the eighteenth century, however, also favored an aqueous origin for granite. They were not about to adopt the igneous view. Hutton, as is well known, attracted considerable opposition to his overall theory of Earth. Some neptunists disputed Hutton's conclusions about the origin of granite, while others rightly found plenty of other ideas objectionable, such as consolidation of sediments by heat. It must be kept in mind, particularly with the question of the origin of granite, that objections to Hutton's and Hall's views did not always emanate from strictly scientific concerns even though the objections may have been couched in scientific terms. Hutton was a hostile opponent of the very idea of Primitive rocks as that term was commonly understood. To Hutton, many geologists envisioned Primitive rocks, with granite being the most primitive, as going all the way back to the original creation of Earth. The Primitive rocks were thought to have been formed in some unknowable manner that fell outside the normal course of nature and outside the bounds of philosophical inquiry, or else they were envisioned to have formed during a singular act of precipitation from a universal ocean. Convinced of the uniform course of nature, however, Hutton also believed that even Primitive rocks had been formed by ordinary, knowable causes that had operated throughout the history of the globe. As a result, by interpreting granites as more recent masses of molten subterranean lava injected into schists and by interpreting schists as altered sediments that had originally formed on Earth's surface, Hutton was clearly implying that the history of the planet was far lengthier than was customarily thought. Hutton's idea about the origin of granite, therefore, was intricately intertwined with a perceived heretical theological concept. Granite played a key part in Hutton's assault on the conception of Primitive rocks, an assault that threatened the contemporary worldview that envisioned Earth as being only a few thousands of years old. In reading scientific rebuttals of Hutton's ideas about granite, one must remember that at least some of his critics, such as Richard Kirwan, were very uncomfortable with the larger view of the world toward which Hutton was moving.

Kirwan (1793) attacked Hutton's theory of granite as it had been enunciated in the 1788 version of "Theory of the Earth" in the last few pages of a lengthy paper entitled "Examination of supposed igneous origin of stony substances." Kirwan (1793, p. 64), for example, assumed that Hutton's famous dictum that "we find no vestige of a beginning" implied that "this system of successive worlds must have been eternal," a concept sufficiently distasteful to Kirwan that he concluded that any such system "must necessarily be false." He found the application of Hutton's system to the formation of granite to be "peculiarly unhappy." Rather than challenge Hutton's field evidence for injection, he mainly disputed the possibility that a granitic liquid could ever produce the mineralogy and texture observed in a granite and challenged any interpretation of melting experiments with glasses that favored an igneous origin of granite. Kirwan claimed that the minerals in granite, having different degrees of fusibility, should naturally run into each other in any heat sufficiently great to melt quartz. Granite also frequently contains mica, but melted mica, he noted, looks quite different from its natural platy structure. Kirwan (1793, p. 77) insisted that granite "can be formed in the *moist* way, but cannot in the *dry*."

Here Kirwan took issue with Thomas Beddoes (1760–1808), another advocate of the igneous theory of granite (R.G., 1885). Beddoes was a medical doctor, a Reader in Chemistry at Oxford until 1791, and the person who introduced Humphry Davy to the world of science. Beddoes (1791) had called attention to the fact that glassy materials could develop textures similar to those of basalt and granite in furnaces if only the rate of cooling were appropriately adjusted. Kirwan wasn't so sure. He countered that such materials did not represent true analogs to rocks because, during the intense heating process, the chemical composition of the glass had been altered as alkalis (salts) were driven off. Of course, its texture would be altered, Kirwan said, but the material does not retain the same chemical composition as before. Moreover, Kirwan (1793, p. 79) maintained that "glasses formed of earthy substances without any salt . . . when once they are perfectly vitrified, a second fusion makes no alteration whatsoever in them, though ever so slowly cooled." For example, he claimed that feldspars, garnets, "shorls," and basalts, once they were converted into glass in a furnace, would remain glass no matter how slowly they were cooled.

Kirwan also argued that the behavior of crystallizing salt solutions and of cooling fused salts should serve as a guide in evaluating the origin of crystalline rocks. Salts of different solubilities crystallize separately from an aqueous solution, he said, but salts never crystallize separately from their fused melts. He argued that it should be the same with rocks. Because granite and basalt are composed of separate crystals, Kirwan reasoned, they are much more likely to be the product of aqueous precipitation than the products of fusion. He even found it easier to visualize aqueous crystallization than crystallization from an igneous fluid on the grounds that in an igneous melt the mineral particles are so crowded together and in contact with each other that the particles would lose their ability to move and unite to produce any separate minerals during solidification. Such motion, interaction, and union of particles to form crystals would be far more likely to occur in a watery fluid. Kirwan (1793, p. 81) concluded his paper with what he probably considered as the key argument, namely, that "granite, recently formed in the moist way, has frequently been found," but he provided only one significant example. A mole constructed on the River Oder in 1722 had been filled with granitic sand that, according to Kirwan, hardened in a short time into a substance so compact that it was impervious to water.

The following year, Kirwan (1794) published the second edition of his *Elements of Mineralogy* and used the occasion to poke fun at Hutton's general theory. Because of the central role of heat in Hutton's view of Earth, Kirwan suggested that Hutton's theory be designated as "the Plutonic theory." Thus, the name now ascribed to the class of plutonic igneous rocks entered the literature by way of ridicule.

As a response to Kirwan's challenges, Hutton set about to elaborate on his theory. The result was the famous landmark book *Theory of the Earth* (Hutton, 1795). Hutton devoted the entire second chapter of his expanded theory to a refutation of Kirwan's paper. In the final pages Hutton specifically addressed Kirwan's comments on granite. Just as Kirwan ignored Hutton's field evidence, so Hutton ignored Kirwan's claims regarding the melting of glasses and salts and the likelihood of crystallization from melt. Instead, Hutton responded only to Kirwan's claim that the mole on the River Oder represented a case in which granite had been formed in the wet way. Hutton prefaced his response by suggesting that the conversion of granitic sand into granite by means of water contradicted chemical knowledge and was "repugnant" to the actual phenomena. He claimed that experience showed that instead of anything "concreting" into granite, exactly the opposite was true, that is, everything is going into a state of decay. Hutton was very skeptical of Kirwan's example. If Kirwan was correct that the sand had hardened, Hutton maintained that the hardening and compaction could have happened simply by introducing a little mud. He charged that insufficient information had been provided and wanted to know how many mineral ingredients made up the supposed granite and whether it was quartz, feldspar, mica, schorl, or a calcareous substance that cemented the grains together.

Two years later, Hutton was dead. The defense of the notion of intrusive igneous granite and other Huttonian distinctives passed into the hands of his popularizer, John Playfair. In his *Illustrations of the Huttonian Theory of the*

Earth, Playfair (1802) suggested that even though masses of granite rarely are incumbent on other rock bodies and even though granite, unlike whinstone, does not alternate with stratified bodies, yet the similarity of origin of whinstone and granite was rather probable. He reasoned for the igneous origin of granite on the grounds of crystal shape, porosity, and orientation of veins.

Like his contemporaries, Playfair was persuaded that the crystals of the minerals in granite provide evidence of the former fluidity of the granite. If one did not accept that inference, he said, then neither "are the figures of shells and of other supposed petrifactions, to be taken as indications of a passage from the animal to the mineral kingdom" (Playfair, 1802, p. 85). But the tough question really concerned the nature of the fluid medium. Playfair reasoned that the texture of the Portsoy granite described by Hutton was not the result of crystallization from solution. In the Portsoy granite, the quartz grains adopted the forms impressed upon them by the earlier crystallized feldspar. But in crystallization from a solution, Playfair (1802, p. 86) claimed, "one kind of crystal ought not to impress another, but each of them should have its own peculiar shape." Playfair also invoked an argument from porosity. He noted that granites lack porosity because all the grains are perfectly consolidated. If granite had crystallized from a solution, he reasoned, it should retain some porosity because once the solids have separated from the solvent, the solvent would still occupy a certain amount of space. Once the residual solvent was removed by evaporation or some other means, he argued, some empty space would be left behind.

Some opponents of Hutton's view had argued that granitic veins, even though younger than the schists, were also younger than the adjacent large masses of granite because they had been formed by the infiltration of water in which the granite was dissolved. Playfair begged the question by refusing to concede the power of water to dissolve granite. More importantly, however, he pointed out that many of the veins intrude upward into the schist from a main body of the granite, an improbability, he thought, on the infiltration hypothesis. As a result, he insisted that the veins and the associated main bodies of granite are coeval. Playfair would also have nothing of the subterfuge that denied that the injected veins were really granite, countering that the force of a fact could not be lessened by changing names. The point, Playfair (1802) urged, is that the vein and the granite mass form one continuous body; it does not matter whether the rocks are termed granite, a rock composed of quartz and feldspar, or syenite, a feldspar-rich rock with relatively little quartz but a substantial amount of hornblende. Nor did it matter to Playfair that a vein might not be exactly the same mineralogically as its host body. He knew of many instances where the vein and its source rocks do have the same mineral content.

A strong objection against aqueous infiltration arose in Playfair's mind from the fact that the granitic veins commonly completely enclose numerous fragments of the schist. He found it hard to conceive how water could sustain such fragments and how water could introduce the fragments into the veins in the first place. He considered it much easier to visualize their transportation by melted granite. Playfair recounted the several examples of intruded granite veins found by Hutton, Hall, and Mackenzie, added some more that he and Lord Webb Seymour had encountered near Loch Chloney in Inverness-shire and at St. Michael's Mount in Cornwall, and mentioned that even the archneptunist Robert Jameson had described intrusive granite veins in the bed of the River Spey. He noted that not just veins but also large granite masses enclosing nodules of gneiss and mica schist had been found. Even Werner, he said, allegedly had a specimen of granite containing gneiss that he considered as proving that the gneiss was older than the granite. Why then, Playfair wondered, would a neptunist like Werner persist in his attachment to aqueous deposition of granite?

> It is impossible to deny that the containing stone is more modern than the contained. The Neptunists indeed admit this to be true, but allege, that all granite is not of the same formation; and that, though some granite is recent, the greater part boasts of the highest antiquity which belongs to any thing in the fossil kingdom. This distinction, however, is purely hypothetical; it is a fiction contrived on purpose to reconcile the fact here mentioned with the general system of aqueous deposition, and has no support from any other phenomenon. (Playfairs, 1802, p. 320)

Playfair discussed graphic granites other than the one at Portsoy and pointed out that in some of them it was the quartz that impressed its crystalline form on the feldspar. Such arrangements, he stated, indicated simultaneous crystallization of the minerals.

Playfair also discussed stratification in granite. He disagreed with Hutton who had come to doubt that it was ever stratified. Playfair accepted the existence of some stratified granite that did not have the character of gneiss, but he suspected that the stratification attributed to granite mentioned by the neptunists was so subtle that it was either an illusion or not true stratification. He accepted the accounts of Pallas, Deluc, and de Saussure of stratified granite but thought it strange that it took them so long to convince themselves of the presence of stratification. Moreover, he charged, the "beds" of granite observed by de Saussure are much too thick and their parallelism is very difficult to ascertain. Such beds, he said, have only the slightest resemblance to the type of beds that water is known to produce. Playfair did acknowledge that there are transitions from gneiss to granite in which the granite may retain a stratified character. Jameson (1800) had reported stratified granite from the island of Arran in his book, *Mineralogy of the Scottish Isles*, but Playfair had seen large tabular masses of granite that were nearly vertical and separated by fissures. These masses struck Playfair as too irregular, too thick, and insufficiently extensive to be regarded as strata. He suggested that no one accept either Jameson's or his own conclusions on granite stratification without further verification. Although conceding that some granitic strata existed, for example, where interlayered with marble and gneiss, Playfair was very doubtful that intrusive granite veins ever proceeded from a body of granite that was really stratified. Still, he suggested that the search for contrary or supporting cases would be worthy of the attention of geologists.

Playfair's support for the Huttonian conception of granite elicited a rapid response from Robert Jameson (1802a) entitled "On granite." Jameson asserted that the structure of primitive granites showed that they had been deposited from chemical solution. Following Werner, Jameson wanted to distinguish primitive granite formations of different ages. Granite, he alleged, is frequently observed to be distinctly stratified. The oldest granite, he said, seldom contains extraneous beds, whereas the frequency of extraneous beds increases in the newer granites. He claimed that wherever granitic veins issue from granite into gneiss and mica slate, the granite mass belongs to a newer formation of granite. Jameson sought to preserve the nonintrusive character of the oldest masses of granite.

But Jameson also responded more explicitly to Playfair's assertion that granitic veins could not have been filled from above. Playfair had claimed that schist fragments in the veins were completely surrounded by granite because they had been transported upward by flowing, melted granite. Jameson countered that these inclusions were only apparently surrounded by the veins. In fact, he said, we can normally see only part of the fragment, and whenever an entire fragment actually can be seen in three dimensions, "we invariably find, that it rests upon another fragment, or upon the sides of the veins" (Jameson, 1802a, p. 232). His own observations on Arran also convinced him that granitic veins are such that both extremes of the vein occur in the same bed of gneiss with which the veins are intimately associated. In other words, Jameson (1802a, p. 232) charged, they are not truly granite veins, because "they are nearly of coeval formation with the gneiss, and have no communication with any formation, older or newer than that in which they occur."

In the same year, John Murray also answered Playfair's defense of the Huttonian theory of granite. Murray (1802, p. 241) was in full agreement with the plutonists that granite was fluid at one time but, he said, the fact that simultaneous crystallization can take place from a solution meant that the structure of granite could "furnish no argument against its aqueous origin." Granite, he thought, afforded the clearest evidence that it had not been formed by fusion. Murray appealed to the laboratory evidence that had been so persuasive to many opponents of the igneous origin of granite, namely, the fusibilities of the constituent minerals. He pointed out that

> felspar is a substance incomparably more fusible than quartz. . . . It is a proposition, therefore, self evident and undeniable, that in the same mass quartz could not be fluid when felspar was solid; and therefore, since in this graphic granite the quartz

is moulded on the crystallized felspar, (and, in the greatest number of granites, the felspar is crystallized while the quartz is not) the fluidity whence both have been consolidated cannot have been fusion from heat." (Murray, 1802, p. 241)

The enclosing of feldspar crystals inside of quartz offered the same proof. So, too, did the fact that "a substance of comparatively easy fusibility . . . is often crystallized in quartz" (Murray, 1802, p. 242). He adduced several instances in which less fusible minerals are included inside more fusible minerals as evidence against origin from melt. Murray (1802, p. 243) was so confident of this line of argument that he asserted that "if they do not prove that these fossils have not been formed by fusion, no conclusion can be established in geology, and we may relinquish every attempt to theorise." Murray indirectly undercut the Huttonian theory of granite by issuing a strong challenge to Hutton's erroneous idea that metalliferous veins had also been intruded as melts. Such veins are presently interpreted as the products of hydrothermal fluids.

A decade later, Robert Jameson (1813) published another book on the geology of the Scottish islands. Little had changed in his attitude about the Huttonian views of granite. He detailed four major objections to Hutton's hypothesis. First, he reiterated an earlier argument that if granite had flowed upward from below, it should have overflowed the neighboring country after it had burst through the strata of mica schist. If Hutton's hypothesis were true, he alleged, Mont Blanc could never have existed. Playfair had already responded to this line of argument in 1802, but Jameson ignored the response and repeated the claim.

Jameson further reasoned that if molten granite had been violently forced upward, then whatever strata cover granite should everywhere be disturbed and broken in every direction at their junction. But geologists had examined vast tracts of granite without ever observing such appearances. The phenomena described by Hutton seem to occur very rarely, he claimed, and, therefore, cannot provide a true indication of the origin of granite. He was also troubled by the fact that many granitic veins occur where no central mass of granite is observed. He repeated the argument that veins are completely confined to a belt of mica schist, and regarded that as a sufficient proof that the veins are not connected with any granite below. He further claimed that granite veins have a character different from allegedly stratiform granite masses, suggesting that the veins were formed at a different period from the granite in strata.

Jameson said that granite grades to gneiss, gneiss grades to schist, and mica schist grades to argillaceous schist. He believed that such rocks were formed in the same manner and nearly at the same time. Veins of granite shooting from the strata of granite into the schist, he thought, could be explained in the same way as common metalliferous or quartz veins. Stratified granite bodies, he alleged, are frequently traversed by veins of granite that have a different grain size from the massive granite. Only with great difficulty, however, could geologists distinguish them from the stratified rock. Fractures appear in closely associated granite, gneiss, schist, and "ardesia," and they are sometimes filled with veins of granite. But, Jameson wondered, if granite fills those rents in granite, why don't we find such veins in all the rocks? Although disputing the plutonic theory, Jameson did not explain where the veins come from.

Ignoring Jameson, Playfair continued his defense of igneous granite. On May 31, 1813, he read a paper before the Royal Society of Edinburgh summarizing observations of the geology of Table Mountain and other features near the Cape of Good Hope at the southern tip of Africa (Hall, 1815). These observations had been sent to Playfair in a series of letters by Basil Hall, a Captain in the Royal Navy, a fellow of the Royal Society, and the son of James Hall. Playfair read excerpts from Hall's letters describing the horizontal beds of sandstone that unconformably overlie granite. What was of particular interest to Hall, however, was the fact that the massive granite sent out veins that penetrated through steeply inclined layers of "schistus" or "killas." Hall (1815, p. 276) concluded that the veins of granite, traversing the "killas" with minute ramifications and insulating fragments of it, "seem to have been in a state of fusion, accompanied by very considerable violence." Playfair added his own view that the evidence Hall presented was "highly favourable to the opinion, that granite does not derive its origin from aqueous deposition" (Hall, 1815, p. 278).

The following year, the Royal Society of Edinburgh heard more on the subject of igneous granite. Playfair had been a little uneasy in defending all of Hutton's assertions without seeing the evidence for them firsthand. By 1802, the year in which he issued his popularization of Hutton's views, he had not been to Glen Tilt. Thus in the autumn of 1807, he and his good friend Lord Webb Seymour paid a visit to the critical locality. Seymour returned for another look the following year. Not until 1814, however, were the results of this investigation presented before the Royal Society by Seymour. Quite possibly Playfair and Seymour were stimulated to make their investigations public because John Macculloch had presented a paper describing the phenomena of Glen Tilt before the Geological Society of London on December 3, 1813. Macculloch described the veining of red granite in schist and limestone in considerable detail, but held back from taking a stance on the origin of granite. Not so Seymour and Playfair. They made a strong case that the veins of granite, which they designated as "sienite," are continuous with a main mass of red syenite and had formed from the same fluid. They ruled out any thought of vein formation by aqueous secretion from surrounding stratified rocks and argued for the role of heat and fusion on the grounds that a series of rocks from trap to "sienite" to granite is present and that the invaded rocks are severely distorted. Inasmuch as trap was already largely accepted as igneous, Seymour and Playfair thought that its gradations into syenite and granite would make the case for the igneous origin of the latter. Moreover, they saw indications of alteration of the stratified rocks by penetrating melt as some of the ingredients in the liquid seeped into the surrounding rocks. Hence, they envisioned a kind of contact metasomatism. They concluded that the sienite, in a state of igneous fusion, was impelled from below, by a violent force, against the strata; that it bent them, broke them, dispersed them, and filled up the intervals, which it now occupies; that the fragments of the strata were in some degree softened by the heated sienite, so as to admit of a mutual action; that, while the whole intermixed mass was still soft, some farther dislocation took place in it, and that all this occurred under a great confining pressure of incumbent matter. (Seymour, 1815, p. 358)

Seymour and Playfair, however, conceded that problems remained for the igneous hypothesis and urged that experiments be performed to obtain knowledge of the properties of the common minerals at high temperatures and pressures.

A sufficient number of examples of apparently intrusive granite having been described, Macculloch (1819) noted that it was not easy to admit the arguments derived from the evidences in favor of the igneous origin of trap but then deny them in the case of granite. Many other geologists accepted an igneous origin for granite (Dean, 1992). Conybeare and Phillips (1822) concurred. Although Boué still had his reservations in 1820, three years later he proclaimed in Jameson's *Edinburgh Philosophical Journal* that Hutton had discovered the true origin of granite. And in 1825 he reported that an increasing number of geologists regarded granite, syenite, and porphyry as igneous. "That granite is identical or analogous with the trap rocks," Macculloch (1826, p. 370) wrote condescendingly, "is another fact established by geologists of a class yet too logical and too philosophical to have made converts of the baser multitudes."

By the 1820s, a large number of geologists were persuaded that igneous activity was not confined to volcanoes. They recognized that igneous activity was also manifested in the presence of cross-cutting veins and dikes of coarse-grained crystalline rocks like granite that had presumably solidified from intrusive bodies of subterranean lava. They recognized the existence of what are now termed plutonic igneous rocks. That recognition, however, was not universal. There still remained the "baser multitudes" of competent geologists who were skeptical of the formation of granite by pure igneous fusion. Although these geologists may no longer have thought in strictly Wernerian terms of granite as an aqueous precipitate from a primitive ocean, they remained convinced that in some way or other water played a very important role in the origin of crystalline granitoid rocks.

The Primitive Era

6 ⁋ WET OR DRY?
The Origin of Granite

As we have just seen, many geologists in the opening decades of the nineteenth century accepted the notion that veins and dikes of coarsely crystalline granitic rocks proceeding from larger masses of granite were formed by the intrusion of subterraneous lava. Some eventually carried the concept of igneous dike intrusion to an extreme. For example, the American geologist Ebenezer Emmons (1799–1863) interpreted intrusive dikes of remobilized marble in the Adirondack Mountains of New York as evidence for their igneous origin (Emmons, 1860). In contrast, numerous geologists throughout much of the century remained skeptical that granite was produced by igneous fusion or that dikes prove igneous injection. Despite the successes of the plutonists, no consensus about the nature of granite emerged for several decades.

The writings of James D. Dana, America's premier nineteenth-century geologist, illustrate the uncertainties about the nature of granite. In his geological report on the Wilkes expedition, Dana (1849) referred to granite and syenite as the products of cooling of melt. In the second edition of *Manual of Geology*, however, Dana (1866) included granite in the category of rocks crystallized by heat without fusion, designated as "metamorphic rocks." He also drew a sharp distinction between "plutonic" rocks and true igneous rocks.

Why did the confused state of affairs about granite and coarsely crystalline rocks persist for so long? Ironically, British geologists did relatively little to advance the discussion about granite after the pioneering efforts of Hutton, Hall, and Playfair. Lyell (1833) did discuss granite and plutonic rocks toward the end of the third volume of his *Principles of Geology*, but, despite his major contributions to the interpretation of geological history, he added little new to the discussion about the origin of granite that had not already been stated by the original plutonists. For the most part, British geologists of the first half of the nineteenth century appear to have regarded the issue of the origin of granite as settled. The neptunist conception of dikes, sheets, or masses of basalt and granite as precipitates from the ocean had largely faded during the 1820s. So, too, had the neptunist conception of granite as exclusively a Primitive formation. Since few Wernerian neptunists remained to challenge the idea of intrusive igneous granite, British geologists saw little need for further debate and turned their attention to other aspects of geology, such as paleontology, stratigraphy, and geomorphology, that proved to be exceptionally fruitful. Between 1830 and 1860, the origin of granite became the concern primarily of continental geologists.

During this period, some advocates of igneous granite, particularly Fournet and Durocher in France, emphasized that granite is the product of "dry igneous fusion." For them, water did not play an essential role in the production of granitic melt. Vigorous opposition to the concept of dry fusion arose, however, on several grounds. The opponents of dry fusion concluded that granite had formed in the "wet way."

Several arguments were leveled against the idea of dry fusion origin of granite. These included the claims that the order of mineral crystallization in granite excludes an origin by fusion; that quartz-bearing volcanic equivalents of granite do not exist; that experimental evidence confirmed the production of quartz only in the "wet way"; that field evidence demonstrates the conversion of metamorphic rocks into granite; and that the presence of "pyrognomic" minerals in granite is inconsistent with their origin by dry fusion.

ARGUMENTS AGAINST THE ORIGIN OF GRANITE BY DRY FUSION

Kirwan, Murray, and other opponents of the views of Hutton and Playfair on the nature of granite had already pointed out that the textures of granite indicate that feldspar crystallized prior to quartz. They reasoned that such an arrangement excludes an igneous origin because feldspar is much more easily melted than quartz. If granite were igneous, the argument went, quartz, the mineral with the very high melting temperature, should have crystallized before the more fusible feldspar. Well-formed crystals of quartz should have imposed their faces on the form of the subsequent feldspar rather than the opposite.

The textural argument was persistently repeated throughout the first half of the nineteenth century. Scipione Breislak, for example, made it in his *Traité sur la Structure du Globe*. Johann Fuchs, a Munich chemist, challenged the "pyrogenic" or igneous origin of granite on textural grounds in 1837 in *Über die Theorien der Erde*. One of the most vigorous advocates of the origin of granite in the wet way was Carl Gustav Christoph Bischof (1792–1870). Of all the participants in the debate, Bischof came the closest to reversion to neptunism. After receiving his doctorate from the University of Erlangen, Bischof began a long tenure as Professor of Chemistry and Technology at the University of Bonn in 1819 (Amstutz, 1970). His early studies of hot springs led him to a vulcanist position, but Bischof soon became much more sympathetic toward transformationist views of granite and similar crystalline

6. Theodor Scheerer (1813–1873). Reproduced by permission of the Bergakademie, Freiberg.

rocks. In the second volume of the English translation of his encyclopedic work on chemical geology, *Elements of Chemical and Physical Geology*, Bischof (1855) rehearsed the long-standing objection that the order of succession of minerals is directly opposed to the idea of the derivation of granite by fusion, and in the third volume he flatly asserted that "crystals of quartz are never found to have interfered with the crystallization of the other minerals" (Bischof, 1859, p. 44).

The textural argument, however, was developed the most thoroughly by Theodor Scheerer (1813–1873) (Figure 6), a chemist and mineralogist at the Bergakademie in Freiberg, in a paper read before the Societé Géologique de France on February 15, 1847. The origin of granite was a very hotly contested topic in France during the course of the 1840s. At successive meetings of the Societé Géologique in late 1846 and throughout 1847, several leading geologists, including Fournet, Durocher, Virlet d'Aoust, Scheerer, Élie de Beaumont, and Delesse, presented papers on the nature of granite. These papers, published in *Bulletin de la Societé Géologique de France* for 1847, provide an excellent cross section of the attitudes toward granite at the time. A more recent granite enthusiast, H. H. Read (1957), referred to 1847 as the "annus mirabilis" and discussed several of the papers in his *The Granite Controversy*.

In his paper entitled "Discussion sur la nature plutonique du granite et des silicates cristallins qui s'y rallient," Scheerer (1847, p. 478) stated that early in his career when he had been unaware of Breislak's objections to igneous granite and before Fuchs had published his, he had been "completely saturated with the most orthodox plutonian theories." In 1833, however, he visited Norway and his earlier beliefs about granite were "completely weakened" as a result of his intense study of Norwegian granites. In 1842, he presented his "altered" views on granite to the Society of Scandinavian Naturalists in Stockholm, confessing that the textural evidence of granite on the island of Hitteröé had made an impression on him. In the Hitteröé granite, he reported, the feldspar had clearly crystallized in an "uninhibited" manner, leaves of mica had been puckered into the contacts with feldspar, and "amorphous" quartz filled in the spaces left over after crystallization of the feldspar and mica. The feldspar in graphic granite, Scheerer said, always won out over the quartz in the "struggle" to crystallize first. On the ordinary plutonic theory, namely, that of pure igneous fusion, he had informed his audience, that textural fact was inexplicable on the grounds that silica, because of its difficulty of fusion, should have solidified well before the silicates of potassium and of magnesium. Heat alone, he judged, had not produced the granite. Scheerer further reported instances of granites in which a host of other minerals such as acmite, garnet, tourmaline, amphibole, orthite, allanite, gadolinite, pyrite, arsenopyrite, cobaltite, and mica, too, had solidified prior to feldspar and, therefore, prior to quartz. Even where the most fusible of these minerals occurs in immediate contact with quartz, Scheerer pointed out, the quartz never prevented them from crystallizing perfectly.

Those who were persuaded by the textural argument were, of course, hampered by a lack of understanding of the internal structure of magma. For the most part, chemists and geologists assumed that individual minerals in a silicate rock like mica, feldspar, and quartz somehow retain their chemical integrity when the rock is melted. Those who used the textural argument probably envisioned that silicate lava consists of domains with chemical compositions identical to those of the minerals that would eventually crystallize upon cooling. The struggle to grasp the internal nature of lava was especially pointed in the writing of Scrope (1825, 1856), who never was convinced that lava is a fused material. Throughout his career he advocated the notion that lava achieved its fluidity because vapor thoroughly penetrated extremely finely comminuted mineral particles. In his view, hornblende, mica, feldspar, and other minerals retained their solid character in lava on an extremely fine-grained scale. Advocates of the textural argument were similarly hampered by the lack of knowledge that magma is a solution in which the phenomenon of freezing-point depression occurs.

The textural argument would have lost some of its force if the presence of crystalline quartz in rocks of known igneous origin, namely lava flows of trachyte, could be confirmed. As a result, many advocates of the textural argument also advanced as an argument the supposed nonexistence of crystalline quartz in volcanic rocks. Scheerer (1847), for example, observed that not even very slowly cooled lava flows erupted from the volcano Jurullo had produced quartz. Noting that although volcanic products like obsidian and pumice have granite-like chemical compositions, he pointed out that they do not contain free quartz. To avoid the implications of the embarrassing absence of quartz in materials of appropriate composition and undoubted igneous origin, the plutonists, in Scheerer's view, were forced to resort to the subterfuge that such silica-rich melts cool much too quickly to form quartz. The plutonists, of course, were right.

In the second volume of his work on chemical geology, Bischof (1855, p. 478) asserted that "quartz crystals have never been found in lava." Any quartz that had been found in lava, he maintained, was essentially a rounded pebble that had been picked up as lava flowed over the surface. He also noted that quartz never crystallized from furnace slags unless they were extremely silica-rich. In volume three, Bischof (1859) eventually conceded that quartz crystals, not just pebbles of quartz, had been found in trachyte, but he took great pains to insist that the quartz crystals were deposited by aqueous solution in cavities or thin seams in the trachyte after its solidification. In Bischof's opinion, if free silica never separated from trachyte, the only rock then considered to contain an excess of silica and of which the igneous origin was certain, there were no grounds for the view that quartz was of igneous origin.

A third argument against the production of granite by dry fusion stemmed from experimental evidence. Opponents of dry fusion insisted that no one had ever produced granite by fusion in a furnace despite the assertion of James Hall that he had recreated substances with stony texture from slowly cooled melts. Nor, opponents alleged, had anyone crystallized quartz from silicate melt. In contrast, several investigators formed quartz from hot water, corroborating the idea that granite was produced in the "wet" way. In 1845, Schafhäutl showed that water vapor heated above 100°C dissolves silica. Upon cooling the vapor, Schafhäutl precipitated hexagonal dipyramidal crystals of quartz. Scheerer (1847) reiterated that no one had yet produced free silica by the slow cooling of a silica-saturated silicate melt, and Virlet d'Aoust (1847) appealed to Schafhäutl's experiments on the solubility of silica to support his contention that high temperature was not necessary to produce granite. Although Bischof (1855) acknowledged the origin of some augite and leucite by fusion, he found from his experiments that all silicate minerals could originate from aqueous solutions without increased temperature and pressure. As far as he was concerned, crystallization from melts was out of the question for several of the minerals commonly found in granite including hornblende, mica, orthoclase, and quartz.

The French experimentalist Daubrée also strongly supported the importance of water in the formation of granite. Gabriel-Auguste Daubrée (1814–1896) studied at the University of Strasbourg and the University of Paris (Chorley, 1971). After several years as a mining geologist, Daubrée was called to the University of Strasbourg in 1848 to serve as Professor of Mineralogy and Geology. There he established an experimental laboratory. After 1861, he became Professor of Mineralogy at the Muséum d'Histoire Naturelle, and from 1872 to 1884 he was also Director of École des Mines in Paris. Although Daubrée's primary contributions focused on metamorphism and meteorites, his studies of silicates in the presence of superheated water demonstrated that many of them crystallized from water at temperatures far below their fusion points. He grew quartz, feldspar, and pyroxene in the wet way.

In experiments on metamorphism, Daubrée (1857) examined the behavior of glass tubes filled with a small quantity of water heated to 400°C for at least a week. The water became charged with alkali silicate. The glass was transformed into an opaque white mass composed of various crystalline substances, one of which was quartz, which lined the tube walls much like a geode. After a month of heating, he produced quartz crystals as much as two millimeters long. He also examined the effect of heated water on obsidian. This rock, he discovered, lost its glassy character and was transformed into a grayish mass that looked like a very fine-grained trachyte and contained abundant microscopic feldspar crystals. Daubrée knew that feldspar had previously been observed in the upper parts of copper smelting furnaces. Rather than conclude that the feldspar crystallized during cooling of melted slag, however, he suggested that the feldspar was deposited on the furnace walls by vapor. After all, he noted, the most skillful chemists had been unable to produce feldspar synthetically by dry fusion, but his experiments on obsidian demonstrated convincingly that feldspar readily formed in the wet way. Even in lavas, Daubrée suggested, copious water vapor was the primary agent in bringing about the crystallization of silicates like feldspar or quartz well below their fusion temperatures. "It is still by this aqueous influence," Daubrée (1857, p. 310) reasoned, "that these same silicates are able to crystallize in a succession that is often opposed to their relative order of fusibility." He maintained that his conclusions were just as applicable to the crystallization of granite as to that of lava.

Heinrich Rose (1795–1864), a Professor of Chemistry at the University of Berlin and older brother of mineralogist Gustav Rose, showed that after fusion, quartz is converted into amorphous silica accompanied by a decrease in specific gravity from 2.6 to 2.2. Rose (1859) argued that the quartz in granitic rocks could not have separated from a dry fused mass and could never have experienced elevated temperature because it always has a specific gravity of 2.6. The American geochemist, T. S. Hunt constantly appealed to the experimental work of Rose and others to support his claim that quartz was never known to be formed in any other way than in the presence of water at temperatures far below those of its fusion temperature.

Igneous granite was also questioned on the basis of field evidence. A number of geologists reported occurrences of gradations of other rock types into granite and thereby concluded that at least some granite was the end result of a metamorphic transformation. This line of argument first appeared in 1825 when Balthazar M. Keilhau (1797–1858), Professor of Mineralogy at the University of Christiania in Norway, suggested that some of the sedimentary rocks of Norway had been transformed into granite.

Another participant in the controversial meetings of the Societé Géologique de France was Théodor Virlet d'Aoust (1800–1894), a mining engineer in Paris. In a paper presented at the Society's meeting of February 15, 1847, Virlet d'Aoust (1847) opposed both the plutonist and Wernerian conceptions of the origin of granite and agreed with much of Scheerer's paper. He noted that red granites at Montabon near Chalon-sur-Saone appear to grade into gneisses. He pointed out that gradations from gneiss to granite had been beautifully preserved in polished slabs of granite from Normandy, such as those that are used as a facing for the sidewalks of Paris. These granites, he said, contain thousands of unmelted, yet modified, fragments of ancient rocks.

Léonce Élie de Beaumont (1798–1874), arguably the preeminent French geologist of his day, also presented a paper on granitic rocks before the Societé Géologique de France in 1847. After studying at École Polytechnique and École des Mines, Élie de Beaumont embarked upon an illustrious career as a mining geologist and leading participant in the effort to produce a geological map of France (Eyles, 1950; Birembaut, 1971). Élie de Beaumont accepted the idea of Virlet d'Aoust that some granites were derived in a cyclic fashion from schists and gneisses that in turn had been derived earlier from metamorphosed granites.

The similarity in composition between many eruptive rocks and slates and graywackes together with the interbedding of granite, gneiss, and sedimentary schists also led Bischof (1855) to agree with Keilhau and Virlet d'Aoust that granite represented altered clay slate.

In the same volume of *Annales des Mines* in which Daubrée had issued his study of experiments on metamorphism, the Professor of Geology at the University of Besançon and Engineer of Mines, Achille Delesse, published a massive three-part series on the metamorphism of rocks. Delesse (1857) distinguished between regional and contact metamorphism and pointed out the differences between the kind of contact metamorphism associated with granites and that associated with lava. He noted that in many cases pelitic sedimentary rocks at the contact with granites had experienced a process of feldspathization to such an extent that they looked very much like granitoid porphyries. From these appearances and other lines of evidence, Delesse concluded that granites were not produced by dry fusion.

James Geikie (1839–1915), younger brother of the renowned Director of the British Geological Survey, Sir Archibald Geikie, was a geologist with the Scottish branch of the Survey and later holder of the Murchison Chair of Geology at the University of Edinburgh. Geikie (1866) contributed a paper on granitic rocks in the southern uplands of Scotland in which he described a series of feldspathic Lower Old Red sandstones that had changed into quartzless syenite with granitoid texture. He also reported graywackes that had apparently been converted into granite along strike. Similarly, A. H. Green described the transformation of other rocks into the Donegal granite in Ireland. On more than one occasion, Geikie referred favorably to the views of Hunt, who also called attention to the gradation of metamorphic rocks into granite

The final argument against a purely igneous origin for granite concerned the presence of what Scheerer (1847) called "pyrognomic" minerals in granite. Pyrognomic minerals were said to give off light spontaneously when heated. Scheerer claimed that pyrognomic minerals such gadolinite, orthite, and allanite become less soluble in acids, their color and transparency are altered, and their density increases during intense heating. How is it, Scheerer asked, that such minerals in their pyrognomic state should be in rocks allegedly formed by igneous fusion? Given that they crystallized prior to quartz and that the quartz allegedly crystallized from melt, Scheerer reasoned that the pyrognomic minerals would have been at a high temperature long after they solidified. How then, he wondered, did they acquire and retain their pyrognomic properties on the igneous theory?

THEORIES OF ORIGIN

Opponents of the origin of granite by dry igneous fusion said that they favored the formation of granite in the wet way. Although their conceptions of what that expression meant commonly lacked rigor and clarity, it was clear enough that they believed that water played a major role in the formation of granite. Perhaps two major ideas about the origin of granite prevailed among the advocates of the wet way. On the one hand were those who perceived granite as the product of metamorphism of preexisting rocks in the presence of pervasive aqueous solutions. On the other hand were those who suggested that granite arose from "aqueo-igneous fusion." For a few like Hunt, the truth, as they saw it, lay somewhere in between.

Those who called attention to the gradations of other rock types into granites typically stressed the metamorphic origin of granite. Virlet d'Aoust (1847), for example, suggested that the transitions to granite represent a more advanced, intense stage of metamorphism than that which produced the gneiss. Such rocks, he said, were derived by transmutation of sedimentary rocks, and he described the processes as "gneissification" and "granitification." He believed that the preservation of stratification or schistosity in the granites of central

France and Brittany indicated that low temperatures had prevailed in some cases of "granitification." C.W.C. Fuchs accepted Bischof's views in a memoir on granite in the Harz Mountains. He regarded the granite as a product of progressive alteration of sedimentary greywacke into hornstone and then into granite by means of water. Geikie (1866) concluded that the crystalline rocks in southern Scotland had formed by the in situ alteration of bedded deposits.

One of the leading proponents of "aqueo-igneous fusion" was Scheerer. Despite his preachments against the plutonists like Durocher, Scheerer (1847) conceded in the end that heat played an important role in the formation of granite. Nevertheless, he pointed out that many of the minerals of granite such as mica, tourmaline, and allanite (he also included gadolinite) contain combined water. Because combined water was presumably present when the granite was still a pasty mass, Scheerer suggested that the granite formed from a paste impregnated by water and heated under strong pressure. Such a melt or paste could form, he argued, at a temperature much lower than an equivalent anhydrous melt. He saw the melting of hydrous salts as an analog. Bemoaning the thought that his conclusion might not be experimentally demonstrable, Scheerer reasoned that water particles helped to separate the atoms in a melt from one another more widely than they would be in an anhydrous melt state at high temperatures. He envisioned that early-forming minerals would be those capable of overcoming the atom-separating effects of water. As they crystallized, water would be concentrated in the residual silica-rich liquid. Thanks to the continuing "augmentation" of water in the increasingly siliceous melt, he thought, quartz would solidify very late. At the time of final granite solidification, water would be released from the melt. In such a way, too, Scheerer maintained, pyrognomic minerals could retain their properties because they would be crystallizing at temperatures much below their points of fusion.

Scheerer also believed that his idea of granite forming from a wet paste or melt could account for the transformation of argillaceous schist into gneiss and granite. He envisioned that a wet, hydrous granitic liquid would be sufficiently fluid that it could easily penetrate between the foliation planes of schist.

In writing on volcanic and metalliferous emanations, Élie de Beaumont (1847) observed that volcanic rocks are dominantly basic and that plutonic rocks are dominantly acidic. Each group is characterized by its own suite of mineralizing emanations, he said. He regarded these emanations as volatile substances, either vapors or solutions, that were capable of transporting silicates and metals. Élie de Beaumont noted that the effect of such emanations was particularly striking in the vicinity of granites such that there is commonly found an "aura" surrounding granite that partakes of some of the characters of granite, including pegmatite, graphic granite, and greisen. While acknowledging the possibility of different origins for granite, he accepted the notion that many granites solidify from melts that are charged with various vapors and mineralizers, particularly water.

The ideas of Hunt fell somewhere between these two extremes. Beginning at least as early as 1858, Hunt, the founder of American geochemistry, wrote a never-ending torrent of papers and delivered an unceasing stream of talks in which he espoused a distinctive view of the chemical history of the Earth that incorporated the idea of the sedimentary origin of granites. Thomas Sterry Hunt (1826–1892) spent a couple of years at Yale, without taking a degree, as an assistant to Benjamin Silliman in performing water analyses (Pumpelly, 1893). In 1847, the Geological Survey of Canada hired Hunt as a chemist and mineralogist, a post he held until 1872. During this time, he simultaneously held professorial posts at Laval University (1856–1862) and McGill University (1862–1868). From 1872 to 1878 he was Professor of Geology at Massachusetts Institute of Technology. It was Hunt who made a motion at the 1876 meeting of the American Association for the Advancement of Science proposing the development of a plan for establishing an International Geological Congress. A brilliant chemist and original thinker whose ideas were insufficiently tempered by careful field work and were typically presented in a controversialist style, Hunt often locked horns with other geologists on many issues. Beginning in 1859, Hunt's radical views appeared on a regular basis in *Geological Magazine*, *Quarterly Journal of the Geological Society of London*, and *American Journal of Science*. In a paper originally read before the American Association for the Advancement of Science meeting in Montreal in August 1857, Hunt (1859) explained his theory of transformation of sedimentary deposits into crystalline rocks. This transformation was not, he argued, simply the result of heating these rocks near the point of igneous fusion. Rather, he claimed that the various silicates in crystalline rocks were formed by the interactions of sedimentary rocks with alkali and earthy carbonates in the presence of silica. Even plutonic rocks, he said, were represented among altered sedimentary strata. Crystalline aggregates of quartz, feldspar, and mica were supposed to display transitions from mica schist through gneiss to stratified granite. Diorite and serpentine, likewise, were said to show such transitions and to have resulted from the alteration of sedimentary rocks rich in magnesia. But metamorphic rocks like granite, diorite, dolerite, serpentine, and limestone, he acknowledged, may appear to be intrusive under certain conditions. As a result, he agreed that the theory of aqueo-igneous fusion applied to granites by Scrope, Scheerer, Élie de Beaumont, and others should be extended to other intrusive rocks. These are in all cases, he insisted, altered and displaced sediments. "The metamorphism of sediments in situ, their displacement in a pasty condition from igneo-aqueous fusion as plutonic rocks, and their ejection as lavas with attendant gases and vapours," Hunt (1859, p. 496) asserted, "are, then, all results of the same cause, and depend on differences in the chemical composition of the sediments, temperature, and the depth to which they are buried." For Hunt, water brought about the transformation of sediments into stratified gneiss containing quartz and ultimately brought about sufficient mobility of the gneiss to result in eruptive behavior. Granite, therefore, was remobilized gneiss that had lost its stratification. Although Hunt eagerly applied Scheerer's name of aqueo-igneous fusion to this water-mobilized mass, it was never clear whether he regarded the mobilized mass as actually molten as Scheerer did.

THE DEFENSE OF DRY FUSION

Subsequent to the labors of the early plutonists like Hutton, Hall, and Playfair, the case for the igneous origin of granite was taken up primarily by French and German geologists, many of whom specifically espoused the idea that granite crystallized from a "dry" melt lacking in water, a melt that they began to designate as "magma." In 1844, the French geologist and chemist, Joseph Fournet (1801–1869) provided the first serious attempt since Playfair to answer the argument from texture against the origin of granite by dry fusion. Fournet graduated from École des Mines in 1822 and became Professor in the Faculté des Sciences at the University of Lyon in 1834, where he specialized in the study of metals and mining geology. Fournet was probably the first geologist to use the term "magma" in its modern sense in reference to igneous melt. In a paper on the circumstances of crystallization in veins, Fournet (1838) noted that "petrosilex" could be chemically distinguished from feldspar by a great excess of silica that is intimately disseminated in a "magma." He also referred to "injected magma." The term "petrosilex" had originally been applied by Alexandre Brongniart to felsites on the mistaken assumption that they were hornstones, that is, very finely crystalline, brittle, flint-like quartz rocks. The rock was considered roughly equivalent to the Scandinavian "hälleflinta." Both rocks eventually proved to contain both quartz and feldspar. Despite the fact that the name felsite was already available, the rock name "petrosilex" persisted for a few decades, at least until the 1860s, and then finally faded from view.

In regard to the texture of granite, Fournet (1844) pointed out that under certain conditions the melting points of some substances appear to occur at higher temperatures than those at which they normally freeze. He suggested that the solidification of quartz in granite might be delayed to a much lower temperature than its fusion point by this phenomenon, which he termed "surfusion." The "surfusion" of quartz, he proposed, accounted for the fact that quartz typically forms later than feldspar. We know now that the equilibrium melting and freezing points of a pure mineral are identical. Before crystallizing, however, many silicate liquids may be greatly undercooled because of their extreme viscosity. Such behavior led Fournet and his contemporaries to believe that the melting and freezing points of a mineral are not necessarily the same. For virtually the entire nineteenth century, therefore, experimentalists made a distinction between the melting point and the solidification point of a mineral.

Fournet's theory of surfusion failed to gain much support from any quarter, even from Durocher, one of the most enthusiastic proponents of dry fusion of granite. Joseph Durocher (1817–1860) was a graduate of École Polytechnique and École des Mines (Joubin, 1900; Troalen, 1943). Durocher participated as an engineer on a scientific voyage to Spitzbergen, the Faeroe Islands, and Finland in 1839 and reported on the mountain structure, glacial erosion, and glacial deposits of those regions. In 1841, he was appointed to the Faculty of Sciences at the University of Rennes and continued service as a mining engineer. His research included investigations of glacial phenomena, structural geology, and metalliferous deposits in Scandinavia as well as igneous and metamorphic rocks throughout the Pyrenees, France, the Alps, and Central America. In a major contribution to the discussion about the origin of granite in which he perpetuated Fournet's use of the term "magma," Durocher (1845) began by noting that the relative arrangement of the minerals (which, at the time, many geologists called "elements") in granite seemed to present an anomaly to the laws of physics in that the arrangement appeared to be incompatible with the easy fusibility of feldspar and mica and the heat-resistant character of quartz. As virtually everyone had recognized, the feldspar commonly imprinted its crystal outlines on the enveloping quartz, indicating that the former had crystallized first. Durocher pointed out examples in which that was indeed the case, but, like Playfair, he also described examples in which quartz crystals are surrounded by feldspar. These reversed situations, he argued, pointed inevitably to the conclusion that the solidification of all the constituent "elements" had to occur at about the same time. How then, he asked, do we explain the nearly simultaneous crystallization of substances whose fusibilities are so different?

Fournet's theory of surfusion was insufficient to account for the observed textures. The problem, said Durocher, is that the temperature difference between the melting or fusion point and the point of solidification of a particular substance is rarely more than 100°C. In his estimation, that was much too small a difference to account for the crystallization of feldspar, tourmaline, or garnet prior to quartz. After all, he pointed out, the difference in fusion points between these minerals and that of quartz is hundreds of degrees. Rather than appealing to Fournet's surfusion theory for support, as was erroneously stated by von Zittel (1899) in his work on the history of geology, Durocher made the far-sighted claim that the component parts of the individual minerals in the molten state of granite were not isolated from one another as they are in the minerals within a rock. In other words, he suggested that melted granite did not consist of molten feldspar domains, molten mica domains, and molten quartz domains, as was commonly envisioned. Rather, he believed that the substances of the various minerals were combined into a homogeneous mass composed of silica, alumina, alkaline and earthy bases, potash, soda, lithia, a little lime and magnesia, oxides of iron and manganese, and even hydrofluoric and boric acids. In effect, Durocher suggested that "le magma granitique" was something like a solution without making the identification explicit as Bunsen (1861) would do later. Durocher proposed to show that the granitic magma would stay fluid and homogeneous while losing its heat until a temperature just above the fusion point of feldspar.

Prior to crystallization of the minerals, Durocher suggested, the silica in the melt was combined with the other silicates in a manner that he regarded as analogous to that of sulfuric acid and alkali sulfate. When silicic acid is united to an alkali silicate or an earthy silicate, he said, heat is produced just as heat is released when sulfuric acid unites with an alkali sulfate. Conversely, he pointed out that separation of these compounds would be accompanied by the absorption of heat. Durocher claimed that when quartz separated from cooling magma, it absorbed heat from the melt, thereby keeping the quartz in a softened condition and also lowering the melt temperature and hastening its solidification. The lengthened state of softness of the quartz would enable it to take the imprint of the crystalline form of the feldspars.

Perhaps not fully confident of his own reasoning, Durocher claimed that even if this temperature-lowering mechanism were insufficient, it still would not invalidate the theory of igneous fusion. With nearly simultaneous crystallization, he argued, some minerals will crystallize sooner than others depending on whether they pass through a viscous state in solidifying or take on the solid state almost immediately. The reason that quartz is generally the last mineral to crystallize, he maintained, is because it normally passes through a viscous condition before it solidifies. Under such conditions quartz may even be pulled into threads. In contrast, he said that feldspar, whose solidification is accelerated by a great tendency to become crystalline, passes much more rapidly from the state of fusion to the solid state. As a result, in the process of consolidating, the quartz remains pasty and soft at the moment when feldspar crystallizes. In Durocher's mind, the fact that quartz leaves an imprint on feldspar in some cases obviously showed that feldspar and quartz both pass more or less simultaneously from the liquid state to the solid state although the amount of time required for the passage would not be the same for both minerals.

Durocher then developed an elaborate argument to demonstrate that mica, feldspar, and quartz, the three "elements" of granite, remained in a combined state until their time of solidification. From the variation in the proportions of quartz, feldspar (orthoclase and albite), and mica in granites and the average chemical compositions of feldspar and mica, Durocher determined average chemical compositions for very feldspathic granites, very micaceous granites, and normal granites. He compared these compositions to those of petrosilex and noted the striking similarity in the composition of petrosilex to the feldspathic and micaceous granites. But, Durocher pointed out, petrosilex would easily melt in the flame of a blowtorch, and its fusibility is only a little less than that of feldspar. From this evidence, he concluded that granites are more fusible than one would suspect initially. The key to solving the problem lay in the fact that when the various silicate minerals were combined, they were much less resistant to melting than when they were isolated. In Durocher's view, feldspar, mica, and quartz associated in the same material possessed a fusibility that is greater than the average fusibility of the minerals considered individually. He saw no reason why granite could not keep all of its "elements" together in the liquid state just prior to solidification if petrosilex could do it.

Durocher drew further evidence from the character of porphyry. Porphyries, he noted, have textures ranging from that of petrosilex to that of granite, and in some cases the textural variants may be seen in the same mass. He was persuaded, too, of the similarities in chemical composition, mineralogy, and density among porphyry, petrosilex, and granite. Moreover, he had found porphyries of the same age as granite, in contradiction to the very common opinion that granites are much older than porphyry. He concluded that:

> All the considerations converge then to justify the combination of granite, of quartz porphyries, and petrosilex in the same class of rocks, unquestionably the most important of all. These three substances, if differing outwardly, constitute the three terms of a series, the granite and the petrosilex being the extreme terms and the quartz porphyries establishing a link between the two. Thus, the granite had to be originally from masses of analogous composition to those of petrosilex; when their cooling had taken place without dividing the elements, they remained in the state of petrosilex; when the separation of elements was incomplete, it formed a quartz porphyry . . . . Finally, when the separation of elements reached its last term, when the igneous mass, which was first of all at the state of a soft paste and homogeneous, was entirely decomposed and resulted in three or four different minerals, it gave birth to granites; the petrosilex rocks then achieved complete development. (Durocher, 1845, p. 1283)

Calling on his experience with metal alloys, Durocher likened this difference in crystallization behavior of granitic liquid to that observed in melted iron when cooled. Under some conditions, he said, some of its carbon is released in crystalline form (black melting), but under other conditions it retains all its carbon in a state of combination (white melting). Cooling circumstances, he suspected, played a crucial role in the formation of the end product, whether petrosilex, porphyry, or granite. A more rapid cooling, he asserted, took place during the formation of petrosilex and porphyry, so that the separation of the constituent minerals was more difficult than in more slowly cooled granites.

Durocher concluded his seminal paper by observing that other presumably igneous rocks had their fine-grained equivalents: aphanite was the equivalent of diorite, melaphyre was the equivalent of basalt, and trachytic porphyry the equivalent of trachyte. "It would be astonishing," Durocher (1845, p. 1284) observed, "that the granites alone . . . never present themselves in the compact state; until the present they seemed to be the exception to the general law, but their combination with the quartz porphyries and the petrosilexes made this exception disappear and came to fill the gap in the natural series of igneous rocks."

Durocher also made a presentation on the origin of granite before the Societé Géologique de France at its meeting of June 7, 1847. Having heard Scheerer's paper at the previous meeting, Durocher (1847) prepared a rebuttal in which he reiterated many of the same lines of argument that he had developed in his 1845 paper. He presented hand specimens and drawings of textures demonstrating that quartz is not always the last mineral to solidify from granite. He reviewed his claim that the minerals in granite crystallized almost simultaneously and that granite, porphyritic trachyte, and petrosilex are closely related. He refuted the argument that quartz does not appear in slowly cooled silica-rich lavas by pointing out that, although presumably not common, there are examples of trachytes from Siebenberge, the Auvergne, and Italy that do contain well-formed quartz crystals. Moreover, he asserted that some of the Italian examples display textural gradations among trachyte, trachyte porphyry, quartz porphyry, and granite. In the eyes of Durocher (1847, p. 1028), these rocks provided "one of the most powerful arguments in favor of the igneous origin of granite." He dismissed Scheerer's argument from the presence of pyrognomic minerals by pointing out analogous situations in which a substance, after having been altered, eventually returned to its original state. Why, he asked, might not a pyrognomic mineral also regain its original characters long after having cooled from a melted state?

Durocher said that he would be the first to accept Scheerer's suggestion that granite might have formed by aqueo-igneous fusion if it were in harmony with facts, but he set out to demonstrate that it was not. On the basis of several chemical analyses he had made, he belabored the point that granites contain only small amounts of combined water, in many cases less than one percent. He knew, too, that where obvious signs of alteration, like kaolinization, are present the water content increases markedly. As a result, Durocher was not fully convinced that the water in seemingly fresh granites might not also in some way be the result of atmospheric contamination. Although he conceded that water in granite might be original, he maintained that the source of that water really had to be regarded as uncertain.

Scheerer had suggested that the melting of hydrated salts at lower temperatures than their anhydrous analogs provided a model for the lowering of the fusion point of granitic liquid containing water. Durocher countered that the analogy was invalid because the melting of a hydrated salt entails little more than increased solubility in its water of crystallization at elevated temperature. What evidence is there, Durocher challenged, of a vastly heightened solubility of quartz and feldspars in water? We know next to nothing about the solubility of quartz in water, he said. Moreover, he believed that there is insufficient water in granite to dissolve the silicate minerals.

Noting the copious volumes of water vapor emitted at volcanic centers, Durocher pointed out that no one had cited any evidence that the water had lowered the fusion temperature of the lava to any significant degree in comparison with an anhydrous silicate melt observed in a furnace. In addition, he said, lavas often contain cavities that are filled with druses as a result of the water present, but granite, as a very compact crystalline rock, presented no such cavities or vesicular structure as lava. Durocher took this lack of vesicularity as one more line of evidence that granitic melts contained very little water. In the end, Durocher said that he would find Scheerer's idea of aqueo-igneous fusion of granite easier to accept if granites contained 12 to 15 percent water and if the minerals of granite were more notably soluble in water. Over the next few years, Scheerer and Durocher offered brief rebuttals of each other's views, but neither moved from his fundamental position.

Many prominent German chemists, geologists, and petrographers also leaned toward a plutonic origin of granite. They did not particularly like the term "igneous" because they felt that it implied the presence of fire in the formation of the rocks. They also believed that metamorphic rocks had been subjected to heat just as much as igneous rocks had. As a result, they much preferred to use the terms "eruptive" or, in some cases, "pyrogenic." For the next several decades, therefore, German geologists characteristically referred to igneous rocks as "Eruptivgesteine" (eruptive rocks), a designation that did not meet with universal approval. England's Joseph Jukes, for example, thought it preferable, at least as regards what we now think of as plutonic igneous rocks, to speak of "intrusive" or "irruptive" rocks inasmuch as such rocks were not erupted onto the surface in a molten condition.

Among the German plutonists were Naumann and von Cotta. Karl Friedrich Naumann (1797–1873) was arguably the leading petrographer of the pre-microscopic era (Burke, 1974b). Naumann was a student of Werner at Freiberg in 1816, and much of his early professional career was spent at the University of Jena and the Bergakademie, first as Professor of Crystallography beginning in 1826, and then as Professor of Geognosy beginning in 1835. At the height of his career, Naumann moved to the University of Leipzig in 1842, where he became Professor of Mineralogy and Geognosy. At Leipzig, he issued a two-volume work, *Lehrbuch der Geognosie* (Naumann, 1850, 1854), probably the most authoritative work on petrography of the mid-nineteenth century. A three-volume second edition appeared between 1858 and 1866.

In the first volume of *Lehrbuch der Geognosie*, Naumann (1850) addressed the issue of the origin of granite. He suggested that Durocher (1845) had satisfactorily answered concerns about the textural character of granite. The appearance of quartz, he concluded, offered no serious objection to the pyrogenic origin of granite. Naumann perpetuated Durocher's references to granitic "magma" by using the term repeatedly in expressions such as "homogeneous molten magma" and "water-free magma." Use of the term "magma" was now beginning to take hold in reference to igneous melts. Naumann was, however, attracted to Scheerer's suggestion that granitic melts contain small amounts of water that contributed to considerable lowering of their solidification temperatures.

Another prominent German petrographer who leaned toward plutonic granite was Carl Bernhard von Cotta (1808–1879). Von Cotta graduated in 1831 from the Bergakademie, where he studied under mineralogist Johann F. A. Breithaupt (1791–1873) and learned neptunian geology from K. A. Kühn (Prescher, 1971). After leaving Freiberg he studied under von Leonhard for a year in Heidelberg. In 1833, he began participation in the Geological Survey of Saxony under the leadership of Karl Naumann and brought the project to completion by 1845. During a part of that time, von Cotta taught at the Tharandt Forestry Academy, but upon the departure of Naumann from Freiberg in 1842 for the University of Leipzig, von Cotta succeeded him as Professor of Geognosy and Paleontology. Von Cotta spent the remaining 32 years of his professional career at the Bergakademie. He first lectured on geognosy, but later added paleontology and perhaps the earliest course in ore deposits. In 1848, he was one of the founders of the German Geological Society (die deutsche geologische Gesellschaft).

In his petrographic work, *Die Gesteinslehre*, von Cotta (1855) accepted Durocher's reasoning regarding the order of crystallization in granite and approved the idea of Bunsen (1861) that granitic melt is a solution. Moreover, von Cotta said that the objection against the igneous origin of granite and some other crystalline rocks from the presence of relatively fusible, but early crystallizing accessory minerals lost its force because some of those same accessory minerals occur in lavas of undoubted igneous origin.

The defense of dry igneous fusion by Fournet and Durocher did not convince everyone. Like Durocher, Scheerer (1847) also disputed Fournet's hypothesis of surfusion, taking him to task for forgetting that the maximum difference in temperature between the fusion point and the solidification point observed for any known substance, namely sulfur, was only about 100°C. Knowing that silica requires a higher temperature to melt than platinum metal, Scheerer estimated the fusion temperature of quartz around 2800°C on the grounds that the melting point of platinum was thought to be a little above 2500°C and that quartz had been melted in a blowtorch flame whose temperature was estimated at 3100°C. He believed that minerals like pyrite probably melt around 1000°C and minerals like amphibole, garnet, and tourmaline have fusion points below 1400°C. As a result, Scheerer maintained, quartz would need to have the property of solidifying anywhere from 1300°C to 1800°C below its point of fusion for the surfusion hypothesis to explain granitic texture. Carl Bischof completely agreed that Fournet's idea of "superfusion" was entirely inadequate to save the case for igneous quartz.

Scheerer also noted that Durocher (1845) had attempted to explain the textural evidence in granite by maintaining that crystallization was essentially simultaneous. If so, Scheerer claimed, one should encounter crystals of feldspar and other more fusible minerals, but not quartz. He believed that, on Durocher's theory, an amorphous silicate something like petrosilex should have solidified in the spaces between feldspar crystals. Durocher's conception, he alleged, should lead to the production of a porphyry that lacks quartz.

FOUNDATIONS FOR A RESOLUTION OF THE DEBATE

For several decades, French and German geologists, the principal participants in the debate about the origin of granite, were unable to reach a consensus. Theories of origin included the injection of anhydrous granitic melts (dry fusion), the injection of granitic melts containing water under pressure (aqueo-igneous or igneo-aqueous fusion), and the conversion of sedimentary rocks into granitic rocks by metamorphic processes aided by water. All these views had several able advocates. Had it not been for the application of the polarizing microscope to the study of rock thin sections beginning in the 1850s and 1860s, theorizing about the origin of granite might have remained in a stalemate for many more years to come. Although not appreciated at the outset, the new technique eventually offered new insights toward a solution of the long-standing knotty problem. The development of the new microscopic methods also lured British geologists back into the fold of thinkers about granite after a long absence.

It was Henry Clifton Sorby (1826–1908) who introduced the polarizing microscope into the discussion about granite. He also led Britain back into the discussion about the nature of granite. Sorby early developed interests in science (Judd, 1908; Higham, 1963). As a man of independent means, he was able to devote himself exclusively to a life of scientific research. After an early interest in the chemical analysis of agricultural materials, Sorby became much intrigued by the processes of sedimentation. As a very young man in the 1840s, Sorby had already begun studying the fossil shells in the Bridlington Crag, which crops out near Scarborough on the North Sea coast of his native Yorkshire, and he was in the habit of using a microscope to study them. It was during this period that he made the acquaintance of William Williamson on a trip from Scarborough to York. Williamson, already skilled in the use of diamond and emery wheels to prepare thin sections of hard materials like teeth, bones, scales, and fossil wood, gave Sorby a lesson in making such slices. Given his interest in rocks, Sorby began to make his own thin slices of rocks and began to publish papers on his discoveries. After a handful of papers on both sedimentary and metamorphic rocks, Sorby (1851, 1853) wanted to see if the new method could shed any light on the debate over granite.

In a paper read before the Geological Society of London on December 2, 1857, and entitled "On the microscopical structure of crystals, indicating the origin of minerals and rocks," Sorby (1858) stated that he would demonstrate that both artificial and natural crystalline substances possess characteristic structures indicating whether they were deposited from aqueous solution or crystallized from a state of igneous fusion. He proposed to approach the question through a study of the fluid inclusions within minerals. Sorby compared the character of fluid inclusions within artificial crystals formed by precipitation from solution, by sublimation, and by fusion with those contained in natural crystals collected from deposits of halite and calcite, quartz veins, metamorphic rocks, volcanic rocks, and granitic rocks.

Sorby found that many of the fluid inclusions in granites display the same characteristics as inclusions in known volcanic rocks. He also found that granitic rocks possess water-filled cavities containing crystals of alkali chlorides and sulfates. He concluded that the range of types of inclusions in granites indicated that they were the product of igneous fusion but specifically fusion in the presence of water under pressure. Sorby, therefore, envisioned the molten silicate rock as dissolving water. Consequently, he found himself in general agreement with Scheerer and Élie de Beaumont concerning the origin of granite. Granites were not, in his view, the product of pure dry igneous fusion. Sorby estimated pressures and temperatures from the inclusions and determined that all the granites that he examined were consolidated under pressures equivalent to those exerted by rock columns 18,000 to 78,000 feet thick. He envisioned a continuum extending from volcanic rocks formed by pure igneous fusion to quartz veins formed by aqueous deposition with granite forming the central link in the series.

According to Dawson (1992), one of Sorby's early studies of thin sections entailed an examination of some sections of metallurgical slags, the products of James Hall's fusion experiments on basalt, and fused samples of Mount Sorrel granodiorite. Sorby (1863) compared thin sections of Hall's fused granodiorite sample with those of the natural rock and concluded that the differences could be attributed to the much slower cooling of the natural granodiorite and to the presence of water under pressure in the granodiorite melt. He maintained that the water acted as a flux and promoted the coarse grain of the granodiorite.

Sorby's fellow Englishman David Forbes (1826–1877) also entered the granite debate in the 1860s (Sorby, 1876; J. M., 1877). A graduate of the University of Edinburgh, Forbes spent much of his career as a mining engineer working with private companies. He developed a wide-ranging knowledge of geology through his extensive experience in such diverse locations as Norway, the South Seas islands, and South America. After returning to London, Forbes became one of the very earliest geologists to be exposed to the possibilities of the new thin-section microscopy. He was aggressively enthusiastic in promoting microscopic investigation and became very impatient with other geologists who failed to incorporate the latest knowledge in chemistry and other sciences into their geological work. His zeal for avant garde science brought Forbes into open conflict with James Geikie and Hunt in regard to the granite question. While agreeing with Geikie that the origin of granite would ultimately be solved by the field observer, Forbes (1867a) insisted that it would be a field observer who had a sound knowledge of chemistry, mineralogy, mathematics, physics, and the new microscopic methods developed by Sorby. Even more than his mentor, Forbes lobbied his fellow countrymen to take full advantage of the insights that could be gleaned through this new technique. While the "mere field observer" might be taken in by Geikie's assertions that stratified beds like graywacke could be converted in situ into granite "by the wondrous agency of hydrothermal action," Forbes (1867a, p. 51) charged, the geologist who possessed even a little knowledge of chemistry would immediately see the incorrectness of the chemistry on points "where even the merest tyro ought not to blunder."

Forbes chided Geikie for failing to provide a chemical demonstration that graywackes had been converted into granite. He also thought it suspicious that graywacke could allegedly be converted into diorite and serpentine, rocks that differed considerably from granite both mineralogically and chemically. He criticized Geikie for failing to determine the kind of feldspar present in the allegedly transformed rock. How then, Forbes asked, can he be sure that the rock is a granite?

Forbes also carried on a running feud with Hunt for several years. A typical response of Forbes (1867b, p. 442) to Hunt charged that it had become fashionable to "'pooh-pooh' the igneous origin of eruptive rocks in general, and of granite in particular." Forbes regarded this attitude as a secession from opinions about granite that, he said, had more or less become universally adopted. He complained that Hunt, like so many geologists and chemists earlier in the century, continued to push the tired claims that quartz, one of the constituent elements of granite, resulted only from secondary processes and that granite, because it contains quartz, must always be a rock of sedimentary origin. Does not Hunt know, Forbes (1867b, p. 442) asked, about the "immense masses of undoubted volcanic rocks scattered all over the surface of the globe which contain abundance of free quartz?" He wondered, too, if Hunt knew about Sorby's microscopic studies showing the identity in structure between volcanic quartz and the quartz of granite and if he knew about Sorby's conclusion that modern volcanic trachytes and old granites have a common igneous origin in which water has played some part.

Within a few years, continued prodding by the microscopists would open the door to much wider acceptance of igneous granite. Another development that ultimately contributed to a recognition of the igneous origin of granite was Bunsen's claim that granitic melt is a solution. Robert W. E. Bunsen (1811–

1899) studied chemistry, physics, and mineralogy at the University of Göttingen (Schacher, 1970). Bunsen (Figure 7) spent much of the 1830s visiting various laboratories and sites of geological interest throughout Europe. In 1832, he studied the mineral and rock collections of the mineralogist C. S. Weiss at the Bergakademie of Freiberg. The following year, Bunsen began his academic career with a series of posts at Göttingen, Kassel, and Marburg. He moved to the University of Heidelberg in 1852, where he remained until retirement. Bunsen's brilliant career as an experimental chemist included the discovery of cesium and rubidium, important work on spectroscopy and galvanic batteries, and the development of instruments such as the famous burner that bears his name.

In 1861, Bunsen published a three-page article entitled "Über die Bildung des Granites." Bunsen (1861) wished to call attention to the long-held erroneous assumption that the igneous origin of granite was negated by the fact that the highly infusible quartz had crystallized last. Bunsen said he was at a loss to understand how that erroneous conclusion had survived for so long and was still being used. Although the earlier efforts of Hall and Durocher to explain granitic texture were moving in the right direction, it was Bunsen who first explicitly pointed out that granitic magma is a solution. "No one," Bunsen (1861, p. 61) lamented, "appears to have thought about the fact that the temperature at which a body solidifies independently is never that temperature at which the body becomes fixed in another body when (the first body) solidifies out of its solutions." One may not object, Bunsen said, that granitic melt cannot be a solution simply because it is hotter than aqueous solutions, and no chemist would arrive at the "widersinninge Idee" (absurd idea) that a solution stopped being a solution just because it is heated a few hundred degrees. Moreover, he pointed out, there is the phenomenon of freezing-point depression that occurs in mixtures such as water and calcium chloride where the temperatures at which ice or calcium chloride crystals form depend on the relative proportions. In any case, Bunsen observed, in these mixtures, the solidification point of water can be lowered 59°C below its freezing point as a pure substance, and the solidification point of calcium chloride can be lowered nearly 100°C below its freezing point as a pure substance. Bunsen also reminded his readers that the sequence of crystallization from solution, whether first water and then salt or first salt and then water, depended on the proportions of water and salt in the solution. For the same reasons, Bunsen insisted, quartz and feldspar will not solidify from their molten granitic solution at their respective melting points. He agreed with the observation of Durocher that in feldspar-rich graphic granite, quartz separates before feldspar, that in other granites quartz crystallizes simultaneously with feldspar, and that in still other granites quartz forms after the feldspars. Bunsen claimed that such phenomena are exactly what one would expect of a solution in which the proportions of dissolved quartz and feldspar varied. Quartz, he concluded, crystallizes from a molten

7. Robert Bunsen (1811–1899). Reproduced by permission of the Archives of the University of Heidelberg.

granitic solution well below its melting point just as it crystallizes from an aqueous solution well below its melting point.

The nineteenth-century debate over the origin of granite largely fizzled out with the dispute between Hunt and Forbes in the 1860s. In the 1870s, the growing acceptance of both the results of microscopic petrography and the solution theory of magma also led to widespread acceptance of igneous, eruptive, magmatic granite. The nineteenth-century furor over the origin of granite had continued for several decades and involved the efforts of some of the greatest geologists of the era. At first, the granite debate primarily involved geologists of France, Germany, and, to a lesser extent, Scandinavia. Only after the late 1850s did geologists from Great Britain and finally America enter the fray. Opinions did not fall out markedly along nationalistic lines. The notion of dry igneous fusion was accepted by French geologists Durocher and Fournet and German geologists Naumann and von Cotta. Aqueo-igneous fusion entailing genuine intrusive granitic melt was accepted by Scheerer and also by Naumann in Germany, Delesse and Élie de Beaumont in France, and Sorby and Forbes in England. Conceptions that entailed the idea of metamorphic transformation in the presence of aqueous fluids were posited by Daubrée in France, Bischof and Rose in Germany, Geikie in England, and Hunt in America. Some geologists, like Durocher, linked their view of granite to a larger conception of primitive magma shells in the interior from which magmas of contrasting composition were derived. Others leaned toward derivation of mobilized granite, whether molten or not, from preexisting rocks within the lower part of the solid crust. Into the 1860s, no consensus had been reached, and the views of geologists about granite played a major role in the manner in which they classified the entire spectrum of rocks.

7 ⟻ CLASSES AND ORDERS

Petrography and Classification in the Early Nineteenth Century

In the early decades of the nineteenth century, most geologists accepted the existence of ancient volcanic rocks and some also accepted the existence of what we now consider to be intrusive igneous rocks. As increasing numbers of geologists investigated more exposures of both volcanic and intrusive crystalline rocks in more localities around the world, it became increasingly evident that such rocks are characterized by considerable variation in mineral content, chemical composition, texture, and geological setting. Geologists who studied real or alleged igneous rocks were forced to develop a language for discussing them with the result that new rock names and textural terms were constantly being coined. During the early years of the debate over the origin of basalt, it was recognized that not all the rocks of presumed volcanic origin are dark-colored basalt. John Strange (1775), for example, detected differences among the columnar rocks in the Euganean Hills, and other investigators recognized lighter-colored flows locally intercalated with the darker ones. To account for the lighter-colored rocks, mineralogist René Haüy had introduced the term "trachyte" in reference to a rock having glassy feldspar crystals in a gray or white feldspathic groundmass with accessory mica.

IGNEOUS ROCK NAMES IN ANTIQUITY

Geologists who participated in the neptunist–vulcanist–plutonist disputes of the latter eighteenth and early nineteenth centuries, however, had already inherited a lengthy list of rock names, including several that referred to rocks now considered as igneous: obsidian, pumice, porphyry, basalt, granite, and syenite. Some of these names had been coined by miners in previous centuries, and still others could be traced to the writers of ancient Greece and Rome. Prior to the recognition of basalt as a volcanic rock, some materials were generally considered as volcanic by both neptunists and their opponents. In his early classification, Werner accepted pumice, lava, and solidified volcanic ash, that is, volcanic tuff, as the only legitimate volcanic rocks (Ospovat, 1971). Of these terms, "pumice" can be traced to classical times when it was known from volcanic terranes. Theophrastus and other Greek writers mentioned pumice. Vitruvius referred to the rock as "lapis spongia" because of its sponge-like appearance. During the Renaissance, Agricola described the characteristics and localities of "pumex" (Bandy and Bandy, 1955). Werner applied the German term "Bimstein" to pumice. Although he associated pumice with volcanism earlier in his career, he ultimately regarded it as an aqueous product to be classified with his newest Flötz-trap unit along with obsidian and pitchstone.

Although Werner did not refer to volcanic ash, or tuff, as tephra, the latter term does stem from classical writers. Pliny the Elder used the word "tephrias" for a rock that was probably of volcanic origin inasmuch as the Greek word "tephra" means "ashes." "Tephrine" was used in the very early nineteenth century by de Lametherie, Brongniart, and Cordier for some volcanic rocks.

Several other names that are presently applied to igneous rocks also date from the classical era. Obsidian, despite its rejection as volcanic by Werner, was regarded by some of his contemporaries as a volcanic rock. According to Harrell (1995), Pliny the Elder said that "obsian" ware, so named from its resemblance to the stone found by Obsius in Ethiopia, was included in the classification of glass. Pliny described the stone as very dark, sometimes translucent, and having a cloudier appearance than glass so that when used as a mirror it reflected shadows rather than sharp images. He had seen gems and statues of Caesar Augustus, elephants, and Egyptian gods made of solid obsidian. Pliny had also seen or heard reports of obsidian from India, Italy, and Spain along the shores of the Atlantic. Whether the Greeks and Romans associated obsidian with volcanoes is not clear.

Whatever the precise etymology of the term "basalt," the word was used by the Romans and Egyptians with reference to rocks used for building monuments. Harrell traced the etymology of the term to the Egyptian "bekhen-stone," a dense, dark graywacke quarried from the eastern desert of Egypt at Wadi Hammamat. He suggested that the Romans used the Greek term "basanites" (touchstone) as a rough equivalent for the "bekhen-stone." Pliny the Elder referred to this rock in his *Natural History*. Harrell maintained that during the Middle Ages a transcription error entered into some of the manuscripts of Pliny's work and that the word "basanites" was transcribed as "basaltes."

Agricola employed the term "basalt" in a section dealing with various types of marble in book VII of *De Natura Fossilium* (Wilsdorf, 1970). Although true marble is usually named after the place where it is found, said Agricola, in some instances it is named after people such as Augustum (Caesar Augustus) and Tiberium (Caesar Tiberius), which latter rock resembled Ethiopian "basaltes" in color and hardness. Agricola further noted that "some marble is iron-gray, for example, the basaltes from Egypt that is found in Ethiopia" (Bandy and Bandy, 1955, p. 154). A similar marble from Misena in Germany was said to be not inferior to "basaltes" either in color or hardness. Agricola noted the existence of angular columns in the "basaltes" in Misena.

The word "basalt" was also used by Conrad Gesner (1516–1565) and passed down into modern European usage for dark crystalline rocks. The Swedish mineralogist Johan Gottschalk Wallerius (1709–1785) spoke of four types of basalt, including a columnar variety, and mentioned the material at Giant's Causeway as an example.

The term "porphyrites" was also coined by Pliny to refer to a red rock with white spots at Gebel Dokhan in the Eastern Desert of Egypt (Harrell, 1995). Pliny reported that statues of the rock were brought from Egypt to Emperor Claudius in Rome. The rock was quarried from about A.D. 50 to 400. The original "porphyrites" is a purplish red andesite with small white to pink plagioclase pnenocrysts. The rock name was derived from the Greek "porphyra" for purple. Agricola made numerous references to this rock.

As we have seen, some geologists at the beginning of the nineteenth century accepted the notion that granite and syenite are rocks that solidified from injected, intrusive masses of subterranean lava. These rock names, too, had a long history. The origin of the term "granite" is obscure. It was reportedly used by Cesalpinus in 1596. Johannsen (1937) speculated that the word was derived from the Italian word "granito" (grained). The term gradually became restricted to coarsely crystalline rocks composed of quartz, feldspar, and mica or hornblende. "Syenite," too, can be traced to ancient Egypt (Harrell, 1995). Syene was the Greek name for an old Egyptian city (1200 B.C.) on the east bank of the Nile in Upper Egypt, formerly known as Swnt and now known as Aswan. Pliny the Elder introduced the term "syenites" in reference to a granite with large red microcline crystals that was quarried at Aswan. Monoliths of the rock were constructed by the kings and dedicated to the Sun god. The term "syenite" was also applied to black granodiorite quarried at Aswan. Agricola spoke of "syenites" as a black rock with reddish spots. The term eventually came to be applied to quartz-poor granitic rocks.

Additional rock names were invented as scientific interest in minerals and rocks increased. During the latter eighteenth and first half of the nineteenth century, the development of a growing nomenclature often occurred in the context of efforts to devise meaningful classifications for all sorts of rocks. Most new rock names, textural terms, and rock classifications of the early nineteenth century emerged from developing German and French petrographic traditions. Lacking the sophisticated instruments available to modern researchers, petrographers of that era were limited to a few very basic tools: a keen, discerning eye, a sturdy hammer, and a hand magnifier. Where possible, hand specimen identification was augmented with blowpipe analysis of separated mineral fragments, chemical analysis of the rock or of mineral separates, and microscopic examination of mineral splinters and slices in ordinary light.

CLASSIFICATION IN GREAT BRITAIN

Despite the leading role of British geologists like James Hutton, James Hall, and John Playfair in making the case for the igneous origin of basalt and granite, a viable petrographic tradition failed to materialize in Great Britain during the first half of the nineteenth century. Although historian John Pinkerton wrote *Petralogy* in 1811, and geologist John Macculloch devised a rock classification in 1821, little attention was paid to igneous rocks in Britain for decades thereafter. For several years after the *Quarterly Journal of the Geological Society of London* began publication in 1845, virtually no articles dealing with igneous rocks were published in that periodical. Lyell and Scrope contributed a couple of papers in which they opposed von Buch's idea of craters of elevation and pondered the internal nature of lava, but no British geologist wrote about igneous *rocks*, whether volcanic or plutonic. Generally, however, rocks and rock structure were subjects to which little very attention had been paid in Britain since the publication of Macculloch's classification. The explanation for this curious deficiency resides in the fact that the early nineteenth-century British geologists like Buckland, Sedgwick, Murchison, Mantell, Lyell, Owen, Conybeare, and Forbes became enamored with stratigraphy, paleontology, and geomorphology, fields to which the geology of England lends itself so successfully. With all the igneous rocks tucked away in the more remote, less accessible reaches of Britain, well up into Scotland or far to the west in Cornwall, common sense dictated that the simple, fossil-rich stratigraphy of the English lowlands around London, Oxford, and Cambridge would receive the lion's share of attention. The spectacular harvest of stratigraphic, paleontologic, and geomorphic results guaranteed further interest. Why trouble oneself with inscrutable crystalline rocks when there were great vertebrate monsters to be discovered, described, and reconstructed in museums for a fascinated public? In time, Great Britain would become a world leader in the study of igneous rocks, but, for now, British enthusiasm for igneous rocks would have to wait until after the polarizing microscope was successfully used in the study of thin sections of igneous rocks in Germany in the 1860s and 1870s. Only then did a British petrographic tradition began.

THE GERMAN PETROGRAPHIC TRADITION

The situation was completely different in Germany. Whereas the English became absorbed by the study of strata and fossils, petrography became a national geological obsession in Germany beginning with the work of Werner. German petrography was overwhelmingly the province of academicians. The tradition of petrographic precision and excellence established by Werner at the Bergakademie in Freiberg, Saxony, was carried on at that institution for decades thereafter by geologists like Naumann and von Cotta. Throughout the nineteenth century, universities in Heidelberg, Berlin, and Leipzig established strong traditions of petrographic research.

Throughout the first half of the century, numerous textbooks on rocks appeared, such as *Charakteristik der Felsarten* (von Leonhard, 1823), *Die Gesteinslehre* (von Cotta, 1855), and *Classification und Beschreibung der Felsarten* (Senft, 1857). Rock classification schemes appeared in general geology texts like *Lehrbuch der Geognosie* (Naumann, 1850, 1854). German writers also contributed numerous articles devoted to the study of specific occurrences of eruptive rocks in geological journals. During the end of the eighteenth century and the beginning of the nineteenth century, geologists had written scientific travel books or contributed to generalized scientific journals, such as the publications of the Royal Society of London and the Academy of Sciences in Paris, or to journals devoted to physics, chemistry, or mining. In the 1830s, however, journals devoted almost exclusively to geology began to emerge. In Germany, K. C. von Leonhard and H. G. Bronn of the University of Heidelberg began to edit *Jahrbuch für Mineralogie, Geognosie, Geologie, und Petrefaktenkunde* in 1830 as an outgrowth of von Leonhard's *Taschenbuch für die gesammte Mineralogie*. Various national geological societies sprang up in England, France, and Germany, and eventually these societies began to publish their own journals, the *Bulletin de la Societé Géologique de France* (1830), *Quarterly Journal of the Geological Society of London* (1845), and *Zeitschrift der deutschen geologischen Gesellschaft* (1849). Within its first six years, von Leonhard and Bronn's *Jahrbuch* included articles on glassy rocks, syenite, felsite, and volcanic eruptions. The first six years of *Zeitschrift der deutschen geologischen Gesellschaft* contained papers on granite, feldspathic rocks, basalt, nepheline-rock, serpentinite, trachyte porphyry, and phonolite. Eruptive rocks, as German writers typically called them, were very much on the minds of German geologists.

Many nineteenth-century German attempts at rock classification were strongly influenced by the successful application of Linnéan methodology to the organic realm. As a result, hierarchies of taxomonic categories derived from biology such as classes, orders, families, and species, not always in the same sequence, were commonly superimposed on rocks by German petrographers.

In addition to the rock names of ancient derivation, several other rock names had come into use in the eighteenth century. According to von Leonhard (1823), the term "Pechstein" (pitchstone) was used as early as 1759 by Schulz and Poetsch for glassy rocks with few crystals, a conchoidal fracture, and a pitchy luster. Werner at first related "Pechstein" to primitive porphyry-slate, and eventually classed it with obsidian and pumice in his newest Flötz-trap unit. "Perlstein" was first employed by Werner, J. E. von Fichtel, and Jens Esmark for glassy rocks with numerous concentric cracks. The term "perlite" was probably first used by the French geologist F. S. Beudant.

Among the "Primitive rocks" of the short classification of 1786, Werner regarded "Porphyrschiefer" (porphyry slate) as a rock with a groundmass "which seems to be intermediate between hornslate and pitchstone, into which feldspar and hornblende are mixed" (Ospovat, 1971, p. 58) These greenish gray rocks were said to possess a dense, coarse-splintery fracture and to form pointed, conical, or jagged mountains. Werner also said that porphyry-slate formations resemble basalt formations because they split into disorderly columns. In later years, Werner included "porphyry-slate" in his Flötz-trap formation, the newest formation among the Flötz strata. He also designated its groundmass as "Klingstein" because of the ringing sound elicited upon striking the rock with a hammer. In 1801, mineralogist Martin Heinrich Klaproth (1743–1817) translated the term "Klingstein" as a Greek word, "phonolite." Once the volcanic nature of such rocks was recognized, "phonolite" became the name of a volcanic rock. In Werner's day, the mineralogical compositions of "Porphyrschiefer" or "Klingstein" were not well known. Studies in the 1820s and 1830s by C. G. Gmelin and Gustav Rose showed that the rock might consist of sanidine and a mixture of nepheline and zeolites. Breithaupt and Rose established the presence of hexagonal prisms of nepheline in phonolites from various locations. Scrope called the rock "clinkstone," a rough translation of Werner's "Klingstein," and considered it to be a rock in which the feldspars are aligned.

Although never contributing a rock classification and not a petrographer as such, Leopold von Buch was extremely interested in volcanic rocks, as evidenced by his studies of the Auvergne, the Canary Islands, the Andes Mountains, and his theory of craters of elevation as an explanation for volcanic structures. During the course of his studies, von Buch (1802) applied the term "domite" to an altered trachyte that forms Puy de Dôme in the Auvergne, and later he referred to a group of rocks from Monzoni and Predazzo in Tyrol as "monzosyenite" (von Buch, 1824). In a paper read before the Akademie Wissenschaftliche Berlin on March 26, 1835, in which he discussed his theory of upheaval craters and volcanoes, von Buch (1836) introduced the term "andesite" in reference to some presumed trachytes in the Andes Mountains. The rock allegedly contained albite and hornblende instead of the hornblende and sanidine of typical trachytes. The term was dropped for awhile when Gustav Rose, mineralogist at the University of Berlin, thought the feldspar might not be albite after all. "Andesite," however, was revived in 1861 by Justus Roth for oligoclase–pyroxene–amphibole rocks once it was established that the feldspar in the Andean rocks was oligoclase.

At this point, I will discuss a series of important rock classification schemes that were devised by German and some French petrographers. Several of these schemes are partially outlined in a succession of tables. Although the major divisions of the respective schemes are presented, finer subdivisions are noted only to the extent that they provide a sense of how what we consider to be igneous rocks were classified by the various petrographers. To that end, the names of igneous rocks in these schemes are italicized. The reader will see that

TABLE 1
Partial Rock Classification of von Leonhard (1823)

Ungleichartige Gesteine (dissimilar, heterogeneous rocks)
  Körnige (granular)
    Granite, syenite, diorite, gabbro
Gleichartige Gesteine (similar, homogeneous rocks)
  Dichte (dense)
    Trachyte, aphanite, serpentine, basalt, pitchstone, obsidian, schists
Trümmer Gesteine (fragmental rocks)
Lose Gesteine (loose rocks)

---

in many instances igneous rock types were lumped together with metamorphic rocks. The tables also convey a sense of the extent to which Linnéan taxonomic categories were employed.

German petrographic investigation underwent a major advance in the hands of K. C. von Leonhard. The most important rock classification scheme after Werner appeared in *Charakteristik der Felsarten* (von Leonhard, 1823). In preparation for his work, von Leonhard thoroughly examined the great rock collection that had been arranged by René Haüy at the Muséum d'Histoire Naturelle in Paris. Von Leonhard developed the first detailed system of rock classification.

Many of the early nineteenth-century rock classification schemes like those of von Leonhard failed to draw sharp distinctions among groups of rocks on the basis of their geological origins as is current practice. Because geologists were not yet in full accord regarding the origin of many rocks, igneous (eruptive) rocks were commonly combined with crystalline rocks of nonigneous origin. Attempts to classify igneous rocks during the nineteenth century typically reflected a struggle to ascertain the relative value of various possible classification factors such as mineral content, structure or texture, geological occurrence, geological age, and chemical composition. Cross (1902) provided an extended and valuable discussion of nineteenth-century rock classifications.

In von Leonhard's system, minerals provided the major distinguishing features of the rocks and, for the first time, texture was considered as important. In contrast to Werner's scheme, geological occurrence was subordinated. Von Leonhard maintained that discussion of the stratigraphic relationships of rocks, so important to Werner, should not proceed until the nature of the rocks had been accurately defined by their mineral content and structure.

Von Leonhard divided rocks into four categories (Table 1): "ungleichartige Gesteine" (dissimilar or heterogeneous rocks), "gleichartige Gesteine" (similar or homogeneous rocks), "Trümmer Gesteine" (fragment rocks), and "lose Gesteine" (loose rocks). Structural characteristics were used for further subdivision into such groups as "porphyritic," "glassy," "dense," "granular," and "schistose." The category of "körnige Gesteine" (granular rocks) among the "heterogeneous rocks" included granite, syenite, diorite, and gabbro. Von Leonhard had difficulty in dealing with the "dichte Gesteine" (dense rocks), a subdivision of the category of "homogeneous rocks" comprising very compact, fine-grained rocks whose composition was not easy to ascertain. "Dense rocks" included trachyte, aphanite, serpentine, basalt, pitchstone, obsidian, and various schists. Many of the rocks that were commonly understood to be of volcanic origin were placed into this group. A revised classification scheme was published in *Lehrbuch der Geognosie und Geologie* (von Leonhard, 1835). The classification had been expanded to include two additional major categories: "scheinbar gleichartige Gesteine" (apparently similar or homogeneous rocks) and "Kohlen" (coals). The former category included lava, trachyte, and "Tonschiefer" (clay slate). Von Leonhard had decided to move at least some volcanic rocks out of the "dense homogeneous" category into the "apparently homogeneous" category.

Although von Leonhard coined no igneous rock names that have persisted, he did introduce fundamental petrographic terms that are still in use. He established a distinction between "wesentliche" (essential) and "beigemengte" (accessory) minerals, the former being those necessarily present within a rock for it to be classified with a certain name, and the latter being widespread minerals that are not crucial to the classification of a rock. Von Leonhard also introduced the term "Grundmasse" (groundmass).

A growing body of chemical data on rocks led Abich to incorporate information about the chemical composition of igneous rocks into classification. Otto Hermann Wilhelm Abich (1806–1886) earned his Ph.D. degree from the University of Berlin in 1831 with a study on the isomorphism of the spinel group minerals (Tikhomirov, 1970). In 1843, Abich was appointed as Extraordinary Professor of Mineralogy at the University of Dorpat (now Tartu, Estonia). Within a few years Abich moved to Russia, where he spent most of his career investigating the geology of the Caucasus Mountains under the auspices of the Corps of Mining Engineers. In his early career, however, Abich concentrated more on mineralogy, petrography, and the chemical aspects of geology. Believing that chemical analyses could provide clues to the nature of the minerals present in fine-grained rocks, Abich (1841) used chemical data in the mineralogical classification of volcanic rocks. He spoke of "Sättigungsgrad" (degree of saturation) of various compounds with silica and regarded volcanic liquids as basic, intermediate, or acid. He also recognized the important role played by the feldspars in the composition of igneous rocks. Noting the wide range of silica content of the feldspars, he concluded that the silica content of a magma determined the composition of the feldspar that crystallized from it. On that basis, he classified volcanic rocks by their feldspars. Abich also employed so-called "oxygen ratios," a chemical method for classifying rocks devised by Karl Friedrich Rammelsberg (1813–1899), Professor of Chemistry at the University of Berlin and an authority on mineral chemistry (Pabst, 1975). In Abich's scheme, the oxygen ratio in a mineral or rock was expressed by the sum of the amount of oxygen in monoxides (RO + R_2O) plus the amount of oxygen in sequioxides (R_2O_3) divided by the amount of oxygen in the silica of a particular mineral or rock, or else its inverse:

$$(O \text{ in } [R_2O + RO + R_2O_3])/(O \text{ in } SiO_2)$$

or

$$(O \text{ in } SiO_2)/(O \text{ in } [R_2O + RO + R_2O_3])$$

Gustav Bischof continued the emphasis on oxygen ratios. On the grounds that the chemical formulas assigned to minerals are nothing more than an expression of the individual views of the chemists who devised them and of no scientific value, Bischof (1855) sought a more objective measure of the chemical constitution of minerals. He asserted that the ratio of the oxygen of the silica to that of all the bases is independent of the partitioning of the bases as monoxides or sesquioxides. The ratio of oxgyen in silica to that of all the bases could be ascertained directly from the chemical analysis, he said. Consequently, he maintained that the best expression for the composition of a mineral is the ratio of oxygen of the silica to that of all the bases. Bischof called this ratio the "Sauerstoffquotient" (oxygen quotient). Bischof conceded that errors in the "oxygen quotient" could arise from the defective determination of the different oxides or iron and manganese, but such errors, he thought, would not be so great as errors in a chemical formula. The failure of the oxygen quotient of a mineral to correspond with that of the ideal, pure mineral was regarded as an indication that some alteration had occurred. In his chapters on the various silicate minerals, Bischof tabulated oxygen quotient values for several analyzed samples. For example, the values for orthoclase and adularia range from 0.310 to 0.374. He took 0.333 as the normal oxygen quotient. Values for plagioclase range from 0.327 (albite) to 1.083 (anorthite).

Lehrbuch der Geognosie (Naumann, 1850, 1854) treated petrographic classification in detail. A three-volume second edition appeared between 1858 and 1866. The first volume of the original edition alone was just short of a thousand pages. After an introductory part of about three hundred pages on the interior and thermal character of Earth, particularly as revealed in the activity of volcanoes and hot springs, Naumann wrote an enormous second section on "Geognosie der festen Erdkruste" (the geognosy of the solid crust) devoted to a statement of the various relationships of the forms, masses, structures, and architecture of the crust. There were four main sections to this part of the text, entitled Morphologie der Erdoberfläche (morphology of Earth's surface), Petrographie, Paläeontologie, and Geotektonik. Particularly noteworthy is the fact that Naumann introduced the term "Petrographie" as a synonym for the more established German term "Gesteinslehre" (rock doctrine, rock science).

TABLE 2
Partial Rock Classification of Naumann (1850)

Class of crystalline rocks
  Order of siliceous rocks
  Order of crystalline silicate rocks
    Family of granite
      Granite
      Syenite
      Miascite
      Gneiss
      Granulite
    Families of *diorite, serpentine, gabbro, diabase, melaphyre, felsite-porphyry, trachyte, basalt, lava,* mica schist
  Order of crystalline haloid rocks
  Order of crystalline ore rocks
Class of clastic rocks
Class of rocks that are neither crystalline nor clastic

By whatever name, Naumann regarded petrography as an extremely important segment of geognosy. With Naumann, mineralogical and textural characters were treated as the leading factors in classification. In the first edition, Naumann divided rocks into three classes (Table 2): crystalline rocks, clastic rocks, and rocks that are neither crystalline nor clastic. Rocks of the first class, the "krystallinische Gesteine" were then divided into four orders: siliceous rocks, crystalline silicate rocks, crystalline haloid rocks, and crystalline ore rocks. Each of the four orders was subdivided into families. It is in the second order that we encounter rocks that Naumann generally considered to be "pyrogene." Here he included families of granite, diorite, serpentine, gabbro, diabase, melaphyre, felsite–porphyry, trachyte, basalt, and lava, but it was also in this order that he included the family of "Glimmerschiefer" (mica schist). Moreover, in the family of granite, Naumann included not only granite, syenite, and miascite, but also gneiss and granulite. Thus, throughout the classification, rocks of different origins were lumped together.

In the second edition, Naumann placed all rocks into just two broad classes: "protogene" (original) and "deuterogene" (derived). The class of "original" rocks incorporated six different orders, one of which was "silicates." He further subdivided these orders into families on the basis of mineral composition. Again, igneous rocks were not clearly distinguished at the outset.

Despite his failure to establish a genuine system, Naumann advanced the general science of rocks and introduced many textural terms that are applicable to igneous rocks, including "hyaline" for glassy materials, "macrocrystalline," "microcrystalline," and "cryptocrystalline" in reference to the general grain size of a crystalline rock, "vesicular," and "saccharoidal." He also coined the important nonigneous terms "metasomatism," "clastic," "pisolitic," and "symplectic." Naumann originated the rock names "felsite porphyry" and "nepheline basalt."

Some further remarks pertaining to Naumann's assessment of the geological position of igneous rocks are in order. The second volume of the first edition of his textbook was fundamentally an excursion through the geological column. Here he looked, in turn, at the various formations that constitute the Primitive, Transition, Trias, Jurassic, and Tertiary divisions. Significantly, Naumann placed a chapter on granitic eruptive formations between his chapters on Primitive and Transition rocks. He also placed chapters on porphyry formations and melaphyre formations between his chapters on Permian formations and Trias formations. And, finally, his next to last chapter was devoted to volcanic formations, sandwiched in between Tertiary and Quaternary formations. The general impression given by the organization of the text is that Naumann perceived granitic rocks to be very old for the most part and volcanic rocks to be rather young for the most part. Rocks like melaphyre, although the chemical and mineralogical equivalent of basalt, also appear to be older than volcanic rocks in Naumann's scheme. Nevertheless, Naumann acknowledged that research on granites from all over the world had exploded the conception that granite is always and everywhere the "Urgestein," the oldest rock of all. Indeed, he noted that most significant granites are eruptive formations that are younger even than the Transition formations and some very much later than that. Naumann implied, therefore, that one must speak of several granite formations.

In his introduction to the discussion of porphyries, Naumann acknowledged that many porphyries occur in much older formations and also that there are porphyries that formed after the Triassic. Thus, he did not claim to restrict porphyries and melaphyres to the Permo-Triassic. Nevertheless, Naumann did claim that the majority of these rocks fell into these two periods. He envisioned the volcanic formations like trachyte and basalt as having been erupted largely throughout the Tertiary and Quaternary periods. Thus, although Naumann had advanced considerably beyond Werner's conception of the relationship between igneous rocks and geological time, he still held to the general Germanic view that the character of igneous activity had undergone significant change throughout geological time.

In 1855, Karl Friedrich Ferdinand Senft (1810–1893) of Eisenach attempted to apply various chemical factors in *Classification und Beschreibung der Felsarten* translated into English two years later as *Classification and Description of Rock Types* (Senft, 1857). Senft regarded rocks as inorganic or organic (Table 3). Inorganic rocks were subdivided into crystalline and clastic rocks. Crystalline rocks were considered to belong to simple or composite classes. Classes were further divided into orders, suborders, and groups. Senft did not think that

TABLE 3
Partial Rock Classification of Senft (1857)

Inorganic rocks
  Division of crystalline rocks
    Class of simple crystalline rocks
      Order of rocks soluble in water
      Order of rocks insoluble or very little soluble in water
        Soluble only in excess of water
        Totally insoluble in water
          Soluble in hydrochloric or sulphuric acid
          Partially soluble in acids
          Insoluble in hydrochloric or sulphuric acid
            Silicate of magnesia
              Amphibolite, *pyroxenite*, talc slate
            Strong silicates
            Silicate of alumina
              Pitchstone, perlite, *obsidian, pumice*
    Class of mixed crystalline rocks
      Order of rocks without labradorite, never containing augite, but generally quartz
        Partially soluble in hydrochloric acid
          Trachyte, *domite, trachytic porphyry, perlite, phonolite, andesite*
        Not decomposable in hydrochloric or sulphuric acids
          Pitchstone-porphyry, felsite-porphyry, granite, syenite, *myascite,* granulite, gneiss
          Mica-slate, calc-mica-slate, talc-mica-slate
          Quartz-slate, itacolumite, greisen, schorl-rock
          Diorite, diorite porphyry
      Order of rocks containing labradorite without orthoclase or quartz
        Mostly decomposable in hydrochloric acid
          Eclogite, *gabbro, hypersthenite*
          Diabase, diabase-porphyry, calc-diabase
          Melaphyre
        Always decomposable in hydrochloric acid
          Dolerite, anamesite, basalt, wacke, *leucite-porphyry, nepheline dolerite, trachyte dolerite*
  Division of clastic rocks (many volcaniclastic rocks were included here)
Organic rocks

structure or mineral composition provided the best basis for arrangement. Because mineral content cannot readily be determined for fine-grained rocks, he concluded that the chemical relationship of rocks, especially crystalline rocks, to certain solvents afforded the best means for classifying rocks. He applied the factor of solubility to the establishment of orders, suborders, and groups. Some rocks, for example, were subdivided into nine groups on the basis of the action of hydrochloric or sulfuric acid on those rocks.

Most rocks now regarded as igneous rocks were placed in the class of mixed crystalline rocks. For example, the first order of this class consisted of rocks without labradorite, never containing augite, but generally quartz. The first suborder of the first order included rocks that are partially soluble in hydrochloric acid. Among these, trachyte and domite, trachytic porphyry, perlite in part, phonolite, and andesite were listed. As another example, the second suborder of the first order of the class of mixed crystalline rocks included rocks that are not decomposable in hydrochloric or sulfuric acids. Pitchstone-porphyry, felsite-porphyry, granite and protogine, syenite and zircon syenite, myascite, granulite, and gneiss were listed. Not only did volcanic and plutonic rocks end up lumped together in this scheme, but so did some high-grade metamorphic rocks! Senft also introduced the term "pyroxenite" for rocks composed essentially of pyroxene.

The beginnings of the modern classification of rocks may be traced to von Cotta. In *Die Gesteinslehre* (von Cotta, 1855), a work that has been described as "the culmination and end of Werner's macroscopic descriptive petrography" (Prescher, 1971, p. 434), von Cotta discussed fourteen different groups of rocks. He regarded mineral content as the leading factor in classification. A second edition of *Die Gesteinslehre* appeared in 1862 and was translated into English as *Rocks Classified and Described: A Treatise on Lithology* (von Cotta, 1866). In the second German and first English edition, von Cotta (1866, p. 123) conceded that "we could not lay down a logically complete system of classification to embrace all rocks, on any principle, whether of ORIGIN, TEXTURE, or COMPOSITION (chemical or mineralogical). . . . There are no rigidly defined classes in nature." He thought that the best scheme was one in which the general features of the scheme coincided with what was known of the origin of the various rocks and which would make use of distinctions based on texture or composition for further subdivision. Von Cotta began his classification by grouping rocks into three major categories based on their geological mode of origin: eruptive (igneous), metamorphic, and sedimentary (Table 4). He considered eruptive rocks to have probably originated through consolidation from molten condition. Along the lines of Bunsen (1851), von Cotta divided eruptive rocks into silica-poor (basic) and silica-rich (acid) types. Each of these two groups was further subdivided into volcanic and plutonic categories.

In the basic volcanic group, von Cotta included dolerite, basalt, and leucite rock. In the basic plutonic group he listed diabase, gabbro, diorite, aphanite, melaphyre, porphyrite, mica trap, syenite, and miascite. His acid volcanic group included trachyte, rhyolite, and phonolite. And in his acid plutonic group he placed granite, granitic porphyry, quartz porphyry, felstone, and pitchstone. Each of these major rock types included several different varieties; for example, gabbro, euphotide, norite, and hyperite were listed as varieties of "gabbro." Andesite, perlite, obsidian, and pumice were listed as varieties of trachyte. Von Cotta also classified an unusual rock type, named tholeite by

CLASSES AND ORDERS

TABLE 4
Partial Rock Classification of von Cotta (1866)

Eruptive rocks
   Class of silica-poor, basic rocks
      Volcanic rocks
         Dolerite, basalt, leucite rock
      Plutonic rocks
         Diabase, diorite, aphanite, melaphyre, porphyrite, mica trap, syenite, miascite
         Gabbro
            Varieties: *gabbro, euphotide, norite, hyperite*
   Class of silica-rich, acid rocks
      Volcanic rocks
         Trachyte
            Varieties: *andesite, perlite, obsidian, pumice*
         Rhyolite
         Phonolite
      Plutonic rocks
         Granite, granitic porphyry, quartz porphyry, felstone, pitchstone
Metamorphic rocks
Sedimentary rocks

Steininger, with either dolerite or basalt. Although he spoke of basic and acid classes, von Cotta departed from the common practice of placing rocks into orders, families, or other taxonomic groups derived from the Linnéan system.

Although proposing no classification scheme until 1868, Ferdinand von Richthofen, an avid student of the volcanic terranes of the Carpathian Mountains in Hungary, coined the term "rhyolite" for quartz-rich trachytes. The term "liparite," named for the volcanic Lipari Islands, was coined for the same kind of rocks in the following year by Roth. Justus Ludwig Adolph Roth (1818–1892) studied in Berlin with mineralogist Gustav Rose and at Tübingen and Jena (Amstutz, 1975). He received his Ph.D. degree from the University of Jena in 1844. Like von Cotta, Roth was one of the founders of die deutsche geologische Gesellschaft in 1848. During his early career, he worked with chemist Eilhard Mitscherlich at the University of Berlin, and in 1867 he was named Extraordinary Professor at the university. He devoted much of his time to the chemical analysis of rocks. From 1879 to 1893, Roth issued a multivolume treatise on general and chemical geology. In 1887, five years before his death, Roth became Professor of Petrography and General Geology at the University of Berlin.

The climactic event of the early era of chemical investigations in igneous geology came with the publication of *Die Gesteins-Analysen in tabellarischer Übersicht und mit kritischen Erläuterungen* (Roth, 1861). This volume summarized in tabular form hundreds of chemical analyses of a wide range of rock types, predominantly igneous, that had been determined by such renowned chemically oriented geologists and geologically oriented chemists as Abich, Bunsen, Bischof, Sartorius von Waltershausen, Durocher, Streng, Kjerulf, Delesse, Rammelsberg, Haughton, von Richthofen, and many others. Roth attempted the first thorough chemical classification of rocks by calculating the mean values of "oxygen ratios" or "oxygen quotients" or "acidity quotients" (the ratio of oxygen in the bases to oxygen in silica) as center points for acid, neutral, and basic families. These families were further subdivided into two subgroups based on the ratio of oxygen in $RO + R_2O$ to that in R_2O_3.

After his efforts, Roth concluded that the chemical groups did not correspond to those in mineralogical classifications. He found, for example, that some rocks having different mineralogical constitutions are chemically similar. Classifying rocks on a purely chemical basis would, therefore, needlessly separate some rocks that are closely related to each other geologically and mineralogically. In the end, Roth decided to arrange rocks on the basis of the kind of feldspar that they contained and on the presence or absence of quartz. He regarded feldspar as a preferable mineralogical discriminator over pyroxene or amphibole because of its greater abundance. Roth's fourfold classification (Table 5) included orthoclase rocks, with and without quartz; oligoclase rocks, with hornblende or with augite; labradorite rocks; and anorthite rocks. As with so many of the classifications of his predecessors, nonigneous rocks like gneiss and granulite were grouped with igneous rocks like granite, syenite, liparite, and phonolite in the category of "Orthoklasgesteine" (orthoclase rocks). In addition to the name "liparite," Roth also coined the names "nephelinite," "augite andesite," and, regrettably, "zobtenite."

Finally, Theodor Scheerer (1862, 1864) also proposed classifying igneous rocks on a chemical basis. Having performed many chemical analyses of rocks, Scheerer was convinced, unlike Roth, that one could set up chemical formulas for igneous rocks. He divided igneous rocks into three groups: plutonites, pluto-vulcanites, and vulcanites (Table 6). He further subdivided each of these groups into an upper, middle, and lower group such that all silicate rocks could be assigned to nine chemical types. Following Bischof and Roth, Scheerer distinguished the nine groups by formulas based on the ratios of oxygen in silica to oxygen in the bases. These values ranged from 4.50 in the upper plutonite to 1.00 in the lower vulcanite. Scheerer accepted the view, widely held in his time, and espoused by contemporaries like Joseph Durocher and Wolfgang Sartorius von Waltershausen, that the chemical character of eruptions had changed throughout geological time from acidic to basic with progressive solidification of Earth. Scheerer believed that the material below the solid crust is density stratified into magma zones with an acid shell lying above a basic shell. On this view, the upper, more siliceous magmas were believed to be first erupted to form granitic plutonites in the crust. These, he thought, would be succeeded in later geological periods by basaltic vulcanites that

TABLE 5
Partial Rock Classification of Roth (1861)

Orthoklasgesteine (orthoclase rocks)
  With quartz or free silica
    Granite
    Gneiss (including granulite and petrosilex)
    Felsite porphyry (including pitchstone)
    Liparite
    Syenite
  Without quartz or free silica
    Quartz-free orthoclase porphyry
    Sanidine trachyte (including pumice)
    Sanidine oligoclase trachyte
    Phonolite
    Leucitophyre

Oligoklasgesteine (oligoclase rocks)
  With hornblende
    Diorite
    Porphyrite
    Amphibole andesite
  With augite
    Oligoclase augite porphyry
    Melaphyre and spilite
    Pyroxene andesite
    Nephelinite
    Hauynophyre

Labradorgesteine (labradorite rocks)
  Labradorite porphyry
  Gabbro
  Hypersthenite
  Diabase
  Dolerite
  Normal pyroxenic rocks
  Basalt

Anorthitgesteine (anorthite rocks)
  With augite
  With hornblende

Various schists and serpentine were treated separately from the above classification. Neptunian rocks were also designated as a separate category.

TABLE 6
Partial Igneous Rock Classification of Scheerer (1864)

Plutonites
   Upper—Rother gneiss
   Middle—Mittlerer gneiss
   Lower—Grauer gneiss

Pluto-vulcanites
   Upper—Quartz-bearing syenite
   Middle—Syenite
   Lower—Melaphyre

Vulcanites
   Upper—Augite porphyry
   Middle—Common basalt
   Lower—Basic basalt

The gneisses in Scheerer's classification are essentially granitoid rocks

would overlie the plutonites in the crust. From this cause, Scheerer believed that the earliest rocks might be supposed to represent the fundamental magmas more nearly than the basic ones, since the latter magmas must have passed through rocks of varied constitution on reaching the surface, and, therefore, suffered much modification through the fusion and assimilation of fragments torn loose on ascending.

THE FRENCH PETROGRAPHIC TRADITION

During the first decades of the nineteenth century, the French also developed a tradition of petrography. Unlike the Germans, French geologists did not write massive petrographic treatises. They did, however, often contribute articles on the geology, mineralogy, and chemical character of igneous rocks. Early issues of *Bulletin de la Societé Géologique de France* included several articles on granite, quartz porphyry, minette, trachyte, basalt, and melaphyre by prominent geologists like Durocher, Delesse, Rozet, Fournet, and Élie de Beaumont. French petrographers were primarily associated with the Muséum d'Histoire Naturelle in Paris, the institution that dominated French igneous petrology throughout the nineteenth century (Limoges, 1980). Quite possibly because of their work in the museum with its exceptional collections of minerals and rocks, French petrographers were less attuned to the geological settings of rocks and more influenced by the tendency to regard rocks as specimens in a drawer or cabinet. As in Germany, so, too, in France, petrographic classification commonly borrowed from the Linnéan system.

The French petrographic tradition began with Abbé René Just Haüy (1743–1822), Professor of Mineralogy at the Muséum from 1802 to 1822 (Kunz, 1918; Hooykaas, 1972). Although Haüy was a crystallographer and mineralo-

TABLE 7
Partial Rock Classification of Haüy (1822)

Class 1. Stony and saline rocks
  Order of phanerogenous rocks
    Genus feldspar
      Species of *granite, syenite, pegmatite, protogine,* gneiss, etc.
    Genus amphibole
      Species of *diorite*
  Order of adelogenous rocks
  Order of conglomerate

Class 2. Combustible nonmetallic rocks

Class 3. Metallic rocks

Class 4. Rocks of an igneous origin according to some, aqueous according to others

Class 5. Volcanic rocks

---

gist rather than a petrographer, he became involved in rock classification. He published *Traité de Minéralogie* (Haüy, 1801), a work of which Cross (1902, p. 340) observed, "this treatise is evidence at once of the advanced state of mineralogy and of the non-existence of anything worthy of being termed a science of rocks, in France, at this time." Haüy's detailed mineralogical system included two brief appendices on rocks. One appendix was entitled, "Agrégates des différentes substances minérales," and the other was entitled, "Produits des volcans." In 1811, Haüy addressed a letter to von Leonhard's magazine, *Taschenbuch für Mineralogie,* indicating that he had come up with an idea for a mineralogical classification of rocks. In view of his specialty and general lack of field experience, he stressed the importance of mineral content over against field occurrence in his classification of rocks. According to German contemporaries, it was essentially a petrography based on cabinet specimens.

In the second edition of *Traité de Minéralogie,* Haüy (1822) put forward a Linnéan scheme of classes, orders, genera, species, and varieties (Table 7). He proposed five different classes of rocks. The fourth class included rocks of an igneous origin according to some but aqueous according to others. The fifth class included volcanic rocks. Some of what we now consider igneous rocks also appeared in a different class as species within genera designated by mineral name. For example, syenite and pegmatite were designated as species within the genus "feldspar" and diorite was designated as a species within the genus "amphibole."

Haüy introduced several petrographic terms that are still in use. He coined the term "diorite," applying it to megascopically granular rocks composed of white feldspar and hornblende. He believed that the white feldspar was orthoclase. "Diorite" was subsequently used for hornblende–feldspar rocks by Brongniart and von Leonhard, who distinguished it from syenite by the predomi-

TABLE 8
Partial Rock Classification of Brongniart (1827)

Class. Roches homogènes ou simples (homogeneous or simple rocks)
  Order. Roches phanérogènes (rocks with distinct known mineral species)
  Order. Roches adélogènes (rocks of unrecognizable constitution)
    Clays, dense schists, *trap, basalt,* etc.
Class. Roches hétérogènes ou composées (heterogeneous or compound rocks)
  Order. Roches de cristallisation
  Order. Roches d'agrégation

nance in the former of hornblende over feldspar. Haüy also introduced the terms "pegmatite," "aphanite," "dolerite," and "eclogite." For many altered gabbros, he used the word "euphotide," a term that was commonly used throughout the nineteenth century but has disappeared from the petrological vocabulary. As noted earlier, Haüy introduced the term "trachyte." Brongniart stated that the rock name "trachyte" had been used by Haüy in his lectures. By the time Haüy (1822) published the term in the second edition of *Traité de Minéralogie*, it was already being used by other French geologists, including d'Aubuisson de Voisins and F. S. Beudant.

Alexandre Brongniart (1770–1847) was Haüy's successor as Professor of Mineralogy at the Muséum d'Histoire Naturelle from 1822 to 1847 (Rudwick, 1970). Well known to geologists for his pioneering stratigraphic studies in the Paris basin with Georges Cuvier, Brongniart also contributed to petrographic classification. His first effort was *Classification minéralogique des roches mélangées* (Brongniart, 1813), but a more mature classification appeared under the title *Classification et Charactères Minéralogique des Roches Homogènes et Hétérogènes* (Brongniart, 1827). Brongniart disliked the geognostic discussions of rock employed by Werner and his followers because they considered rock types so closely in terms of stratigraphic occurrence that if a rock occupied more than one stratigraphic position, and, therefore, formed at different times, then it had to be discussed as many times as the different stratigraphic positions in which it occurred. Brongniart preferred to classify rocks on the basis of mineralogical composition apart from occurrence.

Like so many other early geologists, Brongniart used classes, orders, genera, and species (Table 8). He employed only two classes, however: "homogeneous" or "simple" rocks and "heterogeneous" or "compound" rocks. Some of the fine-grained igneous rocks such as basalt appeared in an order of rocks of unrecognizable constitution in the homogeneous class. The heterogeneous class had two orders: rocks of crystallization and rocks of aggregation. Here Brongniart introduced a genetic factor. Many igneous rocks were included among the former. Finer subdivisions were based almost purely on mineralogy. Brongniart also introduced the terms "diabase," "spilite," and "melaphyre," a term

TABLE 9
Partial Rock Classification of Cordier

Class I. Roches terreuses (earthy rocks)
   First family. Roches feldspathiques (feldspathic rocks)
      Order 1. Phanérogenes (constituents are visible to the unaided eye)
         Genus 1. Agrégés (aggregates)
            Species. Harmophanite, leptynite, gneiss, *pegmatite, granite, syenite*
         Genus 2. Conglomérées (conglomerates)
         Genus 3. Meubles (unconsolidated sands)
      Order 2. Adélogènes (constituents wholly or partially invisible)
   Families of pyroxene, amphibole, garnet, hypersthene, diallage, talc, mica, quartz
   Each family had the two orders and *most igneous rocks belonged to the order of adélogènes*
Class 2. Roches salines ou acidifères nonmétalliques
Class 3. Roches métallifères
Class 4. Roches combustible nonmétalliques

that came to be widely used for basaltic rocks of Carboniferous and Permian age. He also resurrected the ancient terms "tephra" and "basanite."

A contemporary of Haüy and Brongniart was Pierre Cordier, whom Cross (1902) called the master spirit of French petrography during the first part of the nineteenth century. Cordier, Professor of Geology at the Muséum d'Histoire Naturelle, developed a brief classification of volcanic materials as a result of his early studies on crushed lava and volcanic rocks. He divided volcanic rocks into feldspathic and pyroxenic rocks. Further subdivision was based on texture. Cordier continued to work over the classification throughout his career and apply it to the collection of the Muséum d'Histoire Naturelle. Apparently Cordier personally never published his more elaborate classification scheme, but it was included in an article on "Roches" (rocks) by Charles d'Orbigny in *Dictionnaire Universel l'Histoire Naturelle* (1842–8). The classification also appeared in an 1868 volume compiled posthumously by d'Orbigny from Cordier's lectures and manuscripts and entitled *Description des Roches*.

The fundamental idea of Cordier's system was that rocks are species to be determined predominantly on the grounds of mineralogical composition (Table 9). He considered geological origin and occurrence to be of secondary importance in his classification. Cordier's rock species were grouped into genera, orders, families, and classes. The most comprehensive groups, four classes, were based on chemical characters. The four classes were "roches terreuses" (earthy rocks), "roches salines ou acidifères non metalliques" (saline or acid nonmetalliferous rocks), "roches metallifères" (metalliferous rocks), and "roches combustible non metalliques" (combustible nonmetallic rocks). Cordier subdivided the classes into families on the basis of the dominant mineral constituent. As an example, the first class, earthy rocks, was divided into several families based on the predominance of feldspar, pyroxene, amphibole, garnet, hypersthene, diallage, talc, mica, or quartz. The first family of this class, the feldspathic rocks, was based on the predominance of feldspar. Cordier divided the families into two orders, the "phanerogenes," or phaneritic rocks, in which constituents are visible to the unaided eye, and the "adelogenes," or aphanitic rocks, whose constituents are not wholly visible. The phaneritic order of the feldspathic family of the first class consisted of three genera—aggregates, conglomerates, and unconsolidated sands. It is in the first of these genera where at least some of the igneous rocks, "species" like granite, syenite, and pegmatite, were placed along with gneiss. As with so many of the early classifications, igneous rocks were typically included in the same categories as sedimentary and metamorphic rocks.

Cordier's system was used in France for several years after his death in 1861. According to Cross (1902), Cordier's system exerted a retarding influence for several decades on the development of petrography in the country where Brongniart had laid logical foundations, not so much because of the inherent weakness of Cordier's system as in its domination over the thought of his countrymen. Like that of Rene Haüy, Cordier's system was based largely upon the convenient arrangement of cabinet specimens. Both men studied rocks in detail simply as mineral aggregates and ignored their geological relationships, a matter that they regarded as outside the scope of mineralogy.

In 1858, H. Coquand, Professor of Mineralogy and Geology in the College of Besançon, wrote a *Treatise on Rocks* in which he came out with an original system that was very different from that of Cordier. Coquand's scheme was remarkably modern in its outlook. He recognized the existence of three families of rocks: igneous, aqueous, and metamorphic, and he divided igneous rocks into three groups: granitic (granular), porphyritic, and volcanic. His finer subdivisions, however, were less successful.

Petrographic classification in the first half of the nineteenth century was a decidedly unsettled art in which practitioners wrested with finding useful and appropriate ways of organizing rock types. Although many rocks were gradually accepted as igneous and several new igneous rock names were devised, classifications generally remained strictly qualitative, and clear, consistently used criteria continued to elude the grasp of geologists.

8 ❖ BASALT AND TRACHYTE

Early Theories of Diversity

The major questions about igneous rocks during the early decades of the nineteenth century pertained to the origin of basalt, the origin of granite, and classification. Once the criteria for the recognition of the volcanic origin of basalt were established, geologists began to observe considerable variety among volcanic rocks. When some investigators developed criteria for the recognition of the igneous origin of granite, they, too, started to observe variety among crystalline plutonic rocks. As a result, geologists gradually began to ponder the possible causes of the mineralogical and chemical variation observed in rocks that were believed to be igneous. Because there was far greater unanimity regarding the existence of ancient volcanic rocks, the earliest speculations about igneous rock diversity almost invariably addressed the variations among volcanic rocks (Young, 1999a).

EARLY SPECULATIONS

The differences among the early speculations can be attributed to the scientific interests and field experiences of those who proposed hypotheses of diversity, and also to the views that they held concerning the relationship between rock type and geological age. Most of the early hypotheses of diversity were formulated by individuals for whom the diversity of igneous rocks was anything but a burning question. Robert Bunsen and Joseph Durocher came the closest to writing papers, both published in academic journals, that specifically addressed the problem of diversity. In most of the early literature on diversity, however, a generally brief discussion was commonly embedded in a massive expedition report or a broad-based volcanological study.

Early hypotheses of diversity were formulated predominantly by individuals who were not specialists in the study of igneous rocks. Dana, Sartorius von Waltershausen, Durocher, Jukes, and von Richthofen all made important contributions to a wide range of geological fields. Darwin moved from geology to natural history, and Bunsen was primarily a chemist. Although Scrope early became an expert on volcanoes, he spent most of his adult life in the political arena. Von Cotta came the closest to being a specialist in petrography.

Speculation about the causes of diversity of igneous rocks began in England with G. P. Scrope. In his first book, *Considerations on Volcanos*, Scrope (1825) was primarily interested in such matters as eruption phenomena, the nature and behavior of lava, and the structure of volcanoes. He did, however, devote five pages to a discussion of the causes of the variety in mineral composition of lavas. The volcanic rock types recognized by Scrope included trachyte, phonolite, trachytic porphyry, pitchstone, obsidian, basalt, dolerite, and greenstone. In pitchstone and obsidian, Scrope (1825, p. 85) said, the "crystalline particles have been comminuted nearly or quite to the extremest degree." He considered dolerite to be an extremely coarse variant of basalt and greenstone a type of basalt in which hornblende replaced augite. Despite all this variety, Scrope observed that contemporary mineralogists generally classed volcanic rocks into two families based on the prevalence of feldspar or of augite. For Scrope, then, the problem of diversity amounted to finding an explanation for the origin of basalt and trachyte.

Scrope favored the idea that changes in lava occur during its ascent. Deeply affected by his personal experience of the driving power of erupting gases at Mount Vesuvius, Scrope attributed profound abilities to these vapors to produce both textural and compositional variations in lava. He thought that lava was not a homogeneous, completely fused liquid but rather a "compound liquid" like mud, paste, milk, blood, or honey that owed its fluidity to the presence of an aqueous vapor acting as a vehicle for suspended fine particles of minerals, a view that he held throughout his career (Scrope, 1856). He said that solidification of lava occurs because of discharge of the aqueous vapor. He referred to the thorough penetration of crystalline matter by this pervasive vapor as "intumescence." Scrope envisioned the derivation of trachyte, clinkstone (phonolite), or basalt from an original granitic basement rock in terms of the action of these intumescent vapors. In the case of basalt, he suggested that heated vapor volatilized mica and iron-bearing minerals in the granite and separated them from feldspar. Consequently, he maintained that a feldspathic, iron-deficient lava would be left behind in one part of a volcanic chimney and a very iron-rich lava would crystallize in another part. "The subsequent intumescence and protrusion of these lavas," Scrope (1825, p. 146) wrote, "might produce alternate currents of trachyte, clinkstone, compact feldspar, and basalt."

Scrope's emphasis on the role of vapor in altering particle size and separating minerals was shaped by his extensive field experience with volcanic terranes throughout Europe. Because his observations were restricted to continental volcanoes, he saw ample evidence of past and present volatile-driven explosive activity but had little firsthand acquaintance with quiescent eruptions. His frequent references to smoke, vapors, and ash in a list of the world's active volcanoes at the end of his text reinforce the impression that he considered such phenomena to be universally associated with volcanoes. He had observed a wide range of volcanoes throughout Europe, including explosive eruptions of Mount Vesuvius, but lacked firsthand acquaintance with oceanic volcanoes. Impressed by the tremendous driving power of expanding gases, Scrope devised a theory that stressed the role of gaseous constituents.

Scrope's idea, however, was met with considerable skepticism. As one of his biographers observed, it was no wonder that a work professing to account for basalts and trachytes as varieties of lava met with distrust and ridicule in an age when geologists were not yet fully agreed on the volcanic origin of basalt (Anonymous, 1870). After a virtual consensus about the volcanic nature of basalt had finally been achieved, a more cautious proposal about diversity was advanced by Darwin (1844) several years after his voyage on H.M.S. *Beagle*. Unlike Scrope, Charles Darwin (1809–1882) was not a specialist in the study of volcanoes (Bowlby, 1990; Bowler, 1990; Desmond and Moore, 1991). His experience with volcanoes was essentially limited to observations made between 1832 and 1836 on the *Beagle*. During the voyage he observed predominantly dormant and extinct volcanoes in the Andes and on such oceanic islands as the Cape Verde, Fernando de Noronha, Ascension, St. Helena, Tahiti, and the Galapagos. Unlike many of his geological contemporaries and predecessors, Darwin never visited Mount Vesuvius, Mount Etna, or Iceland. Darwin's early career as a geologist has been explored in detail by Secord (1991), Herbert (1991), and Rhodes (1991).

Darwin dismissed Scrope's speculations on the grounds that they lacked positive facts. He also charged that Scrope had overlooked the existence of lighter and heavier minerals in lavas. While on James Island in the Galapagos in 1835, Darwin observed that crystals of glassy albite were much more numerous in the lower, scoriaceous parts of basalt flows. He suggested the sinking of crystals as a mechanism that could throw light "on the separation of the trachytic and basaltic series of lavas" (Darwin, 1844, p. 133). From the specific gravities of common minerals, Darwin concluded that quartz and feldspar should rise and that hornblende, augite, olivine, and iron oxides ought to sink in lava. To explain the anomaly, he attributed the sinking of the albite crystals to a decrease in density of the lavas brought about by their vesicularity.

Darwin also proposed that plutonic rocks might be accounted for by a squeezing mechanism now designated as "filter pressing." Noting the common occurrence of basalt dikes intersecting granite, he wondered if the dikes might not have been formed by more fluid, hornblende-rich parts of cooled granitic and metamorphic rocks oozing into fissures that penetrated the granite. Most plutonic masses, Darwin (1844, p. 140) argued, have been partly drained of "comparatively weighty and easily liquefied elements" that make up basaltic rocks.

It is not surprising that Darwin, a naturalist whose later career was devoted to the causes of variation among plants and animals, turned a finely tuned eye toward variations in the mineral content of rocks. Although beginning his career as a geologist, Darwin already brought with him the eye of the natural historian. Pearson (1996), who discussed Darwin's ideas about igneous rock diversity at length, commented that Darwin's geological writings were full of comments about the gradations between classes of objects. "If anything," Pearson (1996, pp. 54–56) stated, "this was the essence of his scientific method—finding intermediates was the first step toward recognizing a common origin for disparate but related phenomena."

Darwin had an instinct for spotting variation. He still, of course, worked with the most widely accepted categories of "trachyte" and "basalt." Although he saw a nighttime eruption in South America, Darwin witnessed nothing comparable to the cataclysmic explosion of Mount Vesuvius seen by Scrope, and his familiarity with only the products of quiescent oceanic volcanoes prevented him from speculating about "intumescence" or the role of gas as a universal volcanic process. He was content to rely on experimentally verifiable effects owing to contrasts in mineral density. He proposed his ideas as possible explanations for some igneous phenomena, and made no effort to universalize them.

The English-speaking world continued to speculate about igneous rock diversity in the person of the American geologist, J. D. Dana. James Dwight Dana (1813–1899) studied geology under Benjamin Silliman at Yale in the early 1830s (Gilman, 1899; Prendergast, 1978; Stanton, 1971). After a brief stint as a naval instructor, Dana returned to Yale as Silliman's assistant and soon completed his masterwork, *System of Mineralogy*, in 1837. Between 1838 and 1842, Dana participated as the geologist on the U. S. Exploring Expedition to the Pacific under the command of Captain Wilkes. Again he returned to Yale, eventually succeeding Silliman as editor of *American Journal of Arts and Science*. In 1849 he was appointed as Silliman Professor of Natural History at Yale. Dana was widely acknowledged as North America's premier geologist on the grounds of his broad-based expertise on volcanoes, coral reefs, oceanic islands, crustaceans, the structure of mountains, continents, and oceans, and, of course, minerals.

Although far from specializing in igneous petrology, Dana, as an expert in mineralogy, was attuned to mineralogical variations within rocks. During the Wilkes expedition, he visited the active volcanoes of Hawaii and observed the relatively fluid, quiescent eruptions of Kilauea. He noted that the Hawaiian volcanoes consist of basalt and minor trachyte, and that feldspar-rich rocks occur toward the center of most volcanoes, whereas the exterior portions consist of augite-rich basaltic rocks. Given the nonexplosive character of Kilauean eruptions, Dana granted a lesser role to vapor in volcanism than did Scrope. He suggested that lavas slowly rose through the central part of a volcanic conduit as a result of the expansion of the vapors. Accompanying the rising lava was a descending current along the sides of the conduit. Because he believed that augite is more easily fusible than feldspar, Dana suggested that the feldspar would begin to solidify first in the rising, cooling current. The feldspar-depleted liquid, he reasoned, would be driven to the surface by vapor. Having reached the surface, the feldspar-depleted, basaltic liquid was envisioned as descending along the exterior. "If the elements are at hand," Dana (1849, p. 378) concluded, "it requires only different circumstances as regards pressure, heat, and slowness of cooling, to form any igneous rock the world contains."

Dana extended his hypothesis to include coarsely crystalline rocks like syenite and granite. He envisioned the possibility that the cores of some volcanic mountains might be composed of syenite or granite and that such cores might be flanked by hornblende-rich rocks should cooling rates be sufficiently slow. This arrangement "suggested a strong analogy to the trachytes with a circumference of basalt" (Dana, 1849, p. 378). But Dana was ambivalent about granite. As noted in Chapter 6, he later classified granite and syenite as rocks that had crystallized by heat without fusion, that is, as metamorphic rocks. As a geological generalist with wide knowledge of the globe, Dana wanted to propose a generally applicable hypothesis. That hypothesis, however, was shaped by his observations of the distribution of basalt and trachyte in Hawaii, by the relatively placid nature of Hawaiian volcanism, and by his mineralogist's knowledge of mineral fusibilities.

Scrope, Darwin, and Dana all proposed relatively simple theories of igneous rock variation that needed to account primarily for the existence of two major classes of volcanic rock: trachyte and basalt. All three sought for the source of variation either on the surface or within the volcanic edifice. Their familiarity with quite different volcanic terranes accounts for many of the differences among the three hypotheses.

TWO-SOURCE THEORIES

Speculation soon shifted to continental Europe. Here the concept that there had been changes in the chemical and mineralogical composition of volcanic rocks throughout geological time, an idea rooted in Werner's scheme, influenced primarily German, but also French, geologists to seek the causes of diversity in the deep structure of Earth rather than at the site of the volcano. This new approach to igneous rock diversity originated with Robert Bunsen. In 1846, one year after a major eruption of Mount Hekla, Bunsen participated in a Danish expedition to Iceland with the mineralogists Wolfgang Sartorius von Waltershausen and Alfred Des Cloizeaux. Upon his return to Marburg, Bunsen brought his chemical expertise and analytical skills to bear on the problem of diversity by making numerous chemical analyses of eruptive rocks that he had collected. He published the results of his study in 1851.

Bunsen (1851) called attention to aspects of the chemical composition of Icelandic volcanic rocks that, he thought, might provide clues to the origin of eruptive rocks from older geological periods. He believed that Iceland provided an excellent location for evaluating the nature of lava sources because of the absence of chalky and siliceous stratified rocks that could contaminate the lavas. Through chemical analysis, Bunsen concluded that the rocks, despite displaying some gradual transitions, could be separated into two end members that he designated as "normal trachytic" and "normal pyroxenic." He was, of course, analyzing the abundant Icelandic rock types now commonly designated as tholeiitic basalt and rhyolite. He regarded the normal trachytic rocks as silica-rich rocks and the normal pyroxenic rocks as less siliceous and similar to basalt and dolerite. Bunsen calculated hypothetical average compositions for normal trachyte and normal pyroxene rock that he believed to be representative of the compositions of two major rock sources underlying Iceland from which the most acidic and the most basic rock masses were derived. Even though Bunsen essentially accepted the current division of volcanic rocks into trachyte and basalt, he still had to reckon with the existence of rocks of chemically transitional or intermediate character. Neither Scrope nor Darwin nor Dana had specifically addressed the problem of such transitional rocks.

Bunsen, therefore, calculated hypothetical chemical compositions of lavas that would be formed by mixing the two end members in various proportions and compared the analyses of Icelandic rock samples with the calculated intermediate compositions so that he could estimate the relative amounts of the two end members that had mixed together. He believed that the possible compositions of the rocks could be determined in advance. By comparing analyses with theoretically possible compositions, Bunsen (1851, p. 208) concluded that the unmetamorphosed volcanic rocks of Iceland

> are only mixed products of those acidic and basic terminal members or these terminal members themselves, and that the great mineralogical and petrographic variety in which the rocks appear is only a consequence of the one-time mixture relationship and the predominating physical conditions under which the rock acquired its present resting place and form. Of the great number of analyses of Icelandic rocks that have been conducted in my laboratory there is no rock whose composition deviates from the theoretical value calculated in the suggested manner as would be expected from such a figure based on analyses of averages.

Field evidence, he believed, corroborated the conclusion. Numerous intersecting trachytic and basaltic veins were thought to illustrate situations in which the extreme end members had mixed. Chemical analysis of a graded series of samples from trachyte to pyroxenic rock within these intersecting dikes convinced Bunsen that mixing of liquids had occurred.

Bunsen also demonstrated that chemical changes in lava had occurred as a result of melting of wall rock along the contacts of a dike during injection.

He, therefore, allowed for variations in lava composition by contamination. Bunsen's experience with volcanic rocks was predominantly that of a single oceanic volcanic province, Iceland, characterized by a bimodal basalt–rhyolite suite. Working in a terrane in which contamination by various continental lithologies would be minimal, Bunsen postulated the existence of two fundamental lavas derived from two primary magmas.

Despite lacking Scrope's breadth of experience with volcanoes, Bunsen (1851, p. 214) extrapolated his conclusions from the Icelandic data by claiming that the "rock masses of the volcanic period" were everywhere expressed by "great agreement" that prompted the expectation that "these processes of volcanic formation of rocks are not limited solely to Iceland." As evidence for this conjecture, Bunsen pointed out that the volcanic rocks from Armenia chemically resemble those of Iceland. One cannot doubt, he argued, that the volcanic rocks of the Armenian highland and Iceland "flowed from *chemically* identical sources." He speculated that "perhaps all volcanic formations of the Earth's surface had taken their origin from the same sources" (Bunsen, 1851, p. 218). Bunsen even wondered if plutonic rocks might have been melted from the same deep sources.

Long before Bunsen's time, geologists had distinguished basalt and trachyte as the dominant volcanic rock types on the basis of variations in content of augite and plagioclase feldspar. As a chemist, Bunsen demonstrated a definite chemical basis for that distinction. Moreover, his demonstration of a chemical continuum opened the door for the hypothesis of magma mixing. Whereas Scrope, Darwin, and Dana had sought the causes of the separate existence of basalt and trachyte either on the surface or in the volcanic edifice itself, Bunsen traced the cause of that variation to fundamental differences within the interior of Earth.

Bunsen advanced thinking about diversity by considering the variation of eruptive rocks from a chemical point of view, but one of his traveling companions to Iceland advanced the discussion farther by linking igneous rock diversity to the physics of Earth's interior. Wolfgang Sartorius von Waltershausen (1809–1876) was Professor of Mineralogy and Geology at the University of Göttingen for thirty years from 1846 until his death (W. F., 1877). An interest in the physical aspects of the Earth, sparked by studies under Karl Friedrich Gauss during his student days at Göttingen, induced him to conduct measurements of terrestrial magnetism during 1834 and 1835. He later displayed considerable interest in reptilian paleontology and theories of climate change. Volcanoes, however, were his primary concern. Prior to embarking upon his university career, Sartorius von Waltershausen had spent a dozen years exploring several of the volcanic districts of Europe. From the labors of those years emerged a large comparative study of Icelandic and Sicilian volcanoes that was published two years after Bunsen's seminal study (Sartorius von Waltershausen, 1853). As evidenced by his interests in terrestrial magnetism and climate change, Sartorius von Waltershausen thought in global geophysical terms. Consequently, his conception of volcanic activity was intimately linked globally with his conception of the physics of Earth's interior. The broader empirical base of his study and his geophysical leanings also resulted in a theory of volcanic rock diversity of greater universality than that of Bunsen. But Sartorius von Waltershausen's conception of diversity was also shaped by the strong geological tradition, common in Germany and exemplified by such scholars as Pallas, Werner, and Bunsen, that considered granite as the foundational, Primitive rock on which stratified sedimentary, or Flötz, rocks were deposited and that considered volcanism to be a relatively recent phenomenon in Earth history. The conception of the recent, almost accidental character of volcanic activity had originally been closely linked to the notion that basalt was not a volcanic rock. Yet even after basalt was recognized as volcanic, many investigators in the Germanic tradition still considered basalt and other volcanic rocks as belonging predominantly to successions of Tertiary or Recent age, whereas acidic rocks and granite, to the extent that they were regarded as intrusive, magmatic rocks, were typically considered as the product of earlier epochs. Sartorius von Waltershausen, therefore, needed to find a method that would produce acidic rocks like trachyte early in Earth history and basaltic rocks later in Earth history.

To solve the problem, Sartorius von Waltershausen proposed a density-stratified globe. He considered Bunsen's hypothesis of two separate volcanic ovens as essentially "a piece of pure fiction." Bunsen, he believed, had overlooked the most natural hypothesis, namely, the regular disposition of matter in Earth's interior and density increase downward from Earth's surface. Sartorius von Waltershausen maintained that in Earth's outer crust light materials like silica, potash, and soda are particularly abundant, whereas heavier materials like lime, magnesia, alumina, and iron oxides predominate in the deeper layers. Denser materials gradually increase and lighter materials gradually decrease with depth, he argued. At the greatest depths, metallic oxides would increase the specific gravity of the layers still more, until, finally, pure metallic iron, nickel, cobalt, and the like would replace the last oxides.

The layers were considered to have formed from originally liquid layers that cooled downward from the surface. In the cooling process the layers supposedly developed distinctive mineralogical characteristics, with the upper layers containing mainly acidic minerals like feldspars or neutral salts and free silica and the deeper layers containing basic minerals such as augite. Below the depth at which feldspar formation reached its limit, augite was said to predominate, reach a maximum, then decrease and be replaced by magnetite. The magnetite would predominate and then gradually be replaced by the pure metals. When Earth was hotter, the layers would have been more fluid such that a silica-rich magma layer passed downward into an augitic magma layer and lastly into a magnetitic magma. In the view of Sartorius von Waltershausen, these shells of fused material corresponded to the natural succession of volcanic rocks. Trachytic rocks were generally regarded as first ejected in volcanic successions, and the proposed upper position of the acidic shell was surmised to account for their priority of eruption. Similarly, rocks intermediate between acidic trachyte and basic basalt were regarded as derived from intermediate depths at a later time. Recent basaltic rocks were considered to be drawn from the deep basic magma layer. In effect, Sartorius von Waltershausen universalized the two magma sources of Bunsen by envisioning them as two globe-encircling shells.

The conception of two concentric, stratified magma sources within the Earth was further developed by Joseph Durocher. When Durocher (1857) published "Essai de pétrologie comparée," an extensive paper that synthesized knowledge of the "pyrogenous rocks" from the points of view of chemical and mineralogical composition, eruption, and classification, he applied his discussion of diversity to coarse-grained, plutonic igneous rocks to a degree far beyond that of his predecessors. Durocher, who had spent much of his time working with continental rocks, including granite, and was a leading advocate of the origin of granite by dry fusion, recognized that a theory of diversity should account for plutonic as well as volcanic rocks and, therefore, be more complex than the hypotheses of either Bunsen or Sartorius von Waltershausen. He also sought to account for a much larger number of rock types than his predecessors. By 1857, as noted in Chapter 7, increasing refinement of definition had expanded igneous rock nomenclature far beyond the fundamental "trachyte" and "basalt" of Scrope and even Bunsen. At the same time, Durocher, like Bunsen and Sartorius von Waltershausen, was an adherent of the conception that the composition of igneous rocks had changed through time.

Without referring to Sartorius von Waltershausen, Durocher made the strikingly similar claim that all igneous rocks were produced by two magmas, one acid and one basic, each occupying a definite position below the solid crust of the globe. The two magmas, he said, had changed chemically only slightly throughout geological time. Durocher (1857, pp. 220) referred a wide range of rock types to the acid magma, including granite, "the eurites, quartziferous porphyries, and petrosilex, the trachytes, phonolites, perlites, obsidians, pumices, and lavas, with vitreous feldspar." To the basic magma source, he attributed "diorite, ophites, euphotides, hyperites, melaphyres, traps, the basalts, and pyroxenic lavas." Furthermore, he maintained that rocks of an intermediate character like "syenite, the protogenes rich in talc, the trachytes rich in pyroxene and amphibole, and various porphyries that are intermediate between granitic or trachytic porphyries, and amphibolic or pyroxenic porphyries" were derived from a zone of contact between the two magmas (Durocher, 1857, pp. 220–221). He explained the fact that rocks of different mineralogical composition were derived from the same magma source by appealing to differences in pressure and temperature and to the circumstances of the cooling of the rocks.

Durocher regarded the two magmas as forming two distinct layers beneath the solid crust. The upper layer, he said, "is rich in silica, and poor in earthy bases and oxides of iron" (Durocher, 1857, p. 221). Because of its silica-rich composition, this layer was thought to possess the least specific gravity and to be semiliquid or pasty because of its high viscosity. The lighter, more volatile materials, such as the alkali metals, fluorine, and boron, were believed to have collected in the upper layer. The lower layer, Durocher claimed, contained much less silica, was very rich in iron oxides, and was much more dense and fluid than the upper layer.

Because Durocher adopted the view that volcanic rocks are generally more recent than plutonic rocks, he maintained that the compositions of the two magma layers had changed through geologic time. The granitic rocks, he said, were erupted during the Primary or Secondary periods, whereas the trachytic rocks were said to belong to the Tertiary, Quaternary, and modern periods. Durocher believed that during the extended time separating the Primary and the Tertiary periods, the composition of the fluid upper layer from which the granitic and trachytic rocks were derived experienced a decrease in silica and potash and an increase in soda, lime, and iron oxides. The same chemical trends continued since the Tertiary period, he maintained, as indicated by comparison of the compositions of Tertiary trachytes with those of the present. Similarly, Durocher asserted that diorites were ejected more abundantly during the earlier geological periods but were generally replaced by melaphyre, basalt, and dolerite toward the end of the Secondary period and during the Tertiary period. This transition among the basic rocks was also said to be accompanied by a decrease in silica and potash and an increase in soda and lime through time. Even though both magma layers experienced a decrease in silica and potash and an increase in lime and soda, Durocher held that both layers remained distinct. "The trachytic products," he reasoned, "which represent the deeper portions of the siliceous layer differ much less in the whole of the elements from the granites (even the most ancient) than they do from the diorites, or from any other product of the basic layer" (Durocher, 1857, p. 228).

Durocher contended that the first melts to solidify on the surface of the still incandescent globe were granites derived from the uppermost layer. He claimed that the "pyrogenous rocks" found among Azoic slates, Paleozoic rocks, and lower Secondary rocks "are composed almost exclusively of feldspathic and siliceous rocks derived from the upper Magma" (Durocher, 1857, p. 246). These liquids from the upper layer he envisioned as being erupted in great quantities through dislocations in the cooling crust by the force of expanding gases. As a consequence of the ejection of so much acid liquid, however, the conditions of equilibrium of the lower basic magma layer were said to be disturbed in such a way that some basic liquid was drawn into the crust where it formed veins in the zone of granitic rocks. Basic rocks erupted during the Primary geological period, however, were regarded as merely accidental in comparison with the "immense development of the siliceous and feldspathic masses" (Durocher, 1857, p. 248).

By the second half of the Secondary period, the upper layer was said to have thinned as a result of eruptions and partial solidification. The lower basic layer, it was supposed, began to generate great eruptions of magma. Locally, however, the siliceous upper crust allegedly retained sufficient thickness to produce substantial emissions into modern time. In such localized instances, "the outpouring of basalts succeeded large eruptions of trachytic rocks" (Durocher, 1857, p. 249). Generally speaking, however, Durocher envisioned the siliceous upper layer to be almost completely exhausted at present so that more basic rocks predominate.

Although the broad patterns of the diversity and temporal distribution of igneous rocks were accounted for by Durocher in terms of his two-layer magma-source scheme, he proposed additional mechanisms for producing more subtle variations. As a mining engineer who was well acquainted with the properties and behavior of metals and alloys, Durocher proposed a process of "liquation," already known from the cooling behavior of some molten metals. Liquation was said to constitute a division of material derived from one of the magma layers, while in the liquid state, into two batches of liquid of contrasted composition. Durocher was, in essence, referring to liquid immiscibility. He further proposed to account for the sodium enrichment of many modern lavas in terms of mixing of seawater with volcanic liquids. Lastly, he suggested that some volcanic eruptions proceeded from "secondary foci" in the crust by fusion or by what might now be termed assimilation. The eruption products from these foci appeared to Durocher to be composed in part of materials belonging to the surface "which have become ingulfed or incorporated in the secondary foci." In other cases, Durocher explicitly stated that many lavas resulted from the remelting of older lavas.

Despite the resemblance of his theory to that of Sartorius von Waltershausen, on account of a similar commitment to the idea that igneous rocks changed through time, Durocher appreciated the fact that the complexities of igneous rocks might require a theory of petrogenesis that incorporated multiple mechanisms. Not only did his theory attempt to account for chemical, geological, mineralogical, and geophysical factors, but it also introduced considerations pertaining to the Earth's thermal history and the mechanics of igneous eruption.

SUSPICIONS ABOUT MULTIPLE MAGMA SOURCES

The Wernerian concept that igneous lithologies vary with geologic age was prominent among German scholars such as Bunsen and Sartorius von Waltershausen. The French mining engineer Joseph Durocher likewise accepted the notion that igneous rock types had changed through time. Adherents of this view were prone to locate the cause of diversity deep within the fundamental layered structure of the Earth and to attribute diversity to alterations of that structure by long-term cooling of the globe.

Some contemporaries of Bunsen, Sartorius von Waltershausen, and Durocher, however, rejected the view that there exists a correlation between the mineralogy of igneous rocks and their geological age. Coupled with that rejection was skepticism about theories of two magma sources. Among such skeptics was Joseph Beete Jukes (1811–1869), an 1836 graduate of the University of Cambridge where his enthusiasm for geology had been kindled by Adam Sedgwick (J. M. Eyles, 1973b). Jukes had spent more than a year in Newfoundland conducting a geological survey, had been the naturalist on the expedition of the H.M.S. *Fry* to Australia and the East Indies between 1842 and 1846, had mapped structure and studied coal deposits in Wales under the auspices of the Geological Survey of Great Britain, and, since 1850, had been director of the Irish branch of the Geological Survey. The major contributions to geology by Jukes, an outstanding field mapper, were in geomorphology and structure. Igneous petrology was not one of his major interests.

In his textbook, *Students Manual of Geology*, however, Jukes (1857) decided to address the issue of igneous rock diversity. He rejected the commonly held conception that there are "separate deep seated foci or reservoirs for every variety of igneous rock" (Jukes, 1857, p. 294). He attributed the mineralogical differences among igneous rocks to space rather than to time and preferred to think of rocks as "deeply formed" or "superficially formed" rather than "ancient" or "modern." Jukes suspected that depth was a key factor in accounting for igneous rock diversity if only because temperature increases with depth, and substances in molten liquid have different melting temperatures. He envisioned minerals that are difficult to fuse, like pure silica, consolidating at depth within a rising and cooling molten mass, whereas more easily fusible minerals, such as those containing a large quantity of bases, would remain in a molten state all the way to the surface. As he envisioned the process, more readily fusible substances would be successively squeezed out of the same stream of igneous liquid, proceeding from the interior to the surface of Earth, from which the infusible substances solidified.

Jukes (1857, p. 293) saw the associations of such diverse rock types as trachyte and dolerite as due to a cause that tended to "segregate the one from the other out of a generally diffused mass, in which the constituents of both may be equally mingled." Despite his relative lack of exposure to volcanoes,

Jukes was undoubtedly influenced by his experience of the wide range of compositions of various rocks within the continental crust to perceive that an initially uniform magma originating at depth would likely develop differences because it had to pass through different thicknesses of various sorts of rocks. In other words, the magma became contaminated. Along the lines of Dana, he further suspected that diverse rocks developed because of differences in fusibility of substances within the ascending magma as it passed through varying conditions of pressure and temperature. He envisioned that easily fusible substances in the magma separated into "liquid strings and veins from the consolidating rocks below" as the whole mass rose toward the surface.

A year after Jukes' text appeared, a modification of the stratified shell concept was proposed by von Cotta. He voiced the same suspicions as Jukes that no correlation existed between the age of an igneous rock and its mineralogy. Von Cotta (1866) wrote that every geological age produced both acidic and basic igneous rocks. Where a range of igneous rocks had been erupted in the same period, however, he suggested that it was generally the more basic rocks that were emplaced a bit earlier. He proposed that only the pyroxenic shell presently served as a source of volcanic liquids and argued that a hot, molten pyroxenic region lay beneath a solid crust of silica-rich substances. The present diversity of rocks, von Cotta considered, could be accounted for because the liquid from the pyroxenic region would have to pass through the silica-rich crust on its way to the surface and might, therefore, pick up varying amounts of siliceous material. Along the lines of Jukes, he claimed that this rising molten basic material would dissolve variable amounts of the existing siliceous crust, depending on the circumstances. He assumed that the mineralogical diversity and differences in texture could be accounted for by variations in rate of cooling, pressure, and ease of access of water to the rising molten material. In essence, von Cotta retained the two-shell theory of Sartorius von Waltershausen and Durocher but derived magma only from the lower basic shell, thus envisioning basic magma as the only primary magma.

PETROGENESIS COMES TO AMERICA

Despite the long tradition of excellence in petrography in the German universities dating back to Werner, German petrographers, apart from von Cotta, seemed little concerned to address the theoretical question of the origin of igneous rock diversity. They focused on descriptive detail. Surprisingly, it was primarily other German academics who addressed the issue of igneous rock diversity. Bunsen, Sartorius von Waltershausen, von Cotta, and Ferdinand von Richthofen were all members of the German academic establishment.

German petrogenetic theorizing was imported to America when Baron Ferdinand von Richthofen (1833–1905) came to investigate the newly discovered volcanic terranes of the American West and powerfully influenced the thinking of the early American geologist explorers of the western territories, particularly Clarence King (Beckinsale, 1975). Von Richthofen was a geologist who applied a wealth of information from studies of volcanic terranes to the problems of petrogenesis in the years immediately prior to the development of microscopic petrography. In the late 1850s, von Richthofen had gained considerable field experience in the Carpathian Mountains under the auspices of the Austrian Imperial Geological Institute of Vienna. As a result of these studies, von Richthofen coined the term "rhyolite." After a couple of years in southeastern Asia on a mission for the Prussian government, von Richthofen examined the Tertiary and Quaternary volcanic rocks of California and Nevada during a six-year period between 1862 and 1868. The outcome of his research was a massive paper published in the very first volume of *Memoirs of the California Academy of Sciences* and entitled "The natural system of volcanic rocks" (von Richthofen, 1868). After producing that work, von Richthofen returned to Asia and then to Germany where he taught geology and geography at the universities of Bonn, Leipzig, and Berlin until his death in 1905. During the latter part of his career, von Richthofen concentrated on geomorphological studies.

Von Richthofen's "natural system" was still steeped in the view that coarse-grained eruptive rocks are dominated by quartz and feldspar and are predominantly ancient. He considered granites to be Paleozoic or even older rocks, with the major exception being the Mesozoic granites of the Sierra Nevada and South America. In contrast, he viewed volcanism as predominantly recent. He believed that there had also been a transitional epoch in which porphyritic rocks were generated. These relations of age and texture, von Richthofen wrote, were conclusively proved in Europe and seemed to justify the determination that the three classes of eruptive rocks represent three successive and distinct phases of the manifestation of subterranean agencies. Information gleaned from other countries, he thought, appeared to confirm that conclusion.

Von Richthofen's major contribution consisted in his recognition of a general temporal sequence of eruptions within individual volcanic provinces. He postulated that volcanic provinces are characterized by initial eruptions of a rock that he designated as "propylite," later discredited as a distinct rock type when it was recognized as an altered andesite. These eruptions were said to be followed by andesite, trachyte, rhyolite, and, lastly, great outpourings of basalt. Episodes of eruptive activity were seen as beginning with chemically intermediate massive eruptions and proceeding toward increasingly acidic and basic volcanic eruptions.

Although built on the concepts of Bunsen and Sartorius von Waltershausen, von Richthofen's theory was necessarily more complicated than those of his predecessors for at least two reasons. On the one hand, in the American west, von Richthofen was dealing with a volcanic terrane of geological complexity that far exceeds that of Iceland or individual volcanoes like Mount Vesuvius or Mount Etna. In addition, unlike most earlier workers who were concerned primarily about the reasons for the distinction between basalt and trachyte, von Richthofen had to deal with the problem of accounting for the diversity of a range of volcanic rock types that now included such apparently abundant lithologies as andesite, "propylite," and rhyolite. Von Richthofen himself bore much of the responsibility for that expanded nomenclature as originator of the terms "propylite" and "rhyolite." The problem of diversity had become more acute with the recognition that volcanic rock diversity encompassed much more than two fundamental rock types.

Despite his more sophisticated understanding of the patterns of eruption, von Richthofen was still indebted to Bunsen and Sartorius von Waltershausen. He repeatedly insisted that the volcanic rocks obey the law of Bunsen in that they possess chemical compositions lying within the two extremes suggested by the chemist. On that ground alone von Richthofen repudiated the notion that volcanic rocks could be generated by melting deeply buried sedimentary rocks. He reasoned that if eruptive rocks had been generated from sedimentary rocks, then they must be chemically analogous. Eruptive rocks would, in that case, vary within very wide and complex limits that would exceed the range suggested by Bunsen.

Von Richthofen also agreed that the picture of the interior of the Earth with a lighter, acid granitic shell overlying a denser, basic basaltic shell was essentially correct. He concurred, therefore, with Sartorius von Waltershausen that Tertiary volcanism represented the tapping of sources within Earth's deeper, more basic layer. Von Richthofen did, however, suggest that an infinite number of layers passing through infinite gradations of chemical composition and density from the surface of the Earth toward its center would be very likely.

Germanic age-based schemes were distasteful to the British. Although von Richthofen's "natural system" exerted a potent influence on American geologist, Clarence King, the English volcanologist, G. P. Scrope, was singularly unimpressed. In a scathing review of von Richthofen's memoir, Scrope (1869, p. 510) wrote that "these arbitrary classifications and assertions as to the relative age of particular varieties of volcanic rocks are wholly at variance with known facts, and the authority of the most reliable observers of such formations." Scrope concluded his review by wishing that Charles Darwin might someday do properly for the American West what von Richthofen had failed to do. He confessed being saddened that an opportunity for extending knowledge of American volcanic formations and phenomena had been squandered. He thrashed von Richthofen for failing to make good use of his time in California in painstaking examination and accurate description of the volcanic phenomena he observed "instead of rushing into print with a crude classification of rocks founded on imperfect data, and ambitious Cosmical theory, the ideas of which are culled from the works of other geologists, not often referred to, with which the remainder of his memoir is filled" (Scrope, 1869, pp. 515–516).

Having recognized ever greater and greater diversity among at least the volcanic rocks, geologists, chemists, and naturalists of the middle decades of the nineteenth century had begun to speculate on some of the possible causes of that diversity in terms of their own scientific experiences and interests. Many of the processes that are still given currency were proposed in rudimentary form during that period. The present conception of the derivation of basalt from the mantle and granite from the continental crust had its primitive analog in the theory of two magma shells. By one or another geologist, the concepts of mixing of contrasting magmas, contamination by assimilation of crustal rock, partial fusion of crustal rocks, and liquid immiscibility all were placed in the great cafeteria line of petrological ideas. These early views on igneous rock diversity had to account for the origins of just a handful of lithologies. The development of microscopic petrography would put an end to such simplicity.

…
The Microscope Era

9 ⇜ MINUTE OBJECTS, GREAT CONCLUSIONS

The Rise of Microscopic Petrography

Continuous field investigations in Europe and the Americas throughout the early decades of the nineteenth century disclosed the fact that volcanism encompassed rocks with a considerable range of mineral contents. Thanks to the painstaking labors of chemists and geologists a substantial amount of analytical data had been accumulated so that these volcanic rocks were reasonably well characterized in terms of their major-element chemical composition. Geologists also accepted the existence of coarsely crystalline rocks displaying an intrusive, cross-cutting, or "eruptive" character in relation to the surrounding rocks. Despite the fact that considerable controversy swirled around the origin of these coarse-grained rocks, particularly granite, many geologists were convinced that most intrusive, presumably plutonic rocks had crystallized from molten subterranean lava that was gradually being referred to as "magma." As a result, such geologists considered "eruptive" or "plutonic" rocks to be every bit as "igneous" or "pyrogenous" as volcanic rocks. The mineralogical and chemical variation of these rocks, too, was increasingly appreciated. Geologists had even begun to ponder some possible causes for the diversity of igneous rocks, primarily those of volcanic origin.

Although wet chemical and blowpipe analysis and microscopic examination of crushed mineral and rock fragments in ordinary transmitted light contributed some insights into the nature of igneous rocks, the limitations of the hand lens (loupe) and unaided eye in the study of such rocks were clearly evident. Igneous geology was on the verge of stagnation had it not been for a revolutionary change instigated by the application of polarized-light microscopy to the study of rocks and minerals in thin section (Sigsby, 1966; Hamilton, 1982). The development of the polarizing microscope and of thin-section analysis resulted in vastly improved mineral identification, the analysis and interpretation of igneous rock textures, the estimation of the abundances of minerals within rocks, and the determination of the order of crystallization of minerals. Thin-section investigation led to unprecedented advances in igneous rock description and more precise classification. The wealth of new knowledge opened the way for an influx of fresh insights into the genetic interpretation of igneous rocks. In time, more convincing petrologic theories would be constructed on the solid foundation of descriptive petrography that had been laid with the aid of thin-section analysis by the petrographic microscope.

EARLY OPTICAL DISCOVERIES

The first microscope of any kind was invented in the seventeenth century. Robert Hooke (1635–1703) took advantage of the new instrument by examining thin slices of both modern and petrified wood (Drake, 1996). Although Hooke (1665) demonstrated the cellular structure of petrified wood and showed its essential identity to modern wood, no one followed up on the possibilities inherent in Hooke's use of thin slices, at least for another century. The discovery of the phenomenon of double refraction in calcite by Bartholin at about the same time was also of great importance for the nineteenth-century petrographic revolution. These seventeenth-century developments in optics, however, illustrate that the relevance and application of discoveries or advances in one field of science may not become apparent to another field of science for decades or even centuries. Then, too, major advances in geology have sometimes depended not only on creative thinkers but also on fortuitous discoveries in very unlikely places. With a minuscule population because of its remoteness and rigorous climate, Iceland lacked its own scientific establishment. Nevertheless, Iceland provided the stimulus for a major contribution to igneous petrology because of its remarkable volcanic activity that served as the basis for Robert Bunsen's theory of dual magma sources. Arguably, however, Iceland's most significant contribution to igneous petrology stems from a 1668 expedition to the island, then under Danish rule, that discovered a deposit of exceptionally transparent calcite known as Iceland spar. Erasmus Bartholin (1625–1698), a Danish mathematician and astronomer at the University of Copenhagen, made an optical study of the material and discovered that two sets of light rays, which he termed "solita" and "insolita," later to be termed the "ordinary ray" and the "extraordinary ray," are produced by refraction in calcite (Hall, 1970). Bartholin (1669) provided a geometrical construction for determining the position of the extraordinary ray. It remained for the Dutch mathematician, physicist, and astronomer Christiaan Huygens (1629–1695), to discover that both refracted images produced by Iceland spar are polarized when one piece of spar is placed on top of another (Bos, 1972). More than a century later, in 1814, the French mathematical physicist Jean-Baptiste Biot (1774–1862), namesake of the mineral biotite, discovered that the value of the index of refraction, that is, the ratio of the velocity of light in vacuum to the velocity of light in a mineral, associated with the "ordinary ray" is less than the value of the index of refraction associated with the "extraordinary ray" in many minerals, unlike the situation in calcite (Crosland, 1970b). In 1815, Biot also noted that two plates of tourmaline oriented at right angles to one another cause the extinction of light.

Sir David Brewster (1781–1868), Scotland's expert in optics, constantly tried to improve microscopes, explored the phenomena and developed the theory of polarized light and dispersion, and laid the foundations for optical mineralogy with numerous microscopic studies of crushed mineral fragments to ascertain their optical properties (Morse, 1970). By 1819, Brewster had grouped hundreds of minerals into categories based on the number of axes of double refraction. In other words, he established whether the minerals are "uniaxial" or "biaxial." Brewster also made thin slices of minerals as early as 1816. Taking their lead from Brewster, geologists and mineralogists began to examine crushed fragments of rocks under a microscope to make more accurate identifications of the minerals.

In 1828, William Nicol (1768–1851), Lecturer in Natural Philosophy at the University of Edinburgh, reported a method for the construction of a prism for producing plane-polarized light, that is, light constrained to vibrate in a single plane, by gluing together two carefully cut and polished, as well as appropriately oriented, pieces of Iceland spar (Nicol, 1829; Frankel, 1974). Early application of a polarizing microscope was reported by William H. F. Talbot (1800–1877) who had first used slices of tourmaline as a polarizer but abandoned them in favor of the Nicol prism because of the latter's superior "whiteness and transparency" (Talbot, 1834; Jenkins, 1976).

THE DEVELOPMENT OF THIN SECTIONS

The origins of making thin slices of rocks are probably obscured in the misty beginnings of the lapidarist's art. Despite Hooke's and Brewster's preparation of slices of petrified wood and large mineral crystals, respectively, Nicol is commonly credited with development of the rock thin section by perfecting a method of making slices that he learned from George Sanderson, an Edinburgh lapidary. Although he never directly reported on his method, Nicol prepared several thin sections of fossil woods from Carboniferous and Oolitic rocks, some of which he collected himself, for a friend who was compiling a paleontological monograph, *Observations on Fossil Vegetables* (Witham, 1831). Henry Witham (1779–1844), was one of the founders of the Natural History Society of Northumberland, Durham, and Newcastle upon Tyne (Long, 1976). Witham dedicated his book to Nicol and expressed gratitude to his "indefatigable friend" Nicol for providing him with "beautifully prepared sections of plants" (Witham, 1831, p. 4). Witham (1831, p. 2) also announced that "the day is now arrived, when doubts and difficulties as to the class and family to which they are to be referred, must give way under the microscopic examination of the internal structure of these fossil plants." He included a brief description of Nicol's method of making thin sections:

A slice, or thin fragment, is obtained in the usual manner. One side of it is ground and polished, and is then applied to a piece of plate or other glass, by means of a transparent gum or resin. The other side is then ground down, parallel to the glass, and, on being brought to the necessary degree of thinness, polished. By this means, the internal structure may be as distinctly seen as in the slice of a recent vegetable. (Witham, 1831, p. 23)

Witham also provided a much more extended description of the method at the conclusion of his monograph in the hopes that other geologists might be encouraged to examine fossil plants for themselves in this new way. He recorded that a few years earlier, Nicol had recommended the microscopic thin-

MINUTE OBJECTS, GREAT CONCLUSIONS 147

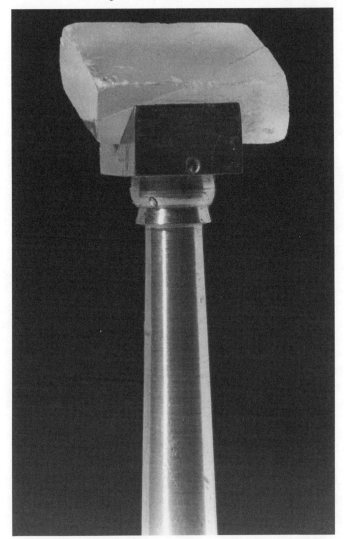

8. Left: thin sections made by William Nicol (1768–1851) for Henry Witham (1779–1844). Right: Nicol's polarizing calcite prism. Reproduced by permission of the National Museums of Scotland.

section method of examination of plants before the York and Newcastle Philosophical and Natural History Societies. Nicol's original thin sections can still be seen in the display of scientific instruments in the Royal Museum of Scotland in Edinburgh (Figure 8). Witham's book was probably not widely read by the more petrographically minded geologists of the era, and, despite the great promise of the thin-slice method for petrography, a few more years passed before further progress was made in igneous rock microscopy. The earliest applications were directed toward paleontology.

Thin sections were later employed in geological studies by William Crawford Williamson (1816–1895), a native of Scarborough, England, who was taught techniques of cutting and polishing by his jeweler–lapidary grandfather (Andrews, 1976). He became curator of the Manchester Natural History Society and established a medical practice in Manchester. After reading Gideon Mantell's *Medals of Creation*, Williamson became fascinated by the notion that chalk consisted of microscopic shells. As a result he undertook microscopic investigation of foraminifera and diatoms and published papers reporting results from a study of thin sections by ordinary, transmitted-light microscopy. Williamson went on to become Professor of Natural History at the newly formed Owens College in Manchester in 1851. John Phillips (1800–1874), Professor of Geology at the University of Oxford, also made microscopic studies of calcareous sedimentary material in thin section (Edmonds, 1974).

The analysis of rock thin sections with a polarizing microscope was developed independently by H. C. Sorby and A. Oschatz. In 1849, Sorby set about to make thin sections of calcareous grits along the coast of his native Yorkshire. In late 1850, he presented the results of his investigation before the Geological Society of London, and his paper appeared the following year. Sorby (1851, p. 1) noted that dissolution of the grit in hydrochloric acid yielded numerous agatized shells and some sandy matter "which without further examination would naturally be thought to be merely sand, and such it has hitherto been considered." He proceeded to show that the sandy matter consists of tiny shells, perhaps of foraminifera, that had been selectively agatized or infiltrated by calcareous matter. Sorby's demonstration entailed examination of thin slices of the rock, about 1/1000 of an inch thick, under a microscope with both transmitted and reflected light. He was able to show that the sandy matter consists of both agate and calcite because of the distinctive behavior of these materials in polarized light. Sorby also made a thin section of an ammonite from the rocks to examine its shelly structure and used his observations to conclude that the tiny bodies had also been shells that were ruptured and subsequently infiltrated. He calculated the approximate abundance of shells and concluded that about two to three million were present in a cubic inch of grit. Sorby's technique had the potential for distinguishing minerals from one another by their interaction with polarized light, for determining the sizes of very small grains, for ascertaining the structure of very small bodies, and for interpreting aspects of the history of objects. Yet Sorby's effort, so impressive in hindsight, went largely unnoticed.

There soon followed a study of the origin of slaty cleavage on the basis of the microscopic structure of slates of North Wales and Devonshire in a paper in which Sorby (1853, p. 142) off-handedly remarked that he had already "prepared and examined thin sections of several hundred specimens of every geological period." Sorby (1853, p. 137) began by observing that microscopic geology "has hitherto attracted little or no attention, though the inspection of two or three thin sections will sometimes solve most important geological problems." In this case he compared thin sections of slate with thin sections of uncleaved sedimentary rocks possessing similar texture and mineral composition. Using a 400× polarizing microscope, Sorby found that he could identify the minerals with certainty even in grains less than 1/1000 inch in diameter. Of special importance was his ability to identify mica grains. He demonstrated from thin sections that had been cut perpendicular to cleavage or to bedding that the micas in cleaved slates display laminae, most of which are strongly oriented parallel, or nearly so, to the cleavage planes, whereas the micas in the sedimentary rocks are much more randomly oriented. He concluded that slates contain evidence of compression and strong elongation and that they had changed position during compression. Again, despite the impressive results of the study, little notice was taken.

Sorby's attention was then directed to the study of fluid inclusions in minerals after he saw the excellent collection of such inclusions owned by Alexander Bryson in Edinburgh. His research resulted in the aforementioned talk to the Geological Society of London and the subsequent paper in which he pointed out the utility of microscopic investigation of the fluid inclusions contained in quartz for determining whether granitic rocks were the results of magma or aqueous precipitation.

Anticipating possible resistance to the impact of his observations, Sorby (1858, p. 497) concluded his talk and paper with an impassioned appeal to fellow geologists to pay attention to the results of microscope work:

> Although with a first-rate microscope, having an achromatic condenser, the structure of such crystals and sections of rocks and minerals as I have prepared for myself with very great care can be seen by good day-light as distinctly as if visible to the naked eye, still some geologists, only accustomed to examine large masses in the field, may perhaps be disposed to question the value of the facts I have described, and to think the objects so minute as to be quite beneath their notice, and that all attempts at accurate calculations from such small data are quite inadmissible. What other science, however, has prospered by adopting such a creed? What physiologist would think of ignoring all the invaluable discoveries that have been made in his science with the microscope, merely because the objects are minute? What would become of astronomy if everything was stripped from it that could not be deduced by rough calculation from observations made without telescopes? With such striking examples before us, shall we physical geologists maintain that only rough and imperfect methods of research are applicable to our own science? Against such an opinion I certainly must protest; and I argue that there is no necessary connexion between the size of an object and the value of a fact, and that, though the objects I have described are minute, the conclusions to be derived from the facts are great.

Despite his impassioned plea, Sorby's work still met with considerable skepticism. His good friend Leonard Horner, then chairman of the Geological Society, said that he did not remember any paper ever having been read that drew so largely on the credulity of the audience. He recalled de Saussure's comment about the impropriety of examining mountains with microscopes. In time, however, Sorby would win over some very effective converts and his method would be vindicated in spectacular fashion.

Before long, as Dawson (1992) pointed out, Sorby (1863) also made thin sections of some of the fused charges of basalt produced by Sir James Hall. He demonstrated conclusively that Hall's fused samples contain the three principal minerals of the original basalt, even though distinctive quench textures are present. Had Sorby's method been in vogue in Hall's day, the basalt controversy might have come to a close much sooner.

The preparation of rock thin sections was developed independently by A. Oschatz of Berlin. Oschatz has received very little recognition for his achievement because so little is known about him. Beginning in 1851 and for several years thereafter, Oschatz annually presented exhibits of his thin sections at meetings of the German Geological Society as reported in *Zeitschrift der deutschen geologischen Gesellschaft* (1851, v. 3, pp. 382; 1852, v. 4, pp. 13–15; 1854, v. 6, pp. 261–263; 1855, v. 7, p. 298; 1856, v. 8, pp. 308, 534). As early as the meeting of September 1851, Oschatz described his method of thin-section preparation with Canada balsam and demonstrated how the structure of minerals could be investigated by use of the microscope. At a meeting in January 1852, he continued to urge upon geologists the extreme importance of undertaking microscopic investigation of minerals in both random-cut and oriented thin sections. He demonstrated sections of various homogeneous materials such as garnet, labrador rock, Bimstein (pumice), and different kinds of glasses, including various obsidian specimens. He pointed out the presence of needle-like crystals or gas bubbles within the obsidians. Some of the gas bubbles were reported to be elongated and oriented. He also showed sections of granite and lapis lazuli and called attention to granular structure in marble, radial structure in malachite, and scaly partings in chrysoprase.

At the meeting of April 1854, Oschatz displayed a more extensive collection and even offered to show thin sections in his home to interested parties. With sections of lazurite and basalt, he demonstrated that color is not so uniform as supposed by examination with the unaided eye but that coloring agents are irregularly distributed and make up a small portion of the material. The following year, Oschatz showed the behavior of thin slices of marble in polarized light.

In March 1856, Oschatz sent a collection of sections that were displayed at a meeting of the society by mineralogist Gustav Rose, the brother of chemist Heinrich Rose. The collection contained labradorite, garnet, obsidian, hypersthene, basalt, dolerite, trachyte, phonolite, porphyry, and other materials.

Oschatz's sections consisted of a thin slice of rock or mineral embedded in Canada balsam and sandwiched between two airtight glass plates. He offered to sell the sections individually or as a collection consisting of 73 pieces. Later in the year, he offered to prepare sections on demand for a fee. At the meeting of March 1856, Oschatz displayed the polarizing characteristics of carnallite and also pointed out its twinning. He then passed from view.

THE FLOURISHING OF THIN-SECTION PETROGRAPHY

The German petrographic tradition provided the most fertile soil in which the new microscopic methods might luxuriate. In the 1850s, a number of German scholars like Gustav Rose and Carl Rammelsberg in Berlin, E. E. Schmid in Jena, Gustav Jenzsch in Dresden, A. Streng in Clausthal, and Gerhard vom Rath in Bonn consistently produced lengthy papers on all sorts of igneous rocks, in some cases using the microscope. Ami Boué (1863) reviewed some of these early microscopic studies. Geologists like Gustav Jenzsch (1830–1877), Gerhard vom Rath (1830–1888), and Max Deiters, who was vom Rath's student at Bonn, began to publish papers in *Zeitschrift der deutschen geologischen Gesellschaft* that included descriptions of thin sections of rocks examined in polarized light and, in some cases, incorporated beautiful diagrams of the rock textures (Jenzsch, 1855, 1856; vom Rath, 1860; Deiters, 1861). An increasing tide of papers based on thin-section work using polarized light appeared in *Zeitschrift* and *Neues Jahrbuch* throughout the 1860s from the pens of A. Knop (Giessen), F. von Hochstetter (Vienna), H. Laspeyres (Bonn), and, particularly, F. Zirkel.

A young mining engineer who had studied under vom Rath and Bischof and received his doctorate in 1861 from the University of Bonn for work on volcanic rocks and ore deposits in Iceland, Ferdinand Zirkel (1838–1912) had already published a microscope-based petrographic paper on the trachytic rocks of Eifel (Zirkel, 1859), but his later enthusiasm for thin-section petrography seems to have been fired largely as result of a chance encounter with Sorby during a cruise on the Rhine River in 1862 (Amstutz, 1976). Zirkel (Figure 9) was serving as a guide on the cruise ship. Upon learning of his interest in geological matters, Sorby, who was vacationing with his mother, struck up a friendship with Zirkel and engaged him in discussions about the volcanic rocks of the Eifel district and their mineral composition. Sorby described the results of his own studies. Upon the conclusion of the voyage and their return to Bonn, he prevailed upon Zirkel to look at some of the sections that he had brought with him. Zirkel was fascinated by the detail of Sorby's sections and became an enthusiast for the new method of rock investigation.

The following winter, Zirkel began to prepare his own thin sections at the Reichsanstalt in Vienna, and shortly thereafter he assumed the position of Associate Professor in the University at Lemberg. On March 12, 1863, he presented

9. Ferdinand Zirkel (1838–1912). Reproduced by permission of the University of Leipzig.

to the Vienna Academy the results of a detailed investigation of the microscopic characteristics of 39 specimens of typical igneous rocks, including granite from Cornwall, England, quartz-bearing trachyte from Iceland, rhyolite from New Zealand, basalt from the Eifel district in Germany, dolerite from Arthur's Seat in Edinburgh, and obsidian from Mexico (Zirkel, 1863). Study of the quartz-bearing trachyte helped resolve the debate over the origin of granite by confirming that quartz does crystallize directly from magma. He focused mainly on the structural and textural characteristics of the rocks and dwelt little upon the identification of the minerals because, at first, he felt that the polarizing microscope held out relatively little promise as a tool for mineral identification. As he put it, "labradorite, oligoclase, and orthoclase, augite and hornblende, minerals whose recognition offers the most important problems in petrography, in most cases cannot be distinguished from one another under the microscope" (Zirkel, 1863, p. 227).

Zirkel (1866) still held the same view when he issued a massive 1200-page, two-volume textbook entitled *Lehrbuch der Petrographie*. Although Zirkel briefly took into account the kinds of things that might be learned through microscope investigation of thin sections of rocks, the book did not contain a single illustration. Lacking such diagrams, the text failed to demonstrate the value of microscope studies as effectively as some of his subsequent publications. In 1868, Zirkel assumed a professorship in Kiel. Two years later he moved to the University of Leipzig where he became the successor to C. F. Naumann. Here Zirkel published a volume on basaltic rocks, dedicated to Sorby, that included considerable material on microscopic petrography. In this book, based on a study of 305 thin sections, Zirkel (1870) described the constituents of basaltic rocks, discussed the general characteristics of their fine structure, and described in detail what he considered to be three main classes of basaltic rocks: feldspar basalts, leucite basalts, and nepheline basalts. One feature of the volume that represented a major advance over his *Lehrbuch* was the inclusion of plates with 79 sketches of microscopic views of sections of different orientations through augite, fluid inclusions, zoned augite crystals, plagioclase twins, olivine, leucite, and a very small number of rock textures.

Three years later, Zirkel published *Die mikroskopische Beschaffenheit der Mineralien und Gesteine*. In this work, Zirkel (1873) discussed the basic principles of the polarizing microscope, the nature of polarized light, and specimen preparation before moving on to a review of the microscopic structure of minerals, the particular microscopic characters of numerous individual minerals, some general comments on the microscopic structure of rocks, and the description of individual rocks. In essence, this work represented the first text of optical mineralogy and petrography. Surprisingly, Zirkel included relatively few diagrams, again failing to convey fully to the reader the advantages offered by the new optical method of studying rocks and minerals. Many of the diagrams that were included had previously appeared in his work on basalts. *Die mikroskopische Beschaffenheit* incorporated diagrams of the microscopic appearance of marble and glassy volcanic rocks but none of granite, syenite, trachyte, phonolite, diorite, gabbro, basalt, gneiss, or mica schist, even though all these rock types were described at great length.

Recognizing the immense value of the work being done by Zirkel, Clarence King requested him to carry out microscopic investigations of the rocks collected on the United States Geological Exploration of the Fortieth Parallel of which King was the leader. Zirkel's impressive memoir on these rocks appeared in 1876. Zirkel continued to publish important papers on igneous rocks, and, toward the latter part of his career, he produced a huge second, thoroughly revised edition of *Lehrbuch der Petrographie* (Zirkel, 1893). This vastly expanded work understandably took considerable advantage of the enormous strides that had been made over the past three decades in the study of rocks with the petrographic microscope. The result was a three-volume, encyclopedic work of 2619 large pages of text with small print, again without a single diagram, and broken only occasionally by a table of chemical analyses! The first volume dealt primarily with general petrographic principles. The final 200 pages of the first volume, the entire second volume, and the first 140 pages of the third volume were devoted to the igneous rocks, which he termed "die massigen Erstarrungsgesteine." Despite the reluctance to use many diagrams, perhaps owing to their production costs, Zirkel provided a powerful impetus to petrographic research through his indefatigable labors of description of innumerable igneous rocks. Zirkel also introduced the rock name "quartz trachyte" and the textural term "ophitic."

A remarkable petrographic triumvirate was established early in Zirkel's career when he was joined by Vogelsang and Rosenbusch. Zirkel's brother-in-law, Hermann Vogelsang (1838–1874), Professor of Mineralogy and Geology at the Polytechnicum in Delft, the Netherlands, shared Zirkel's enthusiasm for the new petrography. In the 1860s, Vogelsang undertook microscope investigations of slags and demonstrated the importance of such studies for the origins of crystalline rocks. Vogelsang (1867) published *Philosophie der Geologie und mikroskopische Gesteinsstudien*, the first geological text whose title signaled the importance of microscopy. He also coined the now widely used terms "felsophyre," "granophyre," "vitrophyre," and "microlite." Unfortunately for the blossoming young science of thin-section petrography, Vogelsang's promising career was cut short by his premature death in 1874.

With the loss of Vogelsang, it was Karl Harry Ferdinand Rosenbusch (1836–1914), the third member of the triumvirate, who was destined to become Zirkel's peer (Judd, 1914; Ramdohr, 1975). Together these two great German petrographers led the scientific world in developing the field of descriptive microscopic igneous petrography. Rosenbusch (Figure 10) had been stimulated to pursue studies in geology and chemistry by the lectures of Robert Bunsen. After receiving his doctorate in 1869, Rosenbusch was appointed Professor Extraordinarius of Mineralogy and Petrography at the University of Strassburg in 1873. In that year, Rosenbusch (1873) issued the first volume of a projected encyclopedic work, entitled *Mikroskopische Physiographie der Mineralien und Gesteine*, that would profoundly influence petrographers through several ever-expanding editions published over five decades. Because of its powerful influence, Rosenbusch's text deserves some detailed attention. The

10. Harry Rosenbusch (1836–1914). Reproduced by permission of the Archives of the University of Heidelberg.

first volume, entitled *Mikroskopische Physiographie der petrographisch wichtigen Mineralien*, remained the bellwether in the field for years to come. This work consisted of a general part devoted to an explanation of the methods of microscopical investigation and a special part devoted to "clear and exact" description of the "species" that are important to the petrography of crystalline rocks.

In the general part, after a brief historical overview and a sketch of methods for preparing thin sections, Rosenbusch reviewed the morphological, physical, and chemical characteristics of minerals. Among the morphological characteristics, he called attention to anomalies of external form like broken crystals or resorption of crystals. He described fluid inclusions, microlites, and twinning phenomena. Among the physical characteristics, Rosenbusch provided a synopsis of optical properties and discussed polarizing instruments, the behavior of thin mineral plates of doubly refracting minerals, interference figures, and pleochroism. Chemical properties and analysis were reviewed in a scant four pages.

The "special part" comprised the bulk of the text. In nearly three hundred pages, Rosenbusch described in detail optical and other characteristics of amorphous bodies like obsidian, Bimstein (pumice), Perlstein (perlite), Pechstein (pitchstone), hyalomelan, tachylyte, palagonite, and opal. He then described ten minerals of the "regular" (isometric) crystal system: magnetite, chromite, spinel, garnet, schorlomite, analcime, sodalite, hauyne, nosean, and pyrite. Ten minerals of the "quadratic" (tetragonal) crystal system, four of which are now known to be different varieties of scapolite, were described. Fifteen hexagonal minerals, including biotite and chlorite, now known to be monoclinic, were described. Sixteen minerals of the "rhombic" (orthorhombic) crystal system, including talc and muscovite, now known to be monoclinic, were discussed. Sixteen minerals of the "klinorhombisches" (monoclinic) system were described, including pyroxenes, amphiboles, epidote, sanidine, and orthoclase. Lastly, Rosenbusch discussed minerals of the "klinorhomboidischen" (triclinic) crystal system. Under this category he described only disthene (kyanite), the triclinic feldspars, and axinite. The text also contained a review of eight homogeneous crystalline aggregates, such as agate, serpentine, kaolin, and sericite. The book concluded with a very extensive list of references and ten plates, each of which consisted of six beautifully colored sketches of textures and minerals.

The second volume, entitled *Mikroskopische Physiographie der massigen Gesteine*, appeared in 1877. In this volume of nearly 600 pages, devoid of illustrations just as many of Zirkel's books were, Rosenbusch (1877a) described seven major categories of igneous rocks in exhaustive detail: (1) orthoclase rocks, (2) orthoclase–nepheline–leucite rocks, (3) plagioclase rocks, (4) plagioclase–nepheline–leucite rocks, (5) nepheline rocks, (6) leucite rocks, and (7) feldspar-free rocks or peridotites. The first category included "families" of granitic rocks, quartz porphyry, felsite pitchstone, syenite, quartz-free porphyry, liparite, acidic volcanic glass, and trachyte. Category two included the families of elaeolite–syenite and phonolite. Category three was subdivided into five major groupings, each of which contained various families. The subdivisions of the third category were plagioclase–mica rocks (families of mica diorite), plagioclase–hornblende rocks (families of dioritic rocks, porphyry, and younger plagioclase–mica and plagioclase–hornblende rocks), plagioclase–augite rocks (families of diabase, diabase porphyrite, melaphyre, augite andesite, and basalt), plagioclase–diallage rocks (family of gabbroic rocks), and plagioclase–enstatite rocks (older members of the plagioclase–enstatite series and younger members of that series). Category four included older plagioclase–nepheline rocks (teschenite), younger plagioclase–nepheline–leucite rocks, and the family of tephrite. Category seven included the older feldspar-free rocks, grainy older peridotite, picrite porphyry, younger feldspar-free rocks, and the family of limburgite. Volcanic ash and keratophyre were included in an appendix. In effect, Rosenbusch provided an elaborate classification of the "massige Gesteine" that dominated thinking about such rocks for the next quarter century. The classification of Rosenbusch went far beyond those of his predecessors in precision and detail. He continued the widespread German practice of making distinctions between older and younger rocks that share a given mineral composition.

In 1878, Rosenbusch assumed the chair of Professor of Mineralogy at the University of Heidelberg, where he remained for the remainder of his academic career. In addition to training young petrographers from all over the world, Rosenbusch spent much of his life in periodically revising his massive books. A second edition of the first volume of *Mikroskopische Physiographie* on the petrographically important minerals appeared in 1885. The value of this new edition was greatly enhanced by the addition of several more diagrams and an interference-color chart, earlier developed in France by Michel-Lévy. The second edition of the second volume on the "massige Gesteine" followed two years later (Rosenbusch, 1887). This edition was improved by the addition of 24 black-and-white photomicrographs. A third edition of *Mikroskopische Physiographie* came out between 1892 and 1896. By the turn of the century, Rosenbusch's magnum opus took on truly monumental proportions as he joined forces with his student E. A. Wülfing in issuing a two-volume fourth edition of the book on the important rock-forming minerals (Rosenbusch and Wülfing, 1904). He continued to author the text on the "massige Gesteine" by himself, but by the time the fourth edition of that work appeared, it, too, consisted of two volumes totaling 1592 pages (Rosenbusch, 1907, 1908). What had begun as a rather sizeable, two-volume work had become a four-volume encyclopedia in the course of three decades. Even after Rosenbusch's death, his co-author Wülfing teamed with O. Mügge to produce a fifth edition of *Mikroskopische Physiographie der petrographisch wichtige Mineralien* in 1927.

Rosenbusch (1898) also found the time to write a more general work on petrography entitled *Elemente der Gesteinslehre*. This textbook ran through four editions, making its final appearance in 1923 under the authorship of another student, A. Osann. In the meantime, the text expanded from 546 to 779 pages and the number of tables and illustrations continually increased. Together Zirkel and Rosenbusch set an extremely high standard for massive, meticulous scholarship in the rapidly burgeoning field of thin-section petrography, particularly with regard to the study of the igneous rocks.

The exhaustive research of these two pioneers, particularly of Rosenbusch, led to the introduction of a host of new terms for textural relationships in igneous rocks that are not observable in hand specimens. Although a few of these terms are no longer in vogue, many of them are permanently embedded in the jargon of modern igneous petrology. Rosenbusch (1877a) formalized the previously used term "holocrystalline" and introduced other textural terms that are familiar to every petrographer: "allotriomorphic-granular," "hypidiomorphic-granular," "panidiomorphic-granular," "idiomorphic," "hypocrystalline," "hyalopilitic," "orthophyric," "pilotaxitic," "miarolitic," and "ocellar."

The development of thin-section petrography made it much easier to recognize subtle differences among rock types that had hitherto escaped attention, thus leading to the irresistible temptation to define entirely new rock types. Rosenbusch named, for example, theralite, vogesite, alnöite, spessartite, camptonite, nepheline syenite, tinguaite, phonolitic trachyte, and harzburgite. In all, he introduced 32 new rock names (Le Maitre, 1989).

The petrographic revolution in Germany was not confined to the two masters. In the final decades of the nineteenth century, professors of mineralogy and/or petrography proliferated throughout German and Austrian universities. Among these scholars, many of whom had studied under either Zirkel or Rosenbusch, were A. von Lasaulx (Bonn), G. Tschermak (Vienna), G. Haarmann (Witten), J.F.E. Dathe (Leipzig), E. Kalkowsky (Leipzig), C. Doelter (Graz), E. Weinschenk (München), A. Osann (Heidelberg), and a host of others. German petrographic texts were also produced by E. Cohen, H. Fischer, and von Lasaulx, who coined the term "troctolite."

Dozens of studies of the volcanic rocks from the Auvergne and from Italy and of alkaline rocks from the Rhine graben filled the pages of *Neues Jahrbuch* and *Zeitschrift der deutschen geologischen Gesellschaft*. German authors also contributed a host of articles on igneous rocks from New Zealand, the Scottish Hebrides, Brazil, and other parts of South America. The established journals that were committed to publishing articles covering the full gamut of geological topics could no longer absorb the plethora of igneous rock studies being produced by the expanding phalanx of German-speaking petrographers. As a result, Gustav Tschermak (1836–1927) of the University of Vienna, and one of the leading mineralogists of the era, founded a new journal devoted more specifically to mineralogical and petrographic topics in 1871, *Mineralogische*

Mitteilungen (Baumgärtel, 1976). Even though articles were exclusively written by German and Austrian authors at first, the journal quickly established a reputation for excellence, and it was not long before American and French authors contributed papers, also in German. In 1878, the name of the journal, so closely associated with the editorship of Tschermak, was changed to *Tschermaks mineralogische und petrographische Mitteilungen*. Its founding editor made his own valuable research contributions to petrology with his demonstration that the plagioclase feldspars form an isomorphous series and by showing how biotite, pyroxenes, and amphiboles may be distinguished from one another in thin section on the basis of their pleochroic properties (Tschermak, 1869). He also coined the rock name "picrite."

THE SPREAD OF THIN-SECTION PETROGRAPHY

Refracted through the prism of the German duo of Zirkel and Rosenbusch, the possibilities latent in polarized light began to shine upon geologists far beyond the boundaries of Germany. Thin-section petrography became cutting-edge science, and bright young geologists, eager to move to the forefront of the glamorous new field, flocked to Germany from France, Scandinavia, England, and America to study under the keen-eyed tutelage of Zirkel and Rosenbusch. Microscopic petrography began to spread around the world as these young scholars returned to their homelands and as other geologists who had not studied under the German masters quickly realized the power of the new method and taught themselves petrographic methods. The early spread of microscopic thin-section petrography has been discussed by Bascom (1927), Hamilton (1982), and Geschwind (1994).

In France, the pioneer of microscopic petrography was Auguste Michel-Lévy (1844–1911), the director of the Service de la Carte Géologique de France from 1887 until his death (Winchell, 1912; Burke, 1974a). Michel-Lévy was also Chief Engineer of Mines after 1883 and Inspector General of Mines after 1907. Although making extremely important contributions to experimental petrology and mineralogy, Michel-Lévy became one of the leading experts in the field of optical mineralogy and petrography. He developed the interference-color chart as well as a method for determining the chemical composition of plagioclase feldspar based on the values of extinction angles. Together with his close friend Ferdinand Fouqué (1828–1904), Professor of Natural History at the Collège de France, Michel-Lévy introduced petrography into France (Burke, 1972). They produced a magnificent text, *Minéralogie Micrographique: Roches Éruptives Françaises* (Fouqué and Michel-Lévy, 1879), that included an extended discussion of general principles of microscopy and more than 300 pages describing French igneous rocks in thin section. The text incorporated a large plate of diagrams of various interference figures. Along with *Les Minéraux des Roches* (Michel-Lévy and Lacroix, 1888), *Minéralogie Micrographique*

11. Alfred Lacroix (1863–1948). Reproduced from the archives of the Sedgwick Museum, University of Cambridge, by permission.

functioned as a streamlined French version of the ponderous German texts of Rosenbusch and Zirkel. Numerous petrographic studies of igneous rocks were undertaken by Michel-Lévy, Fouqué, Barrois, and, particularly, Alfred Lacroix (1863–1948), a student and son-in-law of Fouqué (de Margerie, 1953; Hooker, 1973). Lacroix (Figure 11) carved out a brilliant career as a petrographer while serving as Professor of Mineralogy at the Muséum d'Histoire Naturelle. He pioneered the petrographic study of African rocks, particularly those of Madagascar, and coined 70 new rock names, including the term "oceanite" and dozens of names for obscure alkalic rocks, such as "niligongite," "lustianite," and "kivite" (LeMaitre, 1989).

Also of great importance for the entire petrographic enterprise was the foundational work done in optical mineralogy by des Cloizeaux. Alfred-Louis-Olivier Legrand Des Cloizeaux (1817–1897) had studied with Alexandre Brongniart and worked in the laboratory of Biot at the Collège de France (Taylor, 1971a). He had been to Iceland with Robert Bunsen in 1846 and studied optical mineralogy with Henri de Senarmont. During a career that culminated at the Muséum d'Histoire Naturelle, Des Cloizeaux worked out the optical properties of about 500 minerals, improved polarizing microscopes, developed methods for determining the dispersion and angles of optic axes, and devised the concepts of indicatrix and bisectrix. His textbook of optical mineralogy (Des Cloiseaux, 1862) contains beautiful color diagrams of various interference figures.

Microscopic petrography also slowly found its way into Great Britain at the prodding of the controversial David Forbes. In 1852, Forbes began studying thin sections that had been prepared by Oschatz. He received further training from Sorby and soon began to prepare his own sections. After Sorby became interested in other matters, Forbes became the sole British advocate of the new microscopic methods. Forbes (1867c) repeatedly stressed the tremendous importance of microscopic work in geology in a paper on the microscope in geology. Before long, British geologists began to sense the value of thin-section microscopy. Forbes spoke with some authority, for, by 1867, he possessed a collection of more than 2000 thin sections that he had personally prepared from rocks and minerals collected from the many countries in which he had traveled.

Forbes began his paper by arguing that the advancing state of geology required geologists to avail themselves of all possible means placed at their disposal, including those that could extend their powers of observation beyond the limits of unassisted eyesight. He generously referred to Sorby's work and compared the value of microscope work to that of chemical investigation. After describing his method for making sections, he discussed the kind of results he had obtained for various rocks, driving home his point with sixteen lovely colored diagrams of fields of view. He took pains to point out that one could make confident mineral identifications with polarized light, particularly of translucent, colorless minerals like quartz, leucite, calcite, and feldspar. Likewise, he pointed out, the structure of the rocks, once determined, provided valuable information that might elucidate the mode of formation and origin of the rocks themselves. As a result of Forbes' continued prodding, a generation of eager young British petrographers soon burst upon the scene, including J. C. Ward, J.J.H. Teall, T. G. Bonney, A. Geikie, F. Rutley, J. W. Judd, and F. H. Hatch, along with old-timer Samuel Allport.

British petrographers made great strides in describing the numerous granites and dolerite sills throughout England, Scotland, and Ireland, and, later, the Tertiary volcanic centers of the Hebrides. Articles on petrography in the British Isles regularly appeared in *Geological Magazine*, founded in 1864 as the successor to *The Geologist*, and even the *Quarterly Journal of the Geological Society of London*, for years virtually devoid of any mention of igneous rocks, regularly included important petrographic papers from the 1870s onward.

British petrographers, like their German and French counterparts, perceived the need for petrographic textbooks. The earliest of these was *The Study of Rocks: An Elementary Textbook of Petrology* by Frank Rutley (1842–1904), a petrographer with the Geological Survey of the United Kingdom (Judd, 1904). In his text, Rutley (1879) noted that several manuals of petrology had been published on the continent but that comparatively little had been done in England to supply instruction in the systematic study of rocks. Rutley's book was followed by the classic *British Petrography* of Teall (1888). Jethro Justinian Harris Teall (1849–1924) not only included descriptions of British igneous rocks but also discussed the latest ideas on the genesis of such rocks (Anonymous, 1909). Immediately after publication of his big book, Teall was placed in charge of the petrographic work of the Geological Survey of the United Kingdom until his appointment as Director of the Survey in 1901. In 1891, *An Introduction to the Study of Petrology: The Igneous Rocks*, was published by Frederick H. Hatch (1864–1932) (Spencer, 1933). Although a scant 128 pages, the text of Hatch (1891) was sufficiently successful that it underwent several revisions during his lifetime and has continued to be revised by his successors. The eighth edition of 1926, for example, was co-authored with A. K. Wells and had expanded to 566 pages. Soon after the appearance of Hatch's book, Alfred Harker (1859–1939) published the first edition of *Petrology for Students* (Harker, 1895b). Like the texts of Rutley and Hatch, Harker's book went through numerous editions but remained virtually unchanged through the seventh edition of 1935. Ironically, despite Harker's use of "petrology" in the title, the text was a purely descriptive manual of microscopic petrography with an emphasis on British rocks. The selection of rocks described changed only slightly over the years. The restriction of Harker's text to purely descriptive petrography, unlike those of Rutley and Hatch, is striking in light of the fact that Harker was arguably the prince among the first couple of generations of British interpretive petrographers.

In spite of Harker's descriptive text, the common use of the word "petrology" in the titles of texts and the considerable emphasis on interpretive matters provide one indication that British petrographers were quicker than their continental colleagues to seize upon the implications of the new microscopy for the theoretical side of the study of igneous rocks. Not content with mere description, they realized that thin-section petrography could shed light on the alleged relationship between rock type and geological age and that igneous rocks

MINUTE OBJECTS, GREAT CONCLUSIONS 163

within certain regions possess mineralogical and chemical affinities that distinguish them from similar igneous rocks in other areas.

According to Bascom (1927), thin-section petrography spread to Russia as early as 1867. The Russians were particularly gifted in devising new optical equipment such as the universal stage of Fedorov and various compensators. F. Y. Loewinson-Lessing quickly moved to the forefront as one of the world's premier petrologists. Microscopic petrography was also taken up by mineralogists in other parts of eastern Europe, but these individuals largely remained in the background of the advancing discipline. Petrography flourished in Scandinavia, particularly under the leadership of W. C. Brögger in Norway and J. J. Sederholm, a student of Rosenbusch's, in Finland. Research on specific igneous terranes that utilized the polarizing microscope also began to appear in Italy and Japan during the latter nineteenth century (Bascom, 1927).

Although use of the microscope caught on slowly in the United States, Americans soon rose to the highest ranks of petrographers. A handful of American geologists had received European training in microscopy but found that their skills attracted little interest in the American geological community. Geschwind (1994) showed that it was not until the utility of the polarizing microscope was demonstrated in the solution of thorny stratigraphic correlation problems that the new instrument gained popularity. Clarence King, for example, sensing that the microscope might be useful in unraveling the complex volcanic stratigraphy of the American west, brought in Zirkel to examine hundreds of thin sections. The polarizing microscope helped solve correlation problems of various greenstone units in the Lake Superior mining district. After such applications confirmed the usefulness of the microscope, a considerable number of young American geologists either taught themselves microscopy or trooped to Germany to study with Zirkel, Rosenbusch, or one of the other European masters. H. S. Washington and C. W. Cross studied under Zirkel at Leipzig. Rosenbusch, however, attracted hordes of students from America during the 1870s through the 1890s (for lists of names, see Pirsson, 1918; Bascom, 1927; Servos, 1983; and Geschwind, 1994).

As a result of the importation of microscope skills into America, petrography began to flourish, at first in the United States Geological Survey, and later in American universities. Instruction in petrography was established at Johns Hopkins University in 1883, at Harvard in 1886, at Columbia in 1891, and at Yale in 1892, although G. W. Hawes had provided some instruction at Yale during the 1870s. European texts were commonly used in conjunction with petrography courses, but J. P. Iddings helped to alleviate that situation for American students by translating into English the first volume of the second German edition of Rosenbusch's *Mikroskopische Physiographie* as *Microscopical Physiography of the Rock-Making Minerals* (Iddings, 1888b).

Without question, the contingent of youthful American petrographers acquired the German taste for pure petrographic description, but they did not stop there because Americans matched their British counterparts in ability to combine detailed field study, petrographic description, and interpretive skill. George Huntington Williams (1856–1894) quickly emerged as the early leader among the young practitioners of the petrographic art in America (Iddings, 1894; Clark, 1895; Anonymous, 1896; Pettijohn, 1988). After graduation from Amherst College where he studied with B. K. Emerson, Williams traveled in 1879 to Germany to study at Göttingen, but after the death of one of his professors, he transferred to Heidelberg to work under Rosenbusch. In 1882, he received his Ph.D. degree with a dissertation on the eruptive rocks from the region around Tryberg. Upon his return to America, Williams began petrographic instruction at Johns Hopkins; instituted a systematic program of mapping in the Maryland Piedmont under the auspices of the United States Geological Survey; wrote textbooks; developed an inexpensive polarizing microscope suited to the needs of American students and manufactured in the United States by Bausch and Lomb; conducted important petrographic studies; coined the terms "poikilitic" and "websterite;" and, from 1883 to 1894, supervised the advanced work of 24 students, nine of whom received a Ph.D. degree from Johns Hopkins, the first of the American universities to be built self-consciously on the German pattern of meticulous research. Tragically, Williams' already outstanding career was cut short at the age of 38 when he contracted typhoid fever from drinking contaminated water during a field trip in the Maryland Piedmont.

Kohler (1990) has explored the phenomenon of the dramatic rise in the granting of Ph.D. degrees by American colleges in the latter 1800s as a response to the growing demand for college teachers created by population growth and the proliferation of colleges. Williams' ability to produce a large number of Ph.D. students who went on to serve in the academic world was an indicator that the field of igneous petrology would particpate vigorously in this remarkable American social phenomenon. Among Williams' doctoral students were W. S. Bayley, A. C. Lawson, E. B. Mathews, and Florence Bascom, who became heads of the geology departments at the University of Illinois, the University of California, Johns Hopkins, and Bryn Mawr College, respectively. Among Williams' important petrographic contributions were studies of the Cortlandt Complex in New York (Williams, 1886, 1888), and demonstrations of the igneous origin of Baltimore Gabbro, the volcanic origin of the South Mountain schists in Maryland, and the intrusive igneous nature of American granites. In his history of the geology program at Johns Hopkins, Francis Pettijohn (1988, p. 20) observed that Williams "laid to rest, for all time, the last vestiges of Wernerism," a view that had continued to exert a hold on American geological thinking through the influence of T. S. Hunt. Most significant is the fact that Williams, first and foremost a petrologist, went beyond rock description to emphasize the interpretive analysis of the petrographic data. He employed a dynamic approach of seeking evidence of the instability of the mineral constitution of rocks under changing environmental conditions, and, in so doing, set the tone of American petrographic research to be conducted so ably by Iddings, Pirsson, and their contemporaries.

WOMEN IN PETROGRAPHY

One of the most far-reaching developments of the early petrographic era is that women, too, were swept up into the tremendous growth of the field, particularly in Great Britain and America. Although geology was overwhelmingly the preserve of men throughout its early years, Creese and Creese (1994) have documented that women had played an important role in the geological community as amateur paleontologists throughout the nineteenth century in Britain. From the 1860s onward, they pointed out, British secondary schools for girls commonly offered instruction in the natural sciences in contrast to the more conservative, classically oriented schools for boys. As a result, when Newnham College was established for women at Cambridge in 1871 and universities like Oxford and London were opened to women in the late 1870s, many women were drawn toward a study of scientific subjects, including geology, thanks to their earlier experiences in the sciences.

The first woman to break with the tradition of women as paleontologists and to devote herself to petrography was Catherine Alice Raisin (1855–1945), a protegé of Cambridge geologist T. G. Bonney. Raisin was the first woman to take geology classes at University College, London, in the mid-1870s, just before formal degrees were awarded to women (Reynolds, 1945; Aldrich, 1990; Creese and Creese, 1994). In 1884, she was ultimately granted her baccalaureate degree after which she worked as Bonney's assistant at Cambridge for several years. During this time, she published either alone or with Bonney several petrographic papers in *Quarterly Journal of the Geological Society of London*. In 1890, Raisin was appointed as head of the geology program at Bedford College for Women where she spent the remainder of her career. So highly valued was her work that the Geological Society of London gave her an award from the Lyell Fund in 1893 to pursue research. Ironically, Bonney had to accept the award publicly on her behalf because women were not yet allowed to attend meetings of the Society. In 1919, the Society, after many years of debate, amended its by-laws to allow women to be admitted as fellows, and Raisin was among the first group of British women geologists so admitted. In contrast, Raisin was for 67 years a member of the Geologists' Association, an organization that had never excluded women from membership. In the meantime, Raisin had obtained a D.Sc. degree from the University of London in 1898. During her years at Bedford, she continued her productive output of petrographic papers.

The ranks of women petrographers were doubled a decade later by Florence Bascom (1862–1945), the first American woman to receive a doctoral degree in petrography (1893) and only the second in geology (Ogilvie, 1945; Smith, 1981; Arnold, 1993, 1999, 2000). The daughter of the president of the University of Wisconsin, Bascom went on, after receiving her degree, to found the geology program at Bryn Mawr College, at the time an exclusively women's institution. She headed the department from 1895 to 1932 and was perhaps more instrumental than anyone else in America in introducing women to geology and petrography. She and her numerous students produced a host of studies of the difficult igneous and metamorphic rocks of southeastern Pennsylvania. Together with Britain's Catherine Raisin, Bascom led the way for the introduction of women into the community of igneous petrologists.

10 ⇠ BASALT OR MELAPHYRE?

Igneous Rocks in Time

The application of the polarizing microscope to studies of rocks in thin section resulted in an explosion of knowledge of the mineralogical constitution and textural characteristics of igneous rocks from a wide range of terranes throughout Europe and America. Dozens of new rock types, especially from alkalic terranes, were described and named. The new method also opened the door for an unprecedented understanding of some of the theoretical aspects of igneous petrology. So successful was thin-section microscopy in revealing many of the mysteries of igneous rocks that increasing numbers of geologists began to specialize in the new field. Petrography was introduced into the curricula of a rapidly expanding number of universities in Germany, France, Great Britain, Norway, and the United States. Professorial positions devoted to petrography proliferated in European universities. Petrographers were hired by national geological surveys. Petrographic studies increased so rapidly that existing journals could not keep pace with the influx of papers. New petrographically oriented journals were established to soak up some of the flood of articles, and both elementary and advanced petrographic textbooks abounded in Germany, France, and England.

Although microscopic petrography had a decidedly descriptive flavor, particularly in the hands of its German practitioners, English and American geologists were concerned to use the new technology as a tool in the solution of interpretive problems. One of the most pressing problems concerned whether there exists a distinct relationship or correlation between geological age and rock types. Are specific igneous rock types restricted to specific portions of the geological column or not? Since the days of Werner, a long-standing tradition had been that various epochs in Earth history were in fact characterized by certain lithologies that were not produced to any significant degree during other epochs.

CHAPTER 10

THE CONTINENTAL VIEW OF GEOLOGICAL AGE IN RELATION TO LITHOLOGY

Around 1800, most geologists, whether fully committed to all aspects of Wernerian neptunism or not, believed that there had been changes in rock types formed throughout geological time. Without question, Werner's stratigraphic views had laid the basis for such an opinion. Werner firmly believed that certain rocks were formed during specific periods in Earth's development (Ospovat, 1971). Although he introduced modifications into his scheme from time to time, he persisted in his faith that granite is primitive, that basalt is younger, and that volcanic rocks are very recent. Werner's mature opinion was that granite, gneiss, mica slate, clay-slate, primitive limestone, primitive trap, serpentine, porphyry, syenite, topaz-rock, quartz rock, primitive flinty-slate, primitive gypsum, and white stone belong among the Primitive rocks. Granite, trap, serpentine, porphyry, and syenite, of course, had been denied magmatic origin by Werner. The Flötz formations, according to Werner, were dominated by what are now designated as sedimentary rocks, such as sandstone, limestone, gypsum, chalk, and coal. Werner separated this category into Older and Newest Flötz formations. In addition to the mechanical sedimentary rocks, Werner's Newest Flötz Trap formation consisted of several rock types that would eventually be regarded as volcanic rather than neptunian: basalt, clinkstone porphyry, and trap-tuff, as well as rocks shared with older Flötz formations like greenstone, amygdaloid, pitchstone, obsidian, and pumice. True volcanic materials were still considered to be recent and of minor importance. Werner's stratigraphic work laid the groundwork for the persistent opinion that basalt, clinkstone, greenstone, and the like, rocks that would eventually be regarded as volcanic rocks, are geologically relatively youthful, and that coarse-grained, granitoid rocks are confined to geologically older formations.

Kirwan (1799), for example, stated that granite almost universally served as the base on which the Primitive and Secondary rocks rested. Even in more plutonist-oriented Great Britain, even after Hutton had challenged the concept of primitive granite, as late as 1821, John Macculloch's classification of Primary rocks included granite along with various metamorphic rocks. His Secondary class included mostly sedimentary rocks and pitchstone. "With respect to the order of succession in the primary class," Macculloch (1821, p. 86) wrote, "the claim of granite to the first or lowest place is unquestioned." Without any mention of Hutton's contributions, he listed a number of examples from Great Britain, including Glen Tilt and Arran, where granite was considered to be first in formation. He made no mention of the stratigraphic position of basalt.

In France, Beudant was an important advocate of the generally Wernerian view. François Sulpice Beudant (1787–1850), after 1811, occupied such positions as Professor of Mathematics at the University of Avignon, Professor of

Physics at the University of Marseilles, and, after 1820, Professor of Mineralogy in the Faculty of Science at the Sorbonne (Kendall, 1970). Beudant was appointed by King Louis XVIII as Assistant Director of his cabinet of mineralogy. The French government sent him on a scientific expedition to Hungary, as a result of which he published the three-volume *Voyage Minéralogique et Géologique en Hongrie* (Beudant, 1822). His geological investigations in Hungary led him to the conclusion that all trachytes belong to Secondary strata. From studies of mining districts in volcanic terranes, he concluded that syenite and greenstone are Transition rocks, that andesitic rocks are Secondary, and that basaltic rocks are probably Tertiary. In short, for Beudant, the more coarsely crystalline rocks appeared to be decidedly older than the volcanic rocks.

In later years, Joseph Durocher was another prominent French advocate of the view that coarse granitic rocks are generally older than volcanic rocks. In Germany this notion reappeared in the petrological theories of Bunsen, Sartorius von Waltershausen, Roth, Scheerer, and von Richthofen. Although recognizing that granites are not always Primitive, Naumann (1854) nevertheless structured the second volume of his *Lehrbuch der Geognosie* to convey the strong impression that many igneous rocks prevailed in specific geological time periods.

ZIRKEL AND ROSENBUSCH

German views did not immediately change with the advent of the era of microscopic petrography. In the first edition of *Lehrbuch der Petrographie*, Zirkel (1866) devoted most of his attention to igneous rocks. The vast bulk of what are now regarded as igneous rocks he described in a section entitled "Gemengte kristallinisch-körnige Gesteine" (mixed crystalline-grained rocks). He subdivided this entire category into two large groups: "Ältere Feldspathgesteine" (older feldspar rocks) and "Jüngere Feldspathgesteine" (younger feldspar rocks). At the outset, geological age, broadly speaking, was proposed as a major factor in igneous rock classification. Zirkel's group of older feldspar rocks included quartz-bearing orthoclase rocks like granite; quartz-free orthoclase rocks like syenite, foyaite, and minette; oligoclase rocks like diorite and melaphyre; labradorite rocks like diabase, gabbro, and hypersthenite; and anorthite rocks like eucrite. The group of younger feldspar rocks included crystalline members of the trachyte family such as trachyte, phonolite, and andesite; glassy rocks of the trachyte family like obsidian, perlite, and pumice; a group of nepheline- and leucite-bearing rocks, including nephelinite and leucitophyre; rocks of the basalt family like dolerite, anamesite, and basalt; and feldspar-free rocks such as dunite, lherzolite, and eclogite. The listing makes it plain that Zirkel thought of volcanic rocks as predominantly younger and plutonic rocks as predominantly older.

Noting that similar rocks had already been assigned different names by various writers on the basis of their age, Zirkel commented that he had decided to apply the factor of age in his classification in order to agree with existing usage rather than to eliminate duplicate terms. For example, common usage typically regarded melaphyre and basalt as virtually identical rocks but assigned the name "melaphyre" to older occurrences and "basalt" to younger ones. In his survey of systematic petrography in the nineteenth century, Cross (1902) observed that Zirkel's work was so influential that a lot of the nomenclatural confusion arising from the unnecessary duplication of rock names might have been avoided had Zirkel chosen to clean up the duplication of terminology instead of perpetuating a flawed concept. "The effect of his proposition to apply geological age and the qualitative element of mineral composition as leading factors in the systematic arrangement of the rocks we now term igneous was peculiarly unfortunate," Cross (1902, p. 368) lamented,

> because this invaluable work of reference was issued at the beginning of the era in which petrographers were to be so busily engaged in the microscopical study of rocks that they had no time for systematic work. In the flurry of descriptive literature of the succeeding decade these propositions were adopted almost of necessity. The students from all lands who flocked at this time to Germany to study under Zirkel and other masters, carried the system back to their respective countries, giving it quickly a world-wide usage.

In *Die mikroskopische Beschaffenheit der Mineralien und Gesteine* (Zirkel, 1873), his slight discomfort with the idea of correlation between rock type and geological age became a bit more apparent. In this work, he did not formally recognize age as a classification factor as he had in the *Lehrbuch*. Still, Zirkel did not follow through to the extent of rejecting many of the duplicate terms based on age as a classification factor.

Zirkel never could entirely break free from the old conception, however, if for no other reason than that the age factor had become so entrenched in the naming and description of many igneous rocks. In the second edition of *Lehrbuch der Petrographie*, a text that Cross (1902, p. 483) called "the most comprehensive and complete description of all known rocks ever published" and a work that "represents the present status of the systematic science as a whole, better than any other work," Zirkel (1893) still included geological age as a classification factor. Although he had introduced age as a major factor even before mineralogical composition in the first edition, mineralogy and structure took precedence in the second edition. Age entered as a factor only in the subdivision of some of the mineralogical–structural groups. Zirkel grouped the igneous rocks under seven headings based on mineralogy, such as "rocks with potassium-feldspar and quartz or excess silica" or "rocks with calcium–sodium feldspar without nepheline or leucite." Each of these seven groups was further subdivided into "granular rocks" where Zirkel employed no age distinctions and "porphyritic and glassy rocks" where he did separate the rocks into "pre-Tertiary" and "Tertiary and recent" categories. Cross (1902) suggested that Zirkel's arrangement for the porphyritic and glassy rocks continued to reflect a recognition of contemporary usage of duplicate sets of terms rather than commitment to a definite principle.

Rosenbusch, too, never fully escaped the shackles of geological age as a factor in igneous rock classification. In the first edition of *Mikroskopische Physiographie der massigen Gesteine*, Rosenbusch (1877a) incorporated an age factor. The classification first divided igneous rocks on the basis of mineralogical characteristics: orthoclase rocks, plagioclase rocks, nepheline rocks, and so on. Rosenbusch presented tabular summaries of his classification of these various mineralogical classes. The summary of the "orthoclase rocks" showed a primary subdivision into "ältere" (older) and "jüngere" (younger) groups. Each of those groups was further divided mineralogically into quartz-bearing and quartz-free categories. These categories were then subdivided texturally into "körnig" (granular), "porphyrisch" (porphyritic), and "glasig" (glassy) categories. The older group of orthoclase rocks, therefore, included quartz-bearing rocks such as the families of granite, quartz porphyry, and felsite/pitchstone, and quartz-free rocks such as the families of syenite and quartz-free porphyry. Glassy equivalents were considered to be lacking. The younger group included quartz-bearing rocks such as the families of liparite and of acid glasses like obsidian and pumice, and quartz-free rocks such as the families of trachyte and some glassy rocks similar to those above. "Körnige Gesteine" (granular rocks) were conspicuously absent from the category of younger rocks.

Other mineralogical classes, such as the "plagioclase-nepheline rocks" were also subdivided into older and younger groups with teschenite an example of the former and tephrite an example of the latter. Rosenbusch essentially regarded rocks belonging to pre-Tertiary ages as older and those of Tertiary and recent age as younger.

In the second edition of *Mikroskopische Physiographie*, Rosenbusch (1887) elevated the importance of geological occurrence as a factor in classification. At the outset he divided rocks into "Tiefengesteine" (deep-seated rocks), "Ganggesteine" (dike rocks), and "Ergussgesteine" (effusive rocks). These categories were, in a sense, roughly equivalent to the textural categories, granular, porphyritic, and glassy, of the first edition. Rocks in each of the three categories were then classified mineralogically. Rosenbusch acknowledged that geological age had probably been assigned too important a status in previous classification efforts. Nevertheless, he retained geological age in his classification of the effusive rocks and continued the use of duplicate names for mineralogically very similar rocks of differing ages. Rosenbusch's skepticism about geological age surfaced again in a paper on chemical relationships in igneous rocks (Rosenbusch, 1889), but some consideration was still given to age in the third edition of *Mikroskopische Physiographie der massigen Gesteine* (Rosenbusch, 1896). Indeed, remnants of the concept continued to appear in his early twentieth-century texts.

Zirkel and Rosenbusch were by no means alone in perpetuating the use of geological age as a factor in igneous rock classification. Another of the early petrographic texts was written by Arnold von Lasaulx (1875) of the University of Bonn. *Précis de Pétrographie* made its appearance in both French and German. Von Lasaulx divided all rocks into simple, composite, and clastic groups. Most igneous rocks were placed in a subdivision of the category of "roches composées" (composite rocks) designated as "massive crystalline silicate rocks." These were next divided into a "série ancienne" and a "série récente." Within either of these two age-based series, rocks were considered from a chemical point of view as acid, intermediate, or basic. Within the ancient series, von Lasaulx divided the acid series texturally and then mineralogically. Granite, for example, was classed as a rock of the "série ancienne," in the acid series, having a granitic texture, and composed of orthoclase and quartz. Trachyte was classed as a rock of the "série récente," in the acid series, having a granito-trachytic or trachytic texture, and composed of orthoclase (or sanidine) and quartz. In the scheme of von Lasaulx, the vast majority of the volcanic rocks were included in the "série récente," and the vast majority of coarse-grained, intrusive, igneous rocks were included in the "série ancienne."

The use of age in classification also persisted in France after the onset of the microscopic era. Fouqué and Michel-Lévy (1879), for example, in their *Minéralogie Micrographique*, incorporated age as a factor by making a distinction between pre-Tertiary rocks and those of Tertiary and more recent ages. While acknowledging that the same lithologies occur both before and after the Tertiary, they still maintained that basic volcanic rocks and glassy types predominate in more recent occurrences. The same separation into "ante-tertiare" and "post-tertiare" persisted a decade later in *Structures et Classification des Roches Éruptives* (Michel-Lévy, 1889).

BRITISH VIEWS ON THE ROLE OF AGE IN CLASSIFICATION

Not all German writers were enamored of the traditional view that linked rock type to age. Bernhard von Cotta, for example, in *Die Gesteinslehre*, pointed out that geological age should not be involved in the distinction between the volcanic and the plutonic rocks. The same eruptive mass, he said, may be volcanic at the surface and plutonic below the surface. As a result, plutonic rocks of old ages were bound to have volcanic rocks associated with them at the surface. In contrast, volcanic rocks of younger ages were bound to have plutonic rocks associated with them at depth. Moreover, von Cotta maintained that volcanic agency had been at work throughout Earth history and had always brought forth like products. He conceded that volcanic rocks appear to predominate in younger terranes and that plutonic rocks appear to predominate in older terranes. Rather than assuming that distribution to represent a relic of an original state of affairs, however, he asserted that the older volcanic rocks would be largely removed by erosion, whereas the younger plutonic rocks would not yet be exposed.

In spite of his rejection of the traditional Germanic view, von Cotta was still just a little troubled at the perceived lack of rocks of undoubted basaltic character in pre-Tertiary rocks. He thought that at least some blocks and boulders of genuine basalt should have been encountered in old conglomerates, but he was not aware that any had been found. In the 1878 English version of *Die Gesteinslehre*, the translator took the liberty to comment in a footnote that Joseph Jukes had found pebbles of vesicular trap in Silurian conglomerate and that British geologists had encountered several instances in England, Scotland, and Ireland of basaltic trap rocks that are interstratified with Paleozoic rocks. The footnote raises the question to what extent von Cotta and other continental writers were aware of the work being done in Great Britain that pertained to the question of the relationship between rock type and geological age.

Even before microscopic petrography became established, the idea that there is a correlation between geological age and the character of igneous rocks never engendered a very enthusiastic response in Great Britain. It was in Great Britain and, later, America that geologists used microscopic petrography to produce a continuous stream of compelling evidence against the idea. At the very onset of the microscopical era, Sorby (1858) wrote that the characteristic structures of minerals in ancient trap rocks are analogous or identical to those of modern lavas so that he regarded the purely igneous origin of ancient lavas to be completely established. And Scrope (1862) maintained in the revision of his book on volcanoes that the attempt to limit the production of volcanic rocks of a particular mineral composition to specific periods of Earth history was a great error. "The opinion that mineral composition is any test of age, as respects the pyrogenous rocks," Scrope (1862, p. 128) asserted, "is fast passing away from among geologists."

Although Scrope's assessment of the situation in 1862 was prematurely optimistic, a series of microscopic studies undertaken in Great Britain and America during the succeeding decades would certainly bring Scrope's opinion closer to realization. Among the earliest English petrographers, Allport, Ward, and Judd were especially critical of the continental view. David Forbes, however, got the ball rolling with a paper in which he reported on his examination of thin sections of igneous specimens having quite different appearances and collected from several dikes and bosses at fifteen different localities in Staffordshire. Forbes (1866) discovered that all of these rocks, described by previous authors as basalt, greenstone, trap, feldspathic rock, green rock, white horse, white rock trap, and Rowley Rag, were in reality one and the same rock composed of plagioclase, augite, and ilmenite. In light of the fact that Forbes considered all of these names to be referring to the same rock type, namely dolerite, he took the occasion to deplore the looseness that prevailed in England in the nomenclature of plutonic rocks. By implication, he recognized that a good deal of that looseness could be tightened up with liberal application of thin-section petrography.

Three years later, Allport (1869) expanded on Forbes' contribution with a brief note on basalt in south Staffordshire. Samuel Allport (1816–1897) spent the early years of his life in business and, when serving in South America as a business manager for a firm, became interested in geology (Bonney, 1897). On his return to England, he devoted his spare time to scientific pursuits, all the while making a living as a businessman or college librarian. Despite his location at the periphery of the English professional geological community, Allport was among the earliest to recognize the vast potential of Sorby's thin-section methods and his skills in preparing and analyzing the sections were unsurpassed. He produced several significant papers on the igneous rocks of Arran, and on trap rocks, dolerites, phonolites, and various other volcanic rocks throughout the British Isles. In his note, Allport (1869) called attention to the presence of pseudomorphs after olivine in the Staffordshire basalts that indicated to him various stages in the alteration of that mineral. Allport stressed the utility of the polarizing microscope in detecting alterations that might escape the unaided eye in the examination of a hand specimen of rock and argued that much could be learned about the history of a rock by comparing the altered portions of a rock with its least altered portions. Olivine, he concluded, was an original constituent of the Staffordshire basalts.

Allport (1870) soon published another paper on the composition and microscopic structure of a wide range of so-called "basalts" and "greenstones" from the broader region of the Midland coalfields in which Staffordshire is located. His examination of thin sections led him to conclude that all the basalts and greenstones were fundamentally the same rock. He claimed that the rocks completely correspond to rocks previously designated by such diverse names as trap, basalt, anamesite, dolerite, or melaphyre. Noting that the term basalt was commonly restricted by some geologists to rocks of Tertiary or Recent age, Allport asserted that such a distinction was to be deprecated because no sharp line could be drawn between Tertiary and pre-Tertiary rocks. The only reason that there appeared to be a difference between Tertiary basalts and older rocks of basaltic character, Allport argued, was that the older rocks had been subjected to the chemical action of warm aqueous solutions under pressure over enormous periods of time. He found the occurrence of olivine and its pseudomorphs in the coalfields basalts to be especially interesting. Until recently, he observed, olivine had been regarded as characteristic of younger basalts. That view was no longer tenable, he asserted, because he had discovered olivine to be a constituent of the older rocks. He claimed that his work clearly showed that the original mineral constitution of the Carboniferous dolerites and basalts was precisely the same as that of the Tertiary basalts. The only real difference between the older and younger rocks that he could find was that the older basalts had experienced more alteration. He maintained that the Carboniferous and Tertiary basalts were originally identical.

These conclusions were reinforced in a milestone paper dealing with the microscopic structure and composition of British Carboniferous dolerites. In this article, Allport (1874) noted that he had examined 230 thin sections of Carboniferous basaltic and doleritic rocks collected from 57 different localities throughout England, Scotland, and Ireland. Observing that geologists commonly assigned the name of "melaphyre" or "porphyry" to a rock if it were Paleozoic and "basalt" or "trachyte" if it were Tertiary, Allport (1874, p. 529) charged that "the authors of the existing nomenclature frequently applied it to rocks without any accurate knowledge of their mineralogical composition or structure," thus leading to considerable confusion. He demonstrated again from microscopic studies that the older rocks differ from the younger ones only in their degree of alteration. He proposed dropping several rock names.

Allport provided detailed descriptions of the mineralogy and microscopic structure of trap rocks of undoubted Carboniferous age such as those of Salisbury Crags and Arthur's Seat in Edinburgh and several others from Ireland and England to drive home the point that these rocks are completely like those of Tertiary age. He pointed out that "great variations in texture and composition frequently occur in a single rock, also that they are of the same kind as those which sometimes characterize large rock masses and to which different names have been applied" (Allport, 1874, p. 565). He complained that, although geologists commonly encountered variations in texture or composition within a single quarry of an igneous rock body or even variations from the center to the walls of a single dike, they strangely felt compelled to assign different names where these variant rocks were taken from different localities. Allport (1874, p. 565) concluded that

> all these varieties of composition and structure are common to the Carboniferous and Tertiary dolerites; and an examination of the microscopic structure and physical characters of the individual constituents shows an equally close agreement. The singular varieties of lamination in the triclinic feldspar, the twin structure and lamination of augite, and, in short, all the characters and the mode of occurrence of the original constituents are precisely the same in both series. It will, I think, be evident to any one who will take the trouble to examine and compare a series of Palaeozoic augitic traps with others of Tertiary age, that it is impossible to establish any mineralogical or structure difference between them; in fact the resemblance between individuals of the two groups is often so great that they cannot be distinguished from each other.

This being the case, Allport reasoned, those eruptive rocks composed of the same materials under similar conditions should be placed in one group. The same theme was also emphasized in his papers on phonolite (Allport, 1871) and on the pitchstones and felsites of Arran (Allport, 1872).

Similar studies were undertaken by James Clifton Ward (1843–1880), a graduate of the Royal School of Mines (Bonney, 1899). Ward was employed during his early twenties by the British Geological Survey where he worked on the Millstone Grit and Coal Measures. In 1869, he began eight years of studies in England's Lake District. It was during this period that he undertook microscopic investigations of volcanic rocks. In 1878, just two years before his premature death, Ward was ordained to the curacy of St. John's in Keswick and in 1880 became vicar of Rydal. In a paper read before the Geological Society of London in which he compared the microscopic structure of ancient and modern volcanic rocks, Ward (1875) described examples of the microscopic rock structure in modern lavas, specifically, a trachyte and three leucitic basalts from Italy. He concluded that the order of crystallization had been first magnetite, then feldspar and augite, and, lastly, feldspathoid. Next he described examples of the microscopic rock structure in the lavas and ashes of Wales and of Cumberland. Like Allport, he, too, recognized that these older rocks had experienced extensive alteration. He maintained that microscopic examination could teach geologists about the conditions under which volcanic rocks originated as well as the order of crystallization. In this case, he concluded that the older volcanic rocks were intensely altered to the point where the original structure had been completely obliterated in some samples.

Of greater impact was the early work of John Wesley Judd (1840–1916). Judd was trained as a teacher and began his adult life at a school in Lincolnshire where his geologic interests were stimulated by reading Lyell (Anonymous, 1905). He then studied geology at the Royal School of Mines and, upon graduation in 1864, became employed as a chemist in Sheffield, where he met Sorby, who showed him how to make thin sections and impressed on him the importance of microscopy. Before long, Judd began conducting microscopic investigations of Jurassic sedimentary rocks in Lincolnshire, Northamptonshire, eastern Scotland, and in the western Scottish Highlands and the Hebrides, where his attention was diverted to the interlayered and cross-cutting igneous rocks that are so intimately associated with the sedimentary rocks.

Judd (1874) issued a major paper on the volcanic rocks of the western Scottish Highlands and Inner Hebrides in relation to the surrounding Mesozoic strata. Building on the earlier work of Macculloch, Geikie, Zirkel, and others, Judd demonstrated cogently that the region had been the site of a series of great volcanic centers that had been subjected to varying degrees of erosion and dissection leading to the possibility of detailed three-dimensional study of these former edifices. As he elegantly phrased it,

> In the Hebrides I shall show that we have supplied to us that great geological desideratum—a number of volcanoes so dissected by the scalpel of denudation as to constitute, as it were, a series of anatomical preparations, from which we may learn directly the internal structure of the piles, and obtain bases for reasoning on the causes to which that structure owes its origin. (Judd, 1874, p. 232)

From their relations to the accompanying sedimentary rocks, Judd demonstrated the Tertiary age of the volcanic structures together with all the extrusive and intrusive rocks emanating from them. He determined that the volcanic rocks constituted two distinct, parallel textural series, each of which consists of a highly crystalline rock at one end and a perfect glass at the other. In between he found a complete and insensible gradation between the two extremes. In some instances, he found variations in texture within the same rock mass. He also found that the chemical composition of the rocks within either of the series is nearly identical throughout. Judd discovered, however, that the average chemical composition of the rocks of the two series shows a very marked contrast. These two series were, in essence, the long-recognized trachytic series and basaltic series.

In his studies of Mull, Skye, Ardnamurchan, and the other Hebridean volcanoes, Judd recognized that the highly crystalline gabbros and granites generally constitute large intrusive masses, that dolerites and felsites occur as intrusive sheets or bosses, and that basalts and felstones form narrow dikes and veins that locally pass into tachylyte and pitchstone respectively. He maintained that the relationships of the granitic rocks of the volcanoes to the extensive lava flows could be ascertained from analysis of the varying levels of erosion. At Mull, the relations of the various volcanic products can still be easily traced, he suggested, whereas at Skye the relations are not quite so clear. He also proposed that one might conceive a much more advanced stage of erosion of a volcano where all the eruption products were removed, leaving only a central core of granite or gabbro rising in the midst of a series of stratified rocks.

Judd placed the Hebridean volcanoes in a broader British context. He argued that there had been two distinct periods of volcanic activity in Britain, one from the beginning of the Old Red Sandstone period to the end of the Paleozoic and the other throughout the Tertiary. He found that feldspathic lavas had generally been extruded before basalts in both periods.

In conclusion, Judd argued that if anyone still retained the old view that the highly crystalline character of some igneous rocks provided a criterion of antiquity, then they should consider the Cuillin Hills of Skye. The Cuillins, he said, had long been regarded as of Laurentian (Precambrian) age but now had clearly been shown to be of Miocene age. Having observed so many textural gradations and field transitions among the various rocks at specific volcanic centers, Judd also concluded that the sharp distinction between plutonic, dike (trappean), and volcanic rocks was purely artificial. He regarded them all as products of the same igneous forces operating under different conditions.

At the encouragement of Charles Lyell, Judd decided to test these ideas in the Carpathian Mountains. Although covered by impenetrable forests and lacking the magnificent three-dimensional sections provided by the sea cliffs of the Hebrides, the district surrounding the old mining towns of Schemnitz, Kremnitz, and Königsberg in what is now Slovakia provided Judd with a very convenient opportunity to study the characters and relationships of the various volcanic rocks. Here Judd (1876) found a great tract of volcanic rocks consisting of andesitic lavas, agglomerates, and tuffs. These rocks had been eroded to a broad girdle of isolated mountains overlooking a central depression in which the old mining towns were located. In the central depression, Judd found sedimentary rocks, highly metamorphosed rocks, and intrusive igneous masses, consisting of "greenstone" or "diorite" and rocks that had been previously designated as "syenite" and "granite."

Judd noted that all of the earliest geological interpretations of the district had classed the metamorphic rocks and the granite, syenite, and diorite as Primary and Transition rocks and the volcanic rocks as Secondary and Tertiary. Subsequent geologists began to be uncomfortable with this distinction. In 1848, for example, Charles Daubeny had remarked that many of the interpretive difficulties might disappear if the granites were regarded as the same age as the volcanic rocks. Judd agreed. He established a Miocene age for the andesites of the volcanic girdle, and his microscopic investigations confirmed that the greenstones of the central depression grade insensibly into the andesites. He showed that the syenite is really quartz diorite, and he established several other gradational relationships of the various rocks. As a result, Judd (1876, p. 308) concluded that "all these igneous rocks of the Schemnitz area, 'granites and syenites,' 'greenstone-trachytes,' 'andesitic lavas and tuffs,' are parts of the same great eruptive masses, and are of contemporaneous date."

By the mid-1870s, British geologists had found compelling evidence via microscopic petrography that there exist complete textural gradations between coarsely crystalline rocks and fine-grained volcanic rocks in the very same igneous complex, thereby demonstrating the contemporaneity of plutonic and volcanic igneous rocks. That discovery in and of itself clearly showed that one could not legitimately assign coarse-grained rocks and volcanic rocks to separate portions of the geological time scale. Consistent with Judd's conclusion, British microscopists had already described nearly identical basalt from localities of considerably different geological age. It is small wonder that British geologists had little use for the German and French tendency to interject age as a factor in the classification of the igneous rocks.

A decade later, Judd (1886, p.51) lamented that although his fellow countrymen generally accepted his conclusions, "it would be idle to conceal from myself the fact that they have found but little favour among the petrographers of Europe." Despite some significant support for his views among continental geologists, he took note of the fact that virtually all of the valuable German and French petrographical treatises and monographs continued to insist on the absolute distinction in age between the highly plutonic and the volcanic types of igneous rocks even since the publication of all the English papers. Although Zirkel disagreed with his conclusions, Judd did note that Zirkel and von Lasaulx recognized that the gabbros of the Hebrides had all the characters of true gabbros and thereby named them as such. Consequently, Judd puzzled over the attempts of Rosenbusch and other petrographers to assign different names to the Hebridean gabbros on the sole grounds that they had a Tertiary age. Rosenbusch had, in fact, gone so far as to call Zirkel's designation of the Hebridean rocks as gabbros as "incorrect." Judd would have none of this. As forcefully as he could, he attacked the continental view and defended the British view:

> The geologists and petrologists of this country have always refused to recognize the geological age of a rock as a character upon which its classificatory position and nomenclature ought to be based. They regard such a proceeding as both inexpedient and impracticable—inexpedient, inasmuch as it prejudges the question of the distribution of rocks in time; and impracticable, in that there are many eruptive rock-masses of which it is quite impossible to determine the geological age.
>
> It will be seen, then, that these Hebridean rocks have come to occupy a crucial position in the controversy which has arisen concerning the classification and nomenclature of igneous rock masses. If they are truly gabbros, as maintained by Zirkel, von Lasaulx, and the geologists of this country generally, then it must be admitted that the principle of the classification of rocks by their geological age, having been found inapplicable in this case, has received a severe, and, indeed, a fatal shock. (Judd, 1886, pp. 60–61)

AMERICAN VIEWS ON THE ROLE OF AGE IN CLASSIFICATION

Judd discovered that he was receiving valuable confirmation for his contentions across the Atlantic during the mid 1880s as an outgrowth of the work of geological surveys of the American west. In 1879, Clarence King was appointed as the first director of the brand new United States Geological Survey. Dupree (1957) has summarized the events leading to the formation of the survey. One of King's earliest actions was to hire George Ferdinand Becker (1847–1919), a metallurgist who had graduated from Harvard and did graduate work in Germany with Robert Bunsen (Merrill, 1926). Becker immediately undertook reconnaissance investigations of a number of mining districts in California, Nevada, and Utah. In 1880, he was assigned to a reexamination of the geology of the Comstock Lode in the Washoe district of Nevada that the King party had briefly examined a few years prior during the geological survey along the Fortieth Parallel. Within two years, Becker (1882) had completed the study. He made the important petrographic discovery that von Richthofen's "propylite," allegedly a major rock type in the district, was in reality andesitic lava in an advanced state of decomposition. His examination of propylites from elsewhere showed that they had a similar character to those at the Comstock Lode. He also showed that trachyte was absent in the Comstock lode. Becker, however, did maintain a distinction between Tertiary and pre-Tertiary igneous rocks. After Becker's monograph appeared, a detailed investigation of thin sections from Comstock Lode rocks was initiated by U. S. Geological Survey geologists Arnold Hague and Joseph P. Iddings.

Arnold Hague (1840–1917) entered the Sheffield Scientific School at Yale when King was a senior and graduated in 1863 (Iddings, 1920). After a failed attempt to enlist in the Union Army, Hague went to Germany where he studied with Bunsen in Heidelberg. In 1865, he went to the Bergakademie in Freiberg to spend a year studying with von Cotta. Hague returned to America the next year and was offered a position by King as Assistant Geologist on the Fortieth Parallel survey. At the conclusion of the survey, Hague spent a couple of years in Guatemala and China before returning to America to begin service with the newly organized Geological Survey in 1880. During his career, Hague worked closely with Iddings on many occasions, particularly in the investigation of Yellowstone National Park.

Joseph Paxson Iddings (1857–1920) graduated from Yale in 1877 (Yoder, 1996). He undertook graduate studies at his alma mater and at the Columbia School of Mines. In 1879, Iddings applied to the new United States Geological Survey for a job as a geologist. While waiting for a response to his application, Iddings was presented with an opportunity to see Europe and, as a result, spent several months in Heidelberg studying with Rosenbusch. With newfound expertise under his belt, Iddings encountered Arnold Hague on the ship back to the United States when Hague was returning from his work in China. Iddings had also obtained the position with the Geological Survey and was placed under the supervision of Hague. Although spending much of his professional career with the survey, Iddings was Professor of Petrology at the University of Chicago between 1892 and 1908.

Iddings and Hague worked briefly at the American Museum of Natural History, where King's rock collections had been stored, and also in the Pacific Northwest, before they began their investigation of the Washoe mining district in Nevada. Hague and Iddings (1883) reported that they had been impressed by the textural gradations from glassy rocks to fine-grained granite porphyry among the igneous rocks of the Pacific Northwest, including Mount Rainier. They wanted to find similar situations in the Great Basin and concluded that the Washoe district provided the best opportunity for such a study because of the unparalleled three-dimensional exposure made possible by a spectacular network of shafts, as much as 3000 feet deep, and tunnels, as much as four miles long, that had been excavated during the 25-year mining history of the Comstock Lode. Hague and Iddings (1885) examined about 2000 hand specimens of igneous rocks from the district and about 500 thin sections.

The authors noted that Becker (1882) had agreed with von Richthofen in dividing the igneous rocks of Washoe into Tertiary and pre-Tertiary species, although they had differed as to where to draw the boundaries. Because Becker had access to the deep shafts and galleries of the mines, he found that the rocks obtained from the lower workings closely resembled the so-called pre-Tertiary types in their microscopic structure. As a result, Becker extended the domain of the "older" rocks over a considerable area of great depths below the surface. Upon examination of the same material from the deep locations, Hague and Iddings (1885) concluded that pre-Tertiary and the Tertiary rocks pass by insensible gradations into one another. They found that evidence of a difference in geological age is entirely lacking.

Hague and Iddings also examined hand specimens of numerous rocks that had earlier been classified as either diabase or augite andesite. They undertook a thorough microscopic examination of these rocks and found a complete gradational series in the groundmasses from glassy to coarsely crystalline. The porphyritic crystals, however, were found to be the same throughout the series. They concluded that most of the diabase is identical both microscopically and macroscopically to most of the augite andesite and is transitional between glassy forms at one extreme and the coarsely crystalline forms at the other. Hague and Iddings further demonstrated the existence of complete gradations between porphyritic mica diorite and hornblende andesite and also between pyroxene andesite and hornblende andesite.

The authors ultimately concluded that the degree of crystallization in the igneous rocks was dependent upon the conditions of heat and pressure under which the mass had cooled and was independent of geological time. They also remarked that there are abundant geological and petrographic evidences in the Washoe district to show that all of the igneous rocks are Tertiary in age. Apparently unaware of the important papers that John Judd had written a decade earlier, Hague wrote in the letter of transmittal of their report to the Director of the Geological Survey that, so far as he knew, the results of similar investigations that showed the gradual transition in crystallization under pressure in continuous rock masses had never been published (Hague and Iddings, 1885).

In the 1870s and 1880s, British and American petrographers firmly established that there is no significant correlation between geological time and igneous rock type, a major conceptual advance in igneous petrology that was rendered possible largely by the development of microscopic thin-section petrography. In time, igneous petrologists would become acquainted with the evidence presented by Forbes, Allport, Ward, Judd, and Hague and Iddings and with similar evidence gathered later. When that happened, age as a factor in the naming and classification of igneous rocks finally began to fade from the scene early in the twentieth century. The transition to age-free classification was finally completed when the texts of Rosenbusch went out of print.

11 ❖ PROVINCES AND PLUGS

Igneous Rocks in Space

The petrographic revolution issued in the detailed thin-section reexamination of many classic suites of igneous rocks, such as those from the Auvergne, as well as a plethora of studies of areas never before investigated. From the staggering wealth of petrographic information gleaned from many parts of the world, it became clear that the igneous rocks within certain regions are linked by common petrographic, chemical, and geological features and that the characteristics of the igneous rocks in one region commonly differ from the characteristics of the igneous rocks elsewhere. Although Herman Vogelsang first sensed such relationships in the early 1870s, it remained for J. W. Judd to make explicit some of the geographical relationships of igneous rocks.

THE CONCEPT OF THE PETROGRAPHIC PROVINCE

During the course of his investigations of the volcanic and plutonic rocks of the Hebrides over a number of years, Judd became aware of the mineralogical and textural characters occurring in common at the diverse Hebridean volcanic centers of Skye, Ardnamurchan, and Mull. From the reports of Sartorius von Waltershausen, Zirkel, and others, he further recognized that there are remarkable similarities among the basalts of Iceland, the Faroe Islands, Scotland, and Ireland. The more coarsely crystalline dolerites and gabbros from these diverse locations also appeared to be similar. "The more carefully we study the British and Icelandic types of Tertiary basic rocks," Judd (1886, p. 53) wrote, "the more forcibly are we struck with their essential identity in character." Again and again, he said, we find that certain peculiarities recur in the rocks of both areas. Moreover, he noted that a comparison of the British and Icelandic rocks with rocks of the same age and broadly similar composition from other districts, such as the Auvergne, Bohemia, or Italy, demonstrated that the rocks in each of the areas are marked by highly distinctive groups of characters. In short, the rocks of the diverse regions may be distinguished from one another. These observations led Judd (1886, p. 54) to draw two important conclusions:

The first is that there are distinct *petrographical provinces* within which the rocks erupted during any particular geological period present certain well-marked peculiarities in mineralogical composition and microscopical structure, serving at once to distinguish them from the rocks belonging to the same general group, which were simultaneously erupted in other petrographical provinces.

Judd's second conclusion was that northern Ireland, the Inner Hebrides, the Faroe Islands, and Iceland formed part of the same petrographical province during the Tertiary Period.

In his definition, Judd emphasized the distinctive mineralogical and textural characteristics of a petrographic province, features that, of course, became far easier to identify and describe after the advent of thin-section microscopy. In the case of the British Icelandic rocks, Judd regarded the prevalence of ophitic texture and the absence of porphyritic crystals of augite or hornblende as the salient features of the basic rocks of the province. He was also struck by the absence of nodules of olivine and enstatite embedded in the basic rocks, for such nodules he regarded as common in the provinces of Bohemia, the Eifel, and the Auvergne. Since the days of Judd, the Brito-Arctic Tertiary (Thulean, North Atlantic) Province has been extended to include the rocks of Jan Mayen and some of the rocks exposed along coastal Greenland such as the Skaergaard Intrusion. The province is, of course, also distinguished chemically by the tholeiitic character of its magmas.

The concept of the petrographic province was further developed by Iddings. He introduced chemical composition as one of the distinguishing characteristics of the rocks of an igneous province. Having studied the diverse igneous rocks of Yellowstone National Park in Wyoming since 1883, Iddings came to recognize a strong connection between the mineral composition and the geological occurrence of igneous rocks at Electric Peak and Sepulchre Mountain in Yellowstone. He also described striking similarities in the mineral content, sequence of mineral crystallization, sequence of eruption, and trends in chemical composition of the igneous rock series of the two locations. Iddings found that both series display decreases in CaO, MgO, and FeO and increases in Na_2O and K_2O as the molecular proportions of SiO_2 increase. He concluded that the intrusive rocks of Electric Peak and the volcanic rocks of Sepulchre Mountain were chemically identical and that the entire series of rocks at each location represented "a single, irregularly interrupted succession of outbursts of magma that gradually changed its composition and character" (Iddings, 1892a, p. 202).

In a paper published later that same year, Iddings (1892b) contributed an extensive discussion of the origin of igneous rocks. A major component of his paper was a discussion of the common characteristics within a given district or province which Iddings designated as "consanguinity." Iddings viewed consanguinity as evidence for the common genetic origin of the igneous rocks within the district. Decades of research and speculation, Iddings claimed, had brought petrology to the point where it was now understood that

> there occur in various parts of the world family groups of rocks which, while identical with other groups, are different from many surrounding them and constitute associations of rock groups. The recurrence of these associations establishes a still higher order of relationship or the existence of a more remote common origin. It is to express the idea that all the igneous rocks of any volcanic district have been derived from a common stock—that is, from a common magma—that the writer has applied the term *consanguinity* to the relationship. (Iddings, 1892b, p. 130)

Iddings found mineralogical evidence for consanguinity in the consistency of the color of minerals such as pyroxene, hornblende, or biotite. For example, he noted that in most of the igneous rocks of western America, the augite is pale green and not pleochroic. In contrast, he mentioned that along the eastern flank of the Rocky Mountains, deeper green augite and even aegirine and acmite prevail. In many of the districts of Europe, he pointed out, purplish augite is common. Moreover, he observed that these colors are indicative of variations in the chemical composition of the minerals and, therefore, of the magmas from which they crystallized. He suggested that deeper greens indicate increased sodium content and that purplish augite owes its color to increased titanium and iron-magnesium content. Iddings argued that evidence for consanguinity might also be found in the widespread presence of certain minerals like hornblende within diverse igneous rocks of a particular district.

Iddings (1889, p. 73) had recently coined the term "phenocryst" for the large crystals that "stand out prominently from the mixture of glass and small crystals" in an igneous rock. German writers had commonly employed the term "Einsprenglinge" to refer to phenocrysts. As another example of mineralogical consanguinity, Iddings suggested that orthoclase occurring as "phenocrysts" in silica-poor igneous rocks such as trachyte appeared to be especially characteristic of rocks along the eastern flank of the Rocky Mountains.

Chemical evidence for consanguinity was detected where diverse rocks of a district are characterized by high soda/potash ratios or by high lime content. In the Yellowstone Park area, Iddings noted that the rocks have soda/potash ratios ranging between 3:1 and 2:1. In contrast, he observed that rocks in the vicinity of Mount Vesuvius are much more potassic, with soda/potash ratios from 7:4 to 1:4. Lime, magnesia, and iron oxide were found to decrease considerably from less siliceous to more siliceous rocks at Electric Peak and Sepulchre Mountain in Yellowstone. At Mount Vesuvius, lime and magnesia display a similar decrease, but iron oxide decreases very slightly in abundance in increasingly siliceous rocks and is much more abundant than either lime or magnesia in any of the Vesuvian rocks. Iddings noted that the lack of chemical data in several provinces hampered the search for chemical evidence of consanguinity.

He was also cognizant that the geological literature was clogged with many chemical analyses of low quality as well as many analyses of highly altered material. The confusion brought about by this state of affairs, he acknowledged, had "seriously blocked the progress of petrology for years." Iddings (1892b, p. 141) realized that "the chemical development of the science is still in the future."

Iddings (1892b, p. 143) also noted geological evidence for consanguinity. He suggested that

> the constant recurrence of particular series of rocks often with a certain order of eruption in different localities, and the frequent occurrence of such series at neighboring centers of volcanic activity, sometimes with a repetition of the whole or a part of the series, would be enough to justify the belief that there was a definite connection between the members of a group.

As petrographic, chemical, and mineralogical data continued to accumulate, petrographers identified more and more petrographic provinces: the Auvergne of south-central France; the Eifel district of Germany; the Oslo district of southern Norway; the province of Madagascar and the eastern coast of Africa; the San Juan Mountains province of southwestern Colorado; the potassium-rich Roman "co-magmatic" region of Italy; and the potassic province of central Montana that included the Little Belt, Highwood, Judith, and Bearpaw Mountains. Pirsson (1905a), for example, described the petrographic province of central Montana as characterized by approximately equal percentages of potash and soda in the most siliceous magmas. Moreover, he claimed, the potash/soda ratio persistently increases with decreasing silica and with increasing lime, iron, and magnesia. He also pointed out that the rocks of the province are characterized mineralogically by green augite, strongly pleochroic brown biotite, very little hornblende, and the virtual absence of minerals containing zirconium, thorium, niobium, or the rare earth elements.

In his classic petrology text, *The Natural History of the Igneous Rocks*, British petrologist Alfred Harker (1909, p. 89) wrote that "the doctine of petrographical provinces, supplemented by that which is implied in the use of the term 'consanguinity' in this connection, has thus come to occupy a prominent place in modern petrology, and it is found to be capable of extended application." Indeed, throughout much of his career, Harker attempted to extend the concept of the petrographic province more widely than most other petrologists. Alfred Harker (1859–1939) was the premier British petrologist of the end of the nineteenth century and first decades of the twentieth (Seward and Tilley, 1940). Equally at home dealing with metamorphic or igneous rocks, Harker was an exceptional field worker, petrographer, and theoretical petrologist (Figure 12). He made the acquaintance of J.J.H. Teall when the latter was offering university extension lectures on geology in Hull. Harker's interest in geology was kindled especially by field trips with Teall. He took his undergrad-

12. Alfred Harker (1859–1939). Reproduced from the archives of the Sedgwick Museum, University of Cambridge, by permission.

uate degree in physics and geology at St. John's College, Cambridge, graduating in 1883 and, early in 1884, was appointed as a Demonstrator in Geology at Cambridge. He was promoted in 1904 to University Lecturer, and from 1918 until 1931 he held a Readership in Petrology. Throughout a long and illustrious career at Cambridge, Harker distinguished himself with a series of stimulating papers on the differentiation of igneous rocks, a memoir on the Tertiary igneous rocks of the Isle of Skye, and highly regarded textbooks on petrography, metamorphism, and igneous rocks.

As a field geologist, Harker was especially sensitive to the structural aspects of geology and took pains to stress the connection between igneous activity and tectonic setting. Igneous rocks, he said, are not simply meaningless interpolations into the stratigraphical sequence. Rather, they are "closely bound up with the geological history of the districts in which they occur, and often in intimate relation with folds, faults, unconformities, and other geological accidents" (Harker, 1896b, p. 18). Igneous activity seemed to Harker to be closely related to episodes of important Earth movements such as mountain building. He had been struck by Iddings' observation that relatively alkali-rich rocks are found east of the Rocky Mountains and the Andes Mountains, whereas the rocks of sub-alkalic character are prevalent along the axis of and west of the Rocky Mountains and the Andes Mountains, regions he believed were characterized by compressional tectonics.

Harker became intrigued by large-scale patterns of igneous activity and began to emphasize them. For example, the Brito-Arctic province, Harker suggested, is part of a much larger region that includes the Azores, Canary, and Cape Verde Islands as well as St. Helena, Tristan da Cunha, and parts of the Portuguese and African coasts. This enlarged province he considered as linked tectonically to the great crustal movement of oceanic depression. Harker believed that such areas represented "scattered relics of a belt girdling the Atlantic, analogous to the much better defined 'circle of fire' which surrounds the Pacific Ocean" (Harker, 1896b, pp. 19–20). Harker believed, moreover, that there was evidence that the globe could be divided petrographically into two enormous provinces characterized by two distinct linear series of rock, each of which ranges from extremely silica-rich to silica-poor rocks. The differences between the two series were said to show the greatest divergence in the middle. He further maintained that both series were clearly traceable in either volcanic or plutonic rocks.

To one series Harker referred such rock types as pantellerites, trachytes, phonolites, leucitophyres, nephelinites, leucitites, nepheline basalts, and leucite basalts. He located the dominant chemical distinction of this series in the behavior of its alkalis. Mineralogically, Harker noted that these alkalic rocks are commonly composed of alkali feldspars with little or no quartz, commonly contain feldspathoids such as nepheline, leucite, and sodalite, and are characterized by alkali pyroxenes and alkali amphiboles rather than augite, orthopyroxene, and hornblende among the ferro-magnesian minerals. Harker claimed that such alkalic rocks appear on the eastern slopes of the Rocky Mountains in Montana, Wyoming, Colorado, and Texas. Such rocks, he indicated, are also characteristic farther east in Canada, the New England States, New Jersey, and Arkansas, and also in parts of South America east of the Andes Mountains in Brazil, Argentina, and Paraguay. This alkalic group, Harker said, is developed along the Atlantic slope of the Americas, but he thought that a similar situation also obtained in Europe. Harker designated these areas as an alkalic "Atlantic" province.

To the other, sub-alkalic series of rocks, Harker assigned dacites, andesite, and feldspar basalts, rocks that commonly contain quartz with alkali feldspar and augite, hornblende, and orthopyroxene as the typical ferro-magnesian minerals. The igneous rocks of the sub-alkalic group, he maintained, are located along and to the west of the great continental divide of the two Americas. They are developed along the Pacific slope not only of the America, but indeed around the entire Pacific "circle of fire." Harker assigned these areas to a sub-alkalic "Pacific" province.

Harker became the leading advocate of the concept of two mega-provinces. Not all petrologists were convinced of the existence of these two provinces, and his far-reaching extension of the idea of the petrographic province remained a matter of intense debate in the very early twentieth century.

CHAPTER 11

THE FORMS OF INTRUSIVE IGNEOUS ROCK BODIES

During the final decades of the nineteenth century, geologists also became more intrigued by the spatial relations of individual igneous rock bodies. The heightened interest in igneous rocks led to a growing program of detailed field mapping of igneous rock occurrences and, thereby, to increased awareness that igneous rock bodies occur in a range of sizes and in a variety of forms other than lava flows, dikes, and sheets. The specialists in petrology, however, were so enamored of the petrographic and chemical side of their science that many of the new data and concepts about the sizes and forms of igneous rock bodies were contributed by geologists such as Murchison, Gilbert, Suess, Salomon, and Russell who were primarily interested in the structural or geomorphic aspects of geology rather than in petrology. Reviews of the forms of igneous rock bodies have been provided by Daly (1905), Hunt et al. (1953), and Corry (1988).

Since the early days of the first awareness of the igneous origins of some rocks, the terms "dike" and "vein" or "sheet" and "sill" had been applied to tabular or sheet-like bodies of rock that transect or are conformable to their host rocks, respectively. The origin of these terms is obscure and cannot be attributed to any one individual. It was realized early in the nineteenth century that not all rocks of probable igneous origin occur in tabular dikes or sheets, and more irregularly shaped bodies were typically referred to simply as "masses." Playfair (1802, p. 83), for example, in speaking of granite, wrote that "the fossil now defined exists, like whinstone and porphyry, both in masses and veins, though most frequently in the former."

The only significant refinement in terminology in the first two-thirds of the nineteenth century came from Roderick Murchison. In his synthesis of *The Silurian System* (Murchison, 1839), the famed "King of Siluria" used the term "boss" in reference to irregular, knob-like igneous masses that occur in the sedimentary formations of Shropshire, England.

Nearly forty years later, the next major advance in the understanding of igneous form took place in the desolate, unexplored high plateau of south-central Utah in the American west. During his storied expedition down the Utah portion of the Colorado River in 1869, Major John Wesley Powell (1834–1902) had seen a group of mountains in the distance. Powell deciphered that the Henry Mountains, as he named them in honor of Joseph Henry, the first secretary of the Smithsonian Institution in Washington, D. C., had a domal structure in which Triassic and Jurassic sandstones were upturned along the flanks of masses of dark lava. Powell suspected that these domal structures might represent "craters of elevation."

Ironically, the hypothesis of "craters of elevation" was already largely discredited by Powell's time. The idea, very prominent throughout the first half of the nineteenth century, was originated by Alexander von Humboldt after his study of Jorullo in Mexico (Dean, 1980; Laudan, 1987). Soon thereafter, Leopold von Buch enthusiastically endorsed the elevation-crater concept as a result of his studies in the Auvergne, Italy, and the Canary Islands. Additional advocates included Abich, Élie de Beaumont, Prevost, Daubeny, and many other prominent European geologists. Lyell and Scrope were among its most vigorous antagonists.

The most passionate proponent of the hypothesis, Leopold von Buch (1809), distinguished between true volcanic cones and craters of elevation. He regarded the former as typically small features that had been constructed by the piling up of continuously erupted, viscous, rapidly cooling trachytic lavas and pyroclastic materials. The latter, in contrast, were considered to be much larger and more important structures, typically composed of alternating sedimentary and fluid basaltic strata, generally in the vicinity of the sea. Because of the highly fluid nature of basalt, von Buch did not accept the idea that true volcanic cones could be constructed from such material. He suggested that basalt flows intercalated with sedimentary rocks had issued from a submarine vent on a flat floor without the formation of a volcano. Von Buch attributed the arching of the entire package of volcanic and sedimentary strata to an accumulating subterranean force that eventually broke through to the surface, elevating the beds and creating a crater in which additional volcanic activity might occur. The expansive force thus spent, no further elevation would take place. Throughout successive editions of *Principles of Geology*, Charles Lyell argued against the concept of craters of elevation, reasoning that craters associated with lava flows were the result of volcanic accumulation and, in many cases, marine denudation. Scrope (1859) repeatedly challenged the idea and showed from the eruption of Mount Etna in that year that lava does consolidate on steep slopes, thus confirming the notion that volcanic cones are constructed by successive outpourings of lava.

Very likely, the doming and interstratification of volcanic and sedimentary material in the Henry Mountains suggested the possibility of the presence of "craters of elevation" to Powell. It fell to G. K. Gilbert to show the existence of something very different. Grove Karl Gilbert (1843–1918) graduated from the University of Rochester in 1862 (Pyne, 1980). After working at the Ward Natural Science Establishment in Rochester for five years, Gilbert participated in the Second Geological Survey of Ohio under J. S. Newberry and was a geological assistant on G. M. Wheeler's Geographical and Geological Survey West of the 100th Meridian. In late 1874, Gilbert joined the U. S. Geographical and Geological Survey of the Rocky Mountain Region under Powell, who assigned him to investigate the Henry Mountains, the last group of mountains to be discovered in the area destined to become the conterminous United States. In 1875 and 1876, Gilbert worked on the geology of the region and published his report in 1877. He became a member of the U. S. Geological Survey upon its formation in 1879, and in 1889 he became Chief Geologist. Primarily interested in geomorphology, Gilbert was especially well known for his work on the geomorphology of the Great Basin, the history of Lake Bonneville, fluvial geomorphology, and the idea that the lunar craters are of impact origin.

By his second field season, thanks to the exceptional exposures afforded by the barren terrain of Utah, Gilbert began to realize that the "lavas" of the domal structures in the Henry Mountains are intrusive rocks that had deformed the sedimentary formations that they intruded. At first, he referred in his notebooks to the bubble-shaped intrusives as "lacunes." In his published account, however, Gilbert (1877) used the term "laccolite" to describe these features. Dana (1880) suggested altering the name to "laccolith" in order to avoid confusion with mineral and rock names. For several years thereafter, these bodies were referred to both as "laccolites" and "laccoliths." Eventually, geologists adopted Dana's terminology.

"It is usual," Gilbert (1877, p. 19) wrote,

> for igneous rocks to ascend to the surface of the earth, and there issue forth and build up mountains or hills by successive eruptions.... The lava of the Henry Mountains behaved differently. Instead of rising through all the beds of the earth's crust, it stopped at a lower horizon, insinuated itself between two strata, and opened for itself a chamber by lifting all the superior beds. In this chamber it congealed, forming a massive body of trap. For the body the name *laccolite* (λακκος, cistern, and λιθος, stone) will be used.

Because the concept of lava displacing and lifting overlying rocks was unfamiliar to geologists, Gilbert spelled out five lines of evidence as to why the trachytic rocks had been intruded into, rather than buried beneath, sedimentary materials. The evidences included the absence of trachyte fragments in the overlying strata; the complete lack of vesicularity of the trachyte; an inclination of the arched strata that, in many cases, far exceeded the angle of repose of sediments that might have been deposited on a lava mass; the local cross-cutting nature of some of the trachytic sheets; and the alteration and metamorphism of strata both above and below the trachyte bodies. "In fine," Gilbert (1877, p. 52) concluded, "all the phenomena of the mountains are phenomena of intrusion. There is no evidence whatever of extrusion."

Gilbert's laccolith hypothesis met with a mixed reaction. Several European geologists were skeptical of the idea for many years. Before long, however, American geologists began to discover laccolithic structures over much of the western states and territories of the United States. "Laccoliths" were detected at Shonkin Sag and Square Butte in central Montana, in the La Plata Mountains and the Spanish Peaks, where a radial dike system was also discovered, in Colorado, in the Abajo and La Sal Mountains in Utah, and elsewhere. The "laccolith" became a firm fixture in igneous geology well before the end of the nineteenth century.

Laccoliths, craters of elevation, sills, and other structures associated with volcanoes were reviewed in the massive, multivolume synthesis of global geology, *Das Antlitz der Erde* (Suess, 1885). Eduard Suess (1831–1914), widely known for his work on the tectonics of the Alps, was Professor of Geology at the University of Vienna throughout his career (Wegmann, 1976). He was also very active in Austrian politics, serving in the Vienna city council, provincial diet, and national parliament. In *Das Antlitz der Erde*, Suess described the characteristics of granite masses. He noted that granite bodies commonly are embedded in older stratified rocks, most frequently in schist, and that their form appeared to be that of "*large, irregular loaves or cakes*" (Suess, 1885, p. 167). Granites, he noted, have also produced contact metamorphism in the overlying rocks. In that regard, Suess suggested that although granite masses resemble the trachytic laccoliths of North America, they are far larger. Suess then posed the great "room problem" that has vexed geologists to this day. He asked "how such extraordinarily large cake-shaped masses, with a major axis measuring 10 or 20 kilometers or even more, can have been subsequently intercalated at a definite horizon" (Suess, 1885, p. 167). It was imperative, he argued, that injection of such a hot granitic mass "*should be preceded by the formation of a corresponding cavity*" (Suess, 1885, p. 168). Suess postulated that tangential stresses accumulated during lateral thrusting or folding might create such very large, lenticular spaces in deep-seated zones of schist for the granitic mass to fill and to send apophyses and dikes into adjacent fissures. Suess (1904, p. 168) suggested that

> the magma simply filled the space as far as it extended, and consolidated in it, forming a cake of rock or true *batholite*, which would never be capable of giving rise to any further mountain building, but might in certain cases, in consequence of its great bulk, its rigidity, and configuration, passively influence this process as regards certain subordinate features.

The granite cores of the Pyrenees and the Alps were given as examples of "batholites."

Suess (1895, p. 52) restated his definition of batholith in terms of a theory of intrusion, writing that "a batholith is a stock-shaped or shield-shaped mass intruded as the result of fusion of older formations" and that "either holds its diameter or grows broader to unknown depths." R. A. Daly (1905) noted that the term "batholith" had commonly been used for bodies of intrusive rock of much larger size than stocks or bosses ever since Suess coined the word, but often without the theory of origins invoked by Suess. Daly recommended the continued employment of the term "batholith" for large bodies in a way that freed it from connection to a specific theory. "Stocks, bosses, and batholiths never show a true floor," Daly (1905, p. 506) wrote, "they appear to communicate directly with their respective magma reservoirs."

Both Suess and Daly made reference to "stocks." According to Hunt et al. (1953), Iddings used the term "stock" as early as 1891 in reference to elliptical, circular, or irregular cross-cutting intrusions. And Kemp (1911, p. 256) noted that the term "is an adaptation of the German word for floor or story, and originated in the Saxon tin mines, in which such igneous masses, impregnated with cassiterite, were mined in horizontal slices, like floors." The word was later applied more generally to roughly cylindrical, intrusive bodies in which the stock-work mining system was used.

As mapping of the western United States continued, geologists discovered more and more laccolithic intrusions. As might be expected, however, as increasing numbers of laccoliths were mapped, the likelihood increased that the forms of some of these bodies would be found to deviate to varying degrees from that of the "type" intrusions studied by Gilbert. Enough striking departures from the forms of Gilbert's laccoliths were encountered that geologists decided the time had come to distinguish these "deviant" forms from true laccoliths.

As a result, I. C. Russell (1896a, p. 23) maintained that there were a dozen or so hills in the area of the Black Hills of South Dakota, such as Devil's Tower, Little Missouri Buttes, Bear Butte, and Sun Dance Hill, that "furnish examples of a type of igneous intrusions that does not seem to have been clearly recognized." Israel C. Russell (1852–1906) had graduated from New York University in 1872, after which he accompanied the expedition to New Zealand for the U. S. Transit of Venus in 1874–5 (Willis, 1907). From 1875 to 1877 he was Assistant Professor of Geology at Columbia University. He spent a year with the U. S. Coast and Geodetic Survey and, in 1880, joined the U. S. Geological Survey, where, for several years, he worked on the geology of Pleistocene Lake Lahontan. In 1892, Russell succeeded Alexander Winchell as Professor of Geology at the University of Michigan. Although Russell's interests lay overwhelmingly in the areas of Pleistocene geology, geomorphology, glacial geology, and geography, it was he who coined the term "plug" for these western hills.

Russell claimed that the hills had been formed by the injection of molten rock from below into stratified beds and also that subsequent erosion of the sedimentary strata had exposed an inner core of plutonic rock. The difference from the "laccolites" of Gilbert, however, was said to lie in the fact that the molten rock had not spread out horizontally among the stratified beds as it had in the laccoliths. Russell suggested that these bodies also differed from volcanic necks in that the injected rock did not reach the surface. "As they are composed of igneous matter forced into sedimentary strata and have a plug-like form, it will be convenient to call them *plutonic plugs*," Russell (1896a, p. 25) wrote. He thought that the rock types making up the plugs were rhyolite and sanidine trachyte, although the rocks were later recognized as phonolite.

Russell maintained that the magma in these plugs had cooled slowly, well below the surface under great pressure. As evidence for this mode of solidification, Russell pointed to the geological associations and the coarse-grained or porphyritic structure of the igneous material. As evidence against rapid chilling, he noted the absence of associated obsidian, pitchstone, perlite, scoria, and pumice. He also suggested that the absence of associated dikes or faults indicated injection below the brittle portion of the crust. Russell was uncertain as to how the stratified beds that covered the plugs were displaced.

Iddings was not too sure that he wanted to accept Russell's "plutonic plugs" without more evidence. He thought that many of these bodies might simply represent central eroded remnants of small laccoliths. What he was certain of is that many intrusions are associated with faults. He argued that if vertical displacement by faulting is one of the chief characteristics of an intrusion, then a distinction from a normal laccolithic intrusion needs to be recognized. Iddings (1898a, p. 706) claimed that by vertical displacement of an intrusion along a fault, "the vertical dimension of the intruded mass becomes still greater as compared with the later dimensions, so that its shape is more that of a plug or core. Such an intruded plug of igneous rock may be termed a *bysmalith* (βυσμα = plug, λιθος = stone)." As examples of bysmaliths he proposed the Mount Holmes mass in Yellowstone National Park and Gray Peak in the Gallatin Mountains. Iddings also distinguished "bysmaliths" from "stocks" by limiting the term "stock" to bodies occupying nearly vertical tubes or funnels of indefinite depth in any kind of country rock and maintaining a dike-like, that is, cross-cutting, relationship to them. A stock, Iddings proposed, might represent multiple injections, but a bysmalith resulted from one act of eruption. A year later, Iddings and Hague (1899, p. 16) defined a bysmalith as an injected body filling a "more or less circular cone or cylinder of strata, having the form of a plug, which might be driven out at the surface of the earth, or might terminate in a dome of strata resembling the dome over a laccolith."

Wilhelm Salomon (1868–1941), a German structural and economic geologist at the University of Pavia who was known primarily for his studies of the Adamello region of the Alps, proposed the term "ethmolith" in conjunction with the tonalites of that region (Würm, 1950). Salomon (1903) referred to funnel-shaped plutonic masses that narrow downward and have adjacent strata bent downward as "ethmoliths."

In a review of the classification of intrusive bodies, R. A. Daly (1905) observed that many intrusions have such irregular shapes and that their relationships to the surrounding rocks are so complex that these bodies could not be classified as one of the forms thus far established. Because no generally accepted name had yet been proposed for these irregular intrusions, Daly (1905, p. 499) suggested a name formed from the Greek on the analogy of "laccolith," "bysmalith," and "batholith." It is "chonolith," derived from χωνος, a mold used in the casting of metal, and λιθος a stone. The magma of a "chonolith" fills its chamber after the manner of a metal casting filling the mold. Like a casting, the "chonolith" may have any shape.

Daly regarded a "chonolith" as an igneous body that was injected into any kind of country rock; that is not a true dike, vein, sheet, laccolith, bysmalith, neck, stock, batholith, or ethmolith; and that was composed of magma that had either been passively squeezed into a subterranean orogenic chamber or actively forced apart the surrounding rocks.

MECHANISMS OF EMPLACEMENT

On rare occasions, geologists and physicists such as Reade and Hopkins pondered some of the mechanical problems associated with igneous intrusion, but Gilbert was probably the first geologist to point out clearly that intrusive igneous rocks are capable of deforming their host rocks. In contrast to most geologists of his era, Gilbert did not view laccoliths primarily in terms of chemical composition, differentiation, or geological history. Instead he viewed them from a mechanical point of view by asking how injected magma might lift an overburden of sedimentary rocks. Although petrologists continued to emphasize the mineralogical and chemical aspects of igneous rocks for decades to come, from now on they became more aware of the importance of thinking about the mechanical aspects of magmatism.

Laccoliths posed a twofold problem, Gilbert said. In the first place, laccoliths were emplaced at considerable distances below the ground. The lava did not reach the surface as it does at a volcano. What, he asked, determines the level at which the lava is emplaced? Why do some lavas erupt whereas others are emplaced at depth? And, secondly, once the lava begins to spread horizontally between layers of rock at depth, what prevents its continued lateral spread and leads it to arch the overlying strata?

In answer to the first problem, Gilbert realized that lava acted in accord with the principles of hydrodynamics. On the assumption that the cohesion of the overlying rock did not inhibit the rise of the liquid in the first place, Gilbert thought, the lavas would cease to rise upon reaching a zone in which there is the least hydrostatic resistance to their accumulation. "If this zone is at the top of the earth's crust they build volcanoes; if it is beneath, they build laccolites," said Gilbert (1877, p. 95). But how high a lava might rise in the crust was a matter of the density of the lava in relation to the density of the rocks through which it was attempting to rise. As a result, he maintained that light lavas were more likely to produce volcanoes, whereas heavy lavas were more likely to produce laccoliths.

To test his hypothesis, Gilbert measured the densities of many of the sedimentary rocks of the Henry Mountains and several specimens of trachyte from the laccoliths. On the basis of very limited experimental data, he estimated densities of liquid trachyte and convinced himself that these would be sufficiently great in relation to the densities of the sedimentary rocks through which the liquid needed to rise but not so great as to reach the surface. Hence, it would make sense that the trachytes of the Plateau Province would produce laccoliths, he thought.

Gilbert was less convincing when it came to explaining the common occurrence of basaltic lava flows of the plateau. He recognized that basaltic igneous rocks are considerably more dense than trachyte, and, yet, for basaltic liquids to rise to the surface at volcanoes, they needed to be less dense than trachytic liquids. His suggestion that such might be the case was not so compelling. Gilbert seems not to have considered the effect that very large differences in viscosity between trachytic and basaltic liquids might have had on the ability of these liquids to rise to the levels that they did.

Once the level of injection of the laccolithic magma was determined by relative rock and magma densities, Gilbert reasoned, a thin sheet of lava would intrude along a parting of strata and spread out, extending itself in the direction of least resistance. If the resistances were equal on all sides, the laccolith would take a circular form. Eventually the horizontal extent of the lava sheet would become so great that the lava would overcome the resistance offered by the cover rocks and begin to lift them. At that point the direction of least resistance would be upward so that the reservoir of lava would increase in depth rather than width. With vertical uplift, the sheet became a laccolith. Gilbert maintained that with the continued addition of lava the laccolith grew in height and width until either the supply of lava or the force propelling the lava decreased sufficiently that the lava congealed in its conduit.

Although Gilbert acknowledged that a growing laccolith might have been built by many separate lava flows, he insisted that they must have risen fairly rapidly lest the trachytic magma congeal and clog its conduit. He thought that if the pressure exerted by the lava was just great enough to overcome the resistance to uplift (and Gilbert attempted calculations of the rock pressures to be overcome) then the circumference of the laccolith should be proportional to its depth beneath the surface. "The laccolite in its formation," Gilbert (1877, p. 91) concluded, "is constantly solving a problem of 'least force', and its form is the result."

Two decades later, I. C. Russell (1896b) suggested that igneous rock bodies could be considered to form a kind of graded series from sheets to laccoliths to plutonic plugs to domal uplifts such as those in the Black Hills. He believed that sheets are most common where the invaded strata are nearly horizontal, so that the separation of beds could not be caused by folding. Rather, he attributed the mechanism of separation of beds to injection of highly fluid basaltic

13. Reginald A. Daly (1871–1957). Reproduced by permission of Harvard College Library, Kummel Library of Geological Sciences.

magma close to the surface. The closer to the surface that the intrusion occurred, Russell reasoned, the less the weight of the overlying rocks, and thus more energy would be available for lateral expansion of the magma. In general, he conceived that laccoliths, plutonic plugs, and domal uplifts should be formed at successively deeper levels by increasingly silica-rich, increasingly viscous, and decreasingly fusible magma.

A substantially more controversial and interesting proposal regarding the mechanics of igneous intrusion was advanced by Daly. Reginald Aldworth Daly (1871–1957) graduated from Victoria College in Canada and then pursued graduate work in geology at Harvard University where he earned his Ph.D. degree in 1896 (Billings, 1959; Birch, 1961). During graduate school, he began detailed studies of the Mount Ascutney pluton in Vermont. Daly (Figure 13) spent two years in Europe studying with Rosenbusch and Lacroix and examining classic geological localities. Upon his return to the United States, he became an instructor at Harvard until he resigned in 1901 to undertake the geological investigation of the 49th Parallel for the Geological Survey of Canada. Daly was Professor of Physical Geology at Massachusetts Institute of Technology between 1907 and 1912 and then returned to Harvard as Sturgis-Hooper Professor of Geology. For the remainder of his career, Daly was profoundly interested in the broad topics of tectonics, geophysics, and oceanography but also continued to write about alkalic rocks, the petrology of oceanic islands, and magmatic differentiation. Of particular importance among Daly's writings were his major books, *Igneous Rocks and Their Origin* (Daly, 1914) and its thorough revision, *Igneous Rocks and the Depths of the Earth* (Daly, 1933).

Daly's work on the Mount Ascutney pluton led him to develop the concept of magmatic stoping in a series of papers on the mechanics of igneous intrusion (Daly, 1903a, b, 1908). Thanks to the work of people like Gilbert, Daly granted that geologists were fairly agreed about the intrusion mechanics of dikes, sheets, and laccoliths but reminded his colleagues that the volume of these intrusive masses is rather minor. It was the mechanics of emplacement of the stocks and larger granitic bodies that he regarded as very controversial.

One school of geologists such as Brögger extended the laccolithic idea to most granitic intrusions. But, said Daly, it was clear from a study of the literature that the field relationships such as the cross-cutting character and highly irregular form of the granite bodies and the structural complexity of their invaded rocks rendered implausible the notion that the great majority of granite stocks and batholiths had been injected in the manner of laccoliths. Instead, Daly (1903a, p. 270) argued, the field phenomena indicated that "such magmas actively, aggressively, 'made their way in the world' by the irregular removal of the invaded formations."

Another school of geologists favored the "marginal assimilation" hypothesis that magmas advanced into the overlying rocks by means of a caustic, solvent action on their walls and roofs. Daly suspected that this view appealed more strongly to most geologists who actually worked with granitic bodies in the field, but he maintained that the assimilation theory failed to meet the crucial test of evidence for a chemical and mineralogical relationship between the average intrusive rock and the invaded country rock along their contacts. Daly (1903a, p. 271) pointed out the striking fact that granite reveals "astonishing homogeneity in contact with argillite, limestone, crystalline schists or basic igneous formation—a homogeneity that persists, too, from contact to center of the eruptive." It did not appear to Daly that granitic magmas had assimilated the formations exposed along their contacts.

But, Daly stressed, it is commonplace to find in the inner zone of contact with the roof and walls of a granite intrusion very numerous blocks of the invaded rocks that are characterized by varying size, angular or subangular outlines, sharp contacts with the igneous rock, a normally high degree of crystallinity, and increased density as a result of contact metamorphism. The foreign inclusions become much less common toward the interior, he noted, "until, in the heart of the eruptive area, one may go hundreds of yards or even several miles without discovering any such inclusions" (Daly, 1903a, p. 273).

Daly expressed surprise that there had been no previous adequate discussion and explanation of the presence of foreign blocks within igneous bodies along their contacts and the rarity of blocks toward the centers of the bodies. No one had yet answered the question as to how a block could be suspended in the magma close to its country rock until crystallization of the magma was complete. Nor had anyone answered the question as to whether immersion of a block would result in its floating or sinking in the magma. To answer such questions, Daly discussed at length the densities of various magmas and rock types and summarized which kinds of rocks would sink, which would rise, and which would be suspended in magma of a particular composition and density. He concluded that highly fluid magmas would probably be much more successful in rifting blocks away from the country rocks than viscous ones.

It was Daly's fundamental contention that a magma body gradually worked its way upward through the crust by a process of overhead stoping, a term that he borrowed from the language of mining and that referred to the upward or sideways excavation of ore. Daly envisioned overhead magmatic stoping as entailing the shattering of the rock of roof and walls resulting from the stresses created by temperature differences throughout the country rock. The shattering was said to be accompanied by penetration of fluid magma into the fissures and dislodgment of the fragments of country rock. He also envisioned sunken blocks as having dissolved in the magma at depth rather than near the original contact. The solution of the blocks produced a contaminated or "syntectic" magma, a term that Daly borrowed from Loewinson-Lessing. The entire process Daly referred to as "abyssal assimilation," a theory of magma diversity that he espoused throughout his career. A similar stoping hypothesis was formulated independently by Barrell (1907) for the Marysville stock in Montana, and numerous geologists quickly followed suit in advocating the importance of some sort of stoping mechanism for specific plutons.

At the very beginning of the twentieth century, then, igneous geologists had at their disposal two broad classes of intrusion mechanism. On the one hand, a more active injection of magma produced laccoliths, dikes, and sills by forcing country rock upward or aside. In contrast, a less forceful process of stoping operated in the emplacement of stocks and batholiths whereby magma was envisioned as "eating" its way upward through country rock.

12 ← SPALTUNG UND KERNE
The Emergence of the Theory of Differentiation

The diversity of igneous rocks was generally not considered a problem requiring concerted research effort throughout much of the nineteenth century. Microscopic studies in the 1860s through 1880s, however, brought to light the fact that igneous rocks are far more diverse than had hitherto been recognized. The large contingent of bright young geologists who entered the new glamour field of microscopic petrography were exposed to the extreme diversity of igneous rocks and were confronted with the need to account for that diversity. The new breed of petrographers was able to address the problem with the aid of both microscopy and the developing field of physical chemistry.

KING AND DUTTON

The first speculations on diversity to incorporate the results of investigations with the polarizing microscope were those of Clarence King (Young, 1999a). Clarence Rivers King (1842–1901) was an 1862 graduate of the newly established Sheffield Scientific School of Yale University, where he studied with George Brush and James Dwight Dana (Wilkins, 1988; Goetzmann, 1993). Fresh with enthusiasm after participating in J. D. Whitney's California Geological Survey during the mid-1860s, the youthful King made the brash proposal that the Secretary of Interior of the United States, Edward Stanton, should appoint him to lead a geological and topographical survey of the proposed route for the transcontinental railroad along the Fortieth Parallel. King's proposal was supported by Stanton and the United States Congress, and, by March 1867, King was named United States Geologist in charge of the United States Geological Exploration of the Fortieth Parallel. After sailing to San Francisco, King became acquainted with Baron von Richthofen and engaged in field work in the Washoe, Nevada, region with him during the winter of 1867–1868. King and a large party of topographers, laborers, cooks, photographer, and geological assistants James D. Hague, Arnold Hague, and Samuel F. Emmons crossed California and began work along the eastern slopes of the Sierra Nevada Mountains between 1867 and 1869. King's survey was extended until 1872 after Congress appropriated additional funds.

The King party collected thousands of rock specimens and returned them to New York City, where they were stored for many years at the American Museum of Natural History. Although King lacked expertise with a petrographic microscope, he realized that his geological studies could be enhanced immensely if the microscopic techniques recently developed in Germany could be applied to the rocks. King, therefore, persuaded Ferdinand Zirkel to perform the microscopy. After meeting with King for several weeks to discuss the rocks and their geological settings, Zirkel prepared more than 2500 thin sections for study. As a result of his investigations, Zirkel (1876) issued a handsome memoir for the Fortieth Parallel Survey in which he described propylite, quartz propylite, hornblende andesite, dacite, trachyte, rhyolite, hyaline rhyolite, augite andesite, and basalt as distinct volcanic rock types, thus expanding the range of diversity that needed to be explained. Moreover, he supported von Richthofen's contention that "propylite" is a distinct lithology, claiming that petrographical differences between propylite and hornblende–andesite could no longer be doubted. Zirkel also agreed that von Richthofen's order of succession (propylite, andesite, trachyte, rhyolite, basalt) could be observed over vast stretches of Tertiary eruptive rocks exposed along the Fortieth Parallel.

King's own contribution to the series of massive survey reports was an 800-page volume. King (1878) not only described the volcanic rocks in the vicinity of the Fortieth Parallel in great detail, making the same distinctions as Zirkel, but also theorized at great length about the genesis of volcanic species. Because many of King's petrological views were powerfully shaped by von Richthofen and Zirkel, he assumed the importance of the common German view regarding the age criterion. He also went so far as to question the igneous origin of granitoid plutonic rocks, expressing himself favorably toward T. S. Hunt's views on the subject. Displaying a complete lack of awareness of the research of Allport and Judd, King (1878, pp. 705–706) wrote that

> modern microscopical research ought to have made it forever clear that a sharp line is to be drawn between the so-called Plutonic rocks and the true igneous ones . . . . The adherents of hydrothermal fusion form a class of theorists who, in my belief, fail to appreciate sufficiently the gulf of difference which separates the true igneous rocks from the group embracing granite, syenite, and diorite . . . . They ought, at this stage of microscopic research, to be fully aware that all the volcanic rocks show abundant evidence of fusion in the presence of glass base and glass inclusions, while the group which is typified by granite never shows the slightest trace of the effects of fusion . . . and . . . that every microscopical and macroscopical detail of these rocks allies them in their mode of origin to the crystalline schists which are, beyond all shadow of doubt, the results of low-temperature metamorphism of bedded sediments.

As a result of such opinions, King necessarily restricted his discussion about igneous rock diversity to volcanic rocks. He wrote that all the questions about vulcanicity resolved themselves into the one major problem of the origin of fusion. In his view, any theory of fusion would have to account for the extremely localized character of volcanism, the "non-sympathy of adjacent volcanic regions," and the chemical diversity of volcanic products.

In addressing the question of the sources of magmas, King broke ranks with Sartorius von Waltershausen and von Richthofen. He believed that Sartorius von Waltershausen's scheme of concentric shells of acidic magma passing downward into an augitic and magnetitic magma made sense in the 1850s given the knowledge of the time about the conditions of the natural succession of eruptive rocks on the surface. Because von Richthofen had established a periodic succession of five orders of volcanic rocks, a view that both he and Zirkel accepted with some modification, King maintained that the simple "law of time and depth" of Sartorius von Waltershausen was no longer tenable. Now, because of von Richthofen's work, King (1878, p. 711) judged that "the loci of eruption must have appeared earliest in the basic magma, then risen into the acidic, and lastly been depressed again into lower levels of the basic region." From his own studies on the complex volcanic stratigraphy of the region, King believed that he had further demonstrated the existence of four successive temporal orders of eruptions, each one of which would display the eruptive sequence of five orders of rock types proposed by von Richthofen. If, then, the interior of the Earth were arranged into concentric acid, intermediate, and basic shells, King (1878, p. 711) argued, the theater of eruption would have oscillated through the chemical cycle four times because "the acid and basic products alternate through the whole series of four orders." The difficulties for the Sartorius von Waltershausen and von Richthofen conception of Earth's interior posed by "the remarkable alternation of acid and basic rocks in the full series of successive eruptions" had now become "absolutely insuperable" (King, 1878, p. 711).

In contrast to the theory of multiple magma sources in a series of concentric shells, King located the source of magmas within the solid crust. He assumed Earth to be still a very hot body. He judged that as temperature rapidly increases from Earth's surface downward, a depth would eventually be reached at which the effect of increasing pressure in preventing fusion, that is, the increase of melting temperature as pressure rises, would be at a minimum so that fusion should occur at that level. Below that level, he argued, fusion would become increasingly impossible. King envisioned the existence of both isothermal and isobaric "couches" that mimicked the surface contours of the globe. Probably about forty miles below the surface, he suggested, there is a level that is actually above the temperature of fusion, yet just restrained from fusion by great pressure. That level, he said, must rise beneath the continents, particularly beneath the mountainous regions, and be depressed beneath the ocean basins.

Erosion of mountain material, however, would lead to an adjustment in the levels of the isotherms and isobars. Differences in the rates at which erosion occurred and at which the isothermal levels adjusted would determine whether fusion might occur. During erosion, King argued, the position of the level of fusion temperature would constantly recede toward the center of Earth even as it continued to mimic the surface topography. He maintained that if the rate of erosion and of the lessening of pressure exceeded the rate at which the isotherms receded toward the center of Earth, then local fusion would generate a magma lake. King judged, therefore, that rapid rates of erosion triggered melting. In support of that contention, he maintained that each episode of eruption in western America had followed a rapid period of erosion.

King speculated little about the chemical nature of the crust being fused or about the chemical composition of the magmas generated. Nor did he address in detail the question as to whether or not a variety of magmas could be generated by his fusion process. At most, he suggested that if a fused lake were formed within the acidic parts of Earth, then the resulting lake ought to have a higher proportion of acidic material. He further believed that the lakes of fusion would occur at successively greater depths as the globe gradually cooled. Consequently, he wrote that the secular changes recorded in the subtle petrographical distinctions by which the silica-rich and silica-poor rocks could be distinguished were an expression of depth.

King envisioned that a fused magma lake would eventually be divided into a lighter, silica-rich part and a heavier, silica-poor part. The acidic magma was said to come to the surface before a pyroxenic magma. King suggested two ways in which the chemical separation within the fused lake might occur. On the one hand, he proposed that the heavier constituents of the liquid might concentrate toward the bottom of the lake, leaving a residual layer of lighter liquid at the top. In short, a layer of rhyolitic magma was envisioned as developing on top of a layer of basaltic magma. On the other hand, he accepted the suggestion of Darwin (1844) that crystals that had already formed within the fused lake might separate in accordance with their specific gravities. The fact that some dense minerals do occur in rhyolite, he thought, could be attributed to the fact that the tabular nature of mica and hornblende crystals retarded their tendency to sink. King (1878, p. 716) had personally observed the effects of crystal settling in the lava streams of Kilaeua volcano in Hawaii and had seen that the bottoms of flows were "thickly crowded with triclinic feldspar and augites" whereas the tops were composed of "nearly pure isotropic and acid glass." King (1878, p. 718) concluded that the formation of fusion lakes beneath loci of maximum erosion, the crystallization of minerals within the melted magma, and their separation by specific gravity were sufficient to account for the "periodic succession of volcanic rocks, with their astonishing time-oscillations between the acidic and the basic members."

Similar issues were also addressed by King's American contemporary, Clarence Dutton. Clarence Edward Dutton (1841–1912), an 1860 graduate of Yale, was studying at Yale Theological Seminary when the Civil War broke out (Stegner, 1936, 1971; Longwell, 1958). His decision to enter the war led to a military career. After war service, he was placed in charge of various military arsenals. During his tenure as commander of the Washington arsenal, he made the acquaintance of some of Washington's leading scientists such as Joseph Henry, Ferdinand Hayden, and John Wesley Powell. Dutton's interest in geology was sparked by these contacts, and Powell became sufficiently impressed with Dutton's geological abilities that he persuaded him, like Gilbert, to participate in the United States Geographical and Geological Survey of the Rocky Mountain Region during the late 1870s. Dutton later worked in the newly established United States Geological Survey from 1879 to 1890 and returned to the regular army after 1890.

During his western survey work, Dutton demonstrated a special flair for problems in tectonics, geophysics, and geomorphology. For several years, he wrestled particularly with the general problem of isostasy, a term that he coined as early as 1882, and the complex interrelationships among erosion, sedimentation, uplift, and subsidence. As part of his work with the Survey of the Rocky Mountain Region, Dutton (1880) issued a report on the geology of the high plateaus of Utah. Although he described volcanic rocks and terranes, he admitted that he was loath to engage in speculation regarding the origin of the numerous volcanoes and lavas of the region. Nonetheless, he indulged in such speculation simply because he had been encouraged to do so by John Wesley Powell, his leader on the survey. Dutton (1880, p. 113) confessed that "the origin of volcanic energy is one of the blankest mysteries of science," and he found it odd that phenomena both familiar to the human race for such a long time and so zealously studied were so "utterly without explanation." He stressed that volcanism is a local phenomenon and conceived of each volcano as independently "engaged in its own peculiar business, cooking as it were its special dish, which in due time is to be separately served" (Dutton, 1880, p. 115).

Dutton set himself squarely against the magma-shell view of Sartorius von Waltershausen and Durocher. He considered his field observations that closely spaced volcanic vents erupted totally different kinds of lava and that very different kinds of lava were emitted from the same vent over time to be inconsistent with the idea that lavas represent the "primordial, uncongealed earth-liquid" (Dutton, 1880, p. 116). Rather, he concluded, these facts pointed to the existence of small, relatively shallow reservoirs, each of which contains liquid generated by fusion of the surrounding rocks, a process that occurs within the layered crust. On this point, Dutton sided with King.

Dutton argued that most igneous rocks have the petrographic characteristics to be expected from the fusion of gneisses and schists. "The greater part of the true gneissic rocks," Dutton (1880, p. 118) asserted, "yield by analysis practically the same results as granite, syenite, rhyolite, and acid trachyte." He pointed out that gneisses share abundant feldspar, mica, and hornblende in common with igneous rocks and that textural similarities also exist. Moreover, he argued, there are rock series that present every shade of transition between a wholly crystalline and a wholly amorphous texture.

He did recognize that some metamorphic rocks, such as quartzite, limestone, dolomite, and argillite, have little in common with eruptive rocks. It is, he said, the most refractory metamorphic or sedimentary strata that have no correlatives among the eruptive rocks, and the reason that there are no lava flows having the composition of these materials is that these metamorphic rocks are infusible at the highest volcanic temperatures that are achieved. Of course, he continued, not all eruptive rocks were derived from the fusion of metamorphic rocks because there must, from time to time, also be an input of some primordial lava that can be weathered into sediments and ultimately converted into the metamorphic rocks. Because basalt has been an extremely abundant lava throughout geologic time, and because the weathering of basalt would ultimately yield the materials of sedimentary rocks, Dutton maintained that basalt was that primordial lava. Moreover, he suggested, Earth originally had a homogeneous, primordial basaltic crust that became heterogeneous throughout geological time by weathering and erosion into sediments.

Dutton knew of no facts to justify King's (and Lyell's) idea that a magma would segregate into two or more magmas while in a state of fusion. In addition, he doubted the view of Darwin and King that gravitational separation of crystals would result in a single magma separating into two or more magmas with very different degrees of acidity. He argued that the high silica percentage in a rhyolite, for example, is due not only to the high feldspar content but also to a glassy groundmass that is enriched in silica. In his judgment, crystal settling alone was insufficient to account for the chemical characters of magmas.

Dutton believed that the fusion of crustal metamorphic rocks would be engendered by local increase of temperature in specific horizons, but he confessed that his own efforts to find the reasons for the local temperature increase had been utter failures. He disputed King's contention that relief of pressure brought about by denudation of superincumbent strata was a major cause of melting, for he did not believe that eruptions always occur in localities that had experienced severe erosion. Most existing and recently extinct volcanoes, he said, are situated in regions that have undergone very little denudation in recent geological periods. In fact, he stated that many volcanoes occur in regions of recent deposition.

He also rejected the notion that the mineral melting point depression effected by the presence of water under high pressure is an adequate cause of melting. The amount of water emitted by volcanoes, he maintained, exceeds the water content of the metamorphic rocks from which the lavas were presumed to be melted.

The sequence of erupted volcanic rock types that had been worked out by von Richthofen and modified by King was explained by Dutton in terms of a progressive fusion of the source region from which the liquids were derived. Acid rocks, he and most of his contemporaries incorrectly believed, have the highest melting temperatures, but they correctly understood that basic rocks have the highest specific gravity. As a result, Dutton suggested that, although acid magmas might be of sufficiently low density to erupt at an early stage of the melting process, they would not yet have formed because the temperature was too low. In contrast, basic rock might be melted before an acid rock, but the basic melt was still too dense to reach the surface. He believed, therefore, that the first erupted lava would be of intermediate composition. Dutton drew a melting diagram displaying two intersecting curves, one of which represented the temperature required to render the rocks light enough to rise hydrostatically to the surface and the other of which represented the temperature required to fuse the rocks. From these curves he reasoned that propylite would be in an eruptible condition at a relatively low temperature. With further increase in temperature, both hornblendic andesite and trachyte would become eruptible as the former had passed the fusion point and the latter had passed the density point of eruption. In general, Dutton explained, as the temperature increases, the line of eruptive temperature cuts the two curves at points farther and farther from the lowest point of eruptivity, and these points correspond to rocks which become more and more divergent in their degrees of acidity. One set progresses to the acid extreme, the other to the basic extreme. Dutton suggested that variation on this scheme could be brought about by differences in the depth at which fusion occurred and by differences in the mean density of rocks overlying the locus of fusion.

Although still adhering to von Richthofen's basic ideas about the sequences of eruption of different rock types, Dutton, like King, recognized the shortcomings of the more universal theories of igneous rock diversity of Sartorius von Waltershausen and Durocher. He contented himself with an explanation for the origin of the diversity of local rock suites in terms of local fusion and left to his readers' imaginations the extent to which his proposed mechanism might be applied to volcanic terranes around the world.

THE RISE OF THE THEORY OF DIFFERENTIATION

Before 1880, most ideas about the causes of igneous rock diversity, apart from King's, were proposed without the benefit of microscopic petrography. These early ideas were also formulated without the benefit of a substantial number of chemical analyses of rocks or any of the results of the fledgling science of physical chemistry. The ideas largely preceded the era of specialization in petrology and were conceived primarily by geological generalists like Durocher or Dutton, naturalists like Darwin, and chemists like Bunsen. Early theories of diversity had typically been restricted to volcanic rocks.

Thanks to the wealth of data and the conceptual advances brought about by the microscopic revolution, however, the situation drastically changed throughout the 1880s and 1890s. Because granite and related coarse-grained crystalline rocks were increasingly regarded as magmatic, hypotheses on the origin of diversity began to incorporate plutonic as well as volcanic rocks. Inasmuch as fewer and fewer geologists regarded geological age as relevant to igneous rock classification, theories of multiple magma sources had largely been abandoned. And on account of the host of detailed field studies of igneous rock terranes accompanied by precise petrographic data and a growing volume of chemical analyses, petrologists came to accept the concepts of the petrographic province and of consanguinity. By the 1890s, it was becoming clearer that the diversity of igneous rocks needed to be explained not simply as a generality on a global scale. Rather, it was necessary to account for the diversity of a group of consanguineous, that is, genetically related, igneous rocks produced in the same episode of magmatic activity and occurring within the same petrographic province. The problem of diversity became much more localized. At the same time, the problem of diversity had to be faced repeatedly in numerous petrographic provinces around the globe (Young, 1999b).

Diversity theory was also the beneficiary of several new insights and concepts from the newly developing field of physical chemistry (Servos, 1990). Geologists eagerly attempted to invoke physical mechanisms, such as diffusion, analogous to those observed in aqueous solutions, to account for the diversity within petrographic provinces. Lastly, it must be noticed that, in contrast to the early days of speculating about diversity, those who theorized about diversity were virtually all specialists in petrology and petrography who were reasonably knowledgeable about chemistry (Young, 1999b).

J. P. Iddings (1892b, p. 115) called attention to this theoretical revolution by observing that

> since 1880 the science of petrology has been developing along new lines, and its followers have been busy establishing in great detail the exact mineralogical character of all crystalline rocks, as well as advancing their chemical investigation. A more thorough exploration of their geological occurrence and the finding of more favorable localities for the study of their field relationships have led to a clearer understanding of their true nature and have paved the way for an advance in the theory of their origin.

Four years later, Alfred Harker noted that recent opinion about igneous petrology had revolved about two central ideas, one of which concerned "the production under proper conditions of various rock types from one original rock-magma," an indication of the extent to which a theory of differentiation had begun to occupy center stage as the dominant explanation for igneous rock diversity (Harker, 1896b, p. 14).

Although Charles Darwin in 1844 and Clarence King in 1878 had suggested that crystal settling within a single lava flow produced diverse material, the idea of differentiation emerged in more explicit form during the 1880s within the context of discussions about the nature of magma as a complex silicate solution whose original chemical composition might be modified during cooling. Geologists no longer accepted Scrope's idea that lava is not a true melt, and they had abandoned the notion that individual minerals in some sense retain their integrity in magma as particles having the same chemical composition as the crystalline substances. In other words, granite magma was no longer considered to consist of a mixture of quartz, feldspar, and mica particles or "molecules." Thanks to the ongoing work on aqueous solutions by chemists and to Bunsen's claim that magma is a very hot silicate solution in which the substances of the constituent minerals are somehow dissolved, geologists increasingly observed that magmas possess some of the characteristic properties of solutions. Diversity was viewed as the result of varying the composition of a hot silicate solution by some as yet unverified mechanism or mechanisms.

The earliest notions of differentiation were suggested by Rosenbusch, Lagorio, and Teall. From studies of thin sections in the early 1880s, Rosenbusch worked out the general sequence in which minerals crystallize from cooling magma and proposed that magmas become decreasingly basic during crystallization. In a paper on the chemical relations of rocks, Rosenbusch (1889) reviewed the various proposals of Bunsen and Durocher for generating igneous rock diversity. Although subsidiary to his thesis of two magma shells as the major source of diversity, Durocher's conception of liquation made a strong impression on Rosenbusch. Against the background of Durocher's liquation theory, Rosenbusch put forward the suggestion, as an extension of Durocher's concept, that a chemically homogeneous general magma might undergo a spontaneous "Spaltung" (splitting) or decomposition into two or more chemically different partial magmas. Both Durocher and Rosenbusch were envisioning what today would be called "liquid immiscibility."

Rosenbusch (1889) postulated the existence of a few fundamental chemical constituents in magma that he designated as "Kerne" (nuclei, cores, or kernels). He suggested that there are five major magma types that may be characterized by the presence or dominance of certain molecular groupings or "Kerne." For example, he conceived of the granitic magma type as dominated by a "Kern" having a composition $(NaK)AlSi_2$ as well as free silica with lesser amounts of $CaAl_2Si_4$ and very minor amounts of R_2Si and RSi, where R represented such metals as iron and magnesium. In contrast, the gabbro magma type was said to be dominated by the molecular groups R_2Si, RSi, and $CaAl_2Si_4$. Because he understood these two magma types to contain various "Kerne" in differing proportions, Rosenbusch imagined that under appropriate, but unspecified, conditions, these magmas were capable of further differentiation by spontaneous splitting. This splitting would occur under conditions where the mutual affinities of "Kerne" in a magma for one another had greatly diminished. On the other hand, the foyaite magma type was said to be overwhelmingly dominated by the "Kern" $(NaK)AlSi_2$ with negligible amounts of other groupings, and the peridotite magma type was said to be overwhelmingly dominated by the "Kerne" R_2Si and RSi with negligible amounts of the other groupings. As such, Rosenbusch referred to these two magma types as "pure" magmas, and he regarded them as essentially incapable of undergoing any further differentiation. These magmas, he said, would be unlikely to split spontaneously. Although the idea of splitting or differentiation would gain adherents through the next decade, petrologists were not yet ready to follow Rosenbusch in specifying the composition of the chemical constituents within a magma.

Alexander Lagorio, a Professor of Geology at the Mineralogical Institute of the University of Warsaw, also regarded magma as a complex solution of silicates. Lagorio (1887) compared the chemical analyses of a wide range of volcanic rocks to the analyses of their glassy residua. He interpreted the glass as the chilled liquid remaining after crystallization of early-formed minerals, implicitly suggesting the idea of differentiation. He also worked out a generalized order of separation of minerals from melt but disputed Rosenbusch's idea that magma becomes decreasingly basic during differentiation. Lagorio had discovered that the residual glass in acid and basic rocks commonly has about the same silica content as the whole rock.

In *British Petrography*, J.J.H. Teall (1888) noted that the magmatic origin of volcanic rocks had been confirmed by Fouqué and Michel-Lévy (1882) who had experimentally reproduced the mineralogy and texture of basalt, andesite, and tephrite by igneous fusion. Teall also applied the results of physical chemistry to specific instances of igneous crystallization such as micropegmatite, which he regarded as the product of eutectic crystallization in which a pair of minerals crystallize from a residual melt in fixed proportions at a specific temperature. He also applied solution theory to the question of the order of crystallization of minerals from magma. Teall agreed that Lagorio's studies of glass supported the idea that magmas are solutions and reasoned that intermediate magmas might become "differentiated" into basic and acid magmas.

DIFFERENTIATION IN THE CONTEXT OF FIELD STUDIES

Beginning in 1890, the concept of differentiation was invoked to explain specific suites of diverse igneous rocks that had been the object of detailed field, chemical, and petrographic studies by Brögger and Vogt in Scandinavia, Iddings in the United States, and Harker, Dakyns, and Teall in Great Britain. Waldemar Christopher Brögger (1851–1940) early served as an assistant at the

Mineralogical Institute in Christiania (Oslo) under Theodor Kjerulf and with the Geological Survey of Norway (Andersen, 1941; Tilley, 1941). In 1881, he moved to Sweden and founded a Mineralogical Institute at the Stockholms Høgskola, but in 1890, he returned to his native land to succeed Kjerulf at Christiania. Here he continued his career as Professor of Mineralogy and Geology until retirement in 1916. A man of spectacular talents and energy, Brögger served between 1906 and 1909 as a member of the Norwegian Parliament and between 1906 and 1911 as Rector of the University of Christiania. As a geologist, he established a worldwide reputation with his massive studies on the alkalic igneous rocks of the Christiania area. He also conducted major studies on pre-Cambrian geology, Cambrian and Silurian stratigraphy and paleontology, and Quaternary stratigraphy, tectonics, and paleontology. He was so gifted that he was acknowledged, upon receipt of the Wollaston medal, as a specialist in almost every field of geological science. Nevertheless, Brögger was first and foremost an igneous petrologist, equally at home in field mapping, mineralogy, chemistry, and theory. It was Brögger who gave to petrology the terms "leucocratic" and "melanocratic" as well as 65 new rock names (Le Maitre, 1989).

In the first part of a two-part monograph on syenitic pegmatites in southern Norway, Brögger (1890) described a region of flows, dikes, and massive intrusions with a wide range of mineralogical and chemical compositions and a discernible sequence of emplacement. He demonstrated that the earliest rocks are silica-poor and that the rocks progressively become more enriched in silica. A few years later, Brögger (1894a) issued the first of a series of monographs on the geology of the Christiania region. He described in detail the petrology of the differentiated alkalic grorudite–sölvsbergite–tinguaite series. In the same year, Brögger (1894b) also published a preliminary notice on some basic igneous rocks of Gran in that region. He described a succession of olivine–gabbro–diabase intrusions that are decreasingly basic from north to south. These massive rocks, he found, are cut by a host of camptonite and bostonite dikes. From chemical analyses, Brögger calculated that a composite of average camptonite and average bostonite has virtually the same composition as the average composition of the olivine–gabbro–diabase. He concluded, therefore, that the dike magmas had differentiated from an original olivine gabbro–diabase magma.

Brögger's colleague, Johan Hermann Lie Vogt (1858–1932), undertook graduate study at Stockholm under Brögger and became Professor of Metallurgy at the University of Christiania in 1886 (Oftedahl, 1976). He joined the Technical University of Norway in Trondheim as Professor of Geology, Ore Deposits, and Metallurgy in 1912. After publishing papers on the mineralogy of slags, Vogt (1891) reported on a number of Scandinavian iron–titanium ore deposits associated with basic igneous rocks. He described, for example, a thick dike that grades from mica syenite porphyry in its center to magnetite–biotite–rich kersantite at its margins. In the Ekersund-Soggendal district, he described dikes of ilmenite, hypersthene, and labradorite traversing massive norite. He also described ore deposits in the Taberg district of Sweden that grade outward into a mass of olivine hyperite. Vogt concluded that all the ores resulted from concentration of basic magma into segregations.

In America, the research of J. P. Iddings resulted in several papers on igneous rocks in Yellowstone National Park, including a detailed survey report on the geology of Electric Peak and Sepulchre Mountain (Iddings, 1888a, 1891, 1892a). Iddings stressed the connection between the mineral composition and geological occurrence of the igneous rocks. He reported that Electric Peak consists of sheets of "porphyrite" and dikes and stocks of diorite. The earlier sheets were intruded as a succession that grade into one another mineralogically. The phenocrysts in the porphyrites change from pyroxene to hornblende to biotite to quartz in successively younger sheets. The feldspars become less basic and more abundant, and are associated with increasing amounts of groundmass quartz in the younger sheets. Iddings also found that the proportion of ferromagnesian silicates decreases in the younger sheets. He encountered similar variations in the dioritic dikes. In the massive diorite stock, too, pyroxenes, hornblende, and labradorite were seen to be early-crystallizing minerals whereas biotite, alkali feldspar, and quartz crystallized later. Iddings concluded that the Electric Peak magmas had become more and more siliceous in time.

At Sepulchre Mountain, Iddings encountered similar variations in a succession of andesitic flows and breccias. He concluded that the rock series at Electric Peak and Sepulchre Mountain had commenced with basic magmas and terminated with acidic magmas. The entire series of rocks at each location was viewed as "a single, irregularly interrupted succession of outbursts of magma that gradually changed its composition and character" (Iddings, 1892a, p. 202).

To compare suites of rocks, Iddings compiled chemical analyses and introduced "molecular variation diagrams" on which he plotted molecular proportions of the oxides against the molecular proportions of silica (Figure 14). Both series display decreases in CaO, MgO, and FeO and increases in Na_2O and K_2O as the molecular proportions of SiO_2 increase. Iddings concluded that the intrusive rocks of Electric Peak and the volcanic rocks of Sepulchre Mountain are chemically identical. The different conditions attending the final consolidation of the magmas, he urged, affected their structure and mineralogy. Iddings argued that the mineralogical and chemical consanguinity of these suites of rocks had resulted from chemical differentiation of an intermediate magma in which alkalis typically increased and magnesia, lime, and iron oxide decreased with increase in silica content.

In Great Britain, Teall contributed a landmark field study in which he was joined by John Roche Dakyns (1836–1910), a geologist with the British Geological Survey (Woodward, 1910). Dakyns and Teall (1892) investigated a suite of ultrabasic to acid plutonic rocks that is exposed near the northwest end of Loch Lomond in Scotland. They found that the oldest rocks are peridotites,

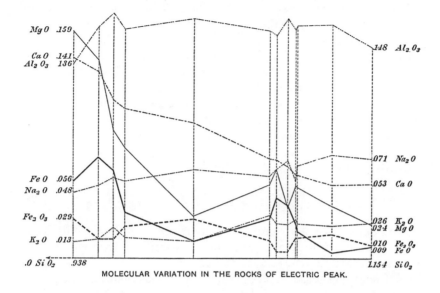

14. Variation diagram, first employed by Iddings (1892), showing molecular variation of oxides in the rocks of Electric Peak, Yellowstone National Park.

followed by epidiorites, tonalites, and granites in order of increasing acidity. They observed that as the rocks become increasingly more siliceous, olivine diminishes and pyroxene increases. Olivine and pyroxene are replaced by hornblende and biotite. Hornblende decreases relative to biotite, and in siliceous veins, the ferromagnesian silicate minerals disappear almost entirely. Dakyns and Teall further observed that plagioclase appears early, to be followed by orthoclase and quartz, and lastly by microcline. They noted that pyroxene, hornblende, biotite, plagioclase, quartz, and microcline typically make their first appearance in the groundmass and only where they become important constituents do they show traces of crystal outlines. The authors also worked out the order of crystallization of minerals within individual thin sections and found that the order strikingly resembles the general distribution of the minerals in the different kinds of rock. Dakyns and Teall also plotted chemical variation diagrams and pointed out that Ca, Fe, and Mg decrease as K, Na, and Si increase. The authors speculated that progressive consolidation of magma yielded increasingly acid liquid.

In a study of composite dikes on the Island of Arran, Judd (1893) concluded that differentiation had occurred in a deep-seated reservoir that fed contrasted magmas into the dikes. Then another British field study supporting the idea of differentiation was undertaken by Alfred Harker (1894b), who published a detailed study of a gabbro intrusion at Carrock Fell in the English Lake district, demonstrating that the abundance of iron ore minerals and the specific gravity of the rocks increases from center to margins. These variations, he said, were related to the boundary of the intrusion and resulted from differentiation that concentrated basic constituents adjacent to the margins. In a sequel paper, Harker (1895a) presented his judgment on the origin of granophyre associated with the gabbro. With a variation diagram, he illustrated how granophyre and gabbro might have differentiated from a common source.

Unlike the speculations about igneous rock diversity earlier in the nineteenth century, these initial proposals advocating differentiation were applied primarily to suites of plutonic rather than volcanic rocks. It became apparent, too, that the differentiation hypothesis could be applied to igneous rock suites occurring in widely separated localities.

MECHANISMS OF DIFFERENTIATION—THE SORET EFFECT

Almost immediately, early proponents of differentiation appealed to the new discipline of physical chemistry as a source of plausible mechanisms to account for the phenomena of diversity in the igneous rocks. As soon as the concept of differentiation began to take hold, the Soret effect quickly emerged as a favored hypothesis.

Charles Soret (1854–1904) was successively Professor of Mineralogy, Professor of Physics, and Rector in the University of Geneva (Anonymous, 1904). He specialized in the optical and thermal properties of crystals and wrote a textbook on physical crystallography, but Soret's contribution to physical chemistry and, ultimately, petrology is found in two papers describing variations in the concentration of dissolved salts along temperature gradients in aqueous solutions. Building on an earlier brief study by Ludwig (1856), Soret (1879, 1881) filled separate glass tubes with solutions of potassium nitrate, potassium chloride, lithium chloride, or sodium chloride and allowed the solutions to come to equilibrium. He placed one end of a solution-filled tube into a hot bath held around 78°C and the other end into a cold bath held between 15°C and 18°C. The solutions were maintained at different temperatures from 9 to 56 days. At the conclusion of each experiment, Soret found that the concentration of the dissolved salt was greater at the cold end of the tube than at the hot end.

The theoretical basis of the diffusion of salts toward the colder region of an aqueous solution was laid by van't Hoff (1887) in terms of the osmotic principle. Van't Hoff calculated that equilibrium concentration of a dissolved salt in dilute solution ought to be inversely proportional to the absolute temperature of the solution.

Early advocates of the Soret effect as a mechanism of differentiation included Lagorio, Teall, Brögger, Vogt, Iddings, and Judd. Lagorio (1887), although not quite ready to abandon the multiple source theory of Sartorius von Waltershausen, saw a role for the Soret action. He reasoned that chemical differences would arise during slow cooling of low-viscosity magma in which the colder, peripheral parts of a dike would become enriched in dissolved salts and the inner part enriched in the solvent. If, however, a magma were viscous or cooled very rapidly, Lagorio suggested that the movement of matter would be hindered and chemical differences would not result.

In *British Petrography*, Teall (1888) proposed that differentiation of intermediate magmas into basic and acid magmas might occur by diffusion of magmatic constituents during cooling in accord with Soret's principle. Four years later, Dakyns and Teall (1892) supposed that the Garabal Hill area represented a vast subterranean reservoir that differentiated during consolidation. The development of early-formed minerals would render the surrounding magma deficient in certain molecular groups, they suggested, but diffusion would tend to restore homogeneity so that the first-formed crystals continued to grow. They appealed to the Soret effect to produce heterogeneity in the original magma.

From his investigations of syenitic pegmatites, Brögger (1890) concluded that differentiation of Na-rich magma had occurred by means of the Soret effect coupled with crystallization. Heavier, dissolved, silica-poor components were said to become enriched at the cooler, outer surface of the reservoir and were the first to be erupted. Succeeding eruptions, he said, became more siliceous until, finally, early crystallized ferromagnesian minerals that had sunk to the bottom of the reservoir were remelted and ready for eruption. Brögger (1894b, 1898) later began to hedge on outright espousal of the Soret effect. He did claim that the basic, camptonite magma at Gran was separated out by diffusion and that the remaining acid magma furnished the material for the bostonite dikes. The differentiation, he suggested, came about before significant crystallization had occurred as the least soluble dissociated and undissociated compounds diffused toward and concentrated at the cooling margin. The matter that diffused to the margins would remain in the liquid state for awhile so that, if disturbances occurred, magma from the walls or from the interior could be squeezed upward to form dikes. In his conclusion, however, Brögger (1894b, p. 37) hesitated discussing "the primary reason for a differentiation of the above-described nature, whether this is to be sought for in Soret's principle, in the effect of chemical affinity, or in other causes."

Because the kersantite margins of the mica syenite porphyry dike studied by Vogt (1891) in Norway were said to represent magma cooled by the escape of heat into the surroundings, he attributed the marginal concentration of basic minerals to the Soret effect. Iddings (1892b) had a very different conception of the internal structure of magma from that of Rosenbusch. Both men regarded magmas as solutions, but rather than adopt the "Kerne" of Rosenbusch such as $(NaK)AlSi_2$ or $CaAl_2Si_4$, Iddings envisioned magma to be a very flexible solution of simple oxide molecules in which no definite compounds or mineral molecules play the part of solvent. He appealed primarily to the Soret effect to account for differentiation in two different directions from an intermediate parent. "The molecular concentration of particular constituents of molten magmas in the cooler parts of inclosed bodies of magma," Iddings (1892b, p. 160) claimed, "is a sufficient cause for their differentiation." Although Iddings suggested that his hypothesis was highly probable, he conceded that experimental facts were not yet available to establish it.

Judd (1893) argued that the deep-seated differentiation producing the magmas feeding the composite dikes of Arran must have occurred before the work of crystallization commenced. Although not ascribing sole potency to the Soret effect, he did acknowledge it as one well-recognized physical principle in accordance with which differentiation would be established in a molten silicate solution.

13 ⟡ PHYSICAL CHEMISTRY AND PETROLOGY

The Mechanism of Differentiation

The assistance afforded to igneous petrologists by physical chemistry almost turned out to be more than they had bargained for. Far from solving the problem of the cause of differentiation, the discipline proved instead to be such a fertile spawning ground for hypothetical mechanisms that igneous petrology was cast into a state of considerable confusion for the next few decades. Despite the early popularity of the Soret effect, petrologists soon found themselves toying with the possibility that such diverse phenomena as liquid immiscibility, gravitational segregation, magnetic attraction, electrical currents, and fractional crystallization might account for the differentiation of igneous rocks.

REJECTION OF THE SORET EFFECT

The Soret effect was the most frequently invoked mechanism of differentiation until 1893, when Bäckström issued a paper attacking it. He was soon joined by Harker, Johnston-Lavis, Michel-Lévy, Becker, Brögger, Loewinson-Lessing, Schweig, and even Teall. Helge Mattias Bäckström (1865–1932) became an assistant at the University of Christiania in 1891, and in 1908 he was appointed Professor of Mineralogy and Petrology at the University of Stockholm (Spencer, 1933). Acknowledging the growing support for differentiation, Bäckström (1893) agreed with many petrologists that differentiation most likely occurred while magma was still fluid. Although noting that several geologists had invoked the Soret effect, Bäckström pointed out that the laws of solutions to which petrologists had appealed applied only to very dilute solutions. Little was known about concentrated solutions like magmas, he observed, particularly in regard to the Soret principle. He insisted that proper application of that principle to petrography required accurate knowledge of both solvent and solute. He pointed out that Vogt, one of the proponents of the Soret effect, had considered that minerals crystallizing during cooling had formerly been dissolved in a magmatic solvent. Bäckström (1893, p. 775) countered that the solvent generally crystallizes first during cooling of a solution until eutectic proportions are attained and argued that on Vogt's reasoning "the more a dilute solution of nitre is diluted with water, so much the more should the water be regarded as the substance dissolved."

In view of the widespread existence of basalt and subordinate rhyolite on Iceland, Bäckström found it incomprehensible why diffusion-driven differentiation never produced an intermediate magma. The Soret theory, he believed, required that rhyolite magma, prior to eruption, would be surrounded by a broad zone displaying all transitions to basalt. Such intermediate magmas should have been erupted at some time, Bäckström (1893, p. 778) argued, but, although rhyolite is "exceedingly subordinate" to basalt, "we know a hundred eruptions of rhyolite but not a single one of andesitic rocks." He preferred the efficacy of liquid immiscibility.

Alfred Harker (1893, 1894b) disputed the Soret effect on different grounds. He concurred with other geologists in the concept of molecular migration of basic constituents toward cool wall rocks but sought to show that Soret's principle is an inadequate explanation for that migration. Noting van't Hoff's demonstration that the concentration of a salt in different parts of a solution is inversely proportional to absolute temperature, Harker reported that the gabbro intrusion at Carrock Fell contains 25 times as much magnetite at the margins as in the center. He seriously doubted that the absolute temperature of the center of the intrusion had been 25 times greater than that at the margins. In addition, he questioned the application of Soret's principle to a nearly saturated solution on the verge of crystallization, reasoning that if differentiation occurred before crystallization began then different varieties of feldspar would have formed in different parts of the Carrock Fell gabbro. Instead, he found, only the amount of feldspar differs. Determining that differentiation must have occurred by diffusion in an essentially saturated magma as the early minerals crystallized, Harker appealed to Berthelot's principle of maximum work to account for the concentration of basic minerals at the margins of the intrusion. He restricted the Soret action to the very earliest stages of crystallization.

A criticism specifically directed at Brögger's advocacy of the Soret effect was advanced by Henry James Johnston-Lavis (1856–1914), a medical doctor who devoted his spare time to geological investigation (Anonymous, 1914). In 1879, Johnston-Lavis moved to Naples to establish a medical practice. Fascinated by Mount Vesuvius, however, he mapped the volcano in detail and published several papers on the volcanoes of southern Italy. In 1893, he was appointed Professor of Volcanology at the Royal University of Naples, but the following year he moved to France, where he spent most of the remainder of his life. Johnston-Lavis (1894) chided Brögger for giving insufficient attention to the evidence of contact metamorphism in the Gran area. No compelling case for differentiation could be made, he maintained, until the extent of migration of constituents from magma into country rock was known. He proposed that there had been an osmotic exchange of material in which silica, alkalis, and alumina had been lost from the magma and basic constituents added, thereby rendering the marginal zone more basic. The evidence, therefore, did not favor Brögger's theory of chemical diffusion.

Not everyone was ready to abandon the Soret principle just yet. In a paper on the Highwood Mountains in Montana, Weed and Pirsson (1895) discussed at length the differentiated Square Butte laccolith. They argued that the distribution of dark, basic shonkinite toward the margins of the intrusion and forming a ring around a central core of more acidic syenitic rock had been produced by differentiation in the liquid state. They claimed that the evidence for differentiation by diffusion of iron and magnesium toward the walls was compelling. Although appreciating Bäckström's espousal of liquid immiscibility, they were unable to accept his critique of Soret diffusion. They wondered if the two mechanisms might not have worked in conjunction. Weed and Pirsson did not address Harker's critique of the Soret effect.

The assault on Soret diffusion, however, resumed when Michel-Lévy (1897) criticized the proposed mechanism in a paper on the classification of igneous rocks. He suggested that volcanic eruption sequences began with acid magma and concluded with basic magma. Such sequences indicated to him that magma chambers consisted of acid magma above and basic magma on the bottom, an arrangement that contradicted the Soret effect whereby, he said, basic magma should concentrate along the cooling top and sides.

The most devastating critique of the Soret effect was leveled in a paper on differentiation by G. F. Becker. Becker (1897a) incorporated the most rigorous attempt yet to apply the principles of physical chemistry to the hypothesis of igneous differentiation. He dismissed the Soret effect on the grounds of the extreme slowness of molecular diffusion. From data on diffusivities of copper sulfate in aqueous solution, Becker calculated the distances to which copper sulfate would diffuse from its source over periods of one year and 1000 years. He calculated, for example, that a concentration of 0.5 percent of that at the source would be attained at a distance of 11.08 meters in 1000 years. Becker suggested that the viscosity of lava would be at least fifty times as great as that of water so that diffusivity through lava would be correspondingly small. Thus, he estimated that in 106 years, "sensible impregnation of the lava" by a diffusing substance would extend only about 49 meters from its source. But heat, he calculated, would diffuse far more rapidly than would dissolved substance. Becker (1897a, p. 31) concluded that

> diffusion of matter in the lava, therefore, takes over 200,000 times as long as the diffusion of heat in solid rock at ordinary temperatures. Even if melted lava conducts heat many thousand times worse than solid rock, so that the conductivity of the fluid

might be neglected, the temperature in an unequally heated mass of melted lava would be sensibly equalized by the conduction of the solid walls of the reservoir before any tendency to molecular flow which difference of temperature might have induced would have had time to produce sensible effects.

Becker concluded that one cubic kilometer of lava would not differentiate by "molecular flow" even if the mass had remained molten since the end of the Archean. And if the bottom of such a mass were hotter than its top, convection would unavoidably prevent segregation by the Soret effect.

For the Seventh International Geological Congress in St. Petersburg in 1897, Russian petrologist F. Y. Loewinson-Lessing presented a monograph on igneous petrology that included nearly one hundred pages devoted to the problems of diversity. Franz Yulevich Loewinson-Lessing (1861–1939) was Professor of Mineralogy at the University of Jurjew (Dorpat) during the 1890s (Meniailov, 1973). In 1902, he became Professor of Mineralogy and Geology at the Polytechnic Institute in St. Petersburg, a position he occupied until 1930. Loewinson-Lessing also became the director of the Petrographical Institute in Moscow, a laboratory that he helped to found. Like Bäckström and Harker, Loewinson-Lessing (1899a) repeated that the Soret effect was applicable only to dilute solutions. He also cautioned that Soret had observed the diffusion of entire salts, not individual ions, in his original experiments. But diffusion in magma, which Loewinson-Lessing regarded as an electrolyte, would likely entail migration of dissociated ions. Comparisons with the Soret mechanisms were, therefore, not likely to be warranted. Moreover, Loewinson-Lessing suggested that, in some cases, MgO and FeO acted more like a solvent than the solutes that were alleged to diffuse.

Former advocates of the Soret hypothesis began to beat a hasty retreat. In the third volume of his series on the eruptive rocks of the Christiania region, Brögger (1898) presented a thorough review of the status of the differentiation question. He questioned Becker's assumption that magma would very severely impede molecular diffusion. He conceded that lavas were sufficiently viscous to retard differentiation phenomena but argued that deep-seated magmas would be far less viscous than surface lavas through the effects of dissolved water and high pressure. Nevertheless, Brögger acknowledged that Bäckström, Harker, and Becker had made a sound case against the efficacy of the Soret mechanism.

Teall and Vogt, both of whom had previously advocated the Soret mechanism, also appreciated the force of the accumulating criticisms. At the turn of the century, Teall was president of the Geological Society of London. His presidential address of 1901 dealt with the evolution of petrological ideas since Hutton. In reviewing differentiation mechanisms, Teall (1901, p. lxxxi) downplayed Soret's principle, conceding that it "will, I fear, help us very little." Unless, he asserted, an effective answer to Becker's arguments could be found,

PHYSICAL CHEMISTRY AND PETROLOGY 219

unaided molecular flow would have to be abandoned as an important factor in the origin of petrographical species. In his monumental two-volume work, *Die Silikatschmelzlösungen*, Vogt (1903, 1904) observed that the distribution of some iron ore segregations implied that diffusion had been toward the central part of a magma body rather than toward the cool margins. He acknowledged that the Soret effect would not yield quantitatively signficant results.

Another major review of the problems of magmatic diversity was published by Martin Schweig of the Geological and Mineralogical Institute at the University of Jena. Schweig (1903) believed that studies of differentiation could benefit from the evidence supplied by glass technology and concluded from his experimental studies that the viscosity of silicate melts is far too great for Soret diffusion to be effective.

ALTERNATIVE DIFFERENTIATION MECHANISMS

Inasmuch as scientists abhor a theoretical vacuum, the demise of the Soret effect was accompanied by the espousal of several alternative differentiation mechanisms, some of which had been previously suggested in conjunction with the Soret action. Many of the alternatives, like the Soret effect, were explicitly borrowed from recent studies in physical chemistry. Despite the repudiation of the Soret effect, many petrologists still envisioned differentiation as primarily a liquid-state process occurring in magma prior to the onset of significant crystallization. These petrologists appealed to the Gouy–Chaperon effect, diffusion along composition gradients in relation to Berthelot's principle of maximum work, or liquid immiscibility.

The French chemists L. G. Gouy and C. Chaperon tried to establish that the density of a standing solution will increase from the top to the bottom as a consequence of gravity. Although Gouy and Chaperon (1887) noted that the effect was too small to be detected experimentally, they derived theoretical equations by which the concentrations of dissolved salts might be calculated from their vapor pressures. They calculated the concentrations at the top and bottom of a column of solution 100 meters high for four different soluble compounds: cadmium iodide, sodium nitrate, marine salt, and sugar. In each case they calculated that the concentration of the solute should be greater at the bottom than at the top, the greatest difference occurring in cadmium iodide, and the least in marine salt. On the basis of the Gouy–Chaperon calculations, some petrologists suggested that magma in a large, quiescent chamber might undergo a spontaneous gradational stratification under the influence of gravity with denser molecules concentrating toward the base of the chamber and less dense molecules rising toward the top.

Judd (1893) appealed to the Gouy–Chaperon effect as one possible mechanism, along with the Soret effect, that produced the diverse magmas injected into the composite dikes of Arran. Brögger (1894b) suggested in his paper on the rocks of Gran that the principle of Gouy and Chaperon might have played a role. Vogt (1891) attributed the Ekersund and Taberg deposits to the Gouy–Chaperon action, arguing that the specific gravity of magma would increase with an increase in the dissolved molecules of iron-bearing minerals like magnetite, ilmenite, pyrite, and ferromagnesian silicates and that molecules of such minerals ought to be more abundant in the lower part of a magma reservoir. Iddings (1892a) also suggested a possible role for the Gouy–Chaperon effect in the Yellowstone rocks.

In opposition to the Gouy–Chaperon concept, however, Harker (1894b) argued that the concentration of basic minerals along both inclined walls of the Carrock Fell gabbro ruled out gravitational segregation of basic constituents in still fluid magma. The Gouy–Chaperon principle was considered to be "even more unsatisfactory" by Teall (1901). Schweig (1903), too, failed to detect any evidence for the Gouy–Chaperon effect in his experiments.

A second mechanism postulated to operate in liquid magma entailed the Berthelot principle. In 1879, Pierre Eugène Marcellin Berthelot (1827–1907), Professor of Organic Chemistry at the Collège de France, enunciated the thermochemical principle of maximum work, a principle that was eventually replaced by the principle of minimum Gibbs free energy (W. R., 1908; Crosland, 1970a). Berthelot (1879) postulated that any chemical change accomplished without the intervention of energy from outside would yield a body or system of bodies producing the most heat. Harker invoked Berthelot's principle, maintaining that cooling at the walls of a magma chamber entailed crystallization of minerals releasing the greatest amount of heat. As Harker (1893, p. 547) stated it:

> In a magma near the point of saturation, whatever promotes crystallization will promote the most rapid evolution of heat. In an unequally heated magma this must be effected by the accumulation of the least soluble ingredients in the part most easily saturated; that is, the concentration of the iron-oxide, etc. in the coolest region, and their crystallization there.

Harker claimed that once precipitation of a constituent began, that constituent would be replenished by diffusion along concentration rather than temperature gradients. With migration of basic constituents toward the crystallizing margins, the magma in the interior would become more acid and more viscous. Diffusion would eventually cease as viscosity increased. Noting Bäckström's objection that magma would initially be too viscous to result in substantial diffusion, Harker appealed to Vogt's experiments on slags indicating that viscosity in basic or water-bearing magmas would be relatively low. Diffusion should, therefore, be feasible in basic magmas, and it was precisely among the basic rocks, Harker said, in which the evidences for differentiation were the most striking. Loewinson-Lessing (1899a) also saw Berthelot's principle as a dominant factor at the onset of crystallization. Just as he had rejected the Soret effect, so, too, did Becker (1897a) reject Harker's idea of molecular flow toward growing crystals because of the "practically infinite" amount of time required for diffusion.

A third suggested mechanism of differentiation entailing only liquid was that of liquation, or liquid immiscibility, first proposed by Durocher (1857) as a secondary process for forming differences in magmas that originated from various magma shells at depth. Once the concept of differentiation supplanted the multiple-shell hypothesis, liquid immiscibility was also suggested as a possible dominant mechanism. Rosenbusch, for example, espoused something like liquid immiscibility with his concept of affinities and lack of affinities among the various "Kerne." Bäckström (1893) enthusiastically endorsed liquid immiscibility. He wished that proponents of differentiation had more seriously considered magmas as analogs of mixtures of aniline and water. He saw no justification for assuming that all chemical compounds forming magma were completely soluble in one another. After all, he noted, immiscible liquid fractions had been recognized in glass manufacture. Many inclusions, Bäckström (1893, p. 778) suggested, might have formed by immiscibility, for

> it is in many cases evident that the inclusions were soft, and then the simplest view is that they were drops, or portions, of a partial magma, which at the temperature, existing immediately before crystallization, could no longer be held in solution by the principal magma, but separated out.

He envisioned that liquid immiscibility adequately explained the bipolar nature of Icelandic lavas. "The acid partial magmas," Bäckström (1893, p. 778) concluded, "were separated out directly from the basic original magma, which by lowering temperature lost its homogeneity."

According to Loewinson-Lessing (1899a), a fundamental phenomenon of differentiation is the tendency toward separation of ferromagnesian magma from siliceous alkalic magma. He regarded such separation as far more likely to result from liquation than the Soret effect.

Liquid immiscibility also came in for its share of criticism. Vogt (1904) was skeptical of Rosenbusch's Kern hypothesis and saw no experimental evidence for liquid immiscibility except in the case of sulfide-silicate pairs. Schweig (1903), likewise, failed to discover any experimental evidence for liquid immiscibility. In his presidential address, Teall (1901) thought that the liquation theory was promising but would be "shrouded in doubt" until it had been established by experiment.

Becker (1897a) granted that immiscibility commonly occurs in aqueous mixtures but wondered how separation into two discrete bodies of magma would take place given the slowness of molecular flow. One means would be for one of the fluids to attach to the walls of its container, but Becker calculated that if half of the material in a spherical mass of an aqueous solution with a radius of 100 meters were to be deposited on the walls by molecular flow, thousands of years would still be required. How much more time would be required, therefore, in the case of viscous magma? An alternative mechanism would involve condensation of one of the liquids as droplets. Even though the droplets might have a different density from the host liquid and, therefore, sink or rise, Becker insisted that the viscosity of the host magma would successfully resist such movement. "In fluids such as lava," Becker (1897a, p. 34) wrote, "it scarcely seems credible that any extensive separation of a precipitated immiscible liquid should occur." For Becker, liquid immiscibility failed to provide an adequate explanation of rock "segregation."

In contrast to the all-liquid differentiation mechanisms, another set of mechanisms was envisioned as occurring primarily during crystallization such that separation of crystals from liquid brought about modifications of liquid composition. Some geologists invoked the significance of eutectic crystallization, whereas others appealed to a process that came to be known as fractional crystallization. The principle of eutectic crystallization was first developed by Frederick Guthrie (1833–1886), a British chemist who received his graduate training at Marburg and then taught in several British institutions (Partington, 1964). He showed that aqueous solutions of many salts experience freezing point depression and that the precipitation of salt or ice upon cooling of the solution depends on the relative proportions of water and salt (Guthrie, 1884). All mixtures enriched in the salt in a proportion greater than a certain fixed ratio would inevitably crystallize the salt. Mixtures with a proportion of the salt less than that of the fixed ratio would crystallize ice. The fixed ratio was designated as the eutectic composition, and it was found that a solution of eutectic composition simultaneously crystallized salt and ice in the eutectic proportions upon cooling. Moreover, Guthrie found that no solution persisted below the temperature at which the eutectic mixture crystallized.

Teall (1888, p. 395) believed that Guthrie's experiments had proved that "fused mixtures are strictly analogous to aqueous solutions." He saw the eutectic crystallization phenomenon described by Guthrie as leading to the formation of porphyritic textures in which early crystallizing compounds formed large crystals and the final eutectic mixture played the role of groundmass. Teall suggested that micropegmatite represented a eutectic mixture and argued that quartz would be the first mineral to crystallize where it is present in excess of a quartz–feldspar eutectic proportion and that orthoclase would be the first mineral to form where it is present in excess of the eutectic proportion. In 1901, Teall still appreciated the applicability of the eutectic principle to igneous rocks, noting that the textures observed in eutectic metal alloys closely resemble those in spherulites and micropegmatites.

In *Die Silikatschmelzlösungen*, Vogt (1903, 1904) also stressed the importance of eutectic crystallization. He concluded that the first mineral to crystallize from magma was a mineral present in excess of eutectic proportions. From a study of slags, he determined the approximate eutectic ratios for several pairs of minerals such as plagioclase–diopside, augite–olivine, and quartz–alkali feldspar. He regarded the groundmass of many rocks as having nearly eutectic compositions and also considered several intergrowths such as hornblende–oligoclase, augite–enstatite, and perthite as eutectic crystallizations. Vogt envisioned the ultimate goal of the differentiation process as leading toward the production of eutectic mixtures.

In his study of the igneous rocks at Magnet Cove, Arkansas, H. S. Washington (1900) also suggested that the differentiation might have resulted from eutectic crystallization. He proposed the subdivision of laccoliths on the basis of their rock types and the nature of their differentiation. Laccoliths of the Henry Mountains type, he suggested, might represent magmas that crystallized as nearly eutectic mixtures.

A second proposed mode of crystal–liquid differentiation entailed the separation of crystals from the liquid, either by sinking or squeezing off of liquid. Teall (1888, pp. 406–407) suggested that heterogeneity in plutonic masses could be referred to "progressive crystallization accompanied by a separation of the first-formed crystals; or possibly to causes of which we are at present totally ignorant." Darwin (1844) and King (1878), of course, had early advocated crystal settling, and, in his overview of the evolution of igneous rocks, Harker (1894a, p. 162) observed that sinking of early, mafic minerals under the influence of gravity "would give rise, in the lower layers of a magma-reservoir, to a relative richness in the more basic minerals." Darwin (1844) had also suggested that oozing of selected portions of partially crystallized granitic liquid into fissures might account for some variations in the mineralogy of dike rocks on James Island. The concept was developed further by George Barrow of the British Geological Survey. Barrow (1893) suggested that small pegmatitic intrusions in the Scottish Highlands formed when late-stage, very K-rich liquid was strained through a mesh of already crystallized quartz, oligoclase, and mica within a larger granitic intrusion. The strained residual liquid was then squeezed to higher crustal levels. Harker (1894a) agreed that Barrow's proposed mechanism, later designated "filter pressing," might indeed serve as one method of differentiation. Both crystal settling and filter pressing would in time be recognized as different modes of the larger process of fractional crystallization.

Becker (1897b) was the first to use the term "fractional crystallization" in a geological context. Interestingly, Becker was not convinced that the existence of originally homogeneous magmas of vast volume was well established and even considered fractional crystallization to be the very opposite of magmatic differentiation, a term that, to him, implied either Soret's principle or liquid immiscibility. Becker (1897b, p. 257) noted that "among the phenomena most often appealed to in support of the theory of magmatic segregation or differentiation is the symmetrical arrangement of material in certain dikes and laccolites." To explain such arrangements he saw no need to resort to the "division of a homogeneous fluid into two or more distinct fluids." He argued that if lava in a dike were "a homogeneous mixture of two liquids of different fusibility, then the crusts which first form upon the walls will have nearly the same composition as the less fusible partial magma." Each time circulating liquid passed the walls it would deposit some of its least fusible components. The liquid composition would tend toward that of the most fusible mixture of its compounds. After that composition was obtained, circulation would cease, and the residual liquid would solidify as a homogeneous material. The dike would, therefore, exhibit gradation in composition from the walls to its center. The same process might occur in laccoliths. Becker (1897b, p. 258) stated that this process

> is one of the most familiar in chemistry and is usually known as fractional crystallization. It has been employed in the purification of compounds ever since chemistry was pursued, and indeed before; for the preparation of salt from sea water or brine depends upon it. It can be and has been employed also to strengthen solutions. A familiar instance is the freezing of weak alcoholic liquids. A bottle of wine or a barrel of cider exposed to a low temperature deposits nearly pure ice on the walls, while a stronger liquor may be tapped from the center. If a still lower temperature were applied the central and more fusible portion would also solidify. Such a mass would be, so far as I can see, a very perfect analogue to a laccolite.

Becker (1897b, p. 259) suggested that the effectiveness of fractional crystallization depended upon convection currents that would be "the mortal enemy of any process of separation involving molecular flow." He believed that because dikes and laccoliths chilled inward, there seemed to be no way to avoid fractional crystallization unless the magma originally possessed either eutectic composition or very high viscosity. On the process of fractional crystallization, he submitted, the agreement between the order of consolidation of minerals recognized under the microscope and the arrangement of minerals in dikes was to be expected. In a final pitch for fractional crystallization, Becker (1897b) argued that the process was well understood, had been demonstrated in hundreds of thousands of experiments, and was sufficiently rapid that it could achieve in a few days diversities of composition that would require centuries through processes dependent on molecular flow.

Schweig (1903) also maintained that the leading factor in differentiation was simply crystallization caused by decrease of temperature or increase of pressure. Unless the magma is too viscous or crystallization occurred too quickly, he said, a separation of crystals from magma would occur in accord with specific weight and differentiation would result.

Additional processes of differentiation were suggested but failed to gain much support. Vogt (1891), for example, had suggested that the mutual magnetic attraction of iron-rich spinels could lead to their segregation within fluid magmas. Rücker, however, pointed out that at the high temperatures of magmas, the magnetic properties of magnetite and other iron-rich spinels would be lost. Barus and Iddings (1892) demonstrated that magmas are capable of carrying electrical currents and must be regarded as electrolytic solutions. As a result, Brögger (1898) suggested that diffusion might be brought about by the passage of electrical currents through magma. For a time, Vogt entertained a similar opinion but soon criticized the proposal on the grounds that an electrical current passing through an electrolyte should result in opposite migration of cations toward a cathode and anions toward an anode. In magmatic differentiation, however, Vogt (1904) said that entire dissolved compounds, cations and anions together, appeared to have diffused in the same direction. Michel-Lévy (1897) attributed much differentiation to the action of circulating mineralizing fluids that were said to dissolve and transport alkalis and other ions from one part of a magma body to another. This view, however, drew little enthusiasm outside of France.

Not every petrologist was convinced that differentiation, whatever the proposed mechanism, was the dominant cause of igneous rock diversity. Partial melting, espoused by Becker and Teall, and assimilation, endorsed by Loewinson-Lessing, were also regarded as important petrological processes. Nevertheless, at the beginning of the twentieth century, differentiation was acknowledged to be the leading source of differences among igneous rocks despite a lack of agreement on the nature of the specific physical mechanisms of differentiation and despite the interest in partial melting, assimilation, or even magma mixing.

One powerful indication of the triumph of the theory of differentiation lies in the fact that it took its place in routine geological instruction. By way of example, examination questions in igneous petrology posed in 1889 may be compared with those posed in 1905 in the Natural Science Tripos at the University of Cambridge. The questions were written by Harker in both instances. In each examination, students were required to describe and identify several hand specimens and thin sections of igneous rocks. In 1889, students were asked to describe textural and structural gradations between granite and rhyolite and between gabbro and basalt and to discuss the bearing of the facts on the question of the conditions of consolidation of igneous rocks. They were also asked to compare granite and basalt in terms of composition, structure, and mode of field occurrence and to draw appropriate conclusions therefrom. Several other questions pertained to rock-forming minerals and the features of igneous rocks. In 1905, however, students were asked the factors that determine the order in which minerals separate from magmas. They were further asked to explain why Rosenbusch's empirical law of increasing acidity of residual liquids held good in many cases. Students were also required to describe three cases of noteworthy variation of composition in different parts of a single intrusive mass, and to state what principles had been applied to the elucidation of the variation. One may reasonably conjecture that a substantial proportion of students quite probably selected Harker's Carrock Fell gabbro as one of the three cases.

Teall reflected on the status of the question of diversity at the turn of the century in his 1901 presidential address concerning the origin of igneous rocks. He revealed that he had adjusted his views since 1888 and achieved some clarity in his own mind by relegating several mechanisms to secondary status. Nevertheless, he indicated that the petrogenetic situation was still sufficiently murky that

> rival theories are struggling for existence, and although it is safe to predict that some will become extinct, that others will be modified, and that natural selection will finally bring about the survival of the fittest, it is impossible to determine, at present, the relative importance of those which claim our attention. (Teall, 1901, p. lxxxvi)

In his judgment, the next great advance in petrological science would

> come about, not so much by adding to our already large store of facts, as by dint of experiment controlled by the modern theory of solutions, and carried out for the express purpose of testing the consequences of that theory and discovering the modifications which may be necessary to adapt it to igneous magmas. (Teall, 1901, lxxxii)

In his call for more petrological experimentation, Teall was not alone. Just a few years later, the experimental revolution in igneous petrology would be underway.

THE PROFESSIONALIZATION OF PETROLOGY AND PETROGRAPHY

The petrogenetic theory of differentiation emerged during an era of increased professionalization and emphasis on detailed specialized research that was typically conducted within organizations devoted specifically to that purpose. Further insights into the relationships among professionalization, research, and universities in the latter nineteenth century have been provided by Veysey (1965), Ben-David (1971), and Shils (1979). German universities had functioned as research institutions since the 1830s, and geologists from other countries commonly looked to Germany for advanced study. As noted earlier, a host of Americans studied petrography with Harry Rosenbsuch in Heidelberg or Ferdinand Zirkel in Leipzig and brought the German ideal of meticulous petrographic investigation back to America. Likewise, new American research-oriented universities established along German lines sprang up toward the end of the nineteenth century, beginning with Johns Hopkins University (1876), Stanford University (1891), and the University of Chicago (1892). American institutions of long standing, like Harvard University, transformed themselves into research-minded schools. Specialized mineralogical research institutes also appeared in many European universities, such as Oslo, Stockholm, Warsaw, and Jena. Opportunity for pure scientific research was further provided within newly established governmental surveys, such as the United States Geological Survey, founded in 1879 as an outgrowth of the surveys of the Fortieth Parallel and of the Rocky Mountain Region.

Evidences of the effect of the broader trend toward scientific professionalization and specialization on petrology may be detected in the occupations, publication outlets, and scientific interests of petrogenetic theorists from 1880 to the beginning of the twentieth century. Those who wrote about the theory of differentiation between 1880 and 1905 were overwhelmingly professional geologists. Apart from an anomaly like Johnston-Lavis, a medical doctor–volcanologist, the days were largely gone when naturalists like Darwin, chemists like Bunsen who did geology on the side, or independently wealthy amateur researchers like Scrope contributed to petrogenetic theory. From 1880 to 1905, virtually everyone who was engaged in igneous petrogenesis, from Rosenbusch to Schweig, worked as a geologist for a government geological survey or a university. J. P. Iddings was one of the first employees of the United States Geological Survey, and when his position was eliminated from the Survey during budget cuts in 1892, he joined the brand new research-oriented University of Chicago as the occupant of the world's first chair of petrology. In Great Britain, theoretical petrology was conducted by Teall and Dakyns at the British Geological Survey and by Harker at the University of Cambridge. In France, the primary theorist, Michel-Lévy, was director of the Service de la Carte Géologique. In eastern and northern Europe, petrogenetic theorizing was overwhelmingly concentrated in academic settings where Rosenbusch, Lagorio, Brögger, Vogt, Bäckström, Loewinson-Lessing, and Schweig all occupied university positions.

Most of the geologists who pondered igneous rock diversity were also specialists in petrology rather than the geological generalists of earlier generations such as Dana, Jukes, or King. Rosenbusch specialized in descriptive microscopic petrography and classification as evidenced by his authorship of exhaustive treatises on optical mineralogy and petrography. Teall was a specialist in igneous and metamorphic rocks whose work included field mapping, descriptive and interpretive petrography, and theorizing about igneous rock diversity. From 1888 to 1901, Teall was occupied primarily with detailed studies of the structure and petrography of schists and gneisses in the Scottish Highlands. His colleague Dakyns conducted field mapping throughout Great Britain but took the greatest delight in mapping the Lower Carboniferous coalfields of England and Wales. He was the least connected to igneous petrology of any of the scientists discussed. Despite occasional contributions to geomorphology, structural geology, and metamorphic rocks, Harker was a specialist in igneous petrology. Throughout his career, Harker, like Iddings, contributed numerous papers on the diversity of igneous rocks. At the same time, he was occupied with field studies, classification, microscopic petrography, the chemistry of igneous rocks, and the question of the distribution of igneous rocks in space and time. Johnston-Lavis, although trained in medicine, became fascinated by Mount Vesuvius, mapped the volcano in detail, and published several papers on volcanoes of southern Italy. In spite of his wide-ranging gifts, geological and otherwise, Brögger was first and foremost an igneous petrologist, equally at home in field mapping, mineralogy, chemistry, and theory. Vogt combined field and petrographic studies with extensive research on the physical chemistry of slags, and he calculated phase diagrams for several silicate mineral pairs from data on melting points, molecular weights, and latent heats of melting. Bäckström was a petrographer with exceptionally strong interests in the mineralogical and crystallographic side of the science. Iddings was equally at home in field studies, chemical petrology, descriptive microscopic petrography, the classification of igneous rocks, and igneous petrogenetic theory. Iddings' U. S. Geological Survey colleague, Becker, discussed differentiation and mapped igneous rock suites, but he was essentially a geophysicist who attempted to apply principles of physics and chemistry to geology as evidenced by his papers on rock mechanics. Michel-Lévy was an expert in optical mineralogy and experimental petrology. Together with Ferdinand Fouqué he experimentally reproduced the texture and mineralogy of several volcanic rock types. Primarily interested in both igneous and metamorphic petrology, Michel-Lévy was skilled in mapping, petrography, and mining geology. Loewinson-Lessing was a specialist in igneous petrology with interests in petrography, field studies, classification, and petrogenetic theory. He wrote numerous papers on volcanic rocks, liparites, anorthosites, on the classification of igneous rocks, on petrography, and on petrogenetic theory.

The tendency toward increasing professionalization in petrology may also be seen in the publication vehicles used during the period of interest. In the earlier part of the century, theories about the diversity of igneous rocks were published in scientific travel books, reports of oceanographic voyages, survey reports, and journals of broad scientific interest. Between 1880 and 1905, petrogenetic theories still appeared in technical monographs and in geological survey reports, but the vast majority of theoretical discussions now appeared in strictly geological journals.

THE ORIGINS OF IGNEOUS PETROLOGY AS A DISCIPLINE

The question arises, however, as to why igneous petrology arose as a separate discipline amenable to professional specialization. At least three reasons may be cited: significant contemporary advances in ancillary sciences, particularly physical chemistry, the proliferation of chemical analyses of igneous rocks, and above all, the application of the polarizing microscope to thin sections of igneous rocks.

Geologists toward the end of the nineteenth century did not conceive of igneous petrology simply as an exercise in field geology aided by the microscope. They freely drew from other disciplines. Becker repeatedly relied on physics to address petrological problems. Insights from metallurgy were applied to igneous petrology by Vogt. Schweig appealed to glass technology. Virtually all petrologists of the era especially looked to the promising new field of the physical chemistry of aqueous solutions for ideas and insight. In particular, they turned to the ideas of Berthelot, Soret, Gouy and Chaperon, and Guthrie to provide a theoretical basis for the nature and behavior of rock magmas and for possible differentiation mechanisms, and they also argued on physical-chemical grounds against some of those mechanisms. Petrologists, too, keenly sensed the need for rigorous, controlled experiments to help solve the petrological problems of order of crystallization, the effects of temperature and pressure, the nature of magmas as solutions, the nature of diffusion, and potential mechanisms of magma behavior. Teall, Loewinson-Lessing, Michel-Lévy, Iddings, and others pleaded for more experiments.

Between 1880 and 1905, no geologist insisted that the problems of petrology had to be solved exclusively in the field. Despite the constant appeals to physical chemistry, no one complained that petrology was in danger of becoming a subfield of another discipline. No one claimed, for example, that field evidence demanded differentiation by molecular diffusion or by liquid immiscibility and that no amount of physico-chemical argument or experimental evidence could persuasively counter the field data. The petrologists of the time, most all of whom were equally at home in the field or peering down a microscope seemed to understand the ambiguities of field evidence and, therefore, enthusiastically welcomed insights offered by chemists, metallurgists, and experimental mineralogists. Although the great zeal shown by petrologists toward physical chemistry, metallurgy, or physics might seem at first glance to run counter to the specialization of the era, it was precisely insights from these other disciplines that showed the way toward providing a sound theoretical basis for igneous petrology.

Another development that contributed to the emergence of igneous petrology as a discipline and especially to the rise of the theory of differentiation was the advance in chemical analytical methods. Among the early petrogenetic speculations on igneous rock diversity, chemical analyses played a major role only in Bunsen's two-source theory as later modified by Sartorius von Waltershausen, Durocher, and von Richthofen. The ideas of Scrope, Darwin, Dana, Jukes, King, and Dutton were unsupported by chemical analyses. In the period under review, however, numerous articles and monographs, for example, those of Lagorio, Brögger, Iddings, Dakyns and Teall, and Harker, contained a wealth of chemical analyses that clearly indicated the gradational chemical character of members in a suite of consanguineous igneous rocks. The gradational chemical character of the rocks, in turn, pointed to the progressive gradational character of the magmas, one of the linchpins in the concept of differentiation.

Perhaps more than anything else, however, the development of thin-section examination with the polarizing microscope created the discipline of igneous petrology. Microscopic petrography disclosed seemingly endless vistas for the study of igneous rocks. Geologists were now able to identify the major minerals in igneous rocks with more confidence than had been possible with examination of hand specimens by hand lens or mineral fragments through a simple microscope. Fine-grained minerals whose existence had not previously been suspected were recognized. More accurate estimates of mineral abundances could be made. Textural relationships among the minerals could be described in great detail. Petrographers could now readily distinguish between primary minerals and those formed by alteration. From the textural relationships, microscopists could also interpret the order in which the various minerals had crystallized. Stimulated by the fruitfulness of the fascinating new method, scores of geologists devoted their careers to use of the microscope for the description of hundreds of rock specimens that had previously been characterized only from hand specimen and chemical studies but also for the description of specimens collected during new detailed field studies.

The development of thin-section petrography led to unprecedented understanding of the mineralogical and textural characteristics of igneous rocks around the world. Major refinements in igneous rock classification were made possible by labors of the great petrographers like Rosenbusch and Michel-Lévy. So fruitful was microscopic petrography that dozens of scholars on both sides of the Atlantic and even in a handful of other parts of the globe became specialists in petrography. The result was a proliferation of petrographic literature in journals and textbooks, the creation of a petrographically oriented journal, the hiring of petrographers by national geological surveys, and the introduction of igneous petrology into the course instruction of a rapidly expanding number of universities in Europe and America.

14 ← THE LANGUAGE OF PETROLOGY

Nomenclature and Classification

As a result of the expanded fund of textural and mineralogical knowledge made available through microscopic petrography, several petrographers tried their hand at the formulation of new classifications of the igneous rocks. From time to time, lexicons of petrographic terminology, such as *Petrographisches Lexikon* of F. Y. Loewinson-Lessing (1893) or Kinahan's *Handbook of Rock Names*, also appeared. In Germany, Zirkel and Rosenbusch refined and modified their original classifications in later editions of *Lehrbuch der Petrographie* (Zirkel, 1893) or *Mikroskopische Physiographie der massige Gesteine* and *Elemente der Gesteinslehre* (Rosenbusch, 1896, 1901, 1910). Von Lasaulx, Roth, Kalkowsky, Osann, Weinschenk, and Lang also contributed classifications. In France, *Minéralogie Micrographique* of Fouqué and Michel-Lévy (1879) and Michel-Lévy's (1889) *Structures et Classification des Roches Éruptives* included classifications. In Great Britain, *The Study of Rocks* (Rutley, 1879), *An Introduction to the Study of Petrology* (Hatch, 1891), and *Petrology for Students* (Harker, 1895b) included classification schemes, and a paper on classification was published by Jevons (1901). In the United States, Iddings (1898b), Cross (1898), and Hobbs (1900) discussed igneous rock systematics, and in Russia the major contributions were those of Loewinson-Lessing (1899a,b, 1900, 1901, 1902).

A host of issues led to divergence in the classifications employed. Should a classification be natural or artificial? Should geological age be regarded as a factor in classification? Should igneous rocks be classified chemically or mineralogically? What role should geological occurrence play in classification? Petrographers would constantly disagree in their answers to these questions. In the end, the petrological community had to contend with a wide spectrum of classifications.

NOMENCLATURE AND CLASSIFICATION BEFORE 1900

Geologists differed as to whether or not a natural classification of igneous rocks is possible. In other words, they disagreed about whether the structure of a classification might be suggested by one or more natural characteristics of the rocks themselves. Despite the wish of many petrologists to discover such a classification, others, suspecting that a legitimate natural classification was improbable, maintained that petrologists needed to superimpose arbitrary boundaries between different rock types.

The use of the age factor in classification died a slow death even after British and American petrographers demonstrated the fallacy of the view that igneous rock types are confined to specific portions of the geologic column. In his earlier work, Rosenbusch made a distinction between paleovolcanic and neovolcanic rocks. Despite gradually retiring the age factor in his formal classification, he still brought up differences in names for identical rocks of different ages as late as the third edition of *Elemente der Gesteinslehre* (Rosenbusch, 1910). The procedure was retained even after Rosenbusch's death in the fourth edition of the textbook (Osann and Rosenbusch, 1922). Kalkowsky (1886) and Roth (1883) continued to use age as a criterion. At least Weinschenk (1912) rejected the age criterion. As late as 1889, Michel-Lévy still made a formal distinction between "ante-tertiares" and "post-tertiares" rocks. In *Einführung in die Gesteinslehre*, A. von Lasaulx (1886) distinguished an ancient series and a recent series of igneous rocks. British and American geologists, of course, did not include age as a factor. Even though his initial scheme resembled that of von Lasaulx in other respects, Hatch (1891) omitted von Lasaulx's separation of igneous rocks into an ancient and a recent series. Harker (1894a, p. 129) spoke for his countrymen when he wrote that the concept of distinctions in rocks based on geological age "is rejected by the British school, and will find no place in the following pages."

Some classifications were based largely on a combination of mineralogy and texture such as those of Zirkel, Rosenbusch, Michel-Lévy, and Harker, whereas other schemes either were purely chemical or else introduced chemical composition as a major factor in subdivision. For example, von Lasaulx (1886), after establishing ancient and recent series of rocks, broke down each of these two groups on chemical terms into an acid series (SiO_2 = 60 to 80 weight percent), intermediate series (SiO_2 = 50 to 70 weight percent), and basic series (SiO_2 = 40 to 60 weight percent). Each chemical series was then subdivided texturally and mineralogically. Hatch (1891) adopted a similar scheme. Loewinson-Lessing also categorized rocks as acid, neutral, basic, or ultrabasic, and then further subdivided them on the basis of the relative proportions of oxides of bases in the chemical analysis of a rock.

Petrographers also disagreed on whether to incorporate geological occurrence as a factor in classification. Rosenbusch consistently proposed three groupings of igneous rocks on the basis of geological occurrence: "Tiefengesteine" (deep-seated rocks), "Ganggesteine" (dike rocks), and "Ergussgesteine" (effusive rocks). In Great Britain, Harker adopted a similar approach. In *Petrology for Students*, Harker (1895b) recognized plutonic, intrusive, and volcanic rocks. In later editions of this text, the term "hypabyssal" replaced "intrusive." In the "hypabyssal" category, Harker included such rocks as porphyry, diabase, and lamprophyre. Von Lasaulx, likewise, proposed three groups but used texture as a discriminant rather than geological occurrence. He distinguished rocks in his acid, intermediate, and basic series on the basis of granitic, porphyritic, or vitreous texture. Rutley (1879) classed rocks texturally into two groups, vitreous and crystalline, and Michel-Lévy (1889) also opposed Rosenbusch's threefold division.

In the early years of the microscopic era, classification schemes could be deciphered only from tables of contents or organizational structures of petrographic textbooks. An important step was taken when Fouqué and Michel-Lévy (1879) published a huge two-page chart of their classification scheme with an abundance of pigeonholes based on mineralogy and structure. Rock names were printed in the appropriate pigeonholes. For the first time, geologists could obtain a convenient overview of a classification scheme without thumbing through pages of text. Michel-Lévy (1889) extended this service to the petrographic community by publishing pigeonhole charts of his latest classification side-by-side with that of Rosenbusch (1887) (Figure 15). Petrographers could now compare the two different classifications at a glance. Pigeonhole classifications would become increasingly popular.

From time to time, geologists made general suggestions for improving classification without necessarily proposing new classifications. W. H. Hobbs (1900) of the University of Michigan, for example, urged the central importance of texture and mineralogy but also stressed the benefit of conveying chemical information by means of diagrams. Both Michel-Lévy and Brögger had previously constructed a type of rose diagram in which eight different oxide components of a chemical analysis might be plotted. Hobbs showed how the shapes of the diagrams might be used to distinguish one igneous rock type from another and to illustrate familial relationships among the rocks (Figure 16). By the beginning of the twentieth century, diagrams increasingly found their way into petrological practice. Reyer (1877, 1888) introduced diagrams for representing relative amounts of rock constituents, either oxides or minerals. Iddings (1892a) originated the variation diagram for his study of Yellowstone igneous rocks, plotting the molecular proportions of oxides in a suite of rocks against the molecular proportions of silica in the same suite of rocks. Variation diagrams found favor in the eyes of petrologists. Before long, Dakyns and Teall (1892), Washington (1900), and Cross (1896) used them in their interpretations of Garabal Hill, the Grecian archipelago, Magnet Cove, and Rosita Volcano in Colorado. Harker (1895a), likewise, employed variation diagrams for

CHAPTER 14

CLASSIFICATION DE M. ROSENBUSCH

ÉLÉMENT FERRO-MAGNÉSIEN	SILICE LIBRE ET FELDSPATH ALCALIN	FELDSPATHS		FELDSPATHIDES			SANS ÉLÉMENT BLANC
		ALCALINS	ALCALINO-TERREUX	AVEC FELDSPATHS		SANS FELDSPATH	
	q et a	a (seuls)	t (avec ou sans q)	a ALCALINS	t TERREUX		
ROCHES DE PROFONDEUR (Un seul temps de consolidation.)							
M	Granites.	Syénites.	Diorites.				
A							
P			Diabases.	Eléolitsyénites.	Théralites.		Péridotites.
H			Gabbros.				
O			Diabases, gabbros.				
ROCHES DE FILONS (Un seul temps.)							
M	Aplites.		Lamprophyres dioritiques. (Kersantites.)				
ROCHES DE FILONS (Deux temps.) On ne considère que les éléments du premier temps, sauf dans les lamprophyres où le feldspath est tout entier du second temps.							
M	Granitporphyres.	Syénitporphyres. Lamprophyres syénitiques.	Dioritporphyrites. Lamprophyres dioritiques.	Eléolitsyénit-porphyres.			
A							
P							
H							
O							
ROCHES D'ÉPANCHEMENTS PALÉOVOLCANIQUES (Deux temps.) On ne considère que les minéraux du premier temps.							
M	Quartz-porphyres.	Porphyres sans quartz. (de première consolidation.)	Porphyrites.				Picrit-porphyrites.
A							
P							
H			Augitporphyrites et mélaphyres.				
O							
ROCHES NÉOVOLCANIQUES (Deux temps.) On ne considère que le premier temps de consolidation, sauf pour le quartz dans les liparites et les trachytes.							
M	Liparites. (Y compris celles qui ne contiennent que du quartz de deuxième consolidation.)	Trachytes. (Non compris ceux qui contiennent du quartz de deuxième consolidation.)	Dacites et andésites.	Phonolites et leucitophyres.	Téphrites et basanites.	Leucitites et néphélinites.	
A							
P							
H							
O			Basaltes.			Leucitbasaltes, Néphélinbasaltes, Mélilitbasaltes.	Limburgites et augitites.

CLASSIFICATION DE MM. FOUQUÉ ET MICHEL LÉVY

ÉLÉMENT FERRO-MAGNÉSIEN	SILICE LIBRE q	FELDSPATHS SEULS				FELDSPATHIDES		l LEUCITE. n NÉPHÉLINE. s SODALITES. h MÉLILITE.	SANS ÉLÉMENT BLANC
		ALCALINS a	ALCALINO-TERREUX			AVEC FELDSPATHS			
			t_1 ANDÉSITIQUES	t_2 LABRADORIQUES	t_3 ANORTHIQUES	ALCALINS $\begin{matrix}l\\n\\s\end{matrix}\Big\}a$	TERREUX $\begin{matrix}l\\n\\s\end{matrix}\Big\{t$	$\begin{matrix}l\\n\\h\end{matrix}$	

ROCHES GRANITOIDES   I

Micas noirs..	M	Granites¹ et microgranites².	Syénites et minettes³.	Kersantites³.		Ditroïtes³.	Teschénites³.		Péridotites³.
Amphiboles..	A			Diorites⁴.		Syénites³ éléolitiques.			
Pyroxènes..	P			Diabases et Gabbros⁴.					
Hypersthènes.	H			Gabbros et norites⁴.					
Olivine....	O			Diabases, gabbros et norites⁴.					

ROCHES PORPHYRIQUES ANTE-TERTIAIRES   II

Micas noirs..	M	Microgranites² et porphyres⁵.	Orthophyres⁶.	Porphyrites⁷.		Porphyres à liébénérite⁶.			Mélaphyrites⁶.
Amphiboles..	A								
Pyroxènes..	P								
Hypersthènes.	H								
Olivine....	O			Mélaphyres⁷.					

ROCHES TRACHYTOÏDES TERTIAIRES ET POST-TERTIAIRES   II

Micas noirs..	M	Microgranites², rhyolites et dacites⁶.	Trachytes⁶.	Andésites⁶.		Phonolites et leucitophyres⁶. (Trachytes à ægirine et haüyne).	Téphrites⁶ (Andésites à haüyne.)	Néphélinites leucitites, mélilit-basaltes⁶.	Augitites, limburgites⁶.
Amphiboles..	A								
Pyroxènes..	P				Labradorites⁷.				
Hypersthènes.	H								
Olivine....	O				Basaltes⁷.				

¹ Comprenant *granites* (α), *granulites* (β), *pegmatites* (γ).
² Comprenant *microgranites* (α), *microgranulites* (β), *micropegmatites* (γ).
³ A structure *grenue* (δ).
⁴ A structures *grenue* (δ) ou *ophitique* (ω).
⁵ A structures *globulaire* (φ) ou *pétrosiliceuse* (π).
⁶ A structure *microlitique* (μ).
⁷ A structures *microlitique* (μ) ou *ophitique* (ω).

15. Comparison of the igneous rock classifications proposed by Auguste Michel-Lévy and Harry Rosenbusch. From Michel-Lévy (1889).

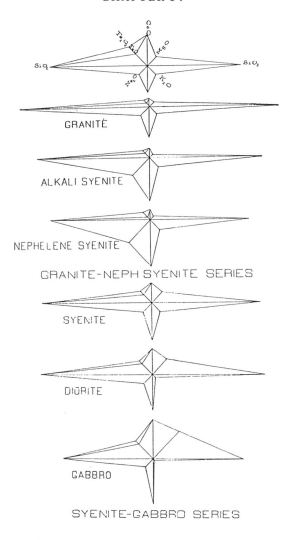

16. Rose diagrams for plotting chemical composition of various rocks by Hobbs (1900). Reproduced from *Journal of Geology* by permission of The University of Chicago Press.

his study of Carrock Fell but plotted atomic ratios of chemical elements instead of molecular proportions of oxides. Later, Harker (1900) suggested the use of weight percentages of the various oxides, including silica, taken directly from the chemical analyses in the construction of variation diagrams. It is probably for this reason that contemporary petrologists typically associate variation diagrams with Harker rather than Iddings, their originator. Iddings (1903) presented an historical overview of the various diagrams in use in petrology after 1877.

Petrographers continually differed over the naming of rocks. Some claimed that there were too many rock names, others not enough. Still others disputed what names to give. Various well-meaning suggestions fell on deaf ears. For example, H. Stanley Jevons, Assistant Demonstrator in Petrology at the Woodwardian Museum of the University of Cambridge and a former student of Rosenbusch, proposed a nomenclatural system designed to take advantage of abbreviations for various minerals and textures. Recognizing that not all granites are alike, Jevons (1901) suggested that one should be able to devise a terminology for distinguishing, say, a hornblende–biotite granite from other types of granites. As the full name "hornblende–biotite granite" is a bit cumbersome, Jevons suggested that it be called a "hornbi-granite." The order of the prefixes horn- and bi- would be used in order of increasing abundance of the mineral whose name was being abbreviated. He suggested that names could also be devised for rocks lacking particular minerals. Hence, he suggested "apyr-magne-peridotite" for schlieren in gabbro and peridotite that are composed of magnetite and olivine but lack the pyroxene. Textural information could be incorporated into a name, Jevons thought, as in "mipegmorhyolite," a name for rhyolite with micropegmatitic texture. Even chemical information might become part of a name; hence, Jevons suggested "ali-syenite" for an alkalic syenite. The prospect of talking about "rhomfels-pyr-alisyenites," "apyr-gabbros," and "hypidiobasalts" evidently left petrographers entirely unenthused, for virutally no one took up Jevons' proposal. Jevons did, however, make the important point that it is not just the existence of minerals in an igneous rock that is important but also the relative proportions of constituents. Regrettably, he pointed out, all too few petrographic descriptions contain any quantitative information about the minerals. To remedy the situation, he urged petrologists to begin including data on mineral abundances in their rock descriptions.

The above observations confirm that igneous rock classification at the beginning of the twentieth century was in a state of confusion with little agreement on the principles on which to construct a classification. Petrographers, however, began to point to the urgent need for quantitative mineralogical data as well as for classifications based on the relative abundances of the various mineral constituents in igneous rocks.

METHODS OF MEASUREMENT OF THE MINERAL PERCENTAGES IN ROCKS

Successful application of a quantitative mineral-based classification of igneous rocks, however, would not be an easy matter because any such scheme required accurate knowledge of the relative abundances of the various minerals present in the rock to be classified and named. Despite Jevons' optimistic call for quantitative data, the determination of such mineral abundances posed formidable problems for petrographers of the nineteenth and early twentieth centuries. Various methods for the determination of abundances had been developed, but none of them was fully satisfactory. All of them were unbelievably tedious.

One early method, developed by Achille Delesse (1847, 1848), involved tracing the outlines of mineral grains within a polished slab of coarse-grained rock onto a piece of transparent oiled paper without the aid of a microscope. The thin piece of paper was then glued to a sheet of metal foil. Each outlined area was labeled in terms of the mineral corresponding to that area. The traced outlines were then cut out, and the pieces representing each mineral were placed into separate piles and weighed. The relative weight of each pile was divided by the density of the mineral it represented to obtain the relative volume of that mineral in the rock. Of course, the results were no more accurate than the tracing, cutting, and weighing. It was also necessary that the paper and metal foil each be of uniform thickness. Delesse's method, too, was severely limited in its ability to yield worthwhile data when applied to fine-grained rocks in which both mineral identification and the drawing of accurate boundaries were far more difficult. After the development of thin-section petrography, the Delesse method was eventually modified by Oxford geologist, W. J. Sollas. Sollas (1892) used a camera lucida apparatus for drawing the outlines of mineral grains in thin sections. John Joly (1903) applied the method directly to photomicrographs.

The next major advance in petrographic measurement was brought about by August Rosiwal (1860–1923), Professor of Mineralogy and Geology at the Technische Hochschule in Vienna and geologist with the Austrian Geological Survey (Spencer, 1924). Rosiwal (1898), who spent much of his career investigating questions of a geotechnical nature, devised a method in which the petrographer followed a series of parallel traverses across a thin section of rock and measured the diameters of each mineral grain encountered along the traverse. The traverse lines were spaced far enough apart that no two lines cut the same grain. If sufficient numbers of grains on a large enough number of traverses were measured, Rosiwal maintained that the relative sums of the diameters of the grains of each mineral should be proportional to the volumes of each mineral in the rock. For his own measurements, Rosiwal drew traverse lines directly onto the coverslips of his thin sections.

Other methods entailed various means of segregating mineral particles from crushed rock. Geologists had long used magnets to remove magnetite from a crushed sample, and Delesse, Fouqué, and Doelter employed electromagnets for more sophisticated separations. Other petrologists made various gravity separations by placing crushed rock fragments into heavy liquids and allowing the minerals to settle or float in accord with their densities. Jigging tables were devised to sort minerals from one another. By means of all these methods, separated minerals could be weighed and their volumes calculated.

During the first couple of decades of the twentieth century, petrographers debated the accuracy of the available methods. Doubts persisted as to whether the areal measurement data of the Delesse-Joly method or the linear measurement data of the Rosiwal method could ever legitimately be converted into volumes. Various investigators also made comparisons among sets of data obtained by chemical analysis, the Rosiwal method, gravity separations, and cutting methods. Although studies suggested that the Rosiwal method did not compare favorably with gravity separations, a study by Lincoln and Rietz (1913) endorsed the value of Rosiwal analysis. A detailed investigation by Johannsen and Stephenson (1919) also showed that Rosiwal analysis compared well with chemical analysis and gravity separation. Lincoln and Reitz determined the number of measurements per thin section that are required in order to obtain accuracy within one percent. They concluded that to do a Rosiwal analysis properly and to obtain the desired accuracy of results, a petrographer needed approximately eleven hours to make the measurements. They suggested that the measurements not be done in eleven consecutive hours, however, to avoid eye strain! In addition, they said, calculations required an additional two hours. As much as two full working days were necessary to determine the mineral percentages in one rock. Under the most favorable circumstances an analysis required at least eight hours, whereas under unfavorable circumstances, as much as eight days of eight hours each might be needed to complete an analysis of one thin section! In light of the conclusions of Lincoln and Reitz (1913), it comes as no surprise that the number of analyses actually performed was not great because of the inherent difficulties and tediousness of the quantitative determination of the minerals in a rock. Had there been quantitative mineralogical classifications available early in the century, relatively few igneous rocks could have been classified accurately anyway.

Felix Chayes (1956), with all the passion that could be summoned up perhaps only by a statistical petrographer of a later generation, lamented the lack of results from "the golden age of petrography" between Delesse and Shand and also the lack of development of instruments that might have overcome the disincentives to quantitative petrography posed by the existing tedious methods. One way around the problem of the lack of quantitative mineralogical classifications was to devise quantitative chemical classifications. Lack of numerical data was not at all a hindrance to this approach in light of the hundreds, if not thousands, of available chemical analyses summarized in successive editions of Roth's *Tabellen*. The most ambitious chemical classification was that of Osann, outlined in a series of very lengthy papers in *Tschermaks mineralogische und petrographische Mitteilungen* (Osann, 1900, 1901, 1902, 1903). C. Alfred Osann (1859–1923) studied at the University of Strassburg and at the University of Heidelberg under Bunsen and Rosenbusch, where he received his Ph.D. degree in 1881 (Spencer, 1924; Johannsen, 1931). From 1884 to 1889, he served as an assistant to Rosenbusch. Osann then went to

Spain and North America, where he worked for the Texas and Canadian geological surveys. In 1893, he returned to Germany to teach at the University of Karlsruhe. In 1897, he became Professor of Mineralogy and Geology at the University of Mühlhausen, and finally, in 1903, Osann joined the faculty of the University of Freiburg, where he was Professor of Mineralogy and Petrography. Osann's contributions were preeminently in the field of petrography. He edited the final editions of some of Rosenbusch's masterworks after his mentor died.

Osann converted the weight percentages of the various oxides in a chemical analysis into molecular proportions by dividing the weight percentages by the molecular weights of the oxides. The molecular proportions were then recalculated to 100 percent. Osann plotted chemical data for igneous rocks on a triangular ACF diagram, where A referred to the sum of the molecular proportions of the alkalis, C referred to the sum of the molecular proportions of the CaO and of any MgO and FeO equivalent to any alumina left over after equivalent amounts of alumina had been allotted to the alkalis and lime, and F referred to the sum of the molecular proportions of FeO and MgO and a possible excess of CaO. The values of A, C, and F were then recalculated to sum to 100 percent in order to plot on the triangular diagram. Osann defined additional parameters such as s, the sum of the molecular proportions of silica, titania, and zirconia, and n, a quantity that essentially measured the soda/potash ratio multiplied by ten. He then established a host of classes of rocks based on the relative proportions of A and C.

THE AMERICAN QUANTITATIVE SYSTEM

Throughout the nineteenth century, igneous rock classification had predominantly been the province of German petrographers, with some contributions from the French and British. At the end of the century, however, a team of four American petrologists, C. Whitman Cross, Joseph P. Iddings, Louis V. Pirsson, and Henry S. Washington (Figure 17), made a remarkable attempt to devise a comprehensive quantitative classification. For a number of years, these four geologists, three of whom were Yale graduates, corresponded in hopes of concocting a classification scheme on which they could all agree. They hoped that a successful team effort might ultimately satisfy a much wider range of petrologists. In 1893, one year after joining the University of Chicago, Iddings wrote to Cross, G. H. Williams at Johns Hopkins, and Pirsson "to obtain their combined judgment on the wisdom of certain generalizations in petrology, in order to meet 'the demands of the classroom'" (Knopf, 1960, p. 239). When Williams died in 1894, he was replaced by H. S. Washington. The four conferred in Washington, D. C., in December 1899, to work out final details.

We have already become acquainted with Iddings in Chapter 10, and now we meet Cross, Pirsson, and Washington. Charles Whitman Cross (1854–1949) graduated from Amherst College in 1875 and undertook advanced study

17. The authors of the American quantitative (CIPW) classification of igneous rocks. Clockwise from upper left: C. Whitman Cross (1854–1949), Joseph P. Iddings (1857–1920), Louis V. Pirsson (1860–1919), Henry S. Washington (1867–1934). Reproduced from the files of the Geophysical Laboratory, Carnegie Institution of Washington.

at Göttingen and then at Leipzig under Zirkel, where he received his Ph.D. degree in 1880 (Larsen, 1958). Upon his return to the United States, he joined the Geological Survey. Stationed in the Denver office, Cross undertook a host of field and petrological studies of numerous regions in Colorado. In later years, he was based in the Washington, D. C., office, where he became Chief of the Section of Petrology. Cross was also instrumental in promoting the establishment of the Geophysical Laboratory of the Carnegie Institution of Washington.

Louis Valentine Pirsson (1860–1919) entered the Sheffield Scientific School of Yale in 1879 and graduated with honors in 1882 after having studied chemistry (Knopf, 1960). He continued graduate study in chemistry at the Sheffield School and served as a teaching assistant for seven years. In 1889, Pirsson was offered a position with the U. S. Geological Survey as a volunteer assistant to Iddings through the efforts of Arnold Hague and George Brush. Although Pirsson knew little geology at the time and did not learn too much from Iddings, who was rather quiet, he found Iddings to be agreeable and decided to go into geology. He began to study crystallography, mineralogy, and petrography under Penfield at Sheffield and did more summer fieldwork with Iddings. In 1891, he went to Heidelberg to learn German and to study mineralogy and petrography with Rosenbusch. Later he studied with Lacroix and Fouqué. In the fall of 1892, he returned to Yale to become Instructor in Mineralogy in the Sheffield Scientific School. He then moved into geology and continued research in Montana where he particularly distinguished himself through his studies of the Highwood Mountains and the differentiated Shonkin Sag and Square Butte laccoliths. Upon his death in 1919, Pirsson held the title of Professor of Physical Geology.

Henry Stephens Washington (1867–1934), the newcomer to the group, graduated from Yale in 1886 after majoring in classics and then pursued graduate study in physics and chemistry (Lewis, 1935). From Yale, Washington went to the University of Leipzig, where he studied under Zirkel and received a Ph.D. degree in 1893. Washington also undertook studies in archeology, and between 1888 and 1894 he spent several months in study at Athens and participated in several excavations in Greece. In 1895, Washington returned to Yale to serve as Instructor in Mineralogy. As he was independently wealthy and teaching was not to his liking, Washington established a private laboratory in his own home, where he undertook investigations of igneous rocks and minerals. In 1912, he took a position with the Geophysical Laboratory and remained there until his death in 1934. Washington was particularly interested in the chemical side of petrology. His studies of the Roman volcanic field were among his most important labors. Perhaps his major contribution, however, was a series of U. S. Geological Survey Professional Papers on the chemical analyses of rocks (Washington, 1903, 1904, 1917). His compilations included detailed evaluations of the analyses as well as calculations of the CIPW norm for each rock. The level of accuracy, precision, and completeness of future chemical analyses was considerably elevated thanks to the prodigious efforts of Washington.

The four American petrologists lamented the fact that igneous rock classification was woefully inadequate because it still labored under the onus of being qualitative. The time had come, they said, for a comprehensive quantitative classification. Their initial classification proposal (Cross et al., 1902) was preceded by a comprehensive review of previous petrographic classifications (Cross, 1902) and succeeded by a book (Cross et al., 1903) and a series of papers through 1910 on improvements to the scheme and on textural terminology in igneous rocks. Cross et al. (1902, p. 556) were convinced that "the best results were to be obtained by the co-operation of several workers who agreed on fundamental principles and who were capable of harmonious collaboration."

Cross, Iddings, Pirsson, and Washington (from this point I refer to them as CIPW) perceived several defects in existing systems. First, they noted the lack of clearly enunciated and generally applicable guiding principles. As a result, they regarded existing systems as largely arbitrary. They perceived a lack of uniformity in the application of existing systems. They maintained that existing systems were, to some extent, founded on theory. They rightly noted that existing systems were largely qualitative rather than quantitative. They maintained that the recognized groups of rocks or rock families were inadequate for expressing known relationships. They found the nomenclature of petrography to be inadequate for expressing the relationships between groups, and they pointed out that there was no guide as to when the use of a new name was necessary or justified. As it stood, each investigator acted as his own judge.

CIPW eventually abandoned all efforts to modify existing systems. Recognizing that glassy rocks and many fine-grained volcanic rocks are not amenable to mineralogical analysis, the four petrologists decided to propose a classification based on chemical composition. To classify a rock completely by their method, it was necessary to know its chemical composition by chemical analysis or by approximation through microscopic methods. Their scheme was very complex and laden with a plethora of new terms. For the reader to grasp why the CIPW system exerted so much influence and aroused such vigorous antipathy in the petrological world, it is needful to describe the system in some detail.

CIPW proposed dividing all igneous rocks into five classes based on the relative abundance of light and dark minerals that they designated as "salic" and "femic," respectively. CIPW, however, excluded such dark ferromagnesian minerals as amphiboles, biotite, and aluminous augite from their "femic" category of minerals. The classification was not strictly chemical, for its initial subdivision into five classes was based on mineral content. How, then, could one use such a classification for a glassy rock? The answer of CIPW was to introduce a set of hypothetical, standard minerals that they termed the "norm." These hypothetical minerals were to serve as the means of classification. The norm was defined as a set of standard, *anhydrous*, chemically simple minerals that were calculated from the chemical analysis. The actual mineral composition of a rock they termed the "mode." They recognized that the norm and the mode of a rock would not be the same. In any case, the relative abundances of salic and femic minerals in the norm placed the rock into one of the five major classes. Salic minerals included normative quartz, feldspar, and feldspathoids, which the authors termed "lenads" from *l*eucite and *n*epheline. Femic minerals included the iron–magnesian minerals like normative olivine, enstatite, and diopside, but not hydrous minerals such as amphiboles or micas or complex minerals like aluminous pyroxene. These minerals were excluded to keep the norm simple.

The five classes, based on the relative abundances of salic versus femic normative minerals, were separated by boundaries located at salic/femic ratios of 7/1, 5/3, 3/5, and 1/7. Class I, for which the salic/femic ratio was greater than 7/1, was termed "persalane," from the adjective "persalic," meaning "extremely salic." Class II, in which the salic/femic ratio was between 7/1 and 5/3, was termed "dosalane" from "dosalic," meaning "dominantly salic." Class III, in which the salic/femic ratio was between 5/3 and 3/5, was termed "salfemane." Class IV, in which the salic/femic ratio was between 3/5 and 1/7, was termed "dofemane" from "dofemic," meaning "dominantly femic." And class V, in which the salic/femic ratio was less than 1/7, was termed "perfemane" from "perfemic," meaning "extremely femic."

CIPW divided these five classes into even more finely discriminated categories that included subclasses, orders, suborders, rangs, subrangs, grads, and subgrads. Each class could possibly be subdivided into subclasses based on the relative proportions of the different salic minerals or the different femic minerals. The subclasses of the three salic-rich classes I–III were based on the relative proportions of salic minerals. Thus subclass 1 of the first three classes had a (quartz + feldspar + feldspathoid)/(corundum + zircon) ratio greater than 7/1. Subclass 2 of the first three classes had a ratio of the same salic minerals between 7/1 and 5/3. The relatively femic-rich classes IV and V were divided into subclasses based on the relative proportions of femic minerals.

CIPW also divided each class into several "orders" based on the relative abundance of the normative minerals within the preponderant group. For example, ratios of quartz/feldspar, feldspar/feldspathoid, pyroxene/olivine, or (pyroxene + olivine)/ore minerals might be used for division purposes. The same boundary lines were used again, and an entirely new set of terms introduced. Thus, order 1 of a salic mineral-rich class or subclass would have a quartz/feldspar ratio greater than 7/1 and would be termed "perquaric," meaning "extremely quartz-rich." Order 2 had a quartz/feldspar ratio between 7/1 and 5/3 and was termed "doquaric," meaning "dominantly quartz-rich." Order 4 had a quartz/feldspar ratio between 3/5 and 1/7 and was termed "quardofelic" to distinguish it from order 6, which had a feldspathoid (lenad)/feldspar ratio between 3/5 and 1/7 and was termed "lendofelic." Orders in the femic-mineral-rich classes were based on the ratio of (pyroxene + olivine)/ore minerals like magnetite, ilmenite, and titanite. Here, for example, order 2 had a (pyroxene + olivine)/ore mineral ratio between 7/1 and 5/3 and was termed "dopolic," meaning "dominantly enriched in pyroxene + olivine."

Suborders could also be devised for classes IV and V on the basis of relative proportions of ore minerals to one another such as, for example, the ratio of magnetite to hematite. To this point, the CIPW classification was based on normative mineralogy. With continued subdivision, however, the classification was based on normative minerals or chemical constituents.

CIPW divided each order into "rangs" (an old term for "rank") on the basis of the relative abundance of bases. In the case of salic-rich rocks, the basis for subdivision was the ratio of alkalis to lime. The same divisional boundaries were used as in the subdivision of classes into orders. As examples, rang 1 of a particular order of the salic classes was one in which the (potash + soda)/lime ratio was greater than 7/1. Such a rock would be "peralkalic." Rang 4 was characterized by a (potash + soda)/lime ratio between 3/5 and 1/7 and was termed "docalcic."

In the case of femic-rich rocks, the rangs were based on the ratio of [(Mg,Fe)O + CaO] to alkalis. As an example, rang 2 would have a ratio between 7/1 and 5/3 and would be termed "domirlic." "Mirlic" was an acronym for magnesia, iron, and lime. Each rang might be further subdivided into subrangs based on either the potash/soda ratio or the magnesia/ferrous oxide ratio.

If further subdivision was required, the next category was the "grad." The grads were based on the proportions of the normative minerals of the subordinate femic or salic groups within classes. In other words, because the dominant minerals in class II were salic minerals, the subordinate group would be the femic group. Hence, for class II, grads were based on the ratio of (pyroxene + olivine)/ore minerals such as magnetite, ilmenite, and titanite. As an example, grad 1 of class II applied where the ratio of (pyroxene + olivine) to ore minerals exceeded 5/3. Such a grad was termed "prepolic" meaning "predominantly pyroxene and olivine." Grad 3 had a ratio of (pyroxene + olivine) to ore mineral less than 3/5 and was termed "premitic" meaning "predominantly magnetite, ilmenite, and titanite." In class IV, however, the dominant minerals were femic and, therefore, the subordinate group was composed of salic minerals. Hence, grads for class IV were based on the relative proportions of the subordinate salic minerals, quartz, feldspar, and feldspathoids. As an example, grad 1 of class IV had a quartz/feldspar ratio in excess of 5/3 and was designated as "prequaric." Still smaller units called "subgrads" could be defined on the basic of chemical ratios such as (magnesia + lime + ferrous iron)/alkalis in the subordinate femic minerals.

Each individual pigeonhole in this elaborate scheme was provided with a magmatic name that was logically derived from a very daunting nomenclatural system. Cross et al. (1902, p. 620) explained their methodology:

> To indicate the relative place of the various divisions in the system, from Class to Subgrad, we have adopted a set of terminations which are to be used invariably with their respective divisions. We have endeavored to select those which are at the same time: mnemonic in suggesting the relative positions of the divisions; euphonious; not in previous use to any great extent; and as far as possible adapted to use in all European languages. The termination -ite is rejected because it is already in use, not only in petrography but in mineralogy, to which latter science, as having a prior claim, we suggest that it be restricted.

Therefore, CIPW proposed the consistent use of the following terminations: for class, -ane; order, -are; rang, -ase; and grad, -ate; and for subclass, -one; suborder, -ore; subrang, -ose; and for subgrad, -ote.

Thus, class II rocks were designated as "dosal*ane*." But a rock belonging to the fourth order in which the quartz/feldspar ratio was less than 3/5 but greater than 1/7 was further designated as "austr*are*." If that rock could be assigned to rang 3, alkalicalcic, in which the ratio of the sum of alkalis to lime was less than 5/3 but greater than 3/5, it was further designated as "tonal*ase*." And if one chose to assign the rock to a subrang, for example, the persodic subrang number 5 in which the potash/soda ratio was less than 1/7, the rock would be designated as "placer*ose*." In short, CIPW proposed to inundate petrographers with a cataclysm of totally new terms such as toscanose, rockallase, kamerunose, baltimorase, baltimoriase, baltimorose, hispanare, germanare, italare, and portugare.

Cross et al. (1902) further proposed adding modifier terms to the name based on textural considerations as well as modal names. They provided a few examples of names from chemical compositions and actual mineralogy. For example, they listed the chemical composition of a typical monzonite from Monzoni studied by Brögger. From the calculated norm, CIPW found that the rock belonged to class II, the dosalanes, the fifth order germanare, the second rang monzonase, the third subrang monzonose, the second grad monzonate, and the second subgrad monzonote. But mineralogical and textural information modified the name, so that a field name of biotitic hornblende–grano dosalane would be applied. Microscopic study of the rock would yield additional information so that the rock could be more precisely depicted as biotitic hornblende–grano–monzonose! Mercifully, for dozens of potentially overwhelmed petrologists around the world, CIPW did allow for the use of traditional field rock names. Once the chemical analysis of a rock became available, however, the new nomenclature became applicable. The latter part of the radical scheme of Cross et al. (1902) laid out the rules and procedures for the calculation of the set of normative minerals from the chemical analysis of a rock and provided examples of the calculations.

The authors combined the two papers of Cross (1902) on the historical review of systematic petrography together with the classification paper (Cross et al., 1902) and published it the following year as a book (Cross et al., 1903). They added numerical tables to assist petrologists in making norm calculations as well as a glossary of the new classification. The reader needs no imagination to appreciate the sense of numbing horror that must have seized the petrological contemporaries of CIPW as they thumbed through the glossary and read such definitions as belcherose—"permagnesic Subrang (1) of the domiric Section (2) of the permirlic Rang (1) of the dopyric Section (2) of hungarare, the perpolic Order (1) of dofemane, class IV. From cortlandtite of Belchertown, Mass." or sagamose—"prepotassic Subrang (1) of bandase, the docalcic Rang (4) of austrare, the quardofelic Order (4) of dosalane, class II. From quartz–diorite of Sagami, Japan." or alferfemphyric—"microscopically porphyritic with both femic and alferric phenocrysts."

Any cynical petrographer tempted to suspect that the entire scheme was nothing more than an elaborate practical joke was quickly faced with cold reality when Iddings and Washington started publishing papers with titles like "Quartz–feldspar porphyry (granophyro liparose–alaskase) from Llano, Tex." (Iddings, 1904) or "Plauenal monzonose syenite" (Washington, 1906). In addition, chemical analyses that appeared in forthcoming publications of the United States Geological Survey were now frequently accompanied by the appropriate CIPW nomenclature and lists of normative minerals. Several writers outside the United States like Lacroix in France and Adams and Barlow in Canada picked up the new terminology. Iddings' (1903) professional paper on diagrams in petrology incorporated a host of chemical data recast in terms of the system, and the compilation of chemical analyses of Washington (1917) also included complete categorization of each analysis in terms of the CIPW scheme. This was no joke.

The attempted petrographic revolution was not completed in 1902. Four years later, CIPW added a detailed paper on textural terms (Cross et al., 1906). They made a careful distinction between crystallinity, granularity, and fabric. They introduced the term "holohyaline" to complement already existing terms like "holocrystalline," "hypocrystalline," and "hypohyaline." The authors also coined the terms "euhedral," "subhedral," and "anhedral." They established size ranges for granularity and developed an incredibly elaborate scheme for working out the textural characteristics of porphyritic rocks that took into account the relative sizes of phenocrysts to groundmass and to one another as well as textural features of groundmass in distinction from phenocrysts. Again, the authors introduced a whole thicket of new terms. For example, they coined the terms "peroikic," "xenoikic," "perxenic," and so on to designate the relative proportions of oikocrysts and xenocrysts! Contemporary petrographers may be grateful that most of these terms have not stuck. Despite its relentless precision, the entire system impressed many petrographers as a case of nomenclatural overkill.

THE REACTION TO CIPW

The American quantitative classification certainly awoke petrographers from a state of classificatory torpor. The most enthusiastic responses predictably came from the U. S. Geological Survey, to which all the authors had very close ties. The Survey made use of the scheme for the next several years. A. C. Lane (1904, p. 83) amply demonstrated that he was not a graduate of the schools of the prophets when he enthused that "whatever modifications the future may have in store, I think that there will be much of their work which will endure, and that such terms as 'persalic' and 'dofemic' will after a little become household words to the petrographers of all countries." A more neutral and cautious reaction came from J. Volney Lewis. In a memorial to H. S. Washington, Lewis (1935, p. 182) observed that

> the Quantitative Classification of Igneous Rocks may not have solved that difficult problem in its entirety, but it stimulated the interest of petrographers and geologists everywhere in the chemical and mineral constitution of these rocks and in the effort to devise a practical scientific method of classification. This interest has been reflected in part in the several alternative systems that have been proposed. Complacent acquiescence in the old condition of chaos in this field has been definitely brought to an end.

On the whole, most petrologists judged that the CIPW norm calculation was a tremendously valuable contribution to the science. As the twentieth century progressed, norms increasingly accompanied chemical analyses. In the end, however, the great majority of igneous petrographers and petrologists rejected the classification scheme itself and the radically new and complex nomenclature with which the scheme was freighted. Unwilling to abandon so many old familiar terms, petrographers kept the old terminology, repudiated the new petrographic language, and salvaged a few useful terms here and there like "euhedral."

Alfred Harker (1903), one of the earliest critics of the CIPW scheme, agreed that the business of nomenclature and classification lay in a state of confusion. He disagreed with their remedy, however. Harker attributed the state of confusion to the fact that microscopic petrography, particularly in the hands of the Germans, had led to so much new descriptive material that synthetic thinkers had been unable to keep up with it. "The pearls are there," Harker (1903, p. 174) said, "but there is no thread of which we can string them." Harker believed that CIPW had given up too easily on the possibility of a natural classification and suggested that petrologists were on the verge of discovering a fundamental principle analogous to the biological principle of descent that might lead to the recognition of a truly natural classification.

To Harker, the new classification was much too rigid, an unyielding framework such that rocks, whether existing or not, had a ready-made pigeonhole. "It follows, of course, since the hard dividing lines have no counterpart in nature, that rocks having the closest affinities must sometimes be divorced," Harker (1903, p. 175) wrote. He favored a more elastic scheme based on facts and not hypothesis but failed to supply a viable substitute. He credited the CIPW classification, despite its apparent complexity, with being straightforward, simple, logical, and precise, but not consistent. He objected, for example, to the subdivision of the first three classes on the basis of the relative proportions of quartz, feldspar, and feldspathoids whereas the last two classes were subdivided on the basis of the relative proportions of femic minerals.

Part of Harker's difficulty lay with the "strange terminology" that CIPW had introduced. It is, he said, "part of their design that each term shall carry its own precise meaning on its face. To this end they have boldly thrown aside the Greek lexicon, and constructed new words, each of which is a frank barbarism, but conveys an explicit meaning on a certain mnemonical plan" (Harker, 1903, p. 176). The nomenclature, he said, had "obvious utility" and yet it also had "a decidedly irritating effect" that would probably prejudice people against the scheme. Again, expressing his preference for the natural as opposed to the artificial, Harker stated that "there is something in human nature which ordains that languages shall grow, and not be produced ready-made; and, whether we call this stubborn conservatism or an instinctive grasp of the evolution philosophy, it insinuates a foreboding for the scheme before us, no less than for its terminology" (Harker, 1903, p. 177).

Harker (1903, p. 178) thought that rather than beginning at the top and dividing down as CIPW had done, we ought to begin at the bottom and build up:

> The unit of classification would be the sufficiently definite entity usually spoken of as a rock-type. The necessity for a given type being once established, and the characters of the type clearly defined, it should receive a name, preferably derived, as is customary, from the type-locality; and thereafter any attempt to alter the definition or to extend the name beyond it, should be firmly resisted. If this proviso could be secured, the often heard objection to the multiplication of names would become merely an objection to the increase of knowledge.

This procedure, he felt, should ultimately lead to clearer "insight into their mutual relations from the genetic point of view .... And a natural system expressing our knowledge of their chemical, geological, and geographical relationships."

Harker was also troubled by the need to do the classification in terms of the norm instead of the mode. Such procedure, he said, belied the fact that this was ultimately a chemical classification. He charged that the norm was only a circuitous device for presenting the chemical composition and wondered why CIPW had not used chemical composition directly as the basis for classification. Why not, he asked, simply begin with five classes divided on the basis of silica percentage instead of salic/femic ratio and then proceed to further divisions based on other chemical constitutents? He thought that procedure would offer advantages without the disadvantages of the present scheme. Furthermore, one of the drawbacks of the chemical approach for Harker was that the quantitative precision of the system required minute accuracy in the diagnosis of every rock specimen and, therefore, a trustworthy chemical analysis of a perfectly fresh specimen. Several of the same criticisms were repeated in *A Natural History of the Igneous Rocks* (Harker, 1909).

Cross (1910) defended the quantitative system against critics like Harker. He pointed out that although the concept of a natural classification is nice in theory, it simply would not work in practice. In fact, Cross (1910, p. 471) argued, despite claims to the contrary, "the systems devised for igneous rocks during the last century are unnatural, and without good reasons for most of their artificial or arbitary features." A classification of natural objects, Cross maintained, ought to be logical and consistent, but most of the igneous rock classifications thus far proposed struck him as unnatural and illogical for two reasons. On the one hand, he said, "what has been found or believed to be true of the rocks of a petrographic province, or of a certain group, has often been too quickly assumed to be true of all igneous rocks" (Cross, 1910, p. 472). As an example, Cross noted that the schemes based on geological age seemed natural to German and French petrographers but not to the British petrographers on the basis of their observations. The second reason for unnatural, illogical classifications, he found, was that many eminent petrographers assumed the truth of certain relationships of rocks that were only approximately in accord with facts.

Cross then reviewed the various grounds on which natural classifications had been attempted. Natural chemical classifications, he said, had been tried on the basis of geographic distribution or of genesis but had failed. The Atlantic and Pacific kindreds so dear to the heart of Harker failed to convince because there are so many exceptions. Rosenbusch's three groups based on geological occurrence appeared unnatural to Cross because there are so many gradations among them. He viewed Rosenbusch's scheme as inadequate in a world of coarse-grained lava flows, fine-grained plutons, and hypabyssal rocks with a wide range of textures.

Nor can we classify by eutectics, Cross argued. Becker thought that the groundmass of lavas closely resembled the products of eutectic crystallization. But a eutectic classification left nonporphyritic rocks out of consideration as well as important constituents of porphyritic rocks. Vogt, too, had suggested the possibility of classification based on eutectics. But Cross responded that the genetic importance of eutexia in magmas remained to be established and defined. Vogt, he said, had talked of anchi-eutectic rocks, but "the range of variation covered by the term 'anchi' (meaning almost), as used by Vogt, is so great as to destroy its significance" (Cross, 1910, p. 487). Cross concluded that classification by eutectics was fundamentally flawed because it rested on hypothesis, did not apply to all rocks, and did not consider the entire magma of most rocks.

Classification by mineral composition or texture was also obviously artificial, Cross continued, because each magma has an infinite range of possible textures depending on conditions of crystallization and cooling. Moreover, Cross (1910, p. 492) asserted that "classification by occurrence, as determining texture, or by texture, as expressing the broad phases of occurrence, is based on long disproved generalizations made from limited observation." He thought that classifying igneous rocks as plutonic, intrusive, and extrusive, or by equivalent terms like Rosenbusch's deep-seated, dike, and effusive rocks, "with the idea that because such a scheme expresses certain geological relationships it must perforce be accepted as 'a natural system', may be compared with dividing vertebrate animals into aquatic, terrestrial, and aerial, or those that swim, walk, and fly, respectively" (Cross, 1910, p. 493). But zoologists, Cross retorted, did not call whales fishes because they lived in water and swam, nor did they call a bat a bird simply because both of them were capable of flying.

Noting that Harker repeatedly insisted that a fundamental principle lying at the root of natural classification was almost within grasp, Cross retorted that Harker was unable to make any definite classificatory proposition even after his exhaustive study of the natural relations of igneous rocks. This failure of Harker's, he felt, confirmed his view that "the laws of magmatic descent operate on such complex solutions and under such a variety of conditions, that factors of necessary simplicity and of clear application in logical system seem to me impossible of realization" (Cross, 1910, p. 497).

Cross then rebutted specific criticisms of the CIPW scheme. To Harker's objection to the use of hypothetical minerals, Cross responded that the norm was compiled from substances that Harker and other petrologists regarded as the principal simple molecules of a magmatic solution. Iddings' participation in the creation of the norm suggests that he had moved away from his earlier idea that magma was a solution consisting of simple oxides to a view of magma as a solution composed of simple mineral molecules.

Harker (1903) had objected to rangs and subrangs based on ratios of alkalis because such a procedure entailed an inconsistent shift from the use of normative minerals to chemicals. Cross responded that the alkali oxide ratios were not based on the total amount of oxides in the bulk rock chemical analysis but rather just on the oxides assigned to normative orthoclase, leucite, albite, nepheline, sodalite, noselite, and anorthite. In effect, Cross maintained, the ratios in rangs are ratios of mineral molecules.

Harker (1909) also did not like the fact that a given designation such as "toscanose" could be applied to so many different kinds of rocks. But Cross indicated that many of the names cited by Harker were inaccurate to begin with, that some trachytes really should have been called quartz latites, that some granites should have been called quartz monzonites, and that many of the rhyolites were inappropriately named because they contain more anorthite than do typical rhyolites. "Instead of showing the heterogeneity of Toscanose," Cross (1910, pp. 501–502) reasoned, "this list of rocks illustrates the lax and inappropriate use of terms under the old system."

Cross' forceful defense of the CIPW system failed to stem the tide of criticism. In his book, *Igneous Rocks and Their Origin*, Daly (1914) contrasted the modal classification of Rosenbusch with the norm-based classification of CIPW and accused the latter of trying to cut the Gordian knot by ignoring the mode. Daly charged that the norm classification was largely founded on assumptions about chemical 'affinities,' and the course of molecule formation in natural magmas. He believed that "the norm system is almost sure to be found too sensitive to the discovery of new *facts* concerning molecular development in silicate melts." He called the system "ingenious but much too delicate" and complained that there was "a great mass of time-consuming labor in calculations and descriptions" that he thought would "turn out to have little permanent value as a system" (Daly, 1914, p. 10). One wonders what Daly's response to the classification would have been if he had had access to a desktop computer!

Like Harker, Daly spoke for other petrologists when he commented that individual subdivisions included rock types that are very different chemically and separated others that are very similar in terms of chemical composition, mineralogy, and genetic origin. As examples, Daly noted that the "camptonose" subrang included both a typical camptonite with 43.98 weight percent silica and a quartz basalt with 54.56 weight percent silica. In contrast, a rhyolite from Colorado plotted in the subrang omeose of the rang liparase in the persalane order brittanare whereas another chemically similar rhyolite plotted in the subrang madeburgose of the rang alaskase in the persalane order columbare.

In summary, Daly charged that the CIPW system disregarded vital principles of scientific classification. Turning Cross' zoological classification analogy against him, he suggested that "from the viewpoint of the geologists," the CIPW system resembled "a zoological system which would place in the same species the Newfoundland cod, the North Sea herring, and the Louisiana alligator, while specifically separating the New England cod from the Labrador cod of slightly different size" (Daly, 1914, p. 11).

Daly also regarded the norm system as having little practical value for the field geologist who had little choice but to use minerals. He claimed that the "quantitative" system did not contain any compelling principle for guiding the field geologist in combining the "species" of the system into larger units that were suitable for the practical geological mapping. Modal classification, however, had worked well in the field, and so, Daly noted, he continued to use Rosenbusch's system even though it still had not been quantified as of the fourth edition of *Mikroskopische Physiographie der Massigen Gesteine* (Rosenbusch, 1908) or the third edition of *Elemente der Gesteinslehre* (Rosenbusch, 1910).

Another extensive review of the CIPW classification was published by Scottish petrologist, George Tyrrell (1914). Tyrrell admired the effort of the four Americans but felt that their scheme failed to meet the everyday needs of working petrographers who really needed a good quantitative mineralogical classification. After all, Tyrrell said with good Scottish common sense, it is difficult to obtain chemical analyses for all the rocks that a petrographer is working on. The chances are far better that mineralogical information can be obtained from the thin sections.

Florence Bascom (1927, p. 54) summed up the general reaction to CIPW when she wrote that the system presented "too radical a break with pre-existing systems of classification to be cordially welcomed by petrographers."

QUANTITATIVE MINERALOGICAL CLASSIFICATIONS

In the end, the magnificent failure of the CIPW classification accomplished two important things for igneous petrology. Despite the abandonment of the overall classification method, petrologists generally recognized the value of norm calculations from the chemical analyses of rocks, especially those that are glassy or very fine-grained. Norms, of course, are calculated for rocks to the present day, and they have served as a basis for chemical classifications of rocks quite different in aspect from that originally proposed by CIPW. The second effect was that the determined effort of Cross and his colleagues to put forward a blatantly quantitative classification stimulated several petrographers to make their own serious attempts at proposing quantitative classifications on a mineralogical rather than a chemical basis.

Consequently, many new quantitative mineralogical classifications appeared during the early decades of the twentieth century. Most of these classifications, including those advanced by Lincoln, Shand, Holmes, and Johannsen, emerged from the English-speaking world, a fact suggesting that the balance of petrographic power was beginning to shift away from Germany. The early masters of petrography, Zirkel and Rosenbusch, died in 1912 and 1914, respectively, and Germany soon found itself embroiled in World War I. The attempt of Osann to update Rosenbusch's *Elemente der Gesteinslehre* clearly showed the effects of the war. In the preface, Osann stated how difficult it had been to gain access to literature published during the war years and apologized for the omissions. The tables of Washington (1917), for example, were not mentioned in Osann's 1922 revision.

Even after the pleas for quantitative classifications, schemes with only a limited quantitative character continued to appear. Hatch (1908) proposed an initial subdivision of rocks on the amount of silica in their chemical analyses. Then he reverted to a loosely semiquantitative mineralogical discrimination. Alexander N. Winchell (1913) proposed an ingenious classification with three coordinates. He expressed appreciation for the fact that the CIPW scheme revealed chemical relationships with fidelity but also saw the great weakness of the classification in the very need to have a chemical analysis available. He recognized that Rosenbusch was able to adapt his classification over the years because he did not put it into diagrammatic form. At the same time, he was puzzled that Rosenbusch (1910) was still paying lip service to geological age in the classification of some groups.

To portray a pigeonhole classification that was based on three factors, Winchell adopted the clever, if less than practical, device of publishing an insert with transparent paper overlays printed with different colored inks. The coordinate axis perpendicular to the overlays related to chemical composition. The rocks to be plotted on the three transparent sheets, therefore, belonged to chemically defined alkali–calcic (alkalcic), alkaline, and peralkaline groups, respectively. As the reader looked through the transparencies, he or she could see within a given pigeonhole the names of three different rocks with the same general mineralogy and mode of occurrence but somewhat different chemical tendency, as, for example, gabbro (alkalcic), essexite (alkaline), and theralite (peralkaline). On a given overlay, the horizontal coordinate related to the mode of occurrence and conditions of formation. In this regard, Winchell followed Rosenbusch and Harker in employing a threefold division of plutonic, hypabyssal, and volcanic. The vertical coordinate on a given overlay related to mineralogy. The categories were based on the dominance of various minerals within the rock.

Noting that quantitative mineralogical classifications were still lacking, the American economic geologist, Francis Church Lincoln, set out to devise one. Lincoln (1913) observed that there were already some potentially quantitative aspects to rock classification. He noted, for example, that Brögger had distinguished leucocratic from melanocratic rocks, had established monzonite as a rock intermediate to syenite and diorite, and had designated quartz monzonite as a rock intermediate to granite and quartz diorite. Lincoln proposed making Brögger's color categories quantitative by establishing boundaries based on volume percentages of leucocratic minerals present in a rock. Thus, he considered a rock with 67 to 100 volume percent of leucocratic minerals as "leucocratic," 33 to 67 volume percent as "mesocratic," and 0 to 33 volume percent as "melanocratic." Each of these three divisions was then subdivided into groups based on the relative amounts of the light minerals in the leucocratic or mesocratic divisions or the dark minerals in the melanocratic division. For rocks in the leucocratic division, for example, Lincoln established five groups, named quartz, quartz–feldspar, feldspar, feldspar–feldspathoid, and feldspathoid. The groups were based on the relative proportions of quartz, feldspar, and feldspathoids within the rock. Thus, for a rock to be placed in the quartz–feldspar group of the leucocratic division, quartz must compose between 33 and 67 volume percent of the light minerals and feldspar must compose between 33 and 67 volume percent of the light minerals. In the feldspar group, quartz lay between 0 and 33 volume percent of light minerals, and feldspar lay between 67 and 100 volume percent of light minerals. Finally, within the individual groups such as the quartz–feldspar group, the rocks were further divided into rock series on the basis of the percentage of feldspar that is orthoclase, or the percentage of feldspathoid that is leucite, or the percentage of ferromagnesian silicate that is olivine. By way of example, the quartz–feldspar group contained three series: granite, with 67 to 100 volume percent of the feldspar being orthoclase; adamellite, with 33 to 67 volume percent of the feldspar being orthoclase; and granodiorite, with 0 to 33 volume percent of the feldspar being orthoclase.

Of more lasting influence was a contribution by S. J. Shand, a Professor of Geology at the University of Stellenbosch in South Africa, a region poised to play a major role in igneous petrology. Because the stunning mineral wealth of South Africa had been recognized during the latter nineteenth century, the region received serious attention from numerous geologists, primarily from Great Britain and the Netherlands, the principal mother countries of the European settlers of South Africa. Geological surveys were in place by 1900. Although a great deal of the early geological effort was directed toward economic concerns, the association of much of the mineral wealth with igneous rocks, for example, platinum and chromium in the Bushveld Igneous Complex, copper in the Phalaborwa alkalic complex, and diamonds in the kimberlites, resulted in the development of considerable interest in igneous petrology in South Africa early in the twentieth century. Wagner, Kynaston, and Molengraaff, early experts on the Bushveld igneous rocks, were employed by geological surveys. Shand was the first of the great academic igneous petrologists of South Africa.

Samuel James Shand (1882–1957) was a native of Edinburgh, Scotland, who studied chemistry and geology at St. Andrews University, from which he graduated in 1905 (Chayes, 1958). He pursued graduate work at the University of Münster and wrote a Ph.D. dissertation on alkalic rocks of Assynt, an accomplishment that launched a life-long interest in silica-undersaturated rocks. In 1910, he also received a D.Sc. degree from St. Andrews. In 1911, Shand went to South Africa to become the chair of the geology program at the University of Stellenbosch, where, for a number of years, he constituted the entire department. At Stellenbosch, Shand undertook investigations of several South African alkalic massifs, a pursuit that consumed him for the rest of his life. In 1937, Shand came to America to occupy the chair of petrology at Columbia University until his retirement in 1950. In addition to studies of alkalic rocks, Shand was particularly interested in the classification of igneous rocks, especially in terms of silica saturation.

Shand (1913) suggested that rock-forming minerals be divided into two classes. The "saturated" minerals, he said, include those that can form in the presence of free silica. Those minerals that do not appear in association with free silica and are believed to be incapable of stable coexistence with it he designated as "unsaturated." Shand then defined "saturated" rocks as those that contain only saturated minerals and "unsaturated" rocks as those that contain only unsaturated minerals. A rock that contains both saturated and unsaturated minerals he designated as "partsaturated." He lumped both the unsaturated and partsaturated rocks into a broader category of "undersaturated" rocks, and a rock containing free silica he called "oversaturated."

While Shand was interested in the petrogenetic implications of the concept of saturation, he believed the concept to be relevant to the issue of rock classification. The concept, he thought, provided "just those natural distinctions between different types, the absence of which petrographers have been accustomed to deplore" (Shand, 1913, p. 512). He suggested that, rather than relying on the silica percentages in rocks, recent classifications like those of Hatch should instead have used mineralogical dividing lines that separated saturated, oversaturated, and undersaturated rocks from one another. Shand (1913, p. 513) claimed that if petrologists subdivided undersaturated rocks on the basis of their leucocratic and melanocratic constituents, and if they also distinguished partsaturated rocks from unsaturated rocks, they would "find themselves in possession of a much more 'natural' classification of rocks, and one which would be vastly simpler to use than any classification which is based upon silica percentages." Petrologists had erred, he asserted, by committing themselves to classifications that were neither strictly chemical nor strictly mineralogical. He pleaded for consistency in classification and maintained that hybrid classifications, like other hybrids, are barren. "Petrographers who shut their eyes to a natural distinction of this kind," Shand (1913, p. 514) rebuked his fellows, "deserve the fate that awaits them at the hands of inventors of arbitrary classifications."

Shand (1914) followed with a paper responding to criticisms and defending a natural classification based on minerals and structure. He agreed that anyone who could produce a system in which chemical and mineralogical characters were correlated satisfactorily would have solved the great problem of petrography. Every attempt to do so thus far had failed, in his view. Osann's method of building a chemical edifice on a mineralogical foundation had not found favor in America, Britain, or France. In his text on *Igneous Rocks*, Iddings (1913) had classified rocks on a simple mineral-textural basis and then appended to each group a discussion of chemical composition, but "to do this he must practically make a separate statement of composition for each rock; no general statements can be made, hence there is no correlation" (Shand, 1914, pp. 488–489). Shand also argued that Hatch's method of correlating feldspars with silica percentage did justice to neither the chemistry nor the mineralogy inasmuch as not all the constituents were taken into account. The classification of Winchell (1913) he regarded as incomplete because it included only purely qualitative expressions of chemical composition.

Shand admitted to having renounced the attempt to produce a single consistent chemical–mineralogical classification. Instead of yearning for the unattainable, he urged, petrographers should strive to agree on one chemical classification and on one mineralogical classification. At least, he thought, petrographers would then have a common language in which to record and compare their observations.

The next year, Shand (1915) proposed a framework for mineralogical classification with five categories: oversaturated rocks, saturated rocks, and three groups of undersaturated rocks. The first group of undersaturated rocks included those in which dyad (divalent) metals are undersaturated, such as olivine-bearing rocks. The second group of undersaturated rocks included those in which monad (univalent) metals are undersaturated, such as feldspathoid-bearing rocks. The third group of undersaturated rocks included those in which both monad and dyad metals are undersaturated, such as those containing both feldspathoids and olivine. For the remainder of his life, Shand ardently advocated the classification of rocks in terms of their degree of silica saturation as reflected by the mineralogy.

In 1917, a brilliant young English petrologist, Arthur Holmes (1890–1965), entered the discussion about classification. Holmes (Figure 18) studied at Imperial College, London, from which he graduated in 1909 at the age of 19 (Reynolds, 1968). During his undergraduate years, stunning discoveries were being made about radioactivity and its bearing on the ages of rocks and minerals. Holmes determined to pursue studies in this field and, upon graduation, spent time working with R. J. Strutt (Lord Rayleigh) on the Pb/U ratios of minerals from Oslo. Holmes' lifelong interest in radioactivity, geochronology, and the age of the Earth began at this time (Lewis, 2000). He received his D.Sc. degree in 1917 from London for his work on geochronology and the geology of Mozambique. He was Demonstrator at Imperial College from 1912 to 1920, then spent three years in Burma with an oil company as chief geologist. In 1924, he began his work at the University of Durham, where he built up the Department of Geology. In 1943, he took up the position of Regius Professor of Geology and Mineralogy at the University of Edinburgh, from which he retired in 1956. Later in life, Holmes married Dr. Doris Reynolds, a leading British advocate of granitization.

During his graduate school years, Holmes (1917) encountered the problem of classification in a very practical manner. He was engaged in rearranging the teaching collection of rocks in the Imperial College and thereby was confronted

18. Arthur Holmes (1890–1965). Reproduced by permission of Grant Institute, University of Edinburgh.

with the need to consider the principles on which igneous rocks should be classified. Holmes agreed with Shand that two systems of classification, one mineralogical and one chemical, were required, but he urged that these systems be arranged as closely as possible along parallel lines.

Holmes voiced many of the objections of others like Harker and Daly to the CIPW classification. He did think, however, that a purely mineralogical classification might proceed along CIPW lines. The terms "leucocratic" and "melanocratic" or "felsic" and "mafic," he pointed out, do correspond to the CIPW "salic" and "femic" classes. He suggested that petrographers could even use CIPW-like terminology and describe rocks as perfelsic, dofelsic, mafelsic, domafic, and permafic rather than persalic, dofemic, and so on. They could even use the same boundary lines as the CIPW classification. In the end, however, Holmes regarded Shand's method of subdivision as more valuable than the color ratio.

Within each of Shand's five groups based on saturation, which Holmes thought corresponded in part to the orders of the CIPW scheme that were based on quartz to feldspar ratios, Holmes suggested a cross classification that would be based on the relative orthoclase content of alkali feldspar versus the soda content of plagioclase. For the orthoclase content of alkali feldspar, Holmes employed the same subdivisions as the CIPW classification, namely 0, 1/7 (12.5 percent), 3/5 (37.5 percent), 5/3 (62.5 percent), 7/1 (87.5 percent), and 100 percent. For the soda content of plagioclase he suggested 0, 15, 50, and 100 percents. He also suggested that feldspathoids, micas, and analcime be recalculated into the equivalent amounts of orthoclase and albite for the classification. Holmes' ongoing interest in matters of classification and nomenclature came to further expression in the publication of his handbook of nomenclature (Holmes, 1920, 1928) as a replacement for Loewinson-Lessing's *Lexicon*. He also found time to produce a major petrographic manual, *Petrographic Methods and Calculations* (Holmes, 1921).

Shand (1917) responded that he was glad Holmes had adopted the degree of saturation as the basis of classification and that he had distinguished potassic from sodic and calcic rocks on the basis of their feldspars. He rejected Holmes' idea of recalculation of feldspathoids and micas into equivalent orthoclase, however, on the grounds that "it takes us away from the actual composition of the rock and partly because in many cases it must be impracticable and its results misleading" (Shand, 1917, p. 465). He preferred the plain truthfulness of the mode in comparison to the "unreliable" norm. He maintained that it was simpler to use the ratio of (orthoclase + leucite)/(albite + nepheline + sodalite) without any modification and with the usual limiting values of 5/3 and 3/5. Shand also favored the use of two groups of textural terms following Zirkel and Iddings as opposed to the three groups used by Rosenbusch, Harker, Winchell, and Holmes. He advocated eliminating the third category of hypabyssal or dike rocks from igneous rock classification on the grounds that there is no special set of textural characteristics belonging to a dike. Shand's classification based on the principle of saturation persisted through the final edition of his textbook, *Eruptive Rocks* (Shand, 1949a).

The most ambitious mineralogical classification of the early twentieth century, however, was devised by Albert Johannsen, Professor of Petrography at the University of Chicago and America's answer to Harry Rosenbusch (Pettijohn, 1963, 1988). An unsurpassed expert on microscopic petrography and author of an infinitely detailed, four-volume work entitled *A Descriptive Petrography of the Igneous Rocks*, Albert Johannsen (1871–1962) was described by Pettijohn (1963, p. 454) as "the greatest and last of the American school of petrographers." After receiving a B.S. degree in architecture from the University of Illinois in 1894, young Johannsen worked at various jobs, one of which was as a field assistant with the United States Geological Survey. He went to the University of Utah for further work and received another B.S. degree in 1898, this time in geology. For graduate work, Johannsen attended Johns Hopkins University, then the recognized American leader in the field of petrography, and received his Ph.D. degree in 1903. At Johns Hopkins, Johannsen studied under E. B. Mathews, himself a pupil of G. H. Williams and Rosenbusch. Johannsen became steeped in the great German tradition of petrography. After a number of years working with the United States Geological Survey in the San Juan Mountains of Colorado and the Black Hills of South Dakota, Johannsen

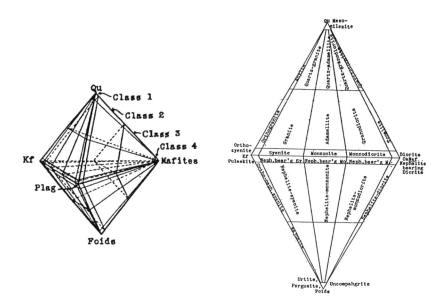

19. Igneous rock classification of Albert Johannsen (1917). Left: subdivisions of the double tetrahedron into classes. Right: the families in class 2 on a quartz-alkali feldspar-plagioclase-feldspathoids (QAPF) double triangle. Reproduced from *Journal of Geology* by permission of The University of Chicago Press.

succeeded Iddings on the geological faculty at the University of Chicago in 1909 but still retained his connections to the survey until 1925. The encyclopedic *A Descriptive Petrography of the Igneous Rocks* (Johannsen, 1931–1937) and *Manual of Petrographic Methods* (Johannsen, 1914) were more than adequate testament to Johannsen's impressive petrographic prowess.

Johannsen (1917, 1920) wanted a strictly quantitative mineralogical classification. Following a modified Linnéan scheme, he proposed dividing igneous rocks into four classes based on the relative abundance of light and dark modal, not normative, minerals. He placed boundaries between the different classes at 5, 50, and 95 percent mafic minerals. He subdivided each of his four classes into four separate orders based on the Na/Ca ratio of the plagioclase feldspar. He located the boundaries between the different orders at 5, 50, and 95 percent anorthite. The orders were then divided into families. Johannsen plotted the classes on a double tetrahedron (Figure 19) whose corners were represented by quartz, K feldspar, plagioclase, mafic minerals, and feldspathoids, which he also termed "foids." The intersections of bounding planes for classes and families divided the double tetrahedron for a particular order into smaller volumes to which numbered rock family names were assigned. For greater clarity of presentation, Johannsen plotted the various families on a series of double triangular diagrams (Figure 19). The boundary lines separating different families in classes 1 to 3 and of Na-rich orders were based on K-feldspar/plagioclase ratios and also quartz percentages. The K-feldspar to plagioclase boundaries were drawn at 5, 35, 65, and 95 percent plagioclase. The quartz boundaries were located at 5, 65, and 95 percent quartz.

Johannsen thus developed an elaborate system of numbered pigeonholes for possible rocks and also provided a numbering system for the rocks. Hence, a rock like granite that belonged to Class two, Order two, and Family seven was designated as rock #227. To fill in the pigeonholes in his scheme, Johannsen coined at least 133 new rock names, making him by far the most prolific contributor to igneous rock nomenclature in the history of igneous petrology (Le Maitre, 1989). Johannsen also laid the basis for the modern system of classification with his creation of the QAPF double triangle. Despite the revisions to his scheme, contemporary systematics owes an enormous debt to the work of Albert Johannsen.

NEW MEASUREMENT TECHNIQUES

The sudden flourishing of quantitative mineralogical classifications during the 1910s and 1920s was accompanied by several technical improvements in obtaining quantitative data. Chayes (1956) regarded Shand's micrometer as the first major improvement on Rosiwal's method. Shand (1916) found a way to reduce some of the tedium of the Rosiwal method by inventing a special movable stage that would do the summing of mineral grain diameters automatically. The biggest problem with the micrometer was that data for only one mineral and for the sum of all the other minerals in the rock could be recorded during one set of traverses of a thin section. C. K. Wentworth (1923) devised an improved recording micrometer that could keep track of at least five minerals at a time. He also abandoned Rosiwal's requirement that no grain should be cut by more than one traverse line. W. F. Hunt (1924) made further refinements to Wentworth's stage.

As a different time-saving effort, Johannsen (1919) devised a planimetric method for measuring the areas of grains. He demonstrated that the measured areas were proportional to volumes provided that the minerals were evenly distributed throughout the rock. He did not regard his planimeter as suitable for the measurement of strongly layered rocks.

Alling and Valentine (1927) compared these various measurement methods. They noted that a complete analysis could be performed on the Shand stage in about 45 minutes. The Johannsen camera lucida method also required about 45 minutes. In contrast, they noted, the standard Rosiwal analysis took several hours and the extreme accuracy method of Lincoln and Rietz (1913) took as much as two days.

In spite of considerable improvement in time efficiency, measurement of the amounts of minerals in igneous rocks continued to lag, even after a handful of quantitative mineralogical classifications had been proposed. Chayes noted that both the Shand micrometer and Johannsen planimeter cost around $25. He bemoaned the fact that "all over the country petrographers who presumably would not afford $25 for an accessory which will permit them to run any number of modes were managing to find and spend at least that much for each chemical analysis they purchased" (Chayes, 1956, p. 37), and this at a time when thousands of complete chemical analyses already existed.

Despite Chayes' later disgruntlement, the means were now at hand, however cumbersome, for making good use of quantitative mineralogical classifications. But which classification to use? Which rock names to use? Throughout the 1920s, petrographers repeatedly expressed exasperation over the growing lack of unanimity regarding classification. Holmes (1920, p. 5), noting the "luxuriant" growth in nomenclature and expressing his wish for more uniformity in terminology, wrote that

> it would, of course, be desirable if definitions of rock-names could be framed by an International Committee endowed with authority to fix meanings finally, and to decide on the validity of new terms at suitable intervals. Unfortunately such a counsel of perfection is not likely to be sought for many years, and even were a powerful committee to be formed, its authority would sooner or later be sapped by disagreement. One such attempt to standardize nomenclature revealed so wide and stubborn a divergency of opinion as to its practicability, and the individual rights of authors to use terms as they choose, that no final decisions were arrived at, and only a few general suggestions and the revised Lexicon of Loewinson-Lessing emerged from the conference.

Tyrrell (1921, p. 494) maintained that

> the question of classification is therefore at the root of our nomenclatorial difficulties; and agreement upon a classification by petrographers, or the enforcement of a classification through the influence of a sufficiently powerful and inspiring teacher, especially in the case of igneous rocks, seems to be a pre-requisite to the successful reform of the nomenclature.

Tyrrell was also of the opinion that, rather than there being too many rock names, there were "ridiculously few," especially in comparison with the sister natural history sciences.

A. K. Wells (1924) conceded that there was little hope of putting rock nomenclature on a firm basis until there was one universally recognized system of rock classification. In contrast to Tyrrell, he griped that there were far too many rock terms, so many of which were named after some obscure location that required a good atlas and gazetteer to locate. "Will not someone in authority," Wells (1924, p. 323) pleaded in exasperation, "lay down the law in this matter, and state dogmatically what differences in composition and texture and considered in what order, shall be of sufficient importance to justify the erection of a new species of rock?" Recognizing that the British had erected a Committee of Nomenclature, he desperately hoped that their recommendations might be adopted.

Shand (1923) also appealed to the British Committee to go beyond "killing off many obsolescent names" and provide further guidance in establishing principles for the construction of new names. The Committee on Petrographic Nomenclature, which all the above writers had in mind, was appointed by the Geological Society of London in 1920 in an effort to bring some order into the nomenclatural chaos of the times. The committee included such luminaries as Teall, Tyrrell, Hatch, and Holmes. The committee issued a report in 1921 in which it recommended the rejection of 27 old names, including diabase and leucitophyre, the reinstatement of trachy-basalt, and the redefinition of basalt in such a way that the term carried no implication as to the presence or absence of olivine. No new names were proposed by the committee, nor was a new classification suggested. The committee also favored the continued use of Rosenbusch's terms for grain shape, "idiomorphic," "hypidiomorphic," and "allotriomorphic," as opposed to Rohrbach's terms "automorphic," "hypautomorphic," and "xenomorphic" or Pirsson's terms "euhedral," "subhedral," and "anhedral."

Both the collaborative effort of the four Americans and the work of the British Committee on Petrographic Nomenclature made it painfully clear to the petrological community that the cooperative efforts so wistfully longed for by many of them would be extremely difficult to organize and would likely yield relatively few results of lasting value. Another half century would elapse before effective cooperative effort in the area of classification and nomenclature would finally take place.

15 ← MELDOMETERS AND THERMOCOUPLES

Early Experimental Petrology

In tracing the story of igneous petrology up to the early twentieth century, I have referred in passing to various individuals who conducted experiments of relevance to igneous petrology. I have mentioned Nicholas Lémery, whose work provided grounds for the notion that volcanic activity might be fueled by the combustion of pyrite, James Hall, whose experiments supported Hutton's conception of granite as a once molten subterranean lava, and Auguste Daubrée, whose investigations lent credence to the view that quartz and, therefore, granite were of aqueous origin. These scientists by no means represented isolated cases in which a glint of experimental light was shed on the problems of igneous petrology, for a substantial amount of petrologically relevant experimental work was conducted throughout the nineteenth century. The value of such efforts was often a matter of debate. Some geologists enthusiastically endorsed the results of experiments to support their views on petrological questions. Such were the geologists who supported their disavowal of the origin of granite by pure igneous fusion with appeals to the work of Schafhäutl, Daubrée, and Heinrich Rose showing that quartz could be produced from water vapor or aqueous solution. On the other hand, Hutton was reluctant to accept experimental support for his ideas about granite because he was skeptical that a furnace could give much insight into the subterranean operations of the globe. There is little doubt, however, that these early labors led to the blossoming of experimental petrology at the beginning of the twentieth century. Before we turn our attention to the petrological revolution associated with the experimental work conducted at the Geophysical Laboratory by N. L. Bowen and his colleagues, we need to gain a sense of the events that made those achievements possible.

EARLY EXPERIMENTAL STUDIES

Even before the well-known experiments of James Hall, several individuals explored the melting of rocks, glass making, and the manufacture of porcelain and other ceramic materials. As early as the seventeenth century, the philosopher Leibniz appreciated the importance of experiments for the interpretation of nature, claiming that we should compare the materials from Earth with those produced in laboratories. He noted that when stones were submitted to fire they yielded glass. In the eighteenth century, Lémery's experiments on the spontaneous ignition of heated mixtures of sulfur and iron led to the view that volcanic activity might be attributed to a similar cause. Buffon showed by his experiments that granite and other crystalline rocks can be converted into glass and maintained that these rocks acquired their crystalline state after a very long period of slow cooling. He also demonstrated that feldspar is much more fusible than the other minerals in granite. Réaumur conducted experimental research on porcelain, and Keir studied the cooling and crystallization of bottle glass.

Spallanzani, one of the most prolific of early experimentalists, subjected a wide range of rock types to fusion. He performed a series of experiments in which he fused lavas and showed the incorrectness of Dolomieu's idea that lavas derive their heat by combustion when they arrive in the atmosphere. D'Arcet, Gerlhard, de Saussure, and Spallanzani all subjected granites to fusion experiments, and Spallanzani observed the relative ease of fusion of feldspar, mica, and quartz and compared their fusibilities against the Wedgwood scale. He also discovered that the degree of fusibility of feldspar is related to variations in its chemical composition and that the ease of fusion of iron-bearing minerals increases as the amount of iron in the mineral increases.

The fusion experiments of James Hall and Gregory Watt lent support to the concept that basalt and lava are fundamentally the same, and Hall produced experimental evidence for the igneous origin of granite. Dartigues experimented with the devitrification of glass. By the early nineteenth century, therefore, a sizeable contingent of investigators across Europe had experimentally examined the melting and cooling behavior of glasses, lavas, and many crystalline silicate rocks and had also developed a rough idea of the relative fusibilities of common igneous minerals.

EXPERIMENTAL SYNTHESIS OF MINERALS

After these early attempts to melt rocks and minerals, experimentalists concentrated on the problem of synthesis of all kinds of minerals. As the synthesis of economically valuable minerals such as galena or sphalerite proved particularly intriguing, considerable attention was devoted to the sulfides. By no means, however, were common rock-forming minerals neglected. The early syntheses were not conducted in a systematic manner. By trial and error, researchers simply wanted to discover the means by which minerals like olivine, quartz, or pyrite could be produced. Although temperatures of synthesis could be estimated, precise determination of melting points of minerals was not possible throughout the first several decades of the nineteenth century. The effects of pressure were also not taken into account.

Throughout virtually the entire nineteenth century, a plethora of investigators succeeded in synthesizing a wide range of common minerals in a variety of ways. The mere synthesis of minerals was a major achievement that provided the first step toward ultimate understanding of the conditions of mineral formation and stability. Numerous investigators such as Senarmont, A. C. Becquerel, Debray, Durocher, Manross, Margottet, G. Rose, and others synthesized dozens of sulfides, oxides, sulfates, and so on. Synthesis of rock-forming silicates, initiated by Mitscherlich, proceeded more slowly.

The German chemist, Eilhard Mitscherlich (1794–1863), entered Heidelberg in 1811 to study Oriental languages (Rose, 1864; Szabadváry, 1974). He continued studies at Paris in 1813, concentrating on the study of Persian in hopes of becoming a member of Napoleon's legation to Persia, but after the fall of Napoleon he returned to Germany. He entered Göttingen in 1817 to read science and medicine, still hoping to reach the Orient. Although he received a doctorate on ancient Persian texts at Göttingen, an interest in chemistry was fueled by Strohmeyer. Mitscherlich then went to Berlin in 1818, where chemical research on phosphates and arsenates induced him to study crystallography. His crystallographic measurements on a variety of crystalline substances led him to recognize the isomorphous nature of many minerals and other crystalline compounds. Indeed, Mitscherlich coined the term "isomorphism."

Between 1819 and 1821, Mitscherlich spent two years in Stockholm with chemist J. J. Berzelius. During that period, he visited mines and the metallurgical works at Fahlun, where he reported on the accidental production of both augite and olivine in the scorias of the copper smelting furnace. He confirmed the identity of the synthetic products with natural minerals by goniometric measurements. He also reported the artificial production of a black mica that resembled biotite. In 1821, Mitscherlich became a professor in the University of Berlin, where he continued his studies of isomorphous substances, synthesized minerals such as melanite garnet and diopside by fusion of mixtures of silica with various metal oxides, and investigated the geology of the volcanic regions of Italy, France, and Germany.

Experimental studies of silicate minerals and rocks especially flourished in France. Pierre Berthier (1782–1861) studied under Berthellot at the École Polytechnique between 1798 and 1801 and then entered upon a career at the École des Mines in 1801 (Kuslan, 1970). From 1816 until his retirement in 1848, he was Professor of Assaying and chief of the laboratory at the École des Mines. While there he analyzed dozens of minerals and ores and discovered several new mineral species, including the Fe-Sb sulfide later named berthierite in his honor. Around 1823, Berthier synthesized augite crystals by melting silica and the appropriate bases in a porcelain furnace. He also obtained olivine and tephroite by purely igneous fusion of mixtures of silica and metal carbonates.

In 1851, Jacques Joseph Ebelmen (1814–1852) reported the synthesis of numerous minerals, including olivine and pyroxene. Ebelmen was a French chemist, mining engineer, and ceramist who was a professor at the École des Mines and director of the Sèvres Porcelain Works (Berner and Maasch, 1996). He produced olivine by melting the constituent elements with acid or alkaline volatile agents at high temperature. For example, he formed olivine and pyroxene crystals by heating mixtures of silica, magnesia, and boric acid or potassium hydroxide. In 1852, Ebelmen reported the synthesis of sphene and perovskite. To produce perovskite, he heated a mixture of titanic acid, alkali silicate, and alkali carbonate. A piece of calcium carbonate was dipped into this melted material and changed entirely to perovskite. He also produced willemite by heating boric acid, silica, and zinc oxide.

One of the major institutions for the prosecution of experimental studies in mineralogy during the nineteenth century was the École Normale Supérieure in Paris. Along with Louis Pasteur, Henri Deville was one of the scholars who helped to increase the prestige of the institution as one of the outstanding centers of scientific education and research in France. Henri Étienne Sainte-Claire Deville (1818–1881) was born in the Virgin Islands of French parents and educated in Paris as a medical student (Leicester, 1971). After earning doctorates in both medicine and science, Deville began his career in 1845 as Professor of Chemistry and Dean of the newly created Faculty of Science at the University of Besançon. In 1851, he moved to the École Normale Supérieure, where he worked on the very high temperature synthesis of minerals with a team that included H. J. Debray, L. J. Troost, H.L.M. Caron, and others. Together with colleagues and students, Deville, particularly in the years 1858 to 1863, synthesized around seventeen minerals, including apatite, corundum, zircon, cinnabar, and greenockite. On his own he successfully synthesized about ten other minerals, including cassiterite and pyrite. In 1861, Deville produced willemite by passing silicon fluoride over melted zinc and red hot zinc oxide. The following year, he synthesized the zeolite minerals lévyne and phillipsite. He formed lévyne by heating a mixture of potassium silicate and sodium aluminate solutions in a sealed tube between 150°C and 200°C, and he synthesized phillipsite similarly at 200°C by heating potassium silicate and potassium aluminate solutions. Deville's assistant, H. J. Debray, synthetically reproduced a host of nonsilicate minerals, such as the phosphates apatite, pyromorphite, vivianite, libethenite, and olivenite, the oxides bunsenite, corundum, hausmannite, and periclase, as well as scheelite, wolframite, and azurite.

One of Deville's prime pupils was Paul Gabriel Hautefeuille (1836–1902) (Nye, 1972). After working as an engineer, Hautefeuille received his doctorate under Deville in 1865 and continued as his assistant at the École Normale Supérieure for many years. He reproduced many minerals by utilizing mineral catalysts. In 1864, Hautefeuille made magnesian pyroxene by heating a mixture of silica and magnesium chloride. The following year, he produced olivine by melting a mixture of silica, magnesia, and magnesium chloride. Hautefeuille became the first experimentalist to synthesize orthoclase. In 1877, he obtained orthoclase by heating a mixture of tungstic acid and potassium alumino-silicate to between 900°C and 1000°C for several days, and he synthesized albite by heating tungstic acid with sodium alumino-silicate for several weeks. He also succeeded in producing orthoclase by heating a mixture of silica, alumina, and potassium tungstate. In 1880, Hautefeuille produced leucite. He also demonstrated the temperature dependence of tridymite and quartz.

G. A. Daubrée grew quartz, feldspar, and pyroxene from aqueous solutions. Among others who succeeded in synthesizing pyroxene and olivine was Étienne Stanislas Meunier (1843–1925), Professor of Geology at the Muséum d'Histoire Naturelle (Spencer, 1927). At the Collège de France, Fouqué and Michel-Lévy not only synthesized a large number of silicate minerals but also reproduced several rocks in which the mineralogy and texture were identical to those of natural volcanic analogs. They were unsuccessful, however, in reproducing the texture of trachyte and rhyolite. They stressed the importance of cooling rate on the grain size of igneous rocks. These lifelong friends and collaborators also published a compilation of the results of experiments on minerals and rocks under the title *Synthèse des Minéraux et des Roches* (Fouqué and Michel-Lévy, 1882).

PYROMETRY

If the artificial production of various minerals was to have any ultimate scientific value in terms of the determination of the physical conditions of mineral and rock formation, geologists would need to determine accurately the temperatures at which minerals melt, crystallize, or react with one another. Such a requirement, of course, demanded the construction of a reliable high-temperature scale and the development of appropriate means for the measurement of high temperatures. To be sure, there had been efforts to measure temperature reliably long before the era of experimental mineral synthesis. Thermometers based on the expansion of liquid had been invented in the early 1600s. These early thermometers, typically based on the expansion of liquid mercury, had arbitrary scales, thus prompting the desire for standardization. The Celsius scale was proposed in the early 1700s, the Fahrenheit scale in 1714, and the Réaumur scale around 1731. Because mercury thermometers are limited by the freezing and boiling points of mercury, other methods needed to be developed for the measurement of the much higher temperatures of interest to igneous petrology.

During the nineteenth century the fusibility of a mineral was one of the physical properties used to identify an unknown. The fusibilities of minerals were pegged to a scale of fusibility analogous to the Mohs' scale of hardness. The scale of fusibility, developed by Franz von Kobell, consisted of several standard minerals, including, in decreasing order of ease of fusibility, stibnite, natrolite, almandine, actinolite, orthoclase, and bronzite, with quartz still higher. The fusibility of an unknown mineral might be determined in comparison with the set of standard minerals by means of a blowpipe. The defect of the fusibility scale was that it was only a relative scale. The minerals in the fusibility scale needed to be calibrated with some standard temperature scale that included precisely determined points at temperatures above the freezing and boiling points of water. How could such higher temperatures be measured accurately and precisely?

In the thick of efforts to extend thermometry to high temperatures was American experimental physicist, Carl Barus (1856–1935). After study at Columbia School of Mines, Barus went to Germany and obtained his Ph.D. degree in physics under Friedrich Kohlrausch at the University of Würzburg in 1879 (Lindsay, 1943). In 1880, Barus joined the U.S. Geological Survey as a geophysicist, worked with George Becker making electrical measurements of ore bodies, calibrated platinum thermocouples for high-temperature measurement, and studied physical and thermal properties of rocks at high temperatures and pressures. In 1892, Barus was dismissed from the Geological Survey along with several other staff members in a budget-cutting action. Then, after several other government jobs of brief duration, Barus was finally appointed in 1895 as Hazard Professor of Physics at Brown University, a position he held for the remainder of his career.

Barus (1889) published a mammoth review of methods of pyrometry, that is, temperature measurement. He listed seventeen different classes of pyrometers. Like the mercury thermometer, many pyrometers were based on the presumed linear expansion or contraction of solids, liquids, or gases with temperature. Among the first pyrometers to be based on the expansion or contraction of solids was the well-known Wedgwood scale of 1782, based on the amount of shrinkage experienced by a small compressed cylinder of clay after exposure to a particular temperature. By 1881, scales appeared that were based on the linear expansion of platinum metal.

Some liquid-based pyrometers employed the thermal expansion of fused metal alloys, mercury, or water under pressure. The difficulty with many expansion-based thermometers was the problem of extrapolating linear expansion beyond the range of the boiling point of water. Several pyrometers utilized the differential expansion of pairs of solids such as platinum and lead. Other pyrometers took advantage of such temperature-dependent properties as the vapor tension of liquids, the dissociation of compounds like calcium carbonate, the fusion, ebullition, specific heat, heat conduction, heat radiation, or viscosities of solids, liquids, or gases. Still others were based on spectrophotometric properties, acoustics, electrical resistance, magnetic moment, and thermoelectrics.

Despite the welter of creative methods of attacking the problem of temperature measurement, Barus (1889) expressed uneasiness and disappointment that the results were in general rather faulty. What he found troublesome was not just that observers using different methods failed to achieve consistent results but that even skilled observers repeating the same method obtained inconsistent data in measuring the boiling point of zinc. Barus found that the published values for that boiling point ranged from 884°C to 1040°C. He suggested that such a wide range of values was hardly likely to inspire confidence.

The two pyrometric methods that had gained the most favor toward the end of the nineteenth century, however, were gas thermometry and thermoelectrics. The method of gas thermometry was based on the assumed linear rate of expansion of a gas with increased temperature. The application of gas thermometry to higher temperatures began in 1828 when Prinsep attempted to measure the melting points of various precious metal alloys. Much of the history of gas thermometry revolved around attempts to find suitable material to use for an inflatable bulb to be filled with a heated, expanding gas during a temperature measurement. Investigators also experimented with different gases with which to fill the bulbs. The volume of the gas-filled bulb, or the gas pressure within a bulb of constant volume, was then regarded as an indicator of the temperature being applied to a substance undergoing melting. Prinsep employed a gold bulb filled with air. In 1837, Pouillet replaced gold with platinum because the melting point of gold was too low to contain extremely hot gas. Becquerel and also Deville and Troost, in turn, substituted thin porcelain bulbs for platinum because of the permeability of platinum to expanding air. They also substituted iodine for air as the gas. For a time, melting-point values of various substances up to 1000°C obtained by Deville and Troost with porcelain bulbs and iodine gas were generally accepted. Investigators eventually abandoned iodine to return to air until Schinz introduced the use of nitrogen. Porcelain bulbs were retained for the rest of the nineteenth century in spite of the fact that unglazed porcelain bulbs proved to be porous to nitrogen, and glazed porcelain bulbs were converted at high temperature into mixtures of uncertain chemical composition that actually released gas. It was not until Holborn and Day, in 1900, returned to use of a platinum bulb filled with nitrogen that the melting point of gold was accurately determined.

The other favored method of temperature measurement entailed the use of thermocouples. The thermocouple was first developed by Thomas Johann Seebeck (1770–1831) in Germany. Seebeck had been experimenting with the effects of temperature on the electrical properties of such metals as bismuth, copper, and antimony. In 1821, he discovered that when two wires composed of different metals were connected to make a complete circuit, an electrical current would flow when one of the connections was heated. He concluded that an electrical potential generated between the hot and cold junctions resulted in the flow of current. By keeping the cold junction at room temperature and using it as a reference, Seebeck found that the higher the temperature of the hot junction, the greater the flow of electrical current through the circuit.

Thermocouples were further developed by Jean Charles Athanase Peltier (1785–1845) and Antoine César Becquerel (1788–1878), grandfather of Henri Becquerel, the discoverer of radioactivity. A. C. Becquerel, a graduate of the École Polytechnique, and a disciple of Alexandre Brongniart, became the first occupant of the chair of physics at the Muséum d'Histoire Naturelle in 1838. He successfully used voltaic cells to synthesize crystals of minerals, particularly sulfides. In 1826, he studied electromotive force (EMF) as a function of temperature, developed thermocouples of platinum and palladium, and proposed that thermocouples could be used for thermometric applications. To that end he attempted to measure the boiling point of zinc.

Henri Deville and Henri Debray were commissioned by the Russian government to investigate the applicability of platinum–iridium alloys to coinage. They discovered that these alloys resist corrosion better than platinum. Debray also studied the electrical and thermal properties of rhodium and various platinum compounds. Building on their work, Henri Le Châtelier (1850–1936) experimented with platinum and platinum-bearing alloys as thermocouples in 1885 (Leicester, 1973). He found that a thermocouple constructed from one wire of pure platinum and another wire composed of an alloy of 90 percent platinum and 10 percent rhodium was more stable at high temperatures than other combinations. As a result, Le Châtelier's platinum–rhodium thermocouple became the thermocouple of choice.

Barus (1889) believed that thermo-electric pyrometry was superior to all other methods and yielded results that were as accurate as any known. To use the thermo-element, however, he said that it was still necessary to determine the temperature equivalent of the EMF for all temperatures of the two thermocouple junctions. That could be done effectively only by calibrating the thermo-elements against a gas thermometer.

Another widely used method of temperature determination entailed the use of so-called pyrometric cones, small triangular ceramic prisms so arranged that, upon heating at a specified temperature at a particular rate, the tip of the cone would bend over to touch the level of the base of the cone. The bending of the cone at a particular temperature was believed to be caused by the formation of viscous liquid within the cone. First used in 1882 in the manufacture of Sèvres porcelain, the pyrometric cones became known as Seger cones when Hermann Seger, Director of the Royal Pottery Works in Berlin, published a book in 1886 in which he described a series of 36 cones having endpoint temperatures ranging from 600°C to 1800°C. The Seger cones were composed of mixtures in various combinations of powdered quartz sand, feldspar, calcium carbonate, kaolin, alumina, ferric iron, sodium carbonate, and boric acid. Seger produced and sold the cones internationally to the ceramic industry.

Because of the general lack of reliable pyrometric methods against which the melting of the cones might be calibrated, Seger did not originally specify their endpoint temperatures. Later, with improvements in gas thermometry and thermo-electrics, Le Châtelier provided the supposed endpoint temperatures of Seger cones.

Still another method of determining melting points of solids was developed by John Joly (1857–1933) at a time when he was an assistant to a physics professor at Trinity College, Dublin (V. A. Eyles, 1973a). Joly (1891) complained that the von Kobell scale of fusibility was beset with errors. By way of example, he pointed out that a filament of fibrous actinolite melts more easily than a compact fragment of orthoclase, and yet, he noted, the melting point of actinolite is more than 100°C higher than that of orthoclase. To determine melting points of minerals more accurately, Joly (1891) devised the "meldometer" method. Joly's meldometer consisted of a thin platinum ribbon about 2 centimeters in length held in tension between two forceps and mounted on a microscope stage. A very fine powder of the mineral whose melting point was to be determined was spread on the ribbon, and a current was passed through the ribbon. The melting of the powdered mineral was observed through the micrsocope, and the temperature of melting was determined from measurements of the linear expansion of the platinum ribbon as the current passed through it. The expansion of platinum had also been calibrated against the melting points of several minerals independently measured by various workers in the field. Joly prepared calibration curves to around 1500°C.

Early determinations of the one-atmosphere melting points of common rock-forming minerals were made by Joly, Cusack, Doelter, Brun, and others. Although Joly measured the melting points of various feldspars, it remained for Ralph Cusack to report the results of melting-point determinations for a wide variety of rock-forming minerals made with Joly's meldometer. Cusack (1896) made melting-point determinations for actinolite, tremolite, hornblende, diopside, diallage, augite, spodumene, enstatite, wollastonite, olivine, almandine, vesuvianite, epidote, zoisite, dioptase, axinite, tourmaline, kyanite, meionite, nepheline, sodalite, leucite, adularia, albite, microcline, labradorite, topaz, titanite, staurolite, andalusite, zircon, and even a few oxides and sulfides. He found differing values for the melting point on different samples of the same mineral and recognized that those differences were probably due to variation in the chemical composition of the samples. He did not regard the variation in the determinations as a major problem, however, because the values obtained generally did not differ by much more than one percent.

Important as Cusack's contribution was, no one did more to determine mineral melting points than Cornelio August Severinus Doelter (1850–1930), one of the major experimentalists of the late nineteenth and early twentieth centuries (Fischer, 1971). After studying with Bunsen at Heidelberg, Doelter spent the majority of his professional career on the faculty of the University of Graz in Austria, where he synthesized nepheline and pyroxenes in 1884, sulfides in 1886, micas in 1888, and zeolites in 1890. He melted and recrystallized rocks and produced augite andesite from a melted eclogite. Doelter also measured volume increases of rocks on melting as well as the viscosity of liquid basalt and granite.

In 1899, Doelter began determinations of the melting points of minerals and mineral mixtures. For his melting-point determinations he constructed a crystallization microscope for observation of run products in the heating ovens manufactured by W. C. Heraeus. Doelter used thermoelectric elements to determine temperatures. He recorded as the melting point the temperature at which the corners or edges of a crystal fragment of a mineral of interest began to show signs of rounding under the microscope. A tireless worker, Doelter measured a staggering number of melting points, including those of the feldspars anorthite, albite, orthoclase, adularia, sanidine, and microcline; the feldspathoids nepheline, haüyne, and sodalite; the micas phlogopite, lepidomelane, biotite, muscovite, lepidolite, and zinnwaldite; the garnets pyrope, grossularite, almandine, and melanite; the olivine minerals monticellite, hyalosiderite, hortonolite, and fayalite; the pyroxenes enstatite, bronzite, hypersthene, diopside, hedenbergite, augite, spodumene, jadeite, aegerine, and acmite; the amphiboles pargasite, hornblende, riebeckite, and arfvedsonite; and also such nonsilicate minerals as chromite, pleonaste, magnetite, hematite, and apatite.

Additional experimental investigations on minerals and rocks were undertaken just prior to the twentieth century by Carl Ötling (1897) of Hamburg and Józef Morozewicz, a student of Lagorio at the University of Warsaw, who subsequently worked with the Russian Geological Survey and taught at the University of Cracow. Morozewicz (1898) provided a valuable review of experimental data on silicate minerals.

Despite clear progress in experimentation, petrological writers around the turn of the twentieth century repeatedly expressed concern about the accuracy of these melting-point measurements. For one thing, different investigators frequently obtained different results for the same minerals. Iddings (1909), for example, incorporated a lengthy table in *Igneous Rocks* comparing the determinations of Doelter with the meldometric measurements of Joly and Cusack. For many minerals, Doelter's determinations differed from those of Joly and Cusack by as much as 50°C. In some instances, Doelter obtained the higher value, but in other cases, Joly or Cusack reported the higher value. Although Iddings believed that Doelter's data were preferable, he recognized the need for further improvement.

It is not surprising that the melting-point determinations were suspect. Because natural rather than synthetic crystals were used, the melting points did not represent the values for pure end-member compositions, and values differed depending on the nature and amount of atomic substitutions in the crys-

20. The original Geophysical Laboratory, Washington, D.C., photographed in 1948 by H. S. Yoder, Jr., on the first day of his employment. From the files of the Geophysical Laboratory, Carnegie Institution of Washington.

tal being melted. In other cases determinations were complicated by the existence of high-temperature polymorphic transitions that were not always recognized by the investigator. The phenomenon of incongruent melting complicated matters, too, as did the persistent problem of the variable kinetics of melting of the different minerals. The sluggishness of some minerals could easily lead investigators to consider the melting points of those minerals to be higher than they actually are. Many of these problems were finally addressed, and further improvement in the temperature scale made, with the establishment of the Geophysical Laboratory.

THE GEOPHYSICAL LABORATORY

The development of the Geophysical Laboratory of the Carnegie Institution in Washington, D. C., arguably stands as the most far-reaching event in the history of experimental igneous petrology. The Geophysical Laboratory (Figure 20), whose formation has been discussed at length by Servos (1983), Yochelson and Yoder (1994), and Yoder (1994), came into existence in 1905 as an outgrowth of the efforts of American geologists who were persuaded of the need for a research laboratory that could bring the results of physics and chemistry to bear upon the solution of some of the major problems in geology. One faction, which included King, Barus, Becker, and T. C. Chamberlin, believed that a research laboratory ought to emphasize geophysics. They maintained that the investigators within such a laboratory should address questions pertaining to the age of Earth and the state of the interior of Earth and that they should engage in the measurement of the physical constants of rocks and minerals such as viscosity, compressibility, and thermal conductivity. Another group, including Iddings and Charles Van Hise of the University of Wisconsin, was equally anxious that a geochemical approach be stressed. They wanted to apply the recently discovered insights of physical chemistry to the problems of igneous and metamorphic petrology and of ore deposits. After an extended period of jockeying for financial support for both sets of interests, the Carnegie Institution of Washington was endowed in 1902 by steel magnate Andrew Carnegie, and the Geophysical Laboratory was established. Despite its name, it was ultimately determined that the laboratory would devote most of its efforts to the geochemical side of geology.

To a considerable degree, the work of the laboratory was initiated by Arthur Louis Day (1869–1960). Day received his Ph.D. from the Sheffield Scientific School at Yale in 1894 (Abelson, 1975). After a brief stint of teaching physics at Yale, Day decided that he preferred research and volunteered to serve as an unpaid assistant in the laboratory of physicist Friedrich Kohlrausch at the Physikalisch-Technische Reichsanstalt (PTR) in Charlottenburg-Berlin, one of the world's premier physics laboratories (Cahan, 1989). The PTR was established in 1887 with Hermann von Helmholtz as the first president. Kohlrausch succeeded him upon his death in 1894. While at the PTR, Day developed an interest in high-temperature physics and thermometry that determined the course of his remaining career. He worked primarily with Ludwig Holborn who became head of the thermodynamics division of the PTR in 1914.

Noting that previous investigators such as Violle, Barus, and Holborn and Wien had obtained diverse results for the value of the melting point of gold (1035°C to 1092°C), Holborn and Day set about to make an accurate determination and to reduce possible sources of error in gas thermometry at high temperatures. Holborn and Day (1899, 1900, 1901) experimented with various bulbs, gases, and designs to improve the gas thermometer. They calibrated the EMFs of thermo-electric measurements against the gas thermometer so that the thermo-electric method could be used independently. They also successfully determined the melting points of gold, silver, copper, lead, zinc, cadmium, and other metals in the range of 300°C to 1100°C. By two different thermo-electric methods, Holborn and Day obtained, by means of EMF measurements, very consistent results for the melting point of gold that yielded an average value of 1063.5°C. As a result of their work and that of others at the Reichsanstalt, a high-temperature scale, based on the accurately measured melting points of cadmium, zinc, silver, and copper (in air and in a reducing atmosphere), became the accepted standard for high-temperature measurement.

In 1900, the United States Geological Survey established a physical laboratory as part of its Division of Physical and Chemical Research under the direction of George Becker. It was intended that the laboratory would conduct high-temperature research on silicate equilibria. Upon his return from Germany late in 1900, Day joined the Geological Survey as a physical geologist. He worked on feldspars and also began an effort to extend the nitrogen gas thermometer scale because, at the time, no reliable gas thermometer measurements had been made at temperatures around and above 1150°C. In setting up the high-temperature laboratory in Washington, Day had to obtain a direct comparison with the gas thermometer at the Reichsanstalt and, in effect, transport the PTR temperature scale to the United States. To accomplish this feat, Day exclusively employed thermo-elements that were cut from the same rolls of wire that had supplied the thermo-elements at the Reichsanstalt. Working with chemist Eugene T. Allen, Day then determined the melting points of exceptionally pure, commercially prepared cadmium, zinc, silver, copper, and antimony in the Survey's laboratory (Day and Allen, 1904). They claimed that temperatures up to 1600°C could be accurately determined by extension of the temperature–EMF relationship of thermo-elements.

When Andrew Carnegie endowed the Carnegie Institution of Washington in 1902, Day received a grant from the new institution to continue the work he was doing at the Survey. Seeing that Day's work was producing results, the Carnegie Institution of Washington decided to establish its own Geophysical Laboratory in 1905 and hired Day as its first director, a position he held from 1906 to 1936.

To increase the absolute accuracy and range of temperature measurements and because the gas thermometer remained the basis on which all other methods of temperature measurement rested, including the thermo-electric method, Day and Holborn, who was still at the PTR, decided to reinvestigate the gas thermometer independently. Hence, the first large piece of apparatus that was installed in the new Geophysical Laboratory was a gas thermometer (Sosman, 1952), but it took Day three years to build the instrument from scratch. He was not able to start making measurements until 1907. Holborn, getting underway in 1904, extended the scale to 1600°C with gas thermometry using a platinum bulb containing 20 percent iridium, but he reported errors up to 10 degrees at very high temperatures. In contrast, Day and Geophysical Laboratory physicist J. K. Clement worked up to only 1200°C to try to reduce further the experimental errors in the lower temperature range. They used a bulb with only 10 percent iridium and eventually concluded that iridium would have to be banished entirely for accurate results at temperatures above 1200°C. Day and Clement (1908) also published revised temperatures for the melting points of pure zinc, silver, gold, and copper in a reducing atmosphere and expressed the hope that they would be able to locate some further calibration points lying between those of zinc (418.5°C) and silver (958.3°C). They expressed satisfaction that they had perfected the constant volume gas thermometer to the point where the aggregate error for measurements between 300°C and 1100°C was no more than 0.5°C.

Robert Sosman, who joined the Geophysical Laboratory in 1908, worked with Day and Allen to extend measurements beyond 1200°C. Day et al. (1910) reported that they had provided a calibration point between that of the melting points of zinc and silver by determining the melting point of antimony (629.2°C); redetermined the melting points of cadmium, zinc, silver, gold, and copper; provided new determinations of the melting points of nickel, cobalt, aluminum, and palladium (1549.2°C) as well as the melting points of the minerals diopside (1391.2°C) and anorthite (1549.5°C); and estimated the melting point of platinum as 1755°C. Thus, by 1911, Day and his colleagues at the Geophysical Laboratory had successfully extended the standard gas thermometer scale, defined in terms of closely spaced melting points of pure substances, well beyond 1200°C. Ludwig Holborn continued his work at the Reichsanstalt and obtained very similar values of temperature for several calibration points on the temperature scale.

SILICATE SYSTEMS INVESTIGATED AT THE GEOPHYSICAL LABORATORY

The establishment of a precise high-temperature scale was not an end in itself. The Geophysical Laboratory wanted to apply the results of such fundamental research to the exploration of geological problems, and under Day's leadership developed a systematic plan for investigating the rock-forming minerals. The plan was first to study the thermal behavior of some of the simple rock-forming minerals by a trustworthy method. Next, the conditions of equilibrium for simple combinations of the rock-forming minerals would be investigated to reach a sound basis for the study of rock formation or magmatic differentiation. Finally, Day and his team hoped to see the day when they could vary both pressure and temperature over a wide range so that their knowledge of the rock-forming minerals would enable them to correlate many geological processes with the appropriate quantitative physico-chemical reactions determined in the laboratory. Now, thanks to the work of Holborn, Day, and their co-workers, the groundwork had been laid for them to embark on that ambitious program of research.

Day chose to initiate the study of the silicate minerals with an investigation of the thermal relations of the plagioclase feldspars on the ground that the lime–soda feldspar series forms the most important group of rock-forming minerals. Tschermak had claimed that plagioclase represents an isomorphous mixture of albite and anorthite, but inasmuch as optical studies had not settled Tschermak's contention, Day hoped that a modern thermal study might resolve the problem. Day et al. (1905) determined the liquidus of the plagioclase ascendant loop diagram for the range of compositions from An_{26} to An_{100} and deduced the form of the remainder of the loop.

Another project began with an examination of the wollastonite/pseudowollastonite inversion (Allen et al., 1906), moved to a determination of the binary phase relations of the lime–silica system (Day et al., 1906), and then expanded to investigation of the binary systems alumina–silica, alumina–lime, and alumina–magnesia (Shepherd et al., 1909). In their study of the lime–silica series of minerals, Day and E. S. Shepherd did the experimental work while F. E. Wright studied the various run products optically. They constructed the phase diagram for lime–silica from results obtained by using a variety of methods, furnaces, and temperature-measuring devices. In these studies, the investigators stressed the extreme importance of using chemically pure starting materials rather than natural minerals if one wanted to produce an accurate phase diagram for an ideal system.

To locate melting points and inversion points on the phase diagram, Day and Shepherd employed the heating-curve method that Day and Allen had previously used in their study of plagioclase and that other investigators had commonly used in the study of minerals at high temperatures. In the heating-curve method, a pulverized mixture of the substances to be investigated, called the charge, was placed in a furnace, the temperature of the furnace was gradually raised, the temperature recorded by the thermocouple was read at regular intervals, and the temperature was then plotted against time on a graph. The temperature of the charge increased regularly if there was no energy change within the charge. If, however, a new phase formed upon continued heating, the rate of heating of the charge would experience a marked change. The magnitude of the change would depend on the amount of energy involved in the phase change. A sudden increase in the rate of heating of the charge would thus produce a break on the plotted heating curve, and the temperature at which the break occurred was regarded as the temperature at which the disappearance of a phase of the system was complete.

Because of the sluggishness of the inversions of the various polymorphs of calcium orthosilicate, the heating curve method yielded unreliable results. Therefore, Day and Shepherd decided to supplement the heating curve runs with measurements that employed a quenching method that had been originally developed by metallurgists for the investigation of the phase relations of metals. The quench method was used more extensively by Shepherd et al. (1909). In this method, the charge was suspended on a porcelain ring through a thin platinum wire that was attached to the ends of thick platinum leads. The charge was then heated to the desired temperature for several hours to a couple of days to insure that equilibrium had been achieved. At the conclusion of a run, a current was passed through the platinum leads until the thin wire broke and the charge vessel dropped into a container of mercury at room temperature. The contents of the charge that had been produced at high temperature were typically frozen almost instantaneously and could then be examined and identified under a microscope.

The investigators at the Geophysical Laboratory soon realized that the increased heating times required by the quenching method were outweighed by distinct advantages. One obtained not only accurate information as to the temperatures at which phases appeared and disappeared, but also, by virtue of microscopic examination, knowledge of the nature of the phases themselves. The experimenter who used the quenching method could detect the existence of phase changes that either took place too slowly or involved too small an energy change to produce an appreciable break on a heating curve. And finally, with the heating-curve method, superheating often occurred for phase changes that involved very viscous substances. Because the rate of heating during employment of the heating-curve method was generally rather rapid, the observed break on the heating curve indicating a phase change was probably located at a temperature higher than the equilibrium transition temperature. The quenching method avoided that problem entirely because the experimenter tried to achieve equilibrium by taking enough time so that all the transformations that could occur would do so. In contrast, the heating-curve method did not always provide an accurate knowledge of the identity of the run products, failed to detect some of the phase changes, and in the case of viscous materials yielded incorrect temperatures for the phase changes.

In spite of the technical achievements in furnace configuration, temperature measurement, and experimental design, the extremely valuable results on melting points and phase equilibria of common silicates were, nevertheless, obtained at atmospheric pressure. Given that magmas are generated and that plutonic igneous rocks crystallize at depth, the effect of pressure on the melting points of minerals and phase equilibria was a question of great interest at the Geophysical Laboratory and elsewhere at the outset. As a result, compressibility studies at Harvard University were supported by the Carnegie Institution of Washington as early as 1903. In 1906, physicist Percy Bridgman began experiments at Harvard University on a range of materials to ascertain the changes in their physical properties with pressure. And, at the Geophysical Laboratory, John Johnston worked on the development of an apparatus that was capable of sustaining both high temperature and high pressure. He achieved a temperature of 400°C at a pressure of 2 kilobars and reported on the melting point changes of some metals with pressure (Adams and Johnston, 1912). The revolution in experimental petrology was under way.

The Experimental Era

16 ← CLEARING THE MISTS

The Theory of Crystallization–Differentiation

The early years of the second decade of the twentieth century witnessed a major turning point in the history of igneous petrology. Between 1910 and 1915, Norman L. Bowen launched his professional career and formulated the initial version of his extremely influential theory of crystallization–differentiation by means of fractional crystallization. Both Bowen's experimental methods and far-reaching theory fomented another revolution in igneous petrology. Before examining Bowen's career and theory, however, a review of the status of igneous petrology at the beginning of the twentieth century is in order.

IGNEOUS PETROLOGY AT THE BEGINNING OF THE TWENTIETH CENTURY

The earlier revolution in microscopic petrography had led to prolific results in the description of igneous rocks. Thanks to hundreds of studies of igneous terranes around the world that incorporated the examination of thousands of thin sections with polarizing microscopes, igneous petrologists discovered the remarkable breadth of mineralogical, textural, and lithological variation that exists among igneous rocks. Descriptive petrography had become a mature science by 1910.

Knowledge of the chemical compositions of the vast range of igneous rocks in terms of their major element contents was also firmly established. Throughout the latter nineteenth century and into the early twentieth century, thousands of chemical analyses of all manner of igneous rocks had been obtained. Now igneous petrologists had an excellent conception of the range of chemical compositions for virtually any important type of igneous rock. Roth (1861, 1869, 1873, 1879, 1884) and Washington (1903, 1904, 1917) had provided the petrological community with detailed summaries of the chemical composition of igneous rocks. Sufficient chemical data had become available that it was possible to gain an overview of the chemical composition of an igneous rock or of a suite of igneous rocks by means of suitable diagrams such as those of Brögger or Iddings. Igneous petrologists had begun to develop a shorthand that would facilitate rapid comparisons among rock samples or suites so that they could more easily discern whether a suite of rocks might be chemically consanguineous.

Because so much information had been accumulated about igneous rocks, petrologists were forced to reevaluate the matter of classification. Igneous rock nomenclature and classification were in a state of chaos in the early twentieth century. The qualitative classifications of Rosenbusch, Zirkel, Michel-Lévy, and other continental geologists had served their purpose for many years, but with the growing need for quantitative classifications, several petrologists tried their hand at meeting that need. The most ambitious attempt was that of Cross et al. (1902). Despite the eventual rejection of their remarkable proposal, the petrological community did recognize the great value of the concept of the norm. Moreover, the CIPW classification stimulated considerable debate about the nature of classification. Because the CIPW classification was chemically based, it prodded petrologists like Shand, Holmes, and Johannsen into developing workable quantitative mineralogical classifications.

In the latter years of the nineteenth century, geologists had also begun to pay attention to the form and mode of emplacement of igneous rock bodies. Whereas geologists at the beginning of the nineteenth century, to the extent that they accepted the idea of plutonic igneous rocks, talked about dikes, sheets, and bosses of coarse-grained igneous rocks, geologists at the beginning of the twentieth century had accepted the concepts of the laccolith and the batholith and to some degree the concepts of the bysmalith, chonolith, and igneous plug. It was clearly understood by most igneous petrologists that igneous rocks in sills and laccoliths had been injected as magmas that forcefully lifted the mass of overlying strata. Daly's concept of magmatic stoping, particularly in stocks and batholiths, had also gained many adherents.

Enormous advances had been made in the experimental study of igneous minerals and rocks. Investigators, particularly chemists and mineralogists in France, had successfully synthesized a wide range of silicate minerals over a range of temperatures by means of a number of chemical reactions. Fouqué and Michel-Lévy had even reproduced the textures of some common volcanic rocks. French investigators such as Deville, Becquerel, and Le Châtelier made technical advances in gas thermometry and thermo-electric pyrometry that led to preliminary approximation of a high-temperature scale. Toward the end of the century, Joly, Doelter, Morozewicz, and others used available pyrometric methods and existing temperature scales to determine melting points of dozens of minerals including rock-forming silicates. Although the results of the different investigators varied considerably, they were sufficiently similar that the values obtained could be regarded as approximate. Moreover, these variations goaded investigators at the Physikalisch-Technische Reichsanstalt in Germany and the Geophysical Laboratory in the United States into making additional improvements. At these institutions, investigators like Holborn and Day made so many refinements in gas thermometry and thermo-electric thermometry that they successfully extended the temperature scale at least to 1550°C and made improved determinations of mineral melting points, particularly after the quenching method of investigation had been developed.

By the early twentieth century, igneous petrologists had made great strides in understanding the causes of the diversity of igneous rocks. As a result of the development of the concepts of the petrographic province and of mineralogical, chemical, and geological consanguinity, and as a result of the abandonment of the concept that igneous rock types are a function of geological age, igneous petrologists rejected the dual-magma-source theories of Bunsen, Durocher, and others that had been so popular in the mid-nineteenth century, and increasingly began to interpret cogenetic suites of diverse igneous rocks in terms of a theory of differentiation of an original parental magma. Competitor theories, such as wholesale assimilation of wall rock, partial fusion of rocks at depth, and mixing of magmas, remained important, but by the early twentieth century, differentiation was the dominant theory of diversity. Petrologists differed widely, however, on the importance of the various possible mechanisms of differentiation, many of which were considered to be consistent with the experimental and theoretical conclusions of physical chemistry. Processes such as Soret diffusion, liquid immiscibility, gravitational segregation of liquid, segregation by means of magnetic or electrical forces, eutectic crystallization, and fractional crystallization all could claim highly respected petrologists as proponents, although Soret diffusion had lost considerable favor by the beginning of the twentieth century.

In the first decade of the century, Daly, Harker, Iddings, Loewinson-Lessing, and Vogt had all addressed the issue of diversity at considerable length. Daly and Loewinson-Lessing had developed the most comprehensive theories of diversity, each invoking substantial amounts of assimilation to go along with differentiation by means of liquid immiscibility, but their views by no means commanded universal assent. There is little doubt that one of the burning questions for igneous petrology when N. L. Bowen embarked on his career was that of the mechanism of differentiation.

The broader institutional and social context of igneous petrology had experienced striking changes by the early twentieth century. Publication outlets had changed dramatically from the early nineteenth century. No longer did geological travelogues, expedition reports, or papers in general scientific journals serve as the major sources of petrological information. Although not absent, these media had largely been supplanted by geological survey publications, petrological textbooks, and professional geological journals, some of which were oriented toward the more mineralogical and petrological aspects of geology. In Germany, *Tschermaks mineralogische und petrographische Mitteilungen* was firmly entrenched as the leading professional journal for the publication of papers pertaining to the experimental and theoretical aspects of igneous petrology.

Doelter, Loewinson-Lessing, Lang, Osann, Rosenbusch, and Morozewicz regularly contributed to its pages. In America, *American Journal of Science*, established in 1818, emerged as the journal of choice for the highly technical reports on high-temperature experimental work emanating from the Geophysical Laboratory. In contrast, *Journal of Geology*, published by the University of Chicago, regularly contained papers dealing with classification, field occurrence, or theoretical igneous petrology. This journal was the favorite outlet, apart from U. S. Geological Survey publications, for J. P. Iddings and his colleagues. More field-oriented studies of specific igneous rock occurrences continued to appear in the journals of the geological societies: *Bulletin de la Societé Géologique de France*, *Zeitschrift der deutschen geologischen Gesellschaft*, *Quarterly Journal of the Geological Society of London*, and *Bulletin of the Geological Society of America*. In the first decade of the twentieth century, several major texts on igneous petrology were published, including the two-volume *Die Silikatschmelzlösungen* (Vogt, 1903, 1904), *Die Petrogenesis* (Doelter, 1906), the first volume of *Igneous Rocks* (Iddings, 1909), and *The Natural History of Igneous Rocks* (Harker, 1909). Further editions of Rosenbusch's petrological works continued to appear, and *Igneous Rocks and Their Origin* (Daly, 1914) would soon appear.

By the early twentieth century, a very subtle shift of leadership in the petrological community toward the English-speaking world, and particularly, toward the United States, was underway. Without question, the German-speaking world continued to make extremely important contributions to petrology, particularly on the experimental side, with the work of Holborn and Doelter. Moreover, Paul Niggli, a contemporary of Bowen, was just beginning his career in Switzerland. At the same time, however, both Rosenbusch and Zirkel died in the early 1910s, and Germany was soon to embroil itself in war.

On the other hand, North Americans like Iddings, Pirsson, Daly, and Day were emerging at the forefront of the science. The Geophysical Laboratory quickly seized leadership from the Reichsanstalt in the application of high-temperature thermometry to the solution of petrological problems. Moreover, as America continued to expand geographically and as its population grew, ever more colleges and universities sprang up. With the need to understand and develop its vast mineral resources and with the need to produce more geology teachers, America's production rate of geologists increased dramatically. American geologists were quick to exploit the technical know-how they had obtained from their graduate studies in Germany. Ben-David (1968) has claimed that during the first half of the twentieth century, America clearly seized leadership in the sciences from continental Europe. He traced the causes of that shift to the flexible, adaptable, entrepreneurial style of the American colleges and universities that enabled them to adjust their research efforts quickly in response to cutting-edge developments within the various disciplines. In contrast, Ben-David maintained, French and particularly German universities retained conservative, inflexible administrative structures that did not allow them to remain at the forefront of scientific investigation as effectively. Although Ben-David's contention has been documented primarily with regard to chemistry and physics, it may also provide insight into the leadership shifts across the Atlantic in the field of igneous petrology.

An important institutional development that affected igneous petrology profoundly was the appearance of the independent research laboratory. As the nineteenth century progressed, igneous petrology was increasingly pursued by professional petrographers who were employed primarily by universities (e.g., Rosenbusch, Zirkel, Fouqué, von Lasaulx, the late Iddings, Harker, Brögger) and secondarily by geological surveys (e.g., Michel-Lévy, Teall, the early Iddings). Academic petrography and petrology, however, with a few exceptions like Doelter, remained oriented toward petrography, chemical analysis, and field work. The experimental side of igneous petrology was given its greatest impetus by means of the Reichsanstalt and the Geophysical Laboratory. There it would remain, particularly in the latter, for the next half century. The profusion of results that emerged from the Geophysical Laboratory profoundly impacted theoretical igneous petrology, and it may be questioned whether igneous petrology in the first half of the twentieth century would have progressed as it did had experimental petrology been done in the context of the academy. The independent research laboratory, with its emphasis on full-time cooperative research programs, dependable flow of funds for the construction of equipment, lack of teaching assignments, and minimum of committee work and other administrative duties, accomplished far more than could ever have been done in the academic universe of the early twentieth century. In the end, it was out of the Geophysical Laboratory, not out of Cambridge, Heidelberg, Paris, or Harvard, that the closest thing that igneous petrology would ever see to a comprehensive, unified theory of igneous rocks would ultimately emerge. Had it not been for the remarkable research opportunity afforded at the Geophysical Laboratory, Bowen's theory of crystallization–differentiation might not have seen the light of day (Young, 2002).

BOWEN'S EARLY CAREER

The Geophysical Laboratory was a perfect match for Bowen's talents and personality. Born in Kingston, Ontario, Canada, Norman Levi Bowen (1887–1956) (Figure 21) completed his collegiate studies in his hometown at Queen's University in 1909 (Young, 1998). During summers, he worked with the Ontario Department of Mines studying diabase sills in the Gowganda Lake district. Ironically, these early field studies of diabase–granophyre sheets led him to reject differentiation as the mechanism of rock diversity. Bowen was eager to pursue graduate study in Norway with Brögger and Vogt but was disappointed to receive a letter from Vogt discouraging him from that course of action. The rejection proved a blessing in disguise, for Bowen headed instead to Massachu-

CHAPTER 16

21. Norman Levi Bowen (1887–1956). From the files of the Geophysical Laboratory, Carnegie Institution of Washington.

setts Institute of Technology, where R. A. Daly took the young scientist under his wing. Bowen was undoubtedly exposed to Daly's petrological lectures that were published a short time later as *Igneous Rocks and Their Origin* (Daly, 1914). At the urging of MIT volcanologist, Thomas Jaggar, Bowen went to the Geophysical Laboratory to do an experimental study for his doctoral dissertation. Director Day suggested that the binary system nepheline–anorthite would be a relatively easy system on which to begin. Bowen obtained data from both the heating-curve and quench methods in constructing the phase diagram for the system but quickly realized the great advantages of the quench method. In his many future phase-diagram determinations, Bowen never again returned to the heating-curve method. He also sensed the tremendous potential of experimental phase-equilibrium studies for the solution of problems in igneous petrology.

Upon receipt of his Ph.D. degree from MIT in 1912, Bowen was besieged by Jaggar, Waldemar Lindgren, and others with attractive offers for employment with the Geological Survey of Canada, the Hawaii Volcano Observatory, and the U. S. Geological Survey. Instead, he chose to continue experimental work as Assistant Petrologist at the Geophysical Laboratory. After determining the phase relations of the system nepheline–kalsilite, the publication of which he delayed for several years, Bowen worked out the phase diagram for plagioclase feldspar, relying exclusively on the quench method. His results led to the publication of his first paper of substantial petrological impact in *American Journal of Science* (Bowen, 1913). Unlike Day et al. (1905), Bowen determined for plagioclase, with an accuracy that has not been superseded, both the liq-

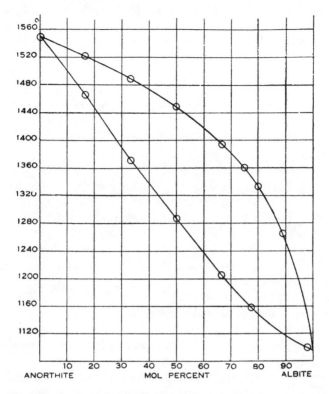

22. Phase diagram of plagioclase feldspars at one atmosphere as determined by Bowen (1913). Reproduced by permission of *American Journal of Science*.

uidus curve, that is, the curve representing liquid compositions at which crystallization begins upon equilibrium cooling, and the solidus curve, that is, the curve representing crystal compositions at which melting begins upon heating. He also clearly demonstrated the ascendant loop, continuous solid-solution behavior of the plagioclase feldspar series (Figure 22). In this paper Bowen first hinted at the possibilities for magmas to experience extreme changes in chemical composition as a result of the separation of crystals from liquid.

Given his success with plagioclase, Bowen next tackled pyroxenes and olivines with Olaf Andersen (1884–1941), a new staff member who, ironically, had received his doctorate at the University of Christiania under Brögger in Norway (Spencer, 1947). Bowen and Andersen (1914) worked out the details of the binary system, $MgO–SiO_2$. This simple system provided information on the melting behavior of both enstatite and forsterite. One of their striking discoveries was that forsterite and enstatite bear a reaction relation, rather than a eutectic relation, to one another by virtue of the incongruent melting character of enstatite. Bowen and Andersen recognized that early-crystallized olivine in systems of appropriate composition would react with residual liquid to precipitate enstatite on cooling. In some instances, olivine that had already crystallized would be resorbed completely and its place taken by enstatite coexisting with an increasingly siliceous liquid. The two authors believed that they had found an explanation for the common occurrence of olivine crystals that are rimmed by pyroxene in igneous rocks. One would not need to explain this textural phenomenon as the result of a sudden change in pressure or of sinking of olivine into a magma of a very different condition. All that was needed was cooling of magma that had previously precipitated olivine.

Bowen next investigated his first ternary system, diopside–forsterite–silica. In this remarkable paper, Bowen (1914) provided a clinic for petrologists in how to read ternary phase diagrams. He demonstrated how paths of liquid composition during cooling and crystallization would be affected by very slow equilibrium crystallization or by the separation of crystals from liquid as a result of sinking of crystals, squeezing off of liquid, or zoning of crystals. With the latter type of crystallization, he pointed out that residual liquids would become sufficiently enriched in silica that pure silica in the form of tridymite would ultimately crystallize.

In this paper Bowen launched his first severe critique of the notion of eutectic crystallization in magmas on the basis of experimental results. Some petrologists, notably Teall and Vogt, had envisioned eutectic crystallization as a significant factor in the differentiation of magmas. The idea was that magmas tend toward ultimate crystallization at a gabbro eutectic, a diorite eutectic, a granite eutectic, and so on. With eutectic crystallization, Bowen made clear, cooling liquids in a specific system would ultimately crystallize at the same temperature to the same set of minerals no matter what the initial composition of the liquid had been. But, he amply demonstrated, that would certainly not happen in this ternary system because of the reaction relation of olivine. He concluded by suggesting that the settling of calcic plagioclase and magnesian pyroxene from a basic magma would be "the dominant control in the differentiation of the ordinary lime–alkali series of igneous rocks." He regarded "fractional crystallization" as "the prime factor in the differentiation of the series mentioned" and spoke of the process as "crystallization–differentiation" (Bowen, 1914, p. 260).

Bowen's next study was an experimental demonstration of the reality of crystal settling in silicate melt (Bowen, 1915a). He took a mixture of diopside and enstatite, melted it thoroughly, cooled it to 1430°C, and held it at that temperature. Olivine crystals formed. At the conclusion of the experiment, Bowen quenched the contents of the crucible and made a thin section of the chilled glass. He discovered that olivine crystals had concentrated toward the bottom of the crucible. In another experiment, he observed the settling of pyroxene crystals from a melt composed of pyroxene and silica. In still another experiment, he took a much more silica-enriched melt in which tridymite was the liquidus phase. At the conclusion of the run, he found that tridymite crystals had begun to concentrate toward the top of the crucible. And in a final experiment, he allowed a sample to crystallize completely. Here he found that the lower part consisted of a mixture of olivine and pyroxene, whereas the upper part consisted of a mixture of pyroxene and silica.

For the first time, Bowen applied his results to a specific igneous body. He noted that the Palisades sill that crops out along the west bank of the Hudson River in northern New Jersey is somewhat enriched in quartz and alkali feldspar toward the top and noteworthy for its distinctive olivine–rich layer a short distance above its base. On the basis of field evidence, Lewis (1907) had already argued for crystal settling of the olivine, and now Bowen, on the basis of his experimental studies, agreed that the Palisades sill offered a prime example of the importance of the vertical movements of crystals in magma under the influence of gravity in bringing about differentiation.

The final experimental paper of Bowen's early career appeared in the same year. In this paper on the ternary system diopside–albite–anorthite, Bowen (1915b) introduced the concept of the "haplobasalt," that is, simple basalt, system (Figure 23). Bowen pointed out the progressive sodium enrichment of residual liquids, illustrated the effect of equilibrium and fractional crystallization on paths of liquid composition, and once more attacked the eutectic theory of differentiation. He made it clear that the search for a gabbro eutectic or a diorite eutectic was futile on the grounds that this ternary system demonstrates that there is neither a eutectic involving diopside and intermediate plagioclase nor any eutectic involving any plagioclase and any pyroxene solid solution.

On the basis of a handful of critical binary and ternary phase diagrams for which he was largely responsible, Bowen believed that the time had come to state a forceful case for the dominant role of crystallization–differentiation in magmas, despite the fact that the systems he had investigated had all been determined at one atmosphere pressure and did not include the extremely important components, iron, potassium, and water. In his third major paper of the year, "The later stages of the evolution of the igneous rocks," Bowen (1915c) published his revolutionary theory in *Journal of Geology*. He disputed the importance of the Soret effect. Despite the sympathy of his doctoral mentor and friend, R. A. Daly, for liquid immiscibility, he dismissed its importance on the grounds that no evidence for the process had been encountered in any of the experiments on silicate systems thus far undertaken at the Geophysical Laboratory. He opposed the idea that magmas can undergo differentiation in the liquid state as a result of purely gravitative differentiation, the so-called Gouy-Chaperon effect. Toward the end of the paper, he questioned whether assimilation played the dominant role accorded to it by such petrologists as Loewinson-Lessing and Daly. Bowen was on a mission to redirect the thinking of petrologists toward the dominance of crystallization–differentiation, that is,

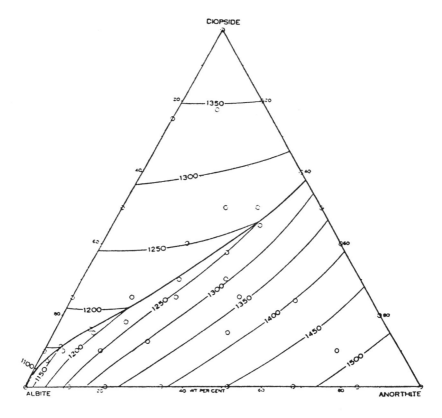

23. Phase diagram of the one-atmosphere liquidus surface in the ternary system albite–anorthite–diopside as determined by Bowen (1914). Reproduced by permission of *American Journal of Science*.

differentiation brought about by the separation of crystals from liquid, namely, fractional crystallization.

In the positive statement of his theory, Bowen urged fellow petrologists to master the principles of physical chemistry and thoroughly reviewed the available phase diagrams, pointing out how different types of crystallization affected the paths of liquid composition and the end points of crystallization. He suggested that crystal settling would play an important role in the early stages of differentiation when a magma body was still largely liquid and that squeezing off of residual liquid would become more important toward the final stages of crystallization. Bowen claimed that a typical basaltic magma would differentiate toward increasingly siliceous compositions. As a result, he regarded granite as the normal major end product of the fractional crystallization of basaltic liquid. With variation in the rate of cooling, however, he suggested that basaltic liquid would differentiate to diorite, granodiorite, or granite. While granting that basalt liquid would not always be the initial magma to undergo differentiation, he argued that in most cases basalt served as the parental magma. In summary, Bowen attributed to fractional crystallization the differentiation of the common, widespread subalkalic igneous rock series, basalt to andesite to rhyolite, or gabbro to diorite to granite.

Bowen (1915c) also intimated that monomineralic rocks like anorthosite and dunite could be the products of early fractionation and settling of plagioclase or olivine. In the most daring part of his paper, he further suggested that alkalic rocks like nepheline syenite had been derived from granitic material by extreme fractional crystallization. In so doing, he challenged another favorite hypothesis of Daly. Soon after Bowen arrived at MIT for graduate study, Daly (1910) advocated the view that alkalic magmas were generated by the assimilation of limestone. In support of this contention, Daly appealed to the very common association of alkalic igneous rocks with carbonate rocks. Daly suggested that when limestone dissolved in subalkalic magma, lime combined with iron oxide, magnesia, and silica to form pyroxenes. The resulting magma was said to be desilicated upon gravitative removal of lime-bearing mafic minerals. For years to come, Daly espoused limestone assimilation as the means for production of alkalic igneous rocks and Bowen disputed the importance of assimilation. In time, Bowen abandoned his early view of alkalic rocks. After the deaths of both Bowen and Daly, the experimental investigation of the effects of CO_2 and H_2O on silicate melts combined with the recognition that many carbonates associated with alkalic rocks are actually igneous carbonatites rather than limestones led to the realization that the limestone assimilation hypothesis was no longer tenable. The history of the limestone assimilation debate was reviewed by Wyllie (1974).

THE REACTION TO BOWEN'S THEORY

Breathtaking for both its simplicity and scope of applicability, Bowen's theory drew both deep appreciation as well as criticism. Arthur Holmes (1916, p. 68) enthused:

> so brilliant is the light which these researches throw on the vexed question of differentiation, that the patient workers to whom they are due deserve not only the gratification that attends success, but also the gratitude of all students of igneous geology. Already they have cleared away some of the mists of speculation. May they be stimulated to penetrate still further the clouds that yet await dispersal!

Holmes, however, was not convinced of Bowen's view of the evolution of all granitic rocks from a basaltic parent. He thought some basalt magmas might give rise to about ten percent of granite provided that they had previously differentiated a similar amount of peridotite. The problem for Holmes was that phonolite and basalt are commonly found together. In Bowen's scheme, however, rhyolite was supposed to be formed between basalt and alkalic material during the course of differentiation. Yet rhyolite is rarely found in alkalic volcanic fields. Why is it, Holmes asked, that volcanic islands of the central Pacific persistently fail to produce rhyolitic lavas? Holmes believed that the basaltic magmas of the Pacific are probably incapable of supplying enough silica to form residual rhyolite (or granite) after ninety percent of the composition of average gabbro has been subtracted from the original average basaltic magma. Instead, he argued, the 10 percent residual liquid would contain a great excess of alumina, alkalis, and iron oxides over that required by granite as well as a marked deficiency in silica. As a result, Holmes suspected that the salic differentiates of basalt ought normally to be markedly alkalic.

Noting that Bowen discounted Daly's magmatic stoping as a leading mechanism of batholithic intrusion, Holmes (1916) also wondered where the rocks are that the batholiths had presumably displaced, given Bowen's idea of derivation of batholithic granite from basalt. Holmes regarded the gigantic masses of granite, always intrusive into quartzo-feldspathic sediments, as the re-fused equivalents of preexisting granite rather than the end result of differentiation. He believed that the processes of stoping, fusion, and assimilation would advance the progress of differentiation in the direction of increasingly siliceous granite. To the extent that basalt magma has been the source of later granites, Holmes claimed, those granites ought to be less siliceous than the earlier granites.

A review of Bowen's theory by Alexander Scott (1916) regarded it as the most satisfactory theory on the diversity of igneous rocks thus far formulated, especially on those points where it did not depend on great extrapolations from the experimental work. "The fact that the theory has a sound experimental basis," Scott (1916, p. 471) wrote, "renders it less liable to criticism than most previous attempts." But Scott had some of the same concerns as Holmes, echoing his complaint that Bowen did not explain what happened to the material into which granitic batholiths were intruded. Scott also indicated that he was much more comfortable with the idea that the granitic massifs had formed by re-fusion of earlier rocks accompanied perhaps by assimilation.

Alfred Harker (1916, p. 554), likewise, deeply appreciated Bowen's work, asserting that his article "will be hailed with satisfaction by all petrologists." Bowen's discussions of courses of crystallization, he said, carried "a great weight of authority." But the general theory of differentiation necessarily introduced an element of hypothesis and, therefore, could no longer command unquestioning acceptance. Harker observed that Bowen had a predilection for the idea of differentiation in situ as opposed to differentiation prior to intrusion. Harker fully agreed that crystal settling did occur after intrusion, but claimed on the basis of his own field experience that it was a relatively rare occurrence. He stated that the clear instances of gravitative differentiation in sills and laccoliths with which he was familiar represented the products of very fluid magmas. On the other hand, as he saw it, the prohibitive viscosity of most magmas in an intrusive body of moderate size would quickly put an end to crystal settling. He could not at all go along with Bowen's explanation of the igneous rocks of Skye as the result of differentiation in place. He said, for example, that the peridotites do not occur at the base of the gabbro as might be expected on Bowen's crystal settling hypothesis but are rather enclosed with the gabbro. Harker's own investigation of Skye led him to believe that differentiation had taken place prior to the intrusion of the igneous masses in some deep-seated reservoir. Nor, predictably, was Harker enamored of Bowen's rejection of his own concept of Atlantic and Pacific provinces of igneous rocks.

Among Bowen's initial American critics were Daly and Grout. In one of his many papers on the genesis of alkalic rocks, Daly (1918) defended his concept of desilication of magma as a result of assimilation of carbonate rocks, challenged Bowen's idea of derivation of alkalic rocks by crystallization–differentiation, and questioned the overwhelming dominance of crystallization–differentiation as the major mechanism productive of igneous rock diversity ascribed to it by Bowen. Daly praised Bowen for his unmatched series of experiments. Bowen, he said, had provided the clearest argument to date on the significance of fractional crystallization. Moreover, Bowen had pointed the way to many new lines of thought and research. Like Harker, however, Daly pointed out that many gabbro sills fail to show the evidence for gravitative fractional crystallization. Daly was also convinced that Bowen had given short shrift to alternative hypotheses, such as gas transfer, assimilation, and liquid immiscibility. Daly wanted to allow a much greater potential role to liquid immiscibility on the grounds that the effects of high pressure, undercooling, and water on possible liquid immiscibility were not yet understood.

Bowen's other early American critic was Frank Fitch Grout (1880–1958). Grout graduated from the University of Minnesota in 1904, received a master's degree from the same institution in 1908, and obtained his Ph.D. degree from Yale University in 1917 (Schwartz, 1959). Grout spent his entire professional career on the faculty of the University of Minnesota but was also actively involved with the Minnesota Geological Survey. As a result, Grout was predominantly a student of the Precambrian igneous geology of that state. Much of Grout's early work centered on the large gabbroic intrusion near Duluth, Minnesota. The Duluth gabbro, for which he coined the term "lopolith," was said to include a thick, lower unit consisting of gabbro, olivine gabbro, troctolite, anorthosite, and peridotite. The lower unit was reported to be overlain by a thinner red rock unit described as a granophyre and having an approximately granitic composition. Grout believed that Bowen's crystallization–differentiation mechanism could be applied to the lower layered gabbroic unit to some degree. He regarded crystal settling alone, however, as inadequate to account for the reversals in density that occurred in the rocks of the unit, and so Grout also invoked the action of convection currents to operate along with crystal settling. On the other hand, he maintained that the distinction between the lower unit and the upper red rock unit was brought about as the result of liquid immiscibility on a large scale.

Grout's most telling objection to Bowen's theory stemmed from the fact that, on the crystallization–differentiation hypothesis, even if the granophyre had been produced by differentation of an originally gabbroic liquid, a substantial volume of rocks intermediate in composition between the lower gabbro and the upper granophyre ought to have been produced. Field evidence, however, disclosed only a very tiny amount of transitional rocks between the two units. A few years later, Grout (1926) calculated the relative amounts of rocks of various compositions that ought to be encountered in a series of rocks produced by fractional crystallization. The virtual lack of rocks of intermediate composition in so many gabbro–granophyre or basalt–rhyolite occurrences was an observation for which Bowen never could provide a satisfactory explanation.

THE EVOLUTION OF THE IGNEOUS ROCKS

With the entry of the United States into World War I, the Geophysical Laboratory diverted its energies from purely scientific work, toward the solution of a pressing national defense problem, namely the manufacture of optical glass for various applications by the armed forces. Bowen and several of his colleagues spent the better part of two years working with private industry in an effort to put American glass production on an efficient and sound scientific basis. After Bowen's war service was completed, he was lured to a teaching position by his alma mater, Queen's University, where he remained throughout 1919 and the first half of 1920. As a result of these digressions, Bowen's experimental petrological studies ground to a halt for several years. While at Queen's, the kind of laboratory facilities he had enjoyed at the Geophysical Laboratory were not available, so he contented himself by completing several optical studies of a few extremely rare minerals he found in the teaching mineral collection. A number of papers on glass technology were also published.

Bowen did manage to write some theoretical papers based on his earlier work including one on the origin of anorthosite (Bowen, 1917). In another paper on crystallization–differentiation in which he responded to the criticisms of Daly and Grout, Bowen (1919) reiterated his opposition to liquid immiscibility and claimed that blotchy or patchy texture would be the result of such a process rather than sharply bounded layers. This article flushed out a youthful H. H. Read, another geologist who became Bowen's chief antagonist in the granite controversy of later years. Read regarded Bowen's theory as a "charming conception" but thought he had overworked the idea of crystallization–differentiation. He pointed out that precisely the blotchy texture to which Bowen had appealed had been found in sills in Scotland and had already been cited as evidence for liquid immiscibility. In short, Read (1920) wanted to make clear that, at least in Great Britain, there was a growing tendency among field workers to believe that liquid immiscibility was an important process in magmas. In response, Bowen (1920, p. 238) retorted that "immiscibility is the petrologic sylph."

In one of the most significant events in the history of igneous petrology, Arthur Day, who himself had been away from the Geophysical Laboratory for some time, returned as director about a year after the conclusion of World War I and made it his top priority to lure Bowen away from Queen's back to Washington. Through a remarkable exchange of letters, Day finally succeeded in enticing a somewhat greedy Bowen back into the fold of the Geophysical Laboratory (Young, 1998). Bowen resumed experimental studies, working on the melting of orthoclase with G. W. Morey and the alumina–silica system with J. W. Greig. He also published the results of some diffusion experiments that he had performed perhaps as early as 1915. In a paper on diffusion in silicate melts, Bowen (1921) described experiments in which he placed a layer of melted diopside beneath a layer of melted plagioclase, quenched the material after a couple of days, and measured the chemical composition of the quenched glasses to determine the extent of diffusion of material from one layer into the next. In support of the arguments of Becker (1897a), Bowen concluded that diffusion of substance in silicate magmas is negligible in comparison with the rate of diffusion of heat. As a result, he now had experimental evidence to back up his rejection of the Soret effect as an important mechanism of differentiation.

Some of Bowen's most significant papers during the 1920s, however, were of a theoretical nature. The Finnish petrologist, Pentti Eskola regarded Bowen's paper on the reaction principle in petrogenesis as the most important petrological paper of the first half of the twentieth century. In this paper Bowen (1922a) proposed the idea of a continuous and discontinuous reaction series (Figure 24). In a continuous reaction series such as that of plagioclase feldspar, he pointed out that crystals are continuously reacting with liquid and thereby constantly adjusting their compositions upon cooling. In contrast, with eutectic crystallization, a crystal, once formed, no longer participates in the equilibrium. Bowen contrasted the flexibility of liquid composition attainable with continuous reaction to its inflexibility in eutectic crystallization. He also indicated the changes in liquid composition made possible in systems with incongruently melting minerals like enstatite. Such reactions he termed discontinuous.

Another landmark paper that year was an enormous article on assimilation in which Bowen (1922b) took pains to bring thermochemical data to bear on the strictly field-based claims about liquids dissolving large quantities of all

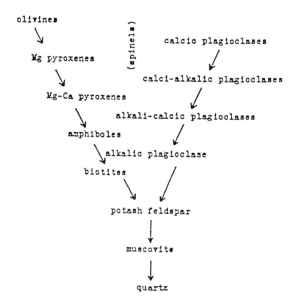

24. The discontinuous and continuous reaction series for subalkaline rocks proposed by N. L. Bowen (1922). Reproduced from *Journal of Geology* by permission of The University of Chicago Press.

manner of country rocks and thereby experiencing substantial changes in chemical composition. Writers like Daly and Shand had assumed that assimilation was an exothermic process that could be sustained for a long time. Bowen showed that many instances of assimilation would be endothermic and would, therefore, tend to cease fairly quickly unless a magma possessed substantial quantities of superheat, that is, existed at a temperature considerably above its liquidus temperature. On grounds such as the widespread presence of phenocrysts in chill zones, Bowen argued that magmas generally do not possess much superheat. He related assimilation to the reaction principle by showing from phase diagrams that basic magmas, crystallizing minerals high in the reaction series, could melt and digest acidic rock fragments and have their compositions modified. In contrast, he maintained that acid magmas, crystallizing minerals lower in the reaction series, would not melt basic rocks. Rather, the magma would simply react with the fragments and precipitate more pyroxene or amphibole or whatever basic minerals had already crystallized. The inability of granitic magmas to dissolve fragments of gabbro, amphibolite, or basalt was one of the points most strongly insisted upon by Bowen. Through his analysis of phase diagrams, he brought to the attention of petrologists the linkage between assimilation and fractional crystallization, a theme to which they returned more than half a century later.

Bowen's contentions about the nature of assimilation brought him into conflict with another prominent geologist, who would become his most vigorous adversary for the next twenty years, his own colleague at the Geophysical Laboratory, Clarence Norman Fenner (1870–1949). Fenner received all his collegiate and graduate education at Columbia University (Wright, 1951). He became a staff member at the Geophysical Laboratory the year in which Bowen became a predoctoral fellow. Fenner (1913) early distinguished himself for his outstanding experimental work on the stability relationships of the silica minerals. His most important contributions, however, resulted from his study of the Katmai volcanic province in Alaska. After the catastrophic eruption of Katmai in 1912, Fenner participated in two major expeditions to the province. He was struck by the important role of volatiles in Katmai magmatism. He was also impressed by the enormous quantities of banded pumice and rhyolite containing abundant streaks and layers of basic material that display evidence of digestion and softening. In a mammoth paper on Katmai volcanism, Fenner (1926) interpreted the basic slabs as having been dissolved by the acid magma and claimed that the magma must have contained sufficient superheat to dissolve the basic material, claims that Bowen denied. Fenner also resisted what he perceived as the too-easy acceptance of fractional crystallization as the dominant mechanism of differentiation by the petrological community. He employed variation diagrams to illustrate his contention that the diverse magmas at Katmai were not necessarily the result of fractional crystallization and that they might have been the expression of some degree of magma mixing.

In the meantime, Bowen was invited to give a series of lectures at Princeton University. These were published a few months later as *The Evolution of the Igneous Rocks* (Bowen, 1928). The book, widely hailed by petrologists for its lucid writing, clarity of thought, and imaginative insight into the nature of igneous rocks, remained a standard petrologic texts for decades thereafter. Much of the book was a compilation of only slightly modified earlier papers, such as those on assimilation, ultrabasic rocks, and the reaction principle. Among the important new information, however, was Bowen's use of the experimental studies of his Geophysical Laboratory colleague, Joseph Greig. Greig (1927) had demonstrated from the phase relations of a host of metal oxide–silica systems that liquid immiscibility does indeed commonly occur in such systems but always in compositional regions far removed from those of natural magmas and also at temperatures much higher than those of natural magmas. Bowen seized on Greig's data to support his ongoing crusade against liquid immiscibility. His arguments had an impact. As indicated by a comparison of *Igneous Rocks and Their Origin* (Daly, 1914) with its thorough revision, *Igneous Rocks and the Depths of the Earth* (Daly, 1933), even Daly had lost much of his earlier enthusiasm for liquid immiscibility as a result of Bowen's withering attack.

Another important development in the book was Bowen's reaction to Fenner's ideas. Surprisingly, he left the chapter on assimilation virtually unchanged from the original article (Bowen, 1922b). In that article, he had not specifically discussed Fenner's claim that acid magmas digest basic rocks, and he saw no need to do so now, feeling perhaps that his theoretical reasoning stood on its own. He did, however, use variation diagrams for the first time to show that Fenner had failed to use them properly. He also included a new chapter on the role of volatiles in magmatism and concluded that although volatiles undoubtedly were of great significance during the final stages of magmatic crystallization in the formation of veins, pegmatite, and ore bodies, they were, nevertheless, of negligible importance during the major course of differentiation from basalt to granitic liquid.

THE END PRODUCT OF CRYSTALLIZATION–DIFFERENTIATION

With the publication of *The Evolution of the Igneous Rocks*, Bowen's reputation as the leading thinker among igneous petrologists of his era was sealed. In effect, Bowen had set the terms of debate in igneous petrology for the first half of the twentieth century. Although other petrologists had proposed theories of diversity and defended them, only Bowen was able to throw the weight of experimental data behind his own theory. For the most part, it was Bowen's theory of crystallization–differentiation that petrologists lined up behind or else challenged. He was acknowledged king of the petrological hill.

While probably any petrologist was uncomfortable with some aspect or other of such an ambitious theory, Fenner continued to be the most outspoken critic. After his Katmai study, his next papers focused primarily on the goal of differentiation. For example, the year after Bowen's book was published, Fenner (1929) issued a paper on the crystallization of basalts. He was perplexed by the fact that, at one time, geologists were unable to find much field evidence to support the concept of gravitative settling of crystals and that all of a sudden, in light of Bowen's new theory, field workers were finding evidence everywhere. Fenner (1929, pp. 225–226) also objected that the new breed of petrologist seemed to be able to

> move crystals of almost any sort at will through the magma body. They cause them to float to the top or to settle to the bottom; move them to the sides or to the center; or they cause them to form layers or schlieren of horizontal or vertical arrangement. Having moved them to the desired position they may allow them to remain, or they may remelt them and drive them forth again as dikes or other intrusive bodies. It is to be feared that this indiscriminate and uncritical advocacy of a process which is probably of great importance will delay the attainment of a full comprehension of all the factors involved in differentiation.

Fenner complained that, although great attention was paid to plagioclase solid solution series as a means for increasing the alkali content of residual liquids, the ferromagnesian solid solution series of pyroxenes and olivines were being ignored as a means of iron enrichment in residual liquids. He described many examples of basalt in which the groundmass is more iron-rich than the early pyroxene phenocrysts. These residua, he said, generally consist of magnetite, pyroxene, and plagioclase and are not particularly enriched in silica. Bowen had repeatedly and consistently argued that crystallization–differentiation resulted in residual liquids enriched in alkalis and silica. That was why he envisioned granitic liquid as the end product of fractionation of basaltic magma. As far as Fenner could tell, however, the actual course of crystal differentiation led to the production of a silica-poor, iron-enriched liquid rather than a granitic liquid. In several more papers, Fenner hammered home the notion that natural basaltic magmas clearly displayed iron-enrichment trends during the course of differentiation.

Another challenge to Bowen's conception of the goal of differentiation came from William Quarrier Kennedy (1903–1979). Kennedy, a 1927 graduate of the University of Glasgow, undertook graduate study and gained his expertise in mineralogy and petrology under the tutelage of Paul Niggli in Zürich (Shackleton, 1970). Kennedy joined the Scotland Office of the Geological Survey and carried out substantive studies of the structure and petrology of the Scottish Highlands. In 1945, Kennedy was appointed Chair of Geology at the University of Leeds. Ten years later Kennedy succeeded in establishing the Research Institute of African Geology at Leeds. For several years Kennedy prosecuted studies of ring complexes, pegmatites, and Precambrian ore deposits in southern Africa.

Throughout his career, Kennedy was highly interested in the source of magmas, as indicated by an early contribution on trends of differentiation in basaltic magmas. In this paper, Kennedy (1933) asserted that there are two different kinds of primary basalt magma. These he termed the olivine–basalt magma type and the tholeiitic magma type. He regarded these as essentially identical to the plateau magma type and nonporphyritic central magma type that E. B. Bailey and his co-workers had described in their classic memoir on the igneous geology of the Isle of Mull in the Scottish Hebrides (Bailey et al., 1924).

Kennedy wrote that the olivine–basalt type contained characteristic olivine, augite, basic plagioclase, and iron ore. He described the pyroxene as titaniferous diopside or augite, and the residual material as alkalic without free quartz. In contrast, he specified that the tholeiitic type contained enstatite–augite (pigeonite), basic plagioclase, and iron ore with residual interstitial material that is acid and dominantly quartzo-feldspathic. The term "tholeiite" did not originate with Kennedy but was coined by J. Steininger in 1840 for a rock that he described as consisting of ilmenite and albite. The rock was found on a hill known as Schaumberg above the town of Tholei between the Saar and Rhine

Rivers. In his *A Descriptive Petrography of the Igneous Rocks*, Johannsen (1931) wrote that "tholeiite" was practically obsolete as a term! Kennedy showed that the tholeiitic type was typically more siliceous than the olivine–basalt type. The tholeiitic type, Kennedy suggested, was well represented by the Deccan traps of India, the Karoo basalts of South Africa, the Paraná sills of South America, and the sills associated with the Sudbury Nickel Irruptive in Ontario, Canada.

The final differentiate of a basaltic magma, Kennedy argued, depended not so much on physical conditions as on the chemical and mineralogical constitution of the parental basalt type. He knew of no instances in which the olivine–basalt parent gave rise upon differentiation to a quartz-bearing pegmatoid. Instead, he claimed, late-stage rhyolites are typically found with tholeiitic basalts and late-stage phonolites generally accompany the olivine–basalt type.

Kennedy argued that the nature of differentiation was controlled by the nature of the pyroxene. In olivine basalt, he said, lime is taken up primarily in pyroxene such that little calcic plagioclase is formed. As a result, alumina in the magma does not combine with lime in the production of calcic plagioclase but is reserved for combination with soda and potash in late-stage feldspathoid minerals. On the other hand, he said, in tholeiitic magmas very little lime is incorporated into pyroxene. As a result, considerable alumina is used up in combination with lime in the formation of basic plagioclase.

Kennedy further noted that there is no evidence to indicate that one primary magma is derived from the other. He observed that it was remarkable that olivine–basalt magma occurs in both continental and oceanic crust whereas tholeiitic magma is always absent in the latter. Tholeiitic rocks, he recognized, seem to be connected with the presence of the granitic crust.

Bowen had, of course, known for years that his sweeping theory of crystallization–differentiation was vulnerable to criticism on the grounds that it was based on a very limited number of experiments. He keenly felt the importance of investigating systems that contained iron, potash, and water as soon as possible. Toward the end of the 1920s, therefore, he began an extended collaborative effort with his close friend and colleague at the Geophysical Laboratory, J. Frank Schairer, to explore the effects of iron and potassium on the crystallization of silicate melts. Their initial efforts focused primarily on the role of iron.

John Frank Schairer (1904–1970) graduated from Yale University with a major in chemistry in 1925 (Yoder 1972, 1995). He earned a master's degree in mineralogy the following year, and in 1928 Schairer (Figure 25) received a Ph.D. degree in physical chemistry. During his graduate school years, Schairer applied to the Geophysical Laboratory, hoping to receive a one-year fellowship to work on his dissertation. Instead, Arthur Day, still the director, proposed that Schairer accept a full-time staff position beginning in the fall of 1927. After completing his experimental work at Yale, Schairer moved to Washington and wrote his dissertation during his first year there. Day encouraged him to inves-

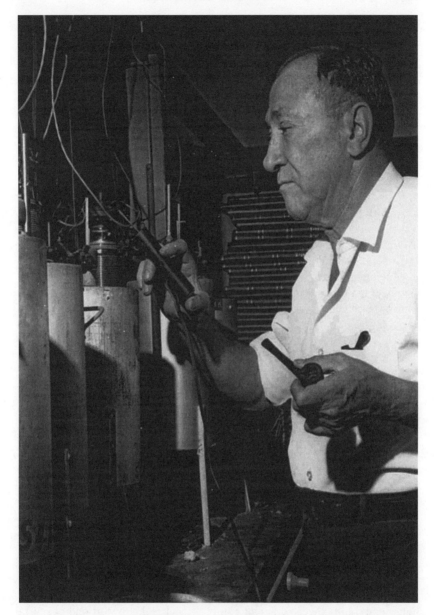

25. J. Frank Schairer (1904–1970). From the files of the Geophysical Laboratory, Carnegie Institution of Washington.

tigate a system with iron oxides, and, consequently, Schairer teamed up with Bowen for the next several years in a sustained assault on iron-bearing silicate systems. Exceptionally skilled and accurate as an experimentalist, Schairer ultimately studied iron-bearing systems beyond what he and Bowen did jointly. One of the important outcomes of that work was Schairer's development of the flow sheet for describing the major pathways of liquid fractionation. During World War II, Schairer helped to develop alloys that minimize gun erosion during firing, and some of these alloys were used subsequently to improve the pressure vessels employed in hydrothermal research at the Geophysical Laboratory. At the conclusion of the war, Schairer resumed his investigations of ternary oxide systems and numerous joins within the basalt system.

In dealing with iron-bearing systems, of course, Bowen and Schairer were faced with the problem of controlling the oxidation state of iron. The problem was relatively easy to solve in their first collaborative venture on the melting relations of acmite (Bowen and Schairer, 1929), a mineral in which all the iron is trivalent. The highly oxidizing conditions required to guarantee the feasibility of acmite synthesis were provided by the atmosphere. Although other workers, including Doelter and Brun, had earlier succeeded in melting acmite and were in reasonable agreement on its melting point somewhere between 950°C and 1020°C, Bowen and Schairer made the discovery that acmite undergoes incongruent melting, a phenomenon that could be recognized only by use of the quenching method. They also made the very important discovery that, at temperatures between 870°C and 850°C, quartz precipitates directly from melts having very silica-rich compositions near a eutectic point at 850°C at which acmite and quartz co-crystallize from residual liquid. Bowen and Schairer had proved the possibility of direct crystallization of quartz from magma. Noting that petrologists had frequently claimed that volatile constituents were necessary to form quartz, a point that was repeatedly made by geologists of the nineteenth century like Daubrée and Rose, Bowen and Schairer pointed out that it was necessary only that the temperature at which silica separates from melt lie within the stability range of quartz in order to precipitate quartz.

Bowen and Schairer next teamed up with a new member at the Geophysical Laboratory, H.M.V. Willems, to continue their experiments in the system Na_2SiO_3–Fe_2O_3–SiO_2. Bowen et al. (1930) wrote that this system well illustrated how thorough the results of fractional crystallization might be in a system involving reaction between liquid and crystals and pointed out that the cooling liquid could follow very different compositional paths toward one of several ternary eutectic points depending on the degree of reaction between early-formed hematite and liquid. The authors pointed out that this system was one in which a low degree of fractionation could lead to an iron-enriched residual liquid and a high degree of fractionation could lead to siliceous residual liquid. They argued that it was precisely in quickly cooled, little fractionated basalts that late crystallization of iron-rich liquid has been urged by petrologists like Fenner. In contrast, they said, late-stage siliceous differentiates had been demonstrated in the more slowly cooled dolerites. The authors had demonstrated that basaltic liquids could proceed in the direction of silica-enrichment and, therefore, granite production, but whereas Bowen had never before talked of iron enrichment and talked as if basalt liquids always moved in the direction of granite, he and his co-workers had to concede that Fenner's iron-enrichment tendency was a possibility. Suddenly, Bowen was championing the reality of alternate fractionation tracks, one iron-rich and one silica-rich, while trying to make it look as if he had done so all along. He was by no means ready, however, to concede that an iron-enrichment trend was of equal importance to a silica-enrichment trend. Fenner had assumed that the iron-enrichment effect shown by ferromagnesian solid solutions would dominate over the silica- and alkali-enrichment effect shown by plagioclase feldspar solid solutions. Bowen and his co-authors argued that one could just as well assume exactly the opposite. They said that the system under investigation made it clear that the lowest melting liquid was not rich in iron but rather rich in silica and alkalis. As far as Bowen and his colleagues were concerned, the experimental evidence on iron-bearing silicates pointed to the ultimate goal of basalt fractionation as most likely being granitic in character.

Fenner was unimpressed and continued his assault on many aspects of the theory of crystallization–differentiation. Bowen and his colleagues continued investigating iron-bearing systems. Their work during the 1930s, culminating in a paper on the investigation of petrogeny's residua system (Bowen, 1937), constituted Bowen's response to Fenner's claim that the natural course of crystallization–differentiation led to residual liquids enriched in iron.

The next goal of Bowen's challenging program of experimentation was to shift from the investigation of systems containing ferric iron to those containing ferrous iron, one of the major constituents of igneous rocks. Bowen and Schairer finally succeeded in overcoming the difficult technical problems encountered in attempting to control the oxidation state of iron. They attempted to create a "neutral" atmosphere by passing a slow stream of purified nitrogen, as deoxidized as possible, through the furnace that contained the charges. Crucibles of electrolytic iron were used. Consequently, all ferrous silicate systems that they investigated were in equilibrium with metallic iron. With this method, Bowen and Schairer found that not all the iron in the melts could be reduced to the ferrous state. The melts always contained just a little Fe_2O_3 for most compositions in their systems, but they found that results could still be portrayed on simplified diagrams for practical purposes.

They began their examination of ferrous oxide–bearing systems with the system $FeO–SiO_2$ (Bowen and Schairer, 1932). Here they established the melting point of fayalite at 1205°C and demonstrated the pseudo-eutectic crystallization of tridymite and fayalite, a point of particular interest because of the coexistence of quartz and fayalite in igneous rocks. This initial paper on a ferrous iron silicate system was followed by three more major studies on the systems Ca_2SiO_4–Fe_2SiO_4 (Bowen et al., 1933a), CaO–FeO–SiO_2 (Bowen et al., 1933b), and MgO–FeO–SiO_2 (Bowen and Schairer, 1935). Bowen's team established relations among metasilicates ($CaSiO_3$–$FeSiO_3$), orthosilicates (Ca_2SiO_4–Fe_2SiO_4), and hedenbergite ($CaFeSi_2O_6$). They recognized that some of the incongruent melting relationships within the system were such that, upon melting, a mineral or solid solution would break down to a silica mineral, namely, tridymite, with concomitant formation of a more basic liquid that was enriched in iron. This discovery would seem to have provided confirmation of Fenner's contention for iron enrichment during the course of fractional crystallization. Bowen maintained that this discovery did not weaken his insistence that silica was normally to be expected as a late product of fractional crystallization but rather simply emphasized the fact that exceptions should be expected. He regarded the formation of an iron-rich residual liquid as just one such exception but by no means the normal course of events.

In their investigation of the system MgO–FeO–SiO_2, Bowen and Schairer (1935) presented a liquidus diagram containing fields of cristobalite and tridymite, of two liquids, and of three series of Mg–Fe solid solutions, the magnesiowüstites, the olivines, and the pyroxenes. No ternary eutectics were found. The fusion surfaces sloped downward toward the FeO–SiO_2 side of the diagram. Bowen and Schairer also explored the orthosilicate join, Mg_2SiO_4–Fe_2SiO_4 and determined solidus and liquidus points of the olivine solid solution ascendant loop at temperatures up to 1500°C. Because the iron crucibles would melt at 1535°C, determinations at higher temperatures were not possible. For compositions more iron-rich than 40% fayalite, in conjunction with the approximate melting point of pure forsterite at 1890°C ± 20°C, they established the solid solution loop of the olivines. This "binary" diagram, now so familiar to petrologists, is in reality a projection of a ternary system that also contains metallic Fe. Moreover, to this day, data points in the magnesian end of the diagram have not been confirmed experimentally because of the limitations of apparatus. Another important join investigated was the metasilicate (pyroxene) join, $MgSiO_3$–$FeSiO_3$. Solidus and liquidus points were determined, and the essentially solid solution loop behavior of the pyroxenes was established as well as the inversion temperatures from clinopyroxenes to orthopyroxenes.

From the results, Bowen and Schairer argued that monomineralic pyroxene or olivine rocks would not likely be the result of crystallization from a magma of pyroxene or olivine composition because of the extremely high magmatic temperatures required. The authors noted that evidence for the presence of volatiles that might lower the liquidus temperatures in the olivine or pyroxene phase diagrams was lacking in dunites or pyroxenites. That left Bowen's favored notion of crystal accumulation of olivine or pyroxene with some special mode of intrusion for a mass of solid crystals as the only viable option for the formation of these monomineralic rocks.

Again, Bowen and Schairer showed that liquids in this system display different compositional trends. Residual liquids, they observed, could become undersilicated or oversilicated depending on the degree of fractionation. Hence, they suggested, some iron-rich basalts might be expected to produce undersilicated, alkalic differentiates, whereas most basalts would likely produce differentiates that would be oversilicated. Despite their efforts to downplay the fact, the experimental studies of iron-bearing systems continued to provide support for the observations of petrologists like Fenner and Kennedy.

Because the important issue of the nature of the ultimate goal of crystallization–differentiation concerned whether residual liquids became enriched in iron or in the constituents of granitic material, namely silica and alkalis, Bowen was understandably eager to begin investigating systems containing both ferrous iron and alkali. Thus he and Schairer began a series of experiments in the system $NaAlSiO_4$–FeO–SiO_2. In an early investigation, Bowen and Schairer (1936) encountered a eutectic point on the albite–fayalite join whose location indicated a much greater enrichment in alkali feldspar than in fayalite. Bowen concluded that the fayalite-bearing rhyolitic and phonolitic lavas of the east African rift valleys were, therefore, the end product of the alleged iron-enrichment trend.

The classic contribution to emerge from this phase of Bowen's work, however, was a paper published under the title "Recent high temperature research on silicates and its significance in igneous geology" (Bowen, 1937). Bowen focused only on the nature of final residual magmas on the basis of studies of silicate equilibrium. He reminded his readers that the plagioclase and diopside–anorthite albite diagrams clearly pointed to soda enrichment of residual liquids. He also showed that low-melting liquids in two potash-bearing systems, leucite–anorthite–silica and leucite–diopside–silica, that he and Schairer had been collaborating on throughout the 1930s, were highly enriched in potash, alumina, and silica and impoverished in magnesia and lime.

Bowen established that residual liquids in iron-bearing systems such as albite–fayalite and $NaAlSiO_4$–FeO–SiO_2 were enriched in soda and depleted in iron. Because of the liquid trends in these systems Bowen argued that the alkali–alumina silicate system was the goal toward which fractional crystallization trends. Then he presented data for what he designated as "petrogeny's 'residua' system," namely, the system $NaAlSiO_4$–$KAlSiO_4$–SiO_2. He pointed out a low-temperature trough within the system and predicted that igneous rocks approaching the diagram in composition should plot within or very close to that thermal trough (Figure 26). He plotted the recalculated normative chemical compositions of forty rhyolites, trachytes, and phonolites that he had collected from the East African rift valley on the diagram. All but three of the

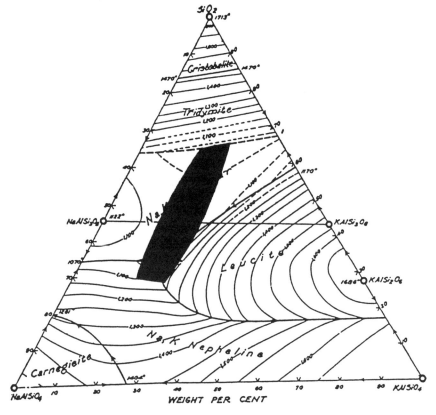

26. Phase diagram of the liquidus surface in "petrogeny's residua system," $NaAlSiO_4$–$KAlSiO_4$–SiO_2 at one atmosphere as determined by Bowen (1937). Blackened area shows the thermal trough. Reproduced by permission of *American Journal of Science*.

analyses plotted within the thermal valley, and the remaining three were very close. The experimental data, he claimed, showed that rhyolites, trachytes, and phonolites were the result of residual liquids derived from the crystallization of complex magmas by fractional crystallization.

The work on iron-bearing silicate systems concluded with a paper on the system $NaAlSiO_4$–FeO–SiO_2 (Bowen and Schairer, 1938). Two important ternary eutectics were encountered within the system. Bowen and Schairer showed that compositions in one part of the system would produce residual liquids relatively poorer in silica and ultimately crystallizing nepheline and that compositions in another part of the system would produce residual liquids relatively richer in silica and ultimately crystallizing tridymite. In both cases, they noted, the eutectic points had chemical compositions that were relatively enriched in alkali–alumina components and relatively poor in the fayalite component.

Bowen and Schairer (1938) believed that study of this system provided a solution to the question as to the dominance of plagioclase crystallization, leading to enrichment of liquid in albite, versus olivine crystallization, leading to enrichment in fayalite. They concluded that the net result of fractional crystallization in the complex liquid of the system would be enrichment in alkali–alumina silicate. Absolute enrichment in iron silicate did not occur, they observed, but enrichment in iron silicate relative to magnesian silicate did. Bowen was confident that the problem of the major goal of crystallization–differentiation had now been solved. Fractional crystallization, he was thoroughly convinced, ordinarily would lead to the formation of alkali–alumina–silica-enriched liquids. Fractional crystallization ordinarily would not lead to the formation of iron-enriched liquids.

Bowen had reached a crucial stage in his career. With the collaboration of Andersen, Morey, Schairer, and other colleagues at the Geophysical Laboratory, he had concluded nearly three decades of groundbreaking experiments on silicate systems at one atmosphere pressure. He had removed the initial weak point of his early set of experiments by performing experiments that included iron and potassium. For the most part, he believed that the theory of crystallization–differentiation that he had laid out in preliminary form in 1915 had been repeatedly confirmed by his subsequent experimental work. He continued to maintain that fractional crystallization was the overwhelmingly dominant mechanism of differentiation in magmas; that fractional crystallization generally entailed crystal settling, squeezing out of residual liquid, or zoning of crystals; that basalt was normally the parental magma; that the course of differentiation could be arrested at various stages depending on rates of cooling; that cooling liquids could follow different compositional paths depending on conditions; and that the predominant compositional path was in the direction of enrichment in silica, alkalis, and alumina. Under normal circumstances, he would continue to insist, granite is the end product of the crystallization–differentiation of basaltic magma.

At this juncture in his career, Bowen expressed a desire to introduce the methods of experimental petrology into the academic world. His friend and mentor, Arthur Day, had just retired from the directorship of the Geophysical Laboratory in 1936. A year later, Bowen left Washington and accepted a position as the Charles L. Hutchinson Distinguished Service Professor of Petrology at the University of Chicago. He became the third in a succession of outstanding igneous petrologists at the university after Joseph P. Iddings and Albert Johannsen. Bowen would achieve renewed prominence in the 1940s when the granite controversy flared up.

CHAPTER 16

ALTERNATIVES TO BOWEN'S HYPOTHESIS

Bowen's theory of crystallization–differentiation remained vulnerable on grounds other than the ultimate goal of differentiation. Several petrologists pointed out that, even if the end product of differentiation of basalt were granitic melt, the total amount that could be generated would not exceed around ten percent, an amount far too small to account for the existence of great granitoid batholiths. And the argument that intermediate rocks such as diorite or andesite that Bowen's theory called for either were lacking entirely or present in minor amounts was made repeatedly. The calculations of Grout (1926) indicated that fractional crystallization of one hundred parts basalt should yield about ten parts diorite, five parts quartz monzonite, and five parts granite. Such volumetric proportions were not consistent with Bowen's claims for crystallization–differentiation.

It was often the virtual lack of intermediate rocks altogether that bothered some petrologists who were, nevertheless, greatly impressed by Bowen's experimental results. Ever since the days of Bunsen, who appealed to two fundamental trachytic and pyroxenic magmas, petrologists had been struck by the bimodal nature of magmatism. Daly (1914) stressed that basalt is overwhelmingly dominant among volcanic products and that granitic material prevails among plutonic rocks, and the strongly bimodal distribution of basalt and granite with its attendant paucity of intermediate types came to be known as the "Daly gap." It was noted that many sills and laccoliths consist of gabbro with minor granitic micropegmatite in their upper parts. In their Mull memoir, Bailey et al. (1924) noted the common occurrence of closely associated mafic and acidic rocks and attempted to account for the association in terms of a modified version of crystallization–differentiation.

Arthur Holmes, however, was not quite so sure about that approach. While expressing deep appreciation for Bowen's experiments and theory in a review of *The Evolution of the Igneous Rocks*, just as he had back in 1916, Holmes dissented from the notion of granite as the end result of crystallization–differentiation and suggested that partial fusion of the crust played an important role. This view came to more pointed expression in several papers that he issued in the 1930s. In an article dealing with the association of acid and basic rocks in the central complexes of Mull and Ardnamurchan, Holmes (1931) rejected the views of Bowen, Harker, and Bailey. Bowen and Bailey invoked varying forms of crystallization–differentiation to account for the diverse rock types, but Holmes argued that differentiating basaltic magmas typically produced trachytic or syenitic rocks rather than granite. And why, he asked as he had in 1916, did not one ever find associated basic and acid rocks in the Pacific Ocean region where basalts are so abundant if acid material results from crystallization–differentiation? Perhaps the lack of sialic crust in the Pacific and its presence in the continents was related in some way to the absence of granitic associates in the former and its presence in the latter, he suspected. Harker had attributed the duality of igneous rocks to separate injections of coexisting siliceous and basaltic magmas that had differentiated in a deeper-seated magma basin. In opposition to these views, Holmes maintained that rising hot basalt magma, gradually stoping its way upward would, upon entry into the upper granitic layer of the crust, possess sufficient heat to fuse that layer, thus generating granitic melt. As a consequence, a cupola with acidic melt resting on basic melt would develop. Any differentiation of the basaltic material would result in small amounts of trachytic, not granitic, liquid. Holmes also allowed for the possibility of some later hybridization of the acidic and basaltic melts to form minor quantities of intermediate rocks. The same hypothesis was also invoked to explain the contrasting acid–basic associations characteristic of the Duluth, Sudbury, and Bushveld lopoliths. Although the details may differ, contemporary petrologists will recognize Holmes' essential idea in current views about the heating and partial melting of upper granitic crust by rising basaltic magmas that have ponded on the underside of that crust. The same arguments were summarized the following year in a paper on the origin of igneous rock (Holmes, 1932).

The association of acid and basic rocks also caught the eye of S.R. Nockolds, then at the University of Manchester. Nockolds (1933) had discussed the nature of contamination in acid magmas, and he attempted to show the tendency of contamination to produce normal rock types. Nockolds (1934) suggested a modification of Bowen's theory of crystallization–differentiation, denying that such differentiation was generally progressive in character. Rather than producing a wide range of rock types, he thought, crystallization–differentiation led to the separation of two fractions, acid and basic, at an early stage. Nockolds referred to his version of crystallization–differentiation as "contrasted differentiation" in opposition to the "progressive differentiation" of Bowen. After the basic material had solidified, the residual acidic melt then reacted with and was contaminated by varying amounts of the basic rock to produce a range of intermediate hybrids. The intermediate rocks, in his view, are not generated in the differentiation stage but rather in a later, contamination stage.

Holmes (1936) vigorously attacked the notion of contrasted differentiation by pointing out from Grout's calculations the small volume of granitic material that could be produced by any form of crystallization–differentiation. He charged that such a small volume of acid residuum could never be separated from the meshwork of early-formed crystals unless intense squeezing had occurred. But such squeezing, brought about by applied stresses, would inevitably result in intense deformation, he thought, and yet indications of planar foliation or cataclastic deformation are characteristically lacking from sills and central complexes where "contrasted differentiation" was alleged to have occurred. What is more, Holmes pointed out that the Whin sill, by way of example, contains its acid residuum, not as a distinct layer but as strictly interstitial material in about the proportions calculated by Grout. The existence of localized patches, streaks, and veins of acidic material in thousands of sills and dikes indicated to Holmes that their occurrence was due to some mechanism other than crystallization–differentiation.

Nockolds (1936) responded to Holmes' denial that the acid and basic magmas were necessarily derived from a common source by arguing that the invariable association of tholeiitic basalts alone, not olivine–basalt, with acidic material rather than alkalic magmas would make no sense apart from their derivation from a common source. He also argued that at least 15 percent of acid residuum could be generated under favorable circumstances, an amount sufficient to be separated from early crystals into a separate layer. Moreover, he thought that the volatile content of the acid residuum would increase its mobility so that it could be squeezed off more readily. He rejected Holmes' claim that an acid residuum could not be produced by differentiation without formation of intermediate liquids that produced intermediate rocks. If Holmes meant rocks of chemically rather than mineralogically intermediate composition, then Nockolds agreed with Holmes. Hornblende-bearing diorites are not present, but, he said, pyroxene–plagioclase–bearing rocks of intermediate composition are.

Holmes (1937) shot back that Nockolds still had not provided clear evidence that the contrasted magma types were generated by differentiation. He agreed with Nockolds on the existence of contrasted acid and basic magma types, but charged that the association "is no more evidence of contrasted differentiation than the observed association of toys and stockings on Christmas morning is evidence of the reality of Santa Claus" (Holmes, 1937, p. 189). Holmes did not feel that Nockolds' claim for a common source had been substantiated. He reiterated that the association could have arisen by partial fusion of acid crust by basic magmas. He also admitted having no objection to the idea of contrasted crystallization–differentiation where the rocks provided evidence of the operation of stress. Even 15 percent interstitial liquid, he said, would still require the application of stress to squeeze it out. Generally, he charged Nockolds with confusing evidence with interpretation.

In general, the experimental work of Bowen, Schairer, and their Geophysical Laboratory colleagues had gone a long way in clarifying the nature of possible processes of differentiation. But to gain lasting geological value, the experimental results had to be applied to the rocks and field contexts of intrusions and volcanic successions. Consequently, further clarity over the nature of differentiation processes had to await the outcome of detailed studies of differentiated sills and large layered intrusions as well as the attainment of some degree of resolution concerning the origin of granite.

17 ❖ AN UNSURPASSED NATURAL LABORATORY

Differentiated Sills and Layered Intrusions

The concept of differentiation was first applied to suites of diverse plutonic rocks interpreted to be genetically related or consanguineous, to use the term of Iddings. In some instances, the concept was also applied to internally heterogeneous, irregular masses of plutonic rock such as those at Garabal Hill and Carrock Fell. Before long, however, geologists began to map and describe thick sills and laccoliths of gabbro that are overlain by more silica-rich material. The internal mineralogical variations within these gabbroic bodies commonly came to be regarded as the product of differentiation. In *Igneous Rocks and Their Origin*, Daly (1914) tabulated a list of 47 localities of sills and laccoliths that he believed demonstrated evidence of differentiation. Of these, he described about 25 as occurrences of sills or sheets. The list included diabase sills in Ontario, some of which had been studied by Bowen (1910), the Moyie River sills of British Columbia described by Daly (1912), the Palisades sill in New Jersey (Lewis, 1907), sills in Sweden, several occurrences of sills in southern Africa, several occurrences of sills in Scotland, and the Electric Peak sheet in Yellowstone National Park (Iddings and Hague, 1899). For the greater number of these sills, most of which are more than thirty meters in thickness, Daly (1914) reported that they consist of diabase or gabbro with an upper facies of micropegmatite or micropegmatitic diabase. He also regarded most of these sheets as products of gravity differentiation and suggested that the different lithologies within the sheets had been assembled by gravity as immiscible liquid fractions. Early in the twentieth century, then, gabbroic sills had become prime exhibits of differentiation. Later on, Bowen applied his concept of crystallization–differentiation to sills, and Nockolds viewed sills as examples of contrasted differentiation.

DIFFERENTIATED SILLS

As soon as Bowen (1915c) put forward the initial statement of his theory of crystallization–differentiation, gabbroic sills, as well as an increasing number of larger gabbroic and ultramafic bodies characterized by pronounced layering, came to be regarded as important sites for testing specific mechanisms of differentiation, in particular, crystal settling. They also became major battlegrounds on which the theoretical war over iron enrichment versus silica enrichment was fought. Precisely because Bowen claimed that basaltic liquid was the typical parent of differentiated suites, basic sills and basic layered lopoliths provided excellent opportunities for determining the end result of the process of differentiation. Intensive study of differentiated gabbroic and ultramafic bodies also opened the way toward an understanding of the dynamic, chemical, and mineralogical processes operating within magma chambers.

One of the most important studies of sills published before Bowen's paper on the later stages of the evolution of the igneous rocks was that of Lewis (1907) on the Palisades sill. The Palisades sill, about 200 to 400 meters thick, crops out in New Jersey and New York as a prominent cliff along the west side of the Hudson River opposite New York City. The sill is related to several similar diabase sills and dikes as well as flows of tholeiitic basalt that crop out throughout the Newark, Hartford, Gettysburg, and other Triassic–Jurassic basins along the eastern side of the Appalachian Mountain chain. The Palisades sill is composed mainly of plagioclase, clinopyroxene, and iron oxides. A prominent layer containing abundant magnesian olivine, commonly known as the olivine ledge, is about 1 to 10 meters thick and is exposed about 10 to 30 meters above the base of the sill. Olivine abruptly disappears at the top of the layer. Near the top of the sill, alkali feldspar and minor quartz are present, in some cases as a micrographic, interstitial intergrowth.

Because of its prominent location near New York City, the sill was always easily accessible, and construction and road building continued to create new exposures so that geological relationships of the sill could readily be observed in a wealth of locations. Hence, the Palisades sill had been examined in reconnaissance for a long time before the first major study was completed by J. Volney Lewis (1869–1969). Lewis worked with the New Jersey Geological Survey from 1892 to 1893, spent a few years at Clemson College in South Carolina as Professor of Geology until 1904, then returned to New Jersey as Professor of Geology at Rutgers University until 1931 (Trager, 1973). He interpreted the sill as the product of a single pulse of magmatic injection and considered the olivine ledge to be the product of crystal settling of early-formed olivine, a conclusion later endorsed enthusiastically by Bowen. Lewis also thought that the material close to the contacts was slightly more basic than the average diabase magma and suggested that the increased basicity had originated through the Soret effect.

One of the most important studies of a differentiated sill to be published immediately after the appearance of Bowen's evolution paper was an investigation of the Lugar sill in southwestern Scotland by George Walter Tyrrell (1883–1961), Professor of Petrology at the University of Glasgow since 1906 (Tomkeieff, 1962). In contrast to the Palisades sill, the Lugar sill is much less accessible, far more poorly exposed, and only about 140 feet thick. The most prominent exposures along Lugar Water, however, are marked by lithologic contrasts. At the upper and lower contacts are fine-grained chill zones. As Tyrrell (1916) described the sill, the top layer is a 20-foot-thick teschenite unit consisting of analcime, plagioclase, Ti-augite, and ilmenite. Below that is a 10-foot-thick, finer-grained theralite unit composed of nepheline, olivine, Ti-augite, and plagioclase. The theralite is underlain by a 90-foot-thick, coarser picrite–peridotite unit composed of olivine, Ti-augite, and the amphibole mineral, barkevikite. The lowest layer is another teschenite unit about 20 feet thick. Between the theralite and picrite layers is a very thin band of lugarite, a coarse-grained nephheline-rich theralite, that Tyrrell regarded as intrusive.

Tyrrell rejected the notion that the Lugar sill was emplaced in a single act of intrusion on the grounds that the chemical composition of the chilled contact-zone rocks differs from the bulk composition of the sill, which he calculated by averaging the analyses of the different units after weighting them by volume. Although perplexed by the lack of xenocrysts, xenoliths, veins, and dikes adjacent to the contacts between allegedly intrusive units, Tyrrell considered the Lugar sill to be a composite sill that resulted from two dominant episodes of intrusion, the first of which introduced the teschenite at the contacts, and the second of which quickly thereafter produced the ultramafic rocks of the interior.

Because of the separate episodes of injection of chemically different magmas, Tyrrell stated that the main differentiation of the Lugar sill had occurred prior to its emplacement. He had originally favored the notion of a differentiation into teschenite and ultrabasic magmas by means of liquid immiscibility prior to injection, but he abandoned that conception following the publication of Bowen's evolution paper. Thanks to the influence of Bowen's work, Tyrrell now became inclined to ascribe the differentiation at the Lugar sill to the sinking of dense olivine crystals through an originally teschenitic magma in a deeper chamber prior to injection into the sill. Tyrrell was also induced to favor gravitational differentiation because he saw evidence for crystal sinking within the ultrabasic magmas that were presumably emplaced by the second injection. He pointed out the presence of 14 volume percent olivine in the theralite unit, 30 volume percent olivine in the picrite, and 65 volume percent olivine in the peridotite, and inferred that early-formed olivine had sunk under the influence of gravity to the lower levels of the ultrabasic unit, a finding that he regarded as in accord with the results of Bowen's experiments on crystal settling of olivine.

Tyrrell illustrated the chemical changes in the sill with a variation diagram in which he plotted the oxide components against the sum of magnesia plus iron oxide. He had decided not to plot the oxide constituents against silica, he said, because the silica content of the sill was relatively constant throughout. He found no tendency toward Bowen's residual granitic material at the Lugar sill.

LARGE LAYERED IGNEOUS INTRUSIONS

Very early in the twentieth century, the petrological community also became aware of the existence of some exceptionally large laccolith-like bodies characterized by pronounced layering of the rocks. Geikie and Teall (1894), for example, described banded gabbros from the Cuillin Hills on the Isle of Skye. Until the mineralogical regularity of the layers was ascertained, they were interpreted as gneisses. The study of Harker (1904) on the geology of the Isle of Skye confirmed that the layered rocks are igneous and also showed that they dip inward toward the center of intrusions. Gilbert (1906) observed alternating light-and-dark banding in some of the granitic intrusions of the Sierra Nevada Mountains in California. N. V. Ussing (1912) described the very well-exposed Ilimaussaq intrusion in Greenland. He regarded this layered body, composed of various alkalic foyaitic and syenitic rocks, as the product of crystallization-controlled differentiation in which both crystal sinking and flotation occurred. Grout (1918b) described the Duluth gabbro and attributed the rhythmic layering of its lower gabbroic units to a combination of crystallization–differentiation and convection. He further assigned the distinction between the lower gabbro and upper red rock granophyre to liquid immiscibility. In the 1920s, H. H. Read mapped a number of rhythmically layered gabbroic bodies in Scotland, such as the Insch and Haddo House intrusions, that were considered as remnants of a much larger gabbroic intrusion. The Great Dyke of southern Africa was also recognized during this era.

Probably no large igneous body with pronounced layering attracted more interest than the enormous Bushveld intrusion near Pretoria in South Africa. The Bushveld Complex is so vast that several years of reconnaissance investigation were required before geologists began to grasp its nature and magnitude. The economic potential of the region certainly contributed to great interest in its geology. Various isolated studies were undertaken beginning in 1872. Soon geologists realized that a variety of igneous rocks are present. The first discussion of the Bushveld Complex as a whole was by Molengraaff in the years between 1894 and 1904. Gustaaf Adolf Frederick Molengraaff (1860–1942) taught at the University of Amsterdam from 1891 to 1897, became State Geologist of Transvaal in 1897 and 1898, and then devoted time to industry in both Indonesia and Transvaal until 1908 (Spencer, 1947). He returned to the Netherlands, where he became Professor of Geology at the Delft Technical

High School. Molengraaff noted an apparent relationship between the position of igneous lithologies in the Bushveld Complex with their specific gravity. He believed that there had been a separation of the Bushveld magma on the basis of density, resulting in an outer basal zone of basic rocks, a middle zone of syenitic rocks, and an upper, central zone of granitic rocks.

In 1902, the Geological Survey of the Transvaal was established and systematic mapping of the Bushveld Complex began under the direction of Herbert Kynaston (1868–1915). Kynaston had been a geologist with the British Geological Survey between 1895 and 1902, became director of the Geological Survey of Transvaal, and later was appointed director of the Geological Survey of the Union of South Africa upon its establishment in 1913 (Woodward, 1916). The most prominent geologists to work with him on the Bushveld Complex were Percy Wagner and Arthur Hall. Important contributions to the study of the Bushveld intrusions were also made by Hatch, Daly, and Shand.

Percy Albert Wagner (1885–1929) studied at South African College in Cape Town and graduated in 1906 from the School of Mines and Technology in Johannesburg (Spencer, 1930). Wagner undertook graduate study at the Bergakademie in Freiberg and studied with Rosenbusch in Heidelberg, where he received a Ph.D. degree with a dissertation on the diamond-bearing rocks of South Africa. He became Professor of Geology at Transvaal University College, worked for DeBeers investigating diamond occurrences in Rhodesia and Southwest Africa, and did consulting work in Johannesburg and Pretoria. From 1914 to 1927, he was a geologist with the Geological Survey of the Union of South Africa. In 1927, he left to become a consulting geologist, but he died prematurely shortly thereafter. Wagner's greatest contributions to Bushveld geology concerned the chromium, nickel, and, particularly, the platinum deposits that were discovered in 1924 and that served as a major spur to further study of the intrusion.

Thanks to such discoveries, the world began to direct its attention toward the Bushveld Complex during the 1920s. In 1922, Harvard University sponsored the Shaler Memorial Expedition to the Bushveld. Daly and Charles Palache from Harvard participated in the expedition along with Frederick E. Wright of the Geophysical Laboratory and Molengraaff, then at Delft. Their guide was Arthur Hall. Later, the 15th International Geological Congress, held in Pretoria in 1929, also focused attention on the Bushveld Complex, particularly by means of a very extended field trip in which petrologists like Bowen participated.

The vast majority of workers on the Bushveld Complex had concentrated on specific areas or specific problems related to the economic geology. Few geologists until Hall had a sense of the larger picture because of the enormous size of the intrusion. It was primarily Hall who established the zonal terminology. Hall defined the main units of the intrusion: a chilled phase, a transition zone, a critical zone, a main zone, and an upper zone. Largely at the urging of

Daly, Hall began work on a massive synthesis of Bushveld data that resulted in the publication of a prodigious memoir, *The Bushveld Igneous Complex of the Central Transvaal* (Hall, 1932). Arthur Lewis Hall (1872–1955) was another of the many British geologists who transferred to South Africa (Haughton, 1956). After an early education in a number of German cities, Hall entered Cambridge from which he graduated in 1899. After a couple of years in teaching, Hall became a geologist with the Geological Survey of the Transvaal, newly reorganized in 1902 under Kynaston after the conclusion of the Boer War. He served for 29 years with the surveys of the Transvaal and of the Union of South Africa. Although Hall investigated the Vredefort dome and also a host of economic deposits throughout South Africa, he was primarily associated with the Bushveld Complex.

In his 1932 memoir, Hall, like Daly and Molengraaff, favored the notion of gravitative differentiation, recognizing that there is an overall decrease in specific gravity of the rocks from the bottom to the top of the intrusion despite local variations and reversals. He was skeptical of the hypothesis of multiple injection of the various zones or of individual units on the grounds that the units such as the chromite layers and the Merensky reef demonstrate remarkable lateral continuity and show no signs of cross-cutting. Thanks to Hall's labors, the Bushveld Complex became one of the world's prime exhibits for the concept of gravitational differentiation. The thick, continuous layers of pure chromite, however, did pose a serious problem for the gravitative differentiation hypothesis, and Hall did not explain why dense monomineralic layers should have settled on top of less dense anorthosite layers.

Although several chemical analyses of individual rocks were included in Hall's memoir, he incorporated surprisingly little information on the chemical composition of individual minerals. In some instances, he noted that plagioclase in this rock is labradorite or that pyroxene in that rock is bronzite, but he included neither chemical analyses of minerals nor optical data for the plagioclase feldspars. Although Hall demonstrated some changes in overall chemical composition within the Bushveld Complex, he had not demonstrated that there is a progressive change in the chemical compositions of minerals from bottom to top. Later detailed studies of the chemical mineralogy by Lombaard and students at the University of Pretoria disclosed the progressive changes in the mineral chemistry from the bottom to the top of the intrusion.

B. V. Lombaard also worked for the Geological Survey of the Union of South Africa. In his study of the felsites and norites in the Lydenburg area of the Bushveld Intrusion, Lombaard (1934) investigated the variation of the chemical composition of plagioclase and pyroxene as determined from their optical properties. He showed that plagioclase generally becomes more sodic upward despite some local reversals. He found that plagioclase compositions range from An_{85} upward to An_{50} in the gabbros, from An_{40} upward to An_{18} in the norite above the gabbro, and from An_{12} to An_0 in the felsites. Lombaard also discovered that the ratio of orthopyroxene to total pyroxene decreases from the bottom to the top of the section. Again, he noted some reversals. From optical properties, Lombaard also estimated changes in the Mg/Fe ratio of pyroxenes, finding that the iron content of both orthopyroxene and clinopyroxene increases toward the top of the intrusion. Lombaard interpreted these chemical variations as the result of a combination of crystal settling and occasional injections of fresh magma of variable composition. Thick monomineralic layers of pyroxenite and anorthosite, for example, he interpreted as the product of crystal setting.

Other major layered mafic and ultramafic intrusions soon came to light. In 1930, under the sponsorship of the United States Geological Survey, J. W. Peoples, A. L. Howland, and Edward Sampson began to study the Stillwater Complex in the Beartooth Mountains of southern Montana. They were accompanied for several field seasons by Harry Hess of Princeton University, and in 1935, Hess began his own detailed mineralogical investigation. These co-workers had originally intended to prepare a joint memoir in which Peoples would write on the general geology, Howland on the floor rocks and sulfide deposits, Sampson on the chromite deposits, and Hess on the mineralogy. Short papers on the chromite deposits, some very brief overviews of the complex, and an abstract on the mineral variation by Hess (1938) were published. World War II, however, delayed the project. Not until 1956 did Hess, who had been an active participant in the war, complete the manuscript of a memoir that was finally published four years later (Hess, 1960). His main ideas on layered intrusions will be examined in Chapter 29.

The Bay of Islands Intrusion on the west coast of Newfoundland was first described by Earl Ingerson (1935) who had worked on the intrusion for his Ph.D. dissertation at Yale University. Ingerson mapped four isolated bodies of layered gabbroic and ultramafic rocks that he considered as four separate laccoliths. Buddington and Hess (1937) visited the occurrence and decided that the four bodies were actually separate parts of one larger lopolith. Presently, the Bay of Islands Complex is regarded as a fragment of uplifted oceanic crust that was affected by transform faulting.

THE SKAERGAARD INTRUSION

Without question, the layered intrusion that has made the greatest impact on igneous petrology is the Skaergaard Intrusion, brought to the attention of geologists in a memoir by Wager and Deer (1939). Taylor and Forester (1979, p. 356) referred to the Skaergaard Intrusion as "the best-studied igneous body on Earth" and "an unsurpassed natural laboratory for the study of igneous petrology and related magmatic and hydrothermal phenomena." In 1930, a massive body of olivine gabbro was discovered on the southeastern coast of Greenland by Laurence Rickard Wager (1904–1965) during a brief exploration

27. Laurence Rickard Wager (1904–1965). Reproduced by permission of Godfrey Argent.

by the British Arctic Air-Route Expedition. Wager graduated in 1926 from Cambridge, where he studied under Harker and Cecil Tilley (Deer, 1967; Brooks, 1981; Hargreaves, 1991; Vincent, 1994). At Cambridge, Wager (Figure 27) became known as one of the premier mountaineers and rock climbers of his generation. These skills contributed to his participation in the original Greenland expedition. In 1929, Wager became a Lecturer in Petrology and Mineralogy at Reading University. Following his initial Skaergaard discovery, Wager returned to Greenland in 1932 under the auspices of the Scoresby Sound Committee's Second East Greenland Expedition to make a preliminary map of the Skaergaard gabbro body. In 1933, Reading University gave him a leave of absence to join an expedition to climb Mount Everest. During the climb, Wager came closer to the summit than anyone ever had without the use of oxygen. He returned to Greenland when the small British East Greenland Expedition 1935–36 spent the period from July 1935 to August 1936 in a thorough study of the Skaergaard Intrusion. In 1939, Wager and one of his collaborators, W. A. Deer of the University of Manchester, compiled the results of their studies into a monumental treatise entitled *The Petrology of the Skaergaard Intrusion, Kangerdlugssuaq, East Greenland* (Wager and Deer, 1939). After

AN UNSURPASSED NATURAL LABORATORY 321

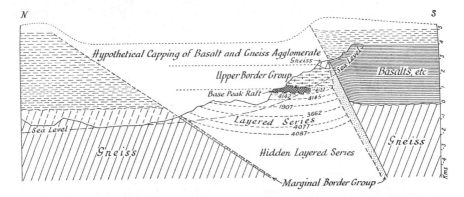

28. North-south cross section of the Skaergaard Intrusion before flexuring and erosion (Wager and Deer, 1939). Reproduced by permission of Danish Polar Center.

publication of his Skaergaard memoir, Wager was commissioned in the Royal Air Force photographic reconnaissance branch for World War II. In 1944, upon release from the Royal Air Force, Wager went to the University of Durham, where he succeeded Arthur Holmes. In 1950, Wager became Professor of Geology at Oxford, where he continued his Skaergaard studies and initiated a program of reinvestigation of the Tertiary volcanic centers of the Scottish islands.

Wager and Deer described a great funnel-shaped layered gabbroic pluton (Figure 28) that displays progressive changes in the chemical composition of the primary mineral phases as well as a wealth of features that they attributed to sedimentation of minerals through the magma onto a gradually rising magma chamber floor. The authors showed that, although the original floor of the intrusion is not exposed, from the exposed stratigraphic base toward the top of the intrusion, the soda content of plagioclase increases and the calcium content of plagioclase decreases, whereas the iron content of both olivine and pyroxene increases and their magnesium content decreases. They attributed these trends in chemical composition of the minerals to gravitational accumulation of crystals that had formed near the roof of the gradually cooling gabbroic magma mass. Just as Grout had proposed for the Duluth gabbro, Wager and Deer also believed that sinking of crystals had been assisted by convection and turbidity currents that swept down the walls of the Skaergaard chamber, flowed across its floor, and arose toward the center of the mass. They explained the prominent rhythmic layering, featuring alternating dark mafic-rich and light plagioclase-rich bands, in terms of variable current and settling velocities in crystal-charged magma sheets flowing across the chamber floor.

Bowen's theory of crystallization–differentiation received a major boost from the Skaergaard investigation because Wager and Deer were convinced that the primary mechanism of differentiation of the Skaergaard magma was fractional crystallization. On the other hand, Bowen's claims about the end products of differentiation received a substantial blow. Wager and Deer (1939) estimated that about 75 percent of the Skaergaard magma had solidified during the early stages of crystallization. This stage, however, resulted only in formation of eucrites and olivine gabbros. It was evident to Wager and Deer that the trend of fractional crystallization in the Skaergaard Intrusion was not one of increasing silica content, for the silica content actually decreases slightly with differentiation. The main changes during the early stage of the Skaergaard differentiation reported by Wager and Deer were increases in the amounts of ferrous and ferric iron, soda, potash, TiO_2, and P_2O_5, and decreases in the amounts of MgO and Al_2O_3. They estimated that another twenty percent of the mass crystallized during the middle stages during which the same chemical trends continued. After 95 percent of the body had solidified, they wrote, the magma contained about 25 percent iron oxides. This middle stage of crystallization was believed to have ended abruptly, however, as the residual magma changed its trend toward a quartzo-feldspathic composition. They attributed the overall increase in iron content and decrease in magnesium content to iron enrichment of the ferromagnesian minerals with minor contributions from iron oxides such as ilmenite and magnetite. Because differentiation had not resulted in silica enrichment, the authors plotted oxides and oxide ratios versus normative albite content of plagioclase rather than silica on variation diagrams as a more accurate indication of trends of differentiation in place of the more traditional Harker diagram. Wager and Deer concluded that, although crystallization–differentiation was a significant petrogenetic process in the Skaergaard Intrusion, the changes in chemical composition were not in the direction insisted on by Bowen. The trend was toward enrichment in iron as advocated by Fenner.

Wager and Deer pointed out that differentiation of Skaergaard magma toward iron-rich rocks precluded them from belonging to the "normal" calc–alkalic series of rocks represented, for example, by the San Juan volcanic rocks of southwestern Colorado or the granitoid rocks of the Sierra Nevada batholith in California. Reasoning that the trend shown by the Skaergaard Intrusion should be regarded as typical for gabbroic magmas, they claimed that some other process in addition to, or in place of, crystal fractionation had been at work in the production of the calc–alkalic series.

Wager and Deer concluded that ferrogabbros, that is, iron-rich gabbros, should be considered the typical residual products of basaltic magma differentiation. The only reason that ferrogabbros are rare, they judged, is that the amount of residual magma of ferrogabbro composition is relatively small. They further suspected that strong, long-continued differentiation of basaltic magma, such as occurred in the Skaergaard Intrusion, was probably relatively rare because fractional crystallization required slow cooling of a large magma body under tranquil external conditions. The relative rarity of layered basic masses indicated to Wager and Deer that the requisite conditions for producing strong fractional crystallization were rather uncommon.

The final stage of Skaergaard differentiation, amounting to about 5 percent of the whole, was described as marked by an abrupt change from iron-rich gabbroic rocks to basic and acid granophyres. The mechanisms by which the granophyres were formed were less clear to Wager and Deer but seemed to involve a combination of gravitative action, filter-press action, and some assimilation. Wager and Deer were now prepared to give Bowen his due. Though noting the uncertainty of the mechanisms of differentiation, Wager and Deer acknowledged that fractional crystallization ultimately produced a magma of general granitic composition, the result expected by Bowen for extreme fractional crystallization of basaltic magma. At least some of the granophyre, they argued, could be attributed to silica enrichment of residual liquid brought about by the olivine-removal process postulated by Bowen, but they were hesitant to assign the same degree of certainty to that claim as to claims about the differentiation of the layered series because of uncertainty about the amount of assimilation of acid gneiss in granophyre production. In any event, the total percentage of granitic material produced was embarrassingly small and provided scant comfort for adherents of Bowen's scheme.

To make clear that the chemical differences between calc–alkalic rocks and strongly differentiated basalts were very striking, Wager and Deer invented the $(K_2O + Na_2O)–FeO–MgO$ (AFM) triangular composition diagram (Figure 29). On this diagram they plotted chemical data from the Skaergaard Intrusion, the Mull Complex, various other dikes, sills, and lavas, and also Daly's average basalt, andesite, dacite, and rhyolite as representative of the calc–alkalic rock series. They adopted the view that the calc–alkalic series, contra Bowen, was generally not the result of fractional crystallization of basalt magma but very probably in the majority of situations the result of magma mixing or assimilation of granitic and basaltic material in varying proportions. They did not deny that, on some occasions, a rock type of the calc–alkalic series might be produced by fractional crystallization of basaltic magma, nor did they maintain that granite could never be produced by such a process. Like other critics of Bowen's ideas, however, they found it difficult to explain by the hypothesis of crystallization–differentiation why the known bulk of intermediate rocks is relatively small in the calc–alkalic series. Moreover, with Daly and Holmes, Wager and Deer suspected that most Phanerozoic granites were due to remelting of the sialic crust.

Not only did the Skaergaard Intrusion provide critical tests of various aspects of the theory of differentiation by fractional crystallization, it also initiated serious thought about the origin of various types of layering as well as magma conditions within a magma chamber. The idea of convection currents within a magma chamber had previously been proposed by Pirsson (1905b) for the Shonkin Sag laccolith in Montana, Grout (1918a,b) for the Duluth lopolith in

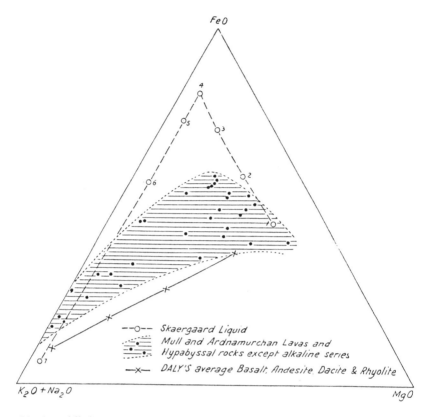

29. AFM (alkalis—FeO—MgO) diagram of Wager and Deer (1939) showing plot of chemical composition of Skaergaard Intrusion liquids, Hebridean lavas and hypabyssal rocks, and Daly's average basalt, andesite, dacite, and rhyolite. Reproduced by permission of Danish Polar Center.

Minnesota, and Holmes (1931) for the British Tertiary plutonic centers. For the Skaergaard Intrusion, Wager and Deer postulated, instead of vertical sinking of crystals, a simple convection system in which currents were conceived to move down the walls, across the floor, and up the center. They suggested that the currents stirred the magma sufficiently to keep it fairly homogeneous and to insure that fractional crystallization would be a relatively uniform process. The crystals, they said, tended to grow in the descending current as a result of increasing hydrostatic pressure. As the liquid flowed horizontally over the pile of crystalline sediment on the chamber floor, the crystals were envisioned as settling through a relatively small thickness of liquid. The currents presumably fluctuated in velocity, thereby producing changes in the proportions of minerals reaching the floor at any one place. The result was rhythmic layering. Turbidity currents were also envisioned as producing graded layers.

Evidence for currents consisted of flow structures in the border group, the orientation of plagioclase crystals parallel to the axis of trough bands suggesting the flow of magma along the trough, concentration of dark, dense, less easily transported minerals near the margins, increase in the size of feldspar crystals toward the center, presumed to be the result of a greater amount of time to grow as the crystals were swept by currents toward the center, and streamlined granophyre xenoliths. Even though feldspars would likely have had a lower density than the surrounding magma, the Skaergaard authors held that they would still be carried by downward currents of dense magma charged with pyroxene and olivine crystals.

Wager and Deer, noting that convection is at an optimum in a body of liquid with a diameter about four times its thickness, postulated that the shape of the intrusion favored the establishment of circulation. They further claimed that the low viscosity of the initial Skaergaard magma favored convection. They believed that the high ferrous iron and low potash contents of the Skaergaard liquid would have given it greater fluidity than that of many basalts.

Early workers on the Ilimaussaq, Duluth, Bushveld, Bay of Islands, and Stillwater Intrusions had reached no consensus as to the origin of the layering. Intermittent crystallization due to gas emanation, relief of pressure, Earth movements, or extrusion of lava had been suggested as possible causes of the alternating light-and-dark layering commonly observed in these bodies. Wager and Deer, however, proposed variation in the velocity of convection currents as the explanation for such layering. If the current and supply of crystals remained constant, they argued, then the proportions of heavy and light minerals deposited at any one place would remain constant, although more of the dense constituents would be deposited near the margin of the intrusion than farther in. If, however, the strength of the currents increased, they said, the heavy minerals would be carried farther than before and would be deposited in greater abundance at a location where they were formerly scarce because the earlier, weak current had not transported them as far. Conversely, they reasoned, a lessening of current velocity would cause the light constituents to be deposited in greater abundance at a location where they had been less abundant because a stronger current had deposited more dense minerals.

Vertical sinking of crystals alone from the top of the magma chamber to its bottom, they thought, would have resulted in a much more marked separation into light and heavy constituents. For that reason, Wager and Deer believed that the convection theory provided a preferable explanation of the major mineralogical variation in the Skaergaard Intrusion.

The work of Wager and Deer on this exceptionally well-exposed layered intrusion was a magnificent achievement, particularly in light of its extreme remoteness. The "great work on the Skaergaard intrusion," said Wager's Oxford colleague E. A. Vincent (1994, p. 128), "was arguably the most important contribution to igneous geology made so far this century." The authors had carried research on layered differentiated intrusions well beyond that of other similar bodies as indicated by their comment that work on the Bay of Islands Complex, the Stillwater Complex, the Great Dyke, and the Sierra Leone Complex was not sufficiently advanced to contribute to the interpretation of the Skaergaard rocks. Within a decade of intensive study, Wager and Deer had demonstrated the progressive changes in chemical composition of bulk rocks and minerals with stratigraphic height in this intrusion; argued for the legitimacy of fractional crystallization; demonstrated the validity of the iron-enrichment trend; lent additional support to Kennedy's concept of the association of diverse differentiation trends with diverse basalt types; brought the role of convection in magma chambers to the forefront; and developed concepts for the formation of igneous layers. The Skaergaard memoir set a very high standard for all future studies of large, differentiated intrusions, studies that continue to the present (see Chapter 29). Although later petrologists have disputed many aspects of the crystallization history of the Skaergaard Intrusion, all have agreed that the prodigious efforts of Wager and Deer resulted in a remarkable understanding of the broad outlines of its crystallization history.

FREDERICK WALKER AND DIFFERENTIATED SILLS

In the decade following the publication of the Skaergaard memoir, advances were also made in understanding processes of differentiation in basaltic magmas through investigations of less spectacular, but certainly more abundant, sills and sheets commonly referred to as diabase or dolerite. One such investigation was carried out by A. B. Edwards on the dolerite sills of Tasmania, Australia, a country on the threshold of great influence in the advancement of igneous petrology in the 1940s. Branagan and Townley (1976) noted that much of the early geological study in Australia was carried out in connection with mining efforts. With the support of regional Bureaus of Mines, the Universities of Melbourne, Sydney, and Adelaide had offered some geological instruction since the late nineteenth century. In the first half of the twentieth century, departments of geology were also established at the Universities of Queensland, Western Australia, and Tasmania. After 1916, the Australian Federal Government became involved in the support of geology. In 1926, the government established a Council for Scientific and Industrial Research, which, a year later, appointed a petrologist. In the late 1920s and throughout the 1930s, therefore, igneous petrology began to emerge as a distinct discipline in Australia. The paper by Edwards was one of the earliest Australian petrology papers of far-reaching import. After his contribution, Australia began to play an increasingly visible role in the development of igneous petrology.

The sills of Tasmania had been examined in rudimentary fashion as early as 1896. Edwards (1942) wanted to know the reasons for the variations that had been observed by earlier workers. He found the commonly exposed chilled margins of the sills to be petrographically and chemically uniform and indicative that the tholeiitic magma was almost completely liquid at the time of its intrusion. Edwards observed that the Tasmanian dolerites bear a striking resemblance to the dolerites of the side of Antarctica facing Australia, but that they could be distinguished from the Palisades sill, the Karoo basalts of southern Africa, and the Whin sill in England by their lower sodium, iron, and titanium contents.

Edwards reported primarily on the Mount Wellington and Mount Nelson sills, each of which exceeds 300 meters in thickness. He sampled rocks at intervals of eight and sixteen meters, obtained chemical analyses of ten bulk rocks, and constructed compositional profiles in which oxide contents were plotted versus height within the sill. The profiles indicated decreases in magnesia and lime content and increases in total iron, silica, soda, potash, and titania from the base of the sills to their tops. Several other sills were also found to be magnesium-rich above their chilled bases and enriched in iron in their residual liquids near their tops. From optical properties and three chemical analyses of pyroxene separates, Edwards demonstrated iron enrichment and magnesium loss in the pyroxenes. He concluded that the chilled bottom of the sills had served as floors on which early-formed magnesium-rich pyroxenes accumulated by sinking from the more slowly cooled parts of the sills.

Edwards speculated that differentiation of a basaltic magma in a chamber of batholithic proportions would result in iron enrichment of ferromagnesian minerals, but that these would sink quickly enough to catch up with, or outdistance, the more slowly sinking plagioclases of an earlier generation. The result, he thought, would be that the residuum at the top of the chamber would tend to become more feldspathic and ultimately siliceous. He envisioned the development of cupolas above the highest points of the chamber with acid-alkalic differentiates concentrating in these protrusions and undergoing further differentiation. Edwards considered the weight of evidence for such a scenario so great that Kennedy's contention that trachytic and phonolitic rocks are the normal end products of the differentiation of undersaturated olivine basalt magma and that silica-rich differentiates are the normal end products of the differentiation of tholeiitic basalt magma should be regarded as established.

Edwards also claimed that some calc–alkalic rocks can be derived from tholeiitic basalt magma by differentiation. He did not think that the Skaergaard trend disposed of that possibility because the Skaergaard Intrusion provided a very special instance of differentiation in a closed chamber. But Edwards, too, realized that the great bulk of calc–alkalic rocks was the main reason for thinking that most were not derived in that manner. He agreed with Wager and Deer that most calc–alkalic suites probably were generated by assimilation of large quantities of sialic material by tholeiitic magma coupled with crystallization–differentiation in a subjacent chamber.

The petrologist who made the greatest contribution to the study of basaltic sills during the 1940s and early 1950s, however, was Frederick Walker (1898–1968). Walker began his professional career at St. Andrews University in Scotland from 1925 to 1939 (Drever, 1969). Like so many of his countrymen, Walker succumbed to the lure of South Africa and moved to the University of Cape Town, where he taught between 1939 and 1956. He then returned to Scotland with a position at University College, Dundee, from 1956 to 1963.

Walker (1930) early distinguished himself with a report on the large crinanite–picrite sill on Garabh Eilean and Eilean an Tighe in the Shiant Isles of the Scottish Hebrides. Walker pointed out that the sill consisted of basal picrite overlain by an olivine dolerite unit and a unit of crinanite, a rock that is essentially an analcime–olivine dolerite. He reported that silica, lime, soda, and potash contents of the rocks increase from the picrite upward to the crinanite, whereas magnesia and iron oxide contents decrease. In addition, Walker noted that the density and total olivine content of the rocks decrease upward and that plagioclase composition becomes increasingly sodic upward. He judged that the crinanite–picrite sill provided an exceptionally clear demonstration of crystallization–differentiation in which the gravitational settling of early-formed olivine crystals played an extremely important role. He noted that all the intrusions that display settling of olivine, with the exception of the Lugar sill, are very thick and concluded that slow cooling appeared to be a sine qua non of this type of differentiation.

Several times during the latter 1930s, Walker visited the Palisades sill and completed petrographic, optical, and chemical work on its rocks in the laboratories at Columbia, Harvard, Cambridge, and Cape Town. The results of his classic reinvestigation of the sill appeared in 1940. From optical studies of the major minerals, Walker (1940) concluded that plagioclase becomes slightly more sodic from the bottom to the top of the sill and that near the upper chill zone it becomes somewhat more calcic. He also determined that augite becomes enriched in the hedenbergite molecule toward the top. On the basis of these optical studies as well as micrometric, chemical, and density analyses, Walker accepted gravitational settling of olivine in stages. He rejected the idea that the olivine layer formed from a subsequent injection of magma on the grounds that the pyroxene to plagioclase ratio within the olivine layer is the same as in the diabase above and below it. Walker attributed scattered olivine near the base of the sill to sinking of crystals formed early in the upper part of sill. He postulated that olivine crystals forming in the middle to lower part of the sill contributed a shower of small olivine grains that settled to form the ledge and that some of the large grains of olivine made their way all the way from the top to settle into the olivine layer. In a comparison between the Palisades sill and other differentiated tholeiitic sills, Walker (1940, p. 1101) was struck that, although many of the features of the Palisades sill had their parallels in the other diabase intrusions, none of those other sills showed "the curious stratum of olivine diabase."

Upon moving to Cape Town in 1939, Walker initiated an ambitious research program on the Triassic Karoo (at the time spelled "Karroo") basalts of South Africa. Although the Karoo basalts had been examined since 1887 by such luminaries as Alexander du Toit, R. A. Daly, and T.F.W. Barth, it was Walker and his graduate student at Cape Town, Arie Poldervaart, who turned the sills into a prime exhibit of the variability of magmatic differentiation in a series of papers. The Karoo project was slowed by World War II-related commitments and occupied the entire decade of the 1940s.

The Karoo system in South Africa is riddled with basaltic dikes and sheets associated with the Drakensberg lavas. Many of the sills are as much as 300 meters thick, and a few such as the Insizwa sheet are around 1000 meters thick. Walker and Poldervaart published about a dozen papers on the Karoo basalts throughout the 1940s, including descriptions of the Hangnest sill (Walker and Poldervaart, 1941) and the New Amalfi sheet (Poldervaart, 1944). The Hangnest sill, about 175 meters thick, they found, is only slightly differentiated. Walker and Poldervaart noted that the only signs of crystallization–differentiation in the sill are the schlieren of dolerite pegmatite, a slight downward increase in the proportion of pyroxene, and an upward increase in the proportion of micropegmatite. The sill is characterized by the almost complete lack of evidence for crystal settling. The authors attributed the lack of evidence for gravitational differentiation to the elongated nature of the early crystallizing minerals, claiming that elongated minerals would not settle as rapidly as more equidimensional crystals because of a greater tendency to interfere with one another. Walker and Poldervaart also observed evidence of a marked tendency for the Hangnest magma to mobilize and react with the associated shales and siltstones. Moreover, they found excellent examples of stoped layers and fragments of country rock at the top and bottom contacts of the sill.

In a later paper, Walker and Poldervaart (1949) reported that the albite content of plagioclase commonly increases toward the top of the various differentiated sheets. Both olivine and pyroxene are characterized by iron-enrichment trends from bottom to top. The authors did find in some sills, however, that olivine disappears in the middle ranges where pyroxene was crystallizing. Walker and Poldervaart included a large number of the "ingenious" AFM diagrams invented by Wager and Deer on which they plotted Daly's average calc–alkalic trend along with analyses of Karoo rocks. They showed that the Karoo trend is similar to that of the Skaergaard Intrusion, namely, a very pronounced differentiation toward iron during the main stages and then toward alkali enrichment in the concluding stages of differentiation. The authors also plotted data on AFM diagrams for the Tasmanian dolerites, the Palisades sill, several Siberian traps, some Scottish tholeiitic rocks, and the Bushveld Complex and concluded that iron enrichment is the principal trend of differentiation in all these cases. Their plot of Hawaiian basalts showed a trend more like that of the calc–alkalic series, however. Despite the Hawaiian example, they wrote that iron enrichment is clearly the normal trend of differentiation during the greater part of the evolution of basaltic magma.

Walker and Poldervaart (1949) believed that differentiation is brought about by crystal fractionation, gravitational effects, segregation of volatile-rich phases as pegmatitic schlieren, acquisition of resurgent volatiles from wall rocks or xenolithic inclusions, and assimilation or metasomatism of sediments. From their field and chemical studies, they also perceived the truth that had been stressed by Bowen nearly two decades earlier on the basis of experimental studies that the differentiation of basaltic magma is dependent on the outcome of a race between the feldspars and the olivines and pyroxenes. Walker and Poldervaart sided with Fenner rather than Bowen regarding the outcome of the race. Fractional crystallization in normal basaltic magma, they said, affects pyroxenes and olivines more than feldspars with the result that iron enrichment is generally more pronounced than enrichment in alkalis. The Karoo magma, they discovered, was no exception to the rule, for here iron enrichment dominated until the final stages of crystallization when alkali enrichment had its opportunity.

Walker (1953) went on to discuss the pegmatitic differentiates of tholeiitic sheets with examples from the Triassic–Jurassic sheets of the eastern United States, the Whin sill, and the Karoo basalts. He paid tribute to Fenner's work on iron enrichment that had "such a profound influence on all recent writings on basaltic crystallization" (Walker, 1953, p. 42). He pointed out that pegmatite commonly occurs in differentiated basaltic sills as lenses or schlieren parallel to the sill contacts, as irregular nests or patches, or as cross-cutting veins. The pegmatites have sharp, unchilled contacts with their host rocks, which have not been altered by the pegmatites. These pegmatites, Walker said, are composed primarily of amphibole, abundant micropegmatite, iron oxides, relatively sodic plagioclase, and iron-rich clinopyroxene. Moreover, Walker found that the pegmatites have lesser lime and magnesia contents and greater iron oxide, titania, and potash contents than the main sill rocks. He plotted the chemical data on an AFM diagram from which he concluded that it was unlikely that the pegmatites had been formed by syntexis of adjacent sedimentary rocks by the hot magma. Instead, he was persuaded that the pegmatite represented layers of residual liquid, rich in iron and volatiles, sitting above the median plane of a sill and generated by far-advanced variation along the liquid line of descent toward iron enrichment. This small amount of low-viscosity residual liquid, he envisioned, must have been squeezed into rifts in the crystal mesh of the sill or trapped as pools between crystal frameworks growing unevenly downward from the upper contact and upward from the lower contact of the sill.

A serious challenge to the adequacy of crystal settling in sills was advanced by J. C. Jaeger, Professor of Geophysics at the Australian National University and an expert on problems of heat conduction in geology. Jaeger and Joplin (1955) claimed that a crystal with a diameter of 2 millimeters would be unlikely to survive for more than one minute in a magma at a temperature above the melting point of the crystal. Assuming the absence of convection in a cooling magma body, they estimated that such a crystal would not fall more than 1 centimeter in one minute, and in this time it would be remelted in the hotter magma below.

Walker, a vigorous advocate of differentiation by sinking of individual crystals in magma, recognized that if conditions were as pictured by Jaeger and Joplin, then the possibility of significant differentiation by the settling of discrete crystals would be eliminated and current ideas on crystallization–differentiation would be demolished at a single stroke. Walker (1956) thought that the fatal objection to Jaeger's remelting hypothesis lay in the idea that there is no hotter magma into which crystals might settle and remelt. Noting that he had examined the chilled contacts of several hundred dolerite intrusions from South Africa, America, and the British Isles, Walker claimed that in every case he had found phenocrysts or microphenocrysts of the mafic minerals, namely, those most likely to settle, an indication that these minerals had begun to crystallize prior to intrusion of the magma. The initial magma was saturated with respect to these minerals, and, therefore, he concluded that they would settle without remelting even though they might react with the magma if not in equilibrium with it.

Walker also disputed the idea of Jaeger and Joplin that blocks of the dense portions of a crystal mush consisting of interlaced crystals and residual liquid broke away from crystallizing material and settled. This mechanism had been suggested to Jaeger and Joplin by magnetic studies that seemed to indicate to them that homogeneous blocks in some sills are separated by narrow regions having very different magnetic properties. Against this view Walker argued that blocks of settled mush should be observed in the field and should be characterized by a high concentration of pyroxenes. He certainly felt that the olivine ledge in the Palisades sill with its smooth top failed to provide evidence for the hypothesis of heaped blocks surrounded by interstitial magma having a different, presumably olivine-poor composition.

By the early 1950s, thanks to the experimental research of Bowen and his colleagues at the Geophysical Laboratory, to the painstaking field, petrographic, and chemical work of several petrologists, particularly Frederick Walker, on thick basaltic sills, and to the similar efforts of several geologists on the recently discovered spectacular layered basic intrusions, in particular the epic work by L. R. Wager and W. A. Deer on the Skaergaard Intrusion in Greenland, the concept of crystallization–differentiation, otherwise known as fractional crystallization, was firmly established in the petrological consciousness as the dominant mechanism of lithologic diversity within an individual sill or lopolith. The idea of multiple injections of magma was typically put aside in favor of in situ differentiation of a single mass of magma. The Lugar sill represented perhaps the major exception to that generality. The notion that differentiation was accompanied by sinking of dense, early-formed individual crystals of minerals such as pyroxene and olivine under the influence of gravity toward the slowly rising floor of a magma chamber was also widely accepted, again thanks to Bowen's experiments and a host of field studies. In the much larger, distinctly layered bodies, such as the Duluth lopolith or the funnel-shaped Skaergaard Intrusion, the adequacy of vertical crystal settling alone was called into question, and a major role for convection of magma in simple cells was invoked as a means for bringing crystals from above toward the chamber floor. Clarence Fenner, who so often seemed to be waging a losing battle in the 1930s against the deductions that Bowen derived from his arsenal of precise, quantitative, and very compelling experiments streaming from the laboratories down the hallway from Fenner's office at the Geophysical Laboratory, received striking vindication for some of his views from investigations of several differentiated tholeiitic intrusions and especially from the Skaergaard Intrusion. Study after study seemed to indicate that the dominant course of differentiation of tholeiitic magmas led toward undeniable iron enrichment rather than toward the silica enrichment so adamantly advocated by Bowen. The same studies further seemed to show that it was only in the final stages of differentiation that small amounts of material approaching granitic composition were generated. Petrologists became more comfortable with the notion that the calc–alkalic series of rocks possesses different chemical characteristics from those of the tholeiitic series. They also became comfortable with the idea that not all of the members of the calc–alkalic series were even produced by differentiation. Concepts of assimilation or partial melting continued to appear in discussions about the calc–alkalic rocks, and serious questions had also arisen in the 1940s as to whether granitoid rocks were really magmatic. Even the paradigm of simple fractional crystallization by crystal sinking in single pulses of magma in sills and layered intrusions would begin to show signs of strain in the 1960s and thereafter.

18 ❖ CONE SHEETS AND CAULDRONS

The Mechanics of Igneous Intrusion

As noted in Chapter 11, geologists learned in the final decades of the nineteenth century that igneous rock bodies occur in a wide diversity of forms on a broad range of scales. To complement the long-standing terms "sill" and "dike," G. K. Gilbert had introduced the concept of the laccolith and Eduard Suess introduced the idea of the batholith. A few other new terms pertaining to the shape of igneous rock bodies also found their way into the pages of the geological literature. Geologists also debated whether certain bodies ought to be designated as "laccolites" or "laccoliths" and others as "batholites," "batholiths," or "bathyliths."

THE CLASSIFICATION OF FORMS OF IGNEOUS BODIES

At the turn of the century, geologists began to ponder classification of the forms of plutonic rock bodies. In the fourth edition of his *Textbook of Geology*, Archibald Geikie (1903) mentioned bosses, sills, veins, and dikes. He regarded stocks as identical to bosses and laccoliths as a variety of sills. He did not use the term batholith. In the second edition of *Geology*, Chamberlin and Salisbury (1909) mentioned dikes, plugs, sills, laccoliths, bysmaliths, and bathyliths. They did not refer to stocks and made no mention of Salomon's "ethmoliths" or Daly's "chonoliths."

The most thorough discussion of the classification of igneous rock bodies in the first decade of the twentieth century appeared in a paper by Daly (1905). According to Daly, geologists were virtually agreed that a dike characteristically has the form of a cross-cutting fissure filling in which the dip angle, that is, orientation, of the fissure is immaterial. He also referred to multiple dikes and composite dikes. Daly further noted the widespread use of the term "sill" and its definition as a sheet of igneous material that had been injected between beds within a succession of sedimentary strata. He called attention to the existence of multiple and composite sills.

In Daly's judgment, the laccolith, as originally described with such clarity and precision by Gilbert, possessed a distinct and unmistakable individuality. He regarded the laccolith as a greatly thickened sill, characterized by a dome-like upper surface, a flat base, and an oval or circular ground plan, that was intruded so as to lift its stratified cover. Daly accepted the idea that many of the bodies in Montana, South Dakota, and elsewhere in the western United States described by Cross, Pirsson, and others might properly be designated as laccoliths, even though differing from those of Gilbert by having a downcurved floor. Daly also accepted the notion of multiple and composite laccoliths. Because of what he considered to be a lack of clarity of definition, Daly did not wish to include the "igneous plug" described by Russell from the Black Hills of South Dakota in his classification of intrusive rock bodies. Iddings' definition of a "bysmalith" was noted, as was the volcanic neck, regarded as the solidified lava filling of a volcanic vent. It was in this paper, too, that Daly introduced his term "chonolith" for irregularly shaped intrusive masses with complicated relationships to the surrounding rocks, bodies that could not be easily classified by one of the other terms.

Daly suggested that bosses and stocks had come to be regarded, at least among English and German geologists, as more or less synonymous and observed that the term "stock" had become the more widely used of the two. He considered bosses or stocks to be cross-cutting bodies with vertical or nearly vertical walls, irregular shape, approximately circular or oval ground plan, and no visible floor. He stated that bodies with diameters in map view ranging from a few hundred feet to several miles had been designated as bosses. Daly proposed an upper limit of 200 square kilometers for the outcrop area of a stock on the grounds that a number of geologists had named certain bodies as stocks that had dimensions approaching that value. Daly also recognized multiple and composite stocks.

In regard to batholiths, originally defined by Suess, Daly recognized that geologists had generally been applying that name for bodies of intrusive rock possessing the general characteristics of stocks but on a much larger scale than is generally attributed to stocks or bosses. No author, however, had yet established a lower limit to the areal dimensions of a batholith or an upper limit to the dimensions of a stock. He also noted that most geologists, unlike Suess, did not attach any particular theory of intrusion to their conception of a batholith. As with so many other intrusive forms, Daly again accepted the notion of multiple and composite batholiths.

PROPOSALS FOR NEW FORMS OF IGNEOUS ROCK BODIES

The urge to coin new terms, of course, did not die with Daly. A year after Daly's paper appeared, C. Burckhardt (1906), in a study of the Sierra de Mazapil in Mexico, felt compelled to apply the term "sphenolith" to an intrusion that wedges out in one direction between steeply dipping beds and is bulgingly cross-cutting in the other direction.

Even Alfred Harker got into the naming act. In *The Natural History of Igneous Rocks*, Harker (1909) offered a very detailed review of the morphology of intrusions. Unlike Daly, Harker thought that intrusive bodies ought to be classified not only with a view to form but also in relation to the different types of crustal movement. As a result, Harker introduced a distinction between plateau-building and mountain-building intrusive bodies. Within each of these categories, he further distinguished concordant and transgressive types.

Among the plateau-building intrusive bodies, Harker considered the sill, laccolite, and bysmalite as concordant, that is, parallel to surrounding rock layers, and the dike as discordant, that is, cutting across surrounding rocks. He preferred to use the suffix "-ite," which he regarded as English usage, in contrast to "-ith" which he considered as a German usage. He stressed that in a laccolite the act of intrusion was the cause rather than the result of the concomitant folding. He did not care for Daly's "chonolite" (chonolith), a term that he believed had not been defined with reference to its geological relationships, and considered it to be an irregular laccolite.

Among the mountain-building bodies, Harker classed transgressive sheets and dikes as discordant and the "phacolite," a term that he freshly coined, as concordant. He claimed that typical sills are not found among bedded strata in mountainous regions because of folding. The crests and troughs of folds, he reasoned, relieve pressure so that magma may find its way between strata to be emplaced along the hinges of folds. He designated igneous bodies emplaced in such a manner as "phacolites." Phacolites were said to differ from laccolites in that they formed as a consequence of folding rather than being the cause of folding. Although Harker did not supply the etymology of his new term, Daly later indicated that it meant "lens rock." Transgressive sheets were said to be typical on the Isle of Skye, where Harker (1904) had mapped numerous "inclined sheets" that cross cut and dip inward toward a common center.

Harker treated "batholites" separately as plutonic intrusions of irregular habit. Batholiths, Harker (1909, p. 82) stated, are "extensive masses of plutonic rocks which have been only partially exposed by erosion, without revealing unequivocal evidence of their true geological relations." He believed that some batholiths have a gigantic laccolithic form but were emplaced by Daly's overhead stoping mechanism in which case the magma would have behaved somewhat passively during intrusion. He suggested that if overhead stoping continued to work toward Earth's surface, an intrusion would gradually acquire a vertical cylindrical form characteristic of a boss.

Daly (1914) took up the matter of the form of intrusions again in *Igneous Rocks and Their Origin*. His list of terms resembled that of the 1905 paper except that now he included the "ethmolith" of Salomon, the "sphenolith" of Burckhardt, and the "phacolith" of Harker. Observation of the actual application of the term "stock" to specific intrusive bodies by geologists since his 1905 paper persuaded Daly that he had set his original upper limit of 200 square kilometers for the outcrop area of a stock on the high side. As a result, he revised that upper limit and, simultaneously, the lower limit of outcrop area for a batholith to 100 square kilometers or about forty square miles. Subsequent geologists have generally accepted Daly's proposal.

For a number of years, Frank Grout had studied the Duluth gabbro in Minnesota, a mass that recent descriptions had commonly referred to as a laccolith. The gabbro has a roof and a floor that are as well defined as those of a laccolith or sill, Grout pointed out, and was intruded along a surface approximately corresponding to a previous structure. As a result, he said, the Duluth gabbro is neither funnel-like nor particularly irregular, thus ruling out terms like "ethmolith" or "chonolith." By a process of elimination, Grout believed, the Duluth gabbro had generally been placed with the laccoliths. In his judgment, however, something was different about the Duluth gabbro and several other large intrusions in comparison to typical laccoliths. Grout suggested that the large intrusions he had in mind have a sunken rather than a domed roof. In effect, he noted, they have the form of great saucers or basins. In all likelihood, he thought, the process of intrusion was quite different from that of a laccolith. Noting that Daly had admitted that these large concavo-convex masses represent a departure from type, even though he still called them laccoliths, and that Joseph Barrell had suggested that they deserved a separate term, Grout proposed referring to them by the term "lopolith" (from λοπος, a basin or a flat earthen dish, and λιθος, a stone). Grout (1918c, p. 518) defined a lopolith as "a large, lenticular, centrally sunken, generally concordant, intrusive mass, with its thickness approximately one-tenth to one-twentieth of its width or diameter." As examples of possible lopoliths, Grout pointed to the Duluth mass, the Sudbury nickel irruptive in Ontario, the Bushveld Complex, an unnamed basin-like mass on the Isle of Skye, and the banded rock of Julianahaab, Greenland. Most of these known lopoliths, he said, consist largely of mafic rocks that had experienced notable differentiation on account of their large size and slow cooling. Grout proposed that the magma of a lopolith had spread along an unconformity. Otherwise, he decided, geologists would be forced to accept the absurd conclusion that the magma knew when to cease its stoping.

A short time later, the German structural geologist, Hans Cloos (1921) coined the name "harpolith" to refer to large sickle-shaped intrusions that had been injected into previously deformed strata and then subsequently stretched horizontally in the direction of maximum orogenic displacement together with the invaded rocks. Then Otto Erdmannsdörffer (1924), a German igneous petrologist, proposed the term "akmolith" for an igneous body that was intruded along a zone of décollement, that is, low-angle overthrust faulting, either with or without extensions into the overlying rock.

Study of the several Tertiary volcanic centers of the Scottish Hebrides and of Ireland proved to be fertile ground for both the recognition of distinctive dike forms and mechanisms of igneous emplacement. One of the most fruitful investigations in this regard was summed up in a memoir of the British Geological Survey on the *Tertiary and Post-Tertiary Geology of Mull, Loch Aline, and Oban* compiled by E. B. Bailey et al. (1924). The project represented the efforts of mapping over a period of twenty years. The leader of the Mull group, Edward Battersby Bailey (1881–1965), graduated from Cambridge in 1902 (Stubblefield, 1965). He immediately joined the Geological Survey and worked in Scotland until 1915, when he was pressed into service in France during World War I. On his return in 1919, Bailey became a District Geologist until the end of 1929, when he resigned the survey to become Professor of Geology at the University of Glasgow. During his survey years, Bailey played a part in the issuance of fourteen Geological Survey memoirs, emerged as an expert in the Tertiary volcanic centers such as Glen Coe and Mull, defined the terms "ring-dyke," "cone sheet," and "screen," and introduced the concept of "magma type." In 1937, Bailey accepted the position of Director of the Geological Survey and Museum, in which capacity he served until 1945. In addition to numerous geological studies, Bailey wrote biographies of Hutton and Lyell and a history of the Geological Survey of Great Britain.

Prior to issuance of the 1924 Mull memoir, arcuate dikes and ring-like structures had been recognized in the Slieve Gullion district of Ireland. T. Thoroddsen had introduced the term "Kreisbrüche" (ring fracture) for similar structures in Iceland. In conjunction with Charles T. Clough and Herbert B. Maufe, Bailey had earlier mapped ring structures at Glen Coe in the Scottish Highlands (Clough et al., 1909), and W. B. Wright had introduced the idea of ring structures during the early stages of the Mull project. J. E. Richey definitively demonstrated the ring structure of the Loch Ba Felsite, and Bailey showed that ring structures dominate in the Glen More district of Mull.

Noting the widespread occurrence of arcuate intrusive bodies on Mull, Bailey et al. (1924, p. 306) defined a ring dike as "a dyke of arcuate outcrop, where there is good reason to believe that its arcuate form is significant rather than accidental. Only in rare instances are ring-dykes so completely developed as to show an entire ring-outcrop." They also defined a screen as "a narrow partition of older rock separating two neighbouring steeply bounded intrusions. Screens separating ring-dykes have arcuate outcrops." Soon thereafter, J. E. Richey described a system of ring dikes in the Ardnamurchan peninsula (Richey and Thomas, 1930) and at Slieve Gullion (Richey and Thomas, 1932). Other ring dike systems came to light at Pilanesberg in South Africa, in the White Mountains of New Hampshire in the United States, and in Nigeria.

The work on Mull also disclosed the existence of a host of Harker's "acid centrally inclined sheets." Bailey and his co-workers expressed some surprise at how infrequently geologists seemed to encounter such sheets at other eruptive centers in light of their profusion at Mull, Ardnamurchan, and Skye. The Mull authors regarded Harker's designation as accurate but "unduly cumbrous." In its place they proposed the concept of the "cone-sheet" inasmuch as "the sheets, viewed as members of a suite, suggest the partial infilling of a number of coaxial cone-shaped fractures with inverted apices united underground" (Bailey et al., 1924, p. 221). At Mull, the average dip of the sheets is about 30° to 40°. Abundant cone sheets were subsequently recognized in the alkalic complexes of Nigeria.

In the second edition of *Principles of Petrology*, George Tyrrell classified intrusive rock bodies in much the same way that Harker had. Thus, Tyrrell (1929) described intrusions in regions of unfolded, gently folded, or tilted strata and intrusions in regions of highly folded and compressed rocks. Within each of these two broad tectonic regimes he listed concordant and discordant types. Among the intrusive bodies of the former tectonic regime, he included sills, laccoliths, and lopoliths as concordant and dikes, cone sheets, volcanic necks, and ring dikes as discordant. Among intrusive bodies of the regime of highly folded rocks, Tyrrell included phacoliths and concordant batholiths as concordant. Stocks or bosses, chonoliths, and discordant batholiths he classed as discordant. Tyrrell mentioned the bysmalith as an extreme case of a laccolith bounded by faults. He did not include such terms as sphenolith, akmolith, ethmolith, and harpolith.

Petrography and Petrology, a textbook by Grout (1932), incorporated the forms mentioned by Tyrrell and also referred to ethmoliths and sphenoliths. Akmoliths and harpoliths were left out of his classification. One may surmise from these omissions either that such terms were not finding general acceptance among petrologists or else that English-speaking petrologists were unaware of the proposals by the German writers Cloos and Erdmannsdörffer.

New designations continued to appear. In a detailed investigation of laccoliths in the Highwood Mountains of Montana (Hurlbut, 1939), David T. Griggs introduced the term "ductolith" to describe horizontal plugs having the shape of a teardrop in cross section. He considered the ductolith as a kind of feeder to laccoliths.

As the twentieth century progressed, the petrological vocabulary was in danger of becoming cluttered with a plethora of obscure names to describe every conceivable variation in shape of an intrusive mass. Mercifully, the number of such terms to be coined dropped spectacularly after Charles Hunt published a tongue-in-cheek definition in his review of the laccoliths of the Henry Mountains (Hunt et al., 1953). With mock seriousness, Hunt described very irregular, crudely cactus-shaped bodies as "cactoliths." "A cactolith," Hunt et al. (1953, p. 151) wrote, "is a quasihorizontal chonolith composed of anastamosing ductoliths whose distal ends curl like a harpolith, thin like a sphenolith, or bulge discordantly like an akmolith or an ethmolith."

Some sense as to which terms found general acceptance, at least among American geologists, may be gleaned from other textbooks published during the 1950s. Ernest Wahlstrom (1950) of the University of Colorado presented a lengthy list of terms in *An Introduction to Theoretical Igneous Petrology*. He included virtually all the types of bodies referred to thus far except ductoliths. Instead of akmoliths, he mentioned "sole injections," and he also noted "stromatoliths" as bodies consisting of intimately and complexly mixed igneous and metamorphic rocks. With considerable wisdom, Wahlstrom urged that the general word "pluton" be employed in place of some of the names that had been invented for bodies of unusual shape. In *Theoretical Petrology*, T.F.W. Barth (1952) did not even bother to discuss the forms of igneous rocks. Likewise, the fourth edition of *Eruptive Rocks* (Shand, 1949a) virtually ignored the form of igneous rocks. In the second edition of *Structural Geology*, Harvard structural geologist Marland Billings (1954) referred to laccoliths, asymmetrical and interformational laccoliths, bysmaliths, lopoliths, phacoliths, sills, dikes, cone sheets, ring dikes, radial dikes, dike swarms, batholiths, and stocks. He used the general term, pluton, and made no mention of ethmoliths, sphenoliths, chonoliths, akmoliths, harpoliths, ductoliths, or even cactoliths! The second edition of *Igneous and Metamorphic Petrology* (Turner and Verhoogen, 1960), like many igneous petrology texts, downplayed the field and tectonic relations of igneous occurrences in comparison to the chemical, mineralogical, petrographic, and thermodynamic side of the discipline. In a very brief compilation, they listed sills, laccoliths, phacoliths, lopoliths, dikes, ring dikes, batholiths, stocks, and plutons as the essential intrusive forms. They showed little interest in the more obscure forms. In general, since the middle decades of the twentieth century, the majority of igneous petrologists would probably have been unable to provide a definition of an ethmolith, bysmalith, chonolith, harpolith, sphenolith, akmolith, or ductolith without looking it up. The useful, all-purpose term "pluton" became well entrenched in the vocabulary of petrologists.

FORCEFUL INTRUSION, STOPING, AND CAULDRON SUBSIDENCE

Soon after the beginning of the twentieth century, geologists typically regarded forceful injection of magma or overhead piecemeal stoping by magma as the two fundamental mechanisms for the emplacement of igneous bodies. Forceful injection was conceived as a process in which magma actively created space for itself. In the case of dikes, the magma was thought to force its walls apart. In the case of sills and laccoliths, the magma was believed to lift or arch the overlying column of strata. In contrast, geologists viewed stoping in more passive terms. Here, the magma was envisioned as gradually "eating" its way upward through the crust by dislodgement and foundering of loosened blocks of country rock. Sunken blocks, in effect, displaced more buoyant magma upward. Daly had made a persuasive case that many stocks and batholiths had been emplaced in such a manner. In the early decades of the century, geologists continued to debate the merits of forceful injection vis à vis stoping, and they sought field criteria that might confirm the dominance of one or the other mechanism.

Some geologists attempted to ascertain the factors that would lead to variations in the shape of sills and laccoliths. Sidney Paige, for example, called attention to the effect of increased viscosity of cooling magma on the form of a laccolith. Paige (1913) argued that a series of shapes from nearly sill-like sheets as at Shonkin Sag to highly bulging, steep-sided, thick lenses as in the Judith Mountains would result from increases in viscosity of magma. Once the cooling mass became very viscous, he thought, it would be possible that faults with a more or less circular configuration would result. Iddings' bysmalith, on Paige's view, would represent the limiting stage in the laccolith-forming process.

MacCarthy (1925) claimed that laccolithic magmas were either intruded into shale horizons that are overlain by competent sandstone or limestone units or else along unconformities with competent, that is, thick and rigid, cover beds. He suggested that rates of intrusion and the thickness and competence of the overlying beds played a larger part in determining the forms of sills and laccoliths than the depth of intrusion. A thick-bedded cover, he maintained, favored the development of domal laccoliths, whereas a thin-bedded cover favored the development of sills. He thought that fissures passing upward into a monoclinal fold or abutting against horizontal beds that rest on an unconformity are especially favorable to the formation of a laccolith. He also suggested that rapid intrusion of magma favored laccolith development and that slow intrusion rates favored the formation of sills. To illustrate his contentions, MacCarthy performed a series of experiments in which he filled a sheet-iron pan with a hole in the bottom with layers of sand and plaster of Paris. With a grease gun he forced a "magma" consisting of colored plaster and water through the hole at varying rates into layered sequences in which he varied the thickness and competence of the layers from one experiment to another. MacCarthy found that rapid injection led to the rupture of competent beds and development of an umbrella-shaped, floorless "stock" that arched the overlying beds. When he reduced the rate of injection and made the layers thicker and more competent, a more laccolith-like body formed. With much thinner layers and a still slower rate of "intrusion," a compound laccolith formed in which the liquid began to inject along bedding. By using competent beds and a very viscous liquid that was slowly injected, MacCarthy produced a steep-sided mass that domed and then split the overlying beds. He had, in effect, modeled a bysmalith. He was also able to reproduce sill-like masses by injecting a very thin liquid.

The discovery of several sets of ring dikes and cone sheets in the Tertiary volcanic centers of Scotland led to the awareness of Daly's magmatic stoping process on a grand scale. In principle, stoping could dislodge blocks of country rock of many sizes and shapes. The realization that stoping of one gigantic block might produce distinctive forms resulted in the concept of cauldron subsidence, a term coined by Eduard Suess but employed to great effect by Clough et al. (1909) in their description of the volcanic center at Glen Coe, Scotland. Here the authors mapped a pile of andesites and rhyolites that had quietly welled upward onto a roughly oval basin of Highland schists and phyllites that is about nine by five miles in area and that is enclosed by arcuate boundary faults surrounded by and locally traversed by Cruachan granite and coarse porphyries. The authors estimated that the central basin had subsided by more than 1000 feet. Evidently, subsidence of a large central block had displaced the basin-filling magmas upward in more than one stage.

In later years, Bailey et al. (1924) recognized cauldron subsidence as the mechanism that produced the Loch Ba Felsite ring dike on Mull, and Richey and Thomas (1932) attributed the ring dikes at Slieve Gullion in Ireland to cauldron subsidence. American geologists began discovering evidence for cauldron subsidence in the ring-dike complexes of New Hampshire. Modell (1936), from a consideration of ring structures at a number of localities, such as the Ossipee Mountains, formulated a set of principles for understanding the patterns of ring dikes. He suggested, for example, that if a subsiding block had the same form in all cross sections and if the subsiding block had dropped directly downward, then a complete ring dike with uniform breadth of outcrop should result; that if a subsiding block had not sunk directly down but had moved over toward one of the walls of the magma chamber, then an incomplete ring dike would likely form; that if the initial fracture dipped much more steeply along some parts of its arc than along others, then the ring dike would vary in thickness; and that a younger intrusion might obliterate an older, complete ring dike.

The mechanical explanation for the formation of ring dikes and cone sheets and the process of cauldron subsidence was developed in preliminary form by E. M. Anderson and presented as a very brief section entitled "dynamics of cone-sheets and ring-dykes" in the Mull memoir of Bailey et al. (1924). Anderson assumed the existence of a magma reservoir having roughly the shape of a paraboloid of revolution several miles beneath a horizontal surface. He further assumed that the density of the magma was equal to that of the country rocks but that the pressure acting on the magma was sufficiently high to raise it to the surface if an outlet were available. Homogeneity of the country rocks was also assumed for simplicity of analysis. Under these conditions, Anderson maintained, the country rock would be under the same horizontal and vertical pressures at every point as the magma. No fractures would develop.

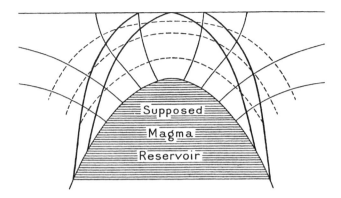

30. Stress trajectories associated with cone sheets and ring dikes proposed by E. M. Anderson. Reproduced by permission of the Royal Society of Edinburgh from *Proceedings of the Royal Society of Edinburgh*, volume 56 (1935–36), pp. 128–157.

Anderson reasoned that an increase in pressure on the magma, however, would result in the formation of a system of stresses in the country rock. Because the magma was pushing upward, the trajectory surfaces of what would now be designated as maximum principal stress were envisioned as having an approximately conical shape with the apices directed toward the top of the magma chamber (Figure 30). The trajectories of what would now be designated as minimum principal stress were envisioned as forming a series of surfaces approximately parallel to the upper surface of the magma chamber. Anderson thought these stress distributions, or what he called tensions, would result in a set of conical fractures along which magma would intrude. Cone sheets, therefore, were regarded as the product of an increase in magma pressure.

In contrast, Anderson proposed that if the pressure of the magma decreased, then the orientation of the trajectories of the maximum and minimum principal stresses would be reversed, resulting in curved tension fractures parallel to the upper surface of the magma chamber and also sets of somewhat paraboloid-shaped shear fractures with gentler dips at the surface and steeper dips at depth. Ring dikes were, therefore, regarded as the product of a decrease in magma pressure. He envisioned that these intersecting sets of fractures might lead to the detachment and sinking of large blocks into the magma and filling of the ring fractures with magma. The farther a block sank into the magma, he noted, the wider the fractures would become with the result that thicker ring dikes would form. Anderson pointed out that the results of his analysis resembled those obtained experimentally by studies of the impingement of curved surfaces on elastic solids or of steel balls on glass. Cone fractures were definitely produced in the latter experiments. Later on, Anderson (1936) developed the mathematical theory behind his analysis of stress distributions established in the two cases that he envisioned. J. E. Richey (1886–1968), however, one of the Mull authors and a long-time geologist with the Geological Survey, acknowledged in a presidential address on the Tertiary ring structures of Great Britain (Richey, 1932) that he had some difficulty in conceiving of the pressure on a magma dropping so much that the roof rocks would be sufficiently subjected to tension as to founder into the magma chamber.

Although geologists accepted the concept of forceful injection in regard to the emplacement of dikes, sills, and laccoliths, they were more inclined to think in terms of stoping and perhaps even cauldron subsidence when it came to stocks and small batholiths. That changed to some degree when the German structural geologist Hans Cloos (1885–1951), of the University of Breslau, undertook an investigation of numerous granite bodies in the cores of German mountains during the 1920s. Previous geologists had regarded these granites as batholithic masses that stoped their way upward. Cloos, however, reinterpreted the relationships between magmatism and tectonics. His predecessors had distinguished older, conformable gneissoid granites with platy structure and oriented inclusions parallel to their foliation from younger, discordant granites proper that lacked foliation and had a circular ground plan. Cloos and his students demonstrated that many of the granite bodies in the second category in fact show a wide variety of subtle structures indicating that they had been forcibly emplaced.

One of Cloos's students, Robert Balk (1899–1955), transported his ideas to the United States. In the 1920s, Balk came to the United States to escape war-ravaged Germany (Pettijohn, 1984). Although living in poverty, he managed to continue his structural studies on igneous rocks. After publication of a Geological Society of America memoir on the structural aspects of igneous plutons (Balk, 1937), he was appointed as a professor at Mount Holyoke College for women. In 1947, he was brought to the University of Chicago but left in 1952 because of its strong emphasis on geochemistry and the downgrading of field geology by Walter H. Newhouse, the autocratic chairman of Chicago's Department of Geology. Balk took a position with the New Mexico Bureau of Mining and was killed in a plane crash outside Albuquerque three years later. In a series of publications, Balk (1925, 1937) pointed out that many granitic plutons contain oriented mineral crystals, clots, schlieren, and inclusions. Members of the Cloos school interpreted such features as linear or planar flow structures formed during differential movement of magma. Balk thought that inclusions would be rotated until their longest axes were aligned parallel to the general flow direction of magma. These flow structures were envisioned as developing in the vicinity of the walls of a pluton.

Cloos and Balk also called attention to the presence of "cross joints" oriented essentially perpendicular to the early-stage flow lineations and foliations. Dikes were seen to fill such cross joints locally. These cross joints developed upon cooling and solidification of the outer shell of a granitic body and were thought to form by yielding of the magma when it had become brittle. Joints and dikes that dip very gently into the intrusive and away from the wall rock were also observed in the vicinity of the contacts of the granites with their wall rocks. These joints, said Balk, are arranged just like the marginal tension joints in a glacier.

All of these structural features, Balk maintained, indicate that many granitic intrusions play a dependent role within Earth's crust in the sense that they are affected by the tectonic conditions of the rocks they invade. Such intrusions appear to penetrate along zones of weakness in the crust and push aside the flexible country rocks into which they are emplaced rather than melt their way upward through the overlying crust.

Balk's insinuation that some intrusive bodies previously attributed to stoping, such as the Mount Ascutney pluton where Daly originally conceived the idea, might actually have been forcefully injected, led the Chapman brothers, Randolph of Johns Hopkins and Carleton of the University of Illinois, to undertake a new study of the Mount Ascutney pluton in light of the newer ideas of forceful injection and cauldron subsidence.

Chapman and Chapman (1940) observed that a north–south-striking quartz sericite schist with a steep easterly dip is generally unaffected by the main Mount Ascutney syenite stock. They also noted that the plunge of crinkles and small folds is consistent around the intrusion. On these grounds they rejected forceful injection as the intrusion mechanism. If the intrusion had been injected forcibly, the low-plunging regional features would have been dragged up along the contacts. But they also questioned whether Daly's preferred mechanism of piecemeal stoping accounted for their observations. They reasoned that piecemeal stoping would have developed more easily parallel to structural planes in the country rocks rather than across them. The intrusive body should have been elongated parallel to the regional structure rather than across it. They also concluded that piecemeal stoping would have produced more irregular and less symmetrical contacts. The Chapman brothers conceded that piecemeal stoping had played some role; after all, they did find small blocks of country rock swimming in the syenite stock. As the dominant mechanism, however, they invoked cauldron subsidence. They believed that two large sunken cylindrical blocks of country rock made space for the main syenite magma and a later granite magma that were emplaced as ring dikes and stocks. They found that volcanic rocks, representing the earliest, mainly arcuate, phases of igneous activity, had been extruded on the ancient surface above the subsiding block but not in the rocks outside the igneous complex. Cauldron subsidence had been vindicated in New Hampshire.

Considerable light was shed on the behavior of ring complexes by the detailed studies of modern calderas and volcanoes, particularly in the Cascade Range of the northwestern United States, conducted by Howel Williams (1898–1980) of the University of California. Williams (1942) answered Richey's concern about the sufficiency of magma pressure reduction to produce tension fractures in the overlying rock by showing that, at Crater Lake, Oregon, such an enormous volume of magma was erupted that magma pressure was drastically reduced and subsidence on a large scale resulted to form the caldera.

Marland Billings and his students at Harvard continued their investigations of the evidences for stoping and cauldron subsidence in the ring dikes, stocks, and small batholiths of the White Mountain Magma Series in New Hampshire. Billings (1945) pointed out that among the more than 36 ring dikes in New Hampshire no gabbro or diorite is present. He found that ring dikes are composed entirely of intermediate and felsic rocks such as monzodiorite, monzonite, quartz monzonite, syenite, nepheline syenite, quartz syenite, and granite. Billings suggested that magmas of these compositions, most being of lesser density than their country rocks, would actively rise, thereby creating the stresses that would lead to the development of cone sheets. Billings also recognized that several of the plutons in New Hampshire, such as the Mount Clough and Kinsman plutons, were probably formed by forceful injection. In these instances, he determined that the schistosity of the country rocks clearly wraps around the allegedly intrusive material, suggesting that it had been shouldered aside by magma forcing its way in.

BATHOLITHS

In the early part of the twentieth century, geologists had greatly advanced their understanding of the emplacement of many plutons. Geologists generally accepted the emplacement of sills, dikes, and laccoliths by forceful injection. Cauldron subsidence had been established for many ring-dike and stock complexes of volcanic centers. Piecemeal stoping appeared to be a valid mechanism for some granitic stocks, and for other smaller granitic plutons the features described by Cloos and Balk pointed to the likelihood of forceful injection. The major problem of emplacement mechanics, however, concerned batholiths. Throughout the first half of the century, geologists constantly debated the geometry and mode of emplacement of these great masses of granitoid rocks.

The investigation of batholiths fell predominantly to geologists from North America, for it is in North America rather than Europe where granitoid bodies of vast scale, such as the Coast Range batholith, the Idaho batholith, the Sierra Nevada batholith, and the Peninsular Range (Southern California) batholith, are located. As a result, North American geologists like Daly, Grout, and Esper Larsen, Jr., moved to the forefront of the discussion about the nature of batholiths.

Many of the difficulties posed by batholiths for early twentieth-century geologists were outlined in a paper by William Richardson (1923). He indicated that several features of batholiths required explanation. Why, he asked, do batholiths generally occur in orogenic belts, typically elongated in the direction of the tectonic axis of the belt? Why do batholiths closely follow a period of orogenic movement? Why do batholiths characteristically have an irregularly domed roof with smooth, steeply inclined walls that enlarge downward? Ever since Suess had postulated that batholiths extend downward "in die ewige Teufe," geologists had assumed that batholiths persist at depth and have no visible floor. Richardson maintained that batholiths had "replaced" the invaded formations rather than "displaced" them as laccoliths had done. Lastly, Richardson pointed to the uniform granitic composition of batholiths as requiring explanation. Ten years later, Daly (1933) essentially seconded Richardson's list of features in need of explanation but also added the problems of the great size, by his definition in excess of forty square miles exposed area, and the generally cross-cutting nature of batholiths. Daly, too, stressed what he called the subjacent nature of batholiths as opposed to their being injected. By "subjacent" he meant the continuity and connectedness of batholithic material at depth. Nearly two decades later, Wahlstrom (1950) recapitulated Daly's list. By mid-century, the batholith problem still required acceptable solutions.

Of course, some of the characteristics assumed by Richardson, Daly, and others to need explanation were themselves in dispute. While there was little debate about the magnitude, composition, and tectonic setting of batholiths, geologists disagreed about whether batholiths were replacive or displacive, and they did not all agree that batholiths are bottomless.

The replacivists included the French petrologists, some British geologists like Grenville Cole, and Daly. Many French geologists and Cole believed that magma could melt its way through the crust and incorporate the melted material into itself. On the other hand, the American geologists Daly, Barrell, and Emmons stressed the idea of stoping of blocks from the roof of the intrusion by unequal heating, veining, and magmatic wedging. Replacivists, however, were confronted by serious objections. Against the melting hypothesis, Tyrrell (1929) charged that the process would require an enormous amount of localized superheat in the magma. But there was no evidence, he thought, for large amounts of superheat to accomplish the melting of wall rocks. Furthermore, Tyrrell pointed out that the composition of batholiths is so remarkably uniform that the idea of wholesale assimilation of wall rock by melting would be improbable in upper levels of the crust. Tyrrell and others conceded that stoping was supported by the irregular, embayed margins of many batholiths and the common presence of xenolithic blocks of all dimensions enclosed within intrusions. Iddings and Cloos, however, regarded stoping as only a subordinate, incidental factor in the emplacement of batholiths. T. C. Chamberlin, too, could not understand why batholiths never worked their way all the way to the surface if stoping had been the primary emplacement mechanism.

Grout (1930) thought that one way to evaluate Daly's stoping hypothesis would be to look for stoped blocks on the floors of the largest known intrusives that do have floors, namely, lopoliths. Since the floor of the Duluth gabbro lopolith provided the closest thing to batholith roots, he claimed that a careful examination of the occurrence and condition of the xenoliths could be instructive. Grout noted that the Duluth gabbro contains abundant xenoliths that are the exact equivalent of the contact rocks along its base. As he found no clear indication that the country rocks along the floor ever existed at the roof, Grout suggested that "underhead" stoping had occurred. In any case, the xenoliths demonstrated some alteration and recrystallization, but he found little evidence to indicate substantial assimilation and concomitant change in the chemical composition of the magma. He concluded that although stoping might occur in batholiths and large blocks might be detached to sink all the way to the floor, the detached blocks would recrystallize and retain their integrity as blocks rather than be assimilated. Grout regarded batholithic assimilation as of rather little quantitative importance.

The displacivists included geologists like Brögger who regarded batholiths as something like large laccoliths. On this view, batholiths were envisioned as pushing the invaded country rocks upward, where they were then eroded. Other displacivists, for example, Cloos and Iddings, maintained that the necessary space for batholithic development had been gained by the shouldering aside and compressing the adjacent rocks. In some instances, wall rocks were also presumed to be partly forced downward concurrently with the folding and thrusting of mountain-building forces. Opponents of the displacive view, however, had difficulty conceiving how a body of magma could shoulder aside the enormous volume of rocks required to make room for a batholith. James Noble (1952) suggested that the seemingly fatal room problem was removed if one regarded batholiths as composite intrusions in which each pluton in a succession of several smaller plutons had made its way by shoving aside a smaller volume of material. The roof pendants of batholiths, he postulated, might not be stoped blocks as suggested by the replacivists but rather compressed and displaced screens of wall rock.

Not all displacivists assumed that batholiths had been emplaced as magma. C. E. Wegmann, for example, introduced the concept of emplacement of granite in the form of diapirs, a term earlier coined by Mrazec (1927). Wegmann (1930) suggested that in situ metasomatism converted deep crustal rocks into a granitic layer. Because of its lower density, he thought that the granitic layer became mobilized during orogenic movements in such a way that plutons moved buoyantly upward in the solid state as diapirs. The same problem remained, however, as to how such large masses could shove aside the large volumes of country rock.

The other main point of dispute concerned the geometry of batholiths. Chamberlin and Link (1927) said that geologists were accustomed to thinking of batholiths as maintaining or enlarging their dimensions with depth. Suess, Daly, Barrell, Grout, Richardson, and many others had assumed that view. Richardson viewed batholiths as the expression of a thickened acidic shell that had been arched under compression. Others denied that the batholiths were necessarily directly connected with a granitic layer at depth but did emphasize their vertical as opposed to their horizontal dimensions. They envisioned batholiths as vast cylinders, teardrops, or lenses.

A number of geologists, however, were beginning to question such ideas. Chamberlin and Link, for example, suggested that batholiths showed a tendency toward horizontal spreading. Batholiths, they suggested, should generally be sheet-like, of great lateral extent, and with relatively little vertical thickness. To support that contention they noted the existence of batholiths in Colorado known to have a pancake-like form. Chamberlin and Link also conducted a series of experiments, similar to those of MacCarthy (1925), in which they injected plaster of Paris, plastic, grease, melted paraffin, and other simulated liquids into artificial strata composed of various materials. They concluded that principles of fracturing determined the migration of liquid. They found that earlier intrusions of simulated liquid resulted in the formation of ring dikes and cone sheets. Simulated liquids injected after the formation of such fractures systems either spread out laterally or moved along inclined planes but did not break through to the surface by way of fractures directly under a central dome. They discovered that the integrity of the roof had been maintained and, therefore, that no stoping had taken place. Lane (1931) and Emmons (1933) agreed with Chamberlin and Link that batholiths do not extend down forever. Lane, in fact, thought that most granite bodies were really interdigitated and had been intruded as a series of interfingering sills.

Perhaps no geologist was more intrigued by batholiths than Grout. As early as 1927, Grout had pleaded for more detailed petrographic work, more chemical study of batholithic rocks, and greater attention to the broad aspects of the batholith problem. He submitted a proposal to the National Research Council to fund a concerted attack on the batholith problem, and, as a result of his initiative, the Council established the Batholith Committee and appointed Grout as chairman. Although the Committee operated between 1928 to 1935, the results were limited. Grout (1945, p. 51) marveled "that eleven geologists of considerable experience could disagree so sharply at so many points." But, since 1932, Grout had also been performing experiments at the University of Minnesota on magma invasion. He placed a matrix of soft wet clay and corn syrup in a glass-sided tank. For scale-model magma bodies, he allowed globules of air, water, and cylinder oil to rise through the matrix. He applied the resulting shapes of the globules to the shapes of batholiths and the structures associated with them. As a result of his experiments and cartoons, the concept of batholiths and plutons rising as teardrop-shaped diapirs penetrating through the crust became entrenched in the consciousness of many petrologists for years to come.

Throughout the twentieth century, it became clearer that the nature of batholiths and the emplacement of granite were conjoined twin problems. One could not be considered apart from the other. To complicate matters further, the issue regarding granite emplacement was no longer simply a matter of forceful injection of magmas versus stoping by magmas or cauldron subsidence into magma chambers. In the 1940s, even as Harvard's Esper Larsen, Jr., was calculating the amount of time required for cooling of the magma of the Southern California batholith, many geologists were questioning whether granites were magmatic at all. They were claiming that batholiths of granite had been emplaced as a result of metasomatic replacement of previously existing rocks. Controversy over the origin of granite had flared up anew.

19 ❖ MAGMA OR EMANATIONS?

The Beginnings of the Granite Controversy

The origin of granite had been hotly contested on two prior occasions. During the latter eighteenth century, Werner's neptunist conception that granite had crystallized from a global ocean prevailed. From cross-cutting relations, however, James Hutton contended that granite was the result of the injection of subterranean lava into preexistent rocks. Hutton's conjecture gained support from James Hall's experiments on the melting of granite. By the 1820s, the neptunist view was collapsing, and the igneous origin of granite gained favor. No sooner was the issue "settled" when a new generation of geologists and chemists, most of whom were by no means Wernerian neptunists, concluded that granite was, after all, produced in the "wet" or "humid" way rather than by pure igneous fusion. Appealing to the textural evidence that quartz in granite typically crystallized after feldspar and other constituents, these writers generally denied an igneous origin to granite on the grounds that quartz, supposedly having a much higher melting temperature than feldspar, ought to have crystallized first rather than last. Moreover, opponents of igneous granite claimed that quartz had been crystallized experimentally from aqueous solution at relatively low temperatures but had not been produced from dry silicate melt. As a result, many geologists argued that granite formed by metamorphic transformation of country rocks, by precipitation of quartz from aqueous solution passing through siliceous rocks, or possibly by crystallization of a granitic melt saturated with water.

In the middle and latter part of the nineteenth centuries, the intrusive magmatic view of granite made a comeback. Sorby's studies of fluid inclusions supported an aqueo-igneous origin for granite. Bunsen undermined the textural argument against the igneous origin of granite by pointing out that one must regard granitic melt as a solution in which the various dissolved constituents mutually affect one another's melting points. Consequently, geologists realized that the melting point of pure quartz would very likely be much higher than that of quartz precipitating from a silicate solution. Lastly, the advent of thin-section petrography made it abundantly clear that quartz in volcanic rocks had crystallized directly from lava and that there is complete textural gradation from silica-rich volcanic rocks to silica-rich plutonic rocks. Thanks to the labors of Rosenbusch and his school, the magmatic origin of granite was once more widely accepted by the end of the century. By the beginning of the twentieth century, the leading petrologists, including Rosenbusch in Germany, Doelter in Austria, Brögger and Vogt in Norway, Teall and Harker in Great Britain, Loewinson-Lessing in Russia, Iddings in America, and Michel-Lévy in France, accepted the igneous origin of granite.

THE FRENCH SCHOOL

Differences existed among magmatists, however. Rosenbusch (1877b) found that the barren contact aureole of the Barr-Andlau granite pluton in the Vosges of northeastern France displays no evidence of metasomatism induced by fluids emanating form the invading magma. In his judgment, the sole effects of metamorphism were induced by heat. He extrapolated these findings to conclude that granitic magmas were generally deficient in dissolved fluids. In contrast, French geologists such as Michel-Lévy typically envisioned granitic magmas as rich in dissolved volatiles that were deemed capable of soaking and altering country rocks upon emplacement of the magma. A half-century later, these divergent views about the behavior of magmatic granite came to play a critical role in the twentieth-century granite controversy. At the beginning of the century, the Germanic view of hot, dry granitic magma intruded from depth was the predominant view, thanks to the influence of the textbooks of Rosenbusch and the fact that so many petrographers had been personally trained by the German master. This state of affairs was recalled near the height of the granite controversy during World War II by British geologist Robert H. Rastall (1871–1950), who was expressing his own sympathies for granitization against the background of his undergraduate training at Cambridge in the first years of the twentieth century. Rastall, the long-time editor of *Geological Magazine* from 1919 to 1950, had studied and taught agriculture before entering Cambridge in 1899 at the age of 28 and graduating in 1903 (Bulman, 1950). He was appointed as Demonstrator in Geology at Cambridge in 1910, and in 1919 he became Lecturer in Economic Geology, a post he held until 1934. A generalist with petrological leanings, he authored several texts in geology. Rastall (1945, p. 19) recollected that

> petrologically I was brought up in the straitest sect of magmatism, with Rosenbusch as its major prophet. All igneous rocks were formed from magmas and all plutonic rocks were bodily intrusions which made room for themselves by shoving aside or lifting the surrounding rocks. Porphyritic structure indicated crystallization in two stages and local variations of composition and associations of cognate rock types in

igneous complexes and petrographical provinces were due to magmatic differentiation.... Altogether the teaching had strong German flavour and the French school was ignored.

Writing on the day that Paris was liberated by the Allies, Rastall (1945, p. 22) commented that "a petrologist could not better celebrate the liberation of France from German tyranny than by reading this literature [i.e., that of the French school] and giving it a fair and open-minded consideration." In light of Rastall's comment, one wonders if emotional factors provoked by the geopolitical situation contributed to the intensity with which the granitization position was held by some British writers during the 1940s. The comment raises the question whether British antipathy to German magmatism and sympathy for French metasomatism was not a subconscious reaction to the rise of German nationalism and militarism in the first half of the twentieth century.

What was this "French school" that so intrigued Rastall and other proponents of granitization in the 1940s? During the debate over the origin of granite in the mid-nineteenth century, many French geologists such as Élie de Beaumont and Fournet had stressed the association of volatile fluids with granitic intrusions and the ability of those fluids to effect striking transformations within wall rocks. Virlet d'Aoust had proposed the "imbibition" or soaking of country rock by granitic material. Delesse, Daubrée, and other experimentalists had pointed out the crucial role of water in the formation of granite. French geologists, however, were by no means unanimous in their devotion to the importance of fluids, for Joseph Durocher was one of the leading advocates of the origin of granite by pure igneous fusion.

The French interest in vapors and fluids flourished again at the end of the nineteenth century. Charles Eugène Barrois (1851–1939), geologist with Service de la Carte Géologique de France, mapped examples of the transformation of large tracts of Precambrian beds into gneisses. Barrois (1884) traced unaltered layers of quartzite, resistant to the process of feldspathization, within successions of argillaceous rocks that had been converted into gneisses. Within porphyritic-looking granites, Barrois encountered enclaves containing large orthoclase crystals, so-called "dents de cheval" (horse's teeth) identical to the phenocryts in the enclosing granites. The various feldspathizing phenomena were regarded as the product of fluids.

Michel-Lévy described a small mass of granite near Flamanville along the Brittany coast of France that invaded Paleozoic sedimentary rocks. The country rocks had been transformed into feldspar-bearing rocks and graded into gneisses. Even granite had developed a schistose texture. Michel-Lévy (1893) attributed the transformation to the agency of "mineralizers" that acted differentially on the various country rocks. In effect, Michel-Lévy's work challenged the ideas of Rosenbusch that feldspathization does not occur at granite contacts and that gneisses are not formed at such contacts. Michel-Lévy suggested that the ascent of granitic magma took place by assimilation or corrosion of the surrounding rocks brought about by the heat and mineralizers coming from the magma. Granites, in his view, were emplaced by the gradual conversion of wall rocks into rocks resembling granite and the incorporation of the converted material into the granitic magma.

In his studies of granitic bodies in the Pyrenees, Alfred Lacroix (1898, 1900) repeatedly recognized the existence of zones of feldspathization adjacent to the contacts of the granites. In these zones, Lacroix traced progressive stages in feldspathization between the feldspathized schists and the granites. Even in the most feldspathized gneisses, Lacroix encountered still recognizable remnants of the various country rocks in intermediate stages of transformation.

Michel-Lévy, Barrois, and Lacroix all regarded magmatic granite as the source of the feldspathizing fluids. In their view, granitic magmas arrived from depth and were responsible for the zones of metamorphism surrounding them. In contrast, Pierre Termier (1859–1930) envisioned the production of granite as the culmination of a granitization process. Termier graduated from the École des Mines in 1883 (Lindgren, 1931). After brief service as a mining engineer in southern France, he became a professor at the mining school at Saint Etienne from 1885 to 1894. Beginning in 1886, he also performed work on behalf of the Service de la Carte Géologique de France. In 1894, Termier returned to Paris as Professor of Mineralogy at the École des Mines, a position he occupied until 1912. In 1911, Termier succeeded Michel-Lévy as Director of the Service de la Carte Géologique. Throughout his career, he was concerned particularly with the structure and metamorphism of the Alpine and Hercynian belts. Termier regarded the appearance of feldspars in schists as undeniable evidence for the transfer of material. He affirmed that there had been an influx of material from depth that expelled some of the old material in a series of crystalline schists. Termier (1904) referred to these rising juvenile fluids as "colonnes filtrantes" consisting of various gases containing dissolved alkali silicates and borates. He envisioned that ascending vapors would raise the temperature of the surrounding rocks and produce intense chemical reaction. Mixtures of minerals with low melting points would be dissolved first. Some excess chemical elements would supposedly be displaced before the advancing "colonne filtrante" to displace still other chemical elements. Termier believed that if a granite body is surrounded by a vast metamorphic aureole, then it has been formed in place by complete fusion of a eutectic mixture while the surrounding rocks were undergoing regional metamorphism.

J. J. SEDERHOLM

When the granite controversy broke out in full force in the 1940s, proponents of granitization like Rastall repeatedly appealed to the earlier work of the French school but also frequently called attention to the labors of Finnish geologist Jakob Johannes Sederholm (1863–1934). After the French geologists turned their attention to other geological problems, such as the theory of nappes, the center for the study of the effects of granitization shifted to Fennoscandia, a region in which such phenomena were believed to be present on a large scale. On a field trip led by Sederholm during the International Geological Congress held in Stockholm in 1910, he applied the term "granitization" to rock assemblages that he showed to trip participants (Backlund, 1946). Sederholm explained the formation of the rocks as the result of penetration of preexisting sedimentary and basaltic rocks by the addition of matter from below without the original rocks having undergone melting. The opposition to Sederholm's field interpretation was almost unanimous. One might have expected that a pure magmatist like Vogt would oppose Sederholm, but even the French geologists who were present, including Barrois, were reluctant to go along with him.

Sederholm studied at the University of Stockholm under Brögger and graduated in 1888. In 1890 and 1891, he also studied in Heidelberg with Rosenbusch and was schooled in the ways of the arch-magmatist. In 1888, Sederholm became a State Geologist for the Geological Survey of Finland, and, in 1893, he was appointed Director of the survey, a post he occupied for forty years. Like Brögger, Vogt, and Bäckström, Sederholm was also involved in politics. After initial studies in surficial geology, Sederholm increasingly devoted his attention to the magnificently exposed Precambrian igneous and metamorphic rocks of Finland and laid the basis for the temporal subdivisions of these rocks.

In his early years, the Rosenbusch-trained Sederholm was an orthodox magmatist. In the summer of 1906, because Sederholm's political duties kept him from conducting his planned field studies, he devoted what spare time he had to an examination of the glacially polished and wave-smoothed rock surfaces of the small islands along the southern coast of Finland. Of these spectacular exposures of complex rocks, Sederholm (1907) wrote that it was here that the problem of gneiss and granite should be solved. In this region Sederholm observed highly folded gneisses intimately intermixed with granitic veins in rock that appeared to have been on the verge of melting. Although Sederholm was uncertain whether the phenomena he observed should be attributed to introduction of heat and material by granitic intrusions or to the subsidence of the entire mass of rock to a depth at which the heat was sufficiently great to begin melting the mass, he leaned toward the idea of a process of formation of granitic magma by the refusion of older rocks with more refractory unmelted portions of gneiss swimming as remnants or fragments in the granite. Sederholm labeled this process of formation of magma by refusion as "palingenesis" or "anatexis" and saw no contradiction between formation of granite at the great depths evidently now exposed in southern Finland and the clearly eruptive intrusive high-level granites so commonly described by the orthodox German magmatists. One was merely observing the end result of differing behaviors of granite and surroundings at different levels within the crust, he thought. He considered the mixed rocks as a kind of stew in which gneissic "shreds of meat" were boiled almost beyond recognition in a granitic "broth." For such rocks of mixed gneiss and granite character, Sederholm coined the term "migmatite." He considered migmatites as rocks that occupy a transitional position between granite and the crystalline schists of partly sedimentary and partly igneous character. To the strong disharmonic folding common in migmatites, he applied the term "ptygmatic." Sederholm regarded his theory of anatectic origin of granitic material as doing justice to both neptunism, in that it accepted the formation of granites from rocks of ultimate sedimentary origin, and plutonism, in that it accepted subterranean heat as the main geological factor in producing granite.

In a number of subsequent papers, Sederholm described more occurrences of migmatites and allegedly anatectic phenomena. By examining terranes that had developed during the severest stages of metamorphism, Sederholm brought into the open the fact that the "life history" of granite could not be confined to the clearly intrusive, magmatic stocks and batholiths of higher levels of the crust that are accompanied by a relatively small volume of metasomatic effects in the surrounding wall rocks. He pointed out that there are regions in the very deep crust where there is a great deal of chemical transfer and mobility. Sederholm (1923) perceptively recognized that the opponents of granitization in his time were petrologists who had relied predominantly on the petrographic microscope for solving problems of rock origins. Moreover, he realized, those petrologists had generally worked on shallow intrusions whereas the field geologists of the granitization school had concentrated their efforts on very deep-seated rocks. Owing to his studies with Brögger and Rosenbusch, Sederholm considered himself as belonging to the Heidelberg school, but, owing to his field studies, he said that he had also become an eager partisan of the theory of granitization. At the same time, Sederholm never divorced granitization and migmatization from granitic magma and never envisioned granitization as the result of diffusion of ions apart from fluids and melts. Moreover, in contrast to P. J. Holmquist, who regarded most pegmatitic veins in gneisses and schists as the product of exudation of material from the surrounding rocks, Sederholm generally regarded such veins as "arterites," namely, pervasive injections of granitic melt into the metamorphic rock. Sederholm was held in high regard by members of the granitization school in later decades, but he remained a magmatist.

Although the "orthodox" German school of granite magmatism, adhering to the ideas that granite plutons were emplaced by forcible injection of magma or piecemeal stoping by magma and that granite bodies are generally not accompanied by feldspathization, injection gneisses, or other significant metasomatic effects, dominated geological thought in the early twentieth century,

French and Scandinavian geologists went about their business describing occurrences exposed from deeper crustal levels in which magmatic granites seemingly are associated with feldspathization, migmatization, and like phenomena. Not until some geologists began to divorce granite magma conceptually from metasomatic granitization effects, however, did the "granite controversy" break out in full vigor.

THE BEGINNINGS OF THE GRANITE CONTROVERSY

According to Norwegian petrologist, T.F.W. Barth (1948, p. 235), Helge Götrick Backlund (1878–1958), Professor of Geology at the University of Uppsala in Sweden, "probably more than anyone else, was responsible for reviving the discussion of the granite problem by propounding a granitization theory disclaiming the primary, magmatic character of the rapikivi granite and similar rocks." Over the next several years, Backlund issued several papers advocating some form of granitization. As an example, Backlund (1938) described sixteen unique features of rapakivi granites that, in his judgment, suggested that they had formed by the replacement of extensive tracts and thick formations of Jotnian sandstones and nearby Archean rocks rather than by injection of magma. He was bothered by the sheet-like morphology of rapakivi masses. Why did not a viscous acid intrusion bulk out beneath its comparatively thin cover of sedimentary rocks? He found that the sheet-like nature of the rapakivi granites was emphasized by horizontal jointing and by the concentration of weathering along planar zones, giving the rocks a horizontally bedded appearance. Backlund suspected that variations in chemical and mineralogical composition of the rapakivi granites are a reflection of the upward variation of their enclosing sedimentary rocks. He conceded that the chemical variations are suggestive of magmatic differentiation but also pointed out that expected variations in specific gravity are lacking as is evidence of early crystallization of mafic minerals, features to be expected in a differentiated sheet. Backlund observed no flow textures or any textures characteristic of laccoliths or batholiths. Rather, he thought that the assemblage of different textural varieties, occurring one inside another with relatively sharp boundaries, contradicted the interpretation of quiet magmatic crystallization under deep-seated conditions. Moreover, he noted that the distinctive ovoids so characteristic of rapakivi granite are also present in the wall rocks. Backlund further stressed that the sheet-like, horizontal bodies of rapakivi granite are completely free of inclusions. He found no evidence of feeder dikes. And he asked why the rapakivi magma did not break through to the surface if, indeed, it had really been a magma. To account for the intrusive relations, Backlund postulated the remobilization of granitized sedimentary materials, referring to them as "rheomorphic."

Backlund attributed the transformation of sandstones and quartzites into rapakivi granite to the addition of alkali aluminates, but he confessed that the problem of the sources of these emanations was very difficult. The emanations, he said, could not have ascended as a regional mass migration from below but must have passed upward through a well-defined and narrowly limited interval so that the resulting rapakivi granite would present sharp boundaries against its roof and against its floor.

Throughout his career, Backlund continued to advocate the large-scale conversion of rocks into granitic material by migration of chemical elements. When he reviewed the granitization problem a few years later, Backlund (1946) noted that Arthur Holmes had emphasized that migrations of chemical elements or oxides in solid matter occur on a small scale and that Doris Reynolds showed that such migrations occur on a medium scale. Why then, he wondered, was there such "grave hesitation" on the part of geologists to extend the concept of chemical element migration to the interpretation of large-scale phenomena? He attributed this perceived hesitation to the decline of the French school of geology. Backlund (1946, p. 106) maintained that the French school had become mute after the turn of the century and just prior to "the orchestral crescendo of discussion relating to magmatic differentiation which at that time was beginning to be claimed as the sole source" of granitic rocks. The conception of granitization, Backlund maintained, had been retarded by two things: first, the triumphal advance of differentiation, "whose most ardent adherents now seem to have completely forgotten that their theory is an extrapolation from exact results acquired exclusively from small scale experiments," and second, "a lack of outcrops in central Europe big enough and sufficiently continuous to be adapted to regional studies on the grand scale."

Backlund attempted to describe what he meant by the "emanations" that were said to produce granitizing transformations. He claimed that in no case did an emanation refer to a circulation of volatiles in the usual sense. Rather, he believed that an emanation implied a migration of ions within crystalline solids by way of structural faults, deformations, and crystal discontinuities. The migration was supposed to be driven by potential differences of lattice energies. Backlund suggested that with complete granitization, volatiles would be vigorously expelled from granitized rock such that the final "granite" would contain only a small quantity of volatiles, primarily H_2O, inasmuch as most volatile constituents cannot be accommodated in the structures of major granitic minerals. As a result, Backlund envisioned a cloud of volatiles to be concentrated in front of a theater of granitization. His conception of emanations as migrations of ions within solids, therefore, placed him in the camp of granitizers known as "dry diffusionists," of which his contemporaries Perrin and Roubault were among the leading spokesmen.

Backlund (1938) had earlier praised the work of René Perrin (1893–1966) and his colleague M. Roubault (1905–1974), a professor at the École de Géologie Appliquée in Nancy, as an "Algerian outlier" of the classical French school. Because of his background in metallurgy, Perrin had been impressed by diffusion phenomena in metals. With Roubault, he attempted to apply ideas about diffusion in the solid state to the textures of granitic rocks in Algeria as early as the 1930s. Throughout the entire granite controversy, Perrin and Roubault persisted in advocating the formation of granites by means of diffusion of ions through solids without the aid of fluids or magmas.

In a later paper, Backlund (1947) observed that the mineral assemblage of granite could not be duplicated in a laboratory furnace. He continued to insist that intense granitization had occurred during each orogeny in the Precambrian terranes of Scandinavia. In these events, sedimentary rocks were said to be replaced by granite, but not by the melting of the sedimentary rocks or invasion of magma from elsewhere. He considered the amount of replaced sedimentary rocks to be much too large and the replacing rocks to be much too homogeneous to be accounted for on magmatist principles. For Backlund, the room problem was one of the major obstacles to the magmatic origin of granite. Granites were formed by replacing emanations. At the same time, Backlund also argued that chemical elements migrated from granites into adjacent rocks, contrary to the view of Rosenbusch. Granite, he alleged, was not simply a passive carrier of heat but an active vehicle conveying both heat and material.

METASOMATISM AND GRANITIZATION

By no means was Backlund alone in the advocacy of granitization by metasomatic processes. Geologists from many parts of the globe reported instances in which sedimentary or metamorphic rocks become increasingly quartzo-feldspathic along strike. Even staunch proponents of magmatic differentiation like Victor Goldschmidt (1922b) claimed that alkali metasomatism was common in silicate rocks because of the high degree of solubility of alkali compounds. He regarded metasomatically generated feldspar and muscovite as widespread (Goldschmidt, 1921). He suggested that the source of the motive energy of metasomatic processes was to be found in all geological agencies that bring rocks of different chemical character into proximity or that set up potentials between different parts of Earth's crust by variations in temperature and pressure. One important source of metasomatic solutions, he argued, was magma. Thus, although Goldschmidt was a magmatist, he left the door open for the possibility of other mechanisms for producing granite.

In Canada, Terence Thomas Quirke (1886–1947), a structural geologist with the Universities of Minnesota and Illinois and the Geological Survey of Canada, also described various instances of granitization. Quirke (1940), for instance, reported on an occurrence near Killarney, Ontario, characterized by a gradation along strike from recognizable Gowganda Formation, a unit of glacially derived conglomerate and mudstones, into a fine-grained porphyritic gneiss and then into a medium-grained microcline gneiss. He further described a migmatite that is transitional between Killarney granite and the porphyritic gneiss. Although insisting that Killarney granite is an intrusive igneous rock, he maintained that the porphyritic gneiss is a highly altered sedimentary rock. The rocks, he said, had been converted by the action of physical and chemical influences on solid rocks. Regardless of their structure, texture, mineralogy, and chemical composition, he alleged, the gneisses did not represent truly igneous rocks.

Even in the United States, generally regarded as a hotbed of orthodox magmatism, examples of granitization were reported. For example, G. H. Anderson (1937) of the California Institute of Technology described an occurrence in the Inyo Mountains of California in which a succession of Precambrian argillite, quartzite, limestone, schist, and altered volcanic rocks had been transformed metasomatically. Anderson claimed that the Pellisier granite, the presumed end stage of the metasomatic transformation, had been produced in situ by fluids. He regarded the texture of the Pellisier granite as so variable that he found it difficult to find a representative sample. The transitions from the granite to argillite, he said, are also characterized by extreme variability. Anderson pointed out that the transitional rocks contain an abundance of relicts of sedimentary material. For Anderson, however, the source of the granitizing fluids was no mystery. Wherever the nearby Boundary Peak granite, which he regarded as plainly magmatic, is in contact with argillite, limestone, or other country rocks, Anderson found zones of silicification, migmatization, or transitions to the metasomatic Pellisier granite. He pointed out that, with increasing distance from the contact of the Boundary Peak granite, there is a gradual decrease in the proportion of magmatic material and in the intensity of alteration. The alteration of the surrounding rocks was effected by fluids that had leaked from the Boundary Peak granite and had reacted with them. Examples of granitization linked to intrusive magmatic granite described by Goldschmidt, Quirke, or Anderson, however, did not bother magmatists as much as did the examples described by Backlund or America's Peter Misch.

Even the magmatist-laden United States boasted its own enclave of advocates of granitization at the University of Washington, where Peter Misch (1909–1987) was perhaps the leading American granitizer. Misch received his Ph.D. degree from the University of Göttingen in 1932 with a dissertation on structure and metamorphism in the Pyrenees (I. McCallum, personal communication). In that year, he participated in an expedition to Nanga Parbat to map the Himalayan foothills. In 1936, Misch escaped from Nazi Germany and ended up in China, where he taught geology in Sun Yat Sen University in Canton. When the Japanese invaded China in 1938, he moved to Yunnan province along with the university, and, in 1940, he joined the national university in Peking. Misch came to the United States in 1946 on a lecture tour and began his career at the University of Washington the following year.

Misch's major work on granitization was a lengthy three-part paper published in 1949. Misch (1949a, p. 210) noted that "it has to be demonstrated that large masses of granitic rocks have as a whole, and not only marginally, formed by metamorphic transformation of solid rocks." He attempted such a demonstration in the Nanga Parbat area in the northwestern Himalaya Mountains. He reported that successions of phyllites, slates, limestones, quartzites, greenstones, tuffs, lava flows, and volcanic breccias grade into mesozonal crystalline schists, then into mesozonal migmatitic augen gneisses and banded gneisses, and ultimately into coarse-grained, light-colored, granitic gneisses or gneissose granites with minor layers of intercalated paragneiss. This succession, he said, is magnificently exposed in continuous sections for dozens of miles over a vertical range of 23,000 feet.

Misch asserted that, despite their igneous appearance, massive augen gneisses revealed their metasomatic origin by the presence of all gradations between augen gneisses and argillite-derived schists and by the interlayering of both kinds of rock. He noted that potash feldspar porphyroblasts occur in a matrix whose mineralogy and texture resembles that of a lower-grade biotite schist. Misch postulated that hot solutions from somewhere below had introduced potassium, but not sodium, into the rocks. Not a single chemical analysis, however, accompanied Misch's article to lend support to his contentions. The banded gneisses, he said, were not the result of metamorphic differentiation. Instead, he attributed the material composing the light-colored layers in the banded gneisses to the infiltration and permeation of solutions along planes of foliation during shearing movements within the country rock. Misch suggested that high-grade regional metamorphism depended on syn-orogenic granitization for its supply of additional heat.

Misch's paper illustrated at least two of the major difficulties that magmatists often encountered with the claims of granitizers. On the one hand, Misch called for an introduction of potash in solution, without the support of chemical data, from a source that was conveniently out of sight. Moreover, he gave no account as to why potash, but not soda, was introduced. The other problem was Misch's use of the term "granitization." The term implies the conversion of rocks into "granite," and yet many of the alleged granites produced by granitization are not that at all. Misch claimed that his massive augen gneiss had an igneous appearance, but the photograph of augen gneiss included in his paper shows a rock with pronounced foliation that looks very much like a gneiss rather than a granite. Had Misch and other proponents of granitization used terms like "gneissification" or "feldspathization" that carry less genetic freight, the controversy might have been less heated.

In his second paper, Misch (1949b) examined the results of alleged static granitization in the Sheku area of the northwestern Yunnan province of China. In this region he described Mesozoic red beds overlying folded and metamorphosed sedimentary and volcanic rocks of Precambrian to Triassic age. Within the upper red beds, Misch recorded irregular bodies of massive granitic rocks that completely lack structures indicative of stress and whose field relations show no indication of magmatic intrusion. The red beds enclosing the granites are undisturbed, and there is no evidence of stoping within the granites. Moreover, some of the red beds are enclosed within the granites, and signs of relict bedding are also found within the granites.

In thin section, the groundmass of the granitic material consists of clastic-appearing, angular and subrounded grains of clear quartz in a matrix of turbid and iron-stained potassium feldspar. "Phenocrysts" of plagioclase appear to have invaded and absorbed the groundmass. Overall the "granite" has a porphyritic appearance. Misch concluded that the groundmass of this "granodiorite porphyry" was not of igneous origin but was derived from the fine-grained clastic red beds by metasomatic processes involving the addition of alkalis. Because of the static conditions, the granitizing solutions were envisioned as spreading irregularly throughout the uniform sedimentary rocks.

In his third paper, Misch (1949c) stated what he perceived to be the relationships between syn-kinematic and static granitization. He regarded syn-kinematic granitization as a fundamental process in geosynclinal orogeny and static granitization as its continuation after the end of an orogeny. As examples of static granitization following syn-kinematic granitization, Misch mentioned Nanga Parbat, West Yunnan, and also north-central Washington west of the Okonogan River. He believed that the majority of large granodioritic and granitic batholiths had probably formed by metasomatic granitization. He had difficulty conceiving of enormous masses of granitic magma rising to shallow levels in the crust without breaking through to the surface. "If the large granitic batholiths were igneous," Misch (1949c, p. 700) wrote, "I should expect that those geosynclinal belts in which they occur, would in post-orogenic time have been drowned in floods of rhyolite, dacites, and related rocks." Misch insisted that magmatic granites with genuine igneous textures for which an intrusive mechanism can be demonstrated are comparatively small.

To account for larger granitic masses, Misch favored the selective mobilization of ions and molecules from nonexposed depths. This mobilization, he thought, would presumably produce an alkali-rich, mobile emanation that would rise into the geosynclinal prism and perform metasomatic granitization by means of infiltration, permeation, and diffusion. Misch could find no magmatic source for the agents that brought about large-scale, regional granitization within the levels of the crust that had been exposed. The birth of granitization, he said, was genetically connected with eugeosynclinal orogeny, and its primary motor is at great depth.

One of Misch's colleagues at the University of Washington from 1919 to 1952 was George Edward Goodspeed (1887–1974). Goodspeed produced a series of very useful papers that acknowledged the existence of both metasomatic and magmatic granitic bodies without indulging in some of the speculative aspects of the debate. He developed sets of field and petrographic criteria for distinguishing magmatic from metasomatic bodies. For example, Goodspeed (1940) distinguished between dilation and replacement dikes. Goodspeed (1948) also distinguished between "xenoliths" as inclusions caught up in truly intrusive magmatic material and "skialiths" as relict inclusions in metasomatized granitic rocks. Later, Goodspeed (1952) developed criteria for the recognition of remobilized, replacement dikes, that is, the "rheomorphic" dikes discussed by Backlund. Finally, he proposed criteria for the recognition of igneous plutonic breccias, replacement breccias, and rheomorphic breccias (Goodspeed, 1953).

THE BRITISH SCHOOL OF GRANITIZATION

Ironically, some of the most enthusiastic proponents of granitization came from Great Britain, the land of Hutton, Hall, and Lyell, the original supporters of magmatic granite, and of Sorby, Judd, Teall, and Harker, all devotées of magmatic granite. The enthusiasm for granitization among British geologists certainly derived in part from acquaintance with field occurrences of granitized rocks. In many cases, however, that enthusiasm also stemmed from an antipathy to the form in which magmatist views about the origin of granite were expressed by the adherents of fractional crystallization, namely, granite magma as the end product of the differentiation of basaltic liquid. The antipathy to the crystallization–differentiation version of the origin of granite magma, espoused by Goldschmidt, Niggli, Eskola, and, above all, Bowen, was also coupled with some degree of suspicion of the results of experimental petrology, viewed by some granitizers as a threat to the insights of the field geologist. In reaction against the concept of the formation of granite by crystallization–differentiation, British granitizers often overstated their case for granitization and minimized the role of granitic magma. Coupled with personality differences, this tendency to overstate only exacerbated an already controversial situation.

The leading advocates of granitization in Great Britain were Arthur Holmes, Doris Reynolds, and H. H. Read. Over the years, Holmes had come to the conclusion that granitic magma was the culmination of a process of granitization, not its cause. Like Rastall, Eskola, and so many others of his generation, Holmes said that as a student he had acquired a highly dogmatic belief in granite magma backed by a great weight of authority but entirely unsupported by evidence (Eskola, 1950). In the light of his own experiences in Uganda and elsewhere and of the experiences of others, he said, he had gradually abandoned the hypothesis that granite resulted from the crystallization of magma. He had completely lost the magmatist faith in which he had been raised.

One reason for Holmes' loss of faith may have been his close relationship with Doris Reynolds (1899–1985) whom he married in 1939. Reynolds had graduated in 1920 from Bedford College, where she was strongly influenced by Catherine Raisin. Between 1921 and 1943, she taught successively at the Queen's University in Belfast, Bedford College, University College, London, and the University of Durham. She received a D.Sc. degree from London in 1937. From 1943 to 1962, Reynolds was an Honorary Research Fellow at the University of Edinburgh. Unlike Catherine Raisin, who pioneered the entry of British women into the field of igneous petrology, Reynolds did not have to operate around the fringes of the geological community nor was she restricted to teaching at a women's college, despite spending a few years at Bedford College. In 1919, the Geological Society, after years of hesitation since first raising the issue in 1880, had voted decisively, 55–12, to admit women as fellows of the society, recognizing that "the work done in the past by a number of women geologists has been of a higher order of merit" (*Geological Magazine*, 1919, p. 193). Reynolds quickly moved to the forefront among contemporary women in petrology. In the United States, Florence Bascom had trained several women petrographers, such as Dorothy Wyckoff and Judith Weiss, and Anna Jonas Stose of the U. S. Geological Survey was also a petrographer, but the reputation of Reynolds as a petrologist surpassed that of all other women.

During the 1930s and 1940s, Reynolds carved out a career examining igneous intrusions and their associated metamorphism with particular emphasis on the Newry igneous complex in Ireland. Reynolds (1946) furthered the cause of the granitizers with a comprehensive paper detailing the sequences of chemical changes that lead to granitization. She compiled chemical data from a very large number of occurrences where granitic rocks have intruded various country rocks and evaluated the chemical changes in the vicinity of the contacts in pelitic, psammitic, carbonate, and basic rocks. For most occurrences she argued that the country rocks farthest from the granite contact had become more basic relative to original, unaltered rock and that the country rocks closer to the granite contact had become more granitized. Likewise, Reynolds found that inclusions of country rock within granite typically have cores of basified rock that are surrounded by granitized rims. From these data she did not conclude that solutions had emanated from the granite magma as, for example, Anderson (1937) had. Similar to the proposal of C. E. Wegmann (1935), who had envisioned an upward advance of migmatite fronts, Reynolds concluded that there had been fronts of basic elements that had been expelled from the rocks being granitized. She envisioned the existence of more or less concentric waves or fronts, first basic and then alkali, attending the emplacement of the granite.

The following year, Reynolds (1947) reviewed the entire granite controversy. She asserted that magmatists like Grout and Tilley misunderstood and persistently misrepresented the views of transformists, that is, granitizers. Grout, she said, had falsely accused transformists of not presenting evidence for their claims. On the contrary, she said, the papers that Grout (1941) had criticized were full of evidence. By analogy with Hutton, who inferred the existence of a source of heat at depth simply on the basis of the field evidence but was not yet able to identify that source, she insisted that transformists should be granted the same liberty regarding the sources of their emanations. Why, she wondered, should transformists not be able to discover the *fact* of granitization of the basis of field evidence even though they did not yet have an explanation for the granitization? She observed that the recognition of granitization and its associated basic fronts led to inferences that implied the operation of unfamiliar physico-chemical processes. Such inferences should, she granted, be readily susceptible to criticism. She complained that Cecil Tilley, rather than offering legitimate criticisms, had simply dismissed the notion of "fronts" as a gross error that he chose not to dwell on. This kind of disdainful dismissal of views, Reynolds (1947, p. 212) regarded as so much "authoritarian bluff" devoid of any scientific interest.

As a transformist, Reynolds maintained that field evidence at many localities sufficiently indicated that the formation of granite could take place by the transformation of preexisting rocks in situ by chemical changes involving both the introduction, particularly of alkalis, and the removal of constituents. Inasmuch as the rocks undergoing transformation to granite are overall generally more basic than granite, the constituents displaced from them during transformation would be unusually rich in Fe, Mg, and Ca. These emigrant constituents could become fixed in a "basic front" of granitization thereby giving rise to other, more basic igneous-looking rock varieties. Contrary to the usual impression conveyed by the granitizers, she insisted that transformists did not deny the existence of granitic magma. If anything, they actively maintained that the transforming processes could lead to the formation of magma given the appropriate conditions of pressure and temperature. She believed that many felsic and intermediate lavas and hypabyssal intrusions could originate in this manner.

Perhaps what came next in Reynolds' 1947 paper was the most troubling to magmatists. While geologists had direct observational knowledge that lava crystallized from a melt, Reynolds claimed, the supposition that plutonic rocks represented the crystallization of magma was pure hypothesis in conflict with many observed phenomena. The strength of the magmatic hypothesis lay in familiarity and usage. She suggested that it was necessary philosophically to add the process of solid diffusion as a possibility concerning the origin of granite. Granites, she said, may have resulted from recrystallization of preexisting rocks in the solid state in response to the changes in chemical composition brought about by solid diffusion and a rise in temperature. Appealing to the phenomenon of diffusion in soft metals, Reynolds noted that as one metal diffuses into the other, a third phase is produced with sharp boundaries representing a diffusion front. By analogy with metals, she suggested that sharp contacts in rocks or minerals might represent limits of diffusion. For example, the contact between a feldspathized zone of quartzite and the original quartzite is remarkably sharp.

Reynolds concluded by charging that magmatists had adopted a defeatist attitude. They were, she alleged, in danger of conceiving of petrology as simply a restricted branch of physical chemistry. In contrast, she said that the transformists wanted to keep petrology as an integral part of geology. They believed that the so-called igneous rocks could not be properly interpreted without paying attention to both the field appearances of the plutonic rocks themselves and their complete geological settings in time and space.

Despite the contributions of Reynolds, Backlund, Holmes, Misch, Perrin, and a host of other eminent transformist geologists, H. H. Read emerged as the most vocal spokesman for the transformist position, a fact that occasionally irritated other transformists who did not agree with all of his pronouncements about granitization. Herbert Harold Read (1889–1970) graduated from the Royal College in 1911, and the M.Sc. degree was awarded to him three years later (Sutton, 1970). During his graduate school years, Read became a staff member of the Department of Geology and began the first of his eight revisions of Frank Rutley's *Elements of Mineralogy*. In 1914, Read joined the Geological Survey of Great Britain, under whose auspices he spent almost two decades mapping complex crystalline terranes throughout Scotland. As a result, Read gained a broad knowledge of the relationships among high-grade metamorphic rocks, migmatite complexes, and granitic rocks. He described the contact relations between basic intrusions and schists and also contributed to an understanding of low-pressure, Buchan-type regional metamorphism.

Read returned to the academic world in 1931 as George Herdman Professor of Geology at the University of Liverpool. In 1939, he transferred to the Imperial College of London, where he continued an active teaching and research program until 1955 when he retired. In the 1940s and 1950s, Read received many honors and frequently filled the offices of several geological organizations. These official positions provided him the opportunity to compose several addresses on the relationship among granites, migmatites, and metamorphic rocks. Over the years, Read had become impressed by the evidence of feldspathization of metasedimentary rocks in the neighborhood of granites. He had also come to appreciate the insights of Scandinavian and French geologists into the nature of the associations of granite, migmatite, and feldspathized metasedimentary rocks. Troubled by what he perceived as a somewhat simplistic approach to the nature of granites espoused by magmatists such as Rosenbusch and especially bothered by the insistence of his own contemporary N. L. Bowen that granite was derived by fractional crystallization of basaltic liquid, Read began in the late 1930s to stress the importance of field relationships pertaining to granite in a series of presidential addresses that gained for Read an international reputation as a spirited, articulate defender of the transformist conceptions of granite. Many of the addresses were later compiled in *The Granite Controversy* (Read, 1957). These addresses amply illlustrated Read's exceptional flair for elegant polemic writing. Indeed, Read's relish for debate and his well-honed literary skill often led him to overreact to his opponents in these addresses and his other articles on granite. According to his protegé, Wallace Pitcher (personal communication), Read was far more judicious and sober on the outcrop and much more ready to concede demonstrations of magmatic granite. But it is for his controversialist literature that Read is remembered by petrologists of later generations.

Like so many geologists of the granitization school, Read (1944) lauded the work of the French geologists. He judged that they had made several major contributions to the study of granitic rocks. First, they had acquainted the geological community with the notion of emanations, mineralizers, and volatiles as extremely important agents in metamorphic and plutonic processes. They had, according to Read, dissected and interpreted the contacts of granite bodies to such a degree of perfection that their notions must still be considered current. The French had insisted that magma does act upon its wall rocks and had demonstrated reciprocal exomorphic and endomorphic operations of great significance. They had pointed out the variability of granite contacts, some of which are sharp and some vague, some of which show little sign of transfer of matter, and some of which contain indications of immense amounts of transfer. Read claimed that the French school, by thus calling attention to the unity of the phenomena at granite contacts, had freed geologists from the illogical nature of the extrapolations of Rosenbusch and his school. He challenged his hearers to compare the beauty and fruitfulness of the French concept of the relations between depth and the character of granite contacts with the barren frigidity of the concept of Rosenbusch. In Read's view, the French had considered the association of gneisses and mica schists with granites as genetically significant rather than as an unnecessary complication. They had realized the significance and importance of feldspathization as a petrogenetic process. They had advanced a mechanism for the emplacement of granites that solved the room problem associated with granite intrusion, and, lastly, Read said, the French school appreciated the fundamental contrasts between the dominantly basic, volcanic association and the dominantly acidic, plutonic association. Read's entire corpus of granitization writings was, in effect, an attempt to vindicate French concepts.

Perhaps more than any other transformist, Read also expressed more forcefully than his fellow granitizers a fundamental philosophical difference that at least some transformists had with magmatists concerning the solution of geological problems. Read was very much a field geologist and was quite reticent about accepting the assistance of allied sciences. Problems of geology, he insisted, needed to be solved geologically. He repeatedly stressed the overwhelming importance of geological evidence in solving all problems of granitization and related processes. What a geologist must not do, Read insisted, is to remain meek in the face of a nongeological argument that was in apparent conflict with geological evidence. Read's overstated antipathy to Rosenbusch's microscopic petrography and Bowen's experimental petrology appears remarkably retrograde, particularly when it is noted that petrologists of earlier generations eagerly sought help in solving petrological problems wherever they could find it. Although their work was primarily field-based, petrologists who worked around the beginning of the twentieth century like Brögger, Iddings, Vogt, and Harker never looked with suspicion on physical chemistry or viewed it as a threat. Quite the opposite; they eagerly sought to apply its insights to their questions. Ironically, too, the advocates of the origin of granite in the wet way who opposed the origin of granite by pure igneous fusion in the mid-nineteenth century were always ready to pounce on the available experimental evidence in support of their ideas about granite. Curiously, a century later, Read, one of the leading skeptics about the idea of magmatic granite, came close to arguing that the problem of the origin of granite could be solved only on the outcrop.

20 ← PONTIFFS AND SOAKS

The Resolution of the Granite Controversy

Upon the death of Rosenbusch, the magmatic view of granite continued in full force. Leading European proponents of magmatic granite included structural geologist Hans Cloos, geochemist Victor Goldschmidt, and petrologists Otto Erdmannsdörffer, Paul Niggli, Pentti Eskola, and Cecil Tilley. In North America, the abundant, clearly intrusive granitic stocks in New England and throughout the western United States coupled with the vast granitoid batholiths of California, the Pacific Northwest, and British Columbia, rendered it a matter of little surprise that the United States would boast such world-renowned petrologists who advocated magmatic granite as A. F. Buddington, Adolph Knopf, Esper Larsen, Jr., Frank Grout, R. A. Daly, and N. L. Bowen. The magmatists saw little reason to doubt that granitic rocks are overwhelmingly magmatic in origin in view of their intrusive relations, textures, chemical similarity to rhyolite, thermal effects on country rocks, and evidence from a growing body of experimental work performed largely by Bowen and his colleagues at the Geophysical Laboratory.

THE MAGMATISTS

While Backlund, Reynolds, Read, and other granitizers were advocating their views, members of the magmatist school were, of course, busy advocating theirs. In a review article on granitization, Frank Grout (1941) reminded his readers that almost all geologists agreed that volatile constituents might emanate from magmas to replace the adjacent rocks and in some places make them look igneous. The trick for the geologist, he said, is to try to discriminate between rocks that are igneous and those that merely look igneous. With such a handy source of fluids available in granitic magmas, Grout maintained that the burden of proof lay on the granitizers to demonstrate the necessity for their mysterious emanations from inaccessible regions at depth.

Grout charged that the granitizers of his generation had been less cautious than Sederholm. Enthusiastic advocates of granitization had carried their ideas too far, he believed, by applying the concept to rock masses several miles across. He noted that the more conservative geologists admitted that granitization had occurred over distances of a few feet and conceded that the truth probably lay somewhere in between the two extremes. The conservatives, he believed, had the weight of evidence on their side so that the burden of proof rested again on the advocates of very extensive granitization.

Grout further charged that transformists were often guilty of the careless use of terminology. He thought that terms such as migmatite, metasomatism, replacement, assimilation, granitization, and even magmatic were often poorly defined. He was much troubled by the characterization of gneisses as granites and by the generally careless definition of rock types. As examples, Grout claimed that it was improper to state that the end result of metasomatism was the production of magma, that assimilation had changed a hornfels into a diorite, or that assimilation had granitized an inclusion. Indeed, one wonders how much controversy there would have been if the term "feldspathization" had been used in place of "granitization."

One of the leading petrologists of his generation was the Swiss geologist Paul Niggli (1888–1953). Niggli graduated in 1907 from the Eidgenössische Technische Hochschule (E. T. H.) in Zürich (Parker, 1954). In 1912, he received his Ph.D. degree from the University of Zürich with a study of chloritoid schists in the Swiss Alps under Ulrich Grubenmann. Niggli spent a year at the Geophysical Laboratory and then returned to Zürich to become a Privatdozent in the university, but in 1914 he took a position in the University of Leipzig as Professor Extraordinarius. In 1918, he became Professor Ordinarius of Mineralogy at the University of Tübingen. Upon Grubenmann's retirement, Niggli succeeded him as Professor of Mineralogy and Petrography at the E.T.H. and the University of Zürich. Throughout his career, Niggli was a prolific writer of articles and books on crystallography, ore deposits, magmas, and volatile constituents.

Niggli (1942) comprehensively reviewed the granite problem, concurred with Grout's contentions, and expressed his magmatist views. A little later, Niggli (1946) acknowledged that not all granites are the same and maintained that each granite body should be interpreted on its own merits. He postulated the existence of three kinds of granite: magmatic granite that had directly crystallized from magma; metagranite that was the product of metamorphic recrystallization of arkose; and migmatite granite that resulted from ultrametamorphism. On the third type of granite, Niggli found some common ground with those transformists who regarded granite magma as the product of extreme granitization.

In all likelihood, Niggli had been influenced by the writings of Pentti Eskola (1883–1964), Professor of Geology and Mineralogy at the University of Helsinki from 1924 to 1953 (Mikkola, 1968). Eskola's lifelong passion was the Precambrian geology of Finland. His early work was in the field of metamorphism and metasomatism. Contact with Goldschmidt in Norway persuaded

Eskola that the behavior of crystalline rocks is determined by the principles of physical chemistry. Thanks to his long friendship with Bowen, begun when at the Geophysical Laboratory, Eskola became an ardent magmatist. Eskola (1932, 1933) had expressed great interest in the idea that many granite magmas were formed by partial re-fusion of the low-melting constituents of silicate rocks deep within the crust in connection with orogenic movements as a result. Years later, although writing freely of metasomatism and soaking and at times sounding like a transformist, Eskola (1950) explicitly denied the possibility of the metasomatism of huge masses of rock. Why should it not be assumed that the granitizing agent has been granitic magma? He could not agree with Read's suggestion that there are "granites and granites" if by that expression Read meant that some granites had come into existence without the mediation of liquid magma while others were magmatic. Eskola's thesis was that all granites are magmatic. Even though he acknowledged that metasomatic granitization had occurred locally, he maintained that such metasomatism had been effected during crystallization of residual magmatic liquids that represented the lower points of the solidus in the polycomponent granite system.

Eskola recognized that the "antimagmatic" theories of granitization had arisen in reaction against the theory of crystallization–differentiation. While not agreeing with all the details of that theory, he regarded it as epoch-making for petrology because, for the first time, that theory placed petrology on a firm physico-chemical foundation that had been laid by Bowen, Niggli, and Goldschmidt. He warned transformists against throwing the baby out with the bathwater in their desire to eliminate magma or replace it with something beyond the possibility of experiment or observation.

Like the transformists, Reginald Daly saw the solution of the granite problem in very broad terms. Its solution, Daly (1949) thought, would require silicate experiments, analogies from experiments on nonsilicates, detailed petrography, seismology, gravity studies, and even input from astrophysics and cosmology. For Daly, the granite problem demanded scholarship of dismaying breadth and thoroughness, not to mention vision regulated by common sense. He vigorously opposed the transformist school. He reminded those who were so fond of appealing to the work of Sederholm, Michel-Lévy, and Lacroix that they all located the source of their granitizing solutions in granitic magma. He did not consider the metasomatists of the his day to be entitled to claim any support from the Fennoscandian and French studies for their idea that metasomatic granite is dominant in the crust. Daly further complained that he could not picture what the metasomatists meant when they discussed the chemistry of emanations. He believed that the argument for extreme metasomatism would be weak in its fundamental premises so long as transformists did not provide a clear statement of the composition of the emanating solutions in terms of all the oxides and elements present.

Transformists, of course, had argued for the metasomatic origin of granitoid batholiths as one way to solve the "room problem" that, they believed, posed an intractable difficulty for advocates of magmatic invasion. Daly still regarded magmatic stoping as the solution to the room problem despite the fact that many fellow magmatists parted company with him on that issue. Granitizers had also argued that if batholiths were truly magmatic, they should have broken through the surface in many instances and thus be associated with large volumes of acid volcanic rocks. In response to their claim that batholiths lack these associated volcanic rocks required for the magmatic view, Daly pointed out that there are many localities in which large volumes of rhyolite, dacite, and obsidian have erupted in the vicinity of syn-genetic granites. In other localities, he noted, erosion had removed large volumes of rhyolitic caps.

THE CONFRONTATION OF READ AND BOWEN

Despite the contribution of an abundance of valuable scientific evidence and pointed rhetoric on all sides of the question by an impressive array of eminent geologists, the evidence and rhetoric of the entire controversy were condensed and concentrated in the writings and addresses of Read and Bowen. Together they formed a mutual irritation society. In the context of their personality clash, which seems to have stemmed back to 1920 when they had a brief but very pointed exchange in *Geological Magazine* over Bowen's views on crystallization–differentiation, the granite controversy flamed to a fever pitch.

Although aspects of the granitization concept had been invoked throughout Bowen's career, its advocates had generally not argued directly against him, perhaps because he had paid little heed to the granitizers. Not until Read began to challenge the applicability of crystallization–differentiation to the origin of granite was Bowen drawn into the growing furor. Earlier in his career, critics of Bowen's scheme of crystallization–differentiation assumed, as he did, the magmatic nature of granite. In the 1940s, however, Bowen's idea about the goal of crystallization–differentiation toward granitic liquid had to include a defense of the very notion that granitic rocks are magmatic.

In an address entitled "Metamorphism and igneous action," Read (1940) uttered his famous assertion that "the best geologist is he who has seen the most rocks." In subsequent addresses, he repeatedly insisted that the granite problem was a problem of field geology and warned lest igneous petrology cease to be geology and become enslaved to the laboratory or to physical chemistry, a comment no doubt bound to annoy Bowen who had been pressing geologists for years to take full advantage of the insights afforded by experiments and by physical chemistry.

In his first meditation on granite, Read (1943) opined that he could discover no warrant for the attitude of those like Bowen who regarded geologists who had doubts that granites are necessarily genetically related to basalts as either mentally deficient, intellectually dishonest, or unpardonably iconoclastic. He thought that Bowen's theory of crystallization–differentiation was just fine for the Geophysical Laboratory, but whether it had actually happened in nature, he thought, should be decided on geological evidence. That evidence, in his mind, clearly opposed Bowen's conception of granite. Read insisted that the gigantic volume of granitic rocks provides an insurmountable objection to Bowen's theory because a vastly greater volume of gabbroic rocks should be associated with the granite than is actually the case. Read reminded geologists, and Bowen, that what has to be explained is what is in the crust, not at the bottom of a little crucible. While hardly discounting the reality of granite magma, Read did take pains to call attention to field evidence for the transformation of rock into granite under extreme metamorphic conditions.

In his second meditation on granite, Read (1944) became even more outspoken in his leanings toward granitization. He acknowledged that detailed petrographic studies had a place in the investigation of granite but insisted that studies of large, clear rock surfaces were more important. He needled experimentalists by stating that plutonic rocks were not emplaced as if they were in platinum crucibles because there must be reaction at the walls. Alfred Lacroix, Read (1957, p. 82) said, in discussing interactions of granite and wall rocks, had appealed to the "only evidence of value, that displayed in the field." Read suggested that the magmatists, and especially the experimentalists who spent their time in the laboratory with tiny little platinum crucibles rather than in the field with huge masses of rock, had become guilty of an unhealthy authoritarianism. More than once, he referred to the views of Niggli as so much "pontificating." Read (1957, p. 82) said that he preferred "to remain in the benighted state of superstition so heretical to Niggli and the other pontiffs of crystallization–differentiation."

Read spelled out what he perceived as some of the weaknesses of Bowen's theory of crystallization–differentiation insofar as it pertained to the production of granite. Granite could not be regarded as the clear-cut end product of Bowen's process in light of the examples of iron enrichment. The calculations of Grout (1926) had undercut Bowen's ideas by showing that only a few percent of granite liquid could be generated from a basaltic parent, a calculation clearly at odds with the field evidence for abundant granite. And he repeatedly hammered on the problem of making space for huge volumes of batholithic granitic liquid.

Read granted that crystallization–differentiation could have played an important role in the development of the basalt–rhyolite volcanic association, but he saw no necessary connection between volcanism and the deep-seated plutonic association of granite, gneiss, and migmatite. He acknowledged the existence of some magmatic granite but increasingly began to suspect that a lot of granite might be the product of transformation of solid metamorphic rocks into granite without passage through a magmatic stage. The distinctions between granite and granitized country rock, he believed, became even more blurred with depth. He began to wonder if magmatic granites of the upper crust were not themselves the product of an extreme granitization or ultrametamorphism at depth that produced magmatic liquid as a result of concentration of fluids and juices in a core of granitization. He repeatedly appealed to J. J. Sederholm's descriptions of migmatites and talk of juices, ichors, and solutions transforming preexisting rocks into granite.

In his own presidential address on magmas, Bowen (1947) not only defended his views on fractional crystallization but also poked fun at the granitization school in one of the most entertaining passages in the corpus of scientific literature. In analyzing the differences between magmatists and transformists, Bowen (1947, p. 264) wrote:

> We can, indeed, for rough purposes, separate petrologists into the pontiffs and the soaks. Yet, among the pontiffs who bear the stigma of magma, there are none who do not believe that magmas contain volatile constituents of which the principal is water, that these may emanate from the magma and give rise to a liquor that pervades the invaded rocks, transforming them at times into igneous-looking rocks. The difference between the pontiff and the soak is that the latter must have his liquor in lavish quantities on nearly all occasions, but the former handles his liquor like a gentleman; he can take it or leave it according to the indications of the individual occasion, he can take it in moderation when it is so indicated, or again he can accept it in copious quantities and yet retain powers of sober contemplation of attendant circumstances.

Bowen teased that the emanations of the granitizers came from nowhere in particular as some sort of Earth sweat. He deplored the vagueness of statements about the alleged processes and agents of granitization such as "waves of solutions."

The granite controversy came to a boil when Read and Bowen squared off at a special symposium on the origin of granite sponsored by the Geological Society of America in Ottawa, Canada, on December 30, 1947 (Gilluly, 1948). Major participants in the symposium included A. F. Buddington, F. F. Grout, and G. E. Goodspeed. More than twenty other geologists, including S. J. Shand, participated in the following discussion. Fittingly, the two main protagonists were separated from one another on the program. Read gave the initial paper. Bowen gave the last. Hatten S. Yoder, Jr. (personal communication), who was present at the conference, recalled that "the sparks flew! What a treat for a graduate student."

In his talk, entitled "Granites and granites," Read (1948) continued to hammer away on his contention that the granite problem was a problem of field geology rather than one of petrography, mineralogy, physical chemistry, or of any other ancillary discipline. Petrology is a branch of geology, he insisted, not a restricted branch of physical chemistry. Read perhaps had some justification for his alarm that petrology might be thought of as a branch of physical chemistry. After all, H. S. Washington (1917, p. 10) had written in the introduction to his renowned tables of chemical analyses that

> it is coming to be recognized by petrologists in general that the study of igneous rocks is in large part the study of silicate solutions and their equilibria, often complicated by the presence of volatile components, and is thus to be regarded as essentially a special branch of physical chemistry.

In reaction to Bowen's humorous classification of petrologists as either "pontiffs" or "soaks," Read (1948, p. 1) replied that "bigots or, if you like, enthusiasts on both sides do a deal of harm, and pontiffs, I suggest to Professor Bowen, while capable of a greater number of good deeds, are also capable of a greater number of bad deeds than the village drunk." The rhetoric was obviously intensifying.

Read reiterated his conviction that the granite problem was a plutonic problem involving complex associations of granite, migmatite, and metamorphic rocks and having little connection with volcanism. He viewed proposed genetic and classification links between rhyolite and granite as "nimble leaps" taken by the likes of Niggli and Bowen that he was unable to take. The association of volcanism and plutonism was purely accidental, he believed. In fact, he would even welcome a return to the two-magma conception entertained by Bunsen (1851) and Durocher (1857). He was willing to accept the validity of crystallization–differentiation for volcanic associations, but he objected to the production of large quantities of granitic magma by that process. The room problem, Read believed, was at the heart of the granite problem. He was unable to conceive how vast batholithic masses could shoulder aside so much rock. Nor did he see field evidence for stoping of batholiths. Granitization seemed to him to avoid the room problem of the magmatists.

Read then reviewed ideas about granitization, noting that some of the more recent granitizers like Wegmann, Kranck, and Backlund had virtually eliminated a major role for magma in the formation of granite. He also reviewed the theory of basic fronts that had been worked out by Doris Reynolds. He further commented that the transformist school seemed to be "moving away from the view that granitization is connected with some kind of soaking by juice or ichor and toward the invocation of ionic migration" (Read, 1948, p. 14). Backlund, Reynolds, Perrin, Roubault, and Scandinavian geologist J.A.W. Bugge had developed the idea that the movements of migmatite fronts correlated with the differing atomic radii of the participating chemical elements. Read thought that this new "dry diffusion" approach to the problem of metasomatism of country rocks to produce igneous-looking rocks held great promise. He concluded that the truly magmatic granites were largely restricted to the upper crust and that, in the plutonic, lower regions, granitic magma was only an end stage of an extreme process of granitization. In essence, then, Read envisioned the existence of a granite series comprising different sorts of granitic bodies, some magmatic and some metasomatic, produced under widely different crustal conditions.

The paper of Bowen (1948) was not a direct response to that of Read, but he had anticipated many of Read's lines of argument and launched into a devastating critique of the theory of granitization. The paper, entitled "The granite problem and the method of multiple prejudices," presented both Bowen's ideas on granite as well as glimpses of his scientific philosophy. While conceding an element of truth to Read's claim that the best geologist was the one who had seen the most rocks, Bowen countered that it was insight rather than sight that was most important. He pointed out that, on numerous occasions, he had taken advantage of the insights gained from physical chemistry or phase diagrams to interpret field evidence more effectively. He suggested that field observation alone was insufficient because it often merely led the petrologist to think of competing hypotheses; as a result, additional resources from the fundamental sciences would often be required to choose among these competing views. Such, he believed, was the case with the origin of granite. For Read the issue at stake was posed as one of the rights of field geology against the pretended rights of an alien nongeological science. In contrast, fully understanding that rocks, minerals, magmas, and fluids are subject to the principles of physical chemistry, Bowen thought that field geology should be done while taking full advantage of the knowledge gleaned from physical chemistry.

As usual, Bowen defended his own conception of granite as the end product of crystallization–differentiation. Of the greatest value in his address were his evaluations of proposed mechanisms of granitization. He examined the views of the "wet granitizers," that is, those who postulated the existence of thin, tenuous fluids, solutions, emanations, or "ichors" that allegedly pervaded the crust and produced colossal masses of granite, especially within thick prisms of geosynclinal sediment. The "ichor," a term originated by Sederholm, was said not only to produce granite by metasomatic replacement but also to cause fusion with the resultant production of granitic magma. Bowen charged that wet granitizers had ignored the energy requirements of granitization. No sufficient supply of energy existed to bring about such large-scale effects, he said. Because the conversion of the low-grade hydrous minerals of geosynclinal sediments like clays, chlorite, and hydromica into the largely anhydrous minerals of granite would be strongly endothermic, he asked where the heat would come from to effect these mineralogical transformations. Bowen believed that it would be a colossal task just for the ichor to raise the temperature of a large mass of rock to the point where low-grade minerals would no longer be stable. To make matters worse, the ichor would have to maintain the higher temperature of the rock mass in the face of the endothermic transformation effects. The prospect was even bleaker than that, he went on, because the ichor also had to dissolve and remove mafic materials not needed for granite. But, he observed, the ichor had presumably already originated in a highly mafic region, namely the lower crust. If the mafic material were more soluble in the ichor than granitic material, he wondered why it had not brought dominantly mafic material along in the first place.

Bowen was most annoyed, however, by the ideas of the "dry granitizers" who were proposing that granitic masses formed by the migration of ions through crystals without the assistance of a pervasive fluid. He concluded that such a process is unlikely on theoretical grounds. Because the rate of ionic diffusion diminishes rapidly with decreasing temperature owing to the decrease in the number of crystal defects, Bowen argued that diffusion would be ineffective below about 0.8 times the melting temperature of the minerals in the rocks that were allegedly undergoing granitization. But, he said, the melting temperature of an assemblage of minerals is typically considerably less than the melting temperature of individual minerals within the assemblage. Hence, the effectiveness of diffusion should be tremendously diminished. In other words, he suggested, the rocks would likely begin to melt before diffusion through the solid minerals could ever become significant. Bowen also argued that the idea that granites were formed by diffusion through solids was incompatible with the fact that granites characteristically contain zoned plagioclase on the grounds that plagioclase zonation would be wiped out by the diffusion process.

One of the leading proponents of dry diffusion was theoretical petrologist Hans Ramberg (b. 1917). While at the Mineralogisk Institutt of the University of Oslo, Ramberg (1944) had postulated that gradients in chemical activity within the crust resulted in the migration or dispersion of light atoms and low density minerals from areas of high activity in the lower crust toward areas of lower activity in the upper crust. In his opinion, potash feldspar, quartz, albite, and minerals with high Fe/Mg ratios have the greatest chemical activities and would migrate the greatest distances. He envisioned this process of dispersion, in conjunction with mechanical movements, as the cause of granitization. He envisioned this net upward movement of granite-making chemical elements as being counterbalanced by predominantly downward mechanical flow of more basic material. As he expressed the concept in *The Origin of Metamorphic and Metasomatic Rocks* (Ramberg, 1952, p. 261), "to compensate for the upper diffusion current, the whole 'crust' sinks bodily, so to speak, through the swarm of slowly rising atoms, ions, and molecules." There was, in his judgment, no need for the element migration to take place with the assistance of volatile substances acting as carriers.

Ramberg's hypothesis was not persuasive to Bowen, who teased that Ramberg and other diffusionists selected for migration precisely those atoms that are needed to make granite without advancing very compelling reasons as to why those particular atoms should be the ones to migrate. Bowen objected that

Ramberg's assumption of an upward diffusion of Fe contradicted his general principle of diffusion because it is Mg that has a lower atomic mass and forms less dense minerals than Fe. At the same time, Ramberg tried to have Fe migrate downward while leaving the high Fe/Mg ratios of granites unexplained. On Ramberg's theory, Si and Al, important constituents in granite, were presumed to rise because of their small size. By the same reasoning, Bowen argued that O and OH, which are also enriched in granite, should lag behind because of their great size. He was perplexed by Ramberg's attempt to rescue the situation by claiming that oxygen migrates upward on the grounds that oxygen atoms are simply too big to remain in high-pressure regions and "up they come." Obviously unconvinced by Ramberg's line of thought, Bowen (1948, p. 84) quipped that "the whole process of solid diffusion is nothing, if not versatile." Ironically, after Bowen left the University of Chicago in 1947, Ramberg became a Research Associate in Geochemistry at the same institution (Fisher, 1963). In 1952, he was promoted to Associate Professor and, the following year, to Professor with responsibility for instruction in petrology. In 1962, Ramberg left Chicago for the University of Uppsala in Sweden.

Bowen concluded his address by plugging his view that granitic magmas, derived by fractional crystallization of basaltic parents, gave off emanations that would accomplish metasomatic replacement including a modest amount of granitization. He also persisted in maintaining that the only really efficient method of having salic material rise and mafic material sink was magmatic crystallization–differentiation. After all, he maintained, granitization does not solve the room problem; it merely evades it because material that rises chemically occupies just as much volume as the same material that rises mechanically as a mass of magma.

AFTER OTTAWA

Bowen's mentor, R. A. Daly (1949) was equally appalled by the claims of solid diffusionists. He insisted that they were duty bound to show how clouds of ions could create extremely thick masses of nearly homogenous granite by diffusing through the crystals that constitute very heterogeneous terranes. He charged that diffusionists had paid inadequate attention to the loss of chemical potential as the ionic clouds worked their way upward, reasoning that the inevitable cooling with its attendant loss of transforming power would win the race against transformism. Daly also regarded the solid diffusionist reliance on tractions due to the differing masses of virtually infinitesimal ions inside each of a multitude of crystals as a *reductio ad absurdum*. He also charged that the solid diffusionist hypothesis was utterly unable to account for the total lack of granite within the sial-free regions of the ocean floors.

The critiques of Bowen, Daly, and others hardly slowed the efforts of the dry diffusionists. By the 1950s, Perrin and Roubault, representing the most radical branch of the transformist school, were now maintaining that virtually all granites had formed in the solid state and questioning the existence of granitic magma. Perrin (1954), for example, appealed to the work of Backlund (1938), Goodspeed (1940), and Misch (1949a) in an effort to dispute the magmatic origin of intrusive dikes, claiming that the evidence from dikes favored rheomorphic, metasomatic origin. He even appealed to the writings of magmatists to the effect that at least some granites had formed by metasomatism and that textural evidence indicated that no granite is purely magmatic. Even magmatist Otto Erdmannsdörffer, he noted, had suggested that pure magmatic textures do not occur in granites, as indicated by evidence of crystal corrosion and presence of metasomatic crystalloblastic fabrics. Although Erdmannsdörffer explained such textures in terms of autometamorphism by residual solutions, Perrin favored the idea of long-distance diffusion. He argued that diffusion was demonstrated by the evidence of corrosion between crystals. Throughout the process of granitization, he maintained, crystals had cleared themselves of inclusions. He suggested that if granitization had occurred as a result of fluid circulation, then the introduction of alkalis should be accompanied by the removal of Fe and Mg in same direction to form a basic front in advance of the granitized rock. But he disputed the existence of basic fronts. On the other hand, he claimed that, if dry diffusion had occurred in a stationary medium, chemical transfers should have taken place in opposite directions. In that case, no basic front should develop. For example, Perrin noted, if the alkalis moved outward and upward, then Fe and Mg ought to move downward on his hypothesis of dry diffusion. Field observation, he thought, should provide the clue to discriminate between these two possibilities. Perrin concluded that alkalis did rise and Fe and Mg did return to depth during diffusion just as Ramberg had explained.

Matt Walton (b. 1915) was perplexed by Perrin's view. Walton was a graduate of the University of Chicago and a recipient of a Ph.D. degree from Columbia University in 1951. In addition to service with the U. S. Geological Survey and the New York State Science Service, he taught geology at Yale University between 1948 and 1965. Walton (1955) did not see why the partitioning of the alkalis and Fe and Mg should lead to long-distance diffusion rather than simply local exchange. Noting that Perrin wanted small, presumably heavy Fe and Mg ions to migrate downward, he inquired if the distribution of Ta, U, Ge, Ga, Be, and Si ions was consistent with Perrin's view. After all, he noted, several of these ions are rather dense. They, too, should all be concentrated downward on Perrin's view but are not. Like Daly, Walton also wondered why, if granitizing countercurrents of ions were set in motion by gravitational potential, the basaltic crust beneath the ocean basins had escaped the effects of granitization throughout geologic time. Of course, in those days, before the development of plate tectonics and deep-sea drilling, neither Walton nor Daly realized that basaltic crust is geologically very youthful, perhaps not sufficiently old to display the effects of ongoing granitization. Walton disputed the tendency of Perrin to reason as if the mere existence of a gravitational potential provided an argument for its realization by the process of ionic diffusion.

For Walton, the basic question in the entire granitization debate concerned how the chemical elements that make up a granite get away from those chemical elements that do not. He claimed that the answer was not yet available. No one, Walton (1955, p. 4) claimed, could "honestly define on the basis of experimental established fact the geologically realizable limiting conditions for the formation of granite magma or granitizing solutions, nor can anyone predict what the final texture of a granite should look like if it has crystallized from a magma under plutonic conditions." While finding small comfort for the solid diffusionists in experimental results, Walton thought that the few experiments on wet diffusion had yielded encouraging results inasmuch as both silica and alkalis showed signs of mobility in the vapor phase, even in Bowen's experiments. Although much of the evidence that had been cited for or against granitization is essentially textural and structural, he observed, virtually no direct knowledge of how textures are influenced by plutonic processes was available.

Walton also questioned the validity of some of the transformist arguments against the magmatist view. He pointed out that a major advantage of the magmatist view is the mere existence of rhyolitic lava. Opponents, of course, downplayed the importance of rhyolite on the grounds of its minor abundance compared with that of basalt. Walton, however, argued that the abundance of rhyolite had very likely been greatly underestimated, reasoning that large amounts of rhyolite had been eroded away because of its occurrence in orogenic belts. He further maintained that a lot of rhyolitic material had been disguised as ash beds and welded tuffs. On other hand, he noted, the amount of true granite had probably been overestimated because granites *sensu stricto* within batholiths are characteristically exceeded in volume by diorites, quartz diorites, and granodiorites.

The room problem that granitizers always urged against the magmatist view Walton regarded as a straw man. He believed that batholiths probably rose buoyantly, diapir-like, through the crust like salt domes rather than by stoping or cauldron subsidence. He pointed out that no one raised the issue of a room problem with salt domes and yet salt domes produced structural dislocations. These dislocations, he noted, were scarcely noticeable beyond a few hundred meters from their contacts and probably would have escaped attention altogether had the rocks involved not been well stratified. If there was a room problem, Walton challenged, then it had to be demonstrated quantitatively in the field. He was also intrigued that bodies of shale sometimes occur in diapir structures, and yet no one ever advocated a process of shalification to avoid a room problem.

Walton was puzzled that transformists so readily accepted the idea of long-range diffusion under low potentials and yet rejected the idea of subsolidus, autometamorphic reorganization of granitic texture by diffusion on the scale of single crystals during the prolonged cooling of a granite that proved by its contact effects that it had been at high temperature.

Despite his litany of objections to many of the claims and arguments of transformists, particularly the dry diffusionists, Walton nevertheless maintained that there was sufficient ambiguity in the evidence in a number of occurrences to suggest that both granitization and granite magma had operated as large-scale processes in the emplacement of granite. Indeed, he thought that geologists were as yet unable to determine with certainty the relative contributions of the two processes for most granites of the world. Although noting a tendency on the part of granitizers to appeal to Backlund's concept of rheomorphism in instances where they were unable to escape the evidence of intrusive relations, Walton acknowledged that rheomorphism might play a major role. After all, he reasoned, why shouldn't a solidified granite body be mechanically highly mobile or even fluid in geological circumstances where accompanying high-grade paragneisses were undergoing strong plastic deformation? He thought that granitic phacoliths in high-grade metamorphic terranes might provide examples of rheomorphism. In the end, Walton concluded that attempts at resolution of the granite problem were premature. The ambiguity of much of the evidence convinced him that granites probably formed in various ways.

Perrin (1956) made no effort to respond to Walton's criticisms of dry diffusion. He simply attacked the magmatic hypothesis again, responding that he had been impressed as early as 1935 by the development of feldspar crystals in country rock that are identical to those in adjacent granites in terms of shape, size, chemical composition, color, and the presence of inclusions of biotite or hornblende. He asserted that there was no probability for such identical feldspar crystals to have developed a few centimeters apart by processes as different as crystallization of magma and crystallization in a solid medium of different texture and composition. He also denied that identical granite portions would have been generated by replacement in the solid state near its borders but by crystallization from magma in its interior. The presence of feldspathized inclusions in the very heart of granite bodies ruled that out.

Perrin, in fact, went so far as to state that he would admit that a particular granite body was magmatic only if convincing facts compelled him to recognize that there was no other possible explanation for its genesis, and he conveyed the impression that that would be nearly impossible to accomplish. Many traditional criteria for magmatic emplacement were unconvincing to him. He doubted that a lack of feldspathization accompanied by sharp contacts was opposed to the origin of the granite by replacement. Because he favored replacement mechanisms, he disagreed with the idea of Tuttle (1952) that perthitic texture in alkali feldspar could be the result of exsolution. Perrin even thought that replacement had occurred in granite bodies containing apparently rotated inclusions. He argued that granitization of a folded migmatite would leave patches of partially granitized material with different orientations. Moreover, he insisted, rotated inclusions are entirely compatible with the rheomorphism of granitized masses.

Vladi Marmo of the Finnish Geological Survey pointed out, as Grout had years before, the importance of precise petrographic definition of the rocks being discussed. Following Eskola (1932), Marmo (1956) believed that careful tectonic discrimination of granitoid bodies as syn-kinematic, late-kinematic, or post-kinematic was essential to a true solution of the granite problem. Syn- and late-kinematic granites, he said, represented two entirely different kinds of rocks structurally, texturally, and stratigraphically, and they should not be lumped indiscriminately under the category of granite. He urged the undertaking of comparative studies of various granitoid bodies involving the labors of experimentalists working in close collaboration with field geologists.

Reynolds (1958) complained again that the granitizers had been the victims of numerous misrepresentations. Eskola, she said, seemed to think that transformists were not interested in applying physico-chemical knowledge to the granite problem. She attributed that misunderstanding to the tendency of magmatists to suppose that Read was the spokesman for the entire transformist school. Read, after all, had repeatedly denigrated any but field evidence and had even professed that he never understood what many of the triangular phase diagrams meant. In fact, Reynolds pointed out, transformists could be as adept at using phase diagrams for their purposes as the magmatists. Barth (1955) and Perrin (1957), she noted, had both shown that Tuttle, in his preliminary experimental work on the granite system, had plotted the compositions only of granites containing 80 percent or more of normative Q + Ab + Or on the granite diagram. In so doing, Tuttle had omitted the most common granitoid rocks of orogenic zones, namely the granodiorites and related types. According to Barth, only relatively high-level granites fell within the area that Tuttle regarded as being appropriate to magmatic granites. All the other granites that Barth plotted showed a nearly perfect dispersal without any relation to the thermal trough.

Reynolds concluded from diagrams plotted by Barth, Eskola, and Perrin that the dispersal of points representing granodiorites and related rocks was incompatible with their having resulted from the crystallization of residual liquids arising from the crystallization–differentiation of basalts or by the crystallization of liquids obtained by partial melting. She concluded that Tuttle's experiments on the granite system actually confirmed her views rather than those of the magmatists.

Another misrepresentation, she said, was that granitizers were not supposed to think that granite was intrusive. Reynolds objected that she had been the first geologist in Britain to map the planar lineations in a granite mass as an indication of flow. But the intrusive movements of granitic materials at high structural levels she attributed to fluidization, the industrial process in which a bed of solid particles expanded by the flowage of gas through it. In the geological counterpart of fluidization, she noted, the temperature was sometimes high enough to cause melting.

Reynolds agreed that granitic melt must be formed at some level within the crust because of its eruption at the surface. The geologist's task, she said, is to discover the evidence of melting and locate the structural level at which it occurs. As Read had done so frequently, she reiterated that the granite problem was really a tectonic problem. She suspected, however, that the correlation between the location of granite diapirs and tectonic structures would be unintelligible if granites resulted from the crystallization of melt derived by crystallization–differentiation of basaltic magma or by differential melting at deep levels.

THE RESOLUTION

Despite the anti-magmatist claims of Barth (1955) and Reynolds (1958), the granite controversy cooled off dramatically after the publication of a Geological Society of America memoir that reported the results of experimental studies in the system $NaAlSi_3O_8$–$KAlSi_3O_8$–SiO_2–H_2O by Tuttle and Bowen (1958). Building on the pioneering experiments on granitic melts by Roy Goranson (1931, 1932, 1938) at the Geophysical Laboratory, Bowen and Orville Frank Tuttle (1916–1983) had been working together for nearly a decade on the granite system (Figure 31). The latest in a long succession of outstanding experimentalists to work with Bowen at the Geophysical Laboratory, Tuttle graduated in 1939 from Pennsylvania State College (Luth, 1987). In the fall of 1940, he began graduate work at MIT. His doctoral work was interrupted by national defense-related activities at both the Naval Research Laboratory and the Geophysical Laboratory. He undertook work on crystal synthesis and in 1946 had already managed to do some experiments at the Geophysical Laboratory in his spare time on silicate–water mixtures and had made some studies on the solubility of water in obsidian. He officially joined the staff of the Geophysical Laboratory at the beginning of 1947. In 1953, Tuttle became Professor of Geochemistry at Pennsylvania State University where he was active in introducing the methods of experimental petrology into the academic world. In 1965, he joined the faculty of Stanford University, where he also helped to developed a laboratory in experimental geochemistry. Among Tuttle's accomplishments was the design of a new type of hydrothermal apparatus that he and Bowen were able to use in their studies of granite and other systems.

Bowen had been moving inexorably toward experimental studies of granite throughout his career. He had already studied dry granitic liquids at atmo-

31. The authors of the "granite memoir" at the Geophysical Laboratory. Left: Norman L. Bowen (1887–1956). Right: Orville Frank Tuttle (1916–1983). From the files of the Geophysical Laboratory, Carnegie Institution of Washington.

spheric pressure with Schairer (Schairer and Bowen, 1935; Bowen, 1937). Further studies were delayed by Bowen's tenure in Chicago and the distractions of World War II. Because Bowen's theory of crystallization–differentiation and his early dry experiments pointed to granites as the ultimate residuum of fractional crystallization, he was eager upon his return to the Geophysical Laboratory in 1947 to extend the experiments in hopes of confirming his ideas and discounting the views of granitizers.

In their memoir, Tuttle and Bowen (1958) presented a diagram showing the positions of the boundary curve between quartz and alkali feldspar for water pressures between 500 and 4000 atmospheres, noting as well the positions of minimum liquidus temperatures on the boundary curves (Figure 32). If granites really are the result of crystal–liquid equilibria, they argued, their compositions ought to cluster around the experimentally determined minima. They also presented a diagram showing contours of the frequency of normative compositions of 571 analyzed granites and syenites and pointed out that the contoured area corresponds closely with the experimentally determined minima, especially at low water contents (Figure 33). They concluded that the clustering of the compositions of granites confirms the hypothesis of their origin by crystallization of the liquid. They also judged that the compositions of granites virtually exclude the diffusion hypothesis.

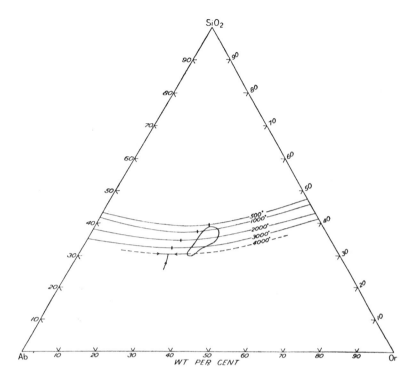

32. Effect of water-vapor pressure on the isobaric minimum in the simple granite system as determined by Tuttle and Bowen (1958). From *Origin of Granite in the Light of Experimental Studies in the System $NaAlSi_3O_8$–$KAlSi_3O_8$–SiO_2–H_2O* by O. F. Tuttle and N. L. Bowen. Reproduced with permission of the publisher, The Geological Society of America, Boulder, Colorado, USA.

Much of the work on the granite project had actually been done by Tuttle because of Bowen's advancing years and failing health, and, in fact, Bowen died in 1956 before publication of the memoir. While in full agreement with Bowen that their experiments corroborated the notion of granite formation from liquids, Tuttle took granite studies in a direction that Bowen had been reluctant to go throughout his career. He presented a persuasive case that large batholithic masses of granite had likely been generated by partial melting of older sedimentary and metamorphic rocks in a zone of melting deep within the crust. In Tuttle's judgment, the room and heat problems would thereby be solved. Production of large masses of granite by Bowen's crystallization–differentiation scheme from a basaltic parent was downplayed.

For the most part, geologists were persuaded by the elegant experiments of Tuttle and Bowen that the magmatists had the better part of the argument. From time to time, a voice in the wilderness would raise questions about

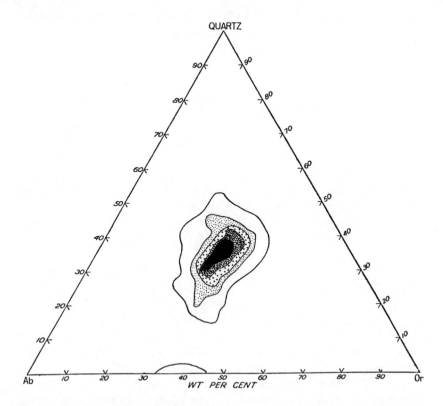

33. Contour diagram illustrating the distribution of normative albite, orthoclase, and quartz in 571 analyzed plutonic rocks from Washington's tables. Tuttle and Bowen (1958) suggested that closeness of the contour plot to the trend of minimum points in the granite system implied that granitic rocks were formed by liquid-crystal equilibria. From *Origin of Granite in the Light of Experimental Studies in the System NaAlSi$_3$O$_8$–KAlSi$_3$O$_8$–SiO$_2$–H$_2$O* by O. F. Tuttle and N. L. Bowen. Reproduced with permission of the publisher, The Geological Society of America, Boulder, Colorado, USA.

certain aspects of granite formation but the controversy had largely dissipated by the early 1960s. The passionate rhetoric of previous decades had spent its force. Bowen, the most passionate of the magmatists, died in 1956, and Read, the most passionate of the transformists, faded from the scene and died in 1970.

The granite controversy, arguably the most spirited and vitriolic controversy in the history of igneous petrology, was more than a simple difference of opinion about scientific evidence. At the root of the controversy were two fundamentally different ways of pursuing the science of geology. On the one hand were those geologists, with Read representing the most extreme proponent of that persuasion, who claimed that the solution to problems of igneous petrology, exemplified by the granite problem, was to be found in the rocks themselves. For them, problems of igneous petrology were geological problems. Read never tired of stating that the granite problem would be solved in the field on outcrops. Granites, he was fond of saying, are very big things that cannot be stuffed into little platinum crucibles. Because he may have wrongly assumed that experimentalists like Bowen never set foot outside the laboratory and into the field, he was very much afraid that some scientists were attempting to convert igneous petrology into a subdivision of physical chemistry. Consequently, Read not only downplayed the insights of phase-equilibrium studies obtained by experiment but even denigrated the work of the micro-petrographical school of Rosenbusch. Read also constantly insisted that the problem of granites was intimately connected with matters of tectonics and intimate associations with high-grade gneisses, schists, and migmatites. Quite rightly, he pointed out that granites differ in different tectonic settings, a fact commonly glossed over by those magmatists who focused on the chemical and mineralogical side of granites.

The other philosophical position, represented by Bowen, Niggli, Eskola, Daly, and others, was not that field evidence is unimportant or even secondary to physical chemistry. Rather they insisted that field evidence is often ambiguous. In such circumstances they felt it necessary to appeal to the insights provided by ancillary sciences on the grounds that experimental studies had, in some instances, successfully discriminated among processes that had some experimental support and hypothetical processes that had no experimental support. Why, then, should we not take advantage of experimental studies to interpret field evidence? Undoubtedly geologists occupied a philosophical continuum between these two poles, and without question, this diversity of philosophy of geology contributed to the misunderstandings and divergences of opinion. The extent to which one accepted the findings of the experimental petrologists, limited at that time to a small coterie of men at the Geophysical Laboratory, inevitably colored the way that one might interpret a given granite occurrence.

Along these lines it is worth observing that the same philosophical differences arose during every eruption of intense interest in the granite problem. During the neptunist–plutonist controversy, it was the experimental work of Hall that persuaded many people to fall in behind Hutton's claim that many granites are intrusive and igneous. The opposing Wernerians remained skeptical of the value of the experiments. Ironically, even Hutton, the prime plutonist, was doubtful of the value of Hall's experimental results. During the subsequent "wet way" versus pure igneous fusion controversy of the nineteenth century, a great deal of experimental data had been accumulated to demonstrate that quartz, a major ingredient of granite, readily crystallized from aqueous solutions but could not be produced by crystallization from silicate melt.

The weight of the experimental evidence lay behind the opponents of magmatic granite. Without questioning the experimental results, the magmatists nonetheless believed that field evidence was sufficiently convincing that the experimental data could not be regarded as decisive. In the twentieth century, however, the tables were turned with the transformists claiming that the field evidence was sufficiently convincing in favor of granitization that the experimental evidence favoring magmatism could not be regarded as decisive. In the end, however, the transformist position gradually gave way in the face of phase diagrams and thermodynamic arguments against large-scale diffusions of ions. The magmatists won the third round of the granite debate. Will there be a fourth round in the twenty-first century in which there will be a resurgence of an anti-magmatic opinion?

Another extremely important aspect of the granite problem concerned the fact that there are so many different kinds of granitoid rocks with varying mineralogical and chemical compositions, textures, forms, and contact relations in so many different geological–tectonic settings. As a result, geologists in different parts of the world were, in effect, like the blind men describing different parts of the proverbial elephant. Geologists working in Germany and North America had an abundance of relatively high-level, clearly intrusive, magmatic granites to examine. North American geologists were quite naturally influenced by their exposure to the great Cordilleran batholiths of the west. In contrast, Scandinavian geologists worked with granitic materials that likely formed by anatexis in the very deepest levels of the crust. French geologists, in turn, studied granitic plutons characterized by an abundance of metasomatic aureoles. It was to be expected that very different views of granite would tend to be associated with national schools of thought. Those geologists, like Read, who took pains to point out the very important fact that there are indeed granites and granites in different structural–tectonic settings all too often failed to gain adherents from among their opponents because they coupled their belief in multiple types of granites with undue insistence on the importance of long-distance diffusion processes for which there was little experimental or theoretical support. In contrast, magmatists were often too busy attacking the proposed mechanisms of granitization to pick up on the more fundamental observation that not all granites are the same and that each one needs to be evaluated on its own. Eskola and Niggli were perhaps the important exceptions in that respect.

One more time, igneous petrology advanced by drawing insights from diverse occurrences of igneous rocks. The resolution of the granite controversy on the side of the magmatists represented an ultimate vindication of the great value of experimental petrology. After the controversy died down in the 1950s and 1960s, igneous petrology routinely incorporated the results of experimental phase-equilibrium studies as much as it had previously been incorporating the results of microscopic petrography. Experimental petrology laboratories soon proliferated in places as far removed geographically from the Geophysical Laboratory as Japan, Australia, Germany, and England. Future generations of igneous petrologists would be drilled in the fine art of reading phase diagrams. No longer would any petrologist, experimentalist or not, claim ignorance, like H. H. Read, as to the meaning of those little triangles.

21 ✦ MODES AND NORMS

Classification in Crisis

The introduction of quantitative igneous rock classifications early in the twentieth century did little to reduce the chaos and confusion that prevailed in the systematics of igneous rocks. In fact, the "granite controversy" of the 1940s and 1950s was exacerbated precisely because of a lack of consensus on the meaning of the term "granite." Discussions about the origin of basalt, too, suffered for lack of consistent terminology. Despite the efforts of Albert Johannsen to devise an acceptable quantitative mineralogical classification around 1920, numerous other petrologists in the succeeding decades continued to muddy the classificatory waters by publishing their own alternative proposals.

The first volume of Johannsen's masterwork, *A Descriptive Petrography of the Igneous Rocks*, was published in 1931. The book included extensive glossaries of rock names and textural terms as well as thorough descriptions of a large number of mineralogically and chemically based igneous rock classifications that had been devised ever since the days of Werner. Johannsen (1931) devoted an entire chapter to an exposition of the classification scheme, now slightly altered, that he had originally developed more than a decade earlier. The publication of the volume, soon to be followed by three more volumes of exhaustive description of virtually all known igneous rocks, stimulated Paul Niggli to respond with his own classification. Ever since the publication of the CIPW "quantitative" classification in 1902, English-speaking authors like Hatch, Shand, Holmes, Cross, Harker, Iddings, and Johannsen had largely dominated the discussion about issues of classification for the first three decades of the twentieth century. The paper of Niggli (1931), however, signaled a renewal of interest on the part of continental European petrologists in matters of classification.

MINERALOGICAL CLASSIFICATION SINCE JOHANNSEN

Like almost everyone else, Niggli (1931) recognized that the CIPW classification was really a classification of magmas and, therefore, not suited to the needs of petrographers. He did, however, laud the norm calculation. He realized the importance of having both quantitative chemical classifications and quantitative mineralogical classifications, but he urged that they be kept strictly distinct from one another. Classifications that attempted to be simultaneously mineralogical and chemical were to be deplored. Niggli approved of Johannsen's mineralogical scheme but thought it a bit artificial in the placement of the boundary lines between different fields of rocks and rock families. Like so many other petrologists, Niggli was still seeking as natural a classification as possible. Although there were many minor differences between his own mineralogical classification and that of Johannsen, Niggli believed that three essential changes in Johannsen's scheme were absolutely necessary.

Johannsen had introduced the QAPF double triangle for the classification of the granitoid and feldspathoidal rocks. To the corner of the double triangle designated A, representing the alkali feldspars, Johannsen had assigned orthoclase, microcline, anorthoclase, sanidine, soda sanidine, and microperthite, including the lamellae of albite. Free, discrete albite grains, however, he excluded, considering them as plagioclase rather than alkali feldspar. Following Rosenbusch, however, Niggli believed that free albite should also be classed as an alkali feldspar. In fact, he was willing to include albite in the broad sense, that is, albite incorporating as much as ten percent anorthite molecule, as an alkali feldspar for classification purposes. Consequently, he restricted minerals to be assigned to the P corner, representing plagioclase, to the more calcic varieties oligoclase through anorthite. Such a move resulted in assignment of different names by Johannsen and Niggli to identical rocks that contained large amounts of both potassium-rich feldspar and albite.

The second point of difference between Johannsen and Niggli concerned the definition of rocks on the basis of the relative proportion of alkali feldspar, however defined, to plagioclase. Niggli agreed with both Rosenbusch and Johannsen that such a criterion of definition is perfectly appropriate for rocks in the QAP triangle on the grounds that quartz is "indifferent" as over against either feldspar. Johannsen had also employed the ratio of alkali feldspar to plagioclase as the criterion for the definition of the feldspathoid-bearing rocks in the FAP triangle. As a result, his two triangles appear somewhat symmetrical (Figure 19). Niggli, however, claimed that such a procedure contradicted knowledge of the paragenesis of feldspathoidal rocks. The feldspathoids, he said, are not analogous to quartz, but are, in effect, undersaturated alkali feldspars. Niggli urged, therefore, that rock definition be based on the relative proportions of the sum of alkali feldspars and feldspathoids to that of plagioclase. In Niggli's scheme this procedure resulted in a somewhat different orientation of boundary lines in the FAP triangle than in the QAP triangle (Figure 34).

The third point of dispute between Johannsen and Niggli concerned the values of ratios to be used for boundary lines between different rock types. Johannsen typically set his boundary lines at ratios of 5:95, 50:50, and 95:5.

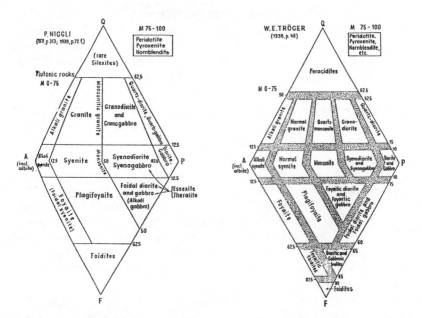

34. Igneous rock classifications of Niggli (1931) and Tröger (1938). Reproduced from Streckeisen (1967) by permission of E. Schweizerbart (http://www.schweizerbart.de).

For example, on the QAP triangle, the boundary line set at 5 percent plagioclase:95 percent alkali feldspar separates the fields of kalisyenite from syenite and kaligranite from granite. In contrast, Niggli preferred to use the octave system employed in the CIPW classification, where boundary lines were set at ratios of 1:7, 3:5, 5:5, 5:3, and 7:1. As a result, Niggli placed a boundary line in the QAP triangle at 12.5 percent plagioclase:87.5 percent alkali feldspar to separate the fields of alkali syenite from syenite and alkali granite from granite. The boundary lines separating granite from syenite also differed in the two systems. Johannsen placed the boundary line at 5 percent quartz whereas Niggli placed it at 12.5 percent quartz. Thus, an alkali feldspar–plagioclase-rich rock with only 5.5 percent quartz was considered a granite by Johannsen. Niggli also placed the upper boundary for granite differently. Johannsen considered only rocks with 50 percent or less of quartz to be granites, whereas Niggli placed the boundary at 62.5 percent quartz.

Throughout his lengthy career, French petrographer Alfred Lacroix, whose petrographic microscope is on display in the mineralogical gallery of the Muséum d'Histoire Naturelle in Paris, had shown his interest in matters of nomenclature by coining 70 new igneous rock names in 36 separate publications (LeMaitre, 1989). Disregarding the warning of Niggli against combined mineralogical–chemical classifications, Lacroix (1933) tried his hand at producing just such a classification. He even attempted to maintain some correspondence with the classes, orders, rangs, and subrangs of the CIPW system. He proposed two major divisions: "roches feldspathiques" and "roches essentiellement feldspathoidiques." The division of feldspathic rocks was split into three subdivisions: "roches quartziques," "roches à silice saturée," and "roches feldspathiques feldspathoidiques." These subdivisions roughly corresponded to the orders of the CIPW classification and were based on the ratio of quartz or feldspathoids to feldspars. The two divisions comprised three sets of families: those with alkali feldspar or with leucite, those with alkali feldspar and plagioclase or with leucite and nepheline, and those with dominant or sole plagioclase or with nepheline. The resulting pigeonholes were further categorized on the basis of the ratio of alkalis to lime (the CIPW rang) and the ratio of potash to soda (the CIPW subrang). By way of example, within Lacroix's Division I, "roches feldspathiques," was the first subdivision, "roches quartziques." Within that subdivision, family B was said to consist of "granites calco-alcalins." This family included "granites potassiques," "granites monzonitiques," "granites akéritiques," and "granodiorites." In Lacroix's system, the name assigned to a rock depended on the relative values of the alkali to lime and potash to soda ratios.

German petrographer Walter Ehrenreich Tröger (1901–1963) proposed yet another mineralogical classification, although he maintained that his scheme was simply the old Rosenbusch system with the geological mode of occurrence suppressed. Tröger (1935b, 1938) joined with Niggli in considering albite as an alkali feldspar (Figure 34). He differed with both Johannsen and Niggli in the placement of the boundary lines, however. For example, in the QAP triangle, Tröger (1935b) placed the boundary between granite and syenite at 10 percent quartz, but shortly thereafter, he adjusted the boundary to 12.5 percent quartz to bring it into agreement with Niggli (Tröger, 1938). His upper boundary for the granite field, however, was set at 50 percent quartz in agreement with Johannsen. Tröger departed drastically from both Johannsen and Niggli by rejecting a boundary between granite and granodiorite, or granogabbro, at 50 percent alkali feldspar and 50 percent plagioclase. Instead, he inserted a field of quartz monzonite between them, as well as a field of monzonite between syenite and syenodiorite, or syenogabbro. The boundaries of these monzonitic fields were placed at 37.5 percent plagioclase and at 62.5 percent plagioclase. As a result, Tröger's field of granite was considerably smaller than those proposed by Johannsen or Niggli. Tröger divided the FAP triangle into more fields than Johannsen or Niggli had done but followed Niggli's general approach. A. Rittmann (1952) essentially applied Tröger's classification to volcanic rocks.

Several additional schemes appeared during the 1950s. S. R. Nockolds (1954) of Cambridge, for example, returned the boundary between granite and syenite to 10 percent quartz (Figure 35). His most radical changes, however, were made in the boundary values assigned to alkali feldspar:plagioclase ratios.

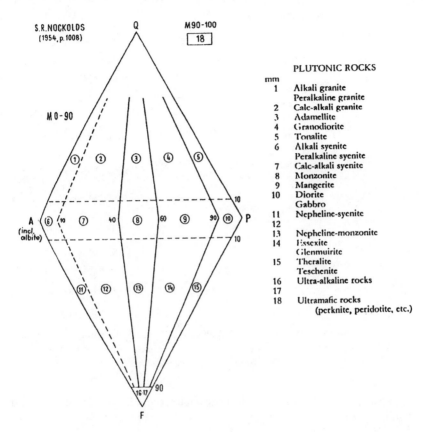

35. Igneous rock classification of Nockolds (1954). Reproduced from Streckeisen (1967) by permission of E. Schweizerbart (*http://www.schweizerbart.de*).

The boundary between alkali granite and granite was placed at 10 percent plagioclase. In place of quartz monzonite, Nockolds used the term "adamellite" and located the boundary between granite and adamellite at 40 percent plagioclase. A boundary between adamellite and granodiorite was located at 60 percent plagioclase. Lastly, he used the term "tonalite" instead of quartz diorite as used by Niggli and Tröger and placed the boundary between tonalite and granodiorite at 90 percent plagioclase. A virtually identical scheme was presented in French by Jung and Brusse (1959) except that some of the rock names differed from those of Nockolds. "Diorite quartzique" was used instead of tonalite, and "granite monzonitique" was used in place of adamellite.

Howel Williams of the University of California did away entirely with distinctions between alkali granite and granite and between alkali syenite and syenite (Figure 36). In Williams et al. (1954), he relocated boundaries between

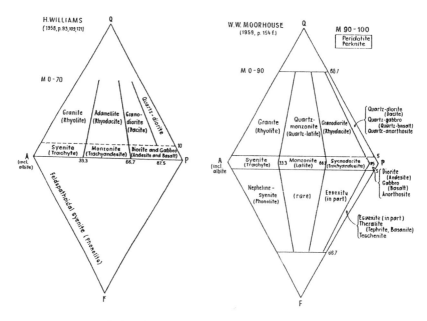

36. Igneous rock classifications of Williams (1958) and Moorhouse (1959). Reproduced from Streckeisen (1967) by permission of E. Schweizerbart (http://www.schweizerbart.de).

granite and adamellite to 33.3 percent plagioclase and between adamellite and granodiorite to 66.7 percent plagioclase. For the boundary between granodiorite and quartz diorite, he inconsistently used the octave system, placing the boundary at 87.5 percent plagioclase. A year later, Walter Wilson Moorhouse (1913–1969), Professor of Geology at the University of Toronto, retained Williams' boundary positions between granite and adamellite, which he called quartz monzonite, and between quartz monzonite and granodiorite (Figure 36). However, Moorhouse (1959) shifted the boundary between granodiorite and quartz diorite from the 87.5 percent plagioclase of Williams to a Johannsenesque 95 percent plagioclase.

To drive this glorious state of confusion farther down the path toward total chaos, several Russian classifications introduced large fields for alaskite, placing them generally where previous writers had placed alkali granite (Kupletsky, 1953; Ginsburg et al., 1962). Some Russian schemes also plotted a field of quartz syenite between syenite and granite. Although they agreed on placing the upper limit of the quartz content of a quartz syenite at 20 percent, they disputed the lower boundary. One of the most striking features of these

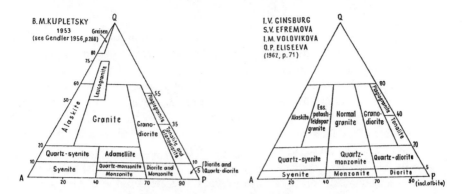

37. Igneous rock classifications of Kupletsky (1953) and Ginsburg et al. (1962). Reproduced by permission of E. Schweizerbart (http://www.schweizerbart.de).

classifications was the location of the fields of adamellite, or quartz monzonite below that of granite rather than alongside it. Kupletsky (1953), for example, plotted a small field of quartz monzonite directly above a small field of monzonite and directly below a field of adamellite, which in turn lay directly below granite (Figure 37). Previously, petrographers had regarded adamellite and quartz monzonite as roughly synonymous. They had also placed them next to granite as rocks with lower ratios of alkali feldspar to plagioclase than granite. Kupletsky saw the difference solely in terms of the abundance of quartz. Gendler (1956) also placed quartz monzonite below adamellite, but essentially shrank the granite field to make room for adamellite, which, like most previous authors, he placed to the plagioclase side of granite. Ginsburg et al. (1962) kept quartz monzonite where Gendler placed it but did away entirely with adamellite (Figure 37). He restricted the field of "normal granite" to rocks with more than 20 percent and less than 60 percent quartz and with feldspar ratios of 40 percent plagioclase on one side and 60 percent plagioclase on the other. The field of "normal granite" of Ginsburg et al. (1962) was so constricted and shifted that it did not even overlap at all with the field of granite proposed by Tröger (1938)! By the same token, the situation regarding the definition of quartz monzonite and adamellite had become nearly hopeless. The series of diagrams presented by Streckeisen (1967, pp. 207–214) helps to convey some sense of the desperate state of petrographic classification and nomenclature in the early 1960s. In the meantime, still other classifications were advanced by the Colorado School of Mines and the American Geological Institute. In matters of characterization of the igneous rocks, the situation well into the 1960s was one in which every petrographer did that which was right in his or her own eyes.

ADVANCES IN QUANTITATIVE MEASUREMENT

While petrographers were struggling to develop appropriate quantitative schemes of classification based on modal mineralogy, modest progress was made in devising improvements in instruments and methods for obtaining reliable quantitative information on mineral abundances in igneous rocks. Since Rosiwal's contribution at the close of the nineteenth century, petrographers had been measuring the total lengths of grains of various mineral species along sets of parallel traverses across a thin section. They assumed that the linear proportions of the minerals along linear traverses would be proportional to the volumetric proportions of constituents. Specialized mechanical stages such as those of Shand and Wentworth improved speed and ease of operation.

A dramatic change in method was introduced in 1933 by Russian petrographer A. A. Glagolev. In the nineteenth century, petrographers measured areas of mineral grains. In the early twentieth century, they measured lengths of mineral grains. Glagolev, however, invented a point-counting method in which the regularly spaced traverses across a thin section were subdivided into a series of equally spaced points. With his procedure, the microscopist identified the mineral lying beneath the intersection of the cross hairs and then pressed a specific key on a tabulator. Depression of a key on the tabulating instrument would then trip a mechanism that translated the counting stage a fixed distance along the line of traverse. The petrographer repeated the procedure until a specified distance had been traversed. The stage was then reset and a new traverse run, and so on, until a sufficient number of points had been traversed and counted. The percentages of points identified as belonging to particular minerals were then regarded as equivalent to the volumetric percentages of the minerals.

The point-counting method did not catch on in other parts of the petrographic world until the late 1940s, thanks in large part to the efforts of Felix Chayes (1916–1993). After a stint with the U. S. Bureau of Mines, Chayes joined the Geophysical Laboratory in 1947 and spent the remainder of his career there as a Systematic Petrographer. Perhaps more than anyone else, Chayes was responsible for calling the attention of petrologists to the importance of statistical treatment of chemical and petrographic data. He compiled a database of more than 16,000 analyses of Cenozoic volcanic rocks and then spearheaded an international effort to develop and maintain a world database for igneous petrology. He also constantly compared different methods for obtaining quantitative data on thin sections and subjected them to statistical tests for precision. For example, Chayes (1946) undertook an investigation of the mineralogical uniformity of thin sections of granite from Woodstock, Maryland, in an effort to ascertain the degree of precision attainable with the Wentworth stage and a stage that had been more recently developed by Cornelius

Hurlbut at Harvard. Both instruments were still used in conjunction with the measurement of the length of traverses across individual mineral grains.

Eventually, Chayes learned of Glagolev's point counting technique. Chayes (1949) reported that he had made nearly 300 thin-section analyses with a mechanical point-counting stage that could be advanced for a fixed distance by hand after each identification had been made. For tabulators he used commercially manufactured blood-cell counters. To evaluate precision and speed, Chayes analyzed 47 thin sections of three different New England granites by measuring each section in two different orientations both by both point counting and by linear analysis. He found that the point-counting method yielded better precision than did linear analysis performed by either the Hurlbut or Wentworth instruments. The point-counter apparatus, he said, was sturdy, inexpensive, and easily assembled. Chayes reported that an hour of work using the Wentworth instrument would normally take half an hour with the Hurlbut instrument but only fifteen minutes with a point counter. He said that he was able with a little practice to make between 75 and 100 counts per minute with a point counter, in spite of delays for identification, changes of focus, and movements of the traverse knobs. A complete thin-section analysis, he enthused, required no more than 15 to 20 minutes. In terms of time and money, the point-counting method brought quantitative modal analysis within the reach of every petrographic laboratory. By the 1950s, then, Rosiwal-style analysis was largely a thing of the past and petrographers increasingly adopted the easy, rapid, and precise procedure of point counting for their thin-section work.

Some years later, Chayes (1956) published a book entitled *Petrographical Modal Analysis: An Elementary Statistical Appraisal.* He noted that some petrographers had been grumbling about his point-counting technique because they had to use both hands during the procedure, one to turn knobs on the mechanical stage and the other to press keys on the tabulator. Despite such minor grousing, Chayes noted that the popularity of point counting was rapidly increasing, and he surmised that, at least in the United States, as many modes were now being determined by point counting as by line integration. Noting that his statistical approach to petrography had also occasioned some criticism, Chayes sought to allay the fears of petrographers that a statistical examination of their results amounted to an indictment of their competence by asserting that the statistician's working assumption was precisely that petrogaphers did not make mistakes. What Chayes' statistical evaluations did do was to indicate that petrographers would generally need to count between 1000 and 2000 points per thin section to obtain accurate results. He again showed that point counting yielded good statistical reproducibility. In support of this contention he noted that in one test he requested five geologists from MIT to do point counts of the same five thin sections. The results, he discovered, were not much different from results obtained when the same petrographer had done

CHEMICAL CLASSIFICATION

all the point counts. Chayes also provided a mathematical demonstration that the ratio of the area occupied by a specific mineral to the area occupied by all the minerals within an area of a thin section being measured provides a consistent estimate of the volume percentage of that specific mineral within the rock.

CHEMICAL CLASSIFICATION

In the wake of the collapse of enthusiastic support for the CIPW system, petrologists still saw the need for various sorts of chemical classifications. Before he published his mineralogical system in response to Johannsen, Niggli (1920) had invented a chemical system, much simpler than that of CIPW, that attracted considerable attention. Niggli proposed the conversion of the chemical analysis of an igneous rock into molecular numbers by dividing the weight percents of the oxides by the respective molecular weights just as was done in calculating a CIPW norm. However, he suggested combining the numbers in a different way, such combinations to be represented by symbols such as *al*, *fm*, *c*, *alk*, *si*, and *mg*. In Niggli's scheme, the symbol *al* represents the sum of the molecular numbers of Al_2O_3, Cr_2O_3, and oxides of the rare earth elements. The symbol *fm* represents the sum of twice Fe_2O_3, FeO, MnO, MgO, NiO, and CuO. The symbol *c* represents the sum of CaO, BaO, and SrO, and the symbol *alk* represents the sum of K_2O, Na_2O, and Li_2O. Niggli then recalculated these four quantities to 100 percent, thus yielding percentages of *al*, *fm*, *c*, and *alk* for the rock.

Additional quantities could also be computed. The molecular number for SiO_2 was reduced proportionally to the sum of the values of the other four quantities to obtain the amount represented by *si*. Similar values for *ti*, *zr*, *p*, and so on were obtained by proportional adjustment of the molecular numbers of TiO_2, ZrO_2, P_2O_5, and other constituents. Two other important numbers in the Niggli system were *k* for the ratio of potash to the sum of alkalis and *mg*, the so-called magnesium number, for the ratio of MgO to the sum of the various constituents that make up *fm*.

Having obtained values for *al*, *alk*, *c*, and *fm* for various rock types, Niggli plotted the data on a series of triangular planes passing through a tetrahedron whose apices are represented by the four Niggli values (Figure 38). He constructed a set of ten planes, all of which contain the *al-alk* edge of the tetrahedron. The planes are equally spaced such that they intersect the *c-fm* edge of the tetrahedron in even tenths. Section I, for example, intersects the *c-fm* edge at *c*5-*fm*95 and section X intersects the edge at *c*95-*fm*5. Individual analyses were plotted on equilateral triangles instead of the actual isosceles triangular sections. The sections were then presented in a series of pairs. Hence, sections I and X, II and IX, III and VIII, IV and VII, and V and VI were presented as attached pairs. Distinct fields for igneous rocks, chemical sediments, limestones, and aluminous sediments appeared on the paired diagrams, and

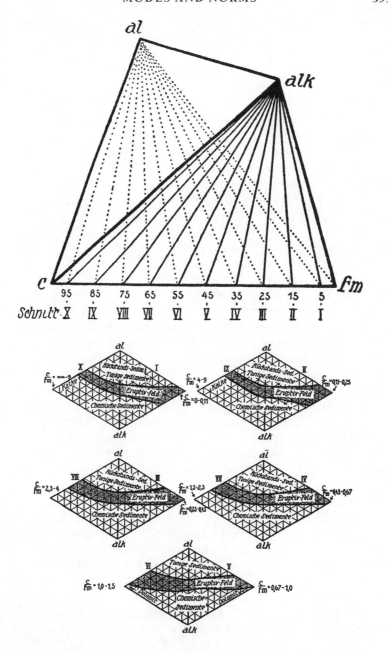

38. Classification scheme using Niggli molecular numbers. Reproduced from Johannsen (1931), *A Descriptive Petrography of the Igneous Rocks*, by permission of The University of Chicago Press.

different igneous rock types also occupied different regions of the diagrams. Niggli further devised a diagram that plotted mg vs. k to display chemical variation among a wide array of igneous rocks. Before long, petrologists were routinely calculating "Niggli numbers" as well as CIPW norms from their chemical analyses.

Niggli (1923) also suggested minor modifications to the CIPW norm, proposing, for example, that normative TiO_2 be substituted for normative ilmenite or titanite and that normative Al_2SiO_5 replace normative corundum. He was also bothered that the norm calculation did not include some of the most important dark minerals, such as biotite and hornblende.

T.F.W. Barth, then at the Geophysical Laboratory, also proposed some modifications to the CIPW norm. Barth (1931) took issue with Niggli's modification of the norm, calling it a "distressing revision." Substitution of Niggli's trifling revisions, he thought, would render direct comparisons of different rock analyses much more difficult. He urged petrologists to stick to the use of the old norm for the sake of making comparisons inasmuch as Washington (1917) had compiled igneous rock analyses up to 1914 and had calculated 8602 different norms! Barth, however, did like Niggli's proposed molecular numbers but thought that the new system should operate side by side with the CIPW norm rather than replace it. He did think that the norm could be modified to reflect the Fe/Mg ratio of rocks and minerals more closely. He proposed, therefore, that normative olivine be discarded in favor of normative fayalite and forsterite in undersaturated rocks. Likewise, on the grounds that diopside is not common in igneous rocks, he suggested that normative pyroxene should be represented, not by the normative diopside and hypersthene of the original CIPW norm, but by the three pyroxene end members, normative wollastonite, enstatite, and hypersthene, where the latter referred to $FeSiO_3$. The pyroxene calculation, Barth suggested, should be done in a manner analogous to that of the feldspars, which were listed as normative orthoclase, albite, and anorthite in the CIPW calculation.

In time, Niggli (1936) proposed a major modification to the CIPW norm calculation. The original method produced a list of normative minerals stated in weight percents. The CIPW norm is, therefore, a weight norm. Niggli proposed instead that the normative minerals resulting from calculation be expressed as mole percents. As a result, his procedure commonly came to be referred to as the Niggli molecular norm. Niggli suggested that instead of calculating molecular proportions of oxides by dividing weight percents of oxides in the chemical analysis by the molecular weights of the oxides, the weight percents in the analysis should be divided by equivalent molecular weights, that is, molecular weights of oxides that are all written such that they all contain just one cation. Thus, SiO_2 is still written as SiO_2 and its weight percent is divided by the molecular weight of SiO_2 to obtain a cation proportion number. In contrast, Na_2O or Al_2O_3 or K_2O would be treated as $NaO_{1/2}$ or $AlO_{3/2}$ or

$KO_{1/2}$ respectively, and their weight percents would be divided by half their molecular weights because the formulas are halved. Thus all oxides contain only one cation. The resulting divisions yield cation proportions that are then recalculated to 100 percent to yield cation percents. The cation percents are next combined in a manner similar to the standard CIPW calculation to yield normative minerals. In calculating normative albite, for example, the molecular norm combines the cation percents of Na, Al, and Si in the ratio 1:1:3 because the formula of albite is $NaAlSi_3O_8$. By comparison, the CIPW norm combines Na, Al, and Si in the ratio 1:1:6 because the formula of albite may be written as $Na_2O.Al_2O_3.6SiO_2$. The sum of the cation percents of Na, Al, and Si yields the mole percent of normative albite in the rock.

The molecular norm calculation of Niggli proved to be somewhat easier to carry out than the CIPW calculation. Its advocates also claimed that the procedure yields values somewhat closer to the actual modal content of igneous rocks than the CIPW weight percent norm. The method also has the advantage that nonstandard normative minerals can more easily be used. Molecular norms are converted more easily into CIPW norms than the other way around. Because of these advantages, Niggli molecular norms became very widely used throughout continental Europe. In recent years, English-speaking petrologists have also begun to take a liking to this scheme.

In the Soviet Union, petrologists employed for many years an elaborate chemical system devised by A. N. Zavaritsky around 1941. Details were translated for English-speaking readers in Zavaritsky (1964). In a manner somewhat analogous to Niggli, Zavaritsky proposed that four "basic numerical characteristics," designated by the letters S, A, C, and B, might be calculated from the chemical analysis of an igneous rock. After weight percents of oxides in the analysis were converted into molecular numbers, the four characteristics were created by appropriate manipulation. The characteristic S was equivalent to the molecular number of the sum of SiO_2 plus TiO_2. The characteristic A was equivalent to twice the sum of K_2O plus Na_2O unless the total alkalis exceeded Al_2O_3. In that case, A represented twice the molecular number of Al_2O_3. In effect, A was an indicator of the relative number of atoms of alkalis appearing in alkali aluminum silicates such as alkali feldspars. The characteristic C was equivalent to the molecular number of Al_2O_3 in excess of the alkalis. It might also be equal to the molecular number of CaO if the amount of Al_2O_3 exceeded the sum of alkalis and lime. But C might also be twice the amount of alkalis in excess of alumina. In essence, then, C was an indicator of the relative number of atoms of calcium appearing in the lime aluminum silicates such as plagioclase. Lastly, the characteristic B referred to various combinations of iron oxides, MnO, and MgO. Either excess CaO or excess alumina might be part of B in certain instances. B was said to be an indicator of the remaining metallic atoms that did not enter into aluminum silicates. The four quantities representing the basic numerical characteristics were then summed and recalculated to

100 percent, much as Niggli had done with *al*, *fm*, *c*, and *alk*. Zavaritsky designated these four percentages as *s*, *a*, *b*, and *c*. He believed that these four basic quantities could be used effectively to discriminate among igneous rocks of normal chemical composition, rocks that are supersaturated with alumina, and rocks supersaturated with alkalis.

In addition, Zavaritsky proposed the use of eight different "supplemental numerical characteristics." These values functioned in a manner somewhat analogous to Niggli's *mg* number. These supplemental numbers typically provided information about ratios involving K and Na or ratios involving Fe and Mg. For example, f' was the ratio of ($2Fe_2O_3$ + FeO + MnO) to *B* multiplied by 100, and m' was the ratio of MgO to *B* multiplied by 100.

For graphically plotting the numerical characteristics obtained from a chemical analysis, Zavaritsky constructed a tetrahedron with *A*, *B*, *C*, and *S* at the four corners. He distorted the base *ASB* into a right triangle with *S* at the right angle and then collapsed the *CSB* side down into the plane of the base to form another right triangle (Figure 39). A point within the tetrahedron representing the chemical composition of the rock was then projected onto the *CSB* and *ASB* triangles to produce two points.

Still another scheme of chemical classification that was widely used in the first half of the century, most notably by Reynolds for her demonstration of basification and granitization, was one devised by Ludwig Ferdinand von Wolff (1874–1952), a former student at the Universities of Leipzig and Berlin under Zirkel and von Richthofen. Von Wolff received his Ph.D. degree from Berlin in 1899. He served as a Privat Dozent at Berlin, as Professor of Mineralogy and Geology at the Technical College of Danzig, and, after 1914, as Professor of Mineralogy and Petrography at the University of Halle. Author of numerous petrological texts, von Wolff (1922) devised a clever method for plotting the chemical analyses of rocks.

In von Wolff's system, the chemical constituents of a rock are lumped into three groupings that are plotted on a triangular diagram. The apex of the triangle represents quartz. A horizontal line drawn halfway up the triangle represents a saturation line such that all rocks plotting above the line contain an excess of silica and those plotting below the line have a silica deficiency. The two ends of the line represent ideal feldspar and augite. An excess of silica above that required for feldspar and augite is indicated by a plus sign and would appear in a rock as quartz. A deficiency of silica is indicated by a minus sign and would appear in a rock as another silicate mineral requiring less silica than feldspar and/or pyroxene.

Von Wolff further subdivided his triangle into a series of smaller triangles or fields by connecting various appropriately located points that represent the compositions of minerals or eutectic points of importance in the genesis of igneous rocks such as olivine, nepheline, leucite, spinel, the albite–diopside

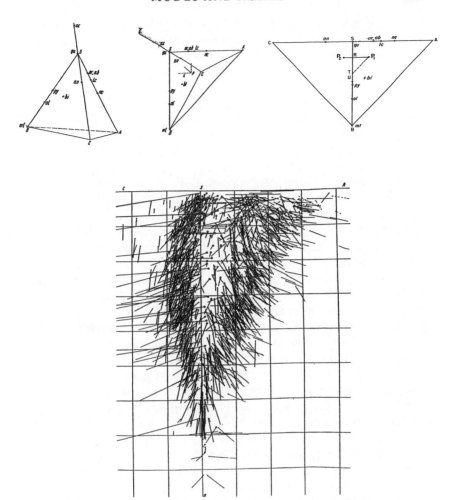

39. Classification scheme of Zavaritsky. Top: A, B, C, S tetrahedron and distortions. See text for description. Bottom: representative plots of chemical data on collapsed ASCB diagram. Reproduced by permission of American Geological Institute.

eutectic, the forsterite–diopside eutectic, and so on. By connecting these various points, von Wolff delineated certain fields in which rocks of definite types typically plot. For example, he discriminated trachytes and andesites from rhyolites, dacites, and pantellerites because they plot in distinct fields. Likewise, nephelinites and nepheline basanites plot in a different field from that of normal basalts. One might, therefore, classify a rock with the von Wolff diagram on the basis of the field in which its recalculated chemical analysis plots. Von

Wolff also believed that the genetic relationships between rocks could be clearly brought out by their positions in the triangle. Even courses of differentiation, he maintained, could be tracked within the diagram.

Some of the more important proposals regarding chemical classification and characterization pertained to variation diagrams. Arguably the most lasting of those contributions was put forward by Martin Alfred Peacock (1898–1950), a petrologist who spent his career at the University of British Columbia, Harvard University, and the University of Toronto. Noting that a division of igneous rock series into alkalic and sub-alkalic groups had long been recognized, Peacock said that some rock series lie in between the two groups, so that they could not be properly described either as alkalic or sub-alkalic.

At British Columbia at the time, Peacock (1931) argued that the character of a rock series is recognizable from ordinary variation diagrams. Because total alkalis invariably increase and lime invariably decreases with an increase in silica content on a variation diagram, the curve for total alkalis inevitably intersects the lime curve at some value of silica content. Peacock suggested that the position of that point of intersection of these two curves is an important index of the character of a rock series. He plotted the variation diagrams of thirteen different rock series and showed the changes in the position of the crossover point from one series to the next (Figure 40). Peacock then divided a variation diagram into four regions in which the crossover point, or "alkali–lime index," might fall. In Peacock's scheme, if the alkali–lime crossover point fell at a value of SiO_2 less than 51 percent, the series was designated as alkalic. If the crossover point plotted between 51 and 56 percent SiO_2, Peacock designated the series as alkali–calcic. If the crossover point plotted between 56 and 61 percent SiO_2, he designated the series as calc–alkalic, and if the crossover plotted at a SiO_2 percent greater than 61 percent, the series was designated as calcic.

To avoid the charge that the boundaries between the four types of series were arbitrarily chosen, Peacock correlated the division boundaries with significant mineralogical characters of the rock series. He suggested that the alkalic series might be distinguished from the alkali–calcic series by the presence of feldspathoid and the absence of quartz in the former and the absence of feldspathoid and presence of quartz in the latter. He distinguished the alkali–calcic series from the calc–alkalic series by the absence of clinopyroxene in the former and its presence in the latter; by the presence of soda pyroxene in the former and the scant presence or absence in the latter; by the absence of hornblende in the intrusive rocks of the former and its presence in the latter; and by the presence of soda amphibole in the intrusive rocks of the former and its absence in the latter. Peacock regarded the mineralogical distinctions between the calc–alkalic and calcic series as more tenuous. He suspected that the difference might rest more on the relative proportions of the characteristic minerals rather than on their kind, although he maintained that a trace of soda pyroxene might be present in the calc–alkalic series but completely absent in the calcic series

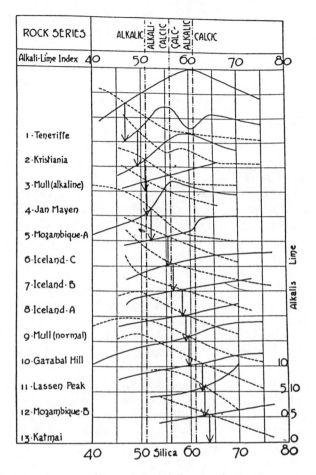

40. Variation diagrams of Peacock (1933) showing distinction among calcic, calc-alkalic, alkali-calcic, and alkalic rock series. Reproduced from *Journal of Geology* by permission of The University of Chicago Press.

and that soda–potash feldspar would normally be present in the former but absent in the latter. Peacock firmly rejected Harker's old proposal regarding alkalic Atlantic and sub-alkalic Pacific provinces, maintaining instead that there is a gradual progression from normal to extreme alkalic types.

From time to time, various petrologists also advocated variations on the standard Harker-type variation diagram to emphasize the degree of fractionation or differentiation in a suite of rocks more effectively. For example, Esper Larsen, Jr. (1879–1961), of Harvard University, believed that SiO_2 content is not the best variable for the abscissa of a variation diagram because, in many otherwise strongly differentiated alkalic, silica-undersaturated suites, the silica content does not vary all that much. Instead, Larsen (1938) suggested that the major oxides of a chemical analysis first be adjusted to sum to 100 and then plotted against a position variable that could have either a positive or negative value. For the position variable, he proposed the quantity ($\frac{1}{3}$ SiO$_2$ + K$_2$O − FeO − MgO − CaO). The variable FeO was obtained from the sum of MnO and total Fe calculated as FeO. BaO and SrO were also added to CaO. Larsen presented plots for several volcanic and plutonic suites that included the San Juan volcanics, the Southern California batholith, the Highwood Mountains of Montana, the Leucite Hills, and the Katmai volcanics.

In a comprehensive study of the geochemistry of igneous rock series, Nockolds and Allen (1953) at Cambridge used a modified version of Larsen's diagram. In place of oxides, Nockolds and Allen used weight percents of chemical elements and plotted them against the index equivalent to ± ($\frac{1}{3}$ Si + K − Ca − Mg). They omitted iron from the differentiation variable to show any iron enrichment in a rock series more satisfactorily. Both the Larsen and Nockolds procedures, however, generally fell from favor among petrologists.

Again, although Harker's variation diagram provided a useful tool for evaluation of calc–alkalic series and the other types of series in Peacock's scheme but not for tholeiitic series, diagrams were developed for the examination of tholeiitic rocks that plot the variation of oxides against the ratio of total iron calculated as FeO divided by the sum of MgO plus FeO. Another similar diagram uses instead a magnesium number, namely, the ratio of the cation percent of Mg to the sum of the cation percents of Mg and Fe^{2+}.

In the 1950s and 1960s several more indices were proposed for determining the extent of differentiation or evolution of an igneous rock. Several Japanese petrologists, led by Hisashi Kuno of the University of Tokyo, proposed a "solidification index." Kuno et al. (1957) defined this index as 100 times the ratio of the weight percent of MgO divided by the sum of the weight percents of MgO, total iron calculated as FeO, Na$_2$O, and K$_2$O. The solidification index, therefore, decreases as the extent of differentiation increases.

At Penn State, Charles Thornton and O. F. Tuttle proposed that weight percents of oxides be plotted against the "differentiation index" as the variable along the abscissa of a variation diagram. Thornton and Tuttle (1960) defined the "differentiation index" as the sum of the weight percentages of CIPW normative quartz, orthoclase, albite, nepheline, leucite, and kalsilite. Because no more than three of these normative minerals can appear in the norm of any specific rock, the differentiation index amounts to the sum of the percentages of just three of the above minerals. They determined the differentiation indices from the average values of several igneous rocks. For example, the differentiation index of average granite is 80, diorite 48, gabbro 30, and peridotite 6, whereas that of rhyolite is 88, andesite 56, basalt 35, and picrite 12. Thornton and Tuttle also presented contour diagrams of the values of various oxides versus differentiation index for 5000 analyses from the tables of Washington (1917).

Arie Poldervaart and Alfred Parker at Columbia University proposed the "crystallization index" as a measure of igneous differentiation in variation diagrams. Poldervaart and Parker (1964) maintained that the Harker diagram and the differentiation index of Thornton and Tuttle both failed to include the effects of the iron-enrichment and alkali-enrichment trends. In particular, they faulted the differentiation index for failing to bring out iron enrichment in some basaltic and extremely alkalic series. They also regarded the diagrams of Larsen and of Nockolds and Allen and the solidification index of Kuno and his colleagues as largely arbitrary. Thus, Poldervaart and Parker defined the crystallization index as the sum of normative anorthite, normative magnesian diopside, normative forsterite plus normative enstatite converted to forsterite, and normative magnesian spinel calculated from normative corundum in ultramafic rocks. This index, they said, measured the progression of magmas or igneous rocks away from the system anorthite–diopside–forsterite. In general, like the solidification index, the crystallization index decreases with increasing degree of differentiation. For example, mafic rocks consisting exclusively of anorthite, of magnesian diopside, of forsterite, or of mixtures of these minerals would have a crystallization index of 100. In contrast, acidic rocks consisting exclusively of quartz, of alkali feldspars, of feldspathoids, or of mixtures of such minerals would have a crystallization index of zero.

While petrologists well into the 1960s continued to search for the most effective means of classifying igneous rocks and characterizing the extent of differentiation in igneous rock suites, a move was underway on the part of several European petrographers to devise a universally acceptable igneous rock classification. The story of that remarkable movement will be reviewed in Chapter 27.

The Geochemical Era

22 ⇿ STRUCTURES AND SPECTRA

The Geochemical Revolution in Petrology

Around the middle of the twentieth century, igneous petrology entered a new phase in which geochemical considerations became paramount. The development of several accurate and precise instrumental analytical methods opened the way for the collection of unprecedented amounts of chemical data on igneous rocks. Trace-element abundances provided a new means for characterizing various igneous rock provinces, for charting the course of differentiation, or for modeling processes of igneous rock origins and history. Likewise, the application of studies of stable and radiogenic isotope ratios to the solution of problems of magma genesis became commonplace as the second half of the century progressed. Chemistry had, of course, played an important role in igneous petrology for much of the previous hundred years. Before looking at details of some of the dominant themes of the geochemical era, therefore, we review some of the significant mileposts in chemistry as applied to igneous petrology in the years leading up to mid-century.

EARLY QUANTITATIVE CHEMICAL ANALYSIS OF ROCKS

A host of chemists, mineralogists, and geologists in the early nineteenth century, such as Berzelius, Élie de Beaumont, Fuchs, Mitscherlich, and Bischof, were fascinated by the chemical compositions of geological materials. Chemical analyses of igneous rocks had been determined by innumerable individuals, and German, French, and Norwegian scientists led the way in rock analysis. Abich, Bischof, Bunsen, Rammelsberg, Sartorius von Waltershausen, Streng, and von Richthofen in Germany, Delesse and Dufrénoy in France, and Kjerulf in Norway were among the most prolific analysts. A great service was done for the geological community when Justus Roth (1861) compiled available rock analyses into his comprehensive *Die Gesteins-Analysen in tabellarischen Übersicht und mit kritische Erläuterungen*. So valuable were Roth's tables of data and interpretive discussion that subsequent compilations were published in 1869, 1873, 1879, and 1884.

Because the importance of chemical analyses of rocks was increasingly recognized throughout the nineteenth century, petrologists sought to obtain as many analyses as possible despite the fact that they were very expensive and extremely time-consuming. A single rock analysis required a lengthy procedure of crushing and digestion of the sample in hazardous chemicals. Exceptional patience and skill were required in preparing and handling solutions and in obtaining and weighing precipitates. Chemical elements present in minor concentrations were determined with very great difficulty or not at all. As reported by Hillebrand (1910, p. 30):

> The question has often been put, "How long does it take to complete an analysis of this kind?" This will depend, of course, on the mineral complexity of the sample and on the personal factor of the individual worker. If there is a competent assistant to do the grinding, and specific gravity determinations are not required, it is quite possible after long experience for a quick worker to learn to so economize every moment of time in a working day of seven hours, with an abundance of platinum utensils and continuous use of air and water or steam baths through the night, as to finish every three days, after the completion of the first analysis, barring accidents and delays, one of a series of rocks of generally similar character, each containing from eighteen to twenty quantitatively determinable constituents excluding, for instance, fluorine, carbon as such, nitrogen, metals of the hydrogen sulphide group, and cobalt. But such an output of work implies an unusual freedom from those occasional setbacks to which every chemist is exposed.

The importance of chemical analysis was finally recognized in the United States when the United States Geological Survey established an analytical laboratory in Denver, Colorado, in 1880, the year after its founding, under the direction of William Francis Hillebrand (1853–1925). Hillebrand had studied in Heidelberg with Bunsen and Kirchhoff in 1872 and at the Bergakademie in Freiberg in 1877 and 1878 (Spencer, 1927). The following year, another Geological Survey laboratory was opened in San Francisco under the direction of W. H. Melville, and in 1883 the Washington office started another analytical laboratory under the direction of F. W. Clarke. In 1885, Hillebrand moved to Washington because the Denver and San Francisco laboratories were closed down and all chemical analytical work done for the Geological Survey was centralized in Washington. Hillebrand published several U. S. Geological Survey bulletins on the principles and methods of rock analysis.

The head of the Washington analytical laboratory, Frank Wigglesworth Clarke (1847–1931), graduated from Harvard University in 1867, then served as Instructor in Chemistry at Cornell and Harvard (Spencer, 1933). He later became Professor of Chemistry and Physics at Howard University, and from 1874 to 1883 he taught at the University of Cincinnati. Clarke served as the Chief Chemist of the United States Geological Survey from 1883 until 1925. He was also an honorary curator of minerals in the U. S. National Museum in

Washington. During his tenure in Washington, he authored several publications on rock analysis. Clarke (1915) observed that nearly 8000 analyses had been performed in the Washington laboratory as well as hundreds more at the other then defunct laboratories. He reported the results of 2789 analyses with 1407 of them being igneous rocks. Many of the analyses included several minor elements reported as ZrO_2, Cr_2O_3, V_2O_3, FeS_2, NiO, MnO, SrO, BaO, CO_2, and TiO_2 in addition to the typical major elements reported as SiO_2, Al_2O_3, Fe_2O_3, FeO, MgO, CaO, Na_2O, and K_2O. The igneous rock analyses were all given names according to the CIPW system.

During the first three decades of the century, two major series of geochemical publications were issued by the U. S. Geological Survey. The first was a series of bulletins published from 1908 through 1924 as five successive editions of *Data of Geochemistry* (Clarke, 1908, 1911, 1916, 1920, 1924). Clarke's work represented a comprehensive compilation of chemical data on minerals, rocks, and natural waters.

The other major series of survey geochemical publications was authored by H. S. Washington. In 1903, he had issued a compilation of chemical analyses of igneous rocks published between 1884, the publication date of the final edition of Roth's table of analyses, and 1900. Washington (1903) compiled 2881 analyses. Of these, 1861 were considered to be superior, a term that Washington applied to analyses that were generally characterized by a summation of oxides fairly close to 100 percent and by completeness in the sense that several minor constituents had been determined along with the major oxides. As was customary U. S. Geological Survey practice after the turn of the century, the analyses were arranged in accord with the CIPW system.

A year later, Washington (1904) issued a supplement to the earlier compilation in which he tabulated superior chemical analyses performed between 1869 and 1884 that he had extracted from Roth's tables. He judged that none of the analyses from the original 1861 tables had been of superior quality in the sense that he defined, largely because few of them contained data on minor elements and were, therefore, incomplete. Washington pointed out that an increasing percentage from each of his successive compilations were of superior analyses, an indication that the thoroughness and quality of analytical work were steadily improving. Much of the credit for that improvement was attributed to the insistence of Hillebrand on the determination of minor elements, particularly if the rocks were known to contain sodalite, zircon, rutile, or apatite.

Several years later, Washington published a massive survey professional paper in which he compiled analyses published from 1884 to 1913. To ensure comprehensiveness, Washington (1917) scoured the geological literature of the world, taking pains to include the work done in countries such as Argentina, Indo-China, Egypt, Romania, and Mexico that were not in the forefront of petrology as well as work done in western Europe and the United States. The end result was a tabulation of 8602 analyses of igneous rocks published since 1884. The quality of analyses continued to improve as more and more of them included minor constituents, although analysts in many countries still neglected to make such determinations. Washington renamed some of the rocks in terms of the CIPW system and assigned a CIPW chemical name to each rock. The tables were formatted systematically in accord with the CIPW scheme. As a result, readers were confronted on the very first page of the tables by the intimidating words: "Class I. Persalane. Rang 1–2 Prealkalic. Dargase."

Despite the difficulties inherent in classical wet chemical analysis, petrological papers published during the latter nineteenth century typically contained several chemical analyses of rocks. Papers like those of Brögger, Harker, and Iddings that advocated the theory of differentiation commonly incorporated considerable analytical data as one of the prime supports for the notion of progressive chemical change in a differentiated rock suite. By the beginning of the twentieth century, their work and that of Lagorio, Rosenbusch, Loewinson-Lessing, and others had made it abundantly clear that differentiated suites are typically characterized by increases in the contents of alkalis and, commonly, in silica and by decreases in the contents of lime, magnesia, and iron oxides. The overall trends in contents of the major elements during differentiation were generally understood. In the 1930s and 1940s, further analytical data provided by Wager and Deer, Walker, and others indicated that the Na/Ca ratio of plagioclase and of bulk rocks and the Fe/Mg ratio of olivine, pyroxene, and bulk rocks also increase during differentiation of magma. The chemical trends detected in the paths of cooling silicate melts in the experiments of Bowen were consistent with the analytical data from natural igneous rocks.

X-RAY DIFFRACTOMETRY AND CRYSTAL STRUCTURES

A scientific breakthrough that would eventually prove to have major implications for the practice of igneous petrology occurred in 1895, when German physicist Wilhelm Konrad Röntgen (1845–1923) discovered a new form of radiation. While experimenting with the production of cathode rays, Röntgen noticed that, upon striking the glass of the cathode ray tube, an electron beam produced radiation of low intensity that excited some nearby fluorescent material to glow in the dark. In time, the new form of radiation was dubbed x-radiation.

In 1911, Max Theodor Felix von Laue (1879–1960), then a physicist at the University of Munich, reasoned that if crystals are composed of regularly spaced atoms then they should act as gratings that would diffract the x-rays. By passing x-rays through a crystal of copper sulfate, he obtained a regular pattern of darkened spots on a photographic plate, supporting his contention that crystals act as diffraction gratings. In 1912, Friedrich and Knipping also observed the diffraction of x-rays by sphalerite crystals. At that time, William Lawrence Bragg (1890–1971) was studying physics at the University of Cambridge (Caroe, 1978). At the age of 22, Bragg reviewed the results of von Laue's experiment and developed a mathematical expression for the condition of x-ray diffraction, the so-called Bragg equation. Bragg's father, William Henry Bragg (1862–1942), was then Professor of Physics at the University of Leeds (Gay, 1974). Together, father and son attacked the problem of solving the crystal structures of simple compounds from the x-ray diffraction patterns they produced. For use in this work they developed an x-ray spectrometer. By 1914, they had successfully deciphered the crystal structures of the very high-symmetry isometric compounds NaCl, KCl, KBr, and KI. Soon thereafter they determined the somewhat more complicated structures of additional isometric minerals, namely, diamond, fluorite, pyrite, cuprite, spinel, and native copper, as well as calcite and corundum. For their work the Braggs were awarded the Nobel Prize in 1915. Lawrence Bragg went on to succeed Ernest Rutherford as Professor of Physics at the University of Manchester from 1919 to 1937, the year in which he published *Atomic Structures of Minerals*. In 1939, Bragg was appointed Cavendish Professor of Experimental Physics at Cambridge and Director of the Cavendish Laboratory, again succeeding Rutherford. During this final phase of his career, Bragg worked on the structures of metals and alloys, the deformation of crystals, and, finally, molecular biology.

At Manchester, Bragg worked on the crystal structures of silicate minerals with colleagues and students such as B. E. Warren and W. H. Taylor. After some preliminary x-ray investigations throughout the 1920s, the heyday of crystal structure determinations of the important igneous rock-forming minerals occurred in the latter 1920s and early 1930s with Manchester at the center of the action. Bragg was involved in determining the crystal structures of olivine, beryl, and diopside. Warren worked on the structure of enstatite, aegirine and other monoclinic pyroxenes, a host of amphiboles, melilite, and several metamorphic minerals such as vesuvianite. Taylor worked on the structure of sanidine and other feldspars as well as zeolite minerals and the alumino-silicate polymorphs. Zachariasen determined the structure of titanite (sphene).

X-ray crystallographers emerged in many locations outside Manchester. As a result, Vegard determined the structure of rutile. Nishikawa and Menzer worked on garnet. Hadding did preliminary studies of feldspars. Binks and Hassel deciphered the structure of zircon. Gottfried determined nepheline. Mauguin and Pauling studied micas, while Jackson and West worked on the structure of muscovite.

Petrologists were quick to seize the benefits of x-ray diffraction studies. Regardless of whether the crystal structure of a specific mineral had been completely determined, the characteristic patterns produced on photographic film by x-rays that had been focused on mineral powders or oriented single crystals provided a very useful tool for the correct identification of unknown minerals. The experimentalists at the Geophysical Laboratory, for example, began to use x-ray diffraction extensively to identify unknown run products. They also used x-ray diffraction to determine the approximate chemical composition of minerals such as olivine that belong to solid-solution series. After the crystal structures of the rock-forming minerals had been determined, petrologists began to alter their conceptions of the structural units within magmas. No longer was it feasible to envision magma as composed of simple oxides as Iddings had once done or to envision a magma composed of simple silicate molecules as the authors of the quantitative classification had done. From now on, petrologists would need to think of magmas in terms of silicate sheets, chains, tetrahedral groupings, and individual ions. They would also need to raise funds to pay for expensive x-ray diffractometers.

CRYSTAL CHEMISTRY

The determination of crystal structure also provided the key to understanding the principles that account for the distribution of chemical elements in the crust generally and igneous rocks particularly. The modern study of chemical element distribution was initiated largely in Russia by Vernadsky and Fersman and in Norway and Germany by Goldschmidt. Vladimir Ivanovich Vernadsky (1863–1945) studied mineralogy at the University of Munich in 1888 with Paul von Groth (1843–1927), the foremost crystallographer of the time, and worked in the laboratories of Fouqué, Le Châtelier, and Curie in 1889 and 1890 before returning to his native Russia (Spencer, 1947; Stadnichenko, 1947). For the next twenty years, Vernadsky taught in Moscow, including a stint from 1898 to 1911 as Professor of Mineralogy and Crystallography at the University of Moscow. Vernadsky spearheaded the development of an independent science of geochemistry by emphasizing the distribution of the chemical elements in time and space. He published his conceptions for a global audience in *La Géochimie* (Vernadsky, 1924).

Although Vernadsky's interests leaned toward the geochemistry of the biosphere, those of his student A. E. Fersman were directed toward the inorganic aspects of geochemistry. Aleksandr Evgenievich Fersman (1883–1945) studied at the University of Moscow under Vernadsky (Spencer, 1947). In 1907, Fersman studied with Lacroix and Rosenbusch, working on the crystallography of diamond, and traveled through Italy and Switzerland. He helped to organize the Peoples' University in Moscow, where he was appointed the first Professor of Mineralogy and gave his first lectures in geochemistry. Subsequently he served as an assistant to Vernadsky and held various posts at the Mineralogical Museum of the Academy of Sciences in Moscow. Fersman was especially interested in pegmatites and the alkalic igneous rocks of the Kola Peninsula, and his curiosity about the distribution of chemical elements was spurred by his pegmatite studies. He regarded atomic structure as the key to element distribution, and he expanded his compilations of data on such distributions to include the entire cosmos.

It was Victor Moritz Goldschmidt (1888–1947), however, who made the greatest advances in geochemistry by spelling out the ways in which the nature of the specific sites in crystal structures and the sizes of ions competing for those sites determine to a considerable degree the manner in which particular elements become concentrated in certain geological settings. Goldschmidt was the son of a chemist who taught at the Universities of Heidelberg and Kristiania, where the Goldschmidt family moved when Victor was only thirteen years old (Mason, 1992). He attended the University of Kristiania after 1905, studied geology, mineralogy, chemistry, and other sciences, and became closely associated with Brögger. He studied for the winter of 1908–1909 at the University of Vienna with Friedrich Becke in order to learn optical mineralogy. His doctoral thesis was published as the classic account of contact metamorphism in the Kristiania area that had been mapped by Brögger (Goldschmidt, 1911). In this paper Goldschmidt proposed the mineralogical phase rule, an idea possibly conceived under the influence of his father, who had worked with van't Hoff, the physical chemist who had applied the phase rule to the study of salt deposits. After receiving his Ph.D. degree in 1911, Goldschmidt worked in Munich with Paul von Groth during the winter of 1911–1912. At about the same time, von Laue was pondering the phenomenon of x-ray diffraction in crystals at the same university. The following year, Goldschmidt returned to Norway to become Docent at the University of Kristiania. When the University of Stockholm attempted to lure Goldschmidt away, the University of Kristiania retaliated by creating a professorship and establishing a mineralogical institute for Goldschmidt, still only 26 years old.

Goldschmidt's earliest professional research concentrated on petrological matters. He investigated regional metamorphism, migmatites, and igneous rocks and authored five monographs on the geology and petrography of the mountainous regions of southern Norway between 1912 and 1921. He coined the igneous rock names "opdalite" and "trondhjemite." In the fourth monograph on the igneous rocks between Stavanger and Trondheim, Goldschmidt (1916) introduced the concept of "Gesteins-Stamm" (rock stem) to refer to a series of genetically related igneous rocks occurring within a geologically limited region. The concept of the rock stem or "rock kindred" was developed further by Tyrrell (1926) to refer to apparently genetically related groups of igneous rocks showing consanguineous chemical and mineralogical characteristics. Tyrrell suggested that the term "kindred" be used for groups in which the degree of consanguinity is somewhat tenuous and that the terms "tribe" and "clan" be used for groups with a greater degree of consanguinity. After concluding his regional petrography monograph series, Goldschmidt (1922a) examined the "Stammestypen" (stem types) of igneous rocks. He proposed a sequence of separation of minerals from cooling magma by fractional crystallization that bore a striking resemblance to the reaction series published by Bowen (1922a) in the same year.

War had temporarily diverted Bowen from his major scientific interest. In Goldschmidt's case, war altered his major scientific focus rather dramatically. During World War I, Norway was cut off from its usual supplies of imported raw materials. Consequently, in late 1917, the government of Norway determined to finance research into the mineral resources of the nation and established a Commission for Raw Materials. Goldschmidt was appointed chairman. Under his leadership, the Commission founded the State Raw Materials Laboratory in the suburbs of Kristiania. One early result of the new laboratory was that Goldschmidt and Brögger mapped and studied a limestone occurrence at Fen with an eye to its economic exploitation. The scientific outcome of the research was the determination that the "limestone" was in reality an intrusion of igneous carbonate. Brögger (1920) described this classic carbonatite occurrence in detail in his monograph *Das Fengebiet in Telemark, Norwegen*. This and other studies on clay and anorthosite as possible sources of aluminum, biotite as a source of fertilizer potassium, ilmenite as a source of titanium for paints, and phosphate deposits stimulated Goldschmidt to a greater interest in the chemical nature of economic deposits and the processes leading to the concentration of chemical elements.

As a result of these developments, Goldschmidt turned his attention to the developing field of crystal chemistry. He got in on the ground floor of the method of crystal-structure determination by x-ray diffraction and turned the new powder diffraction method that had been developed by Debye and Scherrer in 1916 to his advantage. By applying x-ray diffraction to the determination of crystal structure, Goldschmidt and a contingent of bright young research associates, including Ivar Oftedahl, Lars Thomassen, T.F.W. Barth, Gulbrand Lunde, and F.W.H. Zachariasen, worked out the structures of large numbers of oxides and sulfides during the 1920s. From the unit cell dimensions of several simple oxides and halides, Goldschmidt was able to determine the radii of a host of cations of metallic elements as well as the radii of a number of anions. From his studies of the structures of the oxides of the rare earth elements, Goldschmidt discovered the lanthanide contraction. Much of Goldschmidt's crystal chemical work was published in a series of nine articles entitled "Geochemische Verteilungsgesetze des Elemente" (geochemical distribution laws of the elements) issued between 1923 and 1929. Many of the earlier installments of the series reported the results of crystal-structure analyses. The fourth and fifth, for example, respectively described the crystal structures of rare earth element oxides and the isomorphism and polymorphism of rare earth element oxides and the lanthanide contraction (Goldschmidt et al., 1925a, b).

In the seventh article, Goldschmidt (1926) laid out the principles of crystal chemistry and included a table of ionic radii. He established the increase in ionic radius for elements of increasing atomic number within the same vertical group of the periodic table, and he pointed out that cations within a given horizontal period in the periodic table decrease in radius with increasing atomic number. He also confirmed that the radii of elements with multiple valence states decrease from the ions of smaller charge to those of greater charge. It was at this time, too, that he developed the idea that the coordination number of a cation, that is, the number of anions surrounding that cation as nearest neighbors, is largely determined by the radius of the cation divided by the radius of the anion, namely the radius ratio.

Goldschmidt (1926) further proposed the principles of camouflage, capture, and admission. By camouflage he meant the concept that an ion may be "camouflaged" in a mineral or compound containing a chemical element with a similar size and charge, as, for example, the camouflaging of Hf^{4+} in zircon, where it substitutes for Zr^{4+} or the substitution of Ge^{4+} for Si^{4+} in silicate minerals. By capture, Goldschmidt referred to the incorporation of an ion with a similar size but greater charge than the predominant ion in a mineral as, for example, when potassium feldspar "captures" a Ba^{2+} ion in place of the more abundant K^+. And by "admission," he meant the opposite of capture, namely, the substitution of an ion with similar size but lesser charge than the predominant ion in a mineral as, for example, with the "admission" of a Li^+ ion for a Mg^{2+} ion in a pyroxene or olivine.

During the 1920s, Goldschmidt and his team were also accumulating data on the abundances of minor and trace elements, particularly the rare earth elements, in minerals. For analytical purposes, Goldschmidt turned to a new instrument, the x-ray spectrograph, designed in 1921 by his friend Assar Hadding of Sweden. In 1922, Thomassen constructed an x-ray spectrograph for Goldschmidt's laboratory. Goldschmidt successfully detected hafnium with the new technique, concluded that promethium did not exist in nature as a stable element, and discovered the phenomenon of europium anomalies.

During the 1920s, Goldschmidt had received a number of overtures and offers from German universities. During that decade, too, several of his star research associates left. Thomassen departed for the California Institute of Technology, returned for a couple of years, and then went to the University of Michigan in 1929. Barth went to the Technical University of Berlin in 1927 (Dickson and Krauskopf, 1973). Zachariasen, who had married Brögger's daughter, left for Manchester to work with Bragg in 1928 and then went to the University of Chicago in 1930. Lunde, too, departed in 1927. With the dismantling of much of his research staff, Goldschmidt became dissatisfied in Oslo (the former Kristiania). As a result, when the University of Göttingen, at the time renowned for stellar scientists and mathematicians like Born, Prandtl, Hilbert, Courant, and Tammann, offered him a position as Professor of Mineralogy to replace the retiring Otto Mügge (1858–1932), widely recognized for his writing of the fifth edition of Rosenbusch's *Mikroskopische Physiographie der Mineralien und Gesteine*, Goldschmidt accepted. He moved to Göttingen in 1929 and organized another mineralogical institute.

To this point, Goldschmidt had found x-ray spectrography to be satisfactory for his purposes. The method was inadequate, however, for the detection of very low concentrations of rare elements less than 100 ppm (parts per million). He wanted an extremely sensitive analytical method that could detect rare elements present in amounts as little as 1 ppm. He decided to use the optical spectrograph with a carbon arc, an instrument previously used by chemists but never for the quantitative analysis of geological materials. The optical spectrograph was based on the photometric measurement of the intensities of emission lines, first discovered by Bunsen, that are characteristic for the various chemical elements. In the long run, Bunsen's greatest contribution to geology did not come from his Iceland-inspired theory of two magma sources but rather from the spectroscopic discovery that laid the foundation for this powerful analytical tool. Goldschmidt had an optical spectrograph built for his laboratory, and he started making analyses in 1930. Now he was able to achieve a sensitivity of 0.001 percent or even less for most of the chemical elements. The results of his analytical work were published in another long series of papers dealing with the geochemistry of specific chemical elements, including gallium, germanium, scandium, beryllium, boron, selenium, arsenic, precious metals, and the alkali metals.

The rise of the Nazi movement in Germany made Goldschmidt, a member of the Jewish community in Göttingen, very uncomfortable and prompted him to return to Oslo in 1935. After the occupation of Norway by Germany, however, Goldschmidt was imprisoned for a time. By late 1942, he decided to leave Norway, spending a couple of months at the Mineralogical Institute at the University of Stockholm. He declined an offer to replace the retiring Helge Backlund as Professor of Mineralogy at the University of Uppsala. Instead, he spent a few years in Great Britain, returned to Norway in 1946, and died the following year. During his final years, he worked on a comprehensive treatise on geochemistry that included elucidation of the principles of geochemistry and discussions of the distribution of each chemical element. He died before completing the manuscript, but the work was ultimately published posthumously as *Geochemistry* (Goldschmidt, 1954) thanks to the efforts of fellow geochemists and petrologists in fleshing out the text.

THE APPLICATION OF EMISSION SPECTROGRAPHY TO PETROLOGY

Because of the remarkable success of Goldschmidt and his co-workers in obtaining data on elemental abundances in a host of geological materials by optical emission spectrography, the analytical technique gained favor in other laboratories as a means for conducting more detailed examinations of the trace-element compositions of specific suites of igneous rocks. Spectrographic studies of igneous rocks were conducted in chemistry departments at Queen's

University in Canada, the University of Cape Town in South Africa, and the University of Illinois in the United States. By way of example, investigations were made of trace elements in granites from Ontario (Harcourt, 1934) and igneous rocks of the Oliverian magma series in New Hampshire (Chapman and Schweitzer, 1947).

Before long, however, geologists wanted their own emission spectrographs. As a result, the United States Geological Survey began conducting spectrographic analyses in the 1940s, and spectrographs were installed in geology laboratories at MIT, Oxford, and Cambridge. Plans to introduce spectroscopic analysis in the Department of Geology at MIT began in 1935 when economic geologist, Walter Harry Newhouse (1897–1969), who had been on the faculty since 1927, wanted to determine as accurately as possible the amounts of trace elements in individual minerals, mineral deposits, and their associated rocks (Shrock, 1982). Because the very high resolution instrument that Newhouse wanted to construct was too costly to be funded by normal departmental resources, the department persuaded a prominent Boston industrialist, Godfrey L. Cabot, to fund the enterprise. Cabot was interested in the possibility of industrial applications for spectroscopy and agreed to donate the then munificent sum of $4846.50 to build the instrument and another $1000 to provide the salary for a research associate. This acquisition signaled the beginning of a new era in which advances in igneous petrology would become increasingly dependent on sophisticated new technology. Academic departments would need to find substantial sources of funds to acquire and maintain such instruments as x-ray diffractometers and emission spectrographs. Newhouse persuaded Richard Jarrell, a former physics student at MIT who now worked for his father at the Jarrell-Ash Company servicing the quartz spectrographs manufactured in England, to take on the challenge of building the MIT spectrograph.

The spectrograph was operational in 1938, and Newhouse initiated a program of research on trace elements that continued until 1946 when he left to join the University of Chicago (DeVore, 1973). With financial support from several sources, including $2500 from the Carnegie Institution of Washington, six graduate students used the spectrograph to conduct their doctoral investigations. These included studies by Bray (1942) on the igneous rocks from Jamestown, Colorado, and Shimer (1943) on granites and pegmatites of New England.

In 1947, dedicated space for the instrument, darkroom, and offices was set aside as the Cabot Spectrographic Laboratory. A second spectrograph was purchased with funds from the Office of Naval Research, and Harold W. Fairbairn was placed in charge. In the fall of 1948, L. H. Ahrens came to the Cabot Laboratory as a research associate to conduct spectrochemical research. Louis Herman Ahrens (1918–1990), a native of Pietermaritzburg, South Africa, graduated from the University of Natal in 1939 (Shrock, 1977; S. R. Taylor, 1995).

The following year he became a spectrochemical analyst in the Government Metallurgical Laboratory in Johannesburg. He received a D.Sc. degree in chemistry from the University of Pretoria in 1944. Thanks to a postdoctoral research fellowship from the Council of Scientific and Industrial Research of South Africa, Ahrens decided to go to MIT. By 1950, Ahrens had become so involved in the laboratory's activities that Fairbairn relinquished his responsibilities, and Ahrens was appointed Director of the Cabot Spectrographic Laboratory and Assistant Professor of Geology. From 1950 to 1953, Ahrens and his students attracted international recognition for the Cabot Laboratory. Ahrens concentrated on elemental abundances in rocks and on improvements in spectrochemical analysis, publishing the definitive texts *Spectrochemical Analysis* (Ahrens, 1950) and *Quantitative Spectrochemical Analysis of Silicates* (Ahrens, 1954) as well as *Wavelength Tables of Sensitive Lines* (Ahrens, 1951). His doctoral students applied spectrochemistry to newly developing methods in radiometric dating and to meteorite analysis. Ahrens resigned at the end of 1953, however, to accept a Readership in Mineralogy at Oxford University. He was succeeded by William Dennen who ultimately took the spectrography laboratory with him to the University of Kentucky in 1967.

Ahrens left for Oxford thanks to the persuasive efforts of L. R. Wager, who had been called to serve as Head of the Department of Geology in 1950. Wager perceived that geology research at Oxford was lagging badly. After decades in which the focus of Oxford geology had been on stratigraphy, structure, and paleontology, Wager set out to build up a research school in his own mineralogical and petrological image (Vincent, 1994). Rather than bring in an expert in x-ray crystallography, at the time an exciting field that was exceptionally well represented at rival Cambridge, Wager instead proposed to appoint a geochemist. He had good reason to do so.

Wager had already undertaken an impressive spectrographic study of the rocks and minerals of the Skaergaard Intrusion, and he fully intended to continue that work and to make comparisons with similar igneous rock suites. To carry out the work, however, Wager had to enlist the services of Robert L. Mitchell, a spectrographer with the Macaulay Institute for Soil Research in Aberdeen, Scotland. In fact, Mitchell was becoming a rather popular fellow with British petrologists. At Cambridge, S. R. Nockolds had persuaded him to collaborate on a spectrographic study of major and trace elements in rocks and minerals of some Caledonian igneous suites. Frederick Walker and his student Arie Poldervaart also persuaded Mitchell to provide spectrographic analyses of seven rocks from South African sills. Clearly, the demands that other petrologists were putting on Mitchell and the fact that most of his efforts were devoted to work for the Institute for Soil Research made it obvious that petrologists needed their own analytical facilities.

They needed such facilities all the more so because Wager's work on the Skaergaard Intrusion proved to be of exceptional merit and interest. Wager and Mitchell (1943) published a preliminary report on trends in the amounts of trace elements in the Skaergaard layered series. Their major paper, however, included analyses of trace elements in both rocks throughout the intrusion and mineral separates from the border group and the layered series (Wager and Mitchell, 1951). The authors compared their results with the few other investigations that had been done. The paper was a triumph of hard work and valuable scientific results. The mineral separates employed for spectrographic analysis were obtained largely by tedious handpicking under a microscope and without benefit of heavy liquids or magnetic separators. The paper was also groundbreaking in that it appeared in the very first issue of a brand-new journal devoted to geochemistry, *Geochimica et Cosmochimica Acta*, of which Wager was a founding editor.

Wager and Mitchell (1951) found that Ga^{3+}, Sr^{2+}, and Ba^{2+} are concentrated primarily in plagioclase and that the abundances of Sr^{2+} and Ba^{2+} increase dramatically in plagioclase with stratigraphic height in the layered series of the Skaergaard Intrusion. They further discovered that Cr^{3+} is especially concentrated in pyroxene and magnetite toward the bottom of the exposed layered series, that V^{3+} is concentrated in pyroxene, ilmenite, and magnetite toward the bottom of the exposed layered series, that Ni^{2+} and Co^{2+} are especially concentrated in olivine and in lesser amounts in pyroxene and that their abundances diminish strikingly with stratigraphic height, and that Cu^{2+} is present in plagioclase, pyroxene, olivine, ilmenite, and magnetite and tends to increase in abundance upward. It is worth noting that Rb^+ was not detected except from a pyroxene high in the layered series. The values are striking to contemporary eyes in that they are nearly all rounded off to numbers like 80, 200, or 1000 ppm. Emission spectrography represented only a beginning in trace-element analysis, but it was a very important beginning.

Wager and Mitchell were eager to compare their results with other trace-element investigations but were hampered by a relative paucity of data. They managed to compile average values for fifteen trace elements (P, Ga, Cr, V, Ti, Li, Ni, Co, Cu, Sc, Zr, Mn, Sr, Ba, and S) in ultrabasic, basic, intermediate, and acid rocks from the Skaergaard Intrusion as well as average values for the same groups of rocks from the Caledonian igneous suite in Scotland studied by Nockolds and Mitchell (1948). They also compiled values for average ultrabasic, basic, intermediate, and acid rocks in general from the work of Goldschmidt (1934, 1937, 1938) and petrologists such as Tröger (1935a). Although the absolute abundances reported by Wager and Mitchell (1951) differed dramatically for some elements from those of the Caledonian suite or average igneous rocks, the data were generally in agreement that Cr, Ni, and Co decrease progressively from ultrabasic to acid rocks, that V, Sr, Ti, and P are concentrated primarily in basic or intermediate rocks, and that Ba, Zr, and Li tend to increase progressively from ultrabasic to acid rocks. Following

Goldschmidt, crystal chemical explanations based on the sizes and valences of ions were used to account for the entry of those ions into specific minerals.

Wager and Mitchell (1953) again collaborated on a paper on the trace elements in a suite of Hawaiian lavas in an effort to record the variations in trace-element contents and to make a comparison with the suite resulting from fractional crystallization of the Skaergaard Intrusion. They spectrographically analyzed the same elements as in the Skaergaard study and found that Cr, V, Ni, Co, and Cu all decrease in abundance from basalt to oligoclase andesite, that is, with the course of differentiation. They attributed the decrease in abundances of these chemical elements to the fact that, apart from Cu, they are strongly concentrated in early crystallizing ferromagnesian minerals like olivine and pyroxene. Wager and Mitchell further found that Li, Zr, Y, La, Sr, Ba, and Rb generally increase with the course of differentiation toward more siliceous rocks. These elements, they suggested, are those that remain in successive residual liquids because they enter the early crystallizing minerals like Ca-plagioclase, pyroxene, olivine, ilmenite, and magnetite to a very limited extent. With further crystallization to trachyte, Wager and Mitchell found that these trends continue except that Cr and Ni increase slightly, Ba decreases slightly, and Sr decreases abruptly. The authors observed that the magnitudes of the changes in trace-element contents are much greater than for the major elements. They predicted that trace-element geochemistry would "some day provide a powerful weapon for the solution of petrogenetic problems" (Wager and Mitchell, 1953, p. 221).

Although noting that the values obtained from the Skaergaard Intrusion were for successive crystal fractions rather than for residual liquids, the authors still made a comparison of the direction of trace-element content changes between Hawaiian lava suites and the Skaergaard Intrusion. Pointing out the close qualitative, and in some cases the close quantitative, similarity between the trace-element variations in the Hawaiian and Skaergaard series, they maintained that both igneous series had resulted from the same process, namely, fractional crystallization.

Given his passion for geochemistry applied to petrological problems, Wager was undoubtedly overjoyed when Ahrens arrived in Oxford during January 1954. Ahrens quickly ordered a Hilger quartz-prism spectrograph with money from the Higher Studies fund and hired New Zealander Stuart Ross Taylor (b. 1925). Taylor had been stimulated as an undergraduate at the University of New Zealand to pursue geochemistry by Brian Mason, a former student of Victor Goldschmidt. When Mason took a position at Indiana University in the United States, Taylor followed him there to undertake graduate studies (Mason, 1994). At the time he was hired by Ahrens, Taylor had just received a Ph.D. degree in geochemistry from Indiana University, where Neumann, Mead, and Vitaliano had also been developing mathematical expressions for trace-element variation during fractional crystallization (Neumann et al.,

1954). By the end of 1954, Ahrens and Taylor had the new optical spectrograph set up, calibrated, and operating.

Wager was soon to be chagrined, however, when, after only two years, Ahrens accepted the Chair of Inorganic and Physical Chemistry at the University of Cape Town. Ahrens was delighted to return to his homeland. Both salary and laboratory resources were considerably better in Cape Town than at Oxford. Moreover, Ahrens was concerned that, if Wager's tenuous health deteriorated further to the point that he would have to step down as Head, the Department of Geology might revert back to its former focus on classical geology and leave his geochemical interests rather isolated. Ahrens soon became Professor of Geochemistry and Head of the newly formed Department of Geochemistry at Cape Town, where he continued a prolific career of research on meteorites, the distribution of elements in igneous rocks, geochronology, and spectrochemical methods. In the meantime, Ross Taylor continued to head up Oxford's program in spectrographic analysis until he, too, departed for Cape Town in 1958 to work again with Ahrens before ending up at the Australian National University in 1961.

The petrology program at Cambridge had already jumped into spectrographic analysis under the leadership of S. R. Nockolds. Stephen Robert Nockolds (1909–1990) was a 1929 graduate of the University of Manchester, where his interest in igneous petrology was fired (Muir, 1994). He went on to Cambridge for graduate study in the newly established Department of Mineralogy and Petrology under Cecil E. Tilley. Although Harker had recently retired, he was still active, and Nockolds was powerfully influenced by the Cambridge master. He received his Ph.D. degree in 1932 after studying hybridism and contamination in granitic magmas. He returned to the University of Manchester as Assistant Lecturer in Petrology, but then came back to Cambridge in 1937 as a Demonstrator in the Department of Mineralogy and Petrology. In 1945, he was promoted to Lecturer, and in 1957 he became a Reader in Geochemistry.

Early in his career, Nockolds differed from Harker in believing that contamination had played an important role in the evolution of some magmas. One of his early studies on the Glen Fyne–Garabal Hill complex earlier studied by Teall was later extended to a consideration of the geochemistry of several Caledonian plutons. Nockolds and Mitchell (1948) published a paper on the major- and trace-element geochemistry of the Caledonian plutonic rock suites in Scotland. They showed that the trace-element data confirmed the theory that the parental magma was andesitic rather than basaltic. Their work also provided a geochemical support for Bowen's contention that fractional crystallization without substantial contamination was the dominant process in the formation of plutonic rocks. Nockolds' findings also led him to sharp disagreement with transformist conceptions of the origin of granite.

To continue his own geochemical research without having to rely on Mitchell and the Institute for Soil Research, Nockolds set up an emission spectrography laboratory at Cambridge in 1948 with his assistant R. Allen. The major contribution to emerge from Nockolds' geochemical laboratory was a series of three substantial papers on the geochemistry of various igneous rock series. In the first article, Nockolds and Allen (1953) wrote that their object was to study the behavior of the trace elements with respect to the major constituents in the calc–alkalic, alkali, and tholeiitic igneous rock series. They also wished to determine how the behavior of the elements varied from series to series in order to shed light, if possible, on the origin of these three main branches of igneous rock evolution.

The first paper dealt with the calc–alkalic series. After obtaining numerous samples, previously analyzed for major elements, from a host of petrologists, Nockolds and Allen investigated the elemental abundances and trends in the Southern California batholith, the east-central Sierra Nevada batholith, Lassen Peak, the Medicine Lake Highlands of northern California, Crater Lake, the Lesser Antilles, and the Caledonian plutons earlier studied by Nockolds. The results for the major elements were plotted on AFM and Ca–Na–K triangles and smooth curves drawn through the data points (Figure 41). Variation diagrams for major and trace elements were also plotted against the variable, $(\frac{1}{3}\mathrm{Si} + \mathrm{K}) - (\mathrm{Ca} + \mathrm{Mg})$. Nockolds and Allen believed that the plotted curves represented liquid lines of descent. They demonstrated that Ca, Mg, Ni, Fe, Co, Cr, V, Ti, and Sc all gradually decrease in abundance with increasing SiO_2 content in each of the rock suites. Their data indicated that Sr either increases to a maximum and then decreases, or simply decreases, depending on the rock suite analyzed. Where early-formed plagioclase was removed from the cooling magmatic system, Sr content would decrease in the residual liquid, but where little early-formed plagioclase was removed, then Sr content would increase in the residual liquid. They reported that Na content is approximately constant throughout the entire compositional range with a feeble maximum being attained in intermediate members of a rock suite. Li, K, Rb, and Ba were found to increase thoughout any given suite. According to the authors, Zr increases, then becomes constant, and then begins to decrease in the most silica-rich rocks. For most of their samples, they found that Cs, Tl, Pb, Mo, Ce, Ge, Be, and In were below the levels of sensitivity of the spectrographic method.

The second paper examined the chemical composition of alkalic volcanic rock series. Now Nockolds and Allen (1954) reported results from studies of alkali-rich rocks from the Scottish Tertiary volcanic centers, Hawaii, Polynesia, and the east-central Sierra Nevada Mountains. Diagrams identical to those in the first paper were constructed, and the curves again were regarded as representing liquid lines of descent. Nockolds and Allen discovered more variability among these alkali suites than in the calc–alkalic rocks. For the alkali basalt–trachyte suites of the Scottish Tertiary, Hawaii, and Polynesia, they found that

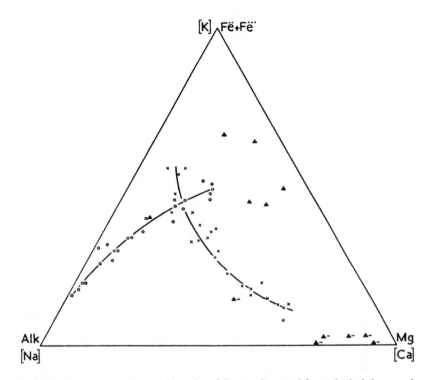

41. Major-element variation in the rocks of the Southern California batholith, one of several diagrams presented by Nockolds and Allen (1953) to show variation in elemental abundances within igneous rock suites. Reprinted from *Geochimica et Cosmochimica Acta*, v. 4, Nockolds, S. R., and Allen, R., The geochemistry of some igneous rock series, pp. 105–142, 1954, with permission from Elsevier Science.

Mg, Ni, Li, Fe, Co, Cr, V, and Ti display similar behavior in all three cases. Mg, Ca, Cr, Ni, and Co decrease in all rock suites with increasing differentiation. In contrast, Li, Na, K, Rb, Y, Zr, and La display increases. Fe, V, Ti, Ba, and Sr increase to maxima and then decrease. For the other alkali rock suites, trends are generally similar, although the location of maxima in abundance curves varies for elements such as Sr and Ba.

Nockolds and Allen (1956) examined tholeiitic rock series in a third paper. Here they compared data for a pair of suites from the Hakone volcano in Japan, two suites of Karoo dolerites from South Africa, two suites of tholeiitic rocks from the British Tertiary province the Northfield sill, and a hybrid rock suite from Satakunta in Finland. They compared results with the Dillsburg diabase sill in Pennsylvania and with the Skaergaard Intrusion. Although the total abundances and rates of increase or decrease with differentiation vary, the authors found that for the various suites, Li, Na, K, Rb, and Ba generally increase with differentiation whereas Mg, Ca, Co, Ni, and Cr characteristically decrease. The behavior of Fe, Sr, and Zr shows more variability.

Regrettably, the fourth paper that promised comparisons among the different series never materialized. In any case, petrologists were now equipped with reasonably large data sets and diagrams illustrating the trace-element characteristics of the chemically distinct tholeiitic, calc–alkalic, and alkalic rock series. Goldschmidt's crystal chemical explanations for element distribution had been reasonably well confirmed, and, for the most part, the trace-element data were considered to be consistent with the operation of fractional crystallization.

THE ACCURACY AND PRECISION OF ANALYSES

By 1950, the attractiveness of spectrochemical methods had become increasingly apparent to petrologists and geochemists. At first, the method had been used primarily for the determination of the abundances of trace elements that were so extremely difficult to analyze by classical wet chemical methods, but analysts believed that emission spectrography might be useful for the determination of major elements as well. As a result, the need for reliable standards for spectrochemical analysis of the major elements became clear. If spectrochemical methods were to be used routinely for major-element analysis, then it was necessary to compare the intensity of a spectral line of an element in a rock being analyzed with the intensity of the line for a known amount of the same element in some standard rock. But to know exactly the elemental contents in standard rocks it was first necessary to select these standards and to conduct interlaboratory analytical comparisons of those standards.

Much of the early work on geochemical standards and comparisons was coordinated by Harold Williams Fairbairn (1906–1994). Fairbairn graduated from Queen's University in 1929 (Shrock, 1977; Hurlbut and Thompson, 1996). After spending a year at the University of Wisconsin, Fairbairn moved to Harvard University, where he was awarded his master's degree in 1931 and a Ph.D. degree in 1932. During his doctoral work, he acquired an interest in structural petrology from reading Bruno Sander's book on the topic. Fairbairn went to Europe to study under Sander, a member of the Hans Cloos school of granite tectonics. He also spent three months at Göttingen with Goldschmidt and three months at the Technische Hochschule in Berlin learning x-ray methods useful for his petrofabric studies. Fairbairn served as an instructor at Queen's from 1934 to 1937 and then began his career at MIT from which he retired in 1972. Fairbairn's interests gradually shifted from petrofabrics and optical mineralogy to spectrochemical analysis and geochronology.

As a first step in the development of geochemical standards, Fairbairn coordinated an interlaboratory analytical study of two rocks, a granite from Westerly, Rhode Island, designated as G-1, and a diabase from Centreville, Virginia, designated as W-1. Approximately thirty analysts from government, industry, and academia participated in analysis of the two standards. The results were published as a U. S. Geological Survey Bulletin (Fairbairn et al., 1951), hailed by some observers as the most important petrological contribution of the century. The wide variation in results of the interlaboratory comparison stunned practicing petrologists. As Vincent (1960, p. 36) noted, "the spread of the results reported for practically every constituent was very much greater than had been anticipated." The gloomy conclusion was that "it was impossible to assess the probable accuracy of any single silicate analysis taken in isolation." It was also discovered that the precision of wet chemical analyses of elements present in low concentrations was significantly poorer that that of elements present in higher concentrations. This fact served as major stimulus to develop improved instrumental methods of analysis.

A few additional laboratories joined the study shortly after publication of the bulletin. The laboratories involved in the wet chemical comparison were located in Australia, Canada, Finland, Germany, Japan, New Zealand, South Africa, Sweden, Switzerland, the United Kingdom, and the United States. No laboratories in France, the Soviet Union, or Norway were involved. The vast majority were in Australia, Japan, the United Kingdom, and the United States. The locations of these laboratories provide just one indication that, by mid-century, interest in igneous petrology and geochemistry had spread far beyond western Europe, North America, and South Africa to the nations of the western Pacific.

The discouraging effort to obtain consistent results for two standard rocks led Fairbairn to assess the quality of the art of rock analysis and to find ways in which analysis might be improved. To that end, Fairbairn (1953) coordinated the very first effort to conduct interlaboratory tests of precision and accuracy. He observed that the precision, that is, the reproducibility of results, of replicate analyses performed by the same experienced analyst working in the same laboratory with the same methods and conditions ought to be very good. The precision of replicate analyses performed by different analysts in the same laboratory, he thought, should not be quite as good. Finally, he suspected that the precision of replicate analyses of the same material obtained by different analysts in different laboratories would be lower still. But just how good or bad would that degree of precision be? Fairbairn used the arithmetic means of the results obtained for the various constituents in G-1 and W-1 analyzed by the 35 laboratories involved for the estimation of precision.

First, a synthetic glass of simple granite composition and lacking Fe, Ti, Mn, H, and trace elements was prepared to serve as a standard for testing the accuracy of the analyses. The glass was analyzed by chemists from twelve laboratories, eleven of which had previously analyzed the standard rocks G-1 and W-1. Comparison of the twelve analyses of the synthetic glass of known composition indicated to Fairbairn that the averages computed for the contents of MgO, CaO, Na_2O, and K_2O analyzed in G-1 and W-1 were probably very close to their true values. He also thought it reasonable to assume that the G-1 value for SiO_2 was about 0.5 weight percent too low and the W-1 value for SiO_2 was about 0.35 weight percent too low. Because the synthetic glass did not contain Fe, Ti, or Mn, Fairbairn tentatively concluded that, since current procedures for the determination of alkalis, lime, and magnesia gave reasonably accurate results, the same also held true for procedures used for the untested constituents.

Fairbairn was satisfied that the test of the analytical data had confirmed a reasonably satisfactory estimate of the chemical compositions of the G-1 and W-1 standard rocks. As a result, these standards would be invaluable for spectrochemical analysis of the major elements. What was still lacking was the evaluation of the precision and accuracy in a single analysis. From comparison with analyses of the synthetic glass, Fairbairn suggested that analyses of silica in rocks were often likely to be reported as less than the true value and alumina as a little higher. He also concluded somewhat pessimistically that the overall picture of the precision of a single analysis of a silicate rock by conventional methods was rather discouraging. He thought that the best that could be done was to assume that the mean values for G-1 and W-1 provided close estimates of their true compositions and that individual analysts should make frequent checks of their work against the standard values. Otherwise, Fairbairn warned, the immense labor of analyses by conventional methods would produce inadequate results much too frequently.

Fairbairn noted that spectrochemical methods offered several advantages, among them increased speed, lower cost, less labor, precision comparable to that of wet chemical methods and improvable with replicate runs, much smaller sample size, and preservation of the results on photographic film so that the analyst could return later to look for additional elements. He concluded with the sobering thought that many unwarranted geological conclusions had probably been drawn from inadequate chemical data, especially where analyses from different laboratories had been used to make correlations. "It should therefore not occasion any surprise," Fairbairn (1953, pp. 155–156) wrote, "if chemical re-study of many suites of rocks fails to confirm earlier findings in all respects."

When the plethora of new analytical methods and instruments was applied to geological problems during the 1950s and 1960s, igneous petrologists would be utterly inundated by an unceasing torrent of analytical data. It would become ever more important to heed Fairbairn's cautions about the reliability of a chemical analysis.

23 ← THE SEA BELOW, THE HEAVENS ABOVE

The Extension of Petrology

Igneous petrology experienced profound intellectual and technological developments during its geochemical era. Not only did a host of new instrumental methods for the analysis of chemical elements and isotopes result in further contributions of data and theory from the geochemical side of petrology, but the basic outlines of the theory of plate tectonics were formulated. The theory soon exerted a dominant influence on virtually every field of geology, including igneous petrology. Moreover, the space race between the Soviet Union and the United States culminated in several manned and unmanned lunar landings that issued in the application of the principles of igneous petrology to rocks on another body in the solar system. Although far-reaching conceptual and technological advances deeply influenced the course of igneous petrology during the middle years of the twentieth century, several interconnected sociological phenomena also began to alter the face of the discipline. The mid-twentieth century was a period of transition in which sweeping changes and trends affected the petrological community in ways not previously seen. Throughout these years, the field of igneous petrology was shaped by increased student enrollments; a proliferation of petrology courses in academic institutions; the availability of steadily growing amounts of funds for petrological research; involvement of increasing numbers of women; the growing internationalization of petrology concomitant with growing Anglicization of the field; an increased tendency to specialize within the discipline; and the formation of new petrological journals. Many, if not most, of these trends and tendencies were not unique to igneous petrology. The trends affecting igneous petrology represented a microcosm of the tide of change sweeping through all of geology and, indeed, all of the natural sciences. Before we resume our review of the leading intellectual developments of the geochemical era, therefore, we pause for a brief look at some of these sociological trends.

CHAPTER 23

GROWTH OF IGNEOUS PETROLOGY
IN ACADEMIA

In the first half of the twentieth century, a few dozen truly outstanding petrologists dominated igneous petrology. Since the 1960s, however, an astounding proliferation of igneous petrologists has been intimately linked with dramatic growth in the number of students enrolled in postgraduate degree programs and in the number of academic institutions offering igneous petrology in their curricula. To my knowledge no study of enrollment trends and of numbers of degrees granted for students specializing in igneous petrology has been undertaken. One indicator of the increase in enrollments, however, may be found in data on numbers of geology students at Massachusetts Institute of Technology compiled by Shrock (1982). He reported that the number of graduate degree candidates in geology had escalated from 27 in 1950 into the 100s during the 1970s. During the same period, the total number of undergraduate geology students declined slightly. In his history of the geology program at the University of Chicago, Fisher (1963) compiled data on Ph.D. degrees granted until 1961 in the top eighteen doctoral programs in geology in the United States. For nearly every program listed, the number of degrees granted in the five-year period between 1957 and 1961 far exceeded the number granted in any previous five-year period.

The number of universities and colleges offering geology programs also increased significantly during mid-century. Undergraduates no longer had to attend the major, prestigious universities to be exposed to igneous petrology. In Great Britain, prior to 1960, Oxford, Cambridge, London, Durham, Manchester, Glasgow, and Edinburgh were the leading academic centers of igneous petrology. In more recent years, however, petrology has been added to the curricula at the universities of Leicester, Bristol, Milton Keynes, Leeds, and elsewhere. In the United States, igneous petrology is no longer the exclusive preserve of schools like Harvard, Yale, MIT, Chicago, and Johns Hopkins. Now petrology is taught in a host of younger state-run institutions. Moreover, undergraduates learn igneous petrology in a multitude of small liberal arts colleges, such as Williams, Franklin and Marshall, Amherst, and Pomona. Several liberal arts colleges have added geology and igneous petrology to their curricula since mid-century, such as Hope (Michigan), Allegheny (Pennsylvania), Dickinson (Pennsylvania), and Carleton (Minnesota).

FUNDING

Some of the growth of interest in igneous petrology during the middle of the century may be traced to the excitement engendered by the granite controversy, to the glamour of experimental petrology, and to the attractions of petrology for physical and analytical chemists. Some of the increased interest may also be linked to population growth. Much of the expansion of petrology, geology, and science in general, however, needs to be attributed to expanded government financial support for the sciences in the universities. Frank Press (1995) has written that basic research in the sciences in America prior to the 1940s was carried out mostly in universities and was funded largely by private philanthropy. Additional research was done in a few government agencies and industrial laboratories. The federal government's involvement in the support of basic science was minor compared to that of the private sector, but Press conceded that government-funded geology has fared relatively well thanks to the existence of the Geological Survey and the Bureau of Mines. The funding balance, however, began to shift dramatically in the 1940s. The concerns for a co-ordinated national policy in the sciences and technology, intensified by the exigencies of World War II, ultimately led to the approval by Congress and President Truman in 1950 of the establishment of the National Science Foundation (Kevles, 1977). Particularly after the Soviet Union launched the Sputnik satellite in 1957, the American federal government began to promote science and technology education vigorously and pumped large amounts of funds into the nation's educational institutions via the National Science Foundation, the National Aeronautics and Space Administration, the Office of Naval Research, the Atomic Energy Commission, and, later, the Department of Energy. Geology programs, and therewith petrology, benefited from the largesse. Geschwind (1995) noted that the Office of Naval Research began awarding contracts for geochemical work such as the construction of mass spectrometers in 1947. Igneous petrology, of course, benefited greatly from the ensuing geochemical research. And beginning in the 1950s, both the Office of Naval Research and the National Science Foundation began the funding of experimental petrology laboratories at Penn State, Chicago, Johns Hopkins, UCLA, and elsewhere. The lunar exploration program proved a great boon to igneous petrologists as the National Science Foundation poured money into departments for the purchase of geochemical analytical equipment and expansion of experimental facilities.

Data from Massachusetts Institute of Technology again help to tell the story (Shrock, 1982). In the academic year 1949–1950, the Department of Geology expended $156,147, of which about one-third came from outside sources. The academic year 1960–1961 was the last year in which internal funds from MIT exceeded the amount of outside funds. From the following year on, external funds spectacularly outstripped MIT funds. In the academic year 1969–1970, the department expended $2,528,131, of which about three-fourths came from outside sources. Over the course of twenty years, the departmental budget grew sixteenfold, and the percentage of outside funds increased from 32 to 75 percent.

In Great Britain, too, governmental support for Earth sciences increased dramatically (Stewart, 1971). During the 1960s, most government money for the Earth sciences came from the Department of Education and Science. Within that department, the University Grants Committee dispensed block grants for salaries, buildings, equipment, and the like to 44 British universities. Stewart reported that in the previous ten years, the number of university staff had increased from about 300 to about 470. Research funding came from the Science Vote, another division of the Department of Education and Science. The Science Vote included several units of which the National Environmental Research Council (NERC) was by far the largest source of government support for research in the Earth sciences in Britain, apart from the University Grants Committee. In 1965–1966, NERC funding for Earth sciences amounted to £4 million and, in 1970–1971, £14 million. The scientific staff of the geological survey of the Institute of Geological Sciences, formerly the Geological Surveys of Great Britain and Overseas, had increased from 244 to 457 in the previous ten years. The number of NERC post-graduate awards in the Earth sciences had also increased by around 50 percent in the seven years previous to 1971. Igneous petrology again participated with all of geology in the flush financial times. Governmental scientific research councils in other countries like Australia, South Africa, and the Scandinavian nations likewise contributed to the dramatic rise in funding.

THE INTERNATIONALIZATION OF PETROLOGY

Mid-century also witnessed increasing diversification within the petrological community. At least in the United States, women made up an increasing percentage of undergraduates, graduates, and professors in igneous petrology as the latter half of the century unfolded. In the first half of the century, igneous petrology among American women was represented to a large extent by Florence Bascom and in Great Britain by Doris Reynolds. By 1996–1997, 31 women were listed specifically as igneous petrologists teaching in American and Canadian universities and colleges in the American Geological Institute's *Directory of Geoscience Departments* (Claudy, 1996).

Increased diversification was also evident in the growing internationalization of the discipline. At the beginning of the twentieth century, igneous petrology was done predominantly in western Europe and the United States at prestigious academic institutions. South Africa, Japan, and Australia became more important participants as the century proceeded. One indication of the trend toward internationalization may be detected in some of the leading petrology journals published during the second half of the twentieth century. For example, the authors of papers in Volume 1 of *Journal of Petrology*, published in 1960, hailed from four countries. Six universities outside the United States were represented, including Oxford, Cambridge, Manchester, and the Imperial College. Three papers were contributed by staff members of the Geophysical Laboratory. In the issues of *Journal of Petrology* published in 1998, however, the authors hailed from nineteen countries and represented 58 universities outside the United States. Only one paper was written by a Cambridge petrologist, none by an Oxford petrologist, and only one by a Geophysical Laboratory staff member. Similar results emerge from *Contributions to Mineralogy and Petrology*. Volume 8 (1961–1962) featured 29 papers, all but three of which were written in German. The four volumes of *Contributions* for 1997 contained 114 papers contributed by authors from seventeen countries, and authors from institutions not typically associated with the academic elite contributed the bulk of the petrological papers. Only one paper was contributed by a staff member from the Geophysical Laboratory and none by an Oxford petrologist. Individuals from many of the nations of Europe that had not previously been heavily involved in petrology, as well as individuals from South America and Asia, began contributing to the discipline. Despite the increased participation of petrologists from many nations around the world, the fact remains that, throughout the twentieth century, igneous petrology was still dominated by Great Britain, the United States, Germany, France, and Scandinavia, with Japan, Australia, and South Africa taking a place within global petrological leadership.

PETROLOGICAL LITERATURE

Because the number of practicing igneous petrologists proliferated so dramatically after the 1950s, the sheer volume of petrological information exploded beyond the point where an individual petrologist could possibly remain abreast of all the important developments. To accommodate the rapidly growing volume of petrological material, several journals devoted almost exclusively to petrology emerged during the middle of the twentieth century. Prior to 1947, few specialized petrology journals existed. These included *Tschermaks mineralogische und petrographische Mitteilungen*, published since 1871, and *Schweizerische mineralogische und petrographische Mitteilungen*, published in Switzerland since 1921. Because articles in these two journals were almost exclusively in German, speakers of other languages resorted to general geological journals such as *Journal of Geology*, *Journal of the Geological Society of South Africa*, or *Bulletin de la Societé Géologique de France* or to various mineralogical journals like *Mineralogical Magazine* as outlets for their papers. In 1947, Germany started production of yet another petrological journal when the University of Heidelberg, under the editorship of Otto Erdmannsdörffer, began to publish *Heidelberger Beiträge zur Mineralogie und Petrographie*. The choice of the title reflects the lasting influence of the German school of descriptive microscopic petrography on the field of igneous petrology. In 1958, the word *Heidelberger* was dropped from the title when Carl Correns of the University of Göttingen took over the editorship. Several papers in English also appeared that year.

The broadening of *Beiträge* and its growing prestige were insufficient, however, to prevent the introduction of *Journal of Petrology* under the leadership of L. R. Wager in 1960. From the beginning, the Britain-based journal published almost exclusively in English. As English gradually became a universal language, *Beiträge* made a concession to that trend in 1966 by adopting a dual title, with the English title *Contributions to Mineralogy and Petrology* appearing above the German title. At the same time, *Petrographie* was changed to *Petrology* to keep in step with the modern emphasis on theory and interpretation within the discipline. Instructions to authors were printed in both languages, and a much higher percentage of papers than ever before were written in English.

A third petrologically oriented journal appeared in 1968 with the publication of *Lithos* under the editorship of T.F.W. Barth at the University of Oslo and the sponsorship of the National Councils for Scientific Research in Denmark, Finland, Norway, and Sweden. Despite its Scandinavian origins, articles were overwhelmingly written in English from the outset. In 1973, *Contributions/ Beiträge* dropped its German title altogether. The Anglicization of the petrological literature continued in 1987 when, after more than 100 years with a German title, *Tschermaks mineralogische und petrographische Mitteilungen* changed its name to *Mineralogy and Petrology* and adopted a policy of publishing only English manuscripts.

All of these journals featured articles on both igneous and metamorphic petrology, but the number of papers on igneous rocks was typically in the majority. The sheer volume of material published also increased spectacularly. In 1960, the first volume of *Journal of Petrology* consisted of 398 pages in three issues. The subscription price was $12. In 1970, three issues of the annual volume amounted to 589 pages and cost $14.50. By 1998, the journal was publishing monthly issues for an annual volume of 2154 pages of larger size than those of 1960. The subscription rate for an individual had soared to $375. The first volume of the forerunner of *Contributions to Mineralogy and Petrology* in 1947 comprised four issues amounting to 721 pages and costing 14.60 Reichmarks. In 1961–1962, the volume of six issues had shrunk a bit to 502 pages but cost 26.80 Deutschmarks. In 1997, the journal was published monthly as four volumes of three issues each annually, and amounted to 1615 pages that are larger than those of earlier years. The annual subscription rate had ballooned to a staggering 3848 Deutschmarks ($2791), well beyond the reach of a typical working petrologist.

Another trend is revealed by the papers in these two leading petrological journals. Since mid-century there has been a distinct movement away from single-author papers to team research and publication, no doubt a reflection of the enormous growth in the literature and the inability of an individual to be sufficiently expert in the various facets of an ever-diversifying field. In the 1960 *Journal of Petrology* 17 papers were written by 23 authors, or 1.35 authors per paper. In 1998, 93 papers were written by 277 authors, or 2.98 authors per paper. The same trend appears in *Contributions to Mineralogy and Petrology*. In 1947, 31 papers were written by 37 authors, or 1.19 authors per paper. The number skyrocketed by 1997 when 114 papers were penned by 298 authors, or 2.61 authors per paper.

The mushrooming expansion of petrological knowledge that began in mid-century has been accompanied by greater specialization within the field. During the first half of the century, igneous petrologists generally were reasonably well versed in most aspects of the discipline. Granted, petrologists had their particular interests: Buddington liked anorthosite, Larsen liked calc–alkalic granitoids, Holmes was interested in radioactivity in igneous rocks, and Bowen was arguably the most specialized igneous petrologist of his era with his single-minded concentration on solving the problem of igneous rock diversity by experimental methods. For the most part, however, igneous petrologists of the first half of the twentieth century were rather well-rounded. During the latter part of the century, however, it became much more commonplace to encounter petrologists and geochemists who were recognized for their abilities in one restricted area. Petrologists became experts on one group of rock types: Roger Mitchell on ultramafic rocks, Wallace Pitcher on granites, Robert Smith on ignimbrites. Other petrologically minded individuals became practitioners of one laboratory method: Donald DePaolo on neodymium isotopes, Hugh Taylor, Jr., on oxygen isotopes, Wilhelm Johannes on experimental petrology of granite, Stephen Sparks on the fluid dynamics of magmas, and Donald Dingwell on the physical properties of magmas. Other geologists became devotees of theoretical thermodynamics such as Mark Ghiorso.

Because igneous petrology became increasingly multi-faceted toward the close of the twentieth century, petrologists of necessity became more specialized, and it made sense for teams of petrologists to join forces and pool their expertise in working to solve complex problems requiring more than one approach.

THE PLATE TECTONICS REVOLUTION

Conceptual and technological developments, of course, continued to shape igneous petrology during the middle of the century. Without doubt, the most far-reaching of the conceptual advances was the formulation of the hypothesis of plate tectonics, an intellectual revolution within the Earth sciences that led to striking new ways of looking at virtually every discipline of geology, including, perhaps especially, igneous petrology. The familiar story of the plate tectonics revolution has been told in many places by many writers (Takeuchi et al., 1967; Hallam, 1973; Wyllie, 1976; Glen, 1982; Menard, 1986). Only the briefest account pertinent to igneous petrology is given here.

Before the middle of the twentieth century, igneous petrology was essentially the igneous petrology of the continents. It was not yet global igneous petrology. Little was known about the geology of the vast ocean basins that comprise approximately two-thirds of Earth's surface. There had, of course, been many expeditions to oceanic islands by geologists like Darwin, Dana, and Daly, and a fair amount was known about islands such as the Cape Verdes, the Azores, Iceland, Hawaii, Gough, Bouvet, St. Helena, Tristan da Cunha, and Ascension. Next to nothing, however, was known about the floors of the ocean basins. Soundings throughout the nineteenth century by rope and wire had yielded a crude picture of the broad topographic features of the seafloor, and echo soundings and seismology, particularly after (and during!) World War II, provided a much more detailed and accurate knowledge of oceanic bathymetry. The ocean ridge and rise system, the abyssal plains, the trenches, and the seamounts were all generally known shortly after the war. During all that time, however, negligible amounts of material had been dredged from the sea bottom. Basalt had been dredged by H.M.S. *Challenger* during its 1872–1876 expedition, and subsequent dredges in the 1930s and 1940s brought some basalt to the surface.

Of particular importance was the 1947 cruise of *Atlantis*, a research vessel of the Woods Hole Oceanographic Institution in Massachusetts under the direction of oceanographer Maurice Ewing. Working near latitude 30°N of the Mid-Atlantic Ridge, *Atlantis* recovered several hundred pounds of boulders from depths as great as 2500 fathoms. Shand (1949b) reported that some limestones were dredged but that the boulders were mostly serpentinite, holocrystalline gabbro, and pillow basalt with and without olivine. He wrote that the *Atlantis* collection confirmed the predominance of basalt on the ocean floor. Daly, who had shown by chemical analyses that basalts from Ascension Island did not differ much from average continental basalt, was persuaded of the relative uniformity of basalt throughout the world. Shand claimed that the basalts of the *Atlantis* collection were consistent with Daly's contention.

The expeditions of Bruce Heezen disclosed the existence of the deep axial rift valleys of the ridge systems (Heezen et al., 1959). More rocks were dredged by R.R.S. *Discovery II* in 1960. Muir and Tilley (1964), summarizing the increased number of descriptions of basalts dredged by the *Discovery II* and other recent voyages, asserted that the Mid-Atlantic Ridge basalts, in fact, did not resemble those of the oceanic islands after all.

Studies of ocean floor basalts took a major step forward in the early 1960s when A.E.J. Engel and his wife, Celeste, turned their attention to the topic.

The Engels, employed respectively by the University of California at San Diego in La Jolla and the United States Geological Survey, took advantage of the opportunity afforded by a close association with the Scripps Institution of Oceanography, a unit of the University of California at San Diego, to participate in voyages and examine dredge samples obtained from the Mid-Atlantic Ridge, the East Pacific Rise, and the Mid-Indian Ocean Ridge. Chemical analyses of the major elements were obtained by Celeste Engel, and the Engels enlisted the assistance of R. G. Havens of the U. S. Geological Survey office in Denver in determining the trace element abundances by emission spectrography.

Engel et al. (1965) summarized their work in a paper on the chemical characteristics of oceanic basalt and the upper mantle. They wrote that recent oceanic studies such as those of Muir and Tilley (1964) suggested that a distinctive tholeiitic basalt is the principal igneous rock in the deeper parts of the ocean basins. These tholeiitic basalts, they said, appear to exist in great volumes along the scarp-like ridges of the oceanic rift zones, on the deeper flanks of the oceanic ridge and rise systems, in the large Hawaiian and Icelandic archipelagic aprons, on the abyssal hills, in the deep oceanic trenches, and even as possible sills beneath the seafloor, as indicated by the core samples obtained from Project Mohole in 1961.

The Engels pointed out that oceanic tholeiite has unique compositional features that include about 49 to 50 weight percent silica, a low ratio of ferric oxide/ferrous oxide in the unaltered rock, and a high ratio of Na/K. They further reported that oceanic tholeiites contain very low abundances of Ba, K, P, Pb, Rb, Sr, Ti, Zr, U, and Th. They noted that oceanic tholeiites differ chemically from continental tholeiitic basalt in that the former contain lower abundances of Si, K, Ba, Cs, Pb, Rb, Sr, Th, U, and Zr than do the latter. They also discovered that continental tholeiitic basalts are characterized by higher isotopic ratios of $^{87}Sr/^{86}Sr$ and $^{206}Pb/^{204}Pb$ than oceanic tholeiites. From these data, the Engels inferred that continental tholeiitic liquids had been contaminated by the chemical elements and isotopes in which they are enriched by scavenging from sialic inclusions and from the walls of their crustal conduits and magma chambers.

The authors further indicated differences between oceanic tholeiites and alkalic basalts found on the higher volcanoes of the eastern Pacific Ocean such as on Guadalupe and the Revillagigedos Islands. They found that alkalic basalts are enriched in Na, K, Ba, Ti, P, U, and Th; have higher ratios of ferric oxide/ferrous oxide; have lower ratios of Na/K; and possess higher isotopic ratios of $^{87}Sr/^{86}Sr$ and $^{206}Pb/^{204}Pb$ than do oceanic tholeiites. Engel and Engel (1964) had earlier suspected that the oceanic alkali-rich basalts were derived from oceanic tholeiites through gravitative differentiation in and below the higher volcanoes in the oceans.

In time, increasing numbers of samples of ocean-floor basalt were obtained thanks to the Deep Sea Drilling Program (DSDP), carried out from the mid-1960s onward by the combined efforts of Scripps Institution of Oceanography, Woods Hole Oceanographic Institution, the Lamont-Doherty Geological Observatory of Columbia University, and the Rosenstiel School of Atmospheric and Oceanic Studies at the University of Miami; to project FAMOUS, the French-American Mid-Ocean Undersea Study that carried out a submersible exploration of portions of the Mid-Atlantic Ridge in 1974; and to the 1983 successor to DSDP, the Ocean Drilling Program (ODP), conducted under the leadership of Texas A & M University. As a result of samples returned by these large-scale cooperative ventures, the relatively simple picture of ocean-floor petrology suggested by Engel et al. (1965) eventually gave way to a more nuanced understanding of the composition and nature of the oceanic crust.

When geophysicist Robert Dietz pondered the ocean floors and petrologist Harry Hess turned from the study of pyroxenes and layered ultramafic intrusions to a consideration of alpine serpentinites and ocean basins, the result was the radical proposal that the ocean floors are not fixed features but, like the continental land masses, are in constant lateral motion. Hess (1962) and Dietz (1961) independently and virtually simultaneously suggested the idea that the seafloor, as a result of convective, conveyor-belt-like motion, is constantly spreading away from the sites of its origin at mid-oceanic ridges and is being destroyed beneath oceanic trenches where it sinks down into the mantle. Menard (1986) has written fascinating recollections of the development, by Harry Hess and Robert Dietz, of the idea of "sea-floor spreading," a term coined by Dietz.

One ultimate effect of the concepts of seafloor spreading and plate tectonics was to precipitate a revolution in the way that petrologists perceived the relationship between magmatism and tectonics. Petrologists had long recognized that island arcs are characterized by calc–alkalic plutons and explosive volcanism featuring andesite and rhyolite. They were also beginning to realize that some island arcs are characterized by parallel belts of igneous activity. Calc–alkalic belts were observed to lie closer to oceanic trenches seaward of island arcs, and more alkalic belts were found to lie farther from the trenches (Miyashiro, 1961). Petrologists had also known that the igneous activity of island arcs is in some way spatially associated with vigorous seismicity. As knowledge of the seafloor increased, they also learned that oceanic trenches are an integral part of that spatial association. Petrologists knew that the continental margins rimming the Pacific Ocean basin are characterized by a virtually continuous belt of mountains extending from the Andes Mountains at the southern tip of Chile all the way up the western coast of the Americas, then through the successive island arc systems of the western Pacific, including the Aleutians, Japan, and the Philippines. The Cordilleran ranges of western North and South

America were known to be characterized by vast composite granitoid batholiths and lesser amounts of calc–alkalic volcanism. Petrologists discovered that many of the circum-Pacific batholith belts are characterized by compositional zoning. Moore (1959), for example, proposed the existence of a "quartz diorite line" that separates more potassic and siliceous granitoids toward the continental interior from more sodic and intermediate granitoids toward the continental margins. Petrologists had also been aware of the close association of the weakly seismic zones of tension and rifting in the interior of continents associated with highly alkalic and even carbonatitic plutonism and volcanism such as that in the Monteregian Hills of Quebec, the Oslo district in Norway, the Rhine graben in Germany, Magnet Cove and Hot Springs in Arkansas, and the great alkalic–carbonatite lava emissions in the rift valleys of eastern Africa.

Plate tectonics opened up new ways of thinking about these and other magmatic-tectonic associations. With the dawn of the plate tectonics era, petrologists needed to understand the generation of the tholeiitic mid-ocean-ridge basalt (MORB) magmas from specific zones in the upper mantle. They needed to understand the generation of calc–alkalic suites either from descending slabs of subducted oceanic crust and its accompanying subducted seafloor sediments; or from a wedge of subcontinental mantle that lay directly above the subducted slab; or from the base of a thickened mass of continental crust; or, finally, from some combination of these sources. Geologists began to think of parallel calc–alkalic and alkalic belts on island arcs in terms of differing depths of magmatic sources at various points along descending slabs, and they began to think of the compositional differences of composite batholiths in similar terms. They needed to understand the generation of alkalic magmas in zones of flexure and tension that may have served as rifts that failed to develop into small, new ocean basins.

After the 1960s, igneous petrologists began to categorize suites of rocks as having been produced at accreting plate margins, for example, mid-ocean-ridge basalts; at collisional destructive plate margins, for example, the great granitoid batholiths; at intraplate localities within the continents, for example, alkalic rock–carbonatite complexes; and at intraplate localities within ocean basins, for example, oceanic islands. They also began to make correlations between igneous rock type and the different sorts of collisional zones. Hence, the great calc–alkalic granitoid batholiths were linked to collisions between continental and oceanic plates. Collisions between two oceanic plates generally were associated with much smaller volumes of magmas, predominantly volcanic, and with more silica-poor to intermediate magmas, such as basalt and andesite and much less rhyolite. Collisions between two continental plates were associated with relatively little magmatic activity or with andesite and small quartz diorite plutons.

The new plate tectonics paradigm also laid to rest a debate over the existence of ultramafic magmas. Harry Hess had accepted the notion that many peridotite plutons, some serpentinized, represent the product of crystallization of an ultramafic magma. In contrast, Bowen doubted the possibility of such high-temperature magmas on experimental grounds and maintained that peridotites and dunites were the result of crystal segregation from silica-poor magmas. These accumulations of olivine and pyroxene crystals, Bowen taught, had been mobilized and injected into orogenic belts in the solid state. After the advent of plate tectonics, petrologists became comfortable with the idea that many alpine-type peridotite–serpentinite bodies represent fragments of oceanic crust and/or upper oceanic mantle, and, in some instances, pieces of seamounts that had been sheared off in subduction zones and emplaced along faults in accretionary wedges. Thus was explained the common association of peridotites with chert and other sea-bottom sedimentary rocks. Many of the igneous features in peridotites were now understood to have been produced at accretion zones where new oceanic crust is being formed rather than during injection or emplacement in the continents.

The effect of the plate tectonics revolution on igneous petrology may be detected from a comparison of the second edition of the classic textbook by Turner and Verhoogen (1960), *Igneous and Metamorphic Petrology*, with its successor, *Igneous Petrology*, by Carmichael et al. (1974). In the 1960 edition, the authors distinguished an "alkaline olivine–basalt association" from a "tholeiitic association." They considered the alkaline association to be widely developed on continents as at Otago, New Zealand, and in the teschenite sills at Lugar and the Shiant Isles in Scotland. They also noted this association in the ocean basins on such islands as Tahiti, Samoa, St. Helena, Ascension, and the Kerguelen archipelago. The tholeiitic association was said to be characteristic of continental flood-basalt provinces like the Deccan plateau of India, the Columbia River plains of the Pacific Northwest of the United States, the Paraná basalts of South America, the Karoo sill swarms of South Africa, and the Triassic dikes and sills of the eastern United States. The authors recognized the close association of both tholeiitic and alkaline olivine–basalt suites in Hawaii and the Tertiary Brito-Arctic province. They said virtually nothing about ocean-floor basalts because hardly anything was known about them in 1960.

Although extremely cautious about drawing too many conclusions concerning the associations of kinds of igneous rocks with certain plate-tectonic regimes, Carmichael, Turner, and Verhoogen distinguished basaltic associations of the ocean basins from continental tholeiitic provinces in 1974. They clearly distinguished between ocean-ridge basalts (abyssal tholeiites) and the basalts of oceanic islands. Moreover, they pointed out the chemical and petrographic uniformity of tholeiitic ocean-ridge basalts. We find, too, that the authors discussed cautiously but in some detail the association of the andesitic volcanism of island arcs and continental margins with subducting plates of ocean-floor crust. After the 1960s, terrestrial igneous petrology was viewed with growing clarity through the lens of plate tectonics.

LUNAR PETROLOGY

A second spectacular scientific achievement of the 1960s was the establishment of American and Soviet programs for the exploration of the lunar surface. In his recounting of the story of lunar exploration, Wilhelms (1993) pointed out that the race between the United States and the USSR to land a person on the moon was motivated much more by nationalistic pride and geo-political considerations than by a thirst for knowledge. Moreover, the effort to reach the moon resembled in many ways a grandiose aeronautical engineering stunt rather than a scientific expedition. Despite the fact that scientific knowledge was not at the top of the list of priorities in reaching for the moon, provision was made for achieving some scientific goals. Scientists, however, were not fully agreed on the specific scientific objectives of the missions. Tensions existed between geologists and the more mathematically inclined community of physicists and astronomers. In the end, the series of lunar landings proved to be an enormous bonanza for igneous petrology.

Although it had long been suspected that much of the lunar surface is composed of lava flows and was profoundly affected by the constant bombardment of meteorites, the landing programs confirmed these conclusions in an impressive way. Prior to the manned landings of the American Apollo series, several preliminary missions were sent to the moon to improve flight procedures and to find appropriate landing sites. In the late 1950s and early 1960s, both American Pioneer and Soviet Luna attempts at lunar flyby, orbiting, or crash landing generally ended in failure. The Soviets achieved a successful lunar crash landing with Luna 2 in 1959 but were unable to make another landing until the 1966 soft landing of Luna 9. The United States initiated its Ranger series of crash landings in 1961, but not until 1964 and 1965 did it achieve success with the landings of Rangers 7 through 9. Just before these Ranger probes self-destructed upon impact with the lunar surface, they transmitted remarkable images to Earth that confirmed the notion that the moon's surface is pockmarked with meteorite impact craters on a wide range of scales.

During 1966 through 1968, the United States undertook a series of seven Surveyor missions, five of which soft landed a probe onto the moon's surface. The Surveyors obtained ground-based images of lunar topography and featured on-board spectrometers that analyzed the chemical composition of lunar soil scooped up by a small trench digger. An approximately basaltic composition was indicated. During the same period, the Soviets launched several successful lunar orbiters and a couple of soft landers of the Luna series, and the

United States initiated its series of five successful Lunar Orbiter missions. The Lunar Orbiter probes were also equipped with spectrometers that measured the chemical composition of the surface. The data indicated that chemical elements of silica-poor igneous rocks and anorthosite are present in great abundance. Lunar orbiters also produced a host of stunning images of a great portion of the moon's features including portions of the backside, previously observed only by one of the Soviet Luna probes. The well-known, magnificent oblique image of the Copernicus crater depicting its great central peaks also resulted from an Orbiter mission.

Between 1969 and 1976, the Soviet Union placed the Lunokhod 1 and 2 rovers on the lunar surface as well as the Luna 16, 20, and 24 sample returners. Between 1969 and 1972, the United States placed a dozen men on the moon's surface during the Apollo 11, 12, 14, 15, 16, and 17 missions. The astronauts landed at and explored a variety of topographic features including flat maria, the edges of lunar highlands, rilles, and aprons of meteorite impact ejecta. The Apollo astronauts retrieved 382 kilograms of lunar rocks and regolith on the six missions. These samples and material retrieved by Luna sample returners have probably been maintained with greater care and analyzed in greater detail than virtually any other geological specimens collected anywhere at any time. The samples returned by the Apollo astronauts were stored at the Johnson Space Center in Houston, Texas, and examined by carefully selected teams of experts from around the world under rigorously controlled conditions. A large number of the world's leading igneous petrologists and geochemists willingly diverted their attention from their ongoing projects to join in a concerted assault on the problem of characterizing the lunar materials and suggesting preliminary interpretations of lunar petrogenesis. Perhaps one of the most eagerly awaited scientific publications ever was the January 30, 1970 issue of *Science*, the "moon issue," published only six months after the landing of Apollo 11 and containing nearly 350 pages of papers on the geology of the Apollo 11 site and samples. The "moon issue" contained several articles on Rb–Sr, U–Th–Pb, and $^{40}Ar/^{39}Ar$ geochronology, a host of trace-element and major element studies, papers on isotopes, studies on magnetic, electrical, and other physical properties, and, of course, studies of mineralogy and petrology. Petrological papers were contributed by large teams from many of the leading petrological centers, including the University of Cambridge, the University of Chicago, the University of Manchester, the Laboratory for Space Physics at the NASA Goddard Space Flight Center, the Geological Survey of Canada, the Institute of Meteoritics at the University of New Mexico, the University of Durham, the Smithsonian Institution, the Grant Institute of Geology at the University of Edinburgh, the Australian National University, the University of Tokyo, the Geophysical Laboratory, Princeton University, the Center for Volcanology at the University of Oregon, the U. S. Geological Survey, and several others. The information gleaned from Apollo 11 was summarized by Mason and Melson (1970) of the Smithsonian Institution.

Investigations of new materials returned from succeeding Apollo missions and ongoing studies of previously collected lunar samples were presented at annual Lunar Science Conferences and successor Lunar and Planetary Science Conferences held in Houston. Petrological knowledge obtained from all the lunar missions was summarized by Ross Taylor in a couple of books (S. R. Taylor, 1975, 1982) and by A. E. Ringwood (1979). Since moving to the Australian National University from Cape Town in 1961, Taylor had developed the technique of spark-source mass spectrometry for the accurate and precise quantitative determination of a vast range of minor and trace elements in geological materials. Given his prodigious expertise in spectrometric methods, Taylor was urged to head up the Preliminary Examination Team that studied the first lunar samples returned to Earth by Apollo 11 in 1969, and he enjoyed the privilege of carrying out the first analysis of a returned lunar sample, albeit by emission spectrography (Lunar Sample Preliminary Examination Team, 1969). He continued as a principal investigator for the National Aeronautics and Space Administration (NASA) for 20 years, in which capacity Taylor was able to develop models of lunar composition, origin, and evolution. In that respect, he had been greatly influenced by L. R. Wager to think about lunar evolution in terms of large-scale Skaergaard-like magma oceans.

From the Apollo and Luna samples, geologists discovered that the moon consists predominantly of igneous rocks. Lacking an atmosphere and oceans, chemical weathering is virtually nonexistent on the moon. As a result, no sedimentary rocks are present. Being a relatively cold body, the moon lacks the internal energy necessary to drive plate tectonics, orogenesis, or compressive tectonics in which the kind of regional metamorphism typical of the terrestrial environment develops. The only significant lunar metamorphism occurs because of the constant bombardment of surface rocks by meteorites. All rocks found on the moon are igneous rocks that display varying degrees of shock metamorphism.

The repeated high-energy impacts on the lunar surface pulverized the surface rocks. The surface soils and breccias, composed of fragments of igneous rocks, are several meters or tens of meters thick. Many impacts fused the surface material so that the soils contain innumerable glass spherules. Surface rocks and soils are commonly covered by a thin crust of glass. Thus, unlike on Earth, where glass is almost invariably the product of rapid chilling of an igneous melt from Earth's interior, lunar glass is the result of shock by high-velocity projectiles.

Study of returned samples has shown that the igneous rocks of the moon are predominantly basalt, gabbro, and anorthosite. The mare regions visited by the astronauts are overlain by blankets of thin, very fluid basaltic lavas. Like terrestrial basalts, the lunar basalts and coarser gabbros are also composed of plagioclase, pyroxene, and olivine. Because they differ from terrestrial igneous rocks in their utter lack of chemical weathering, lunar rocks presented petrographers with almost unbelievably clear and fresh thin sections, a remarkable situation considering that lunar basalts were determined to be 3.2 to 3.9 billion years in age by various radiometric dating techniques. The lunar basalts and gabbros turned out to be considerably different chemically from terrestrial basalts, however, in that they are notoriously deficient in the alkali elements sodium, potassium, and rubidium. Moreover, they are completely lacking in ferric iron and water. Many of them are also strikingly enriched in titanium relative to terrestrial basalts. Lunar basalts and gabbros are also considerably more iron-rich than terrestrial basalts. As a result of these chemical differences, lunar basaltic plagioclase has a much higher anorthite component and a much lower albite component than the typical terrestrial basalt plagioclase. Iron oxide minerals like magnetite and hematite are absent because of the lack of ferric iron. Ilmenite is the common iron-bearing oxide phase. The basalts and gabbros are devoid of common terrestrial minerals like amphibole or biotite because of the lack of water. The chemical composition of lunar basalts also renders them considerably less viscous lava flows than their earthly counterparts.

The Apollo astronauts also returned samples of an unusual basalt that is highly enriched in potassium, the rare earth elements, and phosphorus. This rock was christened with the acronym, KREEP. Petrologists speculated that KREEP basalts might have been generated from a layer of residual ultramafic minerals beneath the anorthosite crust.

Investigators disclosed that anorthosite is the major rock type to compose the vast lunar highland regions. Like terrestrial anorthosite, the lunar counterparts are composed primarily of plagioclase feldspar with minor amounts of pyroxene and olivine. They differ considerably in chemical composition for the same reasons as lunar and terrestrial basalts differ, namely, extreme paucity of alkalis and the lack of ferric iron and water. The plagioclase in lunar anorthosite is extremely anorthite-rich, and the anorthosite lacks magnetite.

The absence of water and ferric iron in lunar samples proved not to be a great surprise to petrologists in view of the fact that no evidence for water on the lunar surface had ever been detected with Earth-based instruments. Moreover, the size of the moon was typically regarded as too small to retain much water during the initial stages of its formation on the grounds that its gravitational field would have been too weak to retain highly volatile hydrogen atoms. The lack of any atmosphere, of course, indicated that oxidation would be extremely unlikely to occur at the surface of the moon. The most surprising geochemical result was the paucity of alkalis. Geochemists also attributed this deficiency to the generally volatile nature of the alkali metals at high temperature and the inability of a relatively small body with a weak gravitational field to retain very large abundances of such volatile substances during the initial stages of its formation.

Our limited acquaintance with the moon has provided but a small glimpse at the igneous petrology of another planetary object whose chemical composition is different from that of Earth in important ways. Although examination of returned samples has continued to the present and has yielded a wealth of valuable information, further substantial developments in lunar igneous petrology have been prevented by the termination of programs of lunar exploration in the 1970s. As the heady years of lunar exploration have receded from view, igneous petrologists have, for the most part, come back down to Earth.

24 ← DELTA AND EPSILON

Stable and Radiogenic Isotopes in Igneous Petrogenesis

Given that igneous rocks are composed of silicate minerals and that the formation of igneous bodies from magma involves reactions and equilibria among complex melts, various crystals, and vapor, it is hardly surprising that the discipline of igneous petrology has been strongly shaped by discoveries in chemistry throughout its history. Even before igneous petrology was acknowledged as a distinct discipline, geologists were curious about the chemical compositions of magmas, igneous rocks, and their constituent minerals. As occasion arose, petrologists applied insights from solution chemistry, phase equilibria, and chemical thermodynamics to the behavior of melts and crystals. Insights from physics, however, have become equally important to igneous petrology. Knowledge of trace-element abundances in igneous rocks, at first acquired largely by spectrography, would not have been possible without the discovery and application of the phenomena of emission spectra produced by atoms excited in a flame. Another development that began to influence igneous petrology especially dramatically in the mid-twentieth century emerged from the field of nuclear physics. This development entailed the application of both stable and radioactive isotope distributions to questions about the formation of minerals and rocks.

THE DISCOVERY OF ISOTOPES

At present, geologists instinctively think of radiometric dating in terms of isotopes like ^{238}U, ^{40}K, and ^{147}Sm. The discovery of the phenomenon of radioactivity, however, preceded the recognition of the existence of isotopes. The earliest attempts to determine the ages of minerals by physicists like Ernest Rutherford, Bertram Boltwood, and R. J. Strutt in the first decade of the twentieth century were based on the understanding that uranium is transformed by some radioactive decay process into lead and helium at measurable rates. These pioneers of radiometric dating, however, had no knowledge at that time of the different forms of uranium or lead. Rutherford and Boltwood obtained ages of 500 Ma (million years) and 410 Ma on allegedly Cambrian uranium-bearing minerals from Glastonbury, Connecticut. Although the dates eventually proved to be inaccurate, they were sufficiently large enough to call into question the estimates of the age of Earth existing at the turn of the century. According to Becker (1908) these estimates ranged from around 24 Ma to around 100 Ma.

In 1907, however, Boltwood separated "ionium," a substance chemically indistinguishable from thorium, another chemical element that was recognized as radioactive, but which possesses a different atomic mass. In 1912, Soddy referred to such apparently different forms of the same chemical element as "isotopes." His conception of isotopes received independent confirmation from the work of J. J. Thomson who produced positively charged rays of ions in a gas discharge tube and passed them through a hole in the tube to be deflected by both electrostatic and magnetic fields. Thomson repeatedly observed multiple parabolic traces of the ion trajectories for various gases, and when he obtained two parabolic traces for ions produced in monatomic neon gas, he realized that two different forms of ionized neon were present. To these forms he assigned the atomic masses 20 and 22.

The investigation of the various possible isotopes of the different chemical elements received substantial impetus when Frederick Aston, working at the Cavendish Laboratory of the University of Cambridge, constructed the first true mass spectrometer in 1919. With the instrument, Aston (1927a) reported that he had been able to distinguish spectral lines produced by atoms with masses differing by about one part in 130. He determined with fair certainty the isotopic constitution of more than fifty chemical elements and demonstrated that the masses of all the atoms except hydrogen could be expressed as integers within one or two parts per thousand on a scale in which the mass of oxygen was set to 16.

Thanks to a generous grant from Great Britain's Department of Science and Industrial Research, construction began in 1921 on a new mass spectrometer that had five times the resolving power and ten times the accuracy of the original one that was dismantled in 1925. Now Aston was able to tackle isotope investigations of the heavier chemical elements. He reported data on isotopes of H, He, B, C, N, F, Ne, P, S, Cl, Ar, As, Br, Kr, Sn, I, Xe, W, and Hg (Aston, 1927a). Soon thereafter, Aston (1927b) further reported that, after repeated failures, he had finally obtained a mass spectrum of ordinary lead that consisted of three principal lines corresponding to ^{206}Pb, ^{207}Pb, and ^{208}Pb. He also noted that many other lead isotopes, including ^{204}Pb, might be present in small proportions.

Aston's spectrometer employed a double focusing of the positive rays to produce sharp lines on a photographic detector. His instrument was eventually replaced by the design of Dempster and Bainbridge who used a large magnet in order to bend the positive ion beams, the different isotopes being bent at different angles depending on their masses. Mass spectrometry was developed further and applied to geological problems by Alfred O. Nier (1911–1994), particularly during the 1930s and 1940s. Nier graduated from the University of Minnesota in 1931 and received a Ph.D. degree in physics from that university in 1936 for mass spectrometric work in which he determined the abundances of Ar and K isotopes and also discovered the existence of ^{40}K (Craig, 1985; Nier, 1985). Nier went on to spend two years at Harvard University with Kenneth T. Bainbridge building a bigger, improved mass spectrometer that could even make quantitative isotopic measurements for heavy elements. At Harvard he studied nineteen chemical elements, discovered several new isotopes, measured the abundances of Pb isotopes, made the first measurement of the ^{238}U/^{235}U ratio, and laid the basis for the mass spectrometric measurement of geological time by means of U–Pb methods.

In 1938, Nier returned to Minnesota, where he served as Professor of Physics until his retirement in 1980. Within five months of his return to Minnesota, Nier had built a mass spectrometer like the Harvard instrument and resumed his isotopic measurements of lead samples. Nier (1985) commented that, in 1939, there were only about half a dozen mass spectrometers in the world capable of making precise isotopic measurements. Part of the difficulty lay in the employment of magnets that weighed as much as two tons. To overcome this problem, Nier designed a much lighter 60° sector magnet for deflecting the ion beams. In 1947, Nier introduced what came to be known as the "Nier machine," a mass spectrometer that incorporated the 60° deflecting magnet, a null method system for precise measurement of isotopic ratios with a double collector, and a capillary-leak, viscous-flow inlet system that was designed to circumvent mass fractionation in gas samples. These technical innovations, together with Nier's generosity in assisting other geochemists, are widely recognized as driving the field of isotope geochemistry by making it feasible for numerous laboratories to build their own mass spectrometers. The Nier-design mass spectrometer fueled the advance of isotope geochemistry much as the polarizing microscope opened up the discipline of igneous petrography.

OXYGEN ISOTOPES

Interest in the application of stable isotope distributions to geological problems began in earnest with the work of Harold Clayton Urey (1893–1981) at the University of Chicago. After doing Nobel prize-winning research on hydrogen isotopes including the discovery of deuterium, Urey began investigating the fractionation patterns of nitrogen, oxygen, carbon, and hydrogen isotopes in nature and their potential use as paleo-thermometers and as tracers of chemical reactions. As early as 1934, he had measured ^{18}O/^{16}O ratios in some meteorites and igneous rocks (Goldsmith, 1991). After World War II, President Hutchins of the University of Chicago hired several outstanding physicists and chemists, including Urey, to work in the newly established Institute of Nuclear Studies (Pettijohn, 1984). Urey had a curiosity about the origin and development of Earth and the other planets and fostered cross-fertilization of ideas with the Chicago geologists. Pettijohn (1984, p. 201) recollected that the "geochemical revolution" taking place at Chicago around 1950 was a time in which the atmosphere was "electric, mutually stimulating, and creative." Urey attracted several bright young researchers to Chicago, including a postdoctoral fellow, Samuel Epstein (1919–2001), who had worked with Harold Thode on the only mass spectrometer in Canada (Epstein, 1997). Urey, Epstein, and their co-workers built a mass spectrometer at the University of Chicago based on Nier's double collector design.

One of the bright young men attracted to Chicago was Peter Baertschi, who had received a grant from the Swiss Foundation to come to the Institute for Nuclear Studies. Urey suggested that he study the isotopic composition of oxygen in silicate rocks. Baertschi (1950) compared the oxygen isotope ratios ($^{18}O/^{16}O$) of his samples with a Precambrian gneissic granite from South Dakota that was used as a standard. The deviations of the oxygen isotope ratios in the samples from the ratio in the standard were measured in per mil, that is, parts per thousand. Seawater, for example, showed a value of −8.5 per mil relative to the standard, indicating that seawater has a smaller $^{18}O/^{16}O$ ratio than the South Dakota standard. Baertschi found that granite (0.0 per mil), basalt (-4.5 per mil), lavas (-1.0 per mil), and obsidian had much lower $^{18}O/^{16}O$ ratio, that is, are much lighter isotopically than sedimentary rocks like quartzite, arkose, shale, flint, and diatomite. He did not know how to account for the slight differences in the oxygen isotope ratios of the various igneous rocks. With the ongoing granite controversy no doubt very much in mind, Baertschi suggested from his limited results that one of the most important applications of accurate oxygen isotope ratio determination in silicate rocks would be the possibility of distinguishing igneous rocks from metasedimentary rocks.

Building on Baertschi's efforts, Sol R. Silverman (b. 1918) continued work on the isotope geology of oxygen for his Ph.D. degree at Chicago with the assistance of Urey, Epstein, and Baertschi. A very important contribution of Baertschi and Silverman was the development of the fluorine-reaction technique. They showed that one could obtain quantitative removal of oxygen gas from nearly all silicate and oxide minerals by reacting them with fluorine gas at 600°C in nickel vessels. The $^{18}O/^{16}O$ ratio of the liberated oxygen gas could then be measured in a mass spectrometer. This basic method continued to be used throughout the next half century for the measurement of $^{18}O/^{16}O$ ratio of minerals and rocks.

For his standard, Silverman utilized pure quartz from a pegmatite at Randville, Michigan. The results from measurements on various samples were then adjusted to the value for Hawaii seawater, which was arbitrarily set at zero per mil. Silverman (1951) analyzed a host of igneous rocks, including

Columbia River basalt, Disco basalt, Mount Etna lava, dunite, olivine basalt, Yellowstone obsidian, Baertschi's standard gneissic granite from South Dakota, nepheline syenite, and Paricutin lava. He also measured several sedimentary rocks, Chicago tap water, Hawaii seawater, and air, but no metamorphic rocks. Like Baertschi, he discovered that the $^{18}O/^{16}O$ ratios of the sedimentary rocks, and especially diatomite, are consistently higher than those of the igneous rocks, implying that the deposition of silica from solution at temperatures of Earth's surface is accompanied by very marked enrichment of ^{18}O in the precipitated phase.

Silverman found that granitic rocks are characterized by higher ^{18}O abundances than basaltic rocks, a phenomenon that he believed to be incompatible with the prevailing idea of fractional crystallization. On the presumption that the heavier ^{18}O isotope would be preferentially incorporated into solid phases upon crystallization, Silverman reasoned that if granites had originated by differentiation from magma, then the earlier crystallizing minerals that form basalts and gabbros ought to be richer in the heavy isotope than the later crystallizing minerals. Silverman also analyzed quartz and feldspar separates from three different granites to see if there is any fractionation of oxygen isotopes between silicates, and he found quartz in every instance to be richer in ^{18}O than coexisting potassium feldspar. Silverman was perplexed by this result. Reasoning that Bowen's reaction principle established an order of crystallization in which potassium feldspar would crystallize earlier than quartz, Silverman expected that the early-forming feldspar would have preferentially extracted the heavier oxygen isotope, leaving behind an isotopically lighter residual melt from which the quartz would ultimately crystallize, but the result was the opposite of what he anticipated.

In retrospect, Silverman wondered if his choice of quartz and potassum feldspar was the most appropriate given contemporary uncertainties about the magmatic origin of granites. He was concerned that his results might be irrelevant to the process of fractional crystallization from a melt, but, he thought, perhaps the concept of diffusion in granitization might be tested with oxygen isotopes. Silverman decided to examine the $^{18}O/^{16}O$ ratios of a group of rocks from Wisconsin that make a transition in wall rocks from granophyre to gabbro over a lateral distance of twenty feet. He found that the rocks become isotopically heavier in passing from the gabbro through the transition zone to the granophyre. Reasoning that ions enriched in the lighter oxygen isotope would have a higher degree of mobility, Silverman concluded that the trends in the isotopic ratios and silicate ion concentrations met the requirements for evidence of solid diffusion. Without providing substantive grounds for thinking that silicate ions should be expected to migrate, he argued that if silicate ions form the migrating particles, then the rocks farthest from the source of migration should be enriched in the lighter isotope relative to material closer to the source. He discounted the effectiveness of Soret diffusion, gravitational forces, assimilation of gabbro by granophyre, mixing of granophyre and gabbro melts, crystal settling within melt, or crystallization from melt as adequate explanations for the transition zone. He concluded that oxygen redistributed itself by some mechanism in solid rock such as permeation, solution, or solid diffusion.

In 1953, California Institute of Technology determined to enter this newly developing field of isotope geochemistry, and hired several prominent researchers from the isotope group at Chicago. The nearly simultaneous departure of Epstein, Heinz Lowenstam, Claire Patterson, Charles McKinney, Harrison Brown, and Gerald Wasserburg for Pasadena left a gaping hole in Chicago's geochemistry program, and in one stroke Caltech moved to the forefront of isotopic geochemistry. Among Samuel Epstein's early graduate students and stellar pupils were Robert N. Clayton (b. 1930) and Hugh P. Taylor, Jr. (b. 1932). A graduate of Queen's University, Clayton received a master's degree from Queen's in 1952 and his Ph.D. degree in chemistry from Caltech in 1955. After teaching for a couple of years at Penn State, he moved to the University of Chicago, where he established the premier laboratory in the world for measuring the temperature dependence of equilibrium $^{18}O/^{16}O$ fractionations among silicates, oxides, carbonates, and H_2O. Clayton's measurements were crucial in the development of the field of oxygen-isotope geothermometry and in utilizing and interpreting all oxygen isotopic data obtained on rocks and minerals. Taylor graduated from Caltech in 1954, and after receiving a master's degree from Harvard in 1955, he returned to Caltech and received his Ph.D. degree in 1959. After teaching for a year at Penn State, Taylor returned to California to pursue a prolific career in the study of oxygen isotopes while serving on the faculty at Caltech. Soon after embarking on his professorial career, Taylor published his Ph.D. dissertation with Epstein (Taylor and Epstein, 1962). This paper represented the first substantial compilation and discussion of oxygen-isotope data on silicate minerals and rocks. They compiled data showing that quartz is the isotopically heaviest silicate mineral in igneous rocks, followed by feldspar, mica, pyroxene and amphibole, olivine, and, lastly, iron oxide minerals like magnetite and ilmenite. They also showed that igneous rocks generally become isotopically heavier from gabbro to granite.

Taylor and Epstein perceived that oxygen isotopes might shed light on petrogenetic mechanisms. The Skaergaard Intrusion had already been established as the world's premier exhibit of fractional crystallization on a spectacular scale. What better place to test the possible relationship between oxygen isotope fractionation and differentiation than the Skaergaard Intrusion? Consequently, Taylor and Epstein (1963) described the results of an oxygen isotope investigation of the intrusion. They employed Potsdam Sandstone from New York as a working standard. During the 1950s, isotope geochemists had begun to employ the Greek letter delta (δ) as a symbol denoting the deviation in parts per thousand (per mil) of the $^{18}O/^{16}O$ ratio of a sample from that of a standard.

The Potsdam Sandstone has a measured δ-value of 15.5 per mil relative to Hawaiian seawater, which they used as the standard for reporting their results. Measuring the oxygen isotope ratios of only two samples from the lower 2000 meters of the exposed layered series of the Skaergaard Intrusion, they obtained a value of 5.9 per mil, very similar to that of normal olivine basalt. From these scant data, Taylor and Epstein concluded that the magmatic liquid had a value of 5.9 per mil during the crystallization interval up to the 2000-meter level, including the lower 60 percent of the intrusion that is not exposed. From rocks at the 2500-meter level, however, they measured a value of 4.3 per mil for fayalite ferrogabbro. They further found that depletion in ^{18}O continues stratigraphically upward to the transgressive granophyre and that the late-stage differentiates of the Skaergaard Intrusion had the lowest δ values then known for any igneous rocks. They reasoned that the differentiation mechanism leading to the extreme iron enrichment had also led to the anomalous oxygen isotope ratios. The lowering of ^{18}O values in the upper differentiates, they suggested, could have resulted from crystallization and settling of minerals that were generally about one per mil heavier than the magma from which they crystallized. Curiously, Taylor and Epstein found in the isotopically anomalous Skaergaard Intrusion the kind of oxygen-isotope distribution that Silverman originally thought would be produced by fractional crystallization.

Taylor continued to accumulate data on oxygen isotopes in various geological materials. In 1968, he published a major research effort that included isotope analyses of 443 samples of igneous rocks and minerals, predominantly from North America. He summarized data on volcanic rocks from Guatemala, Mexico, Easter Island, the East Pacific Rise, the Mid-Atlantic Ridge, Hawaii, and a host of volcanic loci throughout the western United States, including Medicine Lake, Lassen Peak, Clear Lake, and the Mono Craters in California. Taylor (1968) also summarized data on plutonic rocks from several large layered intrusions, such as the Bushveld, Muskox, Kiglapait, Guadalupe, Duluth, and Skaergaard Intrusions; from several anorthosite complexes, including Nain, Egersund, Lac St. Jean, Laramie, and San Gabriel; from the British Tertiary centers at Mull, Skye, and Ardnamurchan; from the Sierra Nevada and Southern California batholiths; and also from the Alta stock in Utah, the Sudbury lopolith, and zoned ultramafic bodies from Alaska. Armed with such a battery of data, Taylor was able to make generalizations about the oxygen isotope contents of igneous rock types and to assess the implications of those data for petrogenesis.

He reported that ultramafic rocks composed predominantly of olivine and pyroxene consistently have δ values between 5.4 and 6.6 per mil, values that are typical of olivine and pyroxene in basalts and gabbros. These values were also found to be similar to those reported for nine meteoritic pyroxene samples. He found that basalt and gabbro δ values range between 5.5 and 7.4 per mil except for the marginal gabbros of the Skaergaard Intrusion, the upper gabbros of the Muskox Intrusion, and gabbros of the British Tertiary centers at Skye and Ardnamurchan. He noted that the mean value of basalt (6.1 per mil) is a little less than the value for gabbro (6.7 per mil), and that values for continental basalt are a little higher than those of oceanic basalt, a phenomenon that he attributed to contamination of basaltic magma by continental crust. No difference in oxygen isotope contents was detected between alkalic olivine basalt and tholeiitic basalt.

Taylor further indicated that andesite, trachyte, and syenite are isotopically indistinguishable from basalt and gabbro in most instances. From the strong coincidence of the narrow range of δ values he concluded that basalts, gabbros, andesites, trachytes, and syenites are all genetically related and that it would be reasonable to assume that andesite, trachyte, and syenite are derived from a parental basaltic magma by a process of magmatic differentiation.

In contrast, Taylor discovered that the granitoid rocks show a very large range of δ values from −6.3 to 17.0 per mil. He divided the granitoid rocks into five groups on the basis of their δ values. He suggested that such rocks with either very high ($\delta > 10.3$ per mil) or very low ($\delta < 5.4$ per mil) δ values are relatively uncommon. Such values he found among late-stage granophyre, pegmatite, and gneissic granite from Precambrian basements. He also indicated that granitoids with intermediate or low δ values ($\delta = 5.5$ to 7.7 per mil) are rather rare and are commonly represented by the so-called hypersolvus granites, that is, those containing essentially one feldspar that is strongly perthitic, as at Mount Ascutney. Some of these granitoids, he thought, might be genetically related to basalt because of their similar isotopic values. The great majority of plutonic granitoids, especially those from large batholiths, Taylor noted, have high δ values in the range of 7.8 to 10.2 per mil. Batholithic granitoids, he said, fall overwhelmingly toward the high end of that range.

Taylor suggested that if batholithic granitoids formed by the differentiation of parental basaltic magma, the process must have occurred at a much lower temperature than the temperatures at which hypersolvus granites formed. He suggested that such lower-temperature differentiation could have been brought about by higher water pressures that would lower liquidus temperatures. On the basis of the well-established fact that the extent of fractionation of isotopes between coexisting phases is a function of temperature, in which the extent of fractionation decreases with increasing temperature, Taylor reasoned that oxygen isotope fractionation between crystals and melt would be sufficiently great at these lower temperatures to produce the much higher $^{18}O/^{16}O$ ratios characteristic of the batholithic granitoids. He also suggested that the isotopic data could be explained if granitoid magmas had been formed by deep-seated fractional melting of ^{18}O-rich metasedimentary rocks and had then experienced varying degrees of mixing or exchange with considerably less ^{18}O-rich gabbroic or andesitic rocks or magmas. Although concluding that the origin of batholithic granitoids was still uncertain in the light of oxygen isotope geochemistry, Taylor did state that it would be essentially impossible for hypersolvus granites with low δ values to have formed by the partial melting of metasedimentary rocks.

Taylor also reviewed oxygen isotope data for individual igneous complexes. Of particular note were his results for the Tertiary volcanic centers in the Scottish Hebrides and at the Skaergaard Intrusion. Taylor found that virtually every analyzed sample from Skye, Mull, and Ardnamurchan is abnormally low in ^{18}O content relative to analogous rocks from other parts of the world. He concluded that the Hebridean igneous rocks must have exchanged oxygen with an external phase that had a very low $^{18}O/^{16}O$ ratio. The only suitable candidate for such a phase, in Taylor's mind, was the meteoric water present in the fractures and joints penetrating the country rocks into which the Hebridean magmas were emplaced. Such water would be derived, he thought, from ^{18}O-depleted rain and snow characteristically precipitated at high latitudes or high elevations. He regarded such a model as consistent with the shallow emplacement of the magmas as ring dikes, cone sheets, and plateau basalts.

A few years earlier, Taylor and Epstein (1963) had reported on the decrease in δ values with stratigraphic height in the Skaergaard Intrusion. They had then regarded the trend as an example of anomalous oxygen isotope fractionation during differentiation. In the meantime, Taylor had become open to the possibility that the anomalously low values, like those in the Hebrides, might be the result of an influx of heated meteoric water into the Skaergaard Intrusion, which would act as a giant heat engine producing a meteoric–hydrothermal convection system that would allow exchange of oxygen in the H_2O with oxygen with the magma in the upper part of the layered series.

At the same time, Taylor (1968) observed that high oxygen isotope ratios accompany fractionation trends toward silica enrichment and low Fe/Mg ratios. In contrast, he noted that depletion in ^{18}O seemed to accompany the ferrogabbro, iron-enrichment trend. While recognizing the probable ^{18}O-depleting role of meteoric water interacting with some magmas, he also proposed to account for these correlations on other grounds. He suggested that the correlation of low $^{18}O/^{16}O$ ratios with the iron-enrichment trends is consistent with the concept that oxygen fugacity strongly influences the course of magmatic differentiation, a proposal put forward earlier by Elburt Osborn at Penn State. Osborn (1957) had accounted for differences between the iron-enrichment and silica-enrichment differentiation trends in terms of early magnetite precipitation or lack thereof, determined by differences in oxygen fugacity brought about by the degree of water concentration in the magma. Because magnetite has by far the lowest ^{18}O content among the common igneous minerals, Taylor argued that changes in its crystallization behavior could affect oxygen isotopic changes occurring during differentiation. Osborn had claimed that abundant early crystallization of magnetite as a result of high water content and oxidizing conditions would lead to a calc–alkalic, silica-enrichment trend. Taylor suggested that abundant precipitation of early-crystallizing magnetite, favored by high oxygen fugacity, would lead to both iron depletion and ^{18}O enrichment in the residual magmas and, therefore, in the later differentiates. On the other hand, he thought, lack of early magnetite crystallization with low oxygen fugacity, might lead to iron enrichment and some ^{18}O depletion in residual liquid, but he was less certain about this possibility, particularly because he suspected that lowered ^{18}O might be the result of interaction with meteoric waters. Early crystallization of quartz, a mineral with high ^{18}O content, would deplete a residual melt in heavy oxygen, but would be unlikely to occur in fractionating basaltic magmas.

Because of the high ^{18}O contents of sedimentary rocks, Taylor regarded any igneous rock with a high ^{18}O content as a candidate for an origin by partial melting of sedimentary rocks. Nevertheless, because the ^{18}O content of igneous rocks is almost always less than that of sedimentary rocks, Taylor argued that melting must have occurred in conjunction with exchange and homogenization of oxygen isotopes at depth. Only igneous rocks with the highest ^{18}O contents would have been the ones most likely to have formed by simple partial melting. He also believed that assimilation, that is, contamination of magma by crustal material, might leave a distinctive isotopic signature. Noting that obsidian samples from the continental interiors are typically richer in ^{18}O than coastal obsidians and that continental gabbros are typically richer in ^{18}O than oceanic basalts, Taylor suggested that assimilation of material from the continental crust by magmas might have produced the observed values. He also regarded any rock with an apparently abnormally high δ value as a plausible candidate for magmatic assimilation.

Taylor continued to be perplexed by the isotope distribution in the Skaergaard Intrusion. His earlier analyses were done on a very limited set of samples obtained from L. R. Wager. Now Taylor decided to see the Skaergaard Intrusion for himself, collect his own samples, and solve the problem of the anomalous isotope trend. In 1971, he participated in an expedition to the intrusion with Alexander McBirney of the University of Oregon. He was accompanied by his graduate student, Richard W. Forester, with whom he had previously conducted studies of the Hebridean complexes. Throughout the 1970s, Taylor analyzed both oxygen and hydrogen isotopes in approximately 400 samples of rocks and minerals obtained during this expedition from the Skaergaard Intrusion and its country rocks. From these data, Taylor and Forester (1979) produced a detailed study of the Skaergaard Intrusion and its 55-million-year-old hydrothermal system. They showed that the oxygen isotope depletions are especially pronounced in those portions of the intrusion that are close to highly permeable fractured country rocks, that is, the overlying Tertiary plateau basalts, and that they are less pronounced adjacent to essentially impermeable country rocks, that is, the underlying Precambrian gneiss. They were also able to establish that the hydrothermal system surrounding the intrusion was established early in the history of the intrusion, but that the hydrothermal waters, despite being in close contact with the magma body throughout its history, were not significantly absorbed by the magma. Rather, they suggested, the hydrothermal system was finally able to interact with the layered gabbroic series only after substantial crystallization had occurred and fracturing of the largely crystallized layered series permitted access of the hydrothermal fluids to the interior of the intrusion. Analysis of deuterium and hydrogen isotopes in the same rocks led Taylor and Forester to confirm that these fluids were, indeed, derived from meteoric ground waters that had originally fallen as rain and snow.

Taylor and his associates continued to compile oxygen and hydrogen isotopic data from a variety of other igneous rocks throughout the 1970s. In his next review of such data, Taylor (1978) had arrived at the view that isotopic studies were useful primarily for understanding the chemical and isotopic exchanges between plutons and their country rocks, the interactions between plutons and aqueous fluids of various origins, and the importance of anatexis or the assimilation of sedimentary and metasedimentary rocks during the evolution of magmas. He asserted that magmas that were derived from the upper mantle have very uniform oxygen and hydrogen isotopic compositions with $\delta^{18}O$ between 5.5 and 7.0 per mil and δD (deuterium) values between -50 and -85 per mil relative to standard mean ocean water (SMOW). On the grounds that isotopic fractionation between silicate liquid and crystals would be rather small at magmatic temperatures, Taylor believed that simple fractional crystallization of magmas from the mantle would not be able of itself to produce major changes in the δ values of oxygen or hydrogen in later differentiates. In particular, Taylor believed that simple fractional crystallization of basaltic magma to produce later granodiorite or granite with $\delta^{18}O$ values higher than 10 per mil could be completely ruled out.

At the same time, Taylor lamented the lack of accurate experimental data on the equilibrium isotopic fractionation factors between coexisting minerals and silicate melts as a function of temperature. Although the calibration of fractionations for mineral pairs involving quartz, alkali feldspar, muscovite, magnetite, calcite, and water had been determined by Taylor, Clayton, James R. O'Neil of the U.S. Geological Survey, and others, there were still virtually no isotope fractionation data on the important OH-bearing minerals biotite and hornblende. Nor were there, at the time, any experimental data on the effects of varying water pressure on oxygen isotope fractionations between minerals and melts. Because of the lack of data, Taylor was not yet ready to rule out the possibility that, under very water-rich, oxidizing conditions, fractional removal of markedly ^{18}O-deficient minerals like magnetite, biotite, or hornblende might produce an enrichment of as much as two to three per mil in ^{18}O in late differentiates. On the other hand, Taylor again expressed reservations about the ability of fractional crystallization alone to produce significant

^{18}O enrichments. He appealed to the fact that in sequences believed to be the result of fractional crystallization, namely, in many well-studied volcanic complexes, there is little or no change in δ^{18}O values from basalt to rhyolite. Consequently, Taylor was inclined to the view that the parental materials of most granitic rocks with relatively high δ^{18}O values, especially those with values in excess of 10 per mil, were formed by partial melting, assimilation, or exchange with parts of the crust that contained a significant amount of sedimentary or volcanic rocks that had become enriched in heavy oxygen at or near the surface as a result of weathering, diagenesis, or hydrothermal alteration. Only granitic rocks with moderately low δ^{18}O values, he thought, could have formed by direct differentiation from a basaltic or andesitic magma. He believed that granitic rocks with δ^{18}O < 6 per mil could not be derived by differentiation from normal basaltic magma either. In his judgment, they must have been formed by the melting of, or by exchange with, preexisting rocks with very low δ^{18}O values, or else must have exchanged oxygen with low δ^{18}O hydrothermal fluids either in the magmatic state or under later subsolidus conditions. Such magmas could also have been derived by differentiation from basaltic magma with an abnormally low ^{18}O content derived as a result of one of those processes. Thus, the results of oxygen-isotope studies had made it reasonably clear that many granite bodies probably had not formed in the manner that Bowen had so vigorously articulated.

The hydrogen in igneous rocks is contained largely in their primary hydrous minerals, the micas and hornblendes, or in the secondary hydrous minerals formed during alteration. Although δD values generally increase from Fe-rich biotite and hornblende to Mg-rich biotite and hornblende, to chlorite, and finally to muscovite, there is much overlap in the δD values of igneous rock types. Taylor believed that the overlap was easily understandable in light of seafloor spreading. He argued that in a slab of subducted oceanic lithosphere, the dominant hydrous minerals, namely, clays, chlorites, and zeolites, would be those present in marine sediments or else those formed by submarine weathering or by hydrothermal alteration of oceanic crust at an ocean-ridge spreading center. Taylor argued that these hydrous minerals would be dehydrated as the slab is subducted and heated and that the released water would be incorporated into any magmas formed during melting of the slab. He suspected that juvenile water in the uppermost mantle would have been overwhelmed by the enormous amounts of subducted water released and cycled throughout the outermost few hundred kilometers of Earth. As a result, he thought that, in terms of its D/H ratio, it would not make much difference whether a magma was derived by anatexis of marine sedimentary and volcanic rocks or whether it formed by the differentiation of a primary basaltic magma derived directly from the upper mantle.

Taylor pointed out that gigantic meteoric–hydrothermal convective circulation systems had been established in the epizonal portions of all batholiths, thus locally producing very low $\delta^{18}O$ values during subsolidus exchange. He claimed that igneous complexes with extreme depletions in heavy oxygen characteristically possess distinctive geological, petrological, and isotopic features. Such intrusions, he said, typically occur in regions of extensional tectonics, commonly as sub-volcanic ring complexes associated with explosion breccias and surrounded by highly fractured, very permeable country rocks. The feldspars of such intrusions are very commonly turbid and are consistently more strongly depleted in ^{18}O than coexisting minerals such as quartz or pyroxene. Taylor found that the primary igneous minerals of these intrusions are commonly altered partly or entirely to uralitic amphibole, chlorite, Fe–Ti oxides, and/or epidote. Granophyric intergrowths of turbid alkali feldspar and quartz are also common in these intrusions. Finally, Taylor claimed that miarolitic cavities are locally present in such intrusions and that veins filled with quartz, alkali feldspar, epidote, chlorite, or sulfides are very common in both the intrusions and their country rocks.

Taylor asserted that one of the most important concepts that had been derived from three decades of stable isotope studies of igneous rocks was that some magmas may be strongly affected by widespread melting or assimilation of hydrothermally altered roof rock above a magma chamber. He noted that the volcanic rocks of most volcanic–plutonic complexes generally have chemical and isotopic properties that are virtually identical to those of the plutonic magmas that had intruded them, because all of these igneous rocks with similar ages are typically related genetically. In such instances, he concluded, few other geochemical criteria are available for distinguishing between primary magmas and magmas that were formed by the melting or large-scale assimilation of their roof rocks. The situation, however, was much different where a substantial meteoric–hydrothermal system had existed. Because the $\delta^{18}O$ and δD values of hydrothermally altered volcanic rocks would practically always be very different from those of primary magmas of deep-seated origin, Taylor argued, it should be much easier to distinguish among the various igneous processes of assimilation versus fractional crystallization.

The work of other isotope geochemists confirmed Taylor's belief that fractional crystallization does not produce very wide variations in oxygen isotope ratios, because the fractionation factors between silicate liquids and crystals are very small at the relatively high temperatures of magmatism. Matsuhisa (1979), for example, reported that $\delta^{18}O$ values increase by approximately one per mil in a Japanese lava sequence from basalt to dacite. Karlis Muehlenbachs, who earned his doctorate with Clayton in the isotopic laboratories at the University of Chicago, and who now teaches at the University of Alberta, analyzed an extremely differentiated suite of volcanic rocks from the Galapagos spreading center and showed that after 90 percent fractionation the residual melt was enriched in ^{18}O by only 1.2 per mil (Muehlenbachs and Byerly, 1982). Similarly, Sheppard and Harris (1985) observed a difference of only 1 per mil in a volcanic suite ranging from alkali basalt to obsidian on Ascension Island.

In addition to the hydrothermal systems at the Skaergaard Intrusion and the Hebrides mentioned above, Taylor and his associates have also observed that igneous rocks with low $\delta^{18}O$ values brought about by hydrothermal alteration are present throughout large areas of the Idaho batholith, the Boulder batholith, the Southern California batholith, the Coast Range Batholith of British Columbia, and the Samail ophiolite in Oman. Such effects are also found in many porphyry copper ore deposits and virtually all epithermal gold–silver deposits and other areas of extensional tectonics, where rift zones provide abundant access for the rise of magma and downward influx of meteoric groundwater and ocean water. These hydrothermal systems basically represent the fossil equivalents of the deep portions of modern geothermal systems such as those in the "black smokers" on mid-ocean rift zones like the East Pacific Rise, or at Wairakei, New Zealand, or Yellowstone National Park.

Many new data have also been accumulated on oxygen isotopes in basalts. A massive compilation by Harmon and Hoefs (1984) demonstrated that basaltic rock from different tectonic environments exhibits characteristic differences in oxygen isotopic composition. These authors showed that the $\delta^{18}O$ values of tholeiitic mid-ocean-ridge basalts vary from 5.4 to 6.6 per mil with an average value of 5.78 per mil; that tholeiitic basalts from oceanic islands have values varying from 5.0 to 5.8 per mil with an average of 5.38 per mil; that alkali basalts from oceanic islands have values of 5.5 to 8.2 per mil with an average of 6.34 per mil; and that oceanic island arc volcanic rocks have values ranging from 5.0 to 9.7 with an average of 6.31 per mil. In contrast, they found that continental basalts are characteristically more enriched in heavy oxygen. Thus, continental intraplate tholeiites have $\delta^{18}O$ of 6.4 to 9.0 per mil and average 6.91 per mil. Continental alkali basalts have values of 5.9 to 8.2 per mil with an average of 6.88, and the basalts of continental margin arcs have values ranging from 5.2 to 8.8 per mil with an average value of 6.71 per mil. The data were obtained from sample sets of 67 to 323 samples per group. Hoefs (1987) noted that the mid-ocean-ridge basalts (MORB) have the lowest average value with the smallest range, consistent with their derivation from a relatively homogeneous source and lack of passage through a contaminating continental crust. Hoefs further suggested that the enrichment in heavy oxygen observed in continental volcanic rocks in comparison with equivalent oceanic volcanics implies either that the subcontinental upper mantle is enriched in ^{18}O by mantle metasomatism or that contamination by ^{18}O-rich continental crust is an important process in the genesis of continental magmas. A comprehensive review of the bearing of oxygen and other isotopes on the origin of igneous rocks was published by Faure (2001).

Since the 1950s, therefore, data on oxygen isotopic distributions in common igneous rocks and minerals have provided a valuable new tool for igneous petrologists. When used in conjunction with field, petrographic, experimental, and chemical information, oxygen-isotope data proved to be an important means for gleaning new insights into the behavior of igneous rocks and for providing constraints on models of their origin.

RADIOGENIC ISOTOPES

The discovery of the various uranium and lead isotopes early in the century meant that many of the early isotopic studies were concerned primarily with geochronological applications. Numerous investigators devised a variety of methods for determining the ages of minerals such as zircon, uraninite, feldspar, and mica from their contents of uranium, thorium, lead, and potassium isotopes. Uranium and lead were employed for determinations of the age of Earth by means of methods proposed by Gerling, Houtermans, and Holmes. In the 1950s, investigators began to use the beta decay of ^{87}Rb into ^{87}Sr as a method for determining the ages of feldspars, micas, and igneous and metamorphic rocks. Among the leading geochronological laboratories was the mass spectrometry laboratory at Massachusetts Institute of Technology headed by Patrick M. Hurley (1912–2000). After his work on petrofabrics and silicate analysis, Harold Fairbairn joined Hurley's program. Together with William H. Pinson, Hurley and Fairbairn conducted a program of geochronological research lasting more than twenty years and generously funded by the Atomic Energy Commission.

In addition to determining the ages of rocks and minerals and devising new methods for obtaining more reliable ages, Hurley's team also developed an interest in the variations in the isotopic abundances of radiogenic isotopes such as ^{87}Sr. In time, the MIT group provided data on the distribution of strontium isotopes in igneous rocks and demonstrated potential petrogenetic applications of such distributions. A groundbreaking study of strontium isotopes in oceanic and continental basalts was done by Gunter Faure (b. 1934), a 1957 graduate of the University of Western Ontario who received his 1961 Ph.D. degree from MIT with a dissertation under Hurley. In 1962, Faure went to Ohio State University and has spent his entire career at that institution. He introduced thousands of students and geologists to the application of isotopes to the solution of geological problems by way of his book, *Principles of Isotope Geology* (Faure, 1977, 1986). The results of his dissertation were published in Faure and Hurley (1963). They proposed that the mode of origin of an intrusive igneous rock now residing in the continental crust might be determined from the ^{87}Sr/^{86}Sr ratio in the rock at the time of its crystallization. They suggested that intrusions derived by assimilation, remelting, or granitization of continental crustal material that had a long crustal history ought to have higher initial strontium isotope ratios than those intrusions that had originated in the upper mantle or in basaltic regions. The reason for the conclusion, they said, was based on geochemical grounds that Rb is an "incompatible" element, that is, an element that is not easily incorporated into the crystal structures of the olivines and pyroxenes that dominate the mantle and oceanic crust. Rb, therefore, is strongly fractionated into the continental crust so that it will have a higher Rb/Sr ratio than the oceanic crust or the mantle. Because Rb includes the radioactive isotope ^{87}Rb, which decays to ^{87}Sr, there should be a greater accumulation of ^{87}Sr in Rb-enriched rocks through time than in Rb-poor rocks. Thus, they reasoned, the Rb-rich continental crust ought to have a ^{87}Sr/^{86}Sr ratio that increases at a greater rate over geologic time than does that ratio in oceanic crust or mantle. As a result, magma being generated in the continental crust should inherit a higher strontium isotope ratio than magma being generated in oceanic crust or mantle.

This elegant concept was not original with Faure and Hurley. In fact, an analogous model had been proposed decades earlier by Arthur Holmes (1932), who had suggested that the diverse origins of igneous rocks might be ascertained from the ^{41}Ca/Ca ratios of rocks. Holmes reasoned that igneous rocks of crustal origin should have higher ^{41}Ca/Ca ratios than those of mantle origin on the grounds that the K/Ca ratio of the continental crust is higher than that of the mantle and that ^{41}K decays into ^{41}Ca. Unfortunately for Holmes' brilliant proposal, it was later demonstrated that ^{41}K is a stable isotope and that ^{41}Ca does not exist!

Before applying their model to igneous rock origins, Faure and Hurley determined the abundance and uniformity of the distribution of ^{87}Sr throughout the source regions of primary basaltic magmas. They measured the strontium isotope ratios of 25 oceanic and continental basalts of Triassic to Recent age, and, where necessary, applied an age correction to obtain the strontium isotope ratio at the time of formation of the rock. This ratio came to be known as the "initial" ^{87}Sr/^{86}Sr ratio. For eleven samples of oceanic basalt from Hawaii, Maui, Samoa, Ascension, the Azores, and the Mid-Atlantic Ridge, they obtained a weighted average initial ^{87}Sr/^{86}Sr = 0.7072. They found the ratio for continental basalts to be slightly higher. The weighted average for fourteen samples from Japan, the Deccan plateau of India, Iceland, Mount Vesuvius, Yellowstone, Squaw Creek (Montana), Columbia River, and Triassic diabase from New Jersey and Connecticut is 0.7082. These values generally agreed with those obtained by Paul Gast of the University of Minnesota for five basalts. Faure and Hurley concluded that the limited variation of the strontium isotope ratio from such widely scattered localities indicated that the source regions from which basaltic magmas are derived must be rather homogeneous in terms of strontium isotopes. The isotopic evidence, they believed, was consistent with the formation of basaltic magma by fractional melting of ultramafic rocks lying in the upper mantle.

Faure and Hurley attempted to estimate the Sr isotope ratio in the continental crust by first making an estimate of the Rb/Sr ratio from values in different rock types. They estimated that the entire continental crust has a Rb/Sr = 0.20 and that Rb/Sr = 0.25 for the upper crust. On that basis, they calculated that the present average continental strontium isotope ratio would be about 0.725, assuming an average age of two billion years for the continents and an initial Sr isotope ratio of 0.704. They found that isotopic analyses of two Paleozoic shale composites yielded values in reasonable agreement with their calculation.

To determine the possible origin of a specific igneous intrusion, however, they noted that it would be necessary to determine its initial strontium isotope ratio. To do that, the present ratio, age, and Rb/Sr ratio would all need to be known. The initial ratio could then be determined by the convergence method in which development lines representing the increases of the strontium isotope ratio through time for at least two whole rock samples from the intrusion having different Rb/Sr ratios would be plotted on a graph of $^{87}Sr/^{86}Sr$ versus time. The point of intersection of the development lines was said to represent the time of crystallization and the initial Sr isotope ratio of the intrusion, provided the rock had not experienced serious metamorphism or diffusion of Sr. This convergence method had already successfully been employed in obtaining the first published whole-rock Rb–Sr age by Schreiner (1958) in South Africa.

Faure and Hurley also suggested that the initial Sr isotope ratio could be obtained by an alternative "isochron" method, first used successfully by Allsopp (1961) in a study of whole rocks and mineral separates from a granite in the central Transvaal. Working at the Bernard Price Institute of Geophysical Research in Johannesburg with L. O. Nicolaysen, H. L. Allsopp plotted present-day strontium isotope ratios against Rb/Sr ratios in a suite of whole rocks and minerals and obtained an excellent straight line through the data points. The intercept of the straight line with the Sr isotope axis, yielding a value of 0.7106, was interpreted as the initial Sr isotope ratio of the granite. From the slope of the line, Allsopp calculated the age of the granite at 3200 my.

After the initial Sr isotope ratio of an intrusion had been ascertained, it would be possible to interpret its origin, according to the model. If the initial ratio of an intrusion or a series of lava flows was significantly greater than the ratio of the source regions of basalt magma, Faure and Hurley claimed, then one could justifiably argue that the rocks had formed by assimilation, fusion, or granitization from sialic crustal material enriched in rubidium. Conversely, the rocks could be considered as having formed by partial melting of a low-Rb source material or else a magmatic derivative of a primary basaltic magma if the initial ratio was the same as, or lower than, that of the source regions of basalt.

Strontium isotope investigations also began to appear elsewhere. While he had been at MIT, Louis Ahrens had supplemented his spectrographic work with investigations into strontium isotopes and radiometric age determinations by Rb–Sr methods. When Ahrens went to Oxford in 1953, therefore, he and Wager, who was anxious to add isotope geology and geochronology to the research mix at Oxford, immediately laid plans for the establishment of a mass spectrometry laboratory (Vincent, 1994). Thanks to several grants from the Department of Scientific and Industrial Research (DSIR), the forerunner to the National Environmental Research Council (NERC), capabilities for undertaking K–Ar, U–Th–Pb, and Rb–Sr projects were put in place. Despite losing Ahrens, Wager successfully created the Oxford Geological Age and Isotope Research Group, composed of geochemists like Stephen Moorbath and Eric Hamilton.

With an ample supply of Wager's Skaergaard samples close at hand, Hamilton (1963) decided to determine whether the $^{87}Sr/^{86}Sr$ ratio, as suggested by Faure and Hurley, could be used to detect contamination of magma by crustal rocks. Hamilton proposed that if the metamorphic rocks into which the Skaergaard magma had been intruded had a markedly different $^{87}Sr/^{86}Sr$ ratio from that of the magma, then assimilation of those rocks by the magma should be detectable from variations in the ratio. For the marginal border group of the Skaergaard Intrusion Hamilton found that the average $^{87}Sr/^{86}Sr$ ratio of three samples is 0.7063. That value, similar to that of other basalts, provided no evidence for contamination. The ratios for nine samples in the layered series range from 0.7044 to 0.7079 and average 0.7066. Likewise, these values provided scant evidence for contamination. From the upper border group, however, where there are inclusions of siliceous rocks, Hamilton obtained an average value of 0.7104, consistent with the idea of some contamination. And three samples of a granophyre sill yielded very high ratios averaging 0.7320 for pink granophyre and 0.7148 for gray granophyre. These high values suggested contamination to Hamilton, but then he found that the $^{87}Sr/^{86}Sr$ ratio of the country rocks is only 0.712, a value much too low for the country rock to be the contaminant. Although acknowledging the problem to be unsolved, Hamilton speculated that the high values might have been the result of absorption of a Rb-enriched facies or solution of biotite from the country rock. The Sr isotope variations in the Skaergaard Intrusion were later studied by Leeman and Dasch (1978), who, using more modern laboratory techniques, found evidence for significant contamination of the marginal border group.

Further studies of radiogenic isotopes in basaltic rocks followed when M. Tatsumoto, Carl Hedge, and A.E.J. Engel of the United States Geological Survey contributed the first Sr isotopic data on dredged seafloor basalt samples. Tatsumoto et al. (1965) pointed out, following the chemical studies of oceanic basalts by Al and Celeste Engel, that oceanic tholeiites seem to be the predominant primary magma erupted from the mantle, and that alkali basalts had been differentiated from them. Given the fundamental nature of tholeiite, it was necessary to determine all the chemical properties of basalt in order to formulate hypotheses regarding heat flow and crustal evolution. Isotopic studies formed a part of that investigation. Tatsumoto and his co-workers analyzed the major heat-producing radioactive elements, K, Rb, U, and Th as well as the Sr isotope ratios of tholeiites from the Mid-Atlantic Ridge and the flanks of the East Pacific Rise. They discovered that the Sr-isotope ratios of oceanic tholeiites are lower than in most other basalts. Although they found the ratio of Mid-Atlantic Ridge basalt to be similar to that of basalts from Ascension and St. Helena, the ratio for East Pacific Rise tholeiite is even lower. Their data agreed with that of Paul Gast that oceanic tholeiites are extremely primitive, characterized by contents of K, Rb, Sr, U, and Th, and Rb/Sr ratios typically lower than those of other basalts and approaching the values for basaltic achondrite meteorites. From analyses of five tholeiite samples from the Atlantic and the Pacific, they obtained an average $^{87}Sr/^{86}Sr$ ratio of 0.702, even lower than that obtained by Faure and Hurley.

A subsequent study of volcanic rocks by Hedge (1966) disclosed that the initial Sr isotope ratios of continental volcanics are much more heterogeneous than those from the oceans. Although he reported that initial ratios of volcanic rocks near the continental margins, such as those in Japan or the Oregon Cascades, are close to those of oceanic basalts, he noted that more radiogenic volcanic rocks occur farther inland where the continental crust has a much longer history. Specifically, he found that, at least in the western United States, volcanic rocks lying to the west of the quartz diorite line of Moore (1959) had initial ratios falling within the range of oceanic basalts, whereas all the more radiogenic volcanics occur to the east of the line.

Hedge also addressed the relationship of tholeiitic basalts to alkali basalts, groups with very different concentrations of Rb and Sr. He concluded that if the two types represented different degrees of partial melting of a similar parent peridotite, then the tholeiite basalt might represent about 10 to 30 percent partial melting, but the alkali basalt a much smaller proportion of melt. On the other hand, if alkali basalt had differentiated from tholeiite as suggested by Engel et al. (1965), a large amount of crystal precipitation would be required to bring about the observed chemical differences. He thought that volatile transfer of Rb and Sr during such differentiation could assist in bringing about the changes. It might be necessary to invoke that process, he said, because known differentiates of tholeiitic basalts do not show the Sr enrichment found in alkali basalts. Hedge suggested that the subalkalic continental basalts might have been derived from old eclogites with high Rb concentrations or else that they were contaminated by large amounts of granitic material. He concluded that a model most consistent with the Sr-isotope and other geological data entailed derivation of magmas from the upper mantle and the subsequent differentiation of the magmas combined with assimilation of crust.

During the late 1960s, a host of geochemists and geochronologists took advantage of the proposal of Faure and Hurley. They analyzed the Rb and Sr isotopes of numerous granitic plutons, plotted isochrons, calculated ages, determined initial Sr isotope ratios, and interpreted the sources of the granite magmas from the initial ratios. Here was a geochemical method for testing Bowen's idea that granites are formed by fractional crystallization of basaltic parents that originated in the mantle against the ideas of Eskola, Niggli, Tuttle, and others that granites are derived by partial melting of crustal rocks. Writing in *Strontium Isotope Geology*, Faure and Powell (1972) gleaned from the literature whole rock Rb-Sr isochron data on 131 granitic bodies from all over the world and tabulated the initial Sr ratios, most of which had been obtained directly from the isochrons. They concluded that about 50 percent of the granitic rocks had initial Sr-isotope ratios consistent with an origin by partial melting of upper mantle or, much more likely, by differentiation of basalt. About 20 percent of the granites showed much higher initial ratios. Faure and Powell interpreted these data to mean that at least a substantial fraction of the material in these granites had previously existed as continental crustal material before the granites themselves had formed. Possibilities included granitization, partial melting, or else derivation from the mantle with a high degree of crustal contamination. The data set also included a sizable percentage of granitoids with initial ratios around 0.707. Such intrusions, they concluded, formed by partial melting of a crustal zone with an appropriate Rb/Sr ratio or else were composed of mixtures of basaltic and sialic materials.

The model of strontium-isotope evolution developed by Faure and Hurley provided a powerful new tool for assessing the origins of magmas of plutons and volcanic rocks. Well after 1972, geochemists continued to determine, and petrologists continued to evaluate and interpret, initial Sr-isotope ratios in igneous rocks. Particularly impressive were some enormous compilations of ages and initial Sr-isotope ratios for the granitic batholiths and plutons of the western United States produced by Ronald Kistler and Zell Peterman of the U.S. Geological Survey (Kistler and Peterman, 1973), Richard Armstrong of the University of British Columbia, and William Taubeneck and Peter Hales of Oregon State University (Armstrong et al., 1977). When contoured on maps of the Cordilleran plutons, the initial Sr-isotope ratios of the granitoids demonstrate a distinct increase from west to east, suggestive of increasing continental crustal components for the magmas eastward. By the 1970s, it was clear that igneous petrogenesis was no longer just a matter of field, petrographic, elemental, and experimental data. Sr isotopes had become an essential component of the petrological tool kit.

Other radiogenic isotopes also proved of interest in the study of igneous rocks. Although U–Th–Pb isotopes have been of inestimable value in the determination of ages of igneous rocks, Bruce Doe (b. 1931) of the U.S. Geological Survey has also pointed out some correlations of Pb-isotope ratios with various igneous rock provinces. Doe (1970) showed that the $^{206}Pb/^{204}Pb$, $^{207}Pb/^{204}Pb$, and $^{208}Pb/^{204}Pb$ ratios of oceanic island basalts are typically higher, that is, more radiogenic than those of oceanic tholeiites. He suggested that the oceanic island volcanic rocks might have been derived from a source that had formed relatively recently. Doe also reported that the Pb-isotope ratios of basalts of island arcs are fairly similar to those of abyssal oceanic tholeiites and suggested that contamination by Pb from subducted oceanic crust and pelagic sediments along Benioff zones might account for this observation. In contrast, continental igneous rocks possess a very wide range of Pb-isotope ratios in comparison with abyssal tholeiites. Doe noted that Pb-isotope ratios for volcanic rocks from the Rocky Mountains are generally lower than those from the Pacific Northwest. In his view, Pb isotopes could provide further clues about sources of magmas.

Because of their similarity in chemical properties, size, charge, and mass, the rare earth elements (REE) had always provided difficult challenges for chemical and isotopic analysis. In time, ion-exchange methods were used for the separation of the different REE, which could then be analyzed by neutron activation. Earlier mass spectrometers, however, lacked sufficient precision for determinations of isotope ratios of some of these elements such as Nd. Thanks to the development at Caltech by Gerald J. Wasserburg of modern mass spectrometric methods employing rapid magnetic field switching and digital data acquisition, the precision of isotope ratio measurements was improved by a factor of 30 (DePaolo, 1988). This improvement in technique made possible the use of the Sm–Nd radiometric dating system. Sm–Nd dating was first applied with great success by G. W. Lugmair to the study of meteorites and lunar samples in the 1970s. The distribution of Sm and Nd in terrestrial rocks and minerals was already generally well understood when the Sm–Nd method was applied to terrestrial rock types.

Donald J. DePaolo (b. 1951) emerged as one of the leaders in the application of Sm-Nd studies to igneous rocks. DePaolo graduated in 1973 from the State University of New York at Binghamton. He pursued his graduate work at Caltech and received his Ph.D. degree in 1978 under Wasserburg. He then began his teaching career at the University of California at Los Angeles (UCLA), and since 1988, he has been on the faculty of the University of California at Berkeley. DePaolo (1988) summarized much of the knowledge of Nd isotopes in his book *Neodymium Isotope Geochemistry*. In addition to valuable geochronological data, Nd–isotope studies have also proved to be an important tracer of igneous rock processes. As with Sr isotopes, the initial Nd–isotope ratio, that is, the original $^{143}Nd/^{144}Nd$ ratio of a magma or rock, is a very important parameter. Values of $^{143}Nd/^{144}Nd$ ratio are commonly reported in relation to a hypothetical "standard" or "reference" value and are indicated by the symbol ε_{Nd}, which gives the deviation of the ratio from the standard in units of 10^4. The initial Nd–isotope ratio at the time of formation of an intrusion is typically compared with the Nd–isotope ratio of a so-called chondritic uniform reservoir (CHUR) at that same time. The value of CHUR is obtained by assuming that Earth, at the time of its formation, had the same Nd–isotope ratio and Sm/Nd ratio as chondritic meteorites.

After a Sm–Nd isochron is obtained for a suite of igneous rocks, its initial Nd–isotope ratio can be determined. These values are typically 0.507 to 0.512.

DePaolo (1988) pointed out that a rock which has $\varepsilon_{Nd} = 0$ would be considered to have crystallized from a magma derived from a source region characterized by the chondritic Sm/Nd ratio from the beginning of Earth to at least the time that the magma was separated from the source region. On the other hand, he noted, if ε_{Nd} were positive, the magma probably had been derived from a source region in which the Sm/Nd ratio had a value greater than the chondritic value during part of the interval from the beginning of Earth to the time of separation of the magma. A negative value of ε_{Nd} would imply the opposite. The inferred Sm/Nd ratio, in turn, indicates the REE concentration pattern in the magma sources.

DePaolo observed that basalts, many of which had been analyzed by O'Nions et al. (1977), possess a wide range of ε_{Nd} values, implying that there exist different domains within the mantle characterized by different Sm/Nd ratios. He suggested that these separate mantle domains could well have persisted since the time of Earth's formation. Because oceanic basalts are less likely than continental basalts to have their ε_{Nd} values modified during passage to the surface, they should be the most reliable indicators of mantle values. DePaolo reported that mid-ocean ridge basalts (MORB) have the highest ε_{Nd} values (~ +10), that is, the greatest displacement from chondritic values among basalts, and a very narrow range. In contrast, oceanic island basalts (OIB) generally have lower ε_{Nd} values (~ +6 for the majority and ranging down to −5). Intra-oceanic arc basalts peak at ε_{Nd} +7 but have values as low as −2.

The Nd data obtained from oceanic basalts implied that the Sm/Nd ratios of the mantle reservoirs were generally greater than the chondritic values, a result that is expected for residual material left behind after repeated extraction of partial melts. In other words, the basalts are derived from a "depleted" source. During any partial melting episode, the slightly lighter Nd should be fractionated into the melt relative to the slightly heavier Sm with the result that the residual mantle would take on a slightly higher Sm/Nd ratio. DePaolo assumed that the mantle originally had a chondritic value but that large portions of the mantle had magmas removed from them in the past, long before the eruption of present seafloor basalts.

Much of the thick continental crust has negative ε_{Nd} values that might be expected to contaminate continental basaltic magmas, yet it was found that at least alkalic continental basalts have values similar to alkalic oceanic basalt, suggesting that mantle reservoirs beneath the continents were isotopically similar to those beneath the ocean basins. DePaolo did report, however, that oceanic tholeiites of the ridges invariably have greater positive ε_{Nd} values than continental tholeiites, suggestive of different tholeiite sources beneath oceans and continents. He postulated that the variability and lower ε_{Nd} values of rocks from continental suites in the Hebrides and Columbia River province might also be the result of contamination by old continental rocks.

DePaolo also summarized data on layered intrusions such as the Skaergaard and Kiglapait Intrusions. He reported that the Archean gneiss wall rocks of the Skaergaard Intrusion have exceptionally low ε_{Nd} values of around −40. The ε_{Nd} values of the intrusion itself are around +5. Because of the very sharp contrast in values, DePaolo maintained that the Nd isotopic composition of the magma should have been quite sensitive to small amounts of assimilation. Although the isotopic variations in the layered series are very small, they seem to correlate with some of the Sr-isotope variations and might be indicative of assimilation. DePaolo noted that the largest isotopic variations were found in the upper border group and also part of the marginal border group. He also thought that lower zone and middle zone rocks of the layered series show the effects of progressive contamination, a trend mirrored in the upper border group. This distribution suggested that there had been good chemical communication, possibly by way of convection, throughout the magma chamber.

A study of the Sr-isotope variations in the Kiglapait Intrusion had been carried out earlier by DePaolo (1985). He found that both the ε_{Nd} and the initial Sr-isotope ratios vary in a regular manner from the bottom to the top of the layered cumulate series, with $^{87}Sr/^{86}Sr$ varying from 0.70408 up to 0.70679 and ε_{Nd} varying from −1.6 to −5.6. He interpreted the data in terms of progressive contamination of the crystallizing magma by small amounts of wall rock. Finding the increase in initial Sr-isotope values to be irregular in the lower part of the intrusion, he suggested that crystallization was interrupted periodically by injections of fresh magma with low initial Sr-isotope ratios. In contrast, he found the increase in initial Sr-isotope ratios in the upper part of the intrusion to be smooth and continuous. He suggested that addition of fresh injections of uncontaminated magma had been unimportant during later stages of crystallization.

DePaolo also reviewed Nd- and Sr-isotopic data for continental magmatic arcs. He noted that ε_{Nd} values and initial Sr-isotope ratios of granitoids become progressively more like those of old Precambrian crustal rocks from west to east across the Cordillera of the western United States. These progressive changes in ε_{Nd}, he pointed out, are interrupted by a discontinuity in central Nevada where ε_{Nd} jumps from −6 to around −14 to −18 in the east. He interpreted the disconntinuity as a reflection of the edge of old Precambrian basement. Batholiths such as the Sierra Nevada batholith of California were said to have isotopic characteristics similar to those of volcanic rocks from continental margins and were interpreted as resulting from the assimilation of variable amounts of crustal material by magmas that had originated in the mantle.

By the late 1970s, then, geochemists had begun to utilize distributions of Nd isotopes, either alone or in conjunction with Sr-isotope data, in assessing possible source regions of magmas as well as the possibility of various amounts of contamination during ascent of large magma volumes through the continental crust or locally by the wall rocks of relatively small plutons such as the

Skaergaard Intrusion. With the advent of Sr-, Nd-, and Pb-isotope geochemistry, igneous petrologists now had powerful new isotopic methods to add to their arsenal for attacking petrological problems, but they also stood in greater need than ever before of collaboration with specialists outside their own field. After the 1960s, petrologists could no longer rely solely on field data, microscopes, and phase diagrams to address their problems. Now they were forced to become acquainted with complex geochemical arguments that required knowledge of crystal chemistry and the equations of radioactive decay. As might be expected, petrologists were not always convinced that the results of the isotope geochemists were compatible with sound geological and petrographic evidence, but they could not afford to ignore the results. Sound petrogenetic theorizing, as it always had, required the application of every scrap of evidence, no matter where it came from.

25 ❖ MATHEMATICAL MODELING

Trace-Element Studies in Igneous Petrogenesis

Sweeping advances in stable and radiometric isotopic geochemistry were made possible by the invention of and continued technical improvements to the mass spectrometer. Only a few institutions, however, possessed the combination of funds and expertise to construct, maintain, and operate mass spectrometry laboratories during the middle of the twentieth century. Most petrologists of the era, although willing to use the data and insights of the isotope geochemists wherever possible, still required the more fundamental data on the chemical compositions of igneous rocks and minerals for their work before they could take advantage of any insights provided by the more esoteric isotopic methods. As a result, the chemists who analyzed the rock and mineral specimens submitted by petrologists for analysis were constantly searching for faster, more accurate, and less expensive methods of determination of the major elements. Despite the breakthrough provided by the emission spectrograph in the analysis of trace elements, measurements obtained with that instrument stood in need of considerable improvement. The race was on to search for the unattainable perfect method or methods that would be very inexpensive, occupy little space, require little maintenance, overrule the mistakes of its operator, entail negligible sample preparation time, and measure a maximum number of chemical elements simultaneously and virtually instantaneously with exceedingly high accuracy, precision, and sensitivity! During the 1950s and 1960s, such important strides were made toward that idealistic goal that trace-element geochemistry experienced another major advance as several new methods of analysis were developed. Before long, igneous petrology would be awash in a sea of trace-element abundance numbers. Petrologists and geochemists also developed a better understanding of the mechanisms that control trace-element distribution, formulated mathematical models that could be used to assess the origin and development of igneous rock suites, and evaluated distribution patterns in ancient igneous rocks as indicators of tectonic environments.

DEVELOPMENTS IN ANALYTICAL METHODS

As noted earlier, classical wet chemical rock analysis, based on the techniques of gravimetric analysis, is a very time-consuming process that requires specific sequences of lengthy and sometimes repeated precipitations. Analysts constantly searched for schemes that would replace gravimetric procedures with simpler, faster techniques. An important breakthrough came in 1952 when Leonard Shapiro and W. W. Brannock of the U. S. Geological Survey published a brief circular outlining a new method for the rapid analysis of silicate rocks (Shapiro and Brannock, 1952). With the stated aim of reducing time and cost, the authors outlined rapid procedures for the determination of silica, alumina, total iron reported as ferric iron, titania, MnO, P_2O_5, lime, magnesia, soda, and potash. They stated that the accuracy and precision of the results obtained by the new rapid methods were probably less than would be obtained by careful, competent analysts using conventional methods. Still, they pointed out, the new methods were widely applicable, rapid, simple, direct, less subjective than conventional methods, reliable, and easily learned and performed by personnel with less training than was typically required for classical wet chemical analysis. They estimated that one analyst using their scheme could analyze approximately sixteen silicate rock samples a week.

Shapiro and Brannock said that most of the analysis could be done with only two solutions, one for silica and alumina, and one for all the rest of the elements. In the rapid scheme, silica, alumina, MnO, titania, P_2O_5, and iron were to be determined with an optical spectrophotometer. In this method the abundance of an element could be determined colorimetrically by measuring the intensity of absorption of characteristic wavelengths. A flame photometer was used for the analysis of sodium and potassium, chemical elements that had been notoriously difficult to determine by standard methods. In flame photometry, a solution is atomized in a flame and the alkali atoms are excited in the process due to transitions of electrons to higher energy levels. During the de-excitation of the alkali atoms when electrons return to lower energy levels, radiation with characteristic wavelengths is emitted. The emission lines are then measured photometrically. Only lime and magnesia were done by titrations with EDTA solutions in the new rapid scheme.

In a later publication, Shapiro and Brannock (1956) reported that they had analyzed 952 silicate rock samples by rapid methods between 1951 and March 1955. They had improved their scheme so that it now included the analysis of water, carbon dioxide, FeO, and Fe_2O_3. They were also using automatic photometric titration for the determination of lime and magnesia. A complete analysis, they said, could be done seven times faster than by conventional methods. Even when an analysis was duplicated, they claimed, results with an accuracy comparable to that obtained by classical methods could still be obtained in less than half the time.

In the meantime, another interlaboratory comparison of analytic results was conducted (Stevens et al., 1960), which showed that rapid methods produce results that are within the ranges of values obtained by analysts using conventional wet chemical methods. A couple of years later, Shapiro and Brannock (1962) reported that, since 1956, they had been analyzing 800 samples per year and that they had added the determination of fluorine and sulfur to their scheme.

New, even more rapid, instrumental methods began to replace the so-called "rapid method," and Shapiro (1967) mentioned in a subsequent report that he was determining lime, magnesia, soda, potash, and MnO by means of atomic absorption spectrometry. He also indicated that he now needed to prepare only one solution, thereby cutting sample preparation time in half and gaining some accuracy. It is possible, he claimed, to obtain determinations on ten chemical elements for thirty samples in just two days. The stage was being set for a vast influx in the amount of major element chemical data that the petrological community would need to digest.

While Shapiro and Brannock were employing atomic absorption spectrometry (AAS) in the determination of some of the major elements, other analysts and geochemists saw the utility of the method for trace-element work. Instrumentation for the new technique had originally been developed in the early 1950s. For analysis, the rock or mineral sample needs to be dissolved, and the solution is then burned in a flame at high temperature. A light source consisting of a noble-gas-filled cathode lamp, in which the cathode is composed of the chemical element to be analyzed, emits radiation with characteristic wavelengths that are absorbed by the atoms of interest in the flame. The intensity of absorption is said to be a function of the concentration of the element in the solution. Although a separate lamp and sets of standard solutions of known concentration are required for each chemical element, the method proved to be very sensitive and highly accurate. Before long AAS was used for the analysis of a wide range of trace elements, including Li, Be, V, Cr, Co, Ni, Cu, Zn, Rb, Sr, Ag, Cd, Ba, and Pb.

Another technique developed during the 1950s that rapidly gained popularity among geochemists and petrologists was x-ray fluorescence spectoscopy (XRF). A major advantage of the method as typically employed by petrologists and geochemists is that the rock or mineral sample does not need to be dissolved, thus saving time and eliminating the need for messy solutions that can easily be spilled. A pellet or disk of pressed rock powder is placed in a beam of primary x-rays generated from a tungsten- or molybdenum-target x-ray tube and passed through an analyzing crystal. The x-ray beam impinging on the sample stimulates the emission of secondary or fluorescent x-rays from the atoms of the various elements at and close to the surface of the sample. A detector then measures the intensity of each wavelength characteristic for particular elements. The method does require careful comparison with standard rocks like G-1 and W-1 as well as elaborate calculations to correct for the effect of absorption of the fluorescent x-rays by the rock matrix itself. Despite the difficulties, the XRF method proved highly reliable in the measurement of major elements with atomic numbers greater than that of Mg and is especially effective in the routine determination of the trace elements Rb, Sr, Y, Zr, and Nb to a few parts per million. X-ray fluorescence spectrometers soon became standard equipment in university geology departments around the world.

The detailed study of the rare earth elements at last became feasible with the development of neutron activation analysis (NAA), a method proposed as early as 1936 and put into practice in the 1940s and 1950s with the development of nuclear reactors that were capable of generating high fluxes of thermal neutrons. Absorption of the neutrons by the target atoms in the sample raises the nuclei of the atoms to an excited state. As the excited nuclei decay into their stable ground states at a known rate of decay, gamma rays with the energies and wavelengths that are specific to each element are emitted. Measurement of the gamma-ray intensity spectrum of the sample by Ge–Li detectors at various time intervals after neutron irradiation leads to the determination of the abundance of the elements. Early applications of NAA in the late 1940s were for the analysis of iron and stony meteorites. NAA proved to be especially valuable for the analysis of a large number of trace elements, including Li, Sc, Cr, Co, Ni, Cu, Ga, Rb, Pd, Sb, Cs, Ba, Hf, Ta, W, Re, Au, Th, and U, in some cases in the range of parts per billion.

NAA had an advantage over most other methods in its greater sensitivity to more elements. During the earlier years of its use, NAA required careful chemical separations of elements prior to analysis. With the development of high-resolution germanium gamma-ray spectrometers in the 1970s, powdered rock samples could be analyzed by instrumental neutron activation analysis (INAA), thus opening the way for the widespread study of the rare earth elements. For instrumental NAA, no chemical processing is necessary before irradiation. Neutron activation analysis also typically permits the simultaneous determination of several chemical elements as well as simultaneous irradiation of several samples. The main disadvantage is that most geology departments do not have access to nuclear reactors at their institutions, so samples must be sent out for irradiation in reactors at other locations.

In the early 1950s, electron probe microanalysis was developed by Robert Castaing and soon applied to petrological problems at Cambridge. This powerful method revolutionized the discipline by providing accurate and precise chemical analysis of individual mineral grains within rocks without the need for tedious, time-consuming, and error-prone mechanical separations of minerals. Indeed, the ability of electron-probe microanalysis to determine in situ mineral compositions that are so useful in the solution of petrological problems precipitated the near abandonment of traditional wet chemical and rapid methods of mineral analysis.

Although not as useful for trace-element determinations as AAS, NAA, or XRF because of lesser sensitivity of the method, microanalysis yields the major-element chemical composition of an extremely small spot (about 1–15 micrometers) on the surface of a single mineral grain in a polished thin section. By moving the electron beam around the grain, dozens of analyses per grain can be obtained to give a picture of compositional zonation within grains. Image maps can be made of chemical variations. An investigator can also determine the consistency of the chemical composition of several grains of the same mineral within a rock. With the data obtained by probe analysis, petrologists can determine the compositions of coexisting phases such as the lamellae of two feldspars or of two pyroxenes within exsolved grains, or two completely distinct mineral grains, such as olivine and pyroxene. Such data have thermometric possibilities and can provide insight into the cooling histories of igneous rocks. In essence, electron probe microanalysis is similar to x-ray fluorescence in that an electron beam targeted onto a spot on the surface of a mineral grain produces secondary, fluorescent x-rays whose intensity is then measured. From intensities of specific wavelengths and corrections for matrix effects, abundances of major elements in the mineral can be calculated. Since the x-ray spectrum of a spot can be obtained with relatively brief irradiation times by the electron beam, several complete chemical analyses of a single grain can be obtained in one hour. The downside of the electron probe is that its cost is prohibitive for all but the major universities and research centers. As a result, petrologists whose institutions could not afford such an instrument needed to obtain funding to have probe analyses made for them at another institution or to travel to that institution to make the measurements themselves.

In the 1970s, ion microprobe analysis, otherwise known as secondary ion mass spectrometry (SIMS), became an increasingly important analytical tool for petrologists. In this technique, an ion beam focused on an area 10–50 micrometers in diameter sputters secondary ions from a mineral grain of interest in a polished thin section. A mass spectrometer analyses the secondary ions and can provide data on both isotopes and trace elements of micron-sized material.

Among the more important new methods to appear in the 1970s were inductively coupled plasma atomic emission spectroscopy (ICP-AES) and inductively coupled plasma mass spectrometry (ICP-MS), methods that simultaneously determine thirty to forty elements within about two minutes. Analysis by some variations of ICP methods require conversion of the sample into solution form. Other variants that employ lasers, however, eliminate the dissolution step and can perform in situ analyses albeit at the price of loss of sensitivity. During the final decades of the twentieth century, ICP increasingly became one of the preferred methods for obtaining rapid, precise, and accurate analyses of minerals and whole rocks for their major, minor, and trace-element contents.

The speed and efficiency of all these analytical techniques was, of course, considerably improved with advancements in computer technology. Increasingly, computers analyzed the data and performed the calculations yielding the elemental concentrations in the samples that decades ago had been done by hand or with the aid of calculators. Computers could also rapidly produce all sorts of graphical analyses of the chemical data. A brief anecdote told by I. D. Muir (1994, p. 315) about S. R. Nockolds conveys some sense of the radical change in igneous petrology brought about by the computer manipulation of the mountain of analytical data being spewed out by the new methods:

> The rapid development in the 1960s of more rapid and accurate physical methods for the determination of trace elements, avoiding the need for tedious mineral separation, overtook his [Nockolds'] laboratory. During the data explosion that followed, Nockolds turned to less demanding activities. Nevertheless, until his retirement his numerous research students continued to make good use of his laboratory. Like many of his generation he never adapted to the computer age. At this time, when Cambridge University had just one computer, a research student calling for advice found him engaged in plotting data sets on a large sheet of squared paper. The student discovered there were 2000 data sets to be plotted and eagerly offered to write a computer programme that would plot them all in an afternoon. 'In an afternoon!' exclaimed Nocky, 'Then what will I do for the rest of my life!'

THEORY OF ELEMENT DISTRIBUTION

The dramatic growth in trace-element analysis first by emission spectrography and then by atomic absorption spectroscopy, x-ray fluorescence spectrography, and neutron activation analysis was accompanied both by increased ability to characterize different igneous rock types by their trace-element abundances, by growing interest in the principles of trace-element distribution, and by development of mathematical expressions for modeling trace-element behavior during various petrological processes.

Goldschmidt had accounted for the trace elements largely in terms of the charges and sizes of the ions such that a trace element might be incorporated into the structure of a growing mineral if its size was sufficiently similar to that of a major element ion that it might replace. Goldschmidt (1937) had also suggested that where a difference in the radii of two ions existed, then, during fractional crystallization, there would be a tendency for the later fractions of the mineral to be enriched in the ion having the larger radius. A classic example was provided by the olivine solid solution series in which the early-formed, high-temperature olivines are enriched in the smaller divalent Mg ions and the later-formed, lower-temperature olivines are enriched in the slightly larger divalent ferrous ion.

During the early 1950s, however, Goldschmidt's principles began to receive some criticism. Among Goldschmidt's critics was Denis Martin Shaw (b. 1923), another product of the geochemical revolution at the University of Chicago. After graduating in 1943 from the University of Cambridge where he had studied with Nockolds and Tilley, and then receiving a master's degree from Cambridge in 1948, Shaw received his Ph.D. degree from Chicago in 1951 with a dissertation on the geochemistry of thallium. In 1948, he joined Canadian isotope geochemist H. G. Thode on the faculty at McMaster University in Hamilton, Ontario. Shaw spent his entire career at McMaster, from which he retired in 1989, and took his place among the leading geochemists of his generation. Shaw (1953) referred to the claim of Goldschmidt enunciated above as the "enrichment principle." He disputed the idea that trace-element distribution in magmatic minerals could be explained solely in terms of the sizes of the ions. While Goldschmidt's enrichment principle might be valid for olivine and some other binary systems, he agreed, it is not valid for the plagioclase feldspars, the alkali feldspars, and several other binary systems. As a result, Shaw was skeptical that Goldschmidt's principle would be valid for complex, multicomponent silicate magmas.

Another modification of Goldschmidt's ideas was put forward by Alfred Edward (Ted) Ringwood (1930–1994). Ringwood was born and educated in Australia and received his Ph.D. degree from the University of Melbourne in 1956 (Mason, 1968). In 1959, Ringwood became a Senior Research Fellow in the Department of Geophysics at the Australian National University, the institution at which he spent his entire professional career. Within a few years, he had advanced to Professor of Geochemistry. Ringwood was known for his research on high-pressure phase transformations, meteorites, planetary interiors, and the genesis of basaltic magmas. It was while he was pursuing his doctoral work at Melbourne, however, that Ringwood proposed a revision of Goldschmidt's idea on enrichment during fractional crystallization by attributing the behavior of trace elements to their electronegativity, that is, the amount of energy that is released upon adding an electron to an atom. Ringwood (1955a) argued that if a trace element has a higher value of electronegativity than the major element it replaces in an igneous mineral, then the trace element tends to form a weaker, more covalent bond in the crystal than the major element. As a result, such an element would generally be rejected by a growing crystal and would be concentrated in residual melt. On the other hand, he suggested, if the trace element has a much lower electronegativity than the major element, it would be preferentially incorporated into the growing mineral and be depleted in the residual liquid. Ringwood thought that Goldschmidt's enrichment principle, based on size, would be generally valid in situations where the differences in electronegativity between trace and major elements are quite small. In a subsequent paper, Ringwood (1955b) further suggested that many trace elements whose ions have a high charge often form complexes with oxygen or hydroxyl in a magma. These complexes, he said, are not easily accepted into crystallizing minerals, with the result that those trace elements are enriched into residual magmas and concentrated to the point at which they can precipitate their own primary minerals in pegmatites.

A few years later, S. R. Nockolds, who had written so extensively on the geochemistry of igneous rock series, noted some difficulties with Ringwood's modification of Goldschmidt's principle in situations where ionic radius and electronegativity act in opposition to one another. Nockolds (1966) thought the problem might be resolved by incorporating charge, radius, and electronegativity into one function. He proposed using bonding energy for that function and developed an expression for the bonding energy between a cation and oxygen. From an examination of calculated bonding energies, he proposed two rules for ionic substitutions. He suggested that, if two cations with the same valence are capable of substitution within a crystal structure, the cation with the greater bonding energy is preferentially incorporated into the crystal. By inference, the ion with lesser bonding energy is concentrated into the residual magma. For the second rule, Nockolds proposed that, if two cations have different valences so that coupled substitutions are involved, the substitution occurs for which the sum of the bonding energies of the coupled ions is greater. Despite providing an explanation for some of Ringwood's discrepancies, a few anomalies still remained even in Nockolds' approach. In the end, Goldschmidt's contention that ionic size and charge are the predominant factors controlling chemical element distribution remained generally valid. Failure of his rules is restricted mainly to special cases, for example, those involving crystal field effects.

TRACE-ELEMENT FRACTIONATION AND MODELING

Whatever the causes of the incorporation or rejection of a cation into a growing crystal might be, however, igneous petrologists and geochemists were still interested in finding ways to relate trace-element abundance variations in igneous rocks to specific processes, such as partial melting or fractional crystallization. Would it be possible, they wondered, to demonstrate from data on trace-element distributions in a suite of igneous rocks that the members of that suite are specifically related to each other by fractional crystallization or by some other process such as assimilation? To that end petrologists and geochemists developed increasingly refined and sophisticated mathematical expressions. Among the earliest mathematical expressions for trace-element distributions were those developed by Heinrich D. Holland and J. Laurence Kulp of Columbia University in a study of the distribution of trace elements in pegmatites (Holland and Kulp, 1949) and Henrich Neumann and his colleagues at Indiana University in a study of trace-element variation during fractional crystallization (Neumann et al., 1954). These expressions typically relate the concentration of a trace element in a crystallizing mineral at a given stage of crystallization; the concentration of that trace element in the coexisting melt at the same stage; the original amounts of melt and of the trace element in the melt, that is, the original concentration of the trace element in the melt; the amount of the solid phase at the given stage of crystallization (which could be related to the fraction of melt remaining); and a "distribution factor." Neumann, Mead, and Vitaliano specifically applied the term "distribution factor" in their mathematical equations to designate a "distribution coefficient" that describes the distribution of a trace element between a mineral and a magma, namely, the ratio of the equilibrium concentration of a trace element in a mineral to the concentration of the element in coexisting melt. The distribution factor, it was understood, is different for each trace element as well as for each mineral and for melts of different chemical compositions and is a function of temperature and pressure. In principle, one could assign values for the distribution factor and the degree of solidification (or fraction of melt remaining) from the mathematical expression and thereby calculate the relative concentration of a trace element in a crystal to that of the trace element in the melt. From several such calculations, Neumann et al. (1954) plotted sets of curves showing the variation in concentration of a trace element in a melt relative to that in crystals of a coexisting mineral with respect to the degree of solidification of the melt for several values of the distribution factor, generally symbolized by the letter K (Figure 42).

During fractional crystallization of magma, however, more than one mineral in general is crystallizing from the melt, and each trace element is partitioned differently between each individual mineral and the melt. As a result, geochemists realized the necessity for developing more complex equations to account for the variation of a trace element with crystallization. Trace-element studies received a significant impetus from the pioneering work of Paul Werner Gast (1930–1973), a 1952 graduate of Wheaton College who received a master's degree in 1956 and his Ph.D. degree in 1957 from Columbia University. Apart from a brief stint on the faculty of the University of Minnesota, Gast was affiliated mainly with Columbia and its Lamont Geological Observatory. Toward the end of his all-too-brief life, Gast became renowned for his contributions to the study of lunar geochemistry, particularly the recognition of KREEP basalts. In the 1960s, Gast developed an equation for evaluating trace-element fractionation in relation to the origin of tholeiitic and alkalic basalt magmas. In essence, Gast (1968) showed that the distribution factor of Neumann et al. (1954) should be replaced by a "bulk distribution coefficient" that is an average of the distribution factors for each individual crystallizing mineral weighted in accord with the proportions in which the mineral phases were crystallizing. Naturally, inasmuch as the proportions of crystallizing phases vary during the course of crystallization and as minerals enter in or drop out as crystallizing phases, the value of the bulk distribution coefficient was difficult to ascertain and was not constant throughout crystallization.

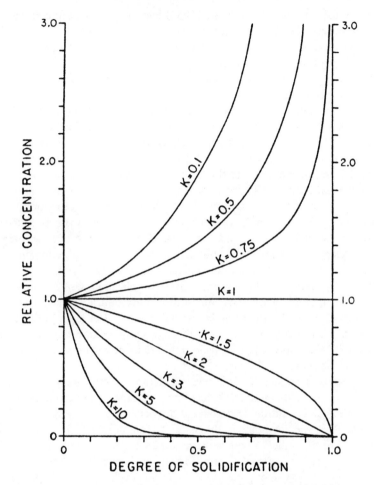

42. Variation in the concentration of a trace element in a melt with degree of solidification for different values of the distribution coefficient as calculated by Neumann et al. (1954). Reprinted from *Geochimica et Cosmochimica Acta*, v. 6, Neumann, H., Mead, J., and Vitaliano, C. J., Trace element variation during fractional crystallization as calculated from the distribution law, pp. 90–99, 1954, with permission from Elsevier Science.

Pointing out that partial melting may be more important than fractional crystallization in determining the trace-element contents of volcanic liquids, Gast also developed an expression for evaluating the variation in trace-element concentrations within increasing amounts of partial melt in equilibrium with a source rock consisting of several mineral phases. He used these mathematical expressions to demonstrate that the high abundances of trace elements with large cations in alkalic basalt could not be consistently accounted for by fractional crystallization, as Engel et al. (1965) had originally advocated. He did argue, however, that trace-element abundances are consistent with the idea that alkalic basalts are produced by about 3 to 7 percent partial melting of accepted upper mantle mineral assemblages.

The power of Gast's analysis unleashed a cascade of similar evaluations of data sets for igneous suites. It also triggered the modern approaches to the study of mantle evolution entailing geochemical and petrological data. Shaw (1970) offered further refinements of the mathematical models for evaluating trace-element variation during partial melting, depending on whether melting was continuous or discontinuous and depending on whether it was modal or nonmodal. In other words, did the minerals melt in the same proportions in which they existed in the source rock or not? Although earlier applications of trace-element modeling had operated with the simplifying assumption that a distribution coefficient, or "partition coefficient," as some workers called it, remains constant during the melting or fractionation process, experimental data had clearly shown that distribution coefficients are variable functions of temperature, pressure, and the bulk chemical compositions of the melts and the mineral phases. To take account of this complication, Greenland (1970) developed a modified expression for fractional crystallization, and Hertogen and Gijbels (1976) did the same for partial melting. The latter authors also dealt with the problem of melting of incongruently melting minerals. Later, Shaw (1978) developed expressions for trace-element variation during partial melting in the presence of a vapor phase.

In the 1970s, leading schools of trace-element modeling emerged in France and Belgium. The Laboratoire de Géochimie et Cosmochimie at the Institut de Physique du Globe and the Département des Sciences de la Terre of the Université de Paris was the home of the research team of Claude J. Allègre, Jean-François Minster, Francis Albarède, Bernard Minster, Michel Treuil, Jean-Louis Joron, and Serge Fourcade. Not confined to trace-element modeling, the versatile French team was among the pioneers in the development and application of the neodymium isotope technique in the1970s. They also studied osmium isotopes in meteorites and terrestrial materials, investigated thorium isotopes in volcanic rocks, discovered ^{129}Xe anomalies in mid-ocean ridge basalts, and successfully employed the ion microprobe for analysis. Of particular importance regarding their theoretical applications of trace-element studies were a series of papers on the systematic use of trace elements in igneous processes (Allègre et al., 1977; Minster et al., 1977; Minster and Allègre, 1978) and a comprehensive review by Allègre and Minster (1978) of most of the quantitative models of trace-element behavior in magmatic processes that had been previously proposed. Allègre and Minster also added some new equations of their own that addressed trace-element variation resulting from mixing processes. Conveying to their readers a sense that the evaluation of a petrological process by means of trace-element modeling is by no means a simple and straightforward task, they presented sets of relatively simple equations to describe trace-element variation for a wide range of petrological processes, including equilibrium (batch) partial melting; disequilibrium partial melting; both continuous and discontinuous fractional melting; simple fractional crystallization; multisequence fractional crystallization; fractional crystallization with simultaneous escape of a fluid phase; fractional crystallization in a convecting magma chamber; fractional crystallization with trapped liquid in the cumulate; simple mixing of two magmas; contamination before or after differentiation; dissolution of wall rock during fractional crystallization; diffusion into melt during fractional crystallization; wall-rock alteration during magma ascent; and replenishment of a magma chamber with fresh magma during fractional crystallization. The trace-element modeling work of the group at the Institut de Physique du Globe was a welcome and important resurgence of French interest in the problems of igneous petrology, an interest that had waned somewhat since the days of Michel-Lévy and Lacroix.

The French were also joined by their neighbors to the north in Belgium where the leading proponents of trace-element modeling were J. Hertogen and R. Gijbels at the Instituut voor Nucleaire Wetenschappen of the Rijksuniversiteit Gent; Jean-Clair Duchesne at the Laboratoire de Géologie, Pétrologie et Géochimie of the Université de l'Etat in Liège; and Daniel Demaiffe at the Laboratoire de Minéralogie et Pétrologie of the Université Libre de Bruxelles. Duchesne and Demaiffe concentrated particularly on the application of trace-element data to modeling of the genesis of anorthosite and related magmas.

Building on the work of Bowen (1922b), who had shown that assimilation generally promotes further crystallization of the minerals that were already crystallizing from the magma, Hugh Taylor (1980) explored the effects of combined assimilation and fractional crystallization (AFC) on the systematics of oxygen and strontium isotope variations in magmas. On the assumption that the rates of wall-rock assimilation and fractional crystallization were inextricably linked, Taylor showed that, by assuming a specific distribution coefficient and a constant ratio of assimilation to fractional crystallization, one could make detailed calculations of the variations in element concentration and isotopic composition as a function of the extent of crystallization. Nd-isotope specialist Donald DePaolo (1981) extended this approach by developing a general mathematical model for any kind of trace elements and isotopes. These AFC models attracted considerable attention among igneous petrologists during the next two decades. DePaolo proposed a general equation for the variation of trace-element concentrations and one for the variation of the δ value or ε value of an isotope. As in Taylor's work, these equations considered the concentration of a trace element as a function of the original concentration, the fraction of melt remaining, and a bulk distribution coefficient, as well as estimates of the relative rates of assimilation and of fractional crystallization. As had been common practice among modelers, DePaolo calculated sets of curves showing variations in the relative concentration of a trace element with fraction of magma remaining for differing values of the bulk solid/liquid distribution coefficient and also various ratios of the relative rates of assimilation and fractional crystallization. Powell (1984) and Roberts and Clemens (1995) later refined these models and issued caveats regarding too-hasty application of AFC modeling to igneous rock suites. Interest in AFC modeling, now also incorporating the concept of magma replenishment, has persisted to the present day (e.g., Spera and Bohrson, 2001; Bohrson and Spera, 2001).

Apart from the glaring fact that all of these mathematical expressions for trace-element variation that were developed since mid-century, no matter how complex, are unavoidably too simplistic to model unerringly the extreme complexity of magmatic processes and are, therefore, open to some degree of criticism and skepticism, the degree of validity and applicability of every one of the trace-element models has been utterly dependent on the accurate assignment of the values of distribution coefficients. In the first place, each mineral crystallizing from a magma upon cooling or each mineral contributing to melt during a partial melting event has a distribution coefficient for each particular trace element, that is, a ratio representing the equilibrium partitioning of the trace element between that mineral and the coexisting melt. That value must be measured in some manner. To make matters worse, the value of a particular distribution coefficient is a function of temperature, pressure, and the compositions of the mineral and particularly the melt. These factors have to be evaluated.

The measurement of distribution coefficients began mainly during the 1960s as serious trace-element modeling was getting underway. Many of the early determinations of distribution coefficients were made by C. C. Schnetzler, John A. Philpotts, and Hiroshi Nagasawa at the Planetology Branch of the Goddard Space Flight Center in Maryland, by Ian S. E. Carmichael, and by R. Berlin and C. M. B. Henderson. The earliest measurements, obtained from coexisting phenocrysts and matrices of natural volcanic rock samples, were made primarily for alkali, alkaline-earth, and the rare earth elements. The Goddard team, for example, determined distribution coefficients for phenocrysts of plagioclase, K-feldspar, clinopyroxene, orthopyroxene, mica, hornblende, garnet, and olivine from a range of volcanic rock types from basalt to rhyodacite. The phenocrysts were mechanically separated, hand picked, and analyzed by mass spectrometric stable isotope dilution techniques.

In the early 1970s, investigators began to determine trace-element distribution coefficients experimentally in the hopes of assessing the effects of experimentally controllable variables as temperature and oxygen fugacity on the partitioning of elements. Numerous workers, for example, examined the partitioning of Ni, Cr, Mn, and REE between olivine and melt; of several transition metals and REE between Ca-pyroxene and melt; and of Sr, Ba, and REE between plagioclase and liquid. Other studies were done on alkali feldspars, garnet, hornblende, ilmenite, and magnetite. Anthony J. Irving (1978, p. 765) of the Lunar and Planetary Institute in Houston summarized the experimentally determined partitioning data and thought there was an "encouraging degree of consistency to the data from different studies." At the same time, he cautioned that many more studies were necessary before usable sets of distribution coefficients suitable for application to petrologic problems could be compiled. Irving warned that selection of distribution coefficients for a specific modeling application must be done with extreme judiciousness. Since that time, dozens of investigators have continued to measure distribution coefficients under carefully controlled conditions. With an ever expanding body of values from which to choose for their modeling calculations, trace-element modelers have had to use great critical judgment. To compound the problem, a sense of uneasiness over the application of trace-element studies on the part of many geochemists has grown with recognition of the strong possibility that equilibrium is not maintained during partitioning processes and by the probability that the equations and models employed do not provide unique solutions to the petrological problems being investigated. And given the uncertainty of the data fed into the modeling equations and the simplicity of the equations themselves in comparison with actual magmatic processes, igneous petrologists have from time to time issued warnings that a trace-element modeling study does not represent the final word in solving a petrological problem but only provides one line of evidence that must be considered in conjunction with field, experimental, isotopic, and other evidence.

RARE EARTH ELEMENTS

Perhaps nowhere did trace-element modeling flourish more than in the study of the rare earth elements (REE), due in part to the fact that REE modeling is less dependent on accurate knowledge of distribution coefficient values than is modeling of other trace elements. After neutron activation analysis had made the study of the rare earth elements feasible on a routine basis, igneous petrologists were quick to seize on the application of such studies to igneous petrogenesis. It became common practice to represent the abundances of the REE graphically. Because a plot of absolute abundances versus atomic number of the REE in virtually any geological material almost inevitably yields a zigzag pattern owing to the fact that even-numbered chemical elements have greater cosmic abundances than the odd-numbered chemical elements adjacent to them in the periodic table, the so-called Oddo-Harkins effect, it soon became common procedure to normalize the REE abundances to some standard to smooth out the jagged nature of the patterns and to plot the REE abundances on a logarithmic scale. Most characteristically, rare earth element abundances in a sample were divided by the abundances in average chondritic meteorites following the procedure established by Coryell et al. (1963).

Three major traits of the semi-logarithmic REE plots came to be regarded as having some petrogenetic significance (Figure 43). One of the most striking features of many plots is the presence of either a peak or a trough at the position of Eu, referred to as a europium anomaly. The second feature is the slope of the plot, either positive, negative, or flat. Lastly, overall REE abundances in the samples being plotted are a major feature. Geochemists attributed europium anomalies to the fact that europium typically exists in the divalent state rather than the trivalent state as is the case with most of the other REE. As a consequence, the ionic radius of europium is considerably larger than that of the other REE and its crystal-chemical behavior is different. Eu is likely to accompany Sr and, therefore, substitute for Ca in feldspars. Thus, Eu is strongly fractionated into feldspars relative to other REE, and a REE plot of feldspar typically displays a very pronounced positive Eu anomaly. Many other minerals like orthopyroxene, garnet, apatite, and hornblende have relative deficiencies in Eu and, therefore, display negative Eu anomalies. The existence of a negative Eu anomaly in an igneous rock might be attributable to derivation of that magma from a feldspar-rich source that retained much of the Eu. On the other hand, precipitation of abundant plagioclase from a magma undergoing fractional crystallization would tend to deplete a magma in Eu relative to other REE and lead to later-formed rocks with negative Eu anomalies. Igneous rocks with positive anomalies could represent plagioclase cumulates as in the case of anorthosites or possibly partial melts from sources that already had very pronounced positive anomalies. Igneous petrologists came to view Eu anomalies as important clues to the nature of magmatic processes.

Jon Steen Petersen of the University of Aarhus in Denmark was but one petrologist who provided an illustration of the use to which Eu anomalies have been put. Petersen (1980) studied the differentiated Kleivan charnockitic pluton in southwestern Norway. He showed that the early charnockites (pyroxene granites) of the pluton are characterized by small positive Eu anomalies. Somewhat more differentiated pyroxene–hornblende granites display negligible Eu anomalies. Negative Eu anomalies appear in the still more differentiated hornblende and hornblende–biotite granites, and spectacularly large negative Eu anomalies are found in the highly differentiated biotite granites. Thus, increasing differentiation was accompanied by transition from a small positive Eu anomaly to a huge negative Eu anomaly. Using distribution coefficients derived from the average phenocryst-melt coefficients for intermediate rocks obtained by Arth (1976), Petersen modeled trace-element variation in the Kleivan pluton, assuming that fractional crystallization was the dominant process and was able to reproduce very closely the observed REE distribution patterns. He attributed the increasingly negative Eu anomaly to crystallization of potash feldspar. Petersen also argued that his attempts to model the REE distribution by partial fusion processes failed.

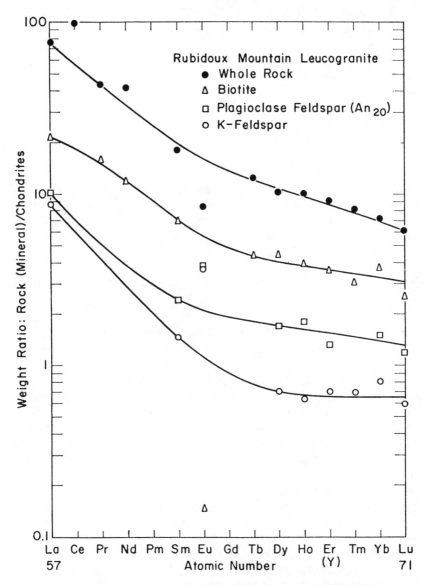

43. Chondrite-normalized rare earth element abundances in Rubidoux Mountain leucogranite, from Towell, D. G., Winchester, J. W., and Spirn, R. V., Rare-earth distributions in some rocks and associated minerals of the batholith of Southern California. *Journal of Geophysical Research*, v. 70, no. 14, pp. 3485–3496, 1965. Copyright © 1965 American Geophysical Union. Reproduced by permission of American Geophysical Union.

The slopes of REE plots, the second main feature of these diagrams, also provide insight into the history of igneous rocks. Because garnet, zircon, orthopyroxene, and, to a lesser extent, hornblende strongly fractionate the heavier REE, an igneous rock plot that is enriched in the light REE, a very common occurrence for granitic rocks, might be interpreted as having formed by melting from a source enriched, for example, in garnet, pyroxene, and zircon. As an example, Joseph G. Arth and Gilbert N. Hanson from the State University of New York at Stony Brook analyzed the REE as well as Rb, Sr, Ba, and K from several quartz dioritic rocks in order to test a proposal by Green and Ringwood (1968) that quartz dioritic magmas are derived by the partial melting of eclogite or amphibolite originally derived from basalt. To do so, Arth and Hanson (1972) examined the Saganaga tonalite and various other trondhjemitic rocks from Minnesota and also the trondhjemite of the Craggy Peak pluton in the Klamath Mountains of northern California. They concluded from the higher abundances in several graywackes analyzed by Larry Haskin of Washington University and his colleagues in 1966 that it is unlikely that the quartz dioritic rocks were derived by partial melting of graywackes. On the other hand, they noted that all the quartz diorites are marked by relatively low contents of the heavy REE. Because garnet is the rock-forming mineral that most strongly retains heavy REE, they postulated that the magmas had been melted from a source in which the residual material contained a high proportion of garnet, a common constituent of mafic rocks under high pressure. They envisioned amphibolite and eclogite as reasonable candidates for such a source rock.

The third feature of REE plots, namely, the overall abundance of the REE, also came to be considered as useful in evaluating magma history. In general, geochemists predicted that, with the exception of Eu, the REE, having large trivalent ions, would tend to remain in a melt during crystallization of early-formed minerals like pyroxene, olivine, and calcic plagioclase. Distribution coefficients of the REE for pyroxene and plagioclase generally support this notion. As a result, REE were expected to concentrate in residual magmas where they would be incorporated mainly in accessory minerals. The study of Petersen on the Kleivan pluton also showed that the total REE abundances generally increase from the less differentiated facies to the more differentiated facies of the pluton. Petersen's fractional crystallization modeling also successfully reproduced the observed abundance patterns.

In a groundbreaking study, L. Peter Gromet of Brown University and Leon T. Silver of California Institute of Technology stressed the extreme importance of paying attention to the accessory minerals in REE studies. They pointed out that the accessory minerals may contain the overwhelming majority of the total rare earth content of intermediate to felsic igneous rocks. Gromet and Silver (1983) had dissected a granodiorite from the Peninsular Ranges (Southern California) batholith into its component minerals and showed that the REE are contained predominantly in accessory titanite (sphene) and allanite with lesser amounts contributed by hornblende, epidote, and apatite. The smallest amounts are contained in the feldspars and biotite. They found that titanite and allanite combined contain 80 to 95 percent of each individual REE and that the major minerals like plagioclase, alkali feldspar, biotite, epidote, and also apatite contain one percent or less of each individual REE, except for Eu, of which plagioclase contains seven percent. They judged that early precipitation of titanite and allanite essentially swept the REE out of the granodiorite magma and left little behind for later crystallizing minerals. A similar conclusion was reached by Mittlefehldt and Miller (1983) who found that monazite crystallization had removed the light REE from residual magma in the Sweetwater Wash pluton in southeastern California.

In general, analysis of granitic rocks has indicated that they are more enriched in REE than basaltic rocks and that ultramafic rocks are rather poor in REE. The abundance of REE presumably ought, in many instances, to increase during fractional crystallization. In principle, it ought to be possible to demonstrate such increases in a suite of rocks known on other grounds to have formed by fractional crystallization as was done by Petersen. Despite uncertainty over the assignment of appropriate values to distribution coefficients and despite the inherent limitations of applying mathematical expressions to extremely complex petrological processes, trace-element modeling took its place alongside isotopic studies during the latter part of the twentieth century as a major geochemical weapon in the assault on petrological problems. In their enthusiasm for the new methods, geochemists and petrologists needed to be reminded from time to time that modeling studies or isotopic studies carried out in isolation from other lines of evidence were sometimes counterproductive. A scientific battle is not won with one weapon alone.

The field of trace-element modeling has been another beneficiary of modern high-speed computers. No longer does a modeler need to plug immense quantities of data into equations and solve for concentrations by hand, and no longer does a modeler need to plot graphical representations of such data by hand.

TRACE ELEMENTS AND PLATE TECTONICS

The large volumes of major-element chemical data accumulated by petrologists over more than a century are, of course, amenable to statistical treatment. From time to time, then, a small cadre of petrologists, like Felix Chayes of the Geophysical Laboratory, attempted to apply statistical methods to igneous rocks in productive ways. Chayes and Velde (1965), for example, believed that they could distinguish circum-oceanic, volcanic arc basalts from oceanic basalts by using a function based on the TiO_2 and MgO contents of the rocks. Chayes also provided a useful service by calling attention to the fact that correlations based on rock analyses must be viewed with a modicum of suspicion because of the closure problem created by the fact that analyses all add up to a constant sum. With the addition of enormous volumes of trace-element abundance data beginning in the second half of the century, statistically-minded petrologists had plenty of material with which to develop tests using various geochemical discriminants. The increased exploration of the seafloors and the advent of the theory of plate tectonics also provided further grounds for developing statistical treatments of geochemical data.

A potentially fruitful line of approach was developed in Great Britain by Julian A. Pearce. After graduating from the University of Cambridge in 1970, Pearce undertook graduate study at the University of East Anglia from which he received his Ph.D. degree in 1973. In his work with marine petrologist Joseph R. Cann, Pearce became particularly interested in whether it is possible to determine the paleo-tectonic environment in which a suite of igneous rocks had formed on the basis of its geochemistry. Together, Pearce and Cann (1971) compared samples from ophiolite complexes to five groups of Cenozoic volcanics on the basis of contents of Ti, Zr, and Y, chemical elements that are commonly presumed to be relatively immobile during weathering and other secondary processes. They were interested in testing the idea that ophiolite complexes might represent fragments of former oceanic crust emplaced onto continental margins by plate-tectonic processes. The five groups of volcanics included ocean-floor basalts, Hawaiian tholeiites, alkali olivine basalts from the Azores, island-arc tholeiites, and island-arc andesites. On a triangular diagram with Ti/100, Zr, and Y at the apices of the triangle, they plotted analyses of these three elements from the five volcanic groups and showed that they fall into distinct regions (Figure 44). A plot of Ti vs. Zr shows similar discrimination of four of the groups. Most of their analyses of ophiolites from Oman, Cyprus, Austria, and Greece were plotted on similar diagrams. The plots showed very close similarities to the ocean-floor basalts. On the basis of discriminant analysis, Pearce and Cann concluded that the ophiolites were probably generated as oceanic crust at spreading plate boundaries and had been thrust into their present positions during continental collisions.

Later, Pearce and Cann (1973) analyzed more than 200 basaltic rocks from different tectonic settings for Ti, Zr, Y, Sr, and Nb. They first distinguished four tectonic settings: ocean-floor basalts, (OFB) erupted at divergent plate margins, volcanic arc basalts erupted at convergent plate margins, oceanic island basalts erupted within an oceanic plate, and continental basalts erupted within a continental plate. Within the volcanic arc category, they further distinguished low-potassium tholeiites (LKT), calc–alkalic basalts (CAB), and shoshonites. On a variety of discrimination diagrams they showed that the analyses plot in distinct regions. For example, on a triangular plot of Ti/100, Zr, and 3Y they discriminated ocean-floor basalts, low-potassium tholeiites, calc–alkalic basalts, and within-plate basalts (WPB). The latter category includes basalts from both ocean-plate and continental-plate interiors, and is especially well distinguished. On a triangular plot of Ti/100, Zr, and Sr/2 they discriminated

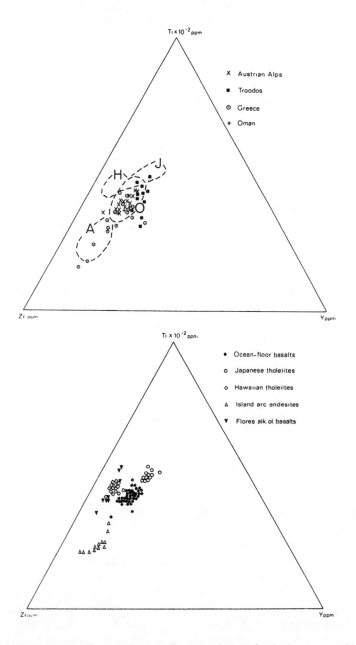

44. Ti–Zr–Y trace-element diagram showing relation of ophiolites to possible parental magma types. Reprinted from *Earth and Planetary Science Letters*, v. 12, Pearce, J. A., and Cann, J. R., Ophiolite origin investigation by discriminant analysis using Ti, Zr, and Y, pp. 339–349, 1971, with permission from Elsevier Science.

ocean-floor basalts, low-potassium tholeiites, and calc–alkalic basalts. Pearce and Cann also suggested the use of Y/Nb as a parameter for distinguishing tholeiitic from alkali basalts.

Pearce also worked on discriminant analysis of the major-element patterns in basalts as part of his doctoral work. In a paper on this work, Pearce (1976) reported that he wanted to test statistically any relationships between major-element patterns in basic volcanic rocks and the tectonic environments of their eruption. He selected a maximum of 75 carefully circumscribed analyses from each of the magma types previously defined by Pearce and Cann (1973). From his discriminant analysis, Pearce claimed that ocean-floor basalts, low-potassium tholeiites from island arcs, calc–alkalic basalts and shoshonites (SHO) from island arcs, and within-plate basalts could be classified correctly about ninety percent of the time by available statistical methods. He did point out, however, that he had been unable to achieve a successful chemical discrimination between within-plate basalts of the oceans, that is, oceanic island basalts (OIB), and the within-plate basalts of the continents. He also cautioned that attempts to classify weathered or metamorphosed ocean-floor basalts were suspect because alteration considerably reduced the success rate of classification.

After leaving East Anglia in 1973, Pearce spent a year as a postdoctoral fellow in Norway and then took a position as Lecturer at the Open University in Milton Keynes, where he worked on ophiolites with Ian Gass. In 1984, he became Lecturer in Geology at the University of Newcastle upon Tyne. Soon thereafter he published work in which he extended his discriminant analysis to granitic rocks. Pearce et al. (1984) statistically evaluated more than 600 high-quality trace-element analyses of granitic rocks from known tectonic settings. Pearce, Harris, and Tindle classified tectonic settings of granites into four groups designated as ocean-ridge granites (ORG), volcanic arc granites (VAG), within-plate granites (WPG), and collision granites (COLG). They achieved the most effective discrimination of granites from these four tectonic settings with diagrams plotting Rb vs. (Y + Nb), Rb vs. (Yb + Ta), Nb vs. Y, and Ta vs. Yb. They also plotted variation diagrams of Y, Yb, Rb, Nb, and Ta against SiO_2 content and obtained very good discriminations of groups of tectonic settings. For example, in a diagram of Yb vs. SiO_2, virtually all WPG and three groups of ORG granites fell above a line at about 3 ppm Yb whereas virtually all VAG, COLG, and one class of ORG granites plot below that line.

Pearce employed another device that had become widely popular with petrologists interested in trace-element geochemistry, namely, the "spider diagram." Various versions of spider diagrams had already been introduced by D. A. Wood, S. S. Sun, and R. N. Thompson in the late 1970s and early 1980s. The typical spider diagram plots the logarithm of the abundances of several trace elements in a rock normalized to their abundances in a standard (Figure 45). In that respect they are like the REE diagrams except that they include a much larger array of elements. As such they are suited for visual evaluation of

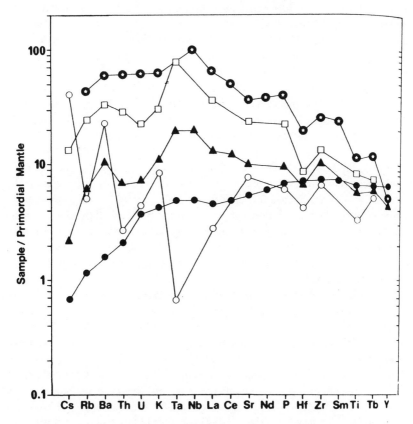

45. Spider diagram originally proposed by Wood et al. (1979) to show element abundances of basic lavas from different tectonic environments. Reprinted from *Earth and Planetary Science Letters*, v. 45, Wood, D. A., Joron, J.-L., and Treuil, M., A reappraisal of the use of trace elements to classify and discriminate between magma series erupted in different tectonic settings, pp. 326–336, 1979, with permission from Elsevier Science.

large amounts of trace-element data. Wood et al. (1979) normalized their data to a hypothetical primordial mantle composition, whereas Sun (1980) and Thompson et al. (1984) normalized their data to average chondritic meteorite abundance. The order in which the elements were arranged on the diagram varied from author to author. The data plotted on a spider diagram generally produce a zigzag pattern. The petrologist can determine visually the degree of similarity or dissimilarity of trace-element patterns in different rock samples. Pearce believed, therefore, that such diagrams have the potential for indicating whether basalts or granites from the same tectonic environment are characterized by similar trace-element patterns. The spider diagrams employed by

Pearce, Harris, and Tindle employ a hypothetical value for ocean ridge granite as the standard (Pearce et al., 1984) They calculated the standard value from average mid-ocean-ridge basalt (MORB) data using distribution coefficients and assuming that the granite had fractionated from MORB. The abundances in their samples divided by those in the hypothetical calculated ORG were then plotted against K_2O, Rb, Ba, Th, Ta, Nb, Ce, Hf, Zr, Sm, Y, and Yb to produce their spider diagrams (Figure 46). The resulting diagrams display notably different patterns for the different tectonic environments. By visual inspection, one can, for example, readily see that the patterns for ocean-ridge granites differ considerably from those for within-plate granites. From their results, Pearce and his colleagues argued that trace-element studies of granites could be a big help in deciphering the tectonic settings of older granites for which the original connection to plate tectonic processes is not immediately evident. At the same time, they warned against application of the trace-element data in isolation from a consideration of geological constraints. Eventually, Pearce became increasingly involved in the field of marine geoscience, and he later held positions at the University of Durham and, beginning in 2000, the University of Cardiff.

In succeeding years, petrologists and geochemists evaluated the geochemistry of all sorts of ancient basalts, granites, and other rock types in an attempt to reconstruct the paleo-plate tectonic settings of the rocks (Wilson, 1989). A number of trace-element diagrams other than those of Julian Pearce were also developed, including several by Thomas H. Pearce (Russell and Stanley, 1989). Some petrologists, however, became concerned about the indiscriminate application of geochemical discriminant diagrams and believed that unjustifiable conclusions about tectonic settings were being drawn in a number of instances. For example, Wang and Glover (1992) of Virginia Polytechnic Institute proposed a tectonics test for several of the most frequently used diagrams. They plotted chemical analyses of the Triassic and Jurassic basalts and diabases from the eastern North America rift basins on a wide array of diagrams to determine whether or not they plot in the "correct" tectonic region of the diagram. Over and over they discovered that the rocks commonly plot in "incorrect" regions. For example, on the triangular diagram of Ti/100-Zr-3Y they found that the rift-basin basalts plot mostly in the ocean-floor and calc–alkalic basalt regions and that the diabases plot in every field except the within-plate basalt field where they should have fallen. Similarly poor results were obtained for several

---

46. Example of spider diagram applied to tectonic discrimination of granitic rocks by Pearce et al. (1984). Representative analyses normalized to ocean ridge granite (ORG). a. Some ocean-ridge granites. b. Volcanic arc granites. c. Within-plate granites. d. Granites in attenuated continental lithosphere. e. Syn-collision granites. f. Post-collision granites. Reproduced from *Journal of Petrology* by permission of Oxford University Press.

MATHEMATICAL MODELING 495

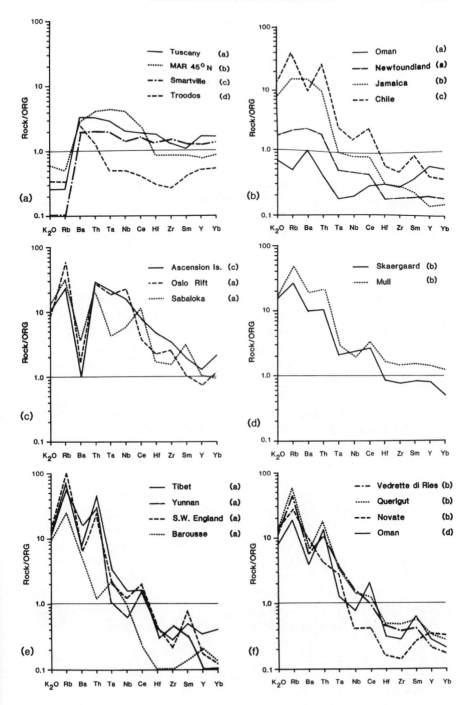

diagrams. Wang and Glover concluded that the various tectonic discriminant diagrams fail to classify the continental basalts correctly. They found that the basalts from eastern North America as well as the Karoo basalts of South Africa, the Deccan traps of India, and the basalts of the Columbia River plateau plot mostly in the volcanic arc and MORB fields. They suggested that large amounts of chemical data had not been available for the continental basalts when the original diagrams had been invented and that had such data been available the misleading nature of the diagrams might have been avoided. Because of the complex nature of the generation of continental basaltic magma, they maintained, it would always be very difficult to create a discriminant diagram that would effectively distinguish continental basalts from other types. They urged petrologists to undertake careful tectonic lithofacies analysis based on field observations, petrographic study, and major-element analysis of basalts of unknown tectonic affinities if they were of a mind to employ chemical discriminant diagrams.

In their earlier attempts to use Pearce diagrams for granites, Förster et al. (1997) from the GeoForschungs Zentrum in Potsdam, Germany, found that Variscan granites from central Europe are not cleanly discriminated but commonly plot on or across field boundaries. This problem led them to seek a geochemical parameter that discriminates more successfully than existing diagrams. In the end, however, they discovered that Pearce's Rb versus (Y + Nb) plot discriminates better than any scheme they tried. Thus they decided to undertake a reevaluation of the tectonic discrimination method based on the Rb versus (Y + Nb) diagram with a new data base of 3500 analyses from more than 250 occurrences of granites worldwide for which the tectonic settings are fairly well known. To minimize the likelihood of inaccurate tectonic assignments to the granites in their data base, they took advantage of the communication possibilities offered by the latest technology by distributing a list of the granitoids in their data set to a global Internet discussion group on granites.

They concluded that oceanic granitoids associated with ophiolite complexes plot in the ORG field only if they formed in a setting unrelated to subduction. In other words, ophiolites that formed at a back-arc ridge, for example, contain granitoids that are indistinguishable from volcanic arc granitoids on the Pearce diagram. They further noted that granitoids from volcanic arcs in an oceanic setting plot in the VAG field without exception. None of them overlap with the WPG field. In contrast to this finding, they discovered that the continental arc setting is much more heterogeneous. Although most of the analyses plot in the VAG field, they detected some overlap with both the WPG and COLG fields. They suspected that some of the overlap is caused by characteristics of the source rocks and argued that a few of the continental volcanic arc granites display COLG characteristics because of the incorporation of pelitic rocks in the melting process.

Förster and his colleagues noted that oceanic within-plate granitoids plot in the WPG field without exception. On the other hand, they found that continental within-plate granitoids occupy different fields. Such granitoids that are associated with mantle plumes producing flood basalt provinces or with major rifting long after previous orogenic processes, however, clearly possess WPG affinities. They concluded that these latter granites and those from oceanic settings would be the most likely to be classified correctly based on geochemistry alone. In contrast, they noted that extensional granitoids that are closely associated in time or space with convergent margins commonly plot in both the WPG and VAG fields and sometimes the COLG field. They maintained that this latter tectonic setting is the most likely to be misclassified on the sole basis of geochemistry.

Förster, Tischendorf, and Trumbull issued the caveat that the correct interpretation of granites from collisional orogens requires knowledge of the stage of collision during which the granite formed, the source rocks involved, if any, and the type of regime that developed after collision. Geochemical information taken in isolation and plotted on a Pearce diagram, they cautioned, would mislead because such granites commonly plot in the VAG, WPG or COLG fields.

The fact that accessory minerals commonly contain substantial abundances of many of the critical trace elements in intermediate to felsic igneous rocks also renders attempts at discrimination of tectonic environments based on trace elements highly susceptible to sampling errors and preparation methods. Despite the complexities and serious pitfalls, however, trace-element studies of various types became increasingly important in petrological thinking during the latter half of the twentieth century as petrologists used trace elements to characterize rock suites, to distinguish rock suites from one another, and to assess the role of various petrogenetic processes involved in the formation of igneous rock suites.

26 ⟐ BOMBS AND BUFFERS

Experimental Petrology after Bowen

Despite their infatuation with isotope and trace-element geochemistry and the expensive analytical instruments like x-ray fluorescence spectrometers and electron-probe microanalyzers that made the geochemical revolution possible, igneous petrologists in the latter twentieth century continued to build on the legacy of experimental petrology established by Bowen, Schairer, and their cohorts at the Geophysical Laboratory. Although the majority of petrologists in the 1950s did not know the difference between a Morey bomb and a Tuttle bomb and may never have seen a furnace, they had become adept at reading phase diagrams under the "distance-learning" tutelage of Geophysical Laboratory scientists through their voluminous output of published descriptions, diagrams, and discussions of silicate systems. Petrologists had also developed an appreciation for the power of phase-equilibrium studies for the interpretation of igneous rocks. Virtually all petrological textbooks of mid-century contained sections on the reading of phase diagrams. Thanks to experimental studies, petrologists of mid-century had a much better awareness of the conditions of stability of igneous rock-forming minerals and of the range of possible hypotheses for magma origins and diversity than did their predecessors. As a result of the growing appreciation for the fruits of phase-equilibrium studies, experimental petrology experienced a profuse flowering during the second half of the century. New experimental technologies and methods were developed, and experimental laboratories sprang up in universities in several countries.

ONGOING WORK AT THE GEOPHYSICAL LABORATORY

Even though Bowen established an experimental laboratory at the University of Chicago after 1937, the Geophysical Laboratory remained the brightest star in the experimental petrology sky for years thereafter. During the 1940s and 1950s, Bowen, who returned to the Geophysical Laboratory from Chicago in 1947, Schairer, Tuttle, Osborn, and others turned out detailed studies of system upon system, join upon join. Although a vast amount of his output was done in collaboration with Bowen, particularly work on ferrous silicate systems and on alkali-alumina-silica systems, J. F. Schairer pursued several other lines of research, often in conjunction with E. F. Osborn, M. L. Keith, H. S. Yoder, Jr., K. Yagi, F. R. Boyd, or D. K. Bailey. Schairer produced studies of several binary joins; the ternary systems pseudowollastonite–åkermanite–gehlenite, $MgO-Al_2O_3-SiO_2$, $FeO-Al_2O_3-SiO_2$, leucite–corundum–spinel, leucite–forsterite–spinel, and nepheline–diopside–silica; and the quaternary systems $CaO-FeO-Al_2O_3-SiO_2$, $CaO-MgO-FeO-SiO_2$, $K_2O-MgO-Al_2O_3-SiO_2$, and $Na_2O-Al_2O_3-Fe_2O_3-SiO_2$. Schairer's studies had important implications for the understanding of a wide range of igneous rock types from melilitic rocks to granites to alkalic rocks. Among Schairer's most valuable contributions was the invention of the "flow sheet," a diagram in which he laid out the major paths of fractionation of liquids in the system $CaO-FeO-Al_2O_3-SiO_2$ (Schairer, 1942). The diagram, however, was potentially applicable to all manner of systems.

In 1948, Hatten Schuyler Yoder, Jr. (b. 1921), who had studied with Bowen as an undergraduate at the University of Chicago and had just received his Ph.D. degree from MIT, joined the staff of the Geophysical Laboratory. Yoder undertook several studies of minerals like quartz, jadeite, and diopside at very high pressure and began a systematic investigation of the micas, particularly muscovite and phlogopite, with H. P. Eugster. He also studied the ternary feldspar system at a water pressure of 5 kilobars with D. B. Stewart and J. R. Smith. To overcome some of the limitations of externally heated pressure vessels, Yoder also constructed an internally heated, gas-media pressure vessel in 1949. The new vessel was capable of sustaining a temperature of 1650°C and a pressure of 10 kilobars for several months.

Arguably, Yoder's greatest achievement was his experimental study of the origin of basaltic magmas in collaboration with Cecil Edgar Tilley (1894–1973), a petrologist from the University of Cambridge. Tilley, a native of Australia, took his undergraduate work at the Universities of Adelaide and Sydney (Chinner, 1974). After service in World War I, he studied with Harker at Cambridge. In 1931, he spent several months working with Schairer at the Geophysical Laboratory and then returned to Cambridge late that year to begin his duties as Professor of Mineralogy and Petrology and also Head of the newly established department of that name. After years of studies on both igneous and metamorphic rocks, Tilley came back to the Geophysical Laboratory in 1955 to work with Yoder on basalt. By the 1950s, extensive experimental experience had already been gained at the Geophysical Laboratory with feldspars and pyroxenes, and the staff believed that the time was right for studying these two major mineral constituents of basalt together. Although the work of systematically determining melting relations in the multicomponent system describing natural magmas had been going on for years, component by component, completion of that ambitious program would have required an unreasonable amount of time before some insights about the relations of basalts could be gleaned. Tilley persuaded Yoder that it was time to move from ideal systems to consider the whole-rock basalt system using natural specimens. Moreover, Tuttle and Bowen (1958) had previously successfully undertaken some systematic experiments using granitic whole rocks. Consequently, Yoder and Tilley decided to gain some preliminary insights into basalt relations by adopting a less rigorous approach in which a series of natural basalts, presumed to have been entirely liquid at one time, would be treated as individual bulk compositions in the multicomponent system. They proposed to examine the courses of crystallization of these basalts by the quenching method.

Yoder and Tilley (1962) conducted melting experiments on 25 natural samples of different kinds of basalts from nine localities, mostly from the Hawaiian Islands. Such diverse types as tholeiite, olivine tholeiite, high-alumina basalt, alkali basalt, and olivine nephelinite were examined, and a few eclogites were also studied. Experiments were conducted over a range of temperatures at pressures from one atmosphere to 40 kilobars and at water pressures from one to ten kilobars. Yoder and Tilley represented their model of the simple basalt system on a tetrahedron with quartz, diopside, nepheline, and forsterite at the corners. The basalt tetrahedron was divided into smaller tetrahedra by two planes connecting the compositions of albite and enstatite. Different basalt types were conveniently represented by compositions in different regions within the tetrahedron (Figure 47). Yoder and Tilley envisioned division of basalts into five groups based on their normative mineral content. For example, tholeiite, containing normative quartz and hypersthene, plots in the smaller tetrahedron bounded by quartz, orthopyroxene (enstatite), plagioclase (albite), and clinopyroxene (diopside). Alkali basalt, containing normative olivine and nepheline, plots in the smaller tetrahedron bounded by nepheline, olivine, plagioclase (albite), and clinopyroxene (diopside). Shortly thereafter, Schairer and Yoder (1964) developed an expanded basalt tetrahedron with nepheline, forsterite, quartz, and larnite at the corners. Not only did the expanded tetrahedron include the simple basalt tetrahedron of Yoder and Tilley, but it also incorporated the compositions of alkalic ultramafic igneous rocks.

Yoder and Tilley found that the two principal basaltic magma types recognized by field geologists, namely, tholeiite and alkali basalt, are separated from one another by a thermal divide at one atmosphere. They also found different thermal divides at elevated pressures and maintained that this change in the position of thermal divides as a function of pressure implies that the same bulk composition undergoing melting could lead to the production of different basalt types depending on the pressure of melting. In general, they concluded that alkali basalt magmas should be generated at high pressure and, therefore, greater depths than tholeiitic magmas. This view was at odds with the claim of Engel and Engel (1964) to be published a short time later that alkali basalt had differentiated from tholeiite. Yoder and Tilley also disputed the contention of Japanese petrologist, Hisashi Kuno, that the chemical variations in basalt

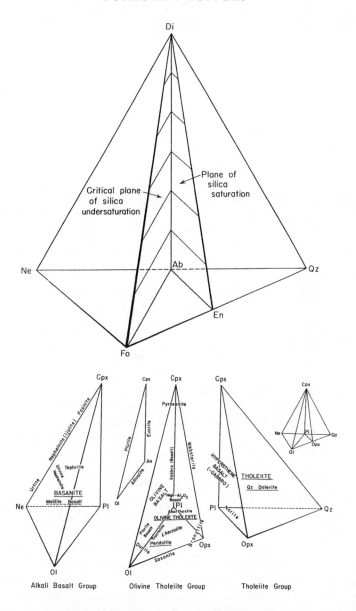

47. The basalt tetrahedron of Yoder and Tilley (1962). Top: schematic representation of the system Di–Fo–Ne–Qz showing plane of silica saturation, Di–En–Ab, and the critical plane of silica undersaturation, Di–Fo–Ab. Bottom: exploded view of simple basalt system with names of rocks whose major normative phases are contained in the tetrahedral. Reproduced from *Journal of Petrology* by permission of Oxford University Press.

magma types may be traced to variations in the chemical composition of their source rocks. Higher pressures and removal of garnet, they concluded, tend to produce alkali basalt, whereas lower pressures and removal of the pyroxene mineral omphacite tend to produce tholeiite. Yoder (1976) continued work on basalt for several years thereafter.

Some of the high-pressure experiments undertaken by Yoder and Tilley on eclogites entailed the use of a new apparatus designed by their fellow workers at the Geophysical Laboratory, Francis R. Boyd and J. L. England. Over the years, a few other researchers had been establishing laboratories for the investigation of high-pressure phenomena independently of the Geophysical Laboratory. These researchers, including Percy Bridgman of Harvard, David Griggs and George C. Kennedy of UCLA, Sydney Clark, Jr., of Yale, and Loring Coes of the Norton Company in Massachusetts, had developed various types of "squeezers" to generate the high pressures that occur in the upper mantle. Boyd and England (1960) modified Kennedy's earlier design and developed a machine that could be internally heated with a thermocouple and that could generate pressures up to 50 kilobars at 1750°C. The way was open for the examination of melting phenomena and phase changes in the upper mantle at depths from 40 to 200 kilometers. During the 1960s, Boyd, Hays, Lindsley, and other scientists at the Geophysical Laboratory determined the P-T melting curves for common igneous minerals such as albite, anorthite, diopside, enstatite, and forsterite, as well as the stabilities of high-pressure metamorphic minerals such as pyrope, jadeite, and coesite. The melting relations of the plagioclase feldspars at 10 and 20 kilobars were also determined.

Ever since Bowen and Schairer began work with iron–silicate systems, experimentalists had been faced with the problem of controlling and measuring the oxidation state of iron during runs. In addition to the technical advances made by Bowen and Schairer in experimental design, further progress had been made by the physical chemists L. S. Darken and R. W. Gurry on the determination of the stabilities of pairs of oxygen buffers (Darken and Gurry, 1945) and by Elburt F. Osborn and Arnulf Muan at Penn State. Without question, however, one of the revolutionary breakthroughs achieved at the Geophysical Laboratory, and one of the crowning achievements of experimental petrology in the second half of the century, was Eugster's invention of the oxygen buffer technique for controlling oxygen fugacity at load pressures greater than one atmosphere. Hans Peter Eugster (1925–1987) was a native of Switzerland who pursued undergraduate studies at the Swiss Federal Institute of Technology in Zürich in engineering geology (Jones, 1988; Gunter, 1990). He received a master's degree in geology in 1948 and then obtained his Ph.D. degree in Zürich under Paul Niggli in 1951. Eugster eventually expected to teach in Niggli's institute, but first he wanted to study optical spectroscopy with Louis Ahrens in the Cabot Laboratory at MIT. In 1952, however, he accepted an invitation to spend time at the Geophysical Laboratory, where Yoder introduced him to the synthesis of micas. Eugster collaborated with Yoder on papers on the synthesis of phlogopite and muscovite. In 1958, Eugster left the Geophysical Laboratory for Johns Hopkins University, where he remained until his death. Although Eugster eventually moved away from experimental igneous petrology to the study of the phase relations of evaporite minerals and other aspects of geochemistry, he exerted a profound influence on igneous petrology by his invention of the oxygen buffer technique and by showing several other young petrologists how to use the method.

During the mid-1950s, Eugster found a way to buffer the oxygen fugacity at a given temperature and pressure in an experiment by using assemblages of minerals such as hematite plus magnetite coexisting at equilibrium. By enclosing a capsule of platinum containing the powdered charge and water with an outer gold capsule containing water and a buffer assemblage, Eugster learned that the oxygen fugacity could be maintained in the fluid phase (water) that was in contact with the charge inside the platinum capsule (Figure 48). Because water within such a gold capsule dissociates, the ratio of hydrogen fugacity to oxygen fugacity is fixed at a specified temperature and pressure, and because the oxygen fugacity is also determined by the buffer assemblage for that temperature and pressure, the hydrogen fugacity within the gold capsule is fixed. In this arrangement, the platinum of the inner capsule has the added benefit of being permeable to hydrogen so that the hydrogen fugacity determined by the buffer assemblage is effectively imposed on the fluid within the platinum capsule, and, in turn, the oxygen fugacity of the fluid is thereby controlled. Eugster (1957) calculated P_{O_2}-T curves for the iron–wüstite, wüstite–magnetite, fayalite–magnetite–quartz, and magnetite–hematite buffer assemblages.

Working in conjunction with David R. Wones, then a fellow at the Geophysical Laboratory while pursuing his Ph.D. degree from MIT, Eugster demonstrated the value of the new oxygen buffer technique in determining the stability of annite, the iron-rich end member of biotite, between 0.5 and 2 kilobars in P-T-P_{O_2} space (Eugster and Wones, 1962). They successfully extended their studies to a wide range of biotite compositions (Wones and Eugster, 1965). The buffer technique was quickly adopted, and in years to come investigators like Wones, M. C. Gilbert, W. G. Ernst, and many others determined the phase relations for biotite, ferriannite, the various end members of hornblende, and numerous predominantly metamorphic minerals. Since the work of Eugster, further data have been obtained for additional oxygen buffer assemblages such as nickel–bunsenite, MnO–Mn_3O_4, graphite–gas, and magnetite–iron, making possible more precise delineation of the fields of stability of iron-bearing minerals.

One of the most valuable studies to emerge from the oxygen fugacity work at the Geophysical Laboratory was that of Lindsley on the stability of Fe–Ti oxides. Donald H. Lindsley (b. 1934), a 1956 graduate of Princeton University, where he had studied with A. F. Buddington, received his Ph.D. degree from

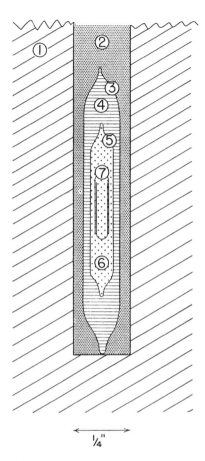

48. Cutaway sketch of the oxygen buffer assembly employed by Wones and Eugster (1962) and designed by Eugster. 1. Pressure vessel. 2. Pressure medium. 3. Sealed gold tube. 4. Oxygen buffer. 5. Sealed platinum tube. 6. Charge. 7. Open silver capsule. Reproduced from *Journal of Petrology* by permission of Oxford University Press.

Johns Hopkins University in 1961. At Johns Hopkins, he made Eugster's acquaintance and learned the oxygen buffer methods. He then spent 1962 to 1970 as Petrologist at the Geophysical Laboratory and became interested in the connection between magnetism and iron–titanium oxides. In his early experiments on oxides at the Geophysical Laboratory, Lindsley recognized that a magnetite–ilmenite solid solution series that had been proposed by Buddington, who had been working on the petrogenetic implications of Fe–Ti oxides for several years, did not exist. Lindsley found that intergrowths between magnetite and ilmenite are not the result of exsolution as Buddington had suggested. He found instead that magnetite and ulvöspinel form a solid solution series and that subsolidus oxidation of ulvöspinel to ilmenite had produced the magnetite–ilmenite intergrowths. When asked to review a paper by Buddington that incorporated his erroneous idea about magnetite–ilmenite exsolution, Lindsley, recognizing that Buddington's manuscript was full of useful analytical data, suggested that they collaborate on a paper using Buddington's analytical and petrographic data and the new experimental information that Lindsley had uncovered. The result was a landmark paper on Fe–Ti oxide minerals and their synthetic equivalents (Buddington and Lindsley, 1964). In effect, Lindsley had calibrated coexisting pairs of Fe–Ti oxides for use as thermometers and oxygen barometers. Because the oxidation state of iron can profoundly influence the differentiation trends of magmas, the newly found ability, developed at the Geophysical Laboratory by Eugster, Lindsley, and others, to measure the fugacity of oxygen by means of common mineral assemblages proved exceptionally valuable to the advance of igneous petrology.

In 1970, Lindsley moved to the State University of New York at Stony Brook where he has spent the remainder of his career. Much of his subsequent experimental work has focused on the determination of phase relations of iron-bearing pyroxenes using oxygen buffers as well as the formulation of thermodynamic models for pyroxenes. In recent years, Lindsley and B. Ronald Frost of the University of Wyoming have studied equilibria among Fe–Ti oxides, pyroxenes, olivines, and quartz and have developed computer programs for calculating temperature, pressure, and oxygen fugacity of magmas from the chemical compositions of coexisting pyroxene–olivine–oxide quartz mineral assemblages in igneous rocks.

An idea of the range of experimental studies accomplished at the Geophysical Laboratory during the 1960s may be gleaned from a perusal of the volume published in 1969 by *American Journal of Science* in honor of J. F. Schairer. The special issue contained, for example, papers by D. K. Bailey and M. C. Gilbert on the stability of acmite; by Ikuo Kushiro on the system forsterite–diospide–silica with and without water at high pressures; by Lindsley and James L. Munoz on the subsolidus relations of iron-rich pyroxenes; by W. C. Luth on albite–silica and potassium feldspar-silica up to 20 kilobars with variable water contents; and by Ian McGregor on the system $MgO-TiO_2-SiO_2$.

THE SPREAD OF EXPERIMENTAL PETROLOGY—PENN STATE

In their book *Igneous Petrology*, Carmichael et al. (1974, p. 219) observed that the type of experimental investigation "associated for all time with N. L. Bowen and J. F. Schairer, has in the last 20 years spread to universities the world over, largely by those associated at one time or another with the Geophysical Laboratory." How was that spread accomplished? After years of groundbreaking experimental studies at the Geophysical Laboratory, Bowen became convinced of the importance of introducing experimental methods into the academic world. To that end, he accepted a position at the University of Chicago in 1937, established a laboratory that was modest in comparison with what he had known in Washington, and supervised a handful of doctoral candidates whose dissertations were published in *Journal of Geology*. Bowen envisioned that experimental petrology would proliferate as his doctoral students took academic positions, set up their own laboratories, and, in turn, trained still more students. The dream initially met with limited success. Half of his students accepted nonacademic positions. Two students who did enter universities contributed relatively little to experimental petrology. Only Julian Goldsmith, who became a faculty member at Chicago, developed a productive career in experimental petrology, and even he diverted his attention from igneous petrology to carbonates and scapolite.

The spread of experimental igneous petrology to universities around the world received its major impetus when Pennsylvania State University (then College) recruited Elburt Franklin Osborn (1911–2001) in 1946 as Professor of Geochemistry and Head of the Earth Sciences Division, one of the three divisions in its School of Mineral Industries (Kerrick, 1987). After receiving his Ph.D. degree in 1938 from Caltech, Osborn developed an interest in research at high temperatures and pressures as a Petrologist at the Geophysical Laboratory between 1938 and 1942. He had also been engaged in wartime research on metallurgical problems and spent a year as a research chemist with Eastman Kodak before coming to Penn State. Osborn's initial experimental work at Penn State was undertaken with his graduate student Rustum Roy, who received his Ph.D. degree in 1948 and spent his entire professional career at Penn State in experimental geochemistry and materials research. After receiving a Ph.D. degree from MIT in 1939, teaching at Queen's University from 1940 to 1947, and serving as a Petrologist at the Geophysical Laboratory from 1947 to 1950, Mackenzie L. Keith (1912–2000) was brought to Penn State in 1950 by Osborn to serve as Associate Professor of Geochemistry. Although Osborn eventually became Dean of the College of Mineral Industries in 1953, he continued research with his graduate student Arnulf Muan, a recipient of a Ph.D. degree in metallurgy in 1955, on iron oxide–silicate systems (Muan and Osborn, 1956). Osborn also worked with other graduate students Robert de Vries and Kenneth Gee in an extensive study of the phase relations in the system $CaO-MgO-Al_2O_3-SiO_2$. The work with Muan, accomplished before Eugster had devised the oxygen buffer method, involved the partial pressure of oxygen as a variable in controlled-atmosphere experiments at one atmosphere total pressure. Those studies led Osborn to formulate his concept of the role of oxygen in the crystallization and differentiation of basaltic magma, as noted in Chapter 24.

Equally important with Osborn's novel idea about the control of differentiation trends by partial pressure of oxygen was the fact that he continued dipping into the Geophysical Laboratory by hiring Bowen's collaborator, O. F. Tuttle, to be Professor of Geochemistry and Chairman of the Division of Earth Sciences in 1953. With Tuttle on board, experimental petrology research at Penn State was extended to higher pressures. While at Penn State, Tuttle completed the manuscript on the landmark granite memoir (Tuttle and Bowen, 1958), continued his studies on the granite system to higher pressures, and began to explore the effect of two volatiles on the granite system. Because commercial apparatus for experimental petrology was not yet available, the various kinds of furnaces and vessels used at the Geophysical Laboratory and at Penn State were constructed entirely by machinists employed by the respective institutions. The expanding experimental program at Penn State, however, drew inquiries from researchers at other institutions who were interested in beginning their own experiments with facilities for research at high temperatures and pressures. As a result, Rustum Roy and Tuttle joined with a graduate student and a post-doctoral research associate to form their own company, Tem-Pres, for the manufacture and sale of high pressure and temperature experimental equipment. Many an experimental petrology laboratory had a modest beginning with Tem-Pres furnaces.

Penn State also hired C. Wayne Burnham in 1955, fresh from Caltech with his Ph.D. degree, as an Assistant Professor of Economic Geology. Although Burnham was particularly interested in ore-forming processes, his work had profound importance for igneous petrology because of his studies of the solubility of water in silicate melts, the composition of fluids in equilibrium with silicate melts, granitic pegmatites, and the internal structure of magmas. So that the chemical composition of volatile phases in equilibrium with melt could be determined more accurately, Burnham and Tuttle subsequently developed a large-volume, internally heated pressure vessel.

Experimental petrology at Penn State took another major step forward with the arrival of Peter John Wyllie (b. 1930) in 1956. Wyllie had done his undergraduate work at St. Andrews University in Scotland and then received his Ph.D. degree in 1958 from the same institution with a study of picritic sills on Skye. His interest in geology had been triggered during his undergraduate years when he joined the British West Greenland Expedition in the summer of 1950. He then spent 1952 to 1954 as a geologist with the British North Greenland Expedition. In 1956, Wyllie came to Penn State as a research assistant under Tuttle to learn the methods of experimental petrology.

A major experimental initiative at Penn State, conducted by Wyllie and Tuttle, concerned the influences of various kinds of volatile substances on the melting of granite. Isolated studies of this sort had been undertaken elsewhere by J. Wyart of Nancy, France, and N. I. Khitarov in Russia. These investigators examined the melting of obsidian or granitic rocks in the presence of alkali or other aqueous salt solutions. Systematic attack on the role of volatiles, however, was done during a three-year period in which Wyllie and Tuttle reported their findings on the effects of CO_2, NH_3, HF, SO_3, P_2O_5, HCl, and Li_2O on the melting temperatures of albite and granite. Early work on carbon dioxide by Wyllie and Tuttle (1959) showed that rates of melting and crystallization of granite and alkali feldspars are not enhanced by the presence of CO_2 under pressure, thus refuting proposals that CO_2 would have an effect similar to that of H_2O. They also found that the melting temperatures are essentially unaffected except insofar as the temperature of melting increases with increases in pressure per se. From their observations, they maintained that CO_2 is less soluble in granitic melts than is water.

In their next major study, Wyllie and Tuttle (1961a) examined the effects of NH_3 and HF on albite and granite. They found that both the curve for the beginning of melting of granite and the albite solidus drastically diminish in temperature with increased weight percent of HF as a second volatile substance. Fluorine, therefore, appeared to be a powerful fluxing agent like water. In contrast, the addition of increased amounts of NH_3 led to very gradual increases in the temperature of the beginning of melting of granite and of albite.

The third major article in the series (Wyllie and Tuttle, 1964) examined the effects of aqueous solutions of H_2SO_4 (SO_3), H_3PO_4 (P_2O_5), HCl, and LiOH (Li_2O) on albite and granite at 2.75 kilobars. The addition of P_2O_5, SO_3, and HCl all led to significant reductions in the beginning of melting of albite with the most pronounced effect being for P_2O_5. For granite, the results were markedly different. P_2O_5 produces lowering of the beginning of melting, SO_3 displays virtually no effect, and HCl produces an increase in the temperature of the beginning of melting of granite. The limited data on Li_2O indicated a very dramatic lowering of melting temperatures of granite, even more than that of HF. Wyllie and Tuttle (1964) concluded that aqueous pore fluids in the crust with appropriate dissolved volatile contents should make it possible for anatexis to occur at higher levels in the crust where temperatures are lower. They also concluded that the Li and F contained in pegmatites are among the constituents that cause pegmatites to remain in molten condition to lower temperatures than ordinary granites. These reconnaissance experiments demonstrated the contrasting behavior of Cl and F in magmas, later quantitatively confirmed by Koster van Groos and Wyllie (1968, 1969) in studies of the systems $NaAlSi_3O_8$–NaF–H_2O and $NaAlSi_3O_8$–$NaCl$–H_2O.

Wyllie became particularly interested in the role of CO_2 in magmatic processes and contributed early experimental studies on the origin of carbonatite, demonstrating that carbonate melts could exist at geologically reasonable temperatures within the continental crust and urging a magmatic origin for intrusive carbonatites. Wyllie and Tuttle (1960) had the honor of contributing the lead article in the very first issue of *Journal of Petrology* on "the system CaO–CO_2–H_2O and the origin of carbonatites." Wyllie spent 1959 to 1961 at the University of Leeds in England but returned to Penn State as a faculty member in 1961. At that point he initiated two avenues of experimentation with his graduate students. One avenue followed from the melting behavior of natural granites to the study of the phase fields intersected by genetically related series of granitic–tonalitic rocks corresponding to a composition line through the whole system. The other avenue involved experiments with various silicate minerals added to the system $CaO-CO_2-H_2O$. Koster van Groos and Wyllie (1963) discovered immiscibility between silicate and carbonate liquids.

The Geophysical Laboratory made a lasting impact on the establishment of experimental petrology in the academic world by contributing Osborn, Keith, and Tuttle to the faculty at Penn State. It is small wonder that Penn State's program, with Burnham and Wyllie added to the mix, rivaled that of the Geophysical Laboratory during the 1950s and early 1960s. This remarkable constellation of experimental petrologists was further reinforced in the early 1960s with the addition, in 1962, of Joseph W. Greig (1895–1977), also from the Geophysical Laboratory, and, in 1960, of Richard Henry Jahns (1915–1983) from Caltech, to serve as Chairman of the Division of Earth Sciences. Although Jahns was widely respected as an excellent field geologist with strong interests in pegmatites and engineering geology, he teamed up with Burnham on experimental studies of pegmatites while at Penn State. He also collaborated with Tuttle and graduate student William Luth in further studies on the granite system.

The memoir of Tuttle and Bowen (1958) had been a watershed in granite studies. Although providing a definitive statement about the origin of granite, to a substantial degree undermining the contentions of the school of granitization, the 1958 granite memoir provided a springboard for an ongoing flood of experimental investigations. Luth, Jahns, and Tuttle extended studies on the liquidus surface in the simple granite system from four to ten kilobars (Luth et al., 1964). Although they found that thermal minima in the liquidus surface at higher pressures are replaced by quaternary eutectic points, these low-temperature points continued the trend of the thermal valley established for lower water pressures and toward higher normative albite and lower normative quartz contents for beginning melts at higher water pressures. They also found that the water content of granitic melts continues to increase and that the temperatures of the liquidus surface as well as the temperatures of quaternary eutectics decrease with increased pressure. In plotting the proportions of normative albite, orthoclase, and quartz for 507 granitic rocks on the anhydrous ternary base of the simple granite system diagram, Luth, Jahns, and Tuttle found that granitic rocks generally plot to the orthoclase rich side of the thermal valley.

The Penn State group also attracted a number of postdoctoral research associates and doctoral students who went on to productive careers as experimental petrologists. Among the research associates were G. M. Anderson from Canada, David L. Hamilton and John Gittins from England, and David H. Eggler from the United States and another product of the Geophysical Laboratory. Among Penn State Ph.D. degree recipients in experimental petrology in the 1960s were

Peter Roeder, Dean Presnall, who later spent 1963 to 1967 at the Geophysical Laboratory, W. C. Luth, and John R. Holloway. These individuals carried experimental petrology with them to other institutions such as the University of Toronto, Queen's University, Stanford University, the University of Texas at Dallas, and Arizona State University. Although August F. Koster van Groos earned his Ph.D. degree from the University of Leiden in the Netherlands, he worked on experimental petrology under Wyllie for his dissertation. In addition, Arthur L. Boettcher, who subsequently changed his last name to Montana, received his Ph.D. degree at Penn State for a field-based petrologic study of a phlogopite deposit but then followed Wyllie to the University of Chicago on a postdoctoral fellowship and there learned experimental petrology from scratch. Boettcher later returned to Penn State and ultimately moved his laboratory to UCLA where he joined experimentalists George Kennedy and W. Gary Ernst who had spent some time at the Geophysical Laboratory.

Although Penn State played a crucial role in the expansion of experimental petrology in North America by means of its doctoral students and postdoctoral research associates, it played perhaps an even greater role in that expansion by reluctantly parting with several of its own faculty stars. The experimental petrology program suffered a major blow in 1965 when Peter Wyllie accepted a position as Professor of Petrology and Geochemistry at the University of Chicago, where his focus on igneous petrology complemented the experimental work of Julian Goldsmith and R. C. Newton. To make matters worse for Penn State, Jahns resigned in the same year to become Dean of the School of Earth Sciences at Stanford. Tuttle went with him. They were joined at Stanford by W. C. Luth in 1967. Thus the Stanford University program in experimental igneous petrology was born in the mid-1960s from the womb of Penn State.

THE SPREAD OF EXPERIMENTAL PETROLOGY—
EUROPE, AUSTRALIA, JAPAN

Experimental petrology laboratories also began to spring up in Europe, again with the assistance of the Geophysical Laboratory. The Geophysical Laboratory had a long tradition of having European petrologists spend time in Washington as guest investigators. Olaf Andersen, Paul Niggli, Cecil Tilley, and T.F.W. Barth, for example, had all spent at least one year at the laboratory during its first few decades. Beginning in the 1950s, investigators from overseas often returned to their native lands to set up their own laboratories. The first experimental petrology laboratory in Europe was established in Great Britain in the 1950s at the University of Manchester by William Scott MacKenzie (b. 1920). MacKenzie had been at the Geophysical Laboratory as a Fellow during 1951–1952 and as a staff petrologist between 1953 and 1956. While there, he worked on the limits of solid solution of nepheline and on the crystalline modifications of albite. From Washington he returned to initiate the experimental program at Manchester, where he often collaborated with W. S. Fyfe. One of his early Ph.D. students was David L. Hamilton. After time at Penn State as a postdoctoral fellow working with Osborn and Burnham on the solubility of water and the role of oxygen fugacity in mafic magmas, Hamilton joined the faculty at Manchester. MacKenzie continued experimental studies on nepheline and the feldspars while Hamilton, together with R. S. James, took granite studies the next step by examining the effect of small amounts of the anorthite component on the phase relationships in the granite system at low water pressures. James and Hamilton (1969) determined the crystal–liquid relations at a water vapor pressure of one kilobar for planes at 3, 5, 7.5, and 10 weight percent normative anorthite in the five component system $NaAlSi_3O_8$–$KAlSi_3O_8$–$CaAl_2Si_2O_8$–SiO_2–H_2O. They showed that residual liquids move toward the quartz-orthoclase sideline with increased anorthite content of the liquid. From a contour plot of the normative quartz, albite, anorthite, and orthoclase contents of 796 volcanic and plutonic rocks from Washington's tables, they demonstrated an even closer coherence between granitic rock compositions and the thermal valley for the anorthite-bearing system in comparison with the haplogranite system, a result attributable to the fact that natural granites do contain some anorthite component (Figure 49). They also showed that the composition of a partial melt would be dependent on the albite/anorthite ratio of the sedimentary rock being melted.

A short time after the establishment of the Manchester laboratory, Michael J. O'Hara (b. 1933) established a laboratory at the University of Edinburgh. O'Hara graduated in 1955 from the University of Cambridge, where he studied under Tilley and Stuart Agrell. He remained at Cambridge for his doctoral work, receiving his Ph.D. degree in 1960 with a dissertation on Lewisian metamorphic rocks in northwestern Scotland. In 1958, O'Hara joined the Grant Institute of Geology at the University of Edinburgh, where he became fascinated with a collection of eclogite and garnet peridotite specimens from South Africa that had been given to Arthur Holmes, who was still present at the university as a professor emeritus. O'Hara's department head, Fred Stewart, was eager to initiate an experimental petrology laboratory at Edinburgh. As a result, O'Hara spent the academic year 1962–1963 at the Geophysical Laboratory, where he was able to pursue his interest in eclogite and peridotite by doing a study of the join diopside–pyrope in collaboration with Schairer and Yoder.

Upon setting up the new laboratory in Edinburgh with funding from the National Environmental Research Council, O'Hara continued experimental work on eclogite, peridotite, and the melting of basaltic magmas. From his own experiments and those of others, he generated a comprehensive scheme for peridotite–basalt petrogenesis. In a pair of reviews on the origin of basalt, O'Hara (1965, 1968) concluded from relevant experimental studies that the chemical composition of magma in equilibrium with mantle assemblages

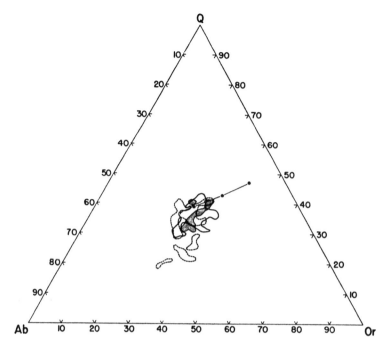

49. Contour diagram illustrating the distribution of normative quartz, albite, and orthoclase for 796 analyzed igneous rocks from Washington's tables that contain 80 percent or more of normative quartz, orthoclase, albite, and anorthite. Line with dots shows path of low-temperature points with increasing anorthite content in granite system. Reproduced from *Contributions to Mineralogy and Petrology*, Phase relations in the system $NaAlSi_3O_8$-$KAlSi_3O_8$-$CaAl_2Si_2O_8$-SiO_2 at 1 kilobar water vapour pressure, James, R. S., and Hamilton, D. L., v. 21, pp. 111–141, figure 10, 1969, copyright © Springer-Verlag 1969.

would vary with increasing pressure. He assumed a mantle consisting primarily of olivine, orthopyroxene, and clinopyroxene (a rock designated as lherzolite) with plagioclase at low pressure, spinel at intermediate pressure, and garnet at high pressure. He maintained that such magmas would be quartz-normative tholeiite at low pressure; high-alumina basalt at somewhat higher pressures around 5 to 8 kilobars; nepheline–normative picritic basalt at eight to 20 kilobars; and hypersthene-normative picritic basalt from 20 to 40 kilobars. O'Hara further argued that the depth of origin of a primary magma exercised little control over final eruption products and that erupted magmas typically do not represent primary magmas inasmuch as they have undergone continuous fractionation, at least of olivine, during their ascent. As a result, he maintained that the abundant oceanic tholeiites are not primary magmas, contrary to the view of Engel et al. (1965). He also suggested that wall-rock contamination probably had not accounted for the variation in magma chemical composition, whereas fractionation of olivine and other phases had. Fractionation trends were envisioned as being controlled by various thermal divides in the basalt system that existed at different pressures. Moreover, O'Hara pointed out, the extent of fractionation of olivine and other materials was strongly influenced by the rate at which magmas ascended toward the surface. He suggested that calc–alkalic magmas most likely developed as high-alumina basalt was hydrated during ascent from dry mantle into a wet orogen and as it fractionated under high oxygen fugacity.

O'Hara's experimental group in Edinburgh also extended related experimental studies to the petrogenesis of lunar rocks. Ironically, the Edinburgh experimental petrology group, then only three years old at the time, was initially the only such laboratory approved for examination of the Apollo 11 samples. Neither the Geophysical Laboratory, nor Penn State, nor Manchester experimental petrologists were involved in this initial look at the Apollo 11 samples. O'Hara eventually left Edinburgh in 1978 and subsequently held positions at the University of Aberystwyth, the Sultan Qaboos University in Oman, and, since 1994, the University of Cardiff. He has continued to focus on experimental studies of basalt origins and lunar petrology.

Among the competitors vying with Edinburgh for supremacy in lunar experimental petrology was the Australian National University in Canberra, because of the fact that high-pressure experimental petrology had found its way into Australia via A. E. Ringwood. In 1963 and 1964, Ringwood had been a fellow at the Geophysical Laboratory working on the petrological constitution of the upper mantle. Thanks to the privileged position enjoyed by the Research School of Earth Sciences at the Australian National University as a full-time research institution, the provision of adequate financial backing by the Australian federal government, and the high quality of its faculty, experimental petrology in Australia quickly enjoyed great success (Green, 1976). Ringwood saw the advantage of examining the phase equilibria in natural rock compositions, following the approaches of Tuttle, Bowen, and Wyllie with granitic rocks and Yoder and Tilley with basalts. By the mid-1960s, this approach had become much more feasible with the development of the electron probe microanalyzer, an instrument that could directly determine the chemical compositions of extremely tiny crystals both in the starting materials and the final run products. Although Ringwood's primary interest concerned phase transformations at pressures on the order of 100 to 250 kilobars, he conducted experimental investigations of the gabbro–eclogite transition and the genesis of basaltic magmas with David H. Green. These experiments provided a solid basis for the later investigation of lunar rocks and related systems.

Green and Ringwood (1967) studied the melting relations in selected olivine-rich basalts with a view to elucidating the nature of basalt fractionation at temperatures as high as 1500°C and pressures as high as 27 kilobars. To estimate directions of fractionation of the magmas, they made electron microprobe determinations of the compositions of olivine, orthopyroxene, clinopyroxene, and garnet that crystallized at high pressures. Unlike that of O'Hara, their mantle model omitted spinel. Green and Ringwood concluded that compositions of magmas are determined by the depth at which magma segregates from residual crystals. They found that basaltic liquid segregating from mantle peridotite, which they termed "pyrolite," at a depth between 35 and 70 kilometers would be alkali olivine basalt with about 20 percent partial melting. With an increase in the amount of partial melting, the liquid would become olivine-rich tholeiite. Magmas segregated at about 30 kilometers, however, were said to be high-alumina olivine tholeiite, similar to those occurring along the mid-oceanic ridge. Quartz-normative melts would develop at shallower levels. The authors suggested that the abundances of incompatible elements in alkali olivine basalt were not consistent with simple crystal fractionation relationships between alkali olivine basalt and other basaltic magma types. In contrast to O'Hara, they maintained that interaction of slowly rising basaltic magma with large volumes of mantle wall rock might account for observed trace-element abundances of erupted basalt. Green continued his experimental investigations of basalt and peridotite in subsequent years at the Australian National University, where he eventually became Director of the Research School of Earth Sciences.

Another important project to emerge from Ringwood's laboratory at the Australian National University, on account of its applicability to the recently formulated hypothesis of seafloor spreading and plate subduction, was an investigation of the genesis of the calc–alkalic igneous rock suite that he undertook with Trevor H. Green, one of his doctoral students. To test several existing models for the genesis of the calc–alkalic suite, T. H. Green and Ringwood (1968) subjected a series of glasses of high-alumina olivine tholeiite, high-alumina quartz tholeiite, basaltic andesite, andesite, dacite, and rhyodacite compositions to temperatures as high as 1535°C and pressures as high as 36 kilobars. They determined the positions of solidus and liquidus curves for each composition in P-T space. On that basis they outlined three possible models of genesis. In one model, they envisioned transformation of piles of oceanic basalt into eclogite composed of garnet, clinopyroxene, and quartz. Then, at a depth of 100 to 150 kilometers, partial melting of the eclogite would give rise to various members of the calc–alkalic suite, depending on factors such as the water content of the eclogite or fractional crystallization of the ascending magmas.

In their second model, T. H. Green and Ringwood envisioned conversion of basalt into amphibolite at the base of a volcanic pile by addition of water. Subsequent partial melting of the amphibolite was said to generate calc–alkalic magmas. For the third model they suggested fractional crystallization of hydrous basalt, derived by partial melting of the mantle, at a depth of 30 to 40 kilometers. The water, they suggested, might have been supplied by contamination with hydrated rocks introduced into the mantle by the sinking limb of a convection cell. This paper tested and supported the possibility that the calc-alkalic rock suites of island arcs and continental magmatic arcs, spatially associated with oceanic trenches and Benioff zones, might originate by partial melting from slabs of oceanic crust descending along subduction zones. Later experiments by Stern and Wyllie (1978), however, demonstrated significant problems with the idea of the formation of andesite in subducted crust. They also redirected attention to the feasibility of partial melting in mantle wedges fluxed by aqueous fluids rising from dehydrated slabs beneath the wedges. Eventually, Trevor Green took a position at Macquarie University in Sydney, where installation of a 50-kilobar piston–cylinder apparatus was made possible in the 1970s with funding by the Australian Research Grants Committee. Similar laboratories were later installed at Monash University and the University of New South Wales.

The Geophysical Laboratory also contributed to the establishment of experimental igneous petrology laboratories in Japan. The roots of experimental petrology in Japan may be traced to Shukusuké Kôzu (1880–1955). Kôzu studied at the Imperial University of Tokyo under B. Koto, the founder of petrology in Japan (Yagi, 1956). After a brief period with the Imperial Geological Survey, during which he made the first discovery of alkalic rocks in Japan, and further periods at the University of Tokyo and the Tohoku Imperial University, Kôzu left Japan early in 1913 for the Geophysical Laboratory, where he conducted optical and thermal studies on Japanese alkalic rocks and befriended A. L. Day, N. L. Bowen, and, particularly, H. S. Washington. In mid-1914, he continued studies in Cambridge and Paris. He returned to Japan in 1916 as Professor of Mineralogy and Petrology at Tohoku University. Because he had been so impressed by the high-temperature experimental work at the Geophysical Laboratory but was unable to obtain equipment for studies of synthetic multicomponent systems, Kôzu instead undertook melting experiments on igneous rocks (Kôzu et al., 1919). When the Institute of Mineralogy, Petrology, and Economic Geology was estblished at Tohoku University in 1921, Kôzu was appointed Head and initiated thermal studies of igneous rock-forming minerals. He retired in 1942.

Kenzo Yagi (b. 1914) became interested in experimental petrology as a student of Kôzu and had the opportunity to spend 1950–1951 at the Geophysical Laboratory, where he worked with J. F. Schairer on nepheline–diopside and the $FeO-Al_2O_3-SiO_2$ system. After returning to Japan, Yagi was appointed as a professor in the newly established Institute of Earth Sciences at Tohoku University. After doing geologic and petrographic studies on Japanese volcanoes for several years, Yagi was eventually able to set up a laboratory for doing experimental work at one atmosphere pressure. In 1962, Yagi moved to the Department of Geology and Mineralogy at Hokkaido University, where he established a high-pressure laboratory and studied the stability relations of minerals and the origin of alkalic rocks along with K. Onuma, his former student at Tohoku, and several graduate students.

Ikuo Kushiro (b. 1934) became one of the most prominent Japanese experimentalists. He became interested in igneous petrology as an undergraduate at the University of Tokyo, where he studied with Hisashi Kuno. Because Kuno encouraged him to study the differentiation of basaltic magma, Kushiro examined several differentiated sills in Japan and became interested in the general problem of the origin of basaltic magma (Kushiro, 2000). He received his Ph.D. degree from the University of Tokyo in 1962 with an investigation of Kuno's hypothesis that tholeiitic magmas are formed at shallower levels in the mantle than are alkalic basalts. Kushiro was encouraged to go to the Geophysical Laboratory to conduct high-pressure experiments relevant to that hypothesis and was appointed as a postdoctoral fellow in 1962. Along with Schairer, he redetermined the system forsterite–diopside–silica originally studied by Bowen in 1914. Then, using the Boyd-England piston-cylinder apparatus, he determined the effects of increased pressure on that system and on the system forsterite–nepheline–silica. Kushiro's work showed that melt composition becomes more silica-undersaturated with increasing pressure. Over the next decades, Kushiro spent the bulk of his career resonating between the University of Tokyo and the Geophysical Laboratory, conducting experiments on the melting of mantle peridotite, the stability of hydrous minerals like phlogopite in the mantle, and the physical properties of magmas. In 1969, Kushiro replaced Kuno, who had just died, in examining Apollo 11 samples. This experience stimulated Kushiro's interest in planetary science, and he subsequently conducted experiments on phase equilibria of silicate systems at extremely low pressures in an effort to simulate conditions in condensing solar nebulae. Experimental petrology has continued to flourish in Japan, thanks in part to the Geophysical Laboratory. Several other Japanese experimental petrologists have also undertaken studies at the Geophysical Laboratory.

The earliest German experimental work in petrology was done at the University of Göttingen by Helmut Winkler and Hilmar von Platen during the late 1950s and early 1960s. Winkler's work was distinctive in that it was directed almost entirely toward the experimental evidence for anatexis of granitoid melts and that only natural rocks were studied (e.g., Winkler and von Platen, 1961). Winkler and von Platen crushed various shale and graywacke samples, adding small amounts of such salts as $CaCO_3$ or $NaCl$, and heated them in the presence of water at a pressure of about 2 kilobars. Von Platen (1965) pointed out that residual granitic melts move increasingly toward the quartz–orthoclase sideline with decreasing albite/anorthite ratios at a water pressure of two kilobars. Hard on the heels of Tuttle and Bowen's work on the simple granite system and of melting experiments of shales by Wyllie and Tuttle (1961b), the studies of Winkler and von Platen made a compelling case for the feasibility of granite genesis by melting of deeply buried sedimentary rocks. In 1966,

Werner F. Schreyer joined the faculty at the Ruhr Universität in Bochum after a stint on the staff at the Geophysical Laboratory and soon established a petrology laboratory. The bulk of Schreyer's work, however, concerned metamorphic petrology. Friedrich Seifert came from Bochum as a guest investigator at the Geophysical Laboratory in 1967–1968, and then returned to Bochum where he collaborated with Schreyer for a time. Eventually Seifert developed a laboratory at Bayreuth. Seifert's laboratory, too, did not strictly focus on igneous applications. In time, an experimental laboratory would also appear at the University of Hannover.

It had been Bowen's dream to see the establishment of the methods of experimental igneous petrology in the academic world. A decade after his death, that dream had probably been fulfilled far beyond what he had dared to envision. By 1966, major centers of experimental research in igneous petrology had been established at Pennsylvania State University, the University of Chicago, Stanford University, the Australian National University, the University of Manchester, the University of Edinburgh, the University of California at Los Angeles, and elsewhere. In most of these instances, the Geophysical Laboratory had played a formative role.

THE SPREAD OF EXPERIMENTAL PETROLOGY— CHICAGO AND STANFORD

The vast majority of studies prior to 1965 concerned phase relations in synthetic systems with pure components. When he left Penn State for Chicago in 1965, Peter Wyllie took some graduate students with him, and they continued for several years thereafter with the Penn State-initiated investigation of the melting behavior of natural granites, and of genetically related series of granitoid rocks. As a basis for comprehension of the phase fields intersected in the rock systems, Wyllie coupled the rock studies with parallel investigation of complex synthetic systems that could be interpreted rigorously in terms of the phase rule.

The crustal pressure experiments were extended to mantle pressures in new piston–cylinder apparatus. Art Boettcher, for example, extended the granite solidus in P-T space to 30 kilobars by heating and squeezing, in the presence of excess water, a natural biotite granite from the Dinkey Lakes pluton in the Sierra Nevada batholith. Boettcher and Wyllie (1968) discovered that the temperature of the water-saturated granite solidus continues to decrease to a pressure around 16 kilobars. At that point the solidus curve develops a positive slope, indicating an increase in the temperature of the solidus with increased water pressure as a result of the breakdown of albite at around 16 kilobars to quartz and the dense phase, jadeite. The positive slope continues all the way to 30 kilobars. Similar results were obtained for tonalite–H_2O and basalt–H_2O by Lambert and Wyllie (1972, 1974).

In several studies on igneous rock series, Wyllie and his students heated crushed samples of rocks from well-mapped and petrographically well-characterized co-genetic igneous rock suites at different load pressures with excess water. By quenching the heated samples at different temperatures, Wyllie's group ascertained the temperatures at which melting began and at which individual mineral phases disappeared completely. From the data obtained at different pressures, curves for the beginning of melting and the disappearance of phases could be plotted on a *P-T* diagram. Although the experiments were run with excess water and oxygen fugacity was not controlled, the oxygen fugacity defined by the apparatus appeared to remain close to that of the natural magmas. The results shed light on the conditions of formation and crystallization of natural igneous rock suites by providing significant parts of the experimental framework that made assessment of these conditions feasible. The range of constraints was progressively extended through the years.

The first such set of experiments was done by Alf Piwinskii, who moved as a graduate student with Wyllie to the University of Chicago in 1965, on granodiorites and tonalites from the zoned Needle Point pluton in the Wallowa batholith of northeastern Oregon (Piwinskii and Wyllie, 1968a). A similar study done at both Penn State and Chicago and submitted for his Ph.D. dissertation at the University of Chicago was that of Piwinskii on granites, granodiorites, and tonalites of the central Sierra Nevada batholith at pressures to 3 kilobars in the presence of excess water (Piwinskii and Wyllie, 1968b). A couple of years later, Piwinskii and Wyllie (1970) returned to Oregon for a look at a late-stage felsic body that intruded the Needle Point pluton. Among other similar studies of rock suites were those by D. L. Gibbon on the Farrington Complex of North Carolina and the Star Mountain rhyolite of Texas (Gibbon and Wyllie, 1969), by J. K. Robertson on the Debouillie stock in northern Maine (Robertson and Wyllie, 1971a), and by S. D. McDowell on the Kungnat syenite complex of southwestern Greenland (McDowell and Wyllie, 1971).

The work of Robertson and Wyllie (1971a) on the Debouillie stock broke new ground because the authors melted members of the rock series with excess water and under water-deficient conditions. They plotted their results on *P-T* diagrams as before and also on diagrams of temperature versus water content (Figure 50). The use of the isobaric temperature–composition diagram was developed further by Robertson and Wyllie (1971b). The use of such diagrams arose primarily because investigators in the latter 1960s more and more stressed the likelihood that melting and crystallization processes occurred under conditions in which $P_{H_2O} < P_{load}$. Virtually all previous hydrothermal experiments, however, had been performed in the presence of excess water so that $P_{H_2O} = P_{load}$ and melts were water-saturated. One method of evaluating phase relationships under water-undersaturated conditions entailed the use of the *T-X* section. Building on his earlier work with such diagrams in dealing with carbonatites, Wyllie decided to extend the approach to the study of silicate

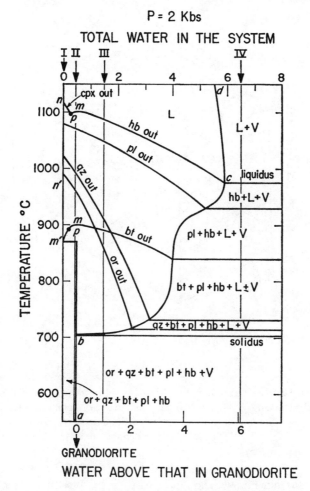

50. Schematic temperature-composition section for the system granodiorite–water determined by Robertson and Wyllie (1971a). Reproduced from *Journal of Geology* by permission of The University of Chicago Press. Copyright (c) 1971. The University of Chicago. All rights reserved.

melts. To illustrate the general principles of such diagrams, Robertson and Wyllie (1971b) constructed from available data a generalized three-dimensional P_{total}-T-X diagram (Figure 51) for the system granodiorite–water as well as schematic T-X sections at a pressure of 2 kilobars. Thus, various fields of coexisting stable phases were displayed as a function of temperature and weight percent water in the system. Robertson and Wyllie identified four types of subsolidus assemblage that could be defined in such diagrams: type I, a water-absent assemblage of anhydrous silicate minerals with no vapor phase

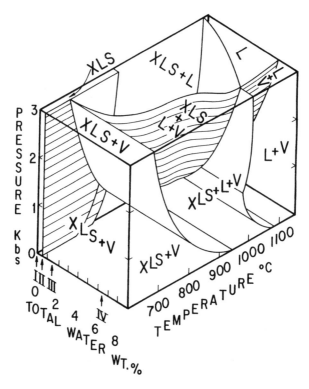

51. Pressure-temperature-water model diagram for the system granodiorite-water from Robertson and Wyllie (1971b). Shaded surface is saturation surface. Below 1 kilobar the hydrous minerals dehydrate below the solidus. Reproduced by permission of *American Journal of Science*.

present; type II, a water-deficient assemblage of silicate minerals, including some hydrous minerals, with no vapor phase present; type III, a water-deficient assemblage of silicate minerals, with or without hydrous minerals, in the presence of a vapor phase; and type IV, a water-excess assemblage of silicate minerals, with or without hydrous minerals, in the presence of a vapor phase, and with sufficient water to saturate the liquid when the mineral assemblage was completely melted at the pressure of the experiment. Robertson and Wyllie pointed out that type I and type IV systems (water-absent and water-excess) had been studied experimentally in detail but that types II and III had received virtually no such attention. From their experimental data and geometrical phase analysis on biotite granodiorite, they produced a *P-T* diagram for the vapor-absent type II system with a solidus curve that is presently called "fluid-absent" or "dehydration melting." In Manchester, Brown and Fyfe (1970) had just published dehydration-melting curves for diorite with added biotite or amphibole.

Sven Maaloe explicitly addressed the problem of deducing the water content of granitic magmas from sequences of crystallization in rocks from experimentally constructed T-X diagrams (Maaloe and Wyllie, 1975). In 1983, Wyllie left for Caltech, where he established a very prolific program of experimental petrology involving both synthetic and natural materials. He and his students continued to produce experimental studies on carbonatites, one of Wyllie's long-standing passions. Extension of carbonate–silicate experiments to mantle pressures has generated results on the effect of CO_2 on mantle melting with applications to kimberlites and alkalic rocks (e.g., Wyllie and Huang, 1975; Wyllie and Lee, 1999).

Although Tuttle and Jahns established experimental igneous petrology at Stanford, Jahns was largely involved with administration and, unfortunately, Tuttle's productivity declined seriously as he gradually lost a battle with Parkinson disease. Nevertheless, studies similar to those of Wyllie's group that took advantage of the T-X diagram began to emerge from the Tuttle–Jahns Laboratory for Experimental Petrology at Stanford as a result of the influence of William Luth, who had been studying the effects of water undersaturation in silicate melts. James A. Whitney, Samuel E. Swanson, and Michael T. Naney all investigated granitic systems under water-undersaturated conditions for their Ph.D. dissertations in 1972, 1974, and 1978, respectively. These studies, however, unlike Wyllie's experiments, returned to the investigation of synthetic materials. Whitney (1975) prepared synthetic gels lacking Fe, Mg, Mn, and Ti that otherwise resembled the average chemical compositions of hornblende–biotite granite, hornblende–biotite adamellite, hornblende–biotite granodiorite, and hornblende–biotite tonalite compiled by Nockolds (1954). To these he added varying amounts of water at pressures between 2 and 8 kilobars. Whitney was able to demonstrate that the order in which the various minerals crystallize would be strongly affected not only by temperature, pressure, and magma composition, but also by water content. Swanson (1979) explored the effect of CO_2 on a single granodioritic bulk composition in the granite system at pressures to 8 kilobars. He found that calcite should be expected as a primary phase in granitic rocks crystallizing in the presence of a CO_2-rich fluid at pressures in excess of three kilobars. From the absence of primary calcite in granitic rocks, Swanson reasoned that, if a CO_2-rich fluid had been present in the magmatic source region, it either remained at the source region or escaped prior to emplacement of the magma. As a result, he concluded that fluids accompanying crystallizing granitic magmas were likely to be H_2O-rich. Naney (1983) took granite studies the next step by investigating the behavior of Fe and Mg in granitic systems. By experimental examination of synthetic granitic and granodioritic compositions modeled after Nockolds' average rocks in the presence of variable amounts of H_2O at pressures between 2 and 8 kilobars and at oxygen fugacities between those of the Ni–NiO and hematite–magnetite buffers, Naney was able to determine the stabilities of several ferromagnesian silicate minerals coexisting with feldspars, quartz, and water-undersaturated granitic melt. He obtained data on the stabilities of hypersthene, augite, hornblende, biotite, and epidote. The studies of Whitney, Swanson, Naney, and others that utilized T-X sections provided a powerful tool for estimating the water contents of magmas from evaluation of the order of crystallization observed in natural igneous rock samples.

THE FURTHER SPREAD OF EXPERIMENTAL IGNEOUS PETROLOGY

Experimental igneous petrology continued at the University of Manchester, where one of the more significant projects was conducted by D.A.C. Manning for his Ph.D. dissertation. Manning (1981) determined the liquidus phase relationships for the haplogranite system in the presence of excess water and an additional 1, 2, or 4 weight percent fluorine (in the form of NaF or KF) at 1 kilobar total pressure. Manning discovered that the quartz–alkali feldspar boundary curve moves away from the quartz apex with increasing fluorine content; that the temperature of the minimum point decreases from 730°C for the purely hydrous melt to 630°C for the system containing 4 weight percent fluorine; and that the position of the minimum moves toward the albite (Ab) corner of the diagram with increased fluorine content. On the basis of these results, Manning maintained that the chemical compositions of fluorine-rich granitic rocks are consistent with an origin by the crystallization of residual melts that had been enriched in fluorine through magmatic differentiation.

New experimental petrology laboratories have also appeared in continental Europe since the 1960s. Wilhelm Johannes (b. 1936), for example, established the Institute of Mineralogy at the University of Hannover in Germany, where he and students such as François Holtz, Andreas Becker, Jagmohan Singh, and Harald Behrens have conducted a steady stream of studies on the melting relations of granitic rocks. Much of their work has been summarized in *Petrogenesis and Experimental Petrology of Granitic Rocks* (Johannes and Holtz, 1996). Johannes has urged caution in the use of phase diagrams of Ca-bearing granitic rocks by pointing out that it has been extremely difficult to demonstrate the attainment of equilibrium in experimental runs for such compositions.

Also among recently founded European laboratories for experimental igneous petrology are those at the Centre de Recherches Pétrographique et Géochimiques in Nancy, France; at the Centre de Recherches sur la Synthèse et la Chimie des Minéraux (CRSCM) of the Centre National de la Recherche Scientifique (CNRS) in Orleans, France; and at the University of Blaise Pascal in Clermont-Ferrand, France. In Nancy, the range of two-volatile studies on granitic systems like those of Manning at Manchester was expanded with the inclusion of boron as one of the volatile phases. Investigation of boron-bearing granitic systems was designed to lead to better understanding of the conditions of occurrence of tourmaline as a common varietal or accessory mineral in granitic rocks. Michel Pichavant, then in Nancy, undertook a study of the effect of boron on water-saturated haplogranite at 1 kilobar total pressure. He discovered that the solidus temperature of haplogranite is progressively lowered with continuing increases in weight percent of added B_2O_3 (introduced in the form of boric acid, H_3BO_2). For example, Pichavant (1981) found a 60°C lowering of the solidus temperature with the addition of 5 weight percent and a 130°C lowering with the addition of 17 weight percent. He also showed that water solubility in silicate melts increases with the addition of boron. In contrast to fluorine, Pichavant found that granitic melts in equilibrium with a boron-rich vapor phase become more potassic.

At the CRSCM in Orleans, Bruno Scaillet was later joined by Michel Pichavant from Nancy and François Holtz from Hannover. In addition to studies on the effect of H_2 on the phase relations of alumino-silicate melts, researchers at the CRSCM conducted melting experiments on natural granites from the Himalaya Mountains and estimated the water contents of their magmas from natural crystallization sequences. They found, for example, that the water content of a natural tourmaline granite may have been in excess of 7 weight percent (Scaillet et al., 1995).

At the University of Blaise Pascal, Daniel Vielzeuf and Jean-Marc Montel have done melting experiments on natural rocks in an effort to determine the most likely fertile source materials for anatectic generation of magmas. For example, Vielzeuf and Montel (1994) conducted experiments between one and 20 kilobars on the fluid-absent melting of a natural quartz-rich aluminous metagraywacke composed of biotite, plagioclase, and quartz. They concluded that such samples represent fertile rocks at temperatures on the order of 800 to 900°C at pressures below 7 kilobars because they obtained between 30 and 60 weight percent silicate melt in their experiments. They also concluded that pelitic rocks would be the most fertile source rocks at pressures in excess of 8 kilobars but that the difference in fertility between pelitic rocks and graywackes would decrease as pressure decreased (Montel and Vielzeuf, 1997). From determination of the compositions of glasses produced in their melting experiments, however, they maintained that anatexis of fertile rock types in the continental crust such as pelites, graywackes, and orthogneisses would produce only peraluminous leucogranites. They noted that more mafic calc–alkalic granites had never been produced in such partial melting experiments. Thus, they argued, whereas peraluminous granites might correspond to liquids formed directly by partial melting of metasediments, calc–alkalic and other granites must have involved some other process such as interaction with mantle-derived material.

One of the major emphases in recent experimental igneous petrology has been the vapor-absent or water-deficient partial melting of various rock types of the type initiated by Rutter and Wyllie (1988) and pursued in detail by Vielzeuf and Montel. At the University of Oregon, for example, A. Dana Johnston, who learned his experimental petrology during a postdoctoral fellowship at Caltech with Wyllie, has carried out studies with his students on the vapor-absent melting behavior of natural metapelites, biotite and hornblende tonalitic gneisses, and metavolcaniclastic rocks. Patiño Douce and Johnston (1991) argued that bulk composition of a source is the overriding factor that controls the amount of melt produced at water-undersaturated conditions around 850°C to 900°C. The amount of melt, they suggested, would be maximized from biotite-rich pelitic source rocks in which the mode coincides with the stoichiometry of a reaction involving biotite, plagioclase, quartz, and an alumino-silicate mineral. Like Vielzeuf they also encountered peraluminous granitic melts. From analysis of dehydration melting of biotite and hornblende in tonalitic gneisses, Skjerlie and Johnston (1993) concluded that, under fluid-absent conditions, such rocks would generally not melt at the temperatures in the continental crust apart from the introduction of very hot, mantle-derived magmas into the lower crust. Such a process, they claimed, might produce alkaline, fluorine-bearing granitic liquids. Skjerlie and Johnston (1996) also performed dehydration-melting experiments on a Precambrian paragneiss of volcaniclastic origin and concluded that the intrusion of hot, mantle-derived magmas along active continental margins could sufficiently heat such rocks to produce melts with compositions ranging from granitic to granodioritic depending on pressure.

After receiving his doctoral degree from the University of Oregon under Johnston in 1990, Alberto E. Patiño Douce joined Whitney and Swanson on the staff of the University of Georgia, where he has continued to conduct similar partial melting investigations of natural materials. He and Kjell Skjerlie, for example, performed water-undersaturated melting experiments on charges that consisted of a layer of sillimanite–bearing metapelite and a layer of garnet amphibolite. Skjerlie and Patiño Douce (1995) found that melt abundances were greater in pelite adjacent to amphibolite than in pelite alone and that melt abundances in amphibolite next to pelite decreased relative to amphibolite alone except near the interface. They attributed the increase in melt fractions in pelite to transfer of Na from amphibolite to pelite. Patiño Douce also conducted melting experiments with James Beard on biotite gneiss and quartz amphibolite (Patiño Douce and Beard, 1995) and on a model metagraywacke composition (Patiño Douce and Beard, 1996). In the former experiments they produced peraluminous granite and strongly peraluminous granodioritic melt respectively. Their experiments on quartz amphibolite suggested that strongly peraluminous cordierite–bearing granites and two-mica granites could even be generated by partial melting of Al-poor source rocks. They found that melting of metagraywacke produced a wide range of melt compositions, depending on pressure and the fugacity of oxygen. Patiño Douce (1996) reiterated that a wide array of melts could be generated from the same source rock depending on compositional factors such as the Fe/Mg and Ca/Fe ratios, water content, and pressure. Dehydration-melting studies have also been done by Stevens et al. (1997) on magnesian metapelites and metagraywackes, and by several other investigators.

During the thirty years since Yoder and Tilley (1962) pioneered with experiments on basalt–H_2O, there had been few additional studies except for those involving excess H_2O. Most experiments on basalt were related to their peridotite source rocks with applications to the petrogenesis of lunar rocks, mid-ocean-ridge basalts, and island arc magmas. With the experimental interest in dehydration melting, however, there came a resurgence of interest in basalt with low H_2O contents. Dry amphibolite was commonly used as a starting material. The initial contributions in this area of research appeared independently in 1991 (Beard and Lofgren, 1991; Rapp et al., 1991; Rushmer, 1991; Winther and Newton, 1991; Wolf and Wyllie, 1991). These investigations were soon followed by additional studies on amphibolite, including those of Sen and Dunn (1994) at the University of New Brunswick in Canada, and of Rapp and Watson (1995) at Rensselaer Polytechnic Institute in New York on four natural amphibolites representing different basalt compositions.

Thanks to the growing number of such dehydration-melting experiments, investigators have been better able to place constraints on their models of the physical conditions, the nature of the source rocks, and the composition of the melts obtained during partial melting in crustal metamorphic terranes and in subducting oceanic plates. They have also been able to determine whether partial melting alone was adequate to account for silica-rich to intermediate rocks or whether or not contributions from mantle-derived magmas were required. Since the publication of Tuttle and Bowen's granite memoir and thanks to early melting experiments on shales and graywackes by Winkler and by Wyllie and Tuttle, petrologists have become increasingly open to derivation of granitoid magmas by pure partial melting of deeply buried metamorphic rocks in the continental crust. The development of plate-tectonic theory opened the possibility for derivation of granitoid rocks by partial melting of subducted sediments and/or the accompanying eclogitic oceanic crust. Further studies of natural rocks showed convincingly that peraluminous granitic melts could indeed be derived by pure anatexis of metasedimentary and metavolcanic rocks. Those same studies, however, also seemed to indicate that most granitoid rocks, if derived by anatexis, also required a significant component of mantle-derived material or of restite for their formation.

Experimental laboratories dedicated primarily to igneous petrology now abound in many parts of the world. Bowen's dream has been magnificently realized thanks to the gradual entry of Geophysical Laboratory staff members and guest investigators into the academic world, to generous funding by governmental agencies in several countries, and to the realization among petrologists that experimental studies have proved their worth in addressing petrological problems. Ironically, coincident with the spread of experimental petrology around the world, experimental phase-equilibrium studies and experimental studies of natural materials of relevance to igneous petrology have faded somewhat in importance at the Geophysical Laboratory. Such studies have largely been displaced by the investigations at extremely high pressures designed to shed light on the structure, composition, and processes of the deep mantle and core.

27 ← CLASSIFICATION SALVAGED?

IUGS to the Rescue

As we have recounted in the immediately preceding chapters, unparalleled advances occurred within igneous petrology during the middle decades of the twentieth century. The discipline took enormous strides thanks to developments in geochemistry. New insights into magmatic processes were afforded by the discoveries of isotope and trace-element geochemistry. Experimental petrology continued to flourish at the Geophysical Laboratory and beyond. Petrological research became increasingly international and collaborative. However, as we saw in Chapter 21, systematics lagged far behind. Although classification schemes abounded, they were typically produced by individuals, and, because so many classifications were available, petrological research evidenced very little uniformity in matters of rock nomenclature and classification. Remarkably, a collaborative approach to systematics and a semblance of agreement on terminology that most petrologists had yearned for so earnestly for decades finally began to take shape by the 1970s.

THE CLASSIFICATION OF VOLCANIC ROCKS

Volcanic rocks posed one of the perennial problems in constructing the numerous strictly mineralogical classifications that had cluttered the petrographic landscape ever since the days of Zirkel and Rosenbusch. These rocks, of course, commonly contain varying amounts of glass and are typically so fine-grained that it is difficult to identify all the minerals accurately without a great expenditure of effort. The problem of identification is exacerbated because very fine-grained volcanic rocks are much more susceptible to alteration than are coarse-grained plutonic rocks. Hence, primary crystallization products in volcanic rocks are commonly destroyed. In some instances the identification of minerals and determination of mineral abundances cannot be made at all. The great success of the CIPW norm calculation and Niggli's modification of it was due largely to the fact that it provided a useful way of characterizing such very fine-grained and glassy igneous rocks. Although the norm provided a quasi-mineralogical categorization of these rocks, volcanic rocks classified by means of the norm were really being classified on the basis of their chemical compositions rather than their mineral constituents.

Throughout the twentieth century, petrographers continued to recognize that a chemical classification provided the most satisfactory means for classifying a significant portion of the volcanic rocks. Arguably the most ambitious chemical classification that was specifically designed for volcanic rocks was proposed by members of the Geological Survey of Canada. In 1966, the Volcanological Subcommittee of the Associate Committee on Geodesy and Geophysics of the National Research Council of Canada asked Neil Irvine and W. Robert A. Baragar of the Geological Survey of Canada along with W. W. Moorhouse to prepare a classification of volcanic rocks that might be recommended for general use in Canada. The subcommittee believed that such a classification was necessary because several institutions in Canada were producing prodigious numbers of chemical analyses of volcanic rocks. A consistent nomenclature was needed to make meaningful comparisons. Moreover, the subcommittee noted, many of the rocks that were being analyzed had been metamorphosed so thoroughly that conventional mineralogical classifications were not applicable to them. Finally, the subcommittee pointed out that there was no existing classification relating such long-standing terms as basalt, andesite, and dacite with newer terms like high-alumina basalt and hawaiite that had become widely used in the petrological literature. After Moorhouse died in 1969, Irvine and Baragar continued the project, distributed a preliminary draft of their report for critical review, and published a final report (Irvine and Baragar, 1971) that was formally accepted by the Volcanological Subcommittee.

Irvine and Baragar proposed two major divisions of volcanic rocks: "subalkaline" and "alkaline." "Peralkaline" rocks were classed in a third minor division. They subdivided the subalkaline rocks into a tholeiitic basalt series and a calc–alkali series. They subdivided the alkaline rocks into an alkaline olivine basalt series and a group of nephelinic, leucitic, and analcitic rocks. They purposely ignored the latter group of feldspathoidal rocks in further classification on the grounds that they are not common in Canada. The alkali olivine basalt series was further split into a sodic series and a potassic series. In several groups of AFM diagrams, ($Na_2O + K_2O$) versus SiO_2 diagrams, cation-normative olivine–nepheline–quartz diagrams, and cation-normative albite–anorthite–orthoclase diagrams, they plotted data from approximately 2500 chemical analyses of volcanic rocks to illustrate the chemical distinctions among tholeiitic, calc–alkaline, and alkaline suites of rocks.

For further subdivision of these three major series, the authors employed a set of diagrams of normative color index versus normative plagioclase composition. These diagrams were broken into several different fields corresponding to specific rock types. Thus, for suites of subalkaline rocks, Irvine and Baragar presented a diagram that shows proposed boundaries for fields of basalt, andesite, dacite, rhyolite, and tholeiitic andesite. For suites of sodic alkaline rocks, they presented a diagram with fields for picrite basalt and ankaramite, alkali basalt, hawaiite, mugearite, benmoreite, trachyte, and nephelinite. And for suites of potassic alkaline rocks, they presented a diagram with fields for picrite basalt and ankaramite, alkali basalt, trachybasalt, tristanite, and trachyte. The intent was that a petrographer might now be able to classify a volcanic rock by calculating the Niggli norm from its chemical analysis and plotting the resulting values for normative color index and normative plagioclase on the appropriate diagram and noting the rock field into which the data point falls.

THE IUGS CLASSIFICATION

The Canadian classification was a welcome contribution to the field of petrographic systematics in that it represented one of the very few collaborative efforts in the history of igneous rock classification. For decades, petrographers had been yearning for cooperative work that might produce a universally, or at least widely, accepted classification. Unfortunately, the very limited previous efforts had yielded correspondingly limited results. The CIPW classification was the product of four American petrologists who evidently worked only among themselves. Although objections were launched on many grounds, the strictly American provenance of the CIPW scheme also worked against its universal acceptance. In the 1920s, British petrologists attempted a cooperative effort but succeeded only in suggesting some changes in rock nomenclature. Their attempts to produce a classification got nowhere. Despite its being a strictly Canadian venture, the Irvine and Baragar classification at least had in its favor the fact that its authors sought constructive feedback from several other petrographers.

The long-awaited cooperative breakthrough, however, finally emerged from continental Europe in the 1960s at the initiative of Albert L. Streckeisen (1901–1998). Streckeisen received his Ph.D. degree from the University of Basel in Switzerland. After seven years as Professor of Mineralogy and Petrology at the Technical Institute of Bucharest in Romania, where he worked on crystalline terranes in the Swiss Alps and the Carpathian Mountains of Romania, Streckeisen became a Lecturer in 1942, and then a Professor in 1954, at the Mineralogic and Petrographic Institute of the University of Bern in Switzerland. When asked to review matters of classification and nomenclature for a new edition of *Tabellen zur Petrographie und zum Gesteinsbestimmen*, a textbook on petrography by Niggli (1939), Streckeisen took it upon himself, in the absence of any international proposals, to review several existing classifications and to submit his own proposals pertaining to the nomenclature and classification of "eruptive rocks" to the international petrographic community (Streckeisen, 1964).

Streckeisen's wise decision to submit questions and proposals regarding classification and nomenclature to a large number of petrologists for their feedback eventually mushroomed into a project of far greater scope than he originally anticipated.

Of particular importance in Streckeisen's paper was a set of fifteen questions to which he invited interested parties to contribute some discussion by the end of 1964. After each discussion question, Streckeisen listed his own proposal. Among the questions, Streckeisen asked whether the boundary for ultramafic rocks should be drawn at a color index (Farbzahl) of 70, 75, 90, or 95. He recommended placing the boundary at a color index of 95. Streckeisen further asked where the boundaries between alkali feldspar and plagioclase should be drawn on a QAPF double-triangle diagram that he included in the paper. He suggested drawing those boundaries at 10, 35, 65, and 90 percent plagioclase. Streckeisen asked how the naming of plutonic rocks in fields 2 and 3 of his proposed QAPF diagram should be handled. He recommended considering these two fields as a double field and naming it "granite" as Rittmann had done. Should the names "monzodiorite" and "monzogabbro," Streckeisen asked, be advocated for field 9? To this query he replied, "yes." Streckeisen asked on what criterion the separation of diorite and gabbro should be based, and he proposed making the separation on the basis of the nature of the plagioclase. Streckeisen further asked on what criterion the separation between andesite and basalt ought to be based. Here he suggested that color index be used as a criterion. Streckeisen also asked how the rock names basalt, dolerite, diabase, and melaphyre should be used. Other questions pertained to the usage of the names phonolite, foyaite, and picrite, to the location of other boundaries, and to the role of light and dark constituents in nomenclature.

Streckeisen received approximately 80 responses from around the world. These responses included both answers to his specific questions as well as other suggestions and proposals. He put together a small team of collaborators to evaluate all the responses. Additional input was received in discussion periods following lectures that he delivered throughout Europe. As a result of the feedback, Streckeisen and a team of petrographers developed a preliminary revised classification proposal that was presented to the Geologische Vereinigung meeting in Strasbourg in 1965. The preliminary proposal was then published in *Geologische Rundschau* (Streckeisen, 1965). In that paper Streckeisen focused on proposals for the classification of igneous rocks with a color index less than 90. A QAPF double triangle, considerably revised from his earlier version, was presented. The feldspathoid-bearing APF triangle was given a symmetrical arrangement, but the alkali feldspar–plagioclase boundaries at 10, 35, 65, and 90 percent plagioclase were retained from his earlier proposal.

A much more thorough version of the revised classification was published by Streckeisen (1967). He referred to this paper as a final report of an inquiry. The report, published in English to make it even more accessible worldwide, listed the names of those who had participated in oral or written discussions about the proposed classifications. Although the vast majority of contributors were from Germany, Switzerland, or the United States, nineteen different countries from six continents were represented. The report also contained a synopsis of previous classification schemes as well as a valuable compilation of diagrams of sixteen earlier schemes. The focus again was the QAPF double triangle with a statement of principles and proposals for naming both plutonic and volcanic rocks in the various fields. Detailed suggestions were provided for the naming of varieties of specific rock types based on color index. Preliminary ideas on the nomenclature of hypabyssal rocks were given. One of the appendices discussed the variation of igneous rocks in the proposed scheme and presented diagrams with plots of rock compositions from various igneous complexes and provinces. The report represented the fruits of the most extensive international petrographic collaboration in the history of igneous petrology. That was only a beginning.

At the 24th International Geological Congress meeting in Prague in 1968, a symposium under the leadership of K. R. Mehnert was planned to discuss the proposals that had been circulated and published by Streckeisen. Unfortunately, the political turmoil in Czechoslovakia (now the Czech Republic and Slovakia) that year prevented the planned discussion. As a result, the International Union of Geological Sciences (IUGS), recently established in 1961, determined to create a Subcommission on the Systematics of Igneous Rocks under the auspices of its Commission on Petrology. The subcommission was charged with deliberating on the various problems of nomenclature, developing a classification system, and presenting some definite recommendations to the IUGS As Streckeisen (1976a, p. 3) observed, this action represented "the first attempt to develop a system through deliberation by a group of geoscientists from all parts of the world." The subcommission began its work in 1969 and also held a working meeting in Bern in 1972. The initial efforts of the subcommission concentrated on plutonic rocks. Recommendations for a system of classification and nomenclature for plutonic rocks were approved by the subcommission in August 1972, during the meeting of the 25th International Geological Congress in Montreal. At that time, the subcommission consisted of 21 members from 19 countries, but several other petrologists helped with the work. Streckeisen served as the chairman of the subcommission. The proposals of the subcommission were published by Streckeisen (1974). This report discussed a refined QAPF double triangle that introduced the terms "quartz syenite," "quartz monzonite," and "quartz monzodiorite." Detailed classification triangles were presented for ultramafic rocks and for gabbroic rocks. The scheme for naming varieties based on color index was expanded and modified from Streckeisen (1967).

The final version of the work on plutonic rocks was published by Streckeisen (1976a). The subcommission retained the QAPF double triangle for quartz–

feldspar and feldspathoidal rocks (Figure 52). They treated free albite with compositions from An_0 to An_5 as alkali feldspar. For the feldspar compositions to serve as boundary markers, the subcommission continued its use of boundaries at 10, 35, 65, and 90 percent plagioclase. Assignment of a rock as granite required that between 35 and 90 percent of the total feldspar be alkali–feldspar and that between 10 and 65 percent of the total feldspar be plagioclase. Assignment as granite also required that between 20 and 60 percent of the sum of quartz and feldspars be quartz. The classification placed quartz monzonite beneath granite rather than beside it or coincident with it as had been done in some earlier classifications. Quartz monzonite was now placed in a row with quartz alkali-feldspar syenite, quartz syenite, quartz monzodiorite, and quartz diorite for rocks containing quartz in an amount between 5 and 20 percent of the sum of quartz and the feldspars. Detailed rules were again provided for applying varietal names and such prefixes as "leuco-" and "mela-" to rock names. The report proposed a tetrahedral classification for ultramafic and plagioclase-rich rocks with the corners represented by olivine, clinopyroxene, orthopyroxene, and plagioclase. The four sides of the tetrahedron could also be opened out into a single plane and represented as a set of four adjoining triangles. Because the classifications, particularly the QAPF double triangle, had by now been before the petrological community in one version or another for some time, they were beginning to gain acceptance. The widely used petrology textbook by Hyndman (1972), for example, incorporated an early version of Streckeisen's QAPF double triangle for felsic rocks.

In the meantime, the IUGS subcommission continued to wrestle with the classification and nomenclature of the volcanic rocks. As a contribution to that discussion, Streckeisen (1976b) also introduced, in the same year as the final report on plutonic rocks, an attempt at a classification of common igneous rocks by means of their chemical compositions. Because of the great difficulties in assigning the orthoclase (Or), albite (Ab), and anorthite (An) molecules to alkali feldspar and plagioclase for plotting on the QAPF double triangle, Streckeisen proposed to abandon such efforts and, instead, to characterize the various rock groups by the relative amounts of the Or, Ab, and An molecules. In particular, he thought that the An/Or ratio would serve as an effective discriminator among the rock groups. Streckeisen proposed first using plutonic rocks as a guide for volcanic rocks. He converted the chemical analyses of about 1600 rocks into the Barth–Niggli cation norms, and subdivided the rocks into three groups: quartz–feldspar rocks containing more than 17 percent normative quartz, feldspar rocks containing less than 17 percent normative quartz and less than 7 percent normative nepheline, and feldspar–foid rocks containing more than seven percent nepheline. The percentages of normative Or, Ab, and An were then plotted on a series of Or–Ab–An diagrams for the three main groups and fields of various plutonic and volcanic rock types such as granite, rhyolite, diorite, granodiorite, gabbro, and trachyte. Although be-

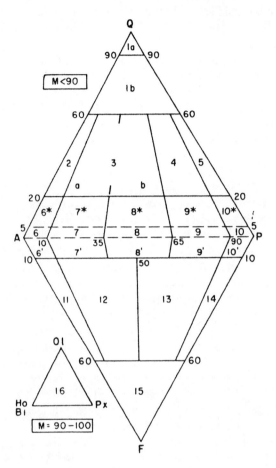

52. Mineralogical classification of plutonic rocks proposed by IUGS from Streckeisen (1976a). 1a. Silexite, 1b. Quartz-rich granitoids, 2. Alkali-feldspar granite, 3. Granite, 4. Granodiorite, 5. Tonalite, 6*. Quartz alkali-feldspar syenite, 7*. Quartz syenite, 8*. Quartz monzonite, 9*. Quartz monzodiorite/quartz monzogabbro, 10*. Quartz diorite/quartz gabbro/quartz anorthosite, 6. Alkali-feldspar syenite, 7. Syenite, 8. Monzonite, 9. Monzodiorite/monzogabbro, 10. Diorite/gabbro/anorthosite, 6'. Foid-bearing alkali-feldspar syenite, 7'. Foid-bearing syenite, 8'. Foid-bearing monzonite, 9'. Foid-bearing monzodiorite/monzogabbro, 10'. Foid-bearing diorite/gabbro, 11. Foid syenite, 12. Foid monzosyenite, 13. Foid monzodiorite/foid monzogabbro, 14. Foid diorite/foid gabbro, 15. Foidolites, 16. Ultramafic plutonic rocks. Reprinted from *Earth Science Reviews*, v. 12, Streckeisen, A., To each plutonic rock its proper name, pp. 1–33, 1976, with permission from Elsevier Science.

lieving that he could achieve a generally good discrimination of rock types by this method, Streckeisen noted that the fields of andesite and basalt overlapped so that the two rock types could not be distinguished by An/Or ratio alone. He admitted that supplementary criteria for classification would be required in that and some other cases. In the end, he decided that his method worked well for leucocratic and some mesocratic rocks in that fair agreement was reached between modal contents and chemical criteria, but he conceded that the method was much less successful in dealing with mafic and ultramafic rocks. "A chemical classification of igneous rocks," Streckeisen (1976b, p. 13) concluded, "has a limited applicability and has to be supplemented and controlled by modal contents."

The subcommission fixed its attention on the systematics of volcanic rocks during meetings at Grenoble, France, in 1975, at the 26th International Geological Congress in Sydney in 1976, and at Prague in 1977. In a publication on volcanic rocks, lamprophyres, carbonatites, and melilitic rocks (Streckeisen, 1978), the subcommission acknowledged the fact that the modal mineral contents of many volcanic rocks cannot be determined accurately. After considering possible chemical and combined chemical and mineralogical schemes, and granting the importance of chemical criteria, the subcommission decided that the classification of volcanic rocks should be consistent with the classification of the plutonic rocks and should, therefore, be based on mineral content. The subcommission further recognized that volcanic rocks had traditionally been named in accord with mineral content. Consequently, the subcommission recommended that volcanic rocks should be named in accord with their position on the QAPF double triangle if the modal mineral content could be established. Proposed names of fields were discussed at length (Figure 53). As examples, the field of rhyolite was considered equivalent to the field of granite, the field of quartz latite was considered equivalent to the field of quartz monzonite, the field of trachyte was considered equivalent to the field of syenite, and the field of basanite was considered equivalent to the field of nepheline gabbro. Suggestions were made for abandonment of the term "melaphyre" and proper usage of such terms as "dolerite," "diabase," "spilite," and "keratophyre."

The subcommission then recommended that, if modal mineral content could not be established, a chemical system be used that correlated with a mineralogical system. In other words, a rock of known mineralogy should be given the same name by both the QAPF diagram and the chemical classification. Without making any recommendation, the subcommission simply took note of existing diagrams in which silica is plotted against alkalis and of Rittmann's method for calculating a mineral assemblage from the chemical analysis of an igneous rock. The paper also provided detailed proposals, including diagrams, for the classification of lamprophyres, carbonatites, and melilite-bearing igneous rocks. Lamprophyres required their own classification, the subcommission determined, because these dike rocks are not simply textural variants

Root names of volcanic rocks
2 alkali(-feldspar) rhyolite (liparite) (see text)
3a
3b } rhyolite (liparite) (see text)
4
5 } dacite (see text)
6* quartz-alkali(-feldspar) trachyte
6 alkali(-feldspar) trachyte
6' foid-bearing alkali(-feldspar) trachyte
7* quartz trachyte
7 trachyte
7' foid-bear. trachyte
8* quartz latite
8 latite
8' foid-bear. latite
9 } andesite, basalt
10 } (see text)
11 phonolite
12 tephritic phonolite
13 phonolitic tephrite (basanite)
14 tephrite, basanite
15a phonolitic foidite
15b tephritic foidite
15c foidite
16 ultramafitite

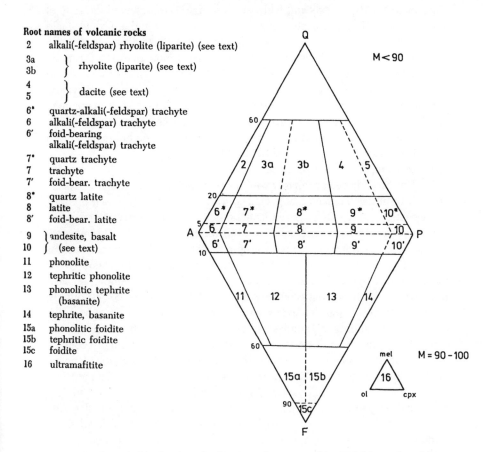

53. Mineralogical classification of volcanic rocks proposed by IUGS Reproduced from Streckeisen (1978) by permission of E. Schweizerbart (http://www.schweizerbart.de).

of plutonic or volcanic rocks but also possess distinctive mineral and chemical compositions.

In conjunction with its labors on volcanic rocks, the subcommission continued to explore possible chemical classifications because of the fine-grained or glassy nature of many volcanic rocks. Streckeisen and Roger W. Le Maitre, a statistical petrographer at the University of Melbourne, advanced a scheme based on the earlier work of Streckeisen (1976b) and tied to the modal QAPF classification already proposed by the IUGS Streckeisen and Le Maitre (1979) developed an orthogonal diagram in which quartz (Q') or feldspathoid (F') ratios were plotted versus a plagioclase ratio (ANOR). Q' represented the ratio of normative quartz to the sum of normative quartz and feldspars; F' represented the ratio of normative nepheline, leucite, and kaliophilite to the sum of normative feldspathoids and feldspars; and ANOR represented the ratio of normative anorthite to the sum of normative orthoclase and anorthite. Field boundaries were drawn from the norms of approximately 2500 chemical analyses. The scheme was then tested against the norms of more than 15,000 analyses that had been retrieved from the CLAIR database compiled by Le Maitre (1973). After discussion of possible sources of error in the method and evaluation of results for specific rock types, Streckeisen and Le Maitre maintained that their method yielded reasonable agreement between plutonic rocks and their volcanic counterparts but cautioned against application of the method to ultramafic rocks and extremely feldspathoid-rich rocks.

Another meeting of the subcommission took place at Padua, Italy, in May 1979. This meeting dealt primarily with the nomenclature of pyroclastic rocks. The subcommission agreed that such a classification should be based on nonpetrogenetic features. The recommendations pertaining to pyroclastic rocks were published by Le Bas and Sabine (1980) and Schmid (1981). The subcommission recommended that individual pyroclasts be designated on the basis of mean diameters of fragments in much the same manner as sedimentary grains are designated by the Udden–Wentworth scale. Thus, for example, an "ash grain" was said to have a diameter between $\frac{1}{16}$ and 2 millimeters (corresponding to sand), a "lapillus" was defined as having a diameter between 2 and 64 millimeters (corresponding to granules and pebbles), and a "bomb" or "block" was said to have a diameter in excess of 64 millimeters (corresponding to cobbles and boulders). Both unconsolidated and consolidated pyroclastic deposits were also named on the basis of the dominant size of clasts. Hence, a pyroclastic rock consisting predominantly of bombs and blocks was termed an "agglomerate" or "pyroclastic breccia," whereas one consisting mainly of ash grains was designated as "coarse tuff." Tuffs and ashes were further subdivided according to the ratios of glass, crystals, and rock among their fragments.

Work on the chemical classification of fine-grained and glassy volcanic rocks continued at the Padua meeting, and also at the 27th International Geological Congress in Paris in 1980, at Cambridge in 1981, and in Granada, Spain, in 1983. After carefully examining many other chemical schemes for volcanic rocks, including the method proposed by Streckeisen and Le Maitre (1979), the subcommission determined, at its Cambridge meeting, to investigate the use of a simple, purely chemical classification based on a total alkali versus silica (TAS) diagram. In arriving at this decision, the subcommission had adopted the principles that a simple classification is to be preferred over a complicated one; that straight-line boundaries are preferable to curved boundaries; and that a rock should be classifiable apart from any knowledge of its locality or associations. At the Granada meeting, a TAS classification was formally approved, and the proposal was presented to the wider petrologic community in publications by Le Maitre (1984) and Le Bas et al. (1986). Ever since Macdonald and Katsura (1964) had originally discovered that a TAS diagram

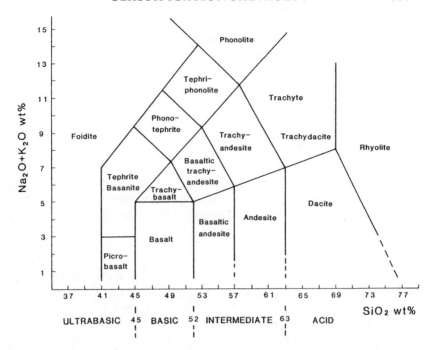

54. Total alkali vs. silica chemical classification diagram for volcanic rocks proposed by IUGS (Le Bas et al., 1986). Reproduced from *Journal of Petrology* by permission of Oxford University Press.

is a useful discriminator between tholeiitic basalt and alkalic basalt, petrologists had widely used the diagram. Some of the fields of volcanic rocks ultimately recommended by the IUGS subcommission had been previously suggested by earlier authors, and the subcommission also refined the positions of rock fields on the diagram by plotting a computer database of about 24,000 chemical analyses of fresh volcanic rocks that had been compiled by Le Maitre. In the final classification, fifteen fields into which volcanic rocks could be plotted were laid out on the diagram of total alkalis versus silica (Figure 54). More than one rock type could be plotted in several of the fields, depending on variations in abundances of different normative minerals.

One such problem with the TAS chemical classification concerned the boundary between the field of the foidites and the field of basanites and tephrites. A difficulty existed because a substantial percentage of rocks designated as nephelinite (the nepheline-rich foidite) on the basis of their mineralogy plotted in the field of basanite on the TAS diagram. On behalf of the subcommission, Le Bas (1989) addressed the question of how to choose the criteria that best distinguish chemically nephelinitic rocks, that is, olivine- or pyroxene-rich rocks in which nepheline is the predominant normative felsic mineral, from basanitic rocks. The proposal was to distinguish basanite, melanephelinite, and nephelinite on the basis of the relative abundances of normative albite and nepheline. Specifically, Le Bas suggested that nephelinite be distinguished from melanephelinite by a CIPW normative nepheline content in excess of 20 percent, and that melanephelinite be distinguished from basanite by a normative albite content of less than 5 percent.

The subcommission had additional meetings at the 28th Geological Congress in Moscow in 1984, in London in 1985, in Freiburg, Germany, in 1986, Copenhagen in 1988, and at the 29th International Geological Congress in Washington, D. C., in 1989. By that time, Le Bas and Streckeisen (1991) observed, 419 petrologists from 49 countries had participated either in the discussions and recommendations made at the various meetings or through discussion papers and questionnaires that were circulated between the meetings. The emerging classifications were, indeed, the fruit of international collaboration. One of the main outcomes was *A Classification of Igneous Rocks and Glossary of Terms: Recommendations of the International Union of Geological Sciences Subcommission on the Systematics of Igneous Rocks* (Le Maitre, 1989), a book summarizing the work of the IUGS subcommission. This volume emerged as a worthy and indispensable successor to earlier petrographic lexicons like those of Loewinson-Lessing and Holmes.

Le Maitre (1989) observed that the single most prolific year in the history of igneous petrology for the introduction of new rock names was 1973 when 54 new names were introduced, mostly as a result of the work of the IUGS subcommission. Thanks, too, to the work of the subcommission, Albert Streckeisen had proposed a total of 87 new igneous rock names by 1989. Despite such "productivity," Streckeisen still wound up in second place far behind Albert Johannsen, who had proposed 133 new names throughout his career in the first half of the twentieth century. Streckeisen was just a bit ahead of third and fourth place holders, Alfred Lacroix (70) and W. C. Brögger (65). Surprisingly, Harry Rosenbusch, of whom petrologists often think first in terms of coining rock names, was a distant fifth with only 32 new names.

Le Bas and Streckeisen (1991) noted that 297 names and terms were used altogether in the IUGS classification. They reported that the subcommission was continuing to debate the best way to classify potassic rocks, lamprophyres, and a few other unusual rock groups. Le Bas and Streckeisen also elucidated ten principles that had guided the subcommission in their work over the years. They stated that: (1) igneous rock nomenclature should be based on descriptive attributes; (2) classification should depend on the actual attributes of rocks and not on interpreted characters; (3) the basic or root name given to a rock should be one that is suitable for all geologists to use; (4) the terms to be used in any classification should follow, as far as possible, those that are currently and widely accepted as being useful terms; (5) classification must consist of classes that are separated by boundary conditions; (6) a classification should have well-established simplicity; (7) any classification of igneous rocks should follow fundamental geological relationships such as plutonic versus volcanic; (8) classification should be based on modal mineralogy as far as possible; (9) if the modal mineralogy of an igneous rock cannot be determined satisfactorily, then chemical analytical parameters should be the next property used; and, finally, (10) all terminology should be internationally acceptable. The authors did concede that using the term "igneous" violated the second principle but argued that the certainty of interpretation of igneous origin justified overlooking that principle in most cases. They also conceded that the division into plutonic and volcanic groups also violated the second principle but argued that the distinction was already so entrenched and accepted that it should be retained.

Most recently, the IUGS subcommission has undertaken further work on the classification of lamprophyres, kimberlites, and other alkalic rocks (Woolley et al., 1996) and on picritic volcanic rocks (Le Bas, 2000).

THE AFTERMATH OF THE IUGS CLASSIFICATION

Although lacking the glamour and prestige brought to igneous petrology by mathematical models of trace-element and isotope distribution, by expensive, technologically advanced analytical instrumentation, or by the extension of experiments on silicate systems to high fluid and load pressures, the classifications of igneous rocks formulated under the auspices of the International Union of Geological Sciences stand as one of the major triumphs within the field of igneous petrology. The remarkable and extended effort of hundreds of petrologists from dozens of countries resulted in a classification that the majority of petrologists were delighted to work with despite its possible shortcomings. At long last, igneous petrology had a widely accepted common standard for its basic petrographic nomenclature. Finally, when an author used the term "quartz monzonite" to describe a rock, most other petrologists now had an understanding of what that author meant. That the new classification of the igneous rocks met with widespread acceptance among petrologists is indicated by the fact that, in the United States, petrology textbooks by Barker (1983), Hyndman (1985), Philpotts (1990), and Winter (2001) adopted the proposals and also by the fact that, in a vast number of recent petrological articles, igneous rocks have been named on the basis of the new classification.

Predictably, the IUGS classification drew some criticism. The textbook on petrography by Williams et al. (1982) stated that the IUGS classification employed two propositions with which they disagreed. On the one hand, they said, the classification presupposed the need to agree upon a single rational and workable system for the naming and classification of igneous rocks. Turner and Gilbert (Howel Williams had died by the time of writing of the revised edition of the textbook), however, countered that unanimity and precision of terminology would be impossible to achieve, and they would not even grant that unanimity and precision are desirable objectives. Secondly, the IUGS classification, they complained, incorporated both "igneous" and "igneous-looking" rocks irrespective of the genesis of the rocks. They charged that grouping the superficially similar products of metamorphism and metasomatism with truly igneous rocks would only confuse an already difficult and complex natural situation by obscuring genetic implications that were otherwise obvious in common igneous assemblages.

The achievement of the IUGS subcommission hardly put an end to alternative or supplemental classification schemes. Among those who were constantly thinking about issues of classification was Eric A. K. Middlemost of the University of Sydney. Middlemost (1971, p. 106) considered petrographic classifications like those that Streckeisen had been working on to be valuable, but, at the same time, he found them to be too "mechanical and inflexible for use in petrogenetic discussions." What was needed, in his eyes, was a flexible genetic classification and still another classification that could be used for studying specific rock complexes. A genetic classification, Middlemost felt, should be based on a coherent model of igneous rock evolution. At the basis of such a model lies the question of the number of primary magmas required to account for that evolution.

Middlemost, therefore, proposed a classification of igneous processes in which the major groups consisted of primary magmas, secondary magmas, syntectic magmas, magmas generated by tectonic activity, and allochthonous igneous bodies. He proposed a division of primary magmas into simple and compound divisions with the former resulting from partial melting in the mantle and the latter resulting from more than one process of magma generation in the mantle. Examples given of simple primary magmas included members of basalt series, gabbros, picrite, and ijolite. Examples of compound primary magmas included kimberlite and carbonatite. Among secondary magmas, Middlemost distinguished high-level differentiates such as rhyolite, trachyte, and phonolite, from cumulates such as peridotite, dunite, pyroxenite, and anorthosite. This classification was clearly intended not as a substitute for, but as a supplement to, the mineralogical classifications then being developed by Streckeisen and the IUGS.

A few years later, Middlemost (1980) continued his consideration of classification by urging the development of complementary classifications applicable to volcanic rocks. He pointed out that because the traditional descriptive petrographic classifications for volcanic rocks were those of natural historians, such classifications were inadequate for the needs of the physical scientist, or experimentalist, who needed to find names for the various materials that were generated, moved, and erupted during the processes of volcano formation in extraterrestrial settings as well as on Earth.

Another petrologist who persistently pondered alternatives to typical classification schemes was Hubert de la Roche. La Roche (1986), for example, perceived that igneous rock systematics was no longer consistent with contemporary knowledge of magmatism. Systematics, he urged, needed to regain that consistency. Although he acknowledged that there had been a long-established tradition for the types of classification that were represented by the QAPF and TAS diagrams adopted by the IUGS, La Roche charged that such classifications were deficient because they divorced mineralogy from chemistry and plutonism from volcanism. To remedy the deficiency, La Roche presented two chemical–mineralogical diagrams derived from the Yoder and Tilley basalt tetrahedron on which he incorporated classifications common to both volcanic and plutonic rocks. A binary diagram, proposed earlier by La Roche et al. (1980), plotted R_2 versus R_1, where $R_2 = 6Ca + 2Mg + Al$ and $R_1 = 4Si - 11(Na + K) - 2(Fe + Ti)$. He obtained the values for Ca, Mg, and the other constituents by dividing the weight percent of a particular oxide by the molecular weight of that oxide based on one cation per formula unit. A second classification was introduced in which he plotted three parameters Rm, Ri, and Rs at the corners of a ternary diagram. In this case $Rm = Al + 6Ca + 2Mg$; $Ri = 2(Fe + Ti) + 7(Na + K)$, and $Rs = 4(Si + Ca - Na - K)$. Classification grids were superimposed on the two diagrams. The parameters calculated from the chemical analysis of a rock could then be plotted on both diagrams to determine its appropriate name. Although both volcanic and plutonic rocks could be plotted on these diagrams by means of chemical composition, La Roche suggested that names could be assigned from mineralogical and modal data as well.

Along similar lines, F. Debon and P. Le Fort, two French petrologists at the Centre de Recherches Pétrographiques et Géochimiques in Nancy who were strongly influenced by La Roche, have wrestled with the problems of typology of igneous rocks. They contended that igneous rock classifications should be quantitatively based on chemical composition in order to be internally consistent and that the classification of individual samples needs to be related to the magmatic associations of which they are a part. Thus, Debon and Le Fort (1983) proposed a classification on the basis of a set of three chemical–mineralogical diagrams. To determine the appropriate name of an individual rock sample, its chemical analysis is plotted on the three diagrams. The first diagram, a modification of La Roche's binary diagram, used two coordinates $Q = Si/3 - (K + Na + 2Ca/3)$ and $P = K - (Na + Ca)$. A grid similar to that of Streckeisen was superimposed on the diagram for classification purposes. A second diagram, called a "characteristic minerals" diagram also consisted of two coordinates $A = Al - (K + Na + 2Ca)$ and $B = Fe + Mg + Ti$. Debon and Le Fort regarded this diagram as providing a measure of the aluminous character of igneous rocks. A third diagram, also adapted from La Roche, used three chemical parameters, Q, B, and $F = 555 - (Q + B)$ to provide an estimate of the proportions of quartz, dark minerals, and feldspars. In the scheme, a rock sample can be characterized further in terms of the balance of the alkalis, Na and K, in terms of the Mg/(Fe + Mg) ratio, and in terms of actual minerals.

By using the same diagrams, Debon and Le Fort then attempted to classify associations of samples that demonstrate a community of characteristics and have come from the same igneous rock body. Associations were categorized as one of three major types, cafemic, alumino-cafemic, or aluminous, from their positions on the "characteristic minerals" diagram. Each of the three associations could then be further divided into subtypes. The cafemic association, for example, includes a calc–alkaline or granodioritic subtype, a subalkaline or monzonitic subtype, an alkaline–peralkaline subtype, and a tholeiitic or gabbroic–trondhjemitic subtype. Among the aluminous associations, Debon and Le Fort distinguished three subtypes based on quartz content, three subtypes based on color index, and three subtypes based on alkali ratio. All the subtypes of the three associations were distinguished from one another by the use of the Q-B-F diagram. The authors claimed considerable success in the application of their scheme to granitoid terranes in south-central Asia. Petrologists in the English-speaking world, however, did not become widely familiar with their diagrams or those of La Roche.

In addition to the various alternative or supplementary classifications such as those just noted, petrologists in recent decades have also devoted considerable attention to the classification of specific groups of igneous rocks, particularly in the case of granitoid rocks. In many instances, proponents of particular granite "typologies" have sought to achieve correlations among such diverse factors as mineralogy, geochemistry, tectonic environment, and source of origin. One scheme of granitoid rock classification that gained considerable favor was first suggested by the Australian geologists Bruce W. Chappell and Allan J. R. White from an investigation of granitoid rocks in the Early Paleozoic Lachlan fold belt in eastern Australia. Chappell and White (1974) described two types of granites with distinct petrographic, geochemical, and field characteristics that they attributed to two different sources. They proposed a distinction between S-type and I-type granitoid rocks. The former, they said, are characterized by high initial Sr-isotope ratios and are rather aluminous rocks, as indicated by a significant normative corundum content and the presence of modal muscovite, garnet, or cordierite. They found that inclusions in the S-type granitic rocks consist primarily of metamorphosed shale or sandstone. They maintained that these characteristics indicated partial melting of metasedimentary source rocks. In contrast, the I-type granitoid rocks were said to be characterized by lower initial Sr-isotope ratios, by normative diopside or extremely low normative corundum, and by modal hornblende and titanite (sphene). I-type granitoid rocks were also said to show a wider range of compositions within a given pluton than is typical of S-type rocks. They observed that inclusions in I-type granites are hornblende–bearing, igneous-looking rocks. They suggested that these rocks had been derived by partial fusion of igneous or metaigneous sources.

After the S-I terminology caught on, additional types were suggested. Loiselle and Wones (1979) proposed an A-type granitoid characterized by high alkalis and high Fe/Mg ratios. A-type rocks were described as more calcium-poor than I-type granites and more aluminum-poor than S-types. In addition, A-type granitoids are much more likely to be restricted to the field of true granite on a QAP diagram. These granites occur in stable continental interiors or rift zones and, therefore, have been intruded into anorogenic environments. White (1979) later added a fourth type of granite, which he designated the M-type. Plagioclase-rich "plagiogranites" occurring in oceanic crust or ophiolite complexes and presumed to be derived from a mantle source were assigned to the M-type.

Despite the very widespread reference to S-type granites, I-type granites, and the like by numerous petrologists, the classification was criticized on various grounds. On the one hand, many examples of granitoids of hybrid S–I or M–I characteristics exist so that it is not always possible to classify a given granite unequivocally. On the other hand, the classification was inconsistent. Whereas the S, I, and M categories were defined in relation to the alleged source of the granite (S for meta-sedimentary sources, I for igneous sources, and M for mantle sources), the A category was defined in relation to tectonic setting (A for *a*norogenic).

A variety of other classifications of granitoid rocks also sprang up in recent decades. In Chapter 25, I discussed the proposal of Pearce et al. (1984) to categorize granites on the basis of trace-element geochemical characters. Pearce discriminated granitic rocks in terms of plate-tectonic environment and recognized volcanic arc granite (VAG), collision granites (COLG), within-plate granites (WPG), and ocean-ridge granites (ORG).

Maniar and Piccoli (1989) advocated a tectonic categorization of granitic rocks on the basis of mineralogy and major-element geochemistry. They defined categories for island-arc granite (IAG), continental-arc granite (CAG), continental-collision granite (CCG), postorogenic granite (POG), rift-related granite (RRG), continental epeirogenic-uplift granite (CEUG), and oceanic plagiogranite (OP). They placed the first three categories into group I, the RRG and CEUG into group II, and the POG into group III. They suggested that these three groups could be distinguished on a variety of chemical diagrams such as plots of alumina versus silica; plots of FeO/(MgO + FeO) versus silica, and AFM and ACF diagrams. OP granites, they stated, could be distinguished from all other granitoids on a plot of potash versus silica.

Among the most ambitious attempts to categorize granitoid rocks have been those of Bernard Barbarin. From approximately twenty different granitoid classifications, Barbarin (1990) attempted a synthesis of mineralogy, chemistry, tectonic environment, and other factors, but he later acknowledged that he had not adequately spelled out the links between different granitoids and their geodynamic environments. To that end, Barbarin (1999) developed a typology that sought to complement recent classifications in terms of chemical, isotopic, field, petrographical, and mineralogical criteria (Figure 55). The granite types that he proposed were said to be indicative of their origin, evolution, and geodynamic environment. Continuing the insidious infatuation of petrologists for abbreviated alphabetical designations, Barbarin proposed seven granitoid types. The muscovite-bearing peraluminous granitoids (MPG) and cordierite-bearing peraluminous granitoids (CPG) were correlated with a crustal origin and a geodynamic environment characterized as continental collision. These granitoids, he said, were emplaced in thickened crust that resulted from continental convergence. MPG were said to be concentrated along transcurrent shear and thrust zones that cut across the thickened crust, whereas CPG were said to be dispersed throughout collisional mountain belts. Amphibole-bearing calc–alkaline granitoids (ACG) and potassium-rich calc–alkaline granitoids (KCG) were linked to a mixed crust and mantle origin. The former were connected to subduction zones, whereas the latter were related to regimes that are transitional between subduction and continental collision. The remaining three types were designated as arc tholeiitic granitoids (ATG), mid-ocean-ridge tholeiitic granitoids (RTG), and peralkaline and alkaline granitoids (PAG). Barbarin attributed a mantle origin to these three types and invoked subduction as the geodynamic environment for the ATG, oceanic spreading for the RTG, and continental doming or rifting for the PAG.

Barbarin viewed granitoid classification as something more than just a device for ease of communication through the use of agreed-upon names. Rather, he believed that, given the importance of granitic rocks in the continental crust, correct typing of granitoids together with clearly defined ages might allow for the construction of well-constrained models of the geodynamic evolution of the continental crust through geologic time. He has shown, as have others before him, that the classification of igneous rocks need not be a dull but necessary task that must be endured before petrologists can get on to the fun of petrogenetic interpretation. Indeed, classification may be a valuable petrogenetic tool in its own right if well conceived.

PARAMETERS	AUTHORS	ORIGIN				
		CRUSTAL		MIXED		MANTLE
FIRST CHEMICAL NOMENCLATURES	SHAND (1927 & 1943)	PERALUMINOUS rocks		METALUMINOUS rocks		PERALKALINE rocks
	LACROIX (1933)	Roches CALCO-ALC. HYPERALUMINEUSES		Roches CALCO-ALCALINES		Roches ALCALINES
PETROGRAPHY	CAPDEVILA & FLOOR (1970) CAPDEVILA et al. (1973)	Granites MESOCRUSTAUX		Granites MIXTES		
	ORSINI (1976 & 1979)		A.M. SUB-ALC. ALUMINEUX	A.M. SUB-ALC. HYPOALUM.	A.M. CALCO-ALC.	
	YANG CHAOQUN (1982)		MM-TYPE	CR-TYPE	MS-TYPE	MD-TYPE
	TISCHENDORF & PALCHEN (1985)	S_i	S_S	I_{KK}	I_{OK}	I_{MT}	I_{MA}
ENCLAVES	DIDIER & LAMEYRE (1969) DIDIER et al. (1982)	C-TYPE (Crustal) ("Leucogranites")		M-TYPE (Mixed or Mantle) ("Monzogranites & Granodiorites")		
MINERALOGY (QAP system)	LAMEYRE (1980) LAMEYRE & BOWDEN (1982)	"LEUCOGRANITES" (Crustal fusion)		CALC-ALKALINE Series (High K, Medium K or Low K)		THOLEITIC Series (PER ALKALINE Series)
MAFIC MINERALS	ROSSI & CHEVREMONT (1987)	A.M. ALUMINOPOTASSIQUE (s.s. ou composites)		A.M. MONZONITIQUE	A.M. CALCOALCALINE	A.M. THOLEITIQUE (PER ALCALINE)
BIOTITE COMPOSITION	NACHIT et al. (1985)	Lignées ALUMINO-POTASSIQUES		Lignées CALCOALCALINES et SUBALCALINES		Lignées ALCALINES et HYPERALCALINES
ZIRCON MORPHOLOGY	PUPIN (1980 & 1985)	TYPE 1	TYPE 2	TYPE 3	TYPE 4 & 5	TYPE 7	TYPE 6
OPAQUE OXIDES	ISHIHARA (1977) CZAMANSKE et al. (1991)	ILMENITE - Series			MAGNETITE - Series	
GEOCHEMISTRY (Major Elements)	CHAPPELL & WHITE (1974 & 1983) COLLINS et al. (1982), WHALEN et al. (1987)		S - TYPE		(I - TYPE)*	M - TYPE	(A - TYPE)*
	LA ROCHE (1986) LA ROCHE et al. (1980)	AK-L M.A.		AK-G M.A.	SA M.A.	CA M.A.	TH M.A.	A-PA M.A.
	DEBON & LE FORT (1983 & 1988)	ALUMINOUS M.A.		ALUMINO-CAFEMIC and CAFEMIC M.A. (Subalkaline, calc-alkaline, tholeiitic, and (per)alkaline)		
	MANIAR & PICCOLI (1989)	CCG		POG	CAG	IAG	OP	RRG	CEUG
GEOCHEMISTRY (Trace Elements)	TAUSON & KOZLOV (1973)	PLUMASITIC LEUCOGR.	ULTRA-MM GRANITES	PALINGENIC GRANITES (Normal and Subalkalines)		PLAGIO-GRANITES	AGPAITIC LEUCOGRANITES
	PEARCE et al. (1984)	COL G - Collision Granites (Syntectonic)		COL G (Post-tectonic)	VAG Volcanic Arc Granites	ORG	WPG Within Plate Granites
ASSOCIATED MINERALIZATIONS	XU KEQIN et al. (1982)	TRANSFORMATION - TYPE (Continental crust)		SYNTEXIS - TYPE (Transitional crust)		MANTLE-DERIVED - TYPE
TECTONIC ENVIRONMENT	PITCHER (1983 & 1987)	HERCINOTYPE		CALEDONIAN - TYPE		ANDINOTYPE	W.PACIFIC TYPE	NIGERIA - TYPE
SUGGESTED SYNTHETIC CLASSIFICATION		**MPG**	**CPG**	**KCG**	**ACG**	**ATG**	**RTG**	**PAG**

55. Compilation of granitoid rock classifications. Reprinted from *Lithos*, v. 46, Barbarin, B., A review of the relationships between granitoid types, their origins and their geodynamic environments, pp. 605–626, 1999, with permission from Elsevier Science.

The Fluid Dynamical Era

28 ❖ RAYLEIGH AND REYNOLDS
The Fluid Dynamics of Magma Chambers

Throughout most of the history of igneous petrology, the chemical and mineralogical aspects of the discipline received more attention than the physical aspects. Most igneous petrologists knew much more about the mineralogical and chemical composition of igneous rocks than they did about the mechanics of intrusion or the physical properties and fluid dynamical behavior of magmas. Cold, hard igneous rocks, after all, are considerably easier to collect and examine than are intensely hot, extremely dangerous lavas. Moreover, direct observation of a subterranean magma chamber has always been out of the question. Necessarily, geologists have concentrated on the chemical characteristics of rocks rather than the motions of inaccessible liquids. More than two centuries ago, geologists realized that igneous rocks are composed of materials that can be chemically analyzed. With the advent of physical chemistry in the latter nineteenth century, petrologists began to regard igneous rocks as the crystallization products of magmatic solutions and to develop ideas about the physico-chemical behavior of rocks and melts in terms of eutectic crystallization, immiscibility, and diffusion. In time, rigorous experimental methods were devised for the determination of the phase relations of igneous materials of very specific chemical compositions under precisely controlled conditions. Textbooks of igneous petrology characteristically emphasized the chemical aspects of petrology.

Despite the strong chemical orientation of most igneous petrologists, however, the physical aspects of the discipline were not ignored entirely. As soon as geologists realized that igneous rocks crystallized from hot melt, they attempted to infer the nature and behavior of that melt within Earth's crust. Geologists like Gilbert at the end of the nineteenth century and like Daly at the beginning of the twentieth century attempted to decipher the mechanics of magmatic intrusion. The formation of dikes and laccoliths and the mechanics of stoping are clearly related to the physical behavior of magma as a fluid, and Gilbert and Daly both realized that they needed to take into account physical properties such as density and viscosity to understand the action of magma. Well into the twentieth century, students of great layered intrusions, particularly Wager and Hess, invoked the concept of large-scale convection cells within magma chambers to explain the movement of crystals from the roof of a chamber to its floor. Geologists like Grout interpreted coexisting masses of igneous rock of contrasted composition as an effect of the segregation of immiscible liquids from one another. Some layered intrusions such as Rum (long spelled as Rhum) were considered by Malcolm Brown, for example, as the result of repeated injections of magma into a chamber. In virtually all of these and in many other cases, petrologists tried to infer the fluid behavior of magma from geological field evidence coupled with mineralogical and chemical information. Only rarely did theoretical or experimental fluid dynamics provide substantial input into the conclusions.

LARGE SILICIC MAGMA CHAMBERS

The field approach unquestionably provided very important insights. Field and chemical investigations of ash-flow deposits have yielded clues to the internal workings of large rhyolitic magma chambers. As an example, Robert L. Smith (b. 1920), a career geologist with the U. S. Geological Survey, concluded that the magma chamber from which the Bandelier Tuff in northern New Mexico erupted was characterized by physical and chemical gradients. Smith and Bailey (1966) pointed out that the upper member of the Bandelier Tuff, associated with the Valles caldera in the Jemez Mountains, consists of several units with distinctive mineralogy, welding characteristics, and inferred crystallization temperatures. They observed systematic upward decreases in the orthoclase content of feldspar, silica content, total alkalis, and FeO/MgO ratio which they interpreted as the result of the magma chamber being tapped at deeper, less fractionated levels during successive eruptive events.

Similar investigations were conducted by several doctoral students of Ian S. E. Carmichael (b. 1930) at the University of California at Berkeley. E. Wesley Hildreth (b. 1938), who received his Ph.D. degree in 1977 and who has been with the U. S. Geological Survey since then, studied the Bishop Tuff, a widespread ash deposit that was erupted from the Long Valley caldera on the eastern side of the Sierra Nevada Mountains in California. Hildreth (1979) discovered that values of paleo-temperature of glass samples determined by means of the Fe–Ti oxide geothermometer increase systematically with stratigraphic position. He regarded the values as an indicator of eruptive progress. He also found that the chemical compositions of several abundant minerals vary progressively with the reported temperatures. The strongest enrichments of elements such as W, Mo, Nb, Sb, Ta, U, Rb, Cs, and Na are present in the earliest erupted materials. Abundances decrease in subsequent ash deposits. In contrast, early deposits are depleted in Mg, Sr, Ba, and Eu and become more strongly concentrated in later eruption products. From these data, Hildreth concluded that the eruptive sequences reflect a systematic, continuous tapping of progressively hotter, presumably deeper, levels within the magma chamber. He believed that the tappings were sufficiently continuous to prevent lengthy time intervals in which magma mixing or phase re-equilibration might take place. The data, he argued, provide no evidence of crystal settling, significant assimilation, liquid immisciblity, or progressive partial melting. The apparent chemical differentiation of the Bishop Tuff magma chamber had presumably occurred in an essentially all-liquid state.

This contention sounded strangely similar to the supposedly discredited ideas propounded by Soret and Gouy and Chaperon a century earlier. Hildreth acknowledged that diffusion rates in rhyolitic liquid are much too slow to permit chemical separations within a reasonable time in a static magma chamber. Nevertheless, he believed that the occurrence of magmas depleted in ^{18}O, the presence of $^{87}Sr/^{86}Sr$ gradients within silicic magma chambers, and water-concentration gradients likely to be present in such chambers all indicated the reality of mass transport in the liquid phase on a substantial scale during a single caldera cycle. While not entirely clear on details of the mechanisms effecting the inferred chemical stratification, Hildreth appealed to the effects of thermogravitation combined with the Soret effect, a suggestion not universally welcomed, and convective circulation. He also believed that water would concentrate in the upper part of the magma chamber. Within the water- and halogen-enriched high-silica roof zone, Hildreth suggested that chemical separations might occur through the combined effects of convective circulation, diffusion, complexing of elements, and wall-rock exchange. The resulting compositional gradients, he maintained, would be linked to gradients in the structure of the melt and would be controlled by the thermal and gravitational fields of the magma chamber.

Similar conclusions were reached by Hildreth's wife, Gail A. Mahood (b. 1951), who received her Ph.D. degree from Berkeley in 1980 and has been on the faculty of Stanford University since 1979. From her studies of the Tala Tuff in Jalisco, Mexico, Mahood (1981) thought it reasonable to assume that a lava flow or an ash flow was tapped from the most differentiated uppermost magma in a chamber during a particular eruption. If so, a sequence of eruptive units would provide a series of "progress reports" on the differentiation mechanisms that operate at depth. As in other voluminous ash-flow localities, she reported progressive chemical variations in the sequence of eruptions, which she interpreted as indicative of chemical zonation in the magma chamber that could not have been produced by crystal settling or partial melting. Mahood, too, attributed the zonation to diffusional processes in the silicate melt, specifically to the transport of trace metals as volatile complexes along thermal and gravitational gradients in a volatile-rich, albeit water-undersaturated magma and also to the migration of trace elements responding to changes in the silicate melt structure that were brought about by decreases in the concentration of network modifiers, dissolved volatiles, and complexing ligands in a relatively dry system.

Since his earlier work on the Bishop Tuff, Hildreth has refined its eruptive stratigraphy (Wilson and Hildreth, 1997). Additional geochemical studies on the Bishop Tuff have been made, for example, by Alfred T. Anderson, Jr. (b. 1937), of the University of Chicago, and his students and associates. Among these studies, Wallace et al. (1999) analyzed melt inclusions within quartz phenocrysts and pumice clasts for volatile constituents and concluded that gradients in exsolved gas constituents had been established prior to eruption. Like Smith, Hildreth, and Mahood, they also concluded that the data were consistent with progressive downward tapping of a compositionally zoned magma body. Anderson et al. (2000) measured the major- and trace-element compositions of melt inclusions in quartz and sanidine phenocrysts. Although they agreed with Hildreth about the absence of convective mixing within the magma chamber, they concluded that there had been some crystal settling of the quartz and sanidine phenocrysts.

ONSET OF THE FLUID DYNAMICS REVOLUTION

As valuable as studies of the Bishop Tuff and other rhyolitic ash deposits had been, and as valuable as the investigations of innumerable plutons like the Skaergaard Intrusion had been for yielding insight into the nature of processes within magma chambers, petrologists still recognized that these processes were being inferred from their end products, namely, cold, immobile rocks that possessed frozen-in, specific textures, structures, mineralogy, and chemical composition. In some ways the situation resembled the days before the development of experimental methods for determining phase equilibria when petrologists tried to deduce crystallization histories and mechanisms of diversity from the rocks alone. As developed at the Geophysical Laboratory, experimental phase equilibria provided a quantitative method for discriminating among competing field-based hypotheses that had been proposed to account for observed igneous rock diversity. The results of experiments led to the widespread conviction among petrologists that some form of fractional crystallization was much more feasible than liquid immiscibility, Soret diffusion, or some other all-liquid differentiation mechanism, field-based ideas that were all very popular among igneous petrologists before the advent of experimental petrology. Thus, the suggestion that the magma chamber of the Bishop Tuff had experienced compositional zoning while in the liquid state led to some uneasiness inasmuch as it seemed to run counter to the conclusions reached by experimental studies.

So, how can petrologists ascertain if convective mixing or the absence thereof is the more likely situation within a magma chamber? How can they know if compositional stratification of the melt in a large rhyolitic magma chamber that produced the Bishop Tuff is a realistic hypothesis? Is crystal settling or crystal flotation to produce distinct compositional layering feasible? Are crystals most likely to grow within a magma chamber at the roof, on the side walls, on the floor, in the middle, or some combination of these possibilities? How are the crystals accumulated into layers or clusters? Can the behavior of fluid masses within a large magma chamber really lead to magmatic differentiation by means of separation of liquid and crystals? When a basaltic magma chamber is replenished by a rhyolitic injection, is it physically realistic to expect the two magmas to mix and produce a hybrid magma?

A new era dawned in the 1960s and 1970s as petrologists realized that, not only phase equilibria, but also theory and experiment in the field of fluid mechanics were required to answer such questions and to make more accurate assessments of hypotheses formulated from field study. To understand magma-chamber processes better, petrologists needed the careful application of theoretical fluid dynamics to well-designed experiments aimed at elucidating the extremely complex behavior of silicate magma in relation to a host of factors, including the location and rate of heat loss from the chamber, rate of heat conduction through the magma, nature of convection within the chamber, location and extent of crystallization, the distribution of crystal sizes and shapes, density contrasts between melt and crystals or between multiple melts, and magma viscosity.

One of the major questions that petrologists set out to resolve with the aid of fluid dynamics concerned the nature and extent of convection in a magma chamber. When heat is lost through the roof, side walls, or floor of a magma chamber, temperature gradients should be established within the magma body. Does heat diffuse conductively through a stagnant magma mass along a temperature gradient? Are density contrasts created by temperature differences sufficient to initiate so-called thermal convection within the magma? If convection does occur, what are the size, shape, and number of convection cells, and how vigorous is the convection? Is convection sufficiently strong to move substantial quantities of crystals from one part of a magma chamber to another thereby promoting differentiation, segregation, or layering? Do conduction and convection occur simultaneously?

As fluid dynamics found its way into the thinking of igneous petrologists, petrological literature began to incorporate references to fluid dynamical concepts and quantities such as the dimensionless Rayleigh number (Ra), the critical parameter that is related to whether a mass of fluid will convect. In its fundamental form, Ra = $g\alpha\,(T - T_o)\,d^3/\,\nu\kappa$, where g is the acceleration due to gravity, a is the thermal coefficient of expansion of the liquid, $(T - T_o)$ is the thermal gradient in the fluid, d is the thickness of the fluid layer, ν is the kinematic viscosity, and κ is the thermal diffusivity. The Rayleigh number essentially measures the balance between propelling and damping forces acting on the fluid, and convection takes place if the propelling forces are sufficiently greater than the damping forces. If the viscosity of the fluid is high, Ra decreases and the likelihood of convection is diminished. If thermal diffusivity is high, heat is conducted more readily through the fluid. If heat is lost through the roof of a chamber and heat can be conducted quickly from magma at the bottom of the chamber, that magma can be cooled, thereby increasing its density so that it remains gravitationally stable at the bottom of the chamber. In other words, the magma is less likely to rise if it can cool. On the other hand, a large temperature difference in a mass of fluid leads to hotter, less dense regions and cooler, more dense regions, and the density difference promotes fluid motion. A large fluid layer depth encourages motion by reducing the retarding effects of walls on circulating fluids. In general, convection occurs in fluids where Ra has values in excess of 10^3. As Ra increases, convection velocity and turbulence increase. For very high Ra, around 10^6 or more, convection may be violently turbulent.

If convection can be driven by differences in density through a mass of magma, then thermal convection, induced by the existence of temperature gradients, is not the only possibility. Within a chemically heterogeneous cooling fluid mass that is undergoing crystallization, local fluid in the vicinity of a growing dense crystal that does not have the same chemical composition as the melt phase may be less dense than the main body of fluid that is away from sites of crystallization. As a result, fluid density differences may develop in response to crystallization. Convection in a chemically complex fluid mass like silicate magma, therefore, may be compositionally driven as well as thermally driven.

Petrologists also began talking about the phenomenon of double-diffusive convection (DDC), a phenomenon that occurs in fluid masses possessing both temperature and chemical gradients. Double-diffusive convection occurs, for example, in masses of fluid that become gravitationally stratified on a local scale in response to temperature and concentration gradients or to multiple concentration gradients as in a solution of salt and sugar. The liquid subdivides into a series of roughly horizontal layers containing small convection cells. If a temperature gradient is present, the system of layers is characterized by an upward decrease in density adjusted to stepwise variations in temperature and chemical composition. If the bottom of the fluid mass is hotter than the top, heat diffuses upward across layer boundaries because of thermal gradients, and there is also some chemical exchange between adjacent layers owing to chemical gradients. The upward diffusion of heat, however, effectively lowers the density of the bottom of the overlying layer and also increases the density of the top of the underlying layer. As a result, local gravitational instability is initiated, thus driving small-scale convection cells within layers.

Important processes also take place when magma already within a chamber is disrupted by an input of a new pulse of magma. Depending on the rate of influx, the relative viscosities of the two magmas, the size of the feeder, and other factors, the input magma may exhibit either laminar (streamline) flow or turbulent behavior. Whether the flow is laminar or turbulent is one factor affecting the extent to which mixing and homogenization of the two magmas occurs. Another dimensionless number, the Reynolds number (Re), relates to laminar versus turbulent behavior in fluids. The Reynolds number is Re = $V \times l/\nu$, where V is the velocity scale, l is the length scale, and ν is the kinematic viscosity. An alternative form of the Reynolds number specifically applicable to the geological case just mentioned is Re = $(g'Q^3)^{1/5}/\nu$, where Q is the volumetric flow rate and g' is reduced gravity ($g\Delta\rho/\rho$), where g is the acceleration due to gravity, ρ is the density of a resident magma and $\Delta\rho$ is the density difference between resident and input magmas. In general, as the Reynolds number of a fluid increases, its flow passes from laminar to turbulent. Increased flow velocity increases the likelihood of turbulent flow. Increased width of the channel in which a flow is confined leads to greater likelihood of turbulence. Thus, basaltic magma flowing rapidly through a fracture that is 1 meter wide is more likely to be laminar than basaltic magma flowing at the same velocity through a fracture that is 10 meters wide. In contrast, increased magma viscosity increases the likelihood of laminar flow. Rhyolite magma is more likely than basaltic magma to display laminar flow because of its great viscosity.

The hope has been that experiments would help petrologists to begin sorting out the effects of such fluid phenomena on magmatic processes. Much of the credit for getting the ball rolling on fluid dynamical studies among geologists goes to Herbert R. Shaw (b. 1930). After receiving his Ph.D. degree from the University of California at Berkeley in 1959, Shaw spent his entire career with the U. S. Geological Survey focusing on the physics of magmas. He experimentally measured the viscosity of obsidian as a function of temperature and water content (Shaw, 1963) and studied the solubility of H_2O in silicate melts (Shaw, 1964). In a trail-blazing study, Shaw (1965) calculated the effects of granitic melt viscosity on rates of settling of different minerals while also evaluating the effects of suspended crystals and gas bubbles on magma viscosity. He concluded that crystals, including those of plagioclase, should be able to settle considerable distances during thousands of years. Shaw also evaluated the conditions of flow during forced convection in dikes and during convection induced by temperature gradients in vertical cylindrical chambers such as stocks or volcanic necks. He concluded that crystal settling might be important during the crystallization of granitic magmas and stated that thermal convection had to be reckoned with in any large pluton in the early stages of its crystallization. In subsequent studies Shaw explored the viscosity of basaltic melts in both field (Shaw and others, 1968) and laboratory investigations (Shaw, 1969), the role of chemical diffusion in both static and convecting magma chambers (Shaw, 1974), and fracture mechanisms of magma transport (Shaw, 1980). In many ways, Shaw's work laid the foundation for much that has been done in the field of geophysical fluid dynamics since the 1970s.

Thanks to the pioneering labors of Shaw, a number of schools and individuals devoted to the application of fluid dynamics to the problems of igneous petrology have emerged, particularly since 1980. Among leaders in geophysical fluid dynamics have been Claude Jaupart, Stephen Tait, Geneviève Brandeis, Anne Davaille, and others at the Institute de Physique du Globe of the Université Paris; Bruce Marsh and his former students such as Katherine Cashman and George Bergantz at Johns Hopkins University; Alexander McBirney of the University of Oregon; Neil Irvine of the Geophysical Laboratory; Frank Spera of the University of California at Santa Barbara; and a large number of researchers affiliated with the University of Cambridge and the Australian National University, including Herbert E. Huppert, R. Stephen J. Sparks, Ross C. Kerr, M. Grae Worster, J. Stewart Turner, Ian H. Campbell, Daniel Martin, and others.

In France, Claude Jaupart initiated a program of fluid mechanical experiments after receiving his Ph.D. degree from MIT in 1981. Jaupart and his doctoral student Geneviève Brandeis, investigated nucleation and crystal growth in cooling magmas, the interaction between convection and crystallization in cooling magmas, and the nature of stagnant layers at the base of convecting magma chambers. Their experiments suggested that convection is likely to occur in magma chambers.

Brandeis and Jaupart (1986) studied the case of thermal convection driven by cooling at a roof and ignored the effects of crystal settling and compositional differences. They found that thermal convection is weak in high-viscosity magmas because the temperature contrast driving convection is very small. In low-viscosity magmas, they found that convective instability of upper boundary layers was characterized by descending crystal-bearing plumes. Bottom crystallization occurred by deposition of crystals from the descending plumes and by nucleation of crystals in place. In a subsequent study, Jaupart and Brandeis (1986) noted that the evolution of a bottom boundary layer of a magma chamber was controlled by cooling through the floor and by penetration of the descending plumes. They cooled silicone oils both from above and below and showed that stagnant layers developed at the bottom that were neither penetrated by convective plumes nor affected by convective mixing. Continued growth of the stagnant layer, however, was impeded by convective erosion. The entire system included well-mixed portions undergoing turbulent convection and stagnant bottom layers. They suggested that stagnant layers within basaltic chambers might attain a thickness of several tens of meters. Brandeis and Jaupart (1987a, b) studied the kinetics of nucleation and growth and calculated nucleation rates for basaltic magma chambers.

Frank Spera investigated the fluid dynamics of magma chambers by means of numerical methods. Spera received his Ph. D. degree from Berkeley in 1977. After teaching at Princeton University, he moved to the University of California at Santa Barbara. Working in collaboration with David A. Yuen of Arizona State University and, later, the University of Minnesota, Spera mathematically evaluated the effects of viscosity contrast on convection, Soret diffusion, and chemical fractionation in boundary layers of silicic magma chambers (Spera et al., 1982). He then numerically examined the dynamics of withdrawal of magma from stratified magma chambers (Spera et al., 1986) and the effects of chemical buoyancy and diffusion where convection occurs along sidewalls of magma chambers (Spera et al., 1989).

After studies at Michigan State University and the University of Arizona, Bruce D. Marsh (b. 1947) received his Ph.D. degree from the University of California at Berkeley in 1974. Since that year he has been on the faculty of Johns Hopkins University and has focused his attention on the physical aspects of magmatism. Marsh (1982, 1984) published major studies on the mechanics of various processes of magma ascent and emplacement such as diapirism, stoping, zone melting, and caldera resurgence. He was also intrigued by the interrelationships among crystallinity, viscosity, and convection within a magma body. For example, Marsh (1981) suggested that when basaltic magma has become more than 50 percent crystallized, it is rheologically impossible to erupt, and the magma ultimately becomes a pluton. Basaltic lava flows, therefore, typically contain less than 50 percent phenocrysts. In contrast, as the silica content of magma increases the "critical crystallinity" decreases. In other words, granitic magma can erupt as rhyolite lava only if its crystal content is much less than that of basalt magma. Rhyolitic lavas ought, therefore, to contain fewer phenocrysts on average than basalts. Beyond the critical crystallinity, a granitic magma could erupt only explosively, creating an ignimbrite. Marsh believed he had found the fundamental explanation for the dichotomy that had long been recognized by petrologists: abundant basalt and granite but subordinate gabbro and rhyolite. Marsh and Maxey (1985), recognizing the central role of the concept of crystal fractionation in igneous petrology, sought to establish the magmatic conditions that favor or discourage the physical separation of crystals from magma by investigating the distribution of crystals in convecting magma. In 1988, Marsh and his graduate student, Katharine Cashman, now on the faculty of the University of Oregon, published a series of papers on crystal-size distribution (CSD). Marsh (1988b) developed the general theory of CSD and showed how the kinetics of crystallization and various physical processes such as crystal fractionation or magma mixing might affect the distribution of crystal sizes within an igneous body. Cashman and Marsh (1988) then applied the theory of CSD to drill core samples taken from the Makaopuhi lava lake that ponded at Kilaeua Volcano on the island of Hawaii in 1965. After measuring the sizes and population densities of plagioclase and ilmenite as the lava lake cooled, they calculated nucleation and crystal growth rates and concluded that crystallization occurred at very small undercoolings. Both Cashman (1992, 1993) and Marsh (1998) have continued to apply textural studies to the interpretation of igneous processes.

THE CAMBRIDGE–ANU GROUP

As much as any individual, J. Stewart Turner was highly influential in his application of fluid dynamics to the processes in magma chambers. Turner's considerable network of colleagues and students associated with the Australian National University and the University of Cambridge has been the most prolific in conducting experimental and theoretical studies relating to the fluid dynamics in magma chambers. Turner graduated from the University of Sydney in 1952 and received a master's degree from his alma mater the following year. He then went to England, where he received his Ph.D. degree from the University of Cambridge in 1957 after studying under renowned fluid dynamicist Sir Geoffrey I. Taylor. Both before and after his doctoral studies, Turner worked in the Cloud Physics group of Australia's Commonwealth Scientific and Industrial Research Organization (CSIRO). Between 1962 and 1966, Turner worked at Woods Hole Oceanographic Institution in Massachusetts, where he investigated mixing processes within the ocean. In 1966, he returned to Cambridge as a faculty member in the Department of Applied Mathematics and Theoretical Physics, where he worked on convection and mixing processes in both the ocean and the atmosphere. He regularly published in the *Journal of Fluid Mechanics* and related journals on topics such as the transport of salt and heat across density interfaces, effects of heating on salinity gradients, and jets and plumes. In 1973, he published *Buoyancy Effects in Fluids*, a text that still exerts wide influence. Turner returned to Australia in 1975 as Professor of Geophysical Fluid Dynamics in the Research School of Earth Sciences at the Australian National University where he served until his retirement in 1995. It was at ANU that Turner's interests turned toward the study of convection and mixing processes in silicate melts.

Turner's colleague at the Australian National University, economic geologist, L. B. Gustafson, realized the importance of applying fluid mechanics to geological problems and inspired Turner's foray into the field of geophysical fluid dynamics. Thus, Turner and Gustafson (1978) reported on investigations of the flow of hot saline solutions from vents on the seafloor and made applications to the formation of ore deposits. They called attention to the importance of compositionally induced density differences in fluids. Of particular interest was the situation in which very salty, hot fluid was injected into less salty, cooler fluid, precisely the situation at seafloor vents. Conducting this work prior to the discovery of black smokers, Turner and Gustafson essentially predicted the existence of such effluents. In such a case, they pointed out that the high salinity and high temperature of the effluent exert opposing effects on the density difference between the exhaled fluid and the surrounding cold sea water. They found that the outflow would separate into a hot, less saline, less dense plume that would rise and a warm, more saline, more dense flow that would spread away from the source as a bottom current, maintaining a sharp boundary with the overlying seawater and mixing with it very little. The authors indicated that the heavier bottom current might flow long distances over the seafloor to settle into depressions where dissolved metals might be deposited. A continuing inflow of the dense fluid into such depressions ought to produce stable stratification. Turner and Gustafson also thought that evidence of density stratification within the Bushveld Igneous Complex hinted at the importance of similar fluid phenomena during the formation of layered igneous intrusions and suspected that chromite, magnetite, and platinum deposits in layered igneous intrusions could be related to the effects of double-diffusive convection.

Earlier in the 1970s, Turner had experimentally investigated double-diffusive convection in fluids, sometimes in collaboration with C. F. Chen of Rutgers University (Turner and Chen, 1974). Turner's research on double-diffusive convection influenced petrologists Alexander McBirney and Neil Irvine, both of whom had a strong interest in fluid dynamics and in magma chambers of layered igneous intrusions. McBirney had already been studying the thermal properties of magmas, and Irvine (1970a) had investigated the cooling and convection of intrusive sheets. McBirney incorporated the concept of double-diffusive convection into a revolutionary paper on the Skaergaard Intrusion (McBirney and Noyes, 1979), and Irvine (1980b) invoked the concept in his study of the Muskox Intrusion. The contributions of McBirney and Irvine to layered igneous intrusions will be examined in more detail in Chapter 29. Knowing that most geologists were still unfamiliar with the concept of double diffusion, Chen and Turner (1980) published an article on crystallization in double-diffusive systems and suggested that geologists begin to apply the concept, as McBirney and Irvine had just done, to the formation of mineral layers in magmatic rocks.

Increasingly addressing problems in igneous petrology, Turner (1980) reported on experiments with aqueous solutions to show how chemical differentiation of a liquid could be brought about by crystallization along the walls of a container. Turner showed that crystal settling, the mechanism of fractional crystallization so dear to the heart of Bowen and a large number of petrologists in the first half of the twentieth century, is not necessarily the only way that differentiation occurs in magmas. He demonstrated that growth of crystals on a side wall boundary released a less dense fluid that rose to the top of its container in a boundary layer flow. In time, a stratified solution was established. Turner argued that upward flow of analogous boundary layers in magma chambers should also be capable of setting up large concentration and density gradients. Having experimentally produced differentiated, stratified aqueous solutions, he urged geologists to compare the consequences of his model with the details of layered igneous intrusions. One of his major conceptual contributions to petrology was to suggest that processes in melt were responsible for large-scale structures in the crystallized solids such as horizontal layering.

Starting in the early 1980s, fluid dynamical studies pertinent to igneous petrology began to flow from Cambridge due to the combined efforts of H. E. Huppert and R.S.J. Sparks. Herbert E. Huppert (b. 1943) graduated from the University of Sydney in 1963. After obtaining his Ph.D. degree from the University of California at San Diego in 1968, Huppert went to Cambridge as a Research Fellow and advanced through the ranks within the Department of Applied Mathematics and Theoretical Physics to the position of Professor of Theoretical Geophysics and Foundation Director of the Institute of Theoretical Geophysics. Huppert met Turner after arriving in Cambridge in 1968 and collaborated with him on a number of papers in oceanography and meteorology. In 1979, Huppert received a telephone call from Stephen Sparks, a volcanologist in the Department of Mineralogy and Petrology at Cambridge, wanting to know if Huppert knew some fluid mechanics. After discussions, Huppert and Sparks quickly realized that joint work applying the concepts of fluid mechanics quantitatively to problems in volcanology would be a very fruitful venture.

Early in their cooperative endeavors, Huppert and Sparks decided to address the fluid dynamical effects of replenishing a magma chamber with a fresh influx of relatively hot and heavy magma. During the 1970s, petrologists had become interested in such replenishment effects. Michael O'Hara, then at the University of Edinburgh, had recently developed a very detailed mathematical model for evaluating the chemical evolution of basaltic magmas in which the magma in a high-level chamber not only underwent continuous fractional crystallization but was also fed periodically with a new batch of parental magma from below. O'Hara (1977) suggested that a new batch of magma would displace some of the residual liquid in the upper part of the chamber as a lava eruption, that the remaining residual liquid would mix with the fresh batch of magma, and that fractionation would then continue. He developed mathematical expressions for tracing changes in the concentrations of the chemical elements during fractional crystallization modified by replenishment.

The Cambridge team, however, was more interested in the fluid dynamical processes accompanying magma replenishment, a situation analogous to the injection of a hot saline flow onto the ocean floor. Prior to teaming up with Huppert, Sparks had already thought about the fluid dynamical processes associated with the injection of a fresh pulse of relatively lighter magma. Following O'Hara, Sparks et al. (1980) suggested that most lava eruptions onto the ocean floor were probably the consequence of influxes of new magma into crustal chambers beneath the ocean floor. The influx of magma into the chamber, they reasoned, caused inflation of the reservoir and an increase in fluid pressure that eventually resulted in a discharge of magma onto the seafloor or onto Earth's surface as at Kilaeua and in the Icelandic volcanoes.

Sparks and his colleagues suggested four possible styles of behavior in basaltic magma reservoirs into which new pulses of magma had been added. These styles of behavior were said to depend on the influx rate, the dimensions of the chamber, and the physical properties of the magmas. They referred to the four styles as laminar plume, semiturbulent plume, fully turbulent plume, and nonbuoyant influx. In the first three styles of behavior, the injected magma would rise to the top of the reservoir. In the laminar plume case, an input magma would ascend and stratify at the top of the reservoir without much mixing. A primitive, low-density magma in the chamber would be displaced and thus erupt. In the cases of semiturbulent and turbulent plumes, mixing would occur, and a hybrid magma would accumulate at the top of the reservoir. The degree of mixing would vary depending on how the buoyant magma rose, on the physical properties of the input and resident magmas, and on the dimensions of the magma chamber.

Finally, Sparks and his colleagues claimed that a nonbuoyant influx of picritic (olivine-rich) magma into the chamber would spread out over the floor of the chamber and would displace some fractionated basalt onto the surface. The picritic influx would not mix effectively with the overlying magma even though convection might occur in both magmas. They used their fluid dynamical model to account for differences in basalt compositions erupted along the East Pacific Rise and the Mid-Atlantic Ridge.

After teaming with Huppert, Sparks could bring the quantitative concepts of fluid dynamics to bear on his contentions. Over the next several years, they conducted a series of experiments on aqueous solutions in which they attempted to model various fluid processes in magma chambers. At first they investigated replenishment phenomena. Huppert and Sparks (1980) examined the fluid dynamics of an influx of hot, dense magma that replenished a reservoir containing less dense, fractionated basaltic magma. From theoretical and empirical models of salt/water systems, they reiterated that an ultrabasic melt would spread over the floor to form an independent layer separated from the overlying magma layer by a relatively sharp interface. Both layers would convect vigorously as heat was transferred from the hot, lower layer to the upper layer at a rate that would exceed the rate at which heat would be lost to the country rocks. Thermal convection would continue until both layers attained the same density. The chemical compositions of the two layers would remain distinct because of the very low diffusivity of mass compared to heat. Huppert and Sparks calculated the temperatures of the two layers as functions of time and evaluated how their cooling rates depended on the viscosities, thermal properties, densities, and thicknesses of the layers. They applied their model to the cyclic ultramafic rock units on the Isle of Rum that had been attributed to successive injections of fresh magma by Brown (1956).

Although Turner left Cambridge for the Australian National University in 1975, he returned to Cambridge from time to time as a visiting scholar. Huppert visited Australia in the same capacity. These exchanges provided opportunity for Turner and Huppert to collaborate. To model the input of dense picritic magma into a chamber, Huppert and Turner (1981) teamed up to experiment with solutions of hot concentrated KNO_3 introduced beneath colder layers of less dense $NaNO_3$ or K_2CO_3. These simple laboratory experiments successfully simulated the theoretical replenishment concept of Huppert and Sparks.

In 1982, Huppert, Turner, and Sparks reported on three experimental variations on their previous work. To continue modeling the effect of injecting pulses of ultrabasic magma into chambers of fractionated basalt, they examined the effects of different rates of injection of various aqueous solutions into both homogeneous and compositionally stratified chambers (Huppert et al., 1982). They rapidly injected a hot, dense fluid into a mass of colder fluid with a vertical composition gradient and found that a series of layers gradually built up from the base of the upper fluid mass as the lower layer of injected fluid cooled, crystallized, and overturned. They also produced a slow influx of hot, dense solution beneath a compositionally homogeneous, cold, less dense layer, and the results were similar to those obtained by Huppert and Turner (1981) on the input of picritic basalt. Finally, Huppert, Turner, and Sparks studied a combination of the first two variations and obtained results similar to those of the case of rapid injection beneath a compositional gradient.

The Cambridge–ANU team continued to develop their replenishment model by considering the effects of viscosity. In previous studies, the pairs of hot, dense fluid and cool, less dense fluid had comparable viscosities. Now Huppert et al. (1983) described laboratory experiments in which the upper layer was a fluid of much higher viscosity than the hot, dense input fluid. In this case, their intent was to model the replenishment of a magma chamber of viscous rhyolite by hot, low viscosity, dense basaltic liquid, a common circumstance in calc–alkalic systems. During cooling and crystallization of the dense input solution, less dense fluid began to ascend in very thin plumes whose tops developed spherical caps. Crystals were carried upward by the plumes but eventually sank. Little mixing occurred between input and high-viscosity resident fluid because of the strong viscosity contrast. From their results, Huppert, Sparks, and Turner suggested that crystallization and possibly volatile exsolution in a hot, dense basaltic liquid injected beneath a cooler rhyolite might result in some entrainment of the less viscous, vesicular mafic magma into the viscous rhyolite. Noting that vesicular mafic inclusions with quench textures are common in silicic volcanic rocks, they proposed that such features be interpreted in terms of entrainment across an interface.

The following year, Sparks et al. (1984) began to stress the importance of convection in the fractionation of magmas. They argued that density differences produced by crystallization would be more effective in promoting convection within magmas than density differences produced by temperature gradients. Maintaining that the velocity of convection within most magma chambers was orders of magnitude greater than the settling velocities of crystals, they disputed the effectiveness of crystal settling as a mechanism for bringing about large-scale fractionation in magmas. Instead, they proposed a concept of "convective fractionation" in which crystal–liquid separation was brought about by convection, not only in fluid basaltic liquids but even in more viscous silicic magmas.

As an example of the influence of crystal-driven convection, Sparks et al. (1985) addressed the phenomenon of adcumulus growth in layered igneous intrusions. Wager et al. (1960) had coined the term "adcumulus growth" to refer to the enlargement of already accumulated crystals, most likely near the floor of a magma chamber, by the addition of material of the same composition as the original crystals from the interstitial (intercumulus) liquid between the accumulated crystals. Noting that adcumulus growth might involve crystallization at the top of a pile of accumulated crystals in direct contact with convecting magma, Sparks and his colleagues proposed that adcumulus growth might involve compositional convection in which low-density liquid released by intercumulus crystallization would be continuously replaced by denser liquid from the overlying magma reservoir. They believed that this process would favor adcumulus growth within the pore space of a cumulate pile.

The authors argued, too, that intercumulus melt could be replaced by more primitive melt emplaced over a cumulate pile during an episode of magma chamber replenishment. A dense, primitive injected magma could sink several meters into the underlying cumulate pile in the form of fingers, replacing the lower density, already differentiated melt. Reactions between the primitive melt and crystalline matrix might lead to changes in mineral composition, mineral textures, and whole-rock isotopic compositions of existing crystals in the cumulate pile.

Thus far, the experiments of Huppert, Turner, and Sparks had concentrated on the effects of nonbuoyant influxes. Consequently, Huppert et al. (1986) performed experiments in which light inputs of freshwater were injected into dense saltwater solutions to observe the laminar or turbulent character of the input plumes. By varying the viscosities of input and resident fluids, they produced similarities or differences in their respective Reynolds numbers, and by varying the input rate they were also able to raise or lower the Reynolds numbers.

Experiments on the effects of fountains and plumes were also conducted when Ian Campbell joined Turner at the Australian National University. Ian Henry Campbell (b. 1942) graduated in 1966 from the University of Western Australia, after which he served as a geologist with Western Mining Corporation, Ltd. During this period, he became acquainted with the layered Jimberlana Intrusion, a body that he investigated for his doctoral dissertation at Imperial College of London. In 1973, he received his Ph.D. degree. After several postdoctoral fellowships, Campbell became Assistant Professor of Geology at the University of Toronto in 1979. He first made Turner's acquaintance in 1982 when he gave a workshop at the University of Western Australia on layered intrusions and ore deposits. Turner showed Campbell how experiments in fluid dynamics might provide analogs of possible mechanisms for forming features observed in layered intrusions. Campbell was promoted to Associate Professor at Toronto in 1984, but shortly thereafter joined the Research School of Earth Sciences at Australian National University, where he began a fruitful collaboration with Turner.

In a series of studies, Campbell and Turner (1989) modeled the replenishment of magma chambers in which dense input magma was injected with sufficient velocity to form fountains or plumes rather than simply flow across the floor. They maintained that fountains might rise hundreds of meters above the floor of a chamber, a result inconsistent with the calculations of Huppert and Sparks (1985). They predicted that hybrid layers would result from mixing of input and residual magmas, and that the hybrid layer on the floor of the chamber would break up into double-diffusive layers. In their judgment, the effects of fountains of injected magma could be detected in the Bushveld Complex, the Great Dyke, and the Jimberlana Intrusion.

These experiments were followed by variable-viscosity studies, the results of which were published much earlier because of unexpected results. Campbell and Turner (1986) modeled the input of low-viscosity magma as a fountain into a reservoir of higher-viscosity magma. The input fluid and host fluid were kept at the same temperature, but the viscosity of the host fluid was varied. Campbell and Turner discovered that turbulence in the fountain resulted in extensive mixing of the fluids if their viscosities were similar. To their surprise no detectable mixing of the two fluids occurred if their viscosities were drastically different even though the fountain was turbulent. The extent of mixing appeared to be related to the ability of the input fluid to distort the contact between the fountaining fluid and the resident fluid. They concluded that if a primitive basaltic magma were injected as a fountain or plume into a chamber of fractionated basalt, the two magmas would readily mix because they have similar viscosities. In contrast, little or no mixing would occur if low-viscosity basaltic magma were injected as a fountain into viscous granitic melt.

The vast majority of petrologically pertinent fluid dynamics experiments had been performed in rectangular tanks. Recognizing that magma chambers are not boxes with vertical walls and perfectly horizontal floors, Huppert et al. (1986) examined the effects of crystallization on an inclined plane, and Turner and Campbell (1986) considered the effects of funnels and inverted funnels on the behavior of fluids. For example, Turner and Campbell showed that if a light fluid is released by crystallization along the walls of a funnel-shaped chamber like the Skaergaard Intrusion, it moves vertically upward to mix convectively with overlying magma. In contrast, if light fluid is released in an inverted funnel, it flows upward along the sloping walls to concentrate at the apex of the chamber. If denser fluid is released during crystallization along the walls of a funnel, the fluid ponds at the floor of the chamber as it flows down the slopes. Denser fluid released by crystallization at the chamber roof descends vertically to mix with magma below. In a reversed funnel, dense fluid flows downward vertically no matter where it forms. Turner and Campbell suggested that stable stratification develops with both geometries but would be better developed in funnels than in inverted funnels.

Both the Cambridge and Australian National University groups also examined the fluid dynamical aspects of melting and assimilation of roof rocks of a magma chamber. Huppert and Sparks (1988a) theoretically and experimentally analyzed the melting of the roof of a chamber that contained turbulently convecting fluid. They considered cases in which the melted material is denser or lighter than the magma in the chamber and recognized that the latter case was relevant to the very important geological situation in which continental crust is melted by the emplacement of hot basaltic magma. In such situations, granitic melt whose viscosity greatly exceeds that of the basalt is produced. The thermal histories were evaluated for the case in which the melt cooled without crystallization and the case in which it crystallized with an increase in viscosity. Huppert and Sparks (1988a, p. 126) also included a mathematical analysis of a case in which the roof did not melt "to evaluate the influence of the overlying solid rock both on the heat transfer from the liquid in a magma chamber and on the state of convection in the liquid." They concluded that the fluid in the chamber would continue to transfer heat to the roof by convection over a long time scale if the roof did not maintain its initial temperature. Given that the notion of granitic magma forming by partial melting of continental crust that had been underplated by basaltic liquid was widely accepted among contemporary petrologists, Huppert and Sparks extended their study of roof melting to examine the specific case of the generation of granitic magma by intrusion of basalt sills into the continental crust (Huppert and Sparks, 1988b). In this paper, they predicted solidification times for basaltic sills and showed how large volumes of overlying convecting siliceous melt would develop.

Campbell and Turner (1987) suggested that simultaneous assimilation and fractional crystallization (AFC) would be a relatively unimportant process in basaltic magma chambers. If granitic liquid formed by melting at the roof of a basaltic chamber were less dense than the magma below, it would remain at the top of the chamber, chemically isolated from the basaltic magma below. The two convecting layers would be separated by a double-diffusive interface. Consequently, they maintained, assimilation would be spatially separated from the crystallization phase within the chamber. Fractional crystallization would occur near the floor whereas assimilation would take place near the roof. Heat required to continue melting the roof would be supplied by the latent heat of crystallization released near the floor and transmitted upward by convection. The heat would then be conducted into the overlying layer of lower-density contaminated magma across a double-diffusive interface between the two magma layers, and, by convection within the upper contaminated magma layer, heat would further be transmitted to the roof. Campbell and Turner envisioned such heat transfer as a response to Bowen's thermal argument against substantial assimilation. Only at a later stage would the contaminated upper layer crystallize. In the eyes of Campbell and Turner, fractional crystallization would follow assimilation rather than occur simultaneously with it in the upper layer because of this chemical isolation of the assimilated melt. "AFC is only an effective process," Campbell and Turner (1987, p. 156) wrote, "if there is extensive mixing between the magma released by fractional crystallization and that released by melting the floor, roof, or walls of the chamber." They suggested that AFC would become important only if a large number of blocks of roof rock fell into the lower fractionating basaltic layer, melted, and then mixed with that layer. Subsequent isotopic studies on the Skaergaard Intrusion by Stewart and DePaolo (1990) were consistent with the models of Campbell and Turner (1987) and Huppert and Sparks (1988b).

A graduate student at the Australian National University under Campbell, Daniel Martin, investigated the phenomenon of convection. From data on the physical and thermal properties of a variety of mafic and ultramafic intrusions, Martin et al. (1987) calculated the heat fluxes through the roofs of these intrusions. From these data, they calculated Rayleigh numbers for thermal convection and for compositional convection caused by the crystallization of dense olivine or pyroxene on a chamber floor. Martin and his co-workers concluded that thermal Rayleigh numbers for homogeneous magma chambers were so large that thermal convection must be highly turbulent. They predicted thermal Rayleigh numbers between 10^{12} and 10^{16} for basaltic magmas and 10^8 to 10^{12} for granitic magmas. Similar predictions were made by Carrigan (1987). They also concluded that compositional Rayleigh numbers for mafic magmas were around 10^{18} to 10^{23} for layer depths between one and ten kilometers. Again, they said, convection must be highly unsteady and disordered in large mafic magma chambers.

Although crystal settling had been a widely accepted mechanism for producing differentiated igneous rock bodies since the days of Bowen, many petrologists had become suspicious of that notion, as will be discussed in the next chapter. Martin and Nokes (1989), however, claimed on theoretical and experimental grounds that crystal settling can occur in magma chambers even where the vertical component of the velocity of convection is greater than the Stokes' law settling velocities of the crystals. That settling is possible was said to be due to the fact that the vertical component of convective velocity is dependent on height within the chamber and diminishes to zero at the chamber boundary. Crystal settling, they concluded, may be an efficient differentiation mechanism. Their study applied to situations where the concentration of crystals in the magma was relatively low, however.

DEBATE ABOUT CONVECTION

Not everyone in the geophysical fluid dynamics community was satisfied with the application of the results of simple experiments to complex magmatic problems by members of the Cambridge–ANU axis. In particular, Bruce Marsh began to question the idea that magma chambers were characterized by vigorous convection and, as a result, touched off a vigorous debate with the Cambridge–ANU group. Noting that "the diversity of igneous rocks gives petrologists their raison d'etre," Marsh (1988a, p. 1720) expressed puzzlement, in light of the common occurrence of fractional crystallization, that just where igneous rock diversity ought to be strong, as, for example, in the vast masses of Hawaiian or ocean ridge basalts, negligible amounts of silicic differentiates were produced. Like many of his predecessors, he was also puzzled that thick diabase sills and gabbroic bodies like the Skaergaard Intrusion had progressed little toward producing eruptible rhyolite. "Why is it," Marsh (1988a, p. 1721) asked, "that the clearest examples of magma chambers, ones we can touch, show little differentiation, whereas the unseen chambers, those spawning suites of lavas, do fractionate?" In essence, he was restating the old questions that had plagued Bowen, Read, Fenner, and other petrologists of earlier generations.

In addressing the problem, Marsh envisioned that a body of magma was already charged with an abundance of phenocrysts at the time of intrusion. He evaluated relative rates of phenocryst settling within a sheet-like magma body in relation to rates of crystal capture by upward and downward growing "solidification fronts." From the calculated rates, he constructed possible profiles of phenocryst distribution with respect to depth in sheet-like igneous bodies (Figure 56). He argued that kinematic shocks within zones of crystal accumulation would lead to the expulsion of fluid and the redistribution of crystals in accord with their buoyancies. Marsh further argued that this process would be modified by convection within the magma between the advancing solidification fronts, but he insisted that convection in the magma would be very slow and sluggish so long as the magma was not superheated. He maintained that the vigor of magmatic convection would be regulated largely by the rate of heat escape through the roof rock and would be strong only at the beginning of the cooling process. In Marsh's view, where heat conduction through the roof was very slow, convection would be suppressed, contrary to the calculations of Huppert and Sparks (1988a). Persuaded of the tranquil character of magma-chamber convection, Marsh objected to the assignment by fluid dynamicists

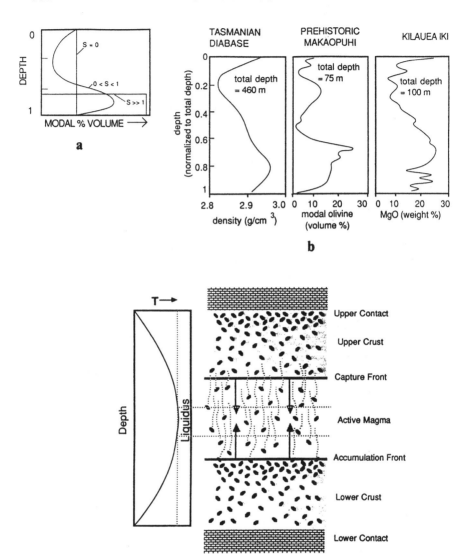

56. Top: a. Idealized modal distribution of phenocrysts in a sheet-like body as a function of rate of crystal settling relative to rate of solidification. b. Distribution of density, modal olivine, and magnesia in three sheet-like basaltic bodies. The profiles have a similar shape to that of the ideal profile. Bottom: Schematic diagram of capture of settling phenocrysts in a sheet-like body by a downward-growing crust and accumulation of phenocrysts in an upward-growing crust. From *Geological Society of America Bulletin*, v. 100, pp. 1720–1737. Marsh, B. D. (1988). Reproduced with permission of the publisher, The Geological Society of America, Boulder, Colorado, USA. Copyright (c) 1988, The Geological Society of America, Inc.

of large values to the Rayleigh number that would imply vigorous convection. Such high values had often been assigned on the assumption of heating of magma from below by replenishment. With cooling of a magma mass, Marsh envisioned that intermittent bursts of tall, narrow plumes would form, collapse, and reform within the magma layer existing between the advancing solidification fronts. These plumes, he said, would contain crystals that would be subject to flow sorting. As the plumes approached the floor, they would slow and spread out in response to a rise in the viscosity associated with the crystallizing crust developing on the floor. More sorting and sedimentation would occur to produce, along with other plumes, a systematically layered series of cumulates.

The foremost conclusion of Marsh was that chemical differentiation by crystal fractionation would be extremely limited in sheet-like gabbroic bodies. He insisted that little andesitic, dacitic, or rhyolitic liquid would be produced because of the presence of the solidification fronts. Inward from the zone of crystal accumulation, possible changes in liquid composition would be drastically limited because of the small degree of crystallinity.

Sparks (1990) objected to Marsh's suggestion that much of what went on in magma chambers concerned the sorting of crystals that were already present upon intrusion, and he also objected to Marsh' claim that convection in magma chambers is lethargic. He responded that convection played a central role in magma chamber differentiation. According to Sparks, the Skaergaard Intrusion lost a lot of heat from its sides as well as its roof by means of the hydrothermal convection model suggested by Norton and Taylor (1979). Hence, convection in the chamber should not be suppressed. Likewise, Sparks pointed out that other studies, such as those of Daniel Martin and Ross C. Kerr, a 1984 Ph.D. student of Huppert, had supported the likelihood of vigorous convection in magma chambers. Such convection could be established by temperature differences of only a few degrees or by crystal contents of just a few percent in a boundary layer that would lead to unstable density contrasts and high Rayleigh numbers.

Marsh (1989) further developed his contention that thermal convection in sheet-like magma chambers, except in the initial stages of rapid cooling, is a transient and sluggish process. He considered convective turbulence to be out of the question as far as magma lacking superheat is concerned. Given that magmas are invariably not superheated upon eruption, Marsh judged that magmas in chambers existed in a state of buffered, transient, small-scale convection characterized by low Rayleigh numbers. The dynamics of an entire magma chamber, he asserted, are largely controlled by the thermal regime of a crystallization front at the roof whose temperature remains close to that of the liquidus. Marsh based his contention for weak convection on the premise of extremely small temperature differences resulting from an absence of superheat in the magma. The reduction of temperature differences to near zero would greatly reduce the Rayleigh number. Presumably the lack of superheat was brought about by its rapid removal during a brief initial stage of convection. Convection would be damped as superheat was lost. Marsh also argued for weak convection on the grounds that the length scale to be used in determining the Rayleigh number is the thickness of a boundary layer rather than the depth of the magma chamber. And he renewed his claim that slow heat conduction through the country rocks would suppress convection.

Huppert, Turner, and Sparks, however, had observed plenty of vigorous convection in their laboratory experiments. Consequently, Huppert and Turner (1991) published a pointed objection to Marsh's view of weak convection. They reminded readers that Huppert and Sparks (1988a) and other writers had shown that internal convection could occur with slow conductive heat loss. They countered that the entire depth of a magma chamber, not just that of a boundary layer, must be considered in calculating the Rayleigh number, and, therefore, in determining the vigor of convection. Both theoretical calculations and experiments, such as those of Kerr et al. (1990) and Worster et al. (1990) showed that in fluids without superheat, convection does occur during crystallization and differentiation. They concluded that in a typical magma chamber, hundreds of meters thick, the density perturbations brought about by differences in temperature, composition, and gas content would "make any experienced fluid dynamicist confident that strong convective motions are highly probable, and definitely *not* out of the question" (Huppert and Turner, 1991, p. 854).

Unconvinced, Marsh (1991) reiterated that convective motion in most magma chambers is weak, at least certainly not turbulent. He charged that Huppert and Turner did not understand the implications of a transient, start-up convective phase, and that they had consistently confused the hard facts of real magma chambers with the results of their laboratory experiments. He again maintained that very efficient heat transfer would rapidly erase significant temperature differences throughout a magma body in an initial phase of convection. The result would be that temperature differences during convection would be reduced to a very small but unknown value for all magma bodies.

Marsh spelled out what he perceived to be the differences between the two approaches. The Cambridge model, Marsh alleged, ignored the realities of magma chamber processes. Their model began with a crystal-free liquid and continued with crystallization occurring simultaneously throughout the system and with all parts of the chamber in close thermal communication. The Cambridge view envisioned crystal nucleation, growth, and transport by convection everywhere throughout the chamber. The Cambridge group, Marsh pointed out, also persistently developed its model on the basis of experiments in which large crystals formed during binary eutectic crystallization. The fact that magmatic crystallization entails crystallization of complex solid solutions and produces small crystals, however, meant that the Cambridge approach was bound to produce misleading results. As one example, Marsh pointed to a study in which Turner et al. (1986) had calculated cooling rates for komatiite flows and found that a komatiite 100 meters thick would cool by 250°C in one month. Marsh noted, however, that the basaltic lava lake at Kilauea Iki in Hawaii had cooled by only 20°C over a ten-year period. He implied that something was not quite right about the Cambridge model.

Marsh said that the contrasting Johns Hopkins model distinguished the fluid dynamical effects of several stages of magmatic evolution and evaluated them separately. Vigorous but short-lived convection was limited to the time of transport before emplacement. The following episode was marked by near-conductive cooling in which crystallization occurred in marginal solidification fronts that advanced toward the interior of the chamber. The chemical evolution of the magma body was controlled by the various chemical and physical processes occurring within the advancing fronts.

Left with a bad taste by the dispute, by perceived misinterpretations of their experiments and calculations, and by the willingness of geological journals to publish what they regarded as incorrect fluid mechanical reasoning, several members of the Cambridge–ANU redirected their interests to what they considered to be more productive research avenues. Both Huppert and Turner gradually became less engaged in issues pertinent to igneous petrology. Huppert continued a vigorous research program relating primarily to particle-driven gravity currents, and Turner returned to his oceanographic studies, retiring in 1995. Sparks, who left Cambridge in 1989 to become Head of the Department of Earth Sciences at the University of Bristol, continued his work in volcanology.

The dispute over convection made it clear that the petrological community needed to exercise very great caution in the application of fluid dynamics to magma chamber problems. Theoretical and experimental studies in petrological fluid dynamics did continue, of course. Some examples follow. Working in conjunction with Huppert and Sparks, M. Grae Worster, a 1983 Cambridge Ph.D. recipient under Huppert, experimentally investigated convection and crystallization in magma chambers that are cooled from above. Worster et al. (1990) concluded that thermal convection coupled with the nonequilibrium kinetics of crystallization could result in preferential crystal accumulation on the chamber floor. In a later paper, Worster et al. (1993) studied the cooling of model "lava lakes" composed of diopside and anorthite. They showed that kinetic undercooling adjacent to a solidification front would lead to thermal convection with large Rayleigh numbers in cooling lava lakes. They also concluded that, in addition to crust formed by cooling at the upper surface of the lake, convection would permit crystallization in the interior of the lake and accompanying crystal sedimentation. Even in a basaltic system with a much higher viscosity than the diopside–anorthite system, they found that convection would still be strong. Claude Jaupart continued his work, much of it with

Stephen Tait and Anne Davaille, who received her Ph.D. degree in 1992 from the Institut de Physique du Globe, where she is now a staff member. Davaille and Jaupart (1993a, b) experimentally studied thermal convection with large viscosity variations as well as convection in lava lakes. Tait and Jaupart (1992) examined the role of compositional convection. Jaupart and Tait (1995) reviewed differentiation in magma chambers from a fluid dynamical point of view. Spera et al. (1995) simulated convection with crystallization in the initially superheated eutectic system $KAlSi_2O_6$–$CaMgSi_2O_6$ (leucite–diopside). Trial and Spera (1990) also studied mechanisms for producing compositional heterogeneity in magma chambers. Marsh has continued his study of convection and crystallization in liquids cooled from above in conjunction with Mattias Hort of the University of Kiel (Hort et al., 1999), and Hort (1997) has explored crystallization in igneous sheets.

FLUID DYNAMICS AND COMPOSITE INTRUSIONS

One of the more interesting applications of fluid dynamics to magmatic problems in recent years related to occurrences of closely associated basic and acid rocks within the same rock body. Such associations puzzled petrologists for a long time. Judd (1893), for example, after noting that Andrew C. Lawson and J.H.L. Vogt had reported similar occurrences, described dikes on the Island of Arran that contain two types of igneous rock of totally different chemical composition. Judd believed that two magmas had been injected into the dikes from the same subterranean reservoir. Such associations were envisioned as evidence of liquid immiscibility by many early workers, but when liquid immiscibility fell into disfavor as a result of the experiments of Greig (1927), the question persisted as to why such contrasted magmas had not mixed and hybridized in the manner that Bunsen (1851) had so long ago envisioned.

One piece of the puzzle was put in place when Wager and Bailey (1953) recognized the chilling of basaltic magma that had been injected into silicic magma. Walker and Skelhorn (1966) claimed that some mixing had occurred. Noting that the basaltic margins of typical composite intrusions invariably contain xenocrysts of minerals that form phenocrysts in the silicic portions, they argued that the xenocrysts had been in the basaltic magma before its emplacement and were derived from a silicic magma. The basaltic margins were said to represent a hybrid magma that formed as a result of thorough mixing of basaltic magma and a porphyritic silicic magma.

Hatten Yoder, Jr., conducted experimental studies on coexisting magmas at the Geophysical Laboratory. From a review of many field examples of coexisting igneous rocks of contrasting composition from both volcanic and plutonic settings, Yoder (1973) maintained that two magmas of greatly contrasting composition, such as basalt–rhyolite, andesite–dacite, alkali basalt–trachyte, or hawaiite–mugearite, were erupted from the same vent or intruded into the same dike at the same time. In view of the field evidence that such magmas had retained their identity with a minimal amount of mixing or chemical diffusion, Yoder posed the question as to how two magmas of such sharp compositional contrast could be generated and maintained at the same place at the same time. Previous solutions to the problem had been advanced. Two magmas might have come from two chambers containing very different liquids as Bunsen had proposed. Basaltic liquid might have invaded solid granite. Or basalt and granite might have solidified from two immiscible liquids. For Yoder, the solution to the problem lay in the probable generation of the two contrasting magmas at, or very near, their own liquidus temperatures. For example, at the temperature at which a water-saturated rhyolitic magma would be just above its liquidus at pressures below 1.5 kilobars, basalt would be completely crystalline. On the other hand, if water-saturated basaltic magma were just above its liquidus in the same pressure range, then coexisting rhyolitic magma would have to be superheated approximately 250°C at atmospheric pressure and 450°C at 1 kilobar water pressure! Because field evidence indicates that both natural magmas contain phenocrysts, the possibility of such excessive superheating was ruled out. Moreover, Yoder noted, as Wager and Bailey first pointed out, natural basaltic magmas have commonly been chilled against rhyolitic magma. The reverse relationship had not been observed in localities of mixed magmas.

To explore the problem, Yoder performed a series of experiments in which he placed cylinders of glass, rock, or powdered rock of sharply contrasted compositions end to end at elevated temperatures. In one set of experiments, he heated cylinders for two hours at a water pressure of 1 kilobar and a temperature of 950°C, above the rhyolite liquidus but below the basalt solidus. In another set of experiments, he heated the cylinders for one hour at 1200°C. No evidence of immiscibility was found. The various cylinders placed end to end developed sharp interfaces with cuspate boundaries. Because the samples showed no obvious homogenization after a couple of hours, Yoder concluded that even water-saturated liquids of greatly contrasting composition could be maintained in contact at least for a limited time. The experiments did show, however, that exchange between the two compositions had already begun. Yoder made electron microprobe analyses across the interface between rhyolite and basalt cylinders from Craignure, Mull, that were heated to 1200°C. Although the contact between rhyolite and basalt appeared to be rather sharp on visual inspection, microprobe examination indicated that the boundary had already started to experience the effects of diffusion. Yoder found a slight rise in TiO_2, MgO, CaO, and FeO from the rhyolite to the basalt and a rise in SiO_2 and K_2O from the basalt to the rhyolite. Yoder's experiments supported results from a field study by Wiebe (1973) that considerable exchange between contrasted magmas occurs rather rapidly but that homogenization is prevented by quick chilling of the magmas.

Robert A. Wiebe (b. 1939) has devoted much of his career to a study of numerous examples of closely associated mafic and felsic igneous rocks. Wiebe, a 1961 graduate of Stanford University, received a master's degree from the University of Washington in 1963 and his Ph.D. degree from Stanford in 1966. Since 1966, he has been Professor of Geology at Franklin and Marshall College in Pennsylvania. Among the earliest examples he examined, Wiebe (1973) discussed the relationships between rocks derived from contrasted magmas in a composite dike along the coast of Cape Breton Island in Nova Scotia. Wiebe calculated that the entire dike solidified in about 28 hours to a temperature of 700°C, an insufficient amount of time for the dense basalt pillows in the composite dike to settle more than about 25 centimeters through the granitic component. Detailed chemical investigation of the typical pillow contacts indicated that there had been considerable chemical exchange between the two magmas but little or no homogenization. Near the contact with the adamellite portion, the basaltic portion is enriched in K and Rb, slightly enriched in Si, Fe, Ti, and Al, and depleted in Na, Ca, Mg, and Sr. Near the contact with the basaltic portion, the adamellite portion is enriched in Ca, Sr, and Na and depleted in K and Rb. Because the pillows chilled in about one day, Wiebe maintained, on the basis of field evidence, that significant exchange of alkalis and alkaline earths is possible between basaltic and granitic magmas within times ranging from one hour to one day. In more recent years, Wiebe (e.g., 1993, 1996) investigated occurrences of commingled basaltic and siliceous magmas in plutons along the coast of Maine.

Eventually, experiments in fluid dynamics were designed in efforts to model the behavior of coexisting magmas of contrasted compositions and to account for field occurrences like those investigated by Wiebe, Sparks, and many other petrologists. Stephen Blake of Auckland University, New Zealand, and Ian Campbell experimentally modeled magma mixing during flow in volcanic conduits by allowing two miscible aqueous fluids of different viscosity to flow through a vertical pipe. The more viscous fluid, analogous to silicic magma, formed a concentric cylinder or ring around a central less viscous fluid, analogous to basaltic magma. Blake and Campbell (1986) found that instability between the two fluids occurred when the Reynolds number of the less viscous fluid had a value of around three, much less than is required for turbulence in the same pipe where all the fluid has the same viscosity. They concluded from their observations that mixing of magma should be possible in dikes that have tapped layers of differing viscosity from a stratified chamber.

Takehiro Koyaguchi of the Earthquake Research Institute at the University of Tokyo and Akira Takada of the Geological Survey of Japan did a similar experimental study on the formation of composite dikes. Koyaguchi and Takada (1994) developed a model for the simultaneous flow of liquid with a silicic core and mafic margins, the reverse of the situation evaluated by Blake and Campbell, along a longitudinal crack adjacent to the chamber wall on the basis of experiments in which both glycerin and dilute glycerin leaked simultaneously from the bottom and side into a crack in gelatin. The viscous glycerin flowed efficiently into the center of the "dike" because it was lubricated by the less viscous dilute glycerin along the margin. The experiments suggested that mafic magma from a lower layer in a magma chamber might leak from a side wall of the chamber and travel faster within a fracture than silicic magma on account of its lower viscosity. After mafic magma reaches a crack tip, Koyaguchi and Takada argued, the rate of dike propagation would be determined by the viscosity of the less viscous mafic magma. Viscous silicic magma would then flow efficiently into the center of the dike, being lubricated by the less viscous mafic magma margins.

Don Snyder and Stephen Tait of the Institut de Physique du Globe (Paris) were concerned with the shapes of inputs of less viscous fluid into more viscous fluid with the intent of modeling intrusion of basaltic magma into silicic magma. In all their experiments using solutions of hydroxyethylcellulose to vary viscosity and NaCl to vary density, Snyder and Tait (1995) found that flows of injected solution quickly developed a viscous gravity current forming a thin wedge-shaped sheet on the floor of the tank. As in the experiments of Turner, Huppert, and Sparks, little or no mixing took place between the input and host fluids. Snyder and Tait observed that the farthest extent of the flow was elevated above the floor of the tank to form a "nose" that trapped a wedge of buoyantly unstable ambient fluid beneath the injected fluid. Only when the input stopped did the nose disappear and the layer become flat. They also noted the development of fingers in the input flow separated by very thin ribbons of the underlying, less dense host liquid that was evidently attempting to rise through the overlying dense input liquid.

Snyder and Tait observed that many intrusions contain similar structures where basaltic liquid had flowed onto the floor of a silicic magma chamber. The basalt in such cases is commonly preserved as fine-grained layers with chilled bases and either chilled tops or else tops that grade continuously upward through a zone of hybrid rocks into a zone of silicic cumulates. Many such flow bottoms are lobate and have been penetrated by thin bodies of silicic material that had frozen into place as it began to rise up through the basalt such as the abundant silicic pipes that have partially or completely penetrated basaltic layers in some of the plutons of coastal Maine described by Wiebe.

The striking similarities between the fingers of dense fluid in their experiments and the field observations of lobes of mafic layers in silicic intrusions led Snyder and Tait to believe that both situations were caused by flow-front instability. They suggested that the veins, ridges, or pipes of silicic rock occurring between fingers observed in the field are analogs of the ribbons of ambient less dense liquid between the fingers produced in the laboratory. They also proposed that many of the mafic enclaves observed in granitic intrusions and in silicic volcanics might have formed when disrupted fingers were sheared off and dispersed throughout the ambient liquid by convection.

Eventually, Wiebe teamed up with the Paris group. Snyder et al. (1997) reported that they knew of no composite dike in which both mafic and silicic rock were in contact with country rock. Dike margins, they noted, are either silicic or mafic, but the features associated with these two classes of composite dikes differ significantly. Composite dikes with mafic margins are common in the Tertiary volcanic regions of Scotland and Iceland and are typically less than 60 meters thick. The mafic margins are symmetrical and thinner than the silicic core. The contacts between the mafic and silicic portions are generally straight, parallel to the dike walls, and relatively sharp or gradational over only a few centimeters. The two lithologies, however, may commonly be intermingled as interpenetrating streaks. Snyder, Wiebe, and their colleagues argued that both lithologies were liquid at same time, presenting as evidence the presence of xenocrysts of quartz and alkali feldspar in the mafic rock and, in some cases, of xenocrysts of augite and plagioclase in the silicic rock. Enclaves of mafic rock are also found in the silicic rock near the contact between the two. In addition to the physical exchange of xenocrysts and xenoliths, substantial chemical exchange had taken place between the two lithologies, so that in some cases the margin rock is intermediate rather than basaltic and becomes more silicic toward its contact with the silicic core of the dike.

Snyder and his team also reported that composite dikes with silicic margins are common in eastern North America as at Mount Desert Island in Maine. In such dikes, the two components are typically not present as distinct parallel sheets. Instead, the central mafic rock is present as discrete pillows that commonly display chilled, crenulate margins with convex-outward lobes that are separated by sharp, inward-pointing cusps. Such pillows are elongated along the strike of the dike, are normally about five to twenty centimeters thick, and have pillow lengths that are roughly proportional to the thickness of the dike. Larger dikes display parallel rows of pillows. The silicic material is variably thick, in many cases only a few millimeters, and the remainder of the silicic material is located between the pillows. The authors believed that the field relationships in the composite dikes of Maine are inconsistent with the assumption that both the mafic and silicic components flowed in the same direction from a single compositionally stratified magma source. They suggested that the two magmas came from distinct sources and may have been flowing in different directions. For the majority of the composite dikes they had seen in Maine, they believed that the magma that formed the margins of a dike encountered a reservoir of the other magma, which, when disrupted, flowed into the center of the dike. In other words, some of the dikes were silicic feeders to composite reservoirs and some of the low-viscosity, high-density basaltic magma in the reservoir flowed down into the viscous, less-dense silicic magma in the dike, where it chilled and developed pillows in the center of the dike. These conclusions were confirmed by further experiments in which liquids of different viscosity and density were intruded into one another at different rates. Snyder and his cohorts always observed that when a liquid was intruded into another of higher viscosity, the flow front became unstable and formed fingers aligned in the direction of flow.

Although petrologists have continued to debate the merits of specific experimental designs and the suitability of specific experiments for appropriately modeling magmatic phenomena and processes such as convection, layer formation, replenishment, or magma commingling and mixing, it is clear that, after the mid-1970s, igneous petrologists discovered that the independent, already highly developed science of fluid dynamics promised great potential for yielding major insights into the behavior of magmas and the fundamental processes underlying differentiation. Fluid dynamics as applied to igneous petrology is still in its early phases. Perhaps, in terms of its stage of development, contemporary fluid dynamics as applied to petrology corresponds to some of the experimental work done on mineral stability prior to the establishment of the phase equilibrium program at the Geophysical Laboratory. Time will tell if another petrologist like Bowen will emerge to bring order out of the results of a growing body of experiments.

29 ✦ PARADIGM LOST, PARADIGM REGAINED?

Cumulate Theory under Fire

Given the importance of large layered intrusions like the Skaergaard, Bushveld, and Stillwater bodies as examples of the products of fractional crystallization and other processes occurring within magma chambers, geochemists in the second half of the twentieth century, going well beyond the pioneering trace-element studies of Wager and Mitchell (1951), began to analyze the distribution of trace elements, stable isotopes, and radiogenic isotopes in such intrusions in detail. During the time in which geochemical approaches requiring complex instrumentation and mathematical analysis became commonplace, further advances in understanding layered intrusions were also made through field, mineralogical, and textural studies.

Insights gleaned from fluid dynamics also contributed to ideas about large layered mafic–ultramafic igneous intrusions. Although early workers on these bodies speculated about crystal settling and convection and applied Stokes' law, more recent students began to apply quantitative experimental and theoretical fluid dynamics more rigorously to the mechanisms of differentiation and the origin of layering. References to dimensionless numbers and to concepts such as double-diffusive convection became frequent occurrences in literature dealing with layered igneous intrusions.

ESTABLISHMENT OF THE CUMULATE PARADIGM

Thanks to the initial work of Wager on the Skaergaard Intrusion, a simple model of crystallization based on crystal settling and convection was proposed for large differentiated intrusions. That model, eventually referred to as the "cumulate paradigm" (McBirney and Hunter, 1995), was further developed throughout the 1960s by Hess, Wager, and others. By the 1970s, the cumulate paradigm was being questioned by numerous investigators, and more complex models of magmatic processes were invoked for large layered intrusions. Such changes in outlook are reviewed in this chapter.

57. Harry Hammond Hess (1906–1969). Reproduced by permission of Archives, Department of Geosciences, Princeton University.

As noted earlier, the Stillwater Igneous Complex of Montana was investigated in preliminary fashion throughout the 1930s by several workers, including H. H. Hess (Figure 57), one of the most versatile petrologists of his era. Harry Hammond Hess (1906–1969), whom we met briefly in Chapter 17, graduated from Yale University in 1927 after studying geology under Adolph Knopf (James, 1971). After spending two years as a geologist in Rhodesia (now Zimbabwe), Hess began graduate study and received his Ph.D. degree from Princeton University in 1932 with a dissertation on an altered peridotite body in Virginia. After a year at Rutgers University, Hess joined the Princeton faculty in 1934, where he carved out a remarkable career until his death from a heart attack in 1969. Hess's wide-ranging geological interests included pyroxenes, ultramafic rocks, the tectonics of the Caribbean region, the Stillwater Igneous Complex, island arcs, and the topography and geophysics of ocean basins. During World War II, Hess was commander of the attack transport ship U.S.S. *Cape Johnson*, and participated in several major landings. During much of the war, Hess continuously operated his echo sounder despite the extreme risk, thereby greatly extending knowledge of Pacific Ocean bathymetry and discovering the flat-topped submarine seamounts that he termed "guyots" in honor of Arnold Guyot, the Swiss-born glaciologist who introduced the study of geology to Princeton. Hess's crowning achievement may have been his seminal paper on the history of ocean basins in which he introduced his concept of seafloor spreading (Hess, 1962).

In his early work on the Stillwater Complex, Hess (1939) postulated that crystals on the surface of a pile that had accumulated on the floor of the magma chamber had grown by the diffusion of appropriate substances from the overlying liquid. In conjunction with such diffusion-fed growth, Hess also envisioned mechanical expulsion of interstitial liquid from between the crystals. The work of Lombaard on the Bushveld Intrusion stimulated Hess to do similar mineralogical studies across sections of the Stillwater, but his work was interrupted by World War II. After the war, Hess took a sabbatical leave during 1949 and 1950 at the University of Cape Town in South Africa, where he had an opportunity to spend several weeks in the company of South African geologists examining the Bushveld Complex and comparing it with the Stillwater Complex. While there, he rewrote his manuscript on the Stillwater Complex, but it was not published until 1960.

In addition to reporting on his detailed mineralogical studies, Hess (1960) also discussed the origin of the Stillwater layering. Hess reported the presence of rhythmic layering consisting of repeated alternations of light plagioclase-rich bands and dark pyroxene-rich bands. He argued that rhythmic layering would result when motions in the magma were not constant in velocity and direction so long as the crystalline phases involved had considerably different settling velocities. The motions were attributed to irregular convection currents. Hess suggested that the very large cycles in mineral variation, represented by thicknesses of crystal accumulations on the order of one thousand feet, might indicate one complete convective overturn of the magma. In contrast, he suggested that the hundreds of smaller cycles in such a great thickness might be the result of minor currents in the magma immediately above the floor. He thought that most of the pronounced rhythmic layering in the Stillwater Complex was probably related to such minor currents. Hess noted that intervals of layered sequences were separated from one another by thicker layers of uniform norite. He interpreted these alternations as indications that currents in the magma immediately above the floor were present for short intervals of time and then became quiescent.

Hess argued that upward currents would exaggerate the settling velocity differences between pyroxene and plagioclase and showed how layering might be produced by variations in current velocity. He reviewed Stokes' law and its effect on settling velocity for spherical crystals of various minerals in terms of their densities and diameters, and he related the size of crystals to the presence or absence of layering. He noted the common association of rhythmic layering with norites or troctolites in which either orthopyroxene or olivine is the only ferromagnesian phase. To explain this association, Hess suggested that if only one ferromagnesian phase was crystallizing it would form relatively large crystals, whereas if two ferromagnesian phases were forming, then about twice the number of nuclei of ferromagnesian crystals would develop to compete for the common ions such as Fe^{2+} and Mg^{2+}. As a result there would be twice as many crystals, each having half the volume. The large single crystals would be the units most likely to settle, although he also accepted the idea that "glomeroporphyritic" aggregates of several crystals might settle as units. For example, he speculated that the small crystals that form chromite layers might have settled as chain-like clusters.

Hess reasoned that containers like the Skaergaard Intrusion, whose horizontal dimensions are slightly larger than their vertical dimensions, favored the development of regular convection cells. In contrast, he thought that lopoliths like the Stillwater and Bushveld are flat lenses with horizontal dimensions on the order of twenty to fifty times their vertical dimensions. He maintained that a mass of liquid of this shape might develop a series of convection cells that were side by side. Because he detected no evidence for a regular convection cell system in the Stillwater or Bushveld Complexes like that of the Skaergaard Intrusion, he postulated magma motions entailing random downward plummeting of denser liquid from near the roof of the magma chambers.

Wager also continued to develop his ideas about processes in differentiating magma chambers. During his tenure at Oxford, Wager accumulated a stellar group of graduate students, like Malcolm Brown and W. J. Wadsworth, set them to work on the Skaergaard, Rum, and other layered intrusions, and began to synthesize their observations. Wager was so struck by the importance of the textures of the rocks in layered intrusions that he co-authored a paper with Brown and Wadsworth on various types of igneous textures. Wager et al. (1960) proposed the term "cumulate," from Latin "cumulus" meaning "a heap," to describe rocks formed by crystal accumulation. In his Skaergaard memoir, Wager had spoken of "primary precipitates" and "interprecipitate liquids" but now advocated replacing them by the terms "cumulus crystal" and "intercumulus liquid," respectively. Cumulus crystals were considered to be discrete crystals that had separated from a magma as a result of their greater density. Wager and his co-authors proposed a distinction between "orthocumulates" and "adcumulates" on the basis of behavior of the intercumulus liquid. Orthocumulates, they said, formed as the intercumulus liquid crystallized slowly through a falling temperature interval to produce new mineral phases between the cumulus phases. They allowed for some possible modification of the earlier cumulus crystals with possible zoning. Adcumulates were said to result from "adcumulus" growth, a process in which the original cumulus crystals were extended by material of the same chemical composition to produce unzoned crystals. Following the proposal of Hess (1939), they said that the adcumulus process gradually reduced the amount of intercumulus liquid by mechanically expelling it, sometimes almost completely. Any part of the intercumulus liquid eventually trapped by further accumulation of crystals would be designated as "trapped liquid." Wager, Brown, and Wadsworth envisioned the adcumulate process as occurring essentially at the same temperature by the diffusion of substance between the intercumulus liquid and the overlying column of magma. They maintained that adcumulates commonly tended toward being monomineralic rocks such as anorthosite if there was only one cumulus mineral.

The authors suggested that "orthocumulates" and "adcumulates" represented two polar extremes. Proposing that rocks produced by an intermediate process be termed "mesocumulates," they advocated use of the term "orthocumulate" for rocks in which adcumulus growth is inconspicious, "mesocumulate" for rocks showing small amounts of pore material, and "adcumulate" for rocks in which the pore material is inconspicuous or absent. Wager, Brown, and Wadsworth reasoned that orthocumulates would be favored by fast bottom accumulation of crystals whereas adcumulus growth would be favored by relatively slow bottom accumulation.

In addition, they defined "heteradcumulates" as rocks in which cumulus crystals are typically enclosed "poikilitically" by one mineral in one place and by a different mineral in a neighboring region. The enclosing unzoned poikilitic crystals were said to have the same chemical composition as the cumulus crystals in the layers above or below. It was suggested that these poikilitic crystals had grown between the cumulus crystals while they were at or near the top of the accumulating pile as a result of the diffusion of material to the relatively rare crystal nuclei that had developed in the intercumulus liquid. They judged that material had diffused from the main bulk of the overlying magma rather than from the trapped liquid because of the unzoned nature of the crystals. They believed that there was a strong likelihood that the textures of cumulate rocks would reflect their origins if they contained a very small number of cumulus minerals.

A short time later, Wager (1963) further developed his ideas about convection in layered intrusions. Rather than attributing rhythmic layering to variations in the velocity of a current, Wager now believed that two types of thermal convection currents existed simultaneously. On the one hand, he envisioned a gentle, fairly continuous convection current with a velocity of about 2 meters per day that deposited the uniform rocks. He also envisioned an intermittent dense current of relatively small volume that periodically descended the walls of the chamber at a velocity greater than 3 kilometers per day and came to rest on the magma-chamber floor and deposited a gravity-stratified layer. He thought these intermittent currents were similar to those postulated by Hess for the Stillwater layered series.

In the slow-moving, continuous convection cells, Wager believed that supersaturated liquid would pass slowly over the magma-chamber floor with a few suspended crystals. It would occasionally deposit some of the crystals and produce adcumulus growth wherein the latent heat of crystallization would pass into the supercooled part of the liquid. Wager noted that the rocks in the observed layered series generally change from adcumulates or mesocumulates downward into mesocumulates and orthocumulates. He speculated that, with increasing depth, the convecting magma might have become increasingly supersaturated, thus causing so much nucleation that no part of the magma would be very far from seed crystals. As a result, less time would be required for the diffusion of heat and substance to produce equilibrium conditions, and little or no adcumulus growth would occur. Yet Wager had to concede that this hypothesis was not applicable to the Bushveld or the Stillwater Intrusions, both of which included many adcumulates apparently formed from bodies of liquid that were several kilometers thick.

Very shortly before his untimely death in 1965, Wager completed two major manuscripts. One was a substantial paper on "rhythmic and cryptic layering in mafic and ultramafic plutons" that Wager (1968) had written for *Basalts: the Poldervaart Treatise on Rocks of Basaltic Composition* (Hess and Poldervaart, 1968). The other was the magnificent volume, *Layered Igneous Rocks*, co-authored with his student and colleague G. Malcolm Brown (Wager and Brown, 1967).

Wager (1968) stated that his estimate of an upward current velocity of more than 1 meter per day was based on the settling velocity of plagioclase crystals in the upper border group having a volume of 0.1 cubic centimeters. He estimated from Stokes' law that the sinking velocity of such crystals in stagnant magma would be around 1 meter per day. Wager assumed that these plagioclase crystals would float to the top of the magma chamber in an upwelling current having a velocity greater than that. The velocity of the intermittent currents was estimated from their ability to carry large blocks of felsic gneiss, occurring as inclusions in the marginal border group, downward from the metamorphic complex surrounding the upper part of the intrusion. Wager maintained that periodic high-velocity currents of 3 kilometers per day would explain why there is so much cross bedding near the outer part of the layered series. He recognized that as the significance of the patterns of layering in the Skaergaard Intrusion became better understood, it would be necessary to modify the simple concept of two different types of convection currents in the magma. He envisioned that additional currents having intermediate velocities and persistence would someday be postulated.

On the basis of a rate of heat loss assumed to be double that in the Stillwater Complex, Wager estimated that the time needed to deposit the Skaergaard layered series was around 12,500 years, or about 20 centimeters per year. Hess had approximated an accumulation rate of about 10 centimeters per year for the Stillwater Complex.

Wager and Brown's masterful synthesis, *Layered Igneous Rocks*, submitted to the publisher just one month before Wager died, represented the culmination of the era of classical studies on layered igneous intrusions with its definitive adoption of the concepts of crystal settling and convection. The text began with Wager's final comprehensive assessment of the Skaergaard Intrusion in which he continued to advance his hypothesis of crystal settling coupled with convection, and he maintained that plagioclase probably remained just a little bit more dense than the liquid on the grounds that plagioclase crystals are among the other cumulus minerals throughout the whole thickness of the layered series. In effect, in Wager's mind, a cumulus mineral was virtually defined as a "settled" mineral. Wager here estimated the time of solidification for the entire Skaergaard Intrusion as about 40,000 years. The book also incorporated numerous lengthy reviews of the geology of other well-known layered intrusions.

Toward the close of Wager's career, several other layered intrusions, some recently discovered, began to attract considerable attention. Among these were the Kiglapait and Muskox Intrusions of Canada. Discovered in 1942 by E. P. Wheeler II to the northeast of the Nain anorthosite in Labrador, the Kiglapait Intrusion was originally thought to be a lobe of that anorthosite body. Field work during 1957 and 1958 by Stearns A. (Tony) Morse in conjunction with his Ph.D. dissertation for McGill University, however, showed that the Kiglapait Intrusion is distinct from the Nain anorthosite. Morse received his degree from McGill in 1962 and spent his career at Franklin and Marshall College and the University of Massachusetts. He supplemented his earlier work with systematic sampling in 1963 and 1964 and eventually produced a memoir on the intrusion (Morse, 1969). He described the Kiglapait Intrusion as a nest of bowls, somewhat resembling the Skaergaard Intrusion in that respect. The intrusion is spectacularly layered and differentiated with a troctolite Lower Zone grading upward through olivine gabbro to a ferrosyenite Upper Zone. The intrusion is capped in its center by an Upper Border Zone consisting of fine-grained olivine gabbro. Flanking the layered series are Inner and Outer Border Zones. Morse showed that the composition of the minerals varies systematically with stratigraphic height as is typical of layered intrusions. The anorthite content of plagioclase decreases upward from An_{67} to An_{10}, and the forsterite content of olivine decreases upward from Fo_{72} to Fo_{10} (Morse, 1979, 1996).

Morse (1969) pointed out that the Kiglapait Intrusion is noteworthy for a host of interesting structural features, including slump breccias, cross bedding, channel scours, drag folding, graded layers, anastomosing layering, rhythmic layering, igneous lamination, and centimeter-scale layering analogous to the inch-scale layering of the Stillwater Complex. He interpreted such features as lamination, cross bedding, channel scours, and down-dip mineral lineation as evidence for crystal settling under the influence of currents. He further believed that adcumulus growth on the floor of the chamber required the renewal of magma by flow. If not, he said, diffusion distances would become impossibly large for sustaining adcumulus growth. Without such renewal by current flow, he maintained that adcumulus growth would quickly lower the concentration of refractory components in the magma overlying the crystal pile, thus resulting in cessation of crystal growth until temperature fell. Because of such abundant evidence for bottom currents throughout the intrusion, Morse argued for the existence of full-depth convection cells throughout most of the history of the Kiglapait crystallization.

Morse suggested that the large extent of many Kiglapait layers and zones of layers indicated a sorting mechanism that operated on an intrusion-wide scale. He believed that sorting at the roof of the intrusion seemed more capable of producing the observed results than sorting along the walls or near the floor, particularly because of the presence of currents with an upward component near the roof. Appealing to Hess's reasoning that changes in the upward velocity of convection effectively initiated new cycles of sorting, Morse suggested that any period of stable current velocity would establish a steady-state difference in the settling rates of olivine and plagioclase, leading to the production of average rock on the floor. He agreed with Hess that increases in upward velocity would retard the sinking of plagioclase behind that of olivine, thus initiating an olivine layer, and that decreases in the upward velocity would allow retarded plagioclase to advance, possibly to form an anorthosite layer. Noting an abundance of anorthosite pods in the layered rocks, Morse suggested that the retardation of plagioclase settling to the point of rafting was a frequent occurrence, if not the normal state of affairs. By way of contrast, Morse reasoned that centimeter-scale layering, because of its extreme delicacy, provided evidence against the existence of appreciable currents during their formation. He attributed their formation to settling very near to the floor. Moreover, he said, the presence of very thin layers with sharp contacts argued against settling of the crystals all the way from the roof.

The Muskox Intrusion was discovered by the Canadian Nickel Company in 1956 in the Coppermine River area astride the Arctic Circle. It was mapped by Charles H. Smith and H. E. Kapp for the Geological Survey of Canada in 1959 and 1960. Smith and Kapp (1963) issued a brief summary of the Muskox intrusion, and that report was followed by detailed studies of Irvine (1970b, 1980b). Smith and Kapp mapped the Precambrian Muskox Intrusion as a 74-mile long dike in the Canadian shield. The dike widens dramatically toward the north. In cross section the intrusion was described as having a funnel shape with a feeder dike. Smith and Kapp also likened the intrusion to a sailing ship with a deep keel plunging gently northward. The feeder, they said, contains bronzite gabbro and picrite units parallel to the nearly vertical walls. The intrusion contains marginal zones as much as 1200 feet thick. The Central Layered Series, approximately 8500 feet thick, was said to consist of 38 main alternating layers of dunite, peridotite, feldspathic peridotite, picrite, olivine clinopyroxenite, websterite, orthopyroxenite, troctolite, olivine gabbro, gabbro, norite, and anorthositic gabbro. The contacts between layers were reported to be sharply defined. An Upper Border Group containing granophyric material was also described. The preliminary studies of Smith and Kapp showed that the forsterite content of olivine in the Central Layered Series decreases upward and toward the walls, leading them to conclude that the ultramafic rocks of the Muskox Intrusion formed during a single cooling process rather than from multiple injections of unrelated magmas.

CHAPTER 29

THE CUMULATE PARADIGM QUESTIONED

During the final decade of Wager's life, a number of features were encountered in layered intrusions that some geologists found difficult to square with the concept of crystal settling and convection acting on a large, single pulse of cooling magma. During the 1950s, Malcolm Brown, W. J. Wadsworth, and others of Wager's students worked on the layered ultramafic complex on the Hebridean island of Rum, formerly spelled "Rhum," just southeast of Skye. Rum was privately owned by Lady Bullough, however, and, therefore, was not normally open to visitors. With the timely help of appropriately connected acquaintances, Wager prevailed on the owner to allow him and his research students to conduct field studies within the strict limits imposed by the management of the deer forest on Rum (Vincent, 1994). Time and again, access to critical rock exposures has proved to be one of the major challenges facing igneous petrologists.

On Rum, Brown (1956) found units about a hundred feet thick that consist of olivine-rich, peridotite layers in the lower part of the complex grading upward into plagioclase-rich, "allivalite" (troctolite) layers in the upper part. Brown believed that olivine crystallized as a primary precipitate during the first stage of development of each rhythmic unit and that plagioclase, often accompanied by augite, crystallized as a primary precipitate during the later stages. To account for such a sequence, Brown suggested that successive pulses of basaltic magma of a critical composition had been injected into the magma chamber. During each pulse, olivine and chromite allegedly settled out, to be followed by plagioclase, and sometimes by augite, after a small interval of cooling. Such an hypothesis, Brown believed, was consistent with the lack of the cryptic layering so typical of other intrusions like the Skaergaard, Bushveld, and Stillwater in which minerals like olivine and plagioclase progressively vary in chemical composition from the bottom to the top of the intrusion.

The Rum Complex also displays features that are difficult to account for on the crystal-settling hypothesis. One such feature is the spectacular development in some regions of prominent olivine growth perpendicular to the layering. This perpendicular olivine growth was originally termed "harrisitic" by Harker (1908) who observed the phenomenon near Harris Bay. Wager et al. (1960) eventually defined the type of rock as a "crescumulate." At Rum, Wager and Brown (1951) attributed the unusual type of crystal development to the upward growth of olivine from the floor of the chamber rather than to settling.

In time, Wager (1959) acknowledged that he had been pondering an alternative hypothesis for some of the rhythmic variation in the Rum Complex. In contrast to the idea that the rhythmic layering was due to differences in crystal density, he proposed that the order of the primary cumulus minerals within the rhythmic units of Rum represented their respective ease of nucleation from a supercooled magma, an order that was dependent on the complexity of their crystal structures. The idea of ease of nucleation goes back at least as far as Harker (1909). Wager also suggested that chromite in the Bushveld Intrusion was the first mineral to nucleate within a macro-unit, followed by bronzite and, lastly, by plagioclase. That sequence of nucleation, he said, is the succession found within the Bushveld macro-units. Wager suspected that minerals with relatively simple structures like chromite and magnetite, minerals with the spinel structure, and olivine, a nesosilicate mineral, should nucleate more easily than complex silicates like the feldspars. For intrusions in which convection currents were not sufficiently vigorous to produce crystal sorting by a winnowing of the sinking crystals, as was the case in the Skaergaard Intrusion, Wager suggested that differences in ease of nucleation might account for the variations in the proportions of minerals in the layers.

Another questioning of crystal-settling orthodoxy appeared in a publication on the textures and mineral associations in the ultramafic zone of the Stillwater Complex by E. D. Jackson. After service in the Marine Corps during World War II, Everett Dale Jackson (1925–1978) graduated from the University of California at Los Angeles in 1950 (Raleigh, 1980). His graduate study at UCLA was interrupted when he was offered a job mapping the Stillwater Complex between 1951 and 1955. Upon returning to UCLA, Jackson used his Stillwater studies as the basis for a Ph.D. dissertation and received his degree in 1961. Soon thereafter, Jackson worked with the National Aeronautics and Space Administration throughout the 1960s in training the Apollo astronauts in geology. He was also active in the investigation of returned lunar rocks and became interested in Hawaiian ultramafic xenoliths and the chronology of the Hawaiian-Emperor chain. His too brief career was cut short at the age of 53.

Jackson (1961) accepted the idea of crystal settling, but with caveats. He noted that the layered rocks composing the Ultramafic Zone of the Stillwater Complex are believed to have formed during crystallization of a single magma by the accumulation of crystal precipitates that fell to the floor of the magma chamber and were enlarged or cemented after deposition. Jackson believed, however, that sorting by currents failed to provide a satisfactory explanation for the rhythmically repeated compositional layers. Specifically, current transportation of crystallization products from the roof to the floor of the intrusion seemed to be incompatible with a lack of current structures and a lack of evidence for hydraulic equivalence in the rocks.

Jackson doubted that magma currents had been important during either the transportation or the deposition of the primary precipitates in the Ultramafic Zone for several reasons. He observed no trough bands. He found the lineation of elongated minerals to be weak or absent. He observed virtually no size-graded bedding. Moreover, he said, in nearly all rocks that contain two coexisting settled phases, the two minerals are not in hydraulic equivalence. In that regard, he pointed out that gravity stratification is properly developed in many olivine chromitite layers in that denser chromites are concentrated at the base of the layers and gradually give way to less dense olivine at the top. But because the size of crystals affects their settling velocities as the square but their density only to a power of one, it turned out that every one of the gravity-stratified layers is hydraulically upside down inasmuch as the olivines are considerably larger than the chromite crystals. Current transportation and deposition of the primary precipitates in the lower part of the Stillwater Complex, Jackson asserted, are incompatible with the textures of the rocks.

Jackson suggested that the most probable explanation was that the olivine–chromite units had formed as a result of changing proportions in the supply of settled olivine and chromite to the deposition surface. He believed that crystallization of the primary precipitates of the Ultramafic Zone had occurred at the bottom of the intrusion. If the settled minerals had formed at the top of the intrusion and then settled through five or six miles of magma, he said, then the rocks would have displayed features of extreme differential settling inasmuch as the time required for a crystal to descend from the roof to the floor would be large in view of the magma viscosity and probable solidification rates of the intrusion. Given the lack of evidence for the kind of sorting to be expected in individual layers and cycles under such circumstances, Jackson maintained that the most reasonable explanation for the observed textures and structures was that the primary precipitates crystallized near the bottom of the intrusion and accumulated by simple crystal settling on the floor as rapidly as they formed. He proposed that crystallization would occur at the base of the intrusion because of the rise in the melting point of minerals with increased pressure. Jackson noted that the difference between the bottom crystallization model proposed by him and by Brown for some aspects of the Rum Complex and the roof crystallization model of Wager and Hess was somewhat analogous to the difference between the formation of chemical sediments and that of detrital sediments.

Further significant challenges to the traditional cumulate model of layered intrusions emerged from studies of the Jimberlana Intrusion of western Australia. Interest in the Jimberlana Intrusion was triggered in 1965, and detailed field work conducted in 1966 became the basis for Ian Campbell's Ph.D. dissertation. Campbell showed that the intrusion had the external form of a dike about 180 kilometers in length and about 2.5 kilometers wide (Campbell et al., 1970). The intrusion was found to exist in doubly plunging synformal, canoe-shaped segments that somewhat resemble the Great Dyke in Zimbabwe (Rhodesia). The internal form of the "dike," however, was described as that of a lopolith containing a layered series of rocks with a vertical thickness of about 1200 meters. The layered series was divided into Upper and Lower Layered Series as well as a Marginal Layered Series. The Upper Layered Series is separated from the Lower Layered Series by a marked unconformity that has been attributed to the entry of a new pulse of magma into the chamber. Both the Upper and Lower Layered Series contain four or five cyclic ultramafic and mafic units. Campbell and his colleagues interpreted these layered rocks as the product of at least three, possibly four, pulses of injected magma.

Later, Campbell (1978) wrote that he had found several features that did not appear to be consistent with the concept of crystal settling. He noted that similar phenomena had also been found in other layered intrusions but complained that the cumulus theory was so widely accepted that petrologists tended to play down the importance of the anomalies. Campbell argued that the layering shape was a significant problem for the crystal-settling hypothesis. He noted that layering is horizontal in the center of the layered series to nearly vertical at the margins with no sharp angular discontinuity at the intersection with the steep marginal layers as is the case in the western part of the Skaergaard Intrusion. Moreover, he noted, the textures in the steeply dipping layers of the marginal zone are those of cumulates. Campbell further reported that the stratigraphic succession at the bottom of the Marginal Layered Series is the reverse of what is normally found in layered intrusions and has many features in common with the Lower Layered Series. Although uncertain of the origin of the Marginal Layered Series, Campbell was certain that the cumulus textures and rhythmic layering in the Marginal Layered Series had not formed by gravity settling.

Campbell also claimed that the distribution of adcumulate and orthocumulate textural types does not accord with standard cumulate theory. He observed that the texture of a layer becomes progressively more orthocumulate as it is traced from the center of the intrusion toward the steeply dipping marginal zone. At the margin, however, the layers are thinner than they are at the center, indicating that the rate of crystal accumulation had been less. Therefore, Campbell deduced that the slower the rate of accumulation had been, the more orthocumulate the textures became. Cumulus theory, however, predicted that slow rates of accumulation would lead to the development of adcumulate textures because more time would be available for the diffusion of substance from the intercumulus liquid.

Campbell observed that the amount of cumulus chromite is inversely proportional to the amount of cumulus bronzite within a layer. Moreover, both olivine and chromite were efficiently separated from bronzite, a phenomenon that he could not explain by recourse to hydrodynamic processes inasmuch as olivine and bronzite have similar sizes and should have settled at similar rates. From Stokes' law and centrifuge experiments, Campbell regarded it as established that layering produced by gravitational settling should be characterized by the concentration of faster settling crystals at the bottom of the layers as well as size sorting of specific minerals such as olivine between the bottom and the top of an individual layer. He suggested that a good example of these phenomena occurred in the Duke Island, Alaska, intrusion. On the other hand, he maintained, crescumulate layering, inch-scale layering, and the hydraulically inverted layering described by Jackson provided examples of in situ crystallization. Campbell agreed with Jackson that the settling distance involved in the production of the hydraulically inverted olivine–chromite layers of the Stillwater Complex had to be small or else the faster sinking olivines would have overtaken the slower sinking spinels. He could envision no mechanical process that could bring small spinel crystals to the bottom of the chamber before the larger olivines. Unlike Jackson, however, who did not question the concept of crystal settling, Campbell maintained that in situ rhythmic crystallization at the floor of the magma chamber from a supercooled melt was a better explanation for the rhythmic layering. Campbell also performed centrifuge experiments on plagioclase buoyancy in basaltic melt and discovered that plagioclase invariably floated rather than sank, a major blow to the conception of gravitationally settled plagioclase cumulus crystals. In the end, Campbell adopted Wager's idea that the order of appearance of minerals in rhythmically layered sequences was a consequence of the order of ease of nucleation of minerals from silicate melt.

The death of Wager in 1965 was followed by a brief hiatus in the investigation of the Skaergaard Intrusion, but studies of the famous body entered a new and exciting phase when A. R. McBirney turned his attention to the mother of all layered intrusions. Alexander R. McBirney (b. 1924) graduated from the United States Military Academy in 1946. After several years in the military, McBirney went to Nicaragua in the early 1950s to develop a coffee plantation (McBirney, personal communication). To support his family, he also worked with a mining company. Through this work, McBirney met Howel Williams, America's premier volcanologist of the mid-twentieth century, who was then studying active volcanoes in Nicaragua. As a result of spending time in the field with Williams, McBirney's interest in geology and volcanology developed. He sold his plantation, went to Berkeley for graduate study, and received his Ph.D. degree from the University of California at Berkeley in 1961 under Williams. In 1965, McBirney established the Center for Volcanology at the University of Oregon, where he spent the remainder of his professional life.

In 1971, McBirney and Hugh Taylor, Jr., an isotope geochemist from Caltech, together with colleagues and students from Oregon and Caltech, embarked on a joint expedition to East Greenland to see the famed intrusion. McBirney led six subsequent expeditions in 1974, 1976, 1979, 1985, 1988, and 1990. His first major paper on the Skaergaard Intrusion, appearing in an issue of *Journal of Petrology* that was devoted entirely to the Skaergaard and Kiglapait Intrusions, was written in conjunction with Richard M. Noyes, an expert in oscillatory chemical systems in the Department of Chemistry at the University of Oregon. The paper was little short of a blockbuster because he took issue with Wager's classical explanation of crystal settling under the influence of gravity as the dominant mechanism of crystal fractionation. While conceding that crystal settling was clearly demonstrable in many cases, McBirney suspected that the process had been overrated, charging that gravitational segregation had been invoked in many cases for want of a better explanation. "For more than half a century," McBirney and Noyes (1979, p. 489) asserted, "the rule of petrology has been 'when in doubt, settle it out.' "

Just what had stimulated the interest of a volcanologist in a layered pluton? McBirney had been struck by the failure, in some cases, of olivine phenocrysts in basaltic lava to sink as predicted by Stokes' law. He noted that abundant olivine phenocrysts in the basaltic Kilauea Iki lava lake of 1959 would have been expected to reach the floor of the lake within a couple of years but had failed to do so. In the Makaopuhi lava lake, olivine settling virtually ceased after a few pyroxene and plagioclase crystals had formed. Similarly, he wondered why dense mafic minerals like olivine are commonly retained in the well-layered rocks under the roofs of layered intrusions rather than sinking. He concluded that the amount of olivine settling was far less than predicted by Stokes' law.

McBirney encountered another difficulty with plagioclase. Wager had assumed that plagioclase, because it formed cumulus crystals in the Skaergaard, must have been slightly heavier than the liquid from which it crystallized. However, Yan Bottinga and Daniel Weill, McBirney's colleagues in the Center for Volcanology, were determining the densities of silicate liquids, including those having the chemical compositions calculated by Wager, and they determined that plagioclase would have floated in Skaergaard magma. Bottinga and Weill (1970) observed that in the Skaergaard or Fenner-type fractionation trend, the liquid became more iron-rich, hence denser, during crystallization, whereas the plagioclase that is being precipitated became more sodium-rich, hence less dense. Although Wager had used a density of 2.58 grams per cubic centimeter for Skaergaard liquid, Bottinga and Weill (1970) considered the density to be much greater. They concluded that gravitative settling of plagioclase was not likely to have been a strong factor in the Skaergaard type of differentiation unless higher H_2O contents were postulated. They thought that their calculated liquid densities were compatible, however, with Jackson's views on the cause of layering in the ultramafic zone of the Stillwater complex.

To confirm the calculations of Bottinga and Weill, Murase and McBirney (1973) directly measured the densities of liquids of the same compositions and compared them with the densities of coexisting minerals over a range of temperatures. They found that plagioclase is less dense than the basaltic liquids throughout most of the range of differentiation. They concluded that the difference in density between plagioclase and melt is sufficiently great that gravitational settling of plagioclase would be a physical impossibility. They also concluded that sinking of "glomeroporphyritic" clusters would not be a satisfactory explanation either because of the regular orientation of the plagioclase crystals in laminated rocks. If plagioclase crystals had been dragged down by mats of denser crystals, they reasoned, then plagioclase should have been concentrated in the lower part of a graded layer along with the mafic minerals rather than in the upper part as it actually is. Centrifuge experiments conducted by Ian Campbell, Peter Roeder, and J. M. Dixon also demonstrated that plagioclase has lower density than iron-rich liquids postulated for the Skaergaard Intrusion and, therefore, should have floated (Campbell et al., 1978).

McBirney and Noyes called attention to the experiments of Herbert Shaw and his colleagues at the United States Geological Survey in which they measured the viscosities of basaltic lava from the Makaopuhi lava lake, formed during the 1965 eruption of Kilauea. Shaw et al. (1968) had demonstrated that magmas exhibit non-Newtonian behavior, that is, they possess a finite yield strength that must be overcome before permanent strain is produced by an applied stress. McBirney noted that his own experiments had supported this notion because he found that a linear relationship between an applied stress and the rate of strain occurred only beyond a certain value of applied stress, namely, the yield strength of the melt. As a result, he believed that crystals of various minerals would have to achieve a minimum diameter in a magma possessing specified yield strength before any movement by viscous flow through the liquid, such as settling, could even begin. For example, as calculated from Stokes' law, an olivine crystal in a basalt magma that possessed a yield strength of 100 dynes per square centimeter would have to have a radius of at least five millimeters to overcome the yield strength of the magma and begin sinking. But Shaw's experiments and those of Murase and McBirney showed that silicate magma yield strengths are considerably higher than predicted so that crystals would need to be even larger to succeed in sinking. Moreover, McBirney observed, the yield strength of a magma will increase during cooling and with increase in the amount of suspended crystals.

McBirney and Noyes reasoned that as heat is lost through the boundaries of a cooling magma, growing crystals exert increasing gravitational stress on the magma in which they are suspended so that they will sink or float at rates governed by the properties of the melt and the sizes and densities of the crystals. In contrast, they pointed out that both falling temperature and increasing amounts of suspended crystals would increase the viscosity and yield strength of the magma. From the combination of effects, McBirney concluded that the gravitational stress exerted by a crystal would increase with falling temperature because the crystal would be growing and its radius increasing. The gravitational stress, however, would not increase as rapidly as the exponential increase in the value of the yield strength over the same interval of decreasing temperature. Therefore, McBirney argued that to escape entrapment and to continue sinking, the crystal must be large and dense enough to sink rapidly before the liquid acquired even a very small yield strength.

The difficulties of sinking crystals from the roof of a magma chamber led McBirney and Noyes to consider crystallization on the floor. Because the effect of increased pressure toward the floor of the chamber should lead to an increase in the liquidus temperature and promote crystallization, they suspected that crystallization might have begun near the floor rather than under the roof, especially if volatile constituents in the magma had migrated toward the top. Higher concentrations of volatiles in the magma near the roof would have lowered liquidus temperatures there even more.

McBirney and Noyes, therefore, proposed boundary-layer crystallization on the floor where the temperature would be the highest. The boundary layer, they said, would consist of a static zone of magma just above the crystalline floor in which most crystallization would be taking place. Just above the static boundary layer would be a layer in which laminar shear was present. They found it somewhat difficult to talk about the process of solidification in terms of the conventional textural terminology that had been introduced by Wager and his colleagues because that terminology implied a mechanism involving crystal accumulation. They argued, instead, for the growth of crystals in situ to explain many of the Skaergaard textures. Upon examination of well-defined graded layers, McBirney found no correlation between modal variation and differences in grain size of minerals, an indication of the lack of hydraulic equivalence stressed by Dale Jackson for the Stillwater Complex. While McBirney could envision possible mechanisms that would bring plagioclase to the base of a settling layer and pin it down beneath mats of more dense crystals, he was unable to account for buoyant plagioclase in proportions of more than 90 percent at the tops of graded layers. Because he did not regard the plagioclase-rich layers as intrusive on the basis of the absence of transgressive graded layers, he insisted that nothing could have prevented the plagioclase in such layers from floating during gravitational sorting.

McBirney and Noyes also maintained that compositional variation within the rhythmic graded layers could not be accounted for by settling. By way of example, they noted that plagioclase becomes more anorthite-rich toward the top of a rhythmic graded layer and that the Mg content of olivine increases abruptly at the base of a layer and then progresses to more iron-rich compositions toward its top. They also found that the trace element contents of oxides and pyroxenes show systematic differences from one layer to the next. They seriously doubted that such systematic compositional variations could have resulted from crystals that grew in different parts of the intrusion and arrived at their present positions by settling from a convecting magma or from density currents. The compositional relations suggested to McBirney that the crystals grew in situ on the floor under conditions that were affected by the proportions of phases, the order and rates of nucleation, and the amounts of equilibration during growth and cooling. In particular, they regarded the compositional differences as best explained as the result of greater nucleation and growth of mafic minerals near the base of a layer and complementary growth of plagioclase from the iron-depleted liquid at the top.

McBirney and Noyes also doubted that density currents could have accounted for most of the rhythmic layering. In a density current, they maintained, crystals would be concentrated and carried along in a zone at some distance above the base so long as the current continued to be swift enough to keep them in suspension. Where an individual flow decelerated, crystals would be deposited. McBirney, however, could not visualize how such a process could produce layers close to the walls and extending across distances of more than one kilometer with nearly uniform compositions and thicknesses. Moreover, he pointed out, he was unaware of any examples in which thick banks of crystals of an unusual size had accumulated in eddies within the currents that would inevitably have developed along the sides of large blocks that would have obstructed density currents. Even around blocks the size of a house, he noted, the layering shows no systematic differences on opposite sides of the blocks. Even a steady convection current, McBirney thought, would probably not produce continuous layers of constant thickness. Rather, crystals would tend to pile up along the margins of the intrusion, and few would be carried to the center of the floor.

In the end, McBirney and Noyes proposed that the rhythmic layering could be explained by oscillatory nucleation and crystallization controlled by the double diffusion of heat and substance next to the very slowly cooling surface of a solidifying magma chamber floor. Gravity was also said to assist in the formation of stratified zones and intermittent layering. They believed that most of the structural features of layered intrusions were explicable by a combination of these processes but also by the action of sporadic currents that eroded the boundary layer and redistributed crystals that had begun to grow there.

CUMULATE THEORY DEFENDED, ADJUSTED, AND CHALLENGED

In the same issue of *Journal of Petrology* in which the paper of McBirney and Noyes appeared, Morse contributed two articles on Kiglapait geochemistry. He agreed that plagioclase did not sink in magma. Nevertheless, he maintained that plagioclase accumulated on the chamber floor, perhaps as a result of nucleation and settling along with olivine crystals. Morse was not yet ready to abandon crystal settling. He had some company.

McBirney's seminal paper was followed by a series of articles on processes in layered intrusions by T. N. Irvine, a petrologist who had joined McBirney on a couple of his Skaergaard expeditions. T. Neil Irvine undertook an investigation of the Duke Island Ultramafic Complex in southeastern Alaska (Irvine, 1963, 1974) for his Ph.D. dissertation at Caltech. After receiving his degree in 1959, Irvine spent some time at McMaster University and the Geological Survey of Canada. Since 1972, he has been Petrologist at the Geophysical Laboratory. Irvine explored the idea of density currents, applied the concept of double-diffusive convection to layered intrusions, and proposed revisions to cumulate terminology. Unlike McBirney, however, Irvine (1980a) was not about to give up so readily on the hypothesis of gravitational magmatic sedimentation. Although granting that plagioclase flotation poses a serious problem for the traditional crystal settling view, Irvine noted that there is still an extensive array of features in many layered intrusions that can be satisfactorily accounted for in terms of gravitational accumulation of crystals, a notion that he felt should not be jettisoned without a thorough reexamination. He stated that he had been impressed by field relations indicating such substantial lateral transport of solids by magmatic currents that the possibility of crystal growth in situ in those instances was effectively ruled out. These field features included the deposition of imbricated rock fragments and the overturning of already deposited layers that had been disrupted by the deposition of a large block. Many of the layering structures in intrusions like the Skaergaard, he said, were strikingly similar to those in the ultramafic, plagioclase-free Duke Island Complex that, he was convinced, were induced by density currents. As a result, he undertook a study to see if there might be some mechanism for depositing minerals on the floor of a magma chamber even when the minerals should float. To that end, he conducted flume experiments on density currents in which masses of glycerin were permitted to flow into silicone. From analysis of the particle motions and of flow patterns, somewhat analogous to those in a sediment-laden turbidity current, Irvine concluded that magma currents charged with suspended crystals were triggered by slumping processes along the walls of the magma chamber from which they swept out across the floor. He maintained that plagioclase would be swept along with the large quantity of mafic minerals in the dense magma flow. Crystals would be deposited, and he suggested that some of the plagioclase would be retained at the depositional surface by the high yield strength of the magma and ultimately buried. Minerals would be deposited from the density current in accord with their sizes and densities, not simply in response to gravity but also in response to accelerations and decelerations in different parts of the descending flow. Although Irvine believed that the repeated rhythmic layering of the Skaergaard Intrusion and the Duke Island Complex could be explained by density currents, he also recognized the role played by bottom crystallization and double-diffusive convection, particularly in the Muskox intrusion.

McBirney and Noyes (1979) had borrowed the idea of double-diffusive convection from fluid dynamicist J. S. Turner and applied it to the Skaergaard Intrusion. Learning of the concept from Turner and McBirney, Irvine was at first skeptical of double-diffusive convection but then applied it to the Muskox Intrusion (Irvine, 1980b). He pointed out that in contrast to the large-scale convection cells that had traditionally been envisioned for layered intrusions, the magma mass had become gravitationally stratified on a local scale such that double-diffusive convection would be very efficient in transferring heat from the bottom of the intrusion to the top and sides as well as in mixing liquids with contrasting chemical compositions.

For the operation of double-diffusive convection in the Muskox Intrusion, Irvine appealed to three lines of evidence. Periodic introductions of fresh, hot magma, represented by a series of at least 25 cyclic units characterized by specific repeated sequences of rock types would have triggered density instabilities in the overlying fractionated residual magma and set up double-diffusive convecting layers. The individual cyclic units were reported as being internally very well differentiated with progressive iron enrichment and nickel depletion in olivines from the bottom to the top of a cyclic unit. From this internal differentiation, Irvine argued that each time a layer of crystals within that layer was fractionated, compositional changes resulting from its removal must have been transmitted through the residual liquid. Finally, Irvine claimed that because the layered series of the Muskox Intrusion terminates against the marginal zones rather than lapping up on them, lateral circulation in the convecting liquid layers removed cooler liquid from the walls of the intrusion and transferred it toward the center. Fractionation was said to be effected as lower density intercumulus liquid, enriched in components not incorporated into crystallizing olivine and chromite and warmed by the release of latent heat of crystallization, streamed upward from piles of crystals to join the overlying convecting layer. In turn, less fractionated, supercooled liquid would move downward to replace the rising liquid. Despite his strong advocacy of density–current transport as the leading mechanism for small-scale layering, Irvine had also come to the conclusion that thickly layered and massive cumulates, like those of the Muskox Intrusion, were probably formed by growth of cumulus minerals in situ. Further adjustments in the composition of cumulus minerals occurred by infiltration metasomatism, a process in which intercumulus liquid was filter pressed upward by compaction of the cumulate pile to react with cumulus olivine. Irvine et al. (1983) later applied the concept of double-diffusive convection to the Stillwater Complex.

Aware that the traditional crystal-settling paradigm was now recognized as inadequate to account for many phenomena in layered intrusions, Irvine (1982) also set out to revise the terminology that had been widely accepted for such stratiform plutons. He recognized that Wager's definitions of "cumulate," "cumulus crystal," and "intercumulus liquid" were genetic expressions incorporating the idea that crystals had accumulated by settling. Inasmuch as such terms had become well entrenched in petrological usage in the past two decades, Irvine chose not to reject the terms but to redefine them in such a way that crystal settling would be acknowledged as a possible, but not essential, process in their formation. Thus, for example, Irvine (1982, p. 131) redefined a "cumulate" as "an igneous rock characterized by a *cumulus framework* of touching mineral crystals or grains that were evidently formed and concentrated primarily through fractional crystallization." The fractionated crystals he called cumulus crystals and said that they were typically subhedral to euhedral.

Alongside proposals to explain the layering by means of density currents, bottom crystallization, and double-diffusive convection were a growing number of claims for the importance of multiple injection. The Bushveld Complex had long been suspected to have experienced injection of several magma pulses. Later, the Rum and Muskox Intrusions were explained in terms of multiple injection. Another blow to the classical crystal-settling hypothesis came when even the Palisades sill, whose olivine layer had been explained by Lewis, Bowen, and Frederick Walker in terms of gravitational settling of olivine crystals, came to be regarded as the product of multiple injections. If Frederick Walker saw the Palisades as a classic example of crystal settling, Kenneth R. Walker did not. Working with Arie Poldervaart at Columbia University and S. R. Taylor at the Australian National University, Walker (1969) undertook a thorough examination of the major and trace elements in both whole rocks and minerals of the Palisades sill. From evaluation of geochemical patterns, Walker concluded that the olivine zone near the base of the sill had crystallized from a second pulse of magma intruded into the sill after about 10 meters of the original liquid had crystallized. He also argued that, on the crystal-settling hypothesis, the olivine layer should be closer to the base of the sill. The layer, however, both rises and falls and thickens and thins, suggesting that it does not represent an accumulation of crystals on the chamber floor. Another study of the Palisades sill was undertaken by David N. Shirley for his Ph.D. dissertation at the University of California at Los Angeles. On the basis of neutron activation analyses of whole rocks, Shirley (1987) discovered that there are several reversals in the overall fractionation trend and concluded that there had been three or four distinct pulses of magma injection. After the final pulse, he believed, the final two–thirds of the sill differentiated by inefficient fractional crystallization in which the cumulate contained as much as 50 percent intercumulus magma. He suggested that late-stage compaction squeezed much of that remaining interstitial melt into a granophyre zone about 255 meters above the base of the sill.

But all was not lost for the hypothesis of crystal settling because alternative views also encountered resistance. Despite his objections to traditional views of the origin of layering in layered intrusions by magmatic sedimentation, McBirney began to have second thoughts about the effectiveness of double-diffusive convection (DDC) as a means for producing layering in the Skaergaard Intrusion. In a reevaluation of the process, McBirney (1985) concluded that DDC would be an important process at the steeply dipping walls of an intrusion where both thermal and compositional gradients are essentially horizontal. Where the gradients are essentially vertical, however, he concluded that DDC was unlikely to occur where ponded tholeiitic or calc–alkalic magmas were cooling and crystallizing beneath a magma chamber roof or on its floor. He concluded that the magnitude of the compositional effect on liquid density was so large that the temperature would need to deviate greatly from the liquidus temperature. In other words, he discovered that convection would be unlikely to occur unless the magma possessed large quantities of superheat or else were drastically undercooled. The Skaergaard layering was attributed to a process that was controlled by differing rates of nucleation, crystal growth, and recrystallization, a view that he would continue to espouse.

Double-diffusive convection in magma chambers was also challenged by Morse (1986). In a detailed analysis of thermal gradients in the vicinity of a cumulate interface, Morse pointed out that the lower density material, which he called "rejected solute," released upon crystallization of mafic cumulates would buoyantly rise and be flushed away by large-scale convection currents, as classical cumulus theory maintained. The rising light rejected solute would cause no double diffusion. On the other hand, felsic cumulates would release denser liquid upon crystallization, in turn producing a stable density gradient because the fluid would tend to pond or stagnate in the cumulate pile. The small amount of latent heat of fusion added from below, he said, would be insufficient to destabilize that density gradient. Morse concluded that double-diffusive convection should not be invoked where adcumulates occur in layered intrusion.

Morse also questioned the viability of some other alternatives to classical cumulate theory. He believed that the process combining filter pressing by compaction and infiltration metasomatism advocated by Irvine (1980b) to account for "adcumulus" growth in the Muskox Intrusion was not a legitimate alternative to adcumulus growth. In reality, Morse insisted, this process led to the production of "mesocumulates" in mafic rocks and "orthocumulates" in felsic rocks. It is, he said, not an adcumulus but an orthocumulus process. Moreover, Morse was skeptical of the importance of compaction on the grounds that depths of crystal mushes on the magma chamber floor on the order of one to ten meters were required. He doubted that such mush thicknesses occurred as a general steady-state condition of large layered intrusions.

Although having retracted his earlier attachment to double-diffusive convection, McBirney continued to explore alternative ways to account for most of the layering in layered intrusions other than magmatic sedimentation of cumulus crystals. After issuing another challenge to the cumulate paradigm (McBirney and Hunter, 1995), he co-authored a pair of papers on the Skaergaard Intrusion in which he distinguished two general types of layering. In the first paper, McBirney and Nicolas (1997) discussed "dynamic" layering, a type that they attributed to magmatic flow. This type of layering was said to consist of contrasting layers with pronounced foliation and linear orientation of elongated mineral grains. Such dynamic layering, they said, is found in association with slumping and cross bedding in a zone that is situated near the margins of the intrusion. McBirney believed that evidence for magmatic flow in the Skaergaard Intrusion is rare toward its interior away from that zone except where masses of material that had fallen from the roof set currents in motion. In the interior of the intrusion, according to McBirney and Alan Boudreau of Duke University, the second type of layering, so-called nondynamic layering, predominates. In the second paper, Boudreau and McBirney (1997) envisioned two end-member processes involved in the production of nondynamic layering. One entailed varying rates of nucleation and crystal growth, and the other involved compaction of rocks formed on the chamber floor, despite the fact that Morse considered compaction of little import. The first end-member process, they thought, was the result of variations in the proportions of simultaneously precipitating minerals brought about by processes such as convective overturn, input of new magma, contamination, or gain or loss of volatiles. The result was generally diffuse layering. Compaction was said to involve solution, reprecipitation and reorientation of minerals and to lead to sharply defined mafic and felsic layers. They rejected Irvine's explanation that considered the mafic–felsic layering to be sediments laid down by turbidity currents sweeping across the floor.

Neil Irvine held fast, however. Building on a lifetime of experience with the Skaergaard Intrusion and numerous other layered intrusions, Irvine, in conjunction with Jens Andersen of Aarhus University in Denmark and Kent Brooks of Copenhagen University, published a definitive study on the field relationships among gabbro, troctolite, and anorthosite blocks and fragments of various dimensions and shapes and the layering in which they are embedded within the Skaergaard Intrusion. Supporting their argument by an extensive array of superb sketches and photographs, Irvine et al. (1998) spelled out in exhaustive and intricate detail the field evidence that compelled acceptance of deposition of the modally graded, rhythmic layers by magmatic currents. Without rejecting the possibility of alternative processes to account for some of the layered features in parts of the intrusion, Irvine nonetheless insisted that sedimentation from magmatic currents had to be an important process. He and his co-workers pointed out that the deformation of layers beneath blocks, many of them presumably detached from the Upper Border Group, indicated that the top of the cumulate pile and the main mass of overlying magma was a sharp interface at the moment of block impact. The relationships also indicated that the cumulates beneath the interface were already coherent and the layer was fully developed up to the interface when the blocks landed on the pile. Thus, Irvine rejected McBirney's ideas that Skaergaard layering had formed by reorganization and recrystallization within the cumulate pile and that the Layered Series had evolved from a crystal mush that was compacting beneath a ponded magma. The evidence from the blocks and layers indicated rather a dynamic interface. Attention was called again to imbricate fragments, draping and thinning of layers over the tops of blocks, and re-deposition of layers by impacting blocks as further evidence of magmatic sedimentation.

Irvine, Andersen, and Brooks suggested two mechanisms for the deposition of graded layers by means of magmatic crystal–liquid suspension currents.

They invoked density surge currents to form graded fragmental layers, imbricate fragments, and layers draping over blocks. Alternating graded and uniform layers were attributed to boundary flow separation and reattachment, a slower process that was said to fractionate liquid away from sorted crystals. The authors also argued that the plagioclase flotation problem for the Skaergaard Intrusion had been exaggerated. On the one hand, they maintained that plagioclase at the top of a modally graded layer would be prevented from floating away because it would be covered immediately by a dense, mafic cumulate. On the other hand, they pointed out that the plagioclase-rich upper parts of the rhythmic layers contained a sufficient number of cumulus mafic grains that the entire package of solids would have a greater density than the liquid.

Throughout his career, Irvine had always delighted his readers with clear, intricate, detailed, and thought-provoking diagrams of field relations, experiments, or hypothetical scenarios of magmatic processes. The paper on blocks and layers perhaps represented the pinnacle of Irvine's artistic achievement and was climaxed with a remarkable sketch (Figure 58) in which he summarized the processes that he and his collaborators envisioned as having occurred in the western half of the Skaergarard magma chamber. If any petrologist ever made it plain that the rigorously scientific field of igneous petrology partakes of a substantial component of the fine arts, it was Neil Irvine.

In a similar vein but on a smaller scale, James Scoates of the Université Libre de Bruxelles in Belgium examined anorthosite and leucogabbro blocks from the Poe Mountain anorthosite in the Laramie Mountains of Wyoming. Scoates (2000) showed from the disrupted and deformed layers beneath the blocks that plagioclase was already accumulating or crystallizing on the floor at the time of block impact and that compositional layering and plagioclase lamination had formed directly at the interface between the crystalline pile and the liquid. The impacts also caused remobilization of dense interstitial melt within the pile. Scoates did reject the idea that plagioclase settled as isolated crystals on the chamber floor and maintained that plagioclase must either have crystallized in situ at the interface between magma and the crystal pile or else arrived on the floor in dense two-phase "packets" of liquid and plagioclase crystals derived from the roof zone of the chamber.

Throughout the 1980s and 1990s, petrologists vigorously debated the relative merits of various hypotheses for the origin of layering in layered igneous intrusions. Compaction, infiltration metasomatism, double-diffusive convection, crystal settling, in situ bottom crystallization, density currents, oscillatory nucleation, and other proposed processes came under intense scrutiny. Petrologists also continued to debate the terminology that should be applied to the rocks and processes of layered igneous intrusions. Major volumes on the origins of igneous layering (Parsons, 1987) and layered intrusions (Cawthorn, 1996) were published during these two decades. Classical cumulate theory, seemingly comatose at one point, was not quite yet dead. The patient proved

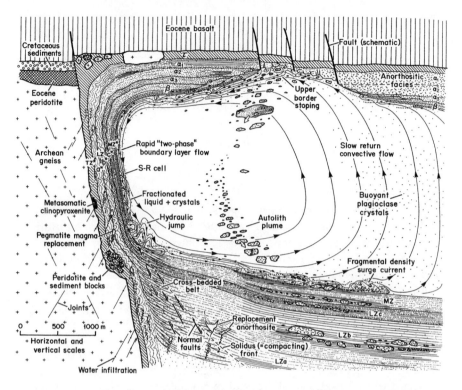

58. Schematic synthesis of magmatic processes in the western part of the magma chamber of the Skaergaard Intrusion, East Greenland, during the Middle Zone stage of differentiation. From *Geological Society of America Bulletin*, v. 110, pp. 1398–1447, Irvine, T. N., Andersen, J.C.Ø, and Brooks, C. K. (1998). Reproduced with permission of the publisher, The Geological Society of America, Boulder, Colorado, USA. Copyright (c) 1998,The Geological Society of America, Inc. (GSA).

to be quite resilient and regained considerable health in the hands of people like Irvine and Morse (1996), who strenuously objected to the "trashing of cumulate theory in recent decades" and maintained that the cumulate theory was alive and well for explaining many features of the Kiglapait Intrusion. At the same time, petrologists learned that the theory of magmatic sedimentation was not universally valid for all layered igneous rocks. If nothing else, the ferment of ideas about processes in layered igneous intrusions heightened the awareness of igneous petrologists about the importance of experimental and theoretical studies in the field of fluid dynamics.

30 ❖ PAST AND FUTURE
Some Concluding Remarks

Because the field of igneous petrology developed so explosively in the last two decades of the twentieth century, the sheer mass of relevant literature and documents utterly prohibits writing a summary of the history of igneous petrology that deals adequately with all of the developments. In addition, our temporal proximity to these developments prevents attainment of a completely satisfactory perspective from which to carry out a meaningful analysis. For that reason, I have chosen not to discuss numerous topics of great interest to contemporary igneous petrologists, a few of which I mention only very briefly later in the chapter. I did, however, single out two developments of the last two decades for discussion in Chapters 28 and 29. The importation of ideas from fluid dynamics into igneous petrology compelled petrologists to hone their mathematical skills and to expand their conceptual arsenals with input from yet another science. The impact of fluid dynamics on how petrologists view magma chamber processes was almost immediate. How well petrologists have understood and properly applied the concepts of fluid dynamics is still being sorted out so that a more satisfactory analysis of the "fluid dynamical era" still lies in the future.

I also addressed some of the recent thinking about the origin of layering in large layered intrusions because petrologists have been wrestling with this long-standing question ever since the publication of Wager and Deer's memoir on the Skaergaard Intrusion. Approaches to this matter have also been heavily influenced by the input of fluid dynamical concepts. This issue, too, is still in ferment. Greater clarity about its resolution must await a later day.

Before concluding the book with a few comments about further research on the history of igneous petrology, I wish to give extended attention to just one final topic, namely, the long-standing "room problem," on which recent studies hold out the promise of a resolution.

GRANITE EMPLACEMENT: DIAPIRS OR DIKES?

Despite the technology-dependent advances in geochemistry and experimental petrology and the applications of fluid dynamics that have revolutionized igneous petrology in recent decades, as reviewed in the preceding chapters, igneous petrology is still ultimately about geological entities, namely, rocks and magmas. Because of the fruitfulness of the new avenues of petrological study, petrologists will need to guard against the disconnection of their research from igneous rock bodies that are walked on, hammered, mapped, measured, admired, and argued about in the field. In the last analysis, igneous rock bodies provide the foundation for the science of igneous petrology. No matter how valuable the input from experiment, instrumental analysis, computer modeling, fluid dynamics, physical chemistry, or other sciences, petrological ideas must constantly be tested against the realities of the outcrop. Moreover, new ideas will continue to emerge from studies of igneous rocks in their natural settings. As one example, new concepts about the nature of granite emplacement have recently evolved to a large extent through fieldwork.

One may argue with some justification that no geologist has done more to advance our understanding of the nature of granitoid batholiths than Wallace Spencer Pitcher (b. 1919). After working as a chemical analyst while taking night courses in chemistry and geology, and serving in World War II, Pitcher studied at and graduated from Chelsea Polytechnic in 1947. Immediately thereafter, H. H. Read employed Pitcher as a Demonstrator at Imperial College, London, where he began research on the Thorr granite in Donegal, Ireland. After advancing to a Lectureship at Imperial College, Pitcher accepted a Readership at King's College in London, where he continued his studies of the granites of Donegal and became impressed with the remarkable variety of intrusive mechanisms that likely produced those bodies, including reactive stoping, diapiric intrusion, cauldron subsidence, and magmatic wedging. In 1962, Pitcher became Professor of Geology and Head of the Department at the University of Liverpool. Shortly after moving to Liverpool, Pitcher initiated a program of research on the granitoid batholiths of Peru to ascertain if the emplacement mechanisms of the Donegal granites had broader applicability. Beginning in 1965, Pitcher put together a coalition of researchers from the Geological Service of Peru and the British Geological Survey along with graduate students at Liverpool to conduct a mapping project on the Coastal Batholith in the northern part of the Lima Province. From time to time the project also incorporated participants from other universities in Peru, Germany, Canada, France, and Great Britain who were engaged in a mapping project to the east under the auspices of the University of Montpellier in France. Pitcher's program received vigorous financial support from the Instituto Geological Mineral y Metalurgica and from the National Environmental Research Council of the United Kingdom.

The exceptional work of Pitcher's team was summarized in the monograph *Magmatism at a Plate Edge: the Peruvian Andes* (Pitcher et al., 1985) and in numerous papers such as Pitcher (1978). Many of Pitcher's insights into batholiths have been incorporated into his book on *The Nature and Origin of Granite* (Pitcher, 1993, 1997). Thanks to exceptional three-dimensional exposures of the Coastal batholith, 1600 kilometers long and as much as 65 kilometers wide, Pitcher and his team have been able to recognize several "superunits" comprising a series of closely related mafic to felsic rocks. These "superunits" consist of more than 800 individual plutons, some of which occupy hundreds of square kilometers in area. The plutons have sharp, cross-cutting contacts and flat roofs at shallow levels. They were originally described as having steep walls that may extend to great depths, but gravity studies have shown that the plutons are thin sheets with an aspect ratio of about 20:1 and a maximum thickness of 5 kilometers (Atherton, 1990). Their general high-level form is that of rectilinear boxes or flat-topped bell jars. Pitcher (1978) maintained that the plutons were so obviously cut out of the crust that emplacement could only have been accomplished by foundering of blocks of crust. Although finding rare instances of impressive piecemeal stoping, he maintained that cauldron subsidence was the dominant process. The cauldron blocks, he suspected, extended far down. Pitcher was very suspicious of diapirism as a mechanism of granitoid emplacement.

Since the days of Wegmann (1930), many geologists had envisioned that granitic plutons were typically emplaced by forceful injection in which a diapir, that is, a roughly teardrop-shaped mass of low-density, buoyant magma was thought to ascend gradually into the crust somewhat like a salt dome, shouldering aside wall rocks as it rose. During the days of the granite controversy, proponents of magmatic batholiths like Grout commonly thought in terms of diapiric emplacement. The granitizers, however, had questioned the concept that batholithic granites are really magmatic on the grounds of the "room problem." They had difficulty envisioning how the room could be created for such enormous masses of liquid. Although the granite controversy was resolved essentially in favor of the magmatists, the mode of emplacement of large masses of magma remained something of an enigma. In the last few years, however, some geologists, including Pitcher, had begun to call the entire diapir concept into question. Many of them suggested that magma ascent takes place in extensional zones and primarily through dikes. D.H.W. Hutton (1982, 1988), for example, noted that room could be provided for granite plutons in the bends and offsets of large transcurrent faults.

Other petrologists followed suit. Allen Glazner of the University of North Carolina at Chapel Hill found a correlation between the obliquity of subduction and the magnitude of plutonism in the Sierra Nevada batholith of California. Glazner (1991) maintained that the most intense episodes of magmatism did not correlate with higher rates of subduction, as advocated by many geologists. Rather, he argued that when the tangential component of plate convergence was large, rates of magmatism increased. That tangential component, he noted, is typically taken up by strike-slip faulting. Hence, Glazner claimed that the correlation between plutonism and oblique convergence in the Sierra Nevada supports the mechanism of emplacement of plutons at releasing bends in strike-slip faults. Tikoff and Teyssier (1992) argued similarly that large volumes of Sierra Nevada magma were intruded during Late Cretaceous strike-slip shearing. They maintained that the orientation of plutons was controlled by emplacement between crustal scale, *en echelon* P-shear zones within larger dextral zones of strike-slip. That hypothesis, they argued, accounted for the consistent elongation of individual plutons approximately 15° counterclockwise off the main trend of the Sierra Nevada batholith, and it accounted for evidence of passive emplacement and of syn-magmatic strike-slip shearing within individual plutons.

The attack on diapiric emplacement and emphasis on the ascent of granitoid magmas through fractures or dikes was promulgated vigorously by J. D. Clemens, then at the University of Manchester, and C. K. Mawer, then at the University of New Mexico. Clemens and Mawer (1992, p. 340) charged that the greatest impediment to understanding the mode of transport of granitic magmas lay in the "ubiquitous cartoons of granitoid diapirs ascending through the crust as inverted tear-drops or prolonged carrots." They asserted that diapirism is not viable on thermal or mechanical grounds and that there is a total absence of field evidence for diapiric rise of magma. The shapes of plutons, they insisted, reveal nothing about the transport of the magma but are only indicators of "arrival phenomena" dictated by local structure, kinematics, and stress states. They claimed that what a field geologist ought to see, if granite magmas rose by diapirism, would be rim synclines, intense local flattening parallel to the margins of a pluton, and cylindrical shear zones around a pluton extending to great depth and displaying pronounced vertical lineations. But such structures are encountered very rarely. Clemens and Mawer asserted, too, that a diapir arising through a softened zone of deformation ought to lose three to ten times as much heat as an identical standing magma body so that contact metamorphism should be much more intense than is observed. Moreover, thermal modeling has shown that a rising diapir should undergo "thermal death," that is, it should solidify and rise no higher than the middle levels of the crust. As a result, they insisted, shallow-level batholithic granitoids could not have ascended by diapirism. In contrast, they considered it highly likely that fractures played a major role and that dikes filled with granitic magma would be catastrophically self-propagating. As an example, they calculated that a single dike 1 kilometer long and 3 meters wide would propagate upward for 20 kilometers in about eight months. Moreover, granitic magma flowing through this dike could inflate a batholith with a volume of 2000 cubic kilometers in less than 900 years. They noted that most recorded felsic dikes are between 1 and 4 meters wide and their lengths are typically as much as 1 kilometer. Clemens and Mawer (1992) concluded that granitoid plutonism is a relatively rapid process with about 100 to 10,000 years required for the emplacement of a typical pluton. They also concluded that most of the crystallization and differentiation of granitoids occurs within the final chamber rather than being inherited from the source region.

Similar conclusions were reached by Nick Petford of the University of Liverpool and, later, the University of Kingston-upon-Thames. Petford and his associates, John Lister at Cambridge and Ross Kerr at the Australian National University, reiterated that many batholiths and plutons, once interpreted as diapirs, are now recognized to be closely associated with large faults and lineaments, often parallel to continental margins and seafloor trenches (Petford et al., 1993). Moreover, the balloon shape of the pluton is now interpreted as the result of final emplacement in dilational jogs in fault systems or other tensional loci rather than as a result of the mode of magma ascent. From a more detailed thermal analysis of heat loss through dike walls and upward heat advection by rising dike magma, Petford and his colleagues showed that viscous magma could rise in narrow conduits without freezing. They determined, for a Peruvian batholith 150 kilometers long and 20 kilometers wide and an estimated volume of 6000 cubic kilometers, that granitoid magma flowing through a dike 10 kilometers long and 6 meters wide would ascend at a rate of about 1 centimeter per second and take about 41 days to rise from its source to the chamber being filled. The entire batholith would be filled in about 350 years. They also recognized that such filling times are probably minimum values and that ascent would be "pulsed" rather than continuous.

Petford et al. (1994) concluded that dike widths would normally be between 2 and 20 meters and that the increased viscosity of a crystal-laden granitoid magma would probably require wider dikes, whereas relatively crystal-free magmas would have lower viscosities and could flow through the narrower dikes. Arriving at the same conclusion as Clemens and Mawer (1992) that batholiths with volumes of several thousand cubic kilometers could be emplaced in a few hundred years, if the flow were continuous, Petford and his colleagues pointed out that the duration of active magma flow and rise would be limited by the amount of magma actually present in the source region. Hence, they noted, as the available melt was drained away from the source by the dike, the rate of flow would diminish and the dike width would decrease until the magma remaining in the dike would freeze. Thus, they envisioned brief episodes in which the dikes contained rapidly ascending magma followed by longer quiescent episodes in which the dikes were closed and new magma was perhaps being generated in the source region. Petford (1996), in a review of the status of the question regarding dikes or diapirs, also noted that assembly of an entire complex batholith might be quite long even though individual plutons might be filled relatively quickly. And more recently, McCaffrey and Petford (1997) have claimed that granitic plutons typically inflate as tabular, sheet-like intrusions and that such geometry is related to the ascent of the magma through vertical dikes.

In summary, in very recent years, petrologists have been approaching the "granite problem" in an entirely new way because of the revised concepts about the ascent and emplacement of granitic magma. The irony is that petrologists had always thought of basaltic magma rising through fractures. At last they realized, thanks not only to field evidence and theoretical considerations, but also to a larger body of experimental data on magma viscosity, that a wide range of magmas could ascend via upward-propagating fractures. Perhaps, after nearly two centuries of puzzlement about granitic plutons, the solution to the long-standing "room problem" is at hand.

MISCELLANY

Before looking at the future of the history of igneous petrology, I also wish to make very brief mention of some other important matters that I did not address earlier. Among the issues of the last two decades that I ignored are the nature and development of so-called "large igneous provinces" represented by the vast outpourings of continental flood basalts in South Africa (Karoo), India (Deccan), Brazil (Paraná), and the northwestern United States (Columbia River); the role of very large magma chambers or magma oceans on the Moon and on Earth during the Archean; the mechanisms of melt generation, migration, collection, and distribution throughout the crust and upper mantle; many aspects of the relationships between tectonism and igneous activity; and the relationships between magmatism and ore generation.

I also passed over the intense effort to measure the physical and thermal properties of multicomponent silicate melts such as density, viscosity, compressibility, thermal expansion, thermal conductivity, enthalpy, and heat capacity. Much labor has also been expended on determination of the solubility of water and other volatile constituents in silicate melts under a range of temperatures and pressures and on measurement of the rates of diffusion of a wide variety of chemical species in silicate melts. Petrologists have also explored the phenomena of nucleation and growth of crystals from silicate melts. Coincident with this recent emphasis on the physical properties of silicate melts has been the effort to understand the internal structure of such melts better. Igneous petrologists have long believed that knowledge of melt structure ought to lead to an improved understanding of diffusion, nucleation, crystal growth, element partitioning between crystals and liquid, and the formation of immiscible liquids. What remains to be seen is how the expanding volume of data on physical properties and the refined knowledge of melt structure will lead to an enhanced understanding of the origin and evolution of igneous rocks.

Marsh (1988a) called the diversity of igneous rocks the igneous petrologist's "raison d'etre." Although I wrote about ideas concerning the causes of diversity, I have also generally ignored ideas concerning the formation and evolution of the individual rock types that compose the full spectrum of that diversity.

The exceptions were granitic rocks and mafic/ultramafic intrusions. One of the trends in igneous petrology during the past half-century has been the specialization in the study of specific igneous rock types. Indicative of that specialization is the proliferation of reference works on specific types. Prior to 1950, igneous petrology had general textbooks like those of Harker and Iddings, petrographic treatises like those of Rosenbusch and Johannsen, and memoirs on specific intrusions or igneous terranes like those of Hall on the Bushveld Complex, Wager and Deer on the Skaergaard Intrusion, or Buddington (1939) on the igneous rocks of the Adirondack Mountains. Despite the publication of *Geologie du Granite* by Raguin (1946) and Geological Society of America Memoir 28 on the origin of granite (Gilluly, 1948), books on individual rock types were few and far between. But since mid-century there has been an eruption of volumes on individual rock types, including some of the more obscure types like orangeites and boninites, rock types whose definition most petrologists would probably need to look up.

The 1950s were the decade of granitic rocks (Read, 1957; Tuttle and Bowen, 1958). During the 1960s, books appeared on carbonatite (Heinrich, 1966; Tuttle and Gittins, 1966), ultramafic rocks (Wyllie, 1967), basalt (Hess and Poldervaart, 1967, 1968), and anorthosite (Isachsen, 1968). During the following decade, books were released that dealt with alkalic rocks (Sørenson, 1974), basalt (Yoder, 1976), carbonatite (Le Bas, 1977), granite (Marmo, 1971; Didier, 1973; Atherton and Tarney, 1979), kimberlite (Boyd and Meyer, 1979a, b), ophiolite (Coleman, 1977), spilite (Amstutz, 1974), and trondhjemite (Barker, 1979). The two volumes on kimberlite edited by Boyd and Meyer represented the proceedings of the Second International Kimberlite Conference held in Santa Fe, New Mexico, in 1977. Students of kimberlite have gathered for an international conference every four or five years and, each time, published two volumes on kimberlite (e.g., Kornprobst, 1984a, b; Ross, 1989a, b; Meyer and Leonardos, 1994a, b). As if the output of the international kimberlite conferences were not enough, other works on kimberlite appeared during the 1980s (Dawson, 1980; Mitchell, 1986). These were joined by several more works on alkalic rocks (Fitton and Upton, 1987; Woolley, 1987; Leelanandam, 1989), andesite (Gill, 1981; Thorpe, 1982), basalt (Morse, 1980; Ragland and Rogers, 1984; Macdougall, 1988; Basaltic Volcanism Study Project, 1981), boninite (Crawford, 1989), carbonatite (Bell, 1989), and komatiite (Arndt and Nisbet, 1982). The flow of literature was unabated throughout the 1990s. Alkalic rocks were discussed by Mitchell (1996) and by Kogarko et al. (1995). Books were also published on anorthosite (Ashwal, 1993), basalt (Floyd, 1991), carbonatite (Bell and Keller, 1995), granite (Didier and Barbarin, 1991; Pitcher, 1993, 1997; Johannes and Holtz, 1996; Clarke, 1992; Brown et al., 1996), kimberlite and orangeite (Mitchell, 1995), lamproite (Mitchell and Bergman, 1991), lamprophyre (Rock, 1991), and ophiolite (Peters et al., 1991; Parson et al., 1992). In addition, there have been several special issues of various journals devoted exclusively to one specific rock type. This compilation is by no means exhaustive, but it does strongly indicate the extent to which igneous petrology has become specialized.

Another issue that I ignored is the overwhelming impact that computers have had on igneous petrology. The topic deserves its own study. During the past two decades, advances in computer technology have occurred at such a breakneck pace that a meaningful historical analysis of the phenomenon may be as feasible as providing a detailed description of a race car as it whizzes by. Nonetheless, just as an observer can still give an impressionistic description of the speeding vehicle, one can also make some impressionistic comments about the computer revolution. Before the early 1960s, few hard-rock geologists ever used a computer. In the 1960s, mineralogists carried out analyses of crystal structure with thick stacks of punch cards and large mainframe computers. Calculations commonly required several hours. Now every petrologist has a desktop computer with vastly more computing power than the old mainframe machines. So accustomed are we to our desktop computers that, in many cases, we become impatient if a graph or the result of a calculation isn't printed in a few seconds. Since 1990, petrologists have had software for calculating CIPW norms; plotting chemical analyses on AFM, REE, TAS, and countless other diagrams; modeling trace-element distributions; or evaluating intensive variables for assemblages of coexisting minerals. Specialists have produced very complex programs for trace-element modeling, manipulating data obtained by spectrometric and other analytical methods, or modeling phase relations in a magmatic system. The MELTS program devised by Mark Ghiorso of the University of Washington, for example, calculates equilibrium phase assemblages and can also be used to model courses of crystallization of melt during either equilibrium or fractional crystallization or to model processes of assimilation and partial melting.

Computers have been a great boon to statistical petrographers. Enormous amounts of quantitative information on igneous rocks are now stored in vast databases, and statistical analyses are readily carried out on selected portions of the databases. Discriminant analyses of petrographic and chemical data have proved valuable in the development of modern classification schemes. Literature databases like GEOREF have rendered the process of searching for technical literature much quicker.

The use of the desktop computer as a word processor has reduced the preparation time required for writing and editing manuscripts. The ability of modern computer systems to produce and print clean, accurate graphs, diagrams, and maps has also greatly reduced the amount of time required in manuscript preparation.

Electronic mail has increased the speed of communication among petrologists. So has the establishment of electronic discussion groups. Unfortunately, the obvious advantages of electronic mail have made the letter an endangered species. Because the internet has facilitated access to information, petrologists should consider including much more biographical data along with their research interests on their web pages.

The electronic revolution has also impacted the dissemination of petrological information. Although the paper journal is not yet in danger of extinction, the staggering increase in the volume of books, reports, and journals has exceeded or threatens to exceed the storage capabilities of many libraries, thus forcing painful decisions about the elimination of subscriptions, the culling of old texts and back issues of journals, or the substitution of on-line electronic journals or CD-ROM versions of journals in place of hard copy. We are still in the midst of a revolution in the mode of dissemination of petrological information of which the outcome is not fully clear. Will the trend toward electronic journals make petrological information more widely accessible or make it more difficult for petrologists in developing countries to keep abreast of recent petrological literature because they may not have access to state-of-the-art technology? Modern computer technology has already exerted, and will continue to exert, for the foreseeable future, a potent impact on the creation, proliferation, and dissemination of petrological knowledge.

THE FUTURE OF THE HISTORY OF IGNEOUS PETROLOGY

The time has come to bring our overview of the history of igneous petrology to a close. The story of igneous petrology is the story of the decipherment of the criteria for the recognition of magmatic rocks, of the discovery of the amazing diversity of such rocks, and of growth in understanding of the possible causes of that diversity. It is the story of the elucidation of the relationships between igneous rocks and their general tectonic environments. It is the story of how petrologists have attempted to figure out the nature of the generation, collection, ascent, emplacement, cooling, crystallization, deformation, and contamination of magma bodies. It is primarily a story that concerned the igneous rocks on Earth, but it has also become the story of the discovery and study of igneous rocks on the Moon and other planetary bodies.

The history of igneous petrology is, of course, the story of the development of scientific ideas about the origin and behavior of a certain class of rocks, but it is more than that because scientific ideas arise, gain acceptance, or are rejected within broader human contexts with geographical, social, and political dimensions. And so, the story of igneous petrology is also the story of its emergence from general geology as a distinct discipline and its gradual professionalization and of its transferral primarily to the academic world. It is the story of the internationalization of a scientific discipline as it spread from western Europe to North America and South Africa and then to Japan, Australia, and gradually around the world. It is also a story of Anglicization in which petrological communication in the English language progressively became dominant.

It is the story of the growing dependence of igneous petrology for further conceptual advances on the data and insights drawn from other geological disciplines, from other sciences such as analytical chemistry, physical chemistry, isotopic chemistry, fluid mechanics, and thermodynamics, and from mathematical disciplines like differential equations, vector algebra, and statistics. It is the story of increased government involvement in the funding of scientific research. And it is the story of the impact of technological developments such as the polarizing microscope, the quenching furnace, the emission spectrograph, the mass spectrometer, the x-ray fluorescence spectrometer, the electron microprobe, the computer, and even the Saturn rocket, tools that extended the senses, aided in calculation, and brought humans into contact with previously unreachable igneous rocks on other worlds.

Because the story of igneous petrology has been told incompletely and imperfectly here, others are encouraged to help tell the story. Inasmuch as the task of exploring and telling the story of igneous petrology has barely begun, a multitude of projects awaits both historians of geology and igneous petrologists who are eager to probe the roots of their discipline. There is, for example, an extreme dearth of detailed scientific biographies of leading igneous petrologists. We do have biographical studies of Bowen, Goldschmidt, Holmes, and Wager, but thorough work on the lives and careers of prominent petrologists like Rosenbusch, Zirkel, Michel-Lévy, Lacroix, Brögger, Vogt, Sederholm, Iddings, Harker, Daly, Grout, Niggli, Eskola, Read, and Hess remains to be done. Iddings, Buddington, and Johannsen have left manuscripts of personal reminiscences that might serve as starting points for biographical work. The archival work necessary to discover other such manuscripts is still necessary. Igneous petrologists, too, need to gain greater appreciation for the significant work of some of the nearly forgotten contributors to petrological progress like Theodor Scheerer, Joseph Durocher, John Judd, J.J.H. Teall, and David Forbes.

Because today's scholars should be the focus of biographical study in another generation or two, contemporary petrologists are strongly encouraged to preserve their correspondence (including e-mail), journals, and notebooks and to write their reminiscences or record some recollections orally. Contemporary petrologists should tell future petrologists and historians of petrology about their education, how they became interested in igneous petrology, the influences on their petrological thinking, their careers, how their ideas developed, and impressions of their contemporaries.

Institutions have been as important as individuals in the progress of igneous petrology. In recent years, the Geophysical Laboratory has been the focus of historical scrutiny. But the role of other institutions central to the development of igneous petrology is a field that is ripe for research. Of particular significance are the Bergakademie at Freiberg, home to Werner, Naumann, and von Cotta; the University of Christiania (Oslo), where Kjerulf, Brögger, Vogt, Goldschmidt, Barth, and Ramberg taught petrology; the University of Chicago, where Iddings, Johannsen, Bowen, Ramberg, and Goldsmith worked and where modern geochemistry received considerable impetus; the University of Heidelberg, where von Leonhard, Bunsen, and Rosenbusch labored; and the Muséum d'Histoire Naturelle in Paris, home to a long line of distinguished mineralogists and petrologists, such as Haüy, Daubenton, Daubrée, Des Cloizeaux, and Lacroix. The role of igneous petrology in the geological surveys of Great Britain, the United States, France, and other nations promises to be a fruitful area of research.

A thorough study of the fortunes of igneous petrology in the academic world would also be welcome. Such a study ought to include data on the beginnings of academic instruction in petrology in various nations, the numerical growth of geology departments and courses in igneous petrology, and student enrollments in such courses, including male, female, and, where relevant, minority enrollments. What factors affected the fortunes of igneous petrology courses at both the undergraduate and graduate levels? How did petrology courses fare in relation to other geological disciplines and why?

A history of igneous petrology written by an American unavoidably suffers from a relative lack of familiarity with German, French, Scandinavian, Russian, eastern European, and Japanese literature. Because the contributions of German, French, Swiss, and Scandinavian petrologists are probably much better known among English-speaking petrologists than is generally the case for contributions from China, Brazil, Japan, Russia, and other eastern European nations, studies of the development of igneous petrology in these nations should yield a more balanced view of its overall history.

Several petrological themes cry out for detailed investigation. Controversial issues are especially interesting and provide good insight into the inner workings of science. It is always worth finding out what factors tilted thinking toward resolution of an issue in one direction or another and how they did so. Controversy, too, often provides some of the more colorful scientific language. The granite controversy of the 1940s and 1950s, a prime example of that very point, begs for historical study, occurring as it did during the height of political and economic tensions throughout the world. So, too, does the nineteenth-century granite controversy in which the feud between David Forbes and T. Sterry Hunt possibly surpassed the more recent dispute between Read and Bowen for its vitriolic rhetoric. More recent disputes, some still ongoing, that would eventually be worth historical scrutiny revolve around the origin of layering in igneous intrusions and around the nature of convection in magma chambers. Participants in these debates would assist future historians of petrology by taking the time to leave lengthy reflections on their roles and that of others in the debates. Without question, the classification of igneous rocks, while lacking the glamour of petrogenetic studies, provided a limitless supply of fuel for debate involving, in many cases, some of the most highly respected petrologists. The entire history of experimental igneous petrology needs close examination. So, too, do the history of ideas about lunar igneous petrology and the history of the impact of the plate tectonics revolution on thinking about igneous rocks.

The field of the history of igneous petrology, although born well after its due date, is now in its extreme infancy. The health and maturation of the youngster remain an open question. Let us hope that the parents of the babe, namely, historians of geology and igneous petrologists, will provide the appropriate nurture for the newborn. Not only will the field of the history of igneous petrology become strong and vigorous if proper attention and care are lavished on it, but those providing the nurture, like any doting parents, will certainly find themselves enriched.

BIBLIOGRAPHY

Abelson, P. H., 1975, Arthur Louis Day. *Biographical Memoirs of the National Academy of Sciences*, v. 47, pp. 27–47.
Abich, H., 1841, *Über die Natur und den Zusammenhang der vulkanischen Bildungen.* Braunschweig, n. p.
Adams, F. D., 1938, *The Birth and Development of the Geological Sciences.* Baltimore, MD, Williams and Wilkins.
Adams, L. H., and Johnston, J., 1912, A note on the standard scale of temperatures between 200° and 1100°. *American Journal of Science*, v. 33 (4th series), pp. 534–545.
Ahrens, L. H., 1950, *Spectrochemical Analysis.* Cambridge, MA, Addison–Wesley.
———, 1951, *Wavelength Tables of Sensitive Lines.* Cambridge, MA, Addison–Wesley.
———, 1954, *Quantitative Spectrochemical Analysis of Silicates.* London, Pergamon.
Aldrich, M. L., 1990, Women in geology, *in* G. Cass-Simon and P. Farnes, *Women of Science: Righting the Record.* Bloomington, Indiana University Press, pp. 42–71.
Allan, T., 1815, An account of the mineralogy of the Faroe Islands. *Transactions of the Royal Society of Edinburgh*, v. 7, pp. 229–267.
Allègre, C. J., and Minster, J. F., 1978, Quantitative models of trace element behavior in magmatic processes, *Earth and Planetary Science Letters*, v. 38, pp. 1–25.
Allègre, C. J., Treuil, M., Minster, J. F., Minster, B., and Albarède, F., 1977, Systematic use of trace elements in igneous processes, I: fractional crystallization processes in volcanic suites. *Contributions to Mineralogy and Petrology*, v. 60, pp. 57–75.
Allen, E. T., White, W. P., and Wright, F. E., 1906, On wollastonite and pseudo-wollastonite, polymorphic forms of calcium metasilicate. *American Journal of Science*, v. 21 (4th series), pp. 89–108.
Alling, H. L., and Valentine, W. G., 1927, Quantitative microscopic analysis. *American Journal of Science*, v. 14 (5th series), pp. 50–65.
Allport, S., 1869, On the basalt of south Staffordshire. *Geological Magazine*, v. 6, pp. 115–116.
———, 1870, On the basaltic rocks of the Midland coal-fields. *Geological Magazine*, v. 7, pp. 159–162.
———, 1871, On the microscopic structure and composition of a phonolite from the 'Wolf Rock'. *Geological Magazine*, v. 8, pp. 247–250.
———, 1872, On the microscopic structure of the pitchstones and felsites of Arran. *Geological Magazine*, v. 9, pp. 1–9, 536–545.
———, 1874, On the microscopic structure and composition of British Carboniferous dolerites. *Quarterly Journal of the Geological Society of London*, v. 30, pp. 529–567.

Allsopp, H. L., 1961, Rb–Sr age measurements on total rock and separated mineral fractions from the Old Granite of the central Transvaal. *Journal of Geophysical Research*, v. 66, pp. 1499–1508.

Amstutz, G. C., 1970, Bischof, Carl Gustav Christoph, in Gillispie, C. C., ed., *Dictionary of Scientific Biography*, v.2. New York, Charles Scribner's Sons, pp. 158–159.

———, ed., 1974, *Spilites and Spilitic Rocks*. New York, Springer-Verlag.

———, 1975, Roth, Justus Ludwig Adolph, in Gillispie, C. C., ed., *Dictionary of Scientific Biography*, v. 11. New York, Charles Scribner's Sons, pp. 560–561.

———, 1976, Zirkel, Ferdinand, in Gillispie, C. C., ed., *Dictionary of Scientific Biography*, v. 14. New York, Charles Scribner's Sons, p. 625.

Andersen, O., 1941, Memorial of Waldemar C. Brögger. *American Mineralogist*, v. 26, pp. 167–173.

Anderson, A. T., Davis, A. M. and Lu, F., 2000, Evolution of Bishop Tuff rhyolite magma based on melt and magnetite inclusions. *Journal of Petrology*, v. 41, pp. 449–473.

Anderson, E. M., 1936, The dynamics of the formation of cone-sheets, ring-dykes, and caldron-subsidences. *Proceedings of the Royal Society of Edinburgh*, v. 56, pp. 128–157.

Anderson, G. H., 1937, Granitization, albitization, and related phenomena in the northern Inyo range of California–Nevada. *Geological Society of America Bulletin*, v. 48, pp. 1–74.

Andrews, H. N., 1976, Williamson, William Crawford, in Gillispie, C. C., ed., *Dictionary of Scientific Biography*, v. 14. New York, Charles Scribner's Sons, pp. 396–399.

Anonymous, 1870, George Poulett Scrope, Esq., F.R.S., F.G.S. *Geological Magazine*, v. 7, pp. 193–199.

Anonymous, 1885, Allan, Thomas, in Stephen, L., ed., *Dictionary of National Biography*, v. 1. London, Smith, Elder, p. 297.

Anonymous, 1896, *George Huntington Williams: A Memorial by Friends for Friends*. Privately Printed.

Anonymous, 1904, Charles Soret (1854–1904). *Mineralogical Magazine*, v. 14, p. 65.

Anonymous, 1905, John Wesley Judd, C. B., LL. D., F.R.S., F.G.S. *Geological Magazine*, v. 2 (5th decade), pp. 385–397.

Anonymous, 1909, Jethro Justinian Harris Teall, M.A., D.Sc., F.R.S., F.G.S. *Geological Magazine*, v. 6 (5th decade), pp. 1–8.

Anonymous, 1914, Henry James Johnston-Lavis, F.G.S. Born July 19, 1856. Died August 10, 1914. *Geological Magazine*, v. 51, pp. 574–576.

Armstrong, R. L., Taubeneck, W., and Hales, P. O., 1977, Rb–Sr and K–Ar geochronometry of Mesozoic granitic rocks and their Sr isotope composition, Oregon, Washington, and Idaho. *Geological Society of America Bulletin*, v. 88, pp. 397–411.

Arndt, N. T., and Nisbet, E. G., eds., 1982, *Komatiites*. London, George Allen and Unwin.

Arnold, L. B., 1993, The Bascom–Goldschmidt–Porter correspondence 1907–1922. *Earth Sciences History*, v. 12, pp. 196–223.

———, 1999, Becoming a geologist: Florence Bascom in Wisconsin, 1874–1887 (Part 1 of 2). *Earth Sciences History*, v. 18, pp. 159–179.

———, 2000, Becoming a geologist: Florence Bascom and Johns Hopkins, 1888–1895. *Earth Sciences History*, v. 19, pp. 2–25.

Arth, J. G., 1976, Behavior of trace elements during magmatic processes—a summary of theoretical models and their applications. *Journal of Research of the U. S. Geological Survey*, v. 4, pp. 41–47.

Arth, J. G., and Hanson, G. N., 1972, Quartz diorites derived by partial melting of eclogite or amphibolite at mantle depths. *Contributions to Mineralogy and Petrology*, v. 37, pp. 161–174.

Ashwal, L. D., 1993, *Anorthosites*. Berlin, Springer-Verlag.

Aston, F. W., 1927a, A new mass-spectrograph and the whole number rule. *Proceedings of the Royal Society of London*, Series A, v. 115, pp. 487–514.

———, 1927b, The constitution of ordinary lead. *Nature*, v. 120, p. 224.

Atherton, M. P., 1990, The Coastal Batholith of Peru. *Geological Journal*, v. 25, pp. 337–349.

Atherton, M. P., and Tarney, J., 1979, *Origin of Granite Batholiths: Geochemical Evidence*. Orpington, UK, Shiva.

B.B.W., 1894, John Murray in Lee, S., ed., *Dictionary of National Biography*, v. 39. London, Smith, Elder, pp. 388–389.

Backlund, H. G., 1938, The problems of the rapakivi granites. *Journal of Geology*, v. 46, pp. 339–396.

———, 1946, The granitization problem. *Geological Magazine*, v. 83, pp. 105–117.

———, 1947, *Om granit och gnejs och jordens alder*. Uppsala, K. Vetenskapssocietetens Arsbok.

Bäckström, H., 1893, Causes of magmatic differentiation. *Journal of Geology*, v. 1, pp. 773–779.

Baertschi, P., 1950, Isotopic composition of the oxygen in silicate rocks. *Nature*, v. 166, pp. 112–113.

Bailey, C., 1947, *De Rerum Natura*, 3 v. Oxford, UK, Clarendon Press.

Bailey, E. B., Clough, C. T., Wright, W. B., Richey, J. E., and Wilson, G. V., 1924, *Tertiary and Post-Tertiary Geology of Mull, Loch Aline, and Oban*. Geological Survey of Scotland Memoir.

Bakewell, R., 1813, *Introduction to Geology*. London, J. Harding.

Balk, R., 1925, Primary structure of granite massives. *Geological Society of America Bulletin*, v. 36, pp. 679–696.

———, 1937, *Structural Behavior of Igneous Rocks*. Geological Society of America Memoir 5.

Bandy, M. C., and Bandy, J. A., 1955, *De Natura Fossilium (Textbook of Mineralogy)*. New York, Geological Society of America.

Barbarin, B., 1990, Granitoids: main petrogenetic classifications in relation to origin and tectonic setting. *Geological Journal*, v. 25, pp. 227–238.

———, 1999, A review of the relationships between granitoid types, their origins and their geodynamic environments. *Lithos*, v. 46, pp. 605–626.

Barker, D. S., 1983, *Igneous Rocks*. Englewood Cliffs, NJ, Prentice Hall.

Barker, F., ed., 1979, *Trondhjemites, Dacites, and Related Rocks*. Amsterdam, Elsevier.

Barrell, J., 1907, *Geology of the Marysville Mining District*. U. S. Geological Survey Professional Paper 57.

Barrois, C. E., 1884, Mémoire sur le granite de Rostrenon. *Societé Geologique Nordique Annales*, v. 12, p. 14.

Barrow, G., 1893, On an intrusion of muscovite-biotite gneiss in the southeastern Highlands of Scotland, and its accompanying metamorphism. *Quarterly Journal of the Geological Society of London*, v. 49, pp. 330–358.

Barth, T.F.W., 1931, Proposed change in calcualtion of norms of rocks. *Mineralogische und petrographische Mitteilungen*, v. 50, pp. 1–7.

———, 1948, Recent contributions to the granite problem. *Journal of Geology*, v. 56, pp. 235–240.

———, 1952, *Theoretical Petrology: A Textbook on the Origin and the Evolution of Rocks*. New York, John Wiley and Sons.

———, 1955, Température de formation de certains granites Précambriens de Norvège méridionale. International Colloquium. Science de la Terre. *Annales de l'École Nationale Supérieure, Nancy*, pp. 119–129.

Bartholin, E., 1669, *Experimento crystalli islandici dis-diaclastici quibus mira et insolita refractio detegitur*. Copenhagen, n.p.

Barus, C., 1889, *On the Thermo-electric Measurements of High Temperatures*. U. S. Geological Survey Bulletin 54.

Barus, C., and Iddings, J. P., 1892, Note on the change of electric conductivity observed in rock magmas of different composition on passing from liquid to solid. *American Journal of Science*, v. 44 (3rd series), pp. 242–249.

Basaltic Volcanism Study Project, 1981, *Basaltic Volcanism on the Terrestrial Planets*. New York, Pergamon.

Bascom, F., 1927, Fifty years of progress in petrography and petrology 1876–1926, *in* Mathews, E. B., ed., *Fifty Years of Progress in Geology 1876–1926*. Baltimore, MD, Johns Hopkins University Press, pp. 33–82.

Baumgärtel, H., 1976, Tschermak, Gustav, *in* Gillispie, C. C., ed., *Dictionary of Scientific Biography*, v. 13. New York, Charles Scribner's Sons, pp. 475–477.

Beard, J. S., and Lofgren, G. E., 1991, Dehydration melting and water-saturated melting of basaltic and andesitic greenstones and amphibolites at 1, 3 and 6.9 kb. *Journal of Petrology*, v. 32, pp. 365–401.

Becker, G. F., 1882, *Geology of the Comstock Lode and the Washoe District*. U. S. Geological Survey Monograph, v. 3.

———, 1897a, Some queries on rock differentiation. *American Journal of Science*, v. 3 (4th series), pp. 21–40.

———, 1897b, Fractional crystallization of rocks. *American Journal of Science*, v. 4 (4th series), pp. 257–261.

———, 1908, Relations of radioactivity to cosmogony and geology. *Geological Society of America Bulletin*, v. 19, pp. 113–146.

Beckinsale, R. P., 1971, Deluc, Jean André, *in* Gillispie, C. C., ed., *Dictionary of Scientific Biography*, v. 4. New York, Charles Scribner's Sons, pp. 27–29.

———, 1975, Richthofen, Ferdinand von, *in* Gillispie, C. C., ed., *Dictionary of Scientific Biography*, v. 11. New York, Charles Scribner's Sons, pp. 438–441.

Beddoes, T., 1791, Observations on the affinities between basaltes and granite. *Philosophical Transactions of the Royal Society of London*, v. 81, pp. 48–70.

Bell, K., 1989, *Carbonatites: Genesis and Evolution*. London, Unwin Hyman.

Bell, K., and Keller, J., eds., 1995, *Carbonatite Volcanism*. Berlin, Springer-Verlag.

Ben-David, J., 1968, *Fundamental Research and the Universities: Some Comments on International Differences*. Paris, Organization for Economic Co-operation and Development.

———, 1971, *The Scientist's Role in Society: A Comparative Study*. Englewood Cliffs, NJ, Prentice-Hall.

Berner, R. A., and Maasch, K. A., 1996, Chemical weathering and controls on atmospheric O_2 and CO_2: fundamental principles were enunciated by J. J. Ebelmen in 1845. *Geochimica et Cosmochimica Acta*, v. 60, pp. 1633–1637.

Berthelot, P.E.M., 1879, *Essai de Mécanique Chimique Fondée sur la Thermochimie*, 2v. Paris, n.p.

Beudant, F.-S., 1822, *Voyage Minéralogique et Géologique en Hongrie pendant l'Année 1818*, 4v. Paris, Verdière.

Biermann, K.-R., 1972, Humboldt, Friedrich Wilhelm Heinrich Alexander von, in Gillispie, C. C., ed, *Dictionary of Scientific Biography*, v. 6. New York, Charles Scribner's Sons, pp. 549–555.

Billings, M. P., 1945, Mechanics of intrusion in New Hampshire. *American Journal of Science*, v. 243A, Daly volume, pp. 40–68.

———, 1954, *Structural Geology*, 2nd ed. Englewood Cliffs, NJ, Prentice-Hall.

———, 1959, Memorial to Reginald Aldworth Daly. *Proceedings of the Geological Society of America for 1958*, pp. 115–122.

Birch, F., 1961, Reginald Aldworth Daly. *Biographical Memoirs of the National Academy of Sciences*, v. 34, pp. 31–64.

Birembaut, A., 1970a, Aubuisson de Voisins, Jean François d', in Gillispie, C. C., ed., *Dictionary of Scientific Biography*, v. 1. New York, Charles Scribner's Sons, pp. 327–328.

———, 1970b, Boué, Ami, in Gillispie, C. C., ed., *Dictionary of Scientific Biography*, v. 2. New York, Charles Scribner's Sons, pp. 341–342.

———, 1971, Élie de Beaumont, Jean-Baptiste-Armand-Louis Léonce, in Gillispie, C. C., ed., *Dictionary of Scientific Biography*, v. 4. New York, Charles Scribner's Sons, pp. 347–350.

Bischof, G., 1854–1859, *Elements of Chemical and Physical Geology*, 3 v. London, Harrison and Sons.

Blake, S., and Campbell, I. H., 1986, The dynamics of magma-mixing during flow in volcanic conduits. *Contributions to Mineralogy and Petrology*, v. 94, pp. 72–81.

Boettcher, A. L., and Wyllie, P. J., 1968, Melting of granite with excess water to 30 kilobars pressure. *Journal of Geology*, v. 76, pp. 235–244.

Bohrson, W. A., and Spera, F. J., 2001, Energy-constrained open-system magmatic processes II: application of energy-constrained assimilation-fractional crystallization (EC-AFC) model to magmatic systems. *Journal of Petrology*, v. 42, pp. 1019–1041.

Bonney, T. G., 1897, Samuel Allport F.G.S. Born January 23, 1816. Died July 7, 1897. *Geological Magazine*, v. 34, pp. 430–431.

———, 1899, Ward, James Clifton, in Lee, S., ed., *Dictionary of National Biography*, v. 59. London, Smith, Elder, p. 319.

Borelli, G. A., 1670, *Historia et meteorologica incendii Aetnaei anni 1669. Responsio ad censuras Rev. P Honorati Fabri contra libram auctoris De vi percussionis*. Regio Iulio in officina Dominici Ferri.

Bos, H.J.M., 1972, Huygens, Christiaan, in Gillispie, C. C., ed., *Dictionary of Scientific Biography*, v. 6. New York, Charles Scribner's Sons, pp. 597–613.

Bottinga, Y., and Weill, D. F., 1970, Densities of liquid silicate systems calculated from partial molar volumes of oxide components. *American Journal of Science*, v. 269, pp. 169–182.

Boudreau, A. E., and McBirney, A. R., 1997, The Skaergaard Layered Series, part III: non-dynamic layering. *Journal of Petrology*, v. 38, pp. 1003–1020.

Boué, A., 1820a, *Essai Géologique sur l'Écosse*. Paris, V. Coucier.

———, 1820b, Short comparison of the volcanic rocks of France with those of a similar nature found in Scotland. *Edinburgh Philosophical Journal*, v. 2, pp. 326–332.

———, 1822, On the geognosy of Germany, with observations on the igneous origin of trap. *Memoirs of the Wernerian Natural History Society*, v. 4, pp. 91–108.

———, 1825, Synoptical table of the formations of the crust of the earth and of the chief subordinate masses. *Edinburgh Philosophical Journal*, v. 13, pp. 130–145.

———, 1863, Über die mikroskopische Untersuchung der Gebirgsarten mit Hilfe ihrer mechanischen Zerreibung, partieller Schleifung und Ätzung. *Sitzungsberichte der mathematisch-naturwissenschaftliche Classe der kaiserlichen Akademie der Wissenschaften, Wien*, v. 47, pp. 457–460.

Bowen, N. L., 1910, Diabase and granophyre of the Gowganda Lake District, Ontario. *Journal of Geology*, v. 18, pp. 658–674.

———, 1913, The melting phenomena of the plagioclase feldspars. *American Journal of Science*, v. 35 (4th series), pp. 577–599.

———, 1914, The ternary system: diopside–forsterite–silica. *American Journal of Science*, v. 38 (4th series), pp. 207–264.

———, 1915a, Crystallization-differentiation in silicate liquids. *American Journal of Science*, v. 39 (4th series), pp. 175–191.

———, 1915b, The crystallization of haplobasaltic, haplodioritic and related magmas. *American Journal of Science*, v. 40 (4th series), pp. 161–185.

———, 1915c, The later stages of the evolution of the igneous rocks. *Journal of Geology*, v. 23, supplement to no. 8, pp. 1–91.

———, 1917, The problem of the anorthosites. *Journal of Geology*, v. 25, pp. 209–243.

———, 1919, Crystallization-differentiation in igneous magmas. *Journal of Geology*, v. 27, pp. 393–430.

———, 1920, Crystallization-differentiation. *Geological Magazine*, v. 7 (6th series), pp. 238–239.

———, 1921, Diffusion in silicate melts. *Journal of Geology*, v. 29, pp. 295–317.

———, 1922a, The reaction principle in petrogenesis. *Journal of Geology*, v. 30, pp. 177–198.

———, 1922b, The behavior of inclusions in igneous magmas. *Journal of Geology*, v. 30, pp. 513–570.

———, 1928, *The Evolution of the Igneous Rocks*. Princeton, NJ, Princeton University Press.

———, 1937, Recent high-temperature research on silicates and its significance in igneous geology: *American Journal of Science*, v. 33 (5th series), pp. 1–21.

———, 1947, Magmas. *Geological Society of America Bulletin*, v. 58, pp. 263–279.

———, 1948, The granite problem and the method of multiple prejudices, in Gilluly, J., ed., *Origin of Granite*. Geological Society of America Memoir 28, pp. 79–90.

Bowen, N. L., and Andersen, O., 1914, The binary system MgO–SiO$_2$. *American Journal of Science*, v. 37, (4th series), pp. 487–500.

Bowen, N. L., and Schairer, J. F., 1929, The fusion relations of acmite. *American Journal of Science*, v. 18 (5th series), pp. 365–374.

———, 1932, The system, FeO–SiO$_2$. *American Journal of Science*, v. 24 (5th series), pp. 177–213.

———, 1935, The system, MgO–FeO–SiO$_2$. *American Journal of Science*, v. 29 (5th series), pp. 151–217.

———, 1936, The system, albite–fayalite. *Proceedings of the National Academy of Sciences*, v. 22, pp. 345–350.

———, 1938, Crystallization equilibrium in nepheline–albite–silica mixtures with fayalite. *Journal of Geology*, v. 46, pp. 397–411.

Bowen, N. L., Schairer, J. F., and Posnjak, E., 1933a, The system, Ca$_2$SiO$_4$–Fe$_2$SiO$_4$. *American Journal of Science*, v. 25 (5th series), pp. 273–297.

———, 1933b, The system, CaO–FeO–SiO$_2$. *American Journal of Science*, v. 26 (5th series), pp. 193–284.

Bowen, N. L., Schairer, J. F., and Willems, H.W.V., 1930, The ternary system: Na$_2$SiO$_3$–Fe$_2$O$_3$–SiO$_2$. *American Journal of Science*, v. 20 (5th series), pp. 405–455.

Bowlby, J., 1990, *Charles Darwin: A New Life*. New York, W. W. Norton.

Bowler, P., 1990, *Charles Darwin: The Man and his Influence*. Oxford, UK, Basil Blackwell.

Boyd, F. R., and England, J. L., 1960, Apparatus for phase-equilibrium measurements at pressures up to 50 kilobars and temperatures up to 1750°C. *Journal of Geophysical Research*, v. 65, pp. 741–748.

Boyd, F. R., and Meyer, H.O.A., eds., 1979a, *Kimberlites, Diatremes, and Diamonds: Their Geology, Petrology, and Geochemistry*. Washington, DC, American Geophysical Union.

———, eds., 1979b, *The Mantle Sample: Inclusions in Kimberlites and Other Volcanics*. Washington, DC, American Geophysical Union.

Branagan, D. F., and Townley, K. A., 1976, The geological sciences in Australia—a brief review. *Earth Science Reviews*, v. 12, pp. 323–346.

Brandeis, G., and Jaupart, C., 1986, On the interaction between convection and crystallization in cooling magma chambers. *Earth and Planetary Science Letters*, v. 82, pp. 345–362.

———, 1987a, Crystal sizes in intrusions of different dimensions: constraints of the cooling regime and the crystallization kinetics. *Geochemical Society Special Publication 1*, pp. 307–318.

———, 1987b, The kinetics of nucleation and crystal growth and scaling laws for magmatic crystallization. *Contributions to Mineralogy and Petrology*, v. 96, pp. 24–34.

Bray, J. M., 1942, Spectroscopic distribution of minor elements in igneous rocks from Jamestown, Colorado. *Geological Society of America Bulletin*, v. 53, pp. 765–814.

Brögger, W. C., 1890, Die Mineralien der Syenitpegmatitgänge der südnorwegischen Augit- und Nephelinsyenite. *Zietschrift für Krystallographie*, v. 16, pp. 1–235.

———, 1894a, *Die Eruptivgesteine des Kristianiagebietes, I: Die Gesteine der Grorudit-Tinguait-Serie*. Kristiania, Jacob Dybwad.

———, 1894b, The basic eruptive rocks of Gran. *Quarterly Journal of the Geological Society of London*, v. 50, pp. 15–38.

Brögger, W. C., 1898, *Die Eruptivgesteine des Kristianiagebietes, III: Das Ganggefolge des Laurdalite*. Kristiania, Jacob Dybwad.

———, 1920, *Die Eruptivgesteine des Oslogebietes, IV: Das Fengebiet in Telemark, Norwegen*. K. Norske Vidensk. Selsk. Skrift, no. 9.

Brongniart, A., 1813, Tableau de la Classification des Roches Mélanges. *Journal des Mines*, v. 34, pp. 31–48.

———, 1827, *Classification et Caractères Minéralogiques des Roches Homogènes et Hétérogènes*. Paris, F. G. Levrault.

Brooks, C. K., 1981, Laurence Rickard Wager, in Gillispie, C. C., ed., *Dictionary of Scientific Biography*, v. 18. New York, Charles Scribner's Sons, pp. 968–970.

Brown, G. C., and Fyfe, W. S., 1970, The production of granitic melts during ultrametamorphism. *Contributions to Mineralogy and Petrology*, v. 28, pp. 310–318.

Brown, G. M., 1956, The layered ultrabasic rocks of Rhum, Inner Hebrides. *Philosophical Transactions of the Royal Society of London*, v. B240, pp. 1–53.

Brown, M., Candela, P. A., Peck, D. L., Stephens, W. E., Walker, R. J., and Zen, E., eds., 1996, *The Third Hutton Symposium on the Origin of Granites and Related Rocks*. Geological Society of America Special Paper 315.

Buddington, A. F., 1939, *Adirondack Igneous Rocks and Their Metamorphism*. Geological Society of America Memoir 7.

Buddington, A. F., and Hess, H. H., 1937, Layered peridotite laccoliths in the Trout River area, Newfoundland (a discussion). *American Journal of Science*, v. 33 (5th series), pp. 380–388.

Buddington, A. F., and Lindsley, D. H., 1964, Iron–titanium oxide minerals and synthetic equivalents. *Journal of Petrology*, v. 5, pp. 310–357.

Bulman, O.M.B., 1950, Obituary R. H. Rastall, 1871–1950. *Geological Magazine*, v. 87, pp. 73–76.

Bunsen, R., 1851, Über die Processe der vulkanischen Gesteinsbildung Islands. *Annalen der Physik und Chimie*, v. 83, pp. 197–272.

———, 1861, Über die Bildung des Granites. *Zeitschrift der deutschen geologischen Gesellschaft*, v. 13, pp. 61–63.

Burckhardt, C, 1906, *Geologie de la Sierra de Mazapit*. IGC Mexico, 10th session, Guide des Excursions, no. 26, p. 33.

Burke, J. G., 1971, Cordier, Pierre-Louis-Antoine, in Gillispie, C. C., ed., *Dictionary of Scientific Biography*, v. 3. New York, Charles Scribner's Sons, pp. 411–412.

———, 1972, Fouqué, Ferdinand André, in Gillispie, C. C., ed., *Dictionary of Scientific Biography*, v. 5. New York, Charles Scribner's Sons, pp. 88–89.

———, 1973, Leonhard, Karl Cäsar von, in Gillispie, C. C., ed., *Dictionary of Scientific Biography*, v. 8. New York, Charles Scribner's Sons, pp. 245–246.

———, 1974a, Michel-Lévy, Auguste, in Gillispie, C. C., ed., *Dictionary of Scientific Biography*, v. 9. New York, Charles Scribner's Sons, pp. 366–367.

———, 1974b, Naumann, Karl Friedrich, in Gillispie, C. C., ed., *Dictionary of Scientific Biography*, v. 9. New York, Charles Scribner's Sons, p. 620.

Cahan, D., 1989, *An Institute for an Empire: The Physikalisch Technische Reichsanstalt, 1871–1918*. Cambridge, UK, Cambridge University Press.

Cameron, H. C., 1952, *Sir Joseph Banks*. London, Batchworth.

Campbell, I. H., 1978, Some problems with the cumulus theory. *Lithos*, v. 11, pp. 311–323.

Campbell, I. H., and Turner, J. S., 1986, The influence of viscosity on fountains in magma chambers. *Journal of Petrology*, v. 27, pp. 1–30.

———, 1987, A laboratory investigation of assimilation at the top of a basaltic magma chamber. *Journal of Geology*, v. 95, pp. 155–172.

———, 1989, Fountains in magma chambers. *Journal of Petrology*, v. 30, pp. 885–923.

Campbell, I. H., McCall, G.J.H., and Tyrwhitt, D. S., 1970, The Jimberlana norite, western Australia—a smaller analogue of the Great Dyke of Rhodesia. *Geological Magazine*, v. 107, pp. 1–12.

Campbell, I. H., Roeder, P. L., and Dixon, J. M., 1978, Plagioclase buoyancy in basaltic liquids as determined with a centrifuge furnace. *Contributions to Mineralogy and Petrology*, v. 67, pp. 369–377.

Carmichael, I.S.E., Turner, F. J., and Verhoogen, J., 1974, *Igneous Petrology*. New York, McGraw-Hill.

Caroe, G. M., 1978, *William Henry Bragg 1862–1942: Man and Scientist*. Cambridge, UK, Cambridge University Press.

Carozzi, A. V., 1969, Rudolph Erich Raspe and the basalt controversy. *Studies in Romanticism*, v. 8, pp. 235–250.

———, 1972, William Hamilton, in C. C. Gillispie, ed., *Dictionary of Scientific Biography*, v. 6. New York, Charles Scribner's Sons, pp. 83–85.

———, 1975, Raspe, Rudolf Erich, in Gillispie, C. C., ed., *Dictionary of Scientific Biography*, v. 11. New York, Charles Scribner's Sons, pp. 302–305.

Carr, J., 1973, Martin Lister, in Gillispie, C. C., ed., *Dictionary of Scientific Biography*, v. 8. New York, Charles Scribner's Sons, pp. 415–417.

Carrigan, C. R., 1987, The magmatic Rayleigh number and time dependent convection in cooling lava lakes. *Geophysics Research Letters*, v. 14, pp. 915–918.

Cashman, K. V., 1992, Groundmass crystallization of Mount St. Helens dacite 1980–1986: A tool for interpreting shallow magmatic processes. *Contributions to Mineralogy and Petrology*, v. 109, pp. 431–449.

———, 1993, Relationship between plagioclase crystallization and cooling rate in basaltic melts. *Contributions to Mineralogy and Petrology*, v. 113, pp. 126–142.

Cashman, K. V., and Marsh, B. D., 1988, Crystal size distribution (CSD) in rocks and the kinetics and dynamics of crystallization, II: Makaopuhi lava lake. *Contributions to Mineralogy and Petrology*, v. 99, pp. 401–405.

Cawthorn, R. G., ed., 1996, *Layered Intrusions*. Amsterdam, Elsevier.

Challinor, J., 1971, Faujas de St. Fond, Barthélemy, in Gillispie, C. C., ed., *Dictionary of Scientific Biography*, v. 4. New York, Charles Scribner's Sons, pp. 548–549.

———, 1975, Playfair, John, in Gillispie, C. C., ed., *Dictionary of Scientific Biography*, v. 11. New York, Charles Scribner's Sons, pp. 34–36.

———, 1976, Whitehurst, John, in Gillispie, C. C., ed., *Dictionary of Scientific Biography*, v. 14. New York, Charles Scribner's Sons, pp. 311–312.

Chamberlin, R. T., and Link, T. A., 1927, The theory of laterally spreading batholiths. *Journal of Geology*, v. 35, pp. 319–352.

Chamberlin, T. C., and Salisbury, R. D., 1909, *Geology*, 2nd ed., 3 v. New York, H. Holt.

Champlin, M. D., 1994, *Raphael Pumpelly: Gentleman Geologist of the Gilded Age*. Tuscaloosa, AL, University of Alabama Press.

Chapman, C. A., and Schweitzer, G. K., 1947, Trace elements in rocks of the Oliverian magma series of New Hampshire. *American Journal of Science*, v. 245, pp. 597–613.

Chapman, R. W., and Chapman, C. A., 1940, Cauldron subsidence at Ascutney Mountain, Vermont. *Geological Society of America Bulletin*, v. 51, pp. 191–211.

Chappell, B. W., and White, A.J.R., 1974, Two contrasting granite types. *Pacific Geology*, v. 8, pp. 173–174.

Chayes, F., 1946, Linear analysis of a medium-grained granite. *American Mineralogist*, v. 31, pp. 261–275.

———, 1949, A simple point counter for thin-section analysis. *American Mineralogist*, v. 34, pp. 1–11.

———, 1956, *Petrographic Modal Analysis*. New York, John Wiley and Sons.

———, 1958, Memorial of Samuel James Shand. *American Mineralogist*, v. 43, pp. 317–324.

Chayes, F., and Velde, D, 1965, On distinguishing basaltic lavas of circumoceanic and oceanic-island type by means of discriminant functions. *American Journal of Science*, v. 263, pp. 206–222.

Chen, C. F., and Turner, J. S., 1980, Crystallization in a double-diffusive system. *Journal of Geophysical Research*, v. 85, pp. 2573–2593.

Chinner, G. A., 1974, Memorial of Cecil Edgar Tilley May 14, 1894– January 24, 1973. *American Mineralogist*, v. 59, pp. 427–437.

Chitnis, A. C., 1970, The University of Edinburgh's natural history museum and the Huttonian–Wernerian debate. *Annals of Science*, v. 26, pp. 85–94.

Chorley, R. J., 1971, Daubrée, Gabriel-Auguste, in Gillispie, C. C., ed., *Dictionary of Scientific Biography*, v. 3. New York, Charles Scribner's Sons, pp. 586–587.

Clark, W. B.,1895, Memorial of George Huntington Williams. *Geological Society of America Bulletin*, v. 6, pp. 432–440.

Clarke, D. B., 1992, *Granitoid Rocks*. London, Chapman & Hall.

Clarke, F. W., 1908, *The Data of Geochemistry*. U. S. Geological Survey Bulletin 330.

———, 1911, *The Data of Geochemistry*. U. S. Geological Survey Bulletin 491.

———, 1915, *Analyses of Rocks and Minerals from the United States Geological Survey 1880 to 1914*. U. S. Geological Survey Bulletin 591.

———, 1916, *The Data of Geochemistry*. U. S. Geological Survey Bulletin 616.

———, 1920, *The Data of Geochemistry*. U. S. Geological Survey Bulletin 695.

———, 1924, *The Data of Geochemistry*. U. S. Geological Survey Bulletin 770.

Claudy, N. H., ed., 1996, *Directory of Geoscience Departments*, 35th ed. Alexandria, VA, Americal Geological Institute.

Clemens, J. D., and Mawer, C. K., 1992, Granitic magma transport by fracture propagation. *Tectonophysics*, v. 204, pp. 339–360.

Cloos, H., 1921, *Der Mechanismus tiefvulkanischer Vorgänge*. Braunschweig, Vieweg und Sohn.

Clough, C. T., Maufe, H. B., and Bailey, E. B., 1909, The cauldron subsidence of Glen Coe, and associated igneous phenomena. *Quarterly Journal of the Geological Society of London*, v. 65, pp. 611–678.

Coleman, R. G., 1977, *Ophiolites: Ancient Oceanic Lithosphere?* Berlin, Springer-Verlag.

Conybeare, W., 1816, Descriptive notes referring to the outline of sections presented by a part of the coasts of Antrim and Derry. *Transactions of the Geological Society of London*, v. 3, pp. 196–216.

Conybeare, W., and Phillips, W., 1822, *Outlines of the Geology of England and Wales*. London, n.p.

Corcoran, T. H., 1971, *Naturales Quaestiones*, 10 v. London, Wm. Heinemann.

Cordier, P. L., 1816, Mémoire sur les substances minérales dites en masse, qui entrent dans la composition des roches volcaniques de tous les ages. *Journal de Physique, de Chimie et d'Histoire Naturelle*, v. 83, pp. 135–163.

Corry, C. E., 1988, *Laccoliths: Mechanics of Emplacement and Growth*. Geological Society of America Special Paper 220.

Coryell, C. D., Chase, J. W., and Winchester, J. W., 1963, A procedure for geochemical interpretation of terrestrial rare-earth abundance patterns. *Journal of Geophysical Research*, v. 68, pp. 555–566.

Craig, H., 1985, Introduction of Alfred O. C. Nier for the V. M. Goldschmidt award 1984. *Geochimica et Cosmochimica Acta*, v. 49, pp. 1661–1665.

Craven, M., 1996, *John Whitehurst of Derby, Clockmaker and Scientist, 1713–88*. Mayfield.

Crawford, A. J., ed., 1989, *Boninites*. London, Unwin Hyman.

Creese, M.R.S., and T. M. Creese, 1994, British women who contributed to research in the geological sciences in the nineteenth century. *British Journal for the History of Science*, v. 27, pp. 23–54.

Crosland, M. P., 1970a, Berthelot, Pierre Eugène Marcellin, *in* Gillispie, C. C., ed., *Dictionary of Scientific Biography*, v. 2. New York, Charles Scribner's Sons, pp. 63–72.

———, 1970b, Biot, Jean-Baptiste, *in* Gillispie, C. C., ed., *Dictionary of Scientific Biography*, v. 2. New York, Charles Scribner's Sons, pp. 133–140.

Cross, W. C., 1896, *Geology of Silver Cliff and the Rosita Hills, Colo.* U. S. Geological Survey Annual Report 17, Part II.

———, 1898, The geological versus the petrographical classification of igneous rocks. *Journal of Geology*, v. 6, pp. 79–91.

———, 1902, The development of systematic petrography in the nineteenth century. *Journal of Geology*, v. 10, pp. 331–376, 451–499.

Cross, W., 1910, The natural classification of igneous rocks. *Quarterly Journal of the Geological Society of London*, v. 66, pp. 470–506.

Cross, W., Iddings, J. P., Pirsson, L. V., and Washington, H. S., 1902, A quantitative chemico-mineralogical classification and nomenclature of igneous rocks. *Journal of Geology*, v. 10, pp. 555–693.

———, 1903, *Quantitative Classification of Igneous Rocks*. Chicago, University of Chicago Press.

———, 1906, The texture of igneous rocks. *Journal of Geology*, v. 14, pp. 692–707.

Cusack, R., 1896, On the melting points of minerals. *Proceedings of the Royal Irish Academy*, v. 4 (3rd series), pp. 399–413.

Dakyns, J. R., and Teall, J.J.H., 1892, On the plutonic rocks of Garabal Hill and Meall Breac. *Quarterly Journal of the Geological Society of London*, v. 48, pp. 104–121.

Daly, R. A., 1903a, The mechanics of igneous intrusion. *American Journal of Science*, v. 15 (4th series), pp. 269–298.

———, 1903b, The mechanics of igneous intrusion. *American Journal of Science*, v. 16 (4th series), pp. 107–126.

———, 1905, The classification of igneous intrusive bodies. *Journal of Geology*, v. 13, pp. 485–598.

Daly, R. A., 1908, The mechanics of igneous intrusion. *American Journal of Science*, v. 26 (4th series), pp. 17–50.

———, 1910, Origin of the alkaline rocks. *Geological Society of America Bulletin*, v. 21, pp. 87–118.

———, 1912, *Geology of the North American Cordillera at the 49th Parallel of Latitude*, 3 parts. Geological Society of Canada Memoir 38.

———, 1914, *Igneous Rocks and Their Origin*. New York, McGraw Hill.

———, 1918, Genesis of the alkaline rocks. *Journal of Geology*, v. 26, pp. 97–134.

———, 1933, *Igneous Rocks and the Depths of the Earth*. New York: McGraw-Hill.

———, 1949, Granite and metasomatism. *American Journal of Science*, v. 247, pp. 753–778.

Dana, J. D., 1849, *Geology. Exploring Expedition During the Years 1838–1842 Under the Command of Charles Wilkes, U.S.N.*, v. 10. New York, G. P. Putnam.

———, 1866, *Manual of Geology*, 2nd ed. Philadelphia, PA, Theodore Bliss.

———, 1880, Gilbert's report on the geology of the Henry Mountains. *American Journal of Science*, v. 19 (3rd series), pp. 17–25.

Darken, L. S., and Gurry, R. W., 1945, The system iron-oxygen, I: the wüstite field and related equilibria. *Journal of the American Chemical Society*, v. 67, pp. 1398–1412.

Darwin, C., 1844, *Geological Observations on the Volcanic Islands*. London, Appleton.

Daubeny, C., 1820, On the volcanoes of Auvergne. *Edinburgh Philosophical Journal*, v. 3, pp. 359–367.

———, 1821a, On the volcanoes of Auvergne. *Edinburgh Philosophical Journal*, v. 4, pp. 89–97.

———, 1821b, On the ancient volcanoes of Auvergne. *Edinburgh Philosophical Journal*, v. 4, pp. 301–315.

———, 1826, *A Description of Active and Extinct Volcanos*. London: W. Phillips.

Daubrée, G. A., 1857, Sur le métamorphisme et recherches expérimentales sur quelques-uns des agents qui ont pu le produire. *Annales des Mines*, v. 12 (5th series), pp. 289–326.

d'Aubuisson de Voisins, J. F., 1814, *An Account of the Basalts of Saxony with Observations on the Origin of Basalt in General*, trans. by P. Neill. Edinburgh, A. Constable & Co.

———, 1819, *Traité de Géognosie*, 2 v. Strasbourg, F. G. Levrault.

Davaille, A., and Jaupart, C., 1993a, Thermal convection in lava lakes. *Geophysical Research Letters*, v. 20, pp. 1827–1830.

———, 1993b, Transient high-Rayleigh-number thermal convection with large viscosity variations. *Journal of Fluid Mechanics*, v. 253, pp. 141–166.

Davies, G.L.H., 1969, *The Earth in Decay: A History of British Geomorphology, 1578 to 1878*. New York, Elsevier.

Dawson, J. B., 1980, *Kimberlites and Their Xenoliths*. Berlin, Springer-Verlag.

———, 1992, First thin sections of experimentally melted rocks: Sorby's observations on magma crystallization. *Journal of Geology*, v. 100, pp. 251–257.

Day, A. L., and Allen, E. T., 1904, Temperature measurements to 1600°C. *Physical Review*, v. 19, pp. 177–186.

Day, A. L., Allen, E. T., and Iddings, J. P., 1905, *The Isomorphism and Thermal Properties of the Feldspars*. Carnegie Institution of Washington Publication 31.

Day, A. L., and Clement, J. K., 1908, Some new measurements with the gas thermometer. *American Journal of Science*, v. 26 (4th series), pp. 405–463.

Day, A. L., Shepherd, E. S., and Wright, F. E., 1906, The lime–silica series of minerals. *American Journal of Science*, v. 22 (4th series), pp. 265–302.

Day, A. L., Sosman, R. B., and Allen, E. T., 1910, The nitrogen thermometer from zinc to palladium. *American Journal of Science*, v. 29 (4th series), pp. 93–161.

Dean, D. R., 1980, Graham Island, Charles Lyell, and the craters of elevation controversy. *Isis*, v. 71, pp. 571–588.

———, 1992, *James Hutton and the History of Geology*. Ithaca, NY, Cornell University Press.

———, ed., 1997, *James Hutton in the Field and in the Study*. Delmar, NY, Scholars' Facsimiles and Reprints.

———, 1999, *Gideon Mantell and the Discovery of Dinosaurs*. Cambridge, UK, Cambridge University Press.

De Beer, G., 1957, Gregory Watt's tour on the continent, 1801. *Annals of Science*, v. 13, pp. 127–136.

———, 1962, The volcanoes of Auvergne. *Annals of Science*, v. 18, pp. 49–61.

Debon, F., and Le Fort, P., 1983, A chemical–mineralogical classification of common plutonic rocks and associations. *Transactions of the Royal Society of Edinburgh: Earth Sciences*, v. 73, pp. 135–149.

Deer, W. A., 1967, L. R. Wager. *Biographical Memoirs of Fellows of the Royal Society*, v. 13, pp. 359–385.

de Goër, A., Boivin, P., Camus, G., Gourgaud, A., Kieffer, G., Mergoil, J., and Vincent, P. M., 1994, *Volcanology of the Chaîne des Puys*. Clermont-Ferrand, Parc Naturel Régional des Volcans d'Auvergne.

Deiters, M., 1861, Die Trachydolerite des Siebengebirges. *Zeitschrift der deutschen geologischen Gesellschaft*, v. 13, pp. 99–135.

Delesse, A., 1847, Procédé méchanique pour déterminer la composition des roches. *Comptes Rendus Hebdomadaires des Séances de l'Académie des Sciences*, v. 25, pp. 544–545.

———, 1848, Procédé méchanique pour déterminer la composition des roches. *Annales des Mines*, v. 13, pp. 379–388.

———, 1857, Études sur le métamorphisme. *Annales des Mines*, v. 12 (5th series), pp. 705–772.

della Torre, G. M., 1755, *Storia e Fenomeni del Vesuvio*. Napoli, G. Raimondi.

de Margerie, E., 1953, François Antoine Alfred Lacroix. *Obituary Notices of Fellows of the Royal Society*, v. 8, pp. 193–205.

den Tex, E., 1996, Clinchers of the basalt controversy: empirical and experimental evidence. *Earth Sciences History*, v. 15, pp. 37–48.

DePaolo, D. J., 1981, Trace element and isotopic effects of combined wall rock assimilation and fractional crystallization. *Earth and Planetary Science Letters*, v. 53, pp. 189–202.

———, 1985, Isotopic studies of processes in mafic magma chambers, I: The Kiglapait intrusion, Labrador. *Journal of Petrology*, v. 26, pp. 925–951.

———, 1988, *Neodymium Isotope Geochemistry: An Introduction*. Berlin, Springer-Verlag.

Des Cloizeaux, A., 1862, *Manuel de Minéralogie*, 2 v. Paris, n.p.

Desmarest, N., 1774, Mémoire sur l'origine et la nature du basalte. *Mémoires de l'Académie Royale des Sciences pour l'Année 1771*, pp. 705–775.

Desmond, A. J., 1975, The discovery of marine transgression and the explanation of fossils in antiquity. *American Journal of Science*, v. 275, pp. 692–707.

———, 1982, *Archetypes and Ancestors: Palaeontology in Victorian London, 1850–1875*. Chicago, University of Chicago Press.

Desmond, A. J., and J. Moore, 1991, *Darwin*. New York, Warner Books.

De Vore, G. W., 1973, Memorial of Walter Harry Newhouse December 13, 1897–September 21, 1969. *American Mineralogist*, v. 58, pp. 380–382.

Dickson, F. W., and Krauskopf, K. B., 1973, Memorial of Tom F. W. Barth May 18, 1899—March 7, 1971. *American Mineralogist*, v. 58, pp. 360–363.

Didier, J., 1973, *Granites and Their Enclaves: The Bearing of Enclaves on the Origin of Granites*. Amsterdam, Elsevier.

Didier, J., and Barbarin, B., eds., 1991, *Enclaves and Granite Petrology*. Amsterdam, Elsevier.

Dietz, R. S., 1961, Continent and ocean basin evolution by spreading of the sea floor. *Nature*, v. 190, pp. 854–857.

Doe, B. R., 1970, *Lead Isotopes*. New York, Springer-Verlag.

Doelter, C., 1906, *Die Petrogenesis*. Braunschweig, F Vieweg und Sohn.

Dolman, C. E., 1975, Lazarro Spallanzani, in C. C. Gillispie, ed., *Dictionary of Scientific Biography*, v. 12: New York, Charles Scribner's Sons, pp. 553–567.

Dolomieu, D., 1788, *Mémoire sur les Iles Ponces, et Catalogue Raisonné des Produits de l'Etna*. Paris, Cuchet.

———, 1794, Mémoire sur les roches composées en général et particulièrement sur les pétro-silex, les trapps et les roches de corne, pour servir à la distribution méthodique des produits volcaniques. *Journal de Physique*, v. 44, pp. 175–200.

———, 1796, Lettre sur la nécessité d'unir les connaissances chimiques à celles du minéralogiste; avec des observations sur la différente acception que les auteurs allemands et français donnent au mot chrysolithe. *Journal des Mines*, an. V., pp. 365–376.

Drake, E. T., 1996, *Restless Genius: Robert Hooke and His Earthly Thoughts*. New York, Oxford University Press.

Drever, H. I., 1969, Frederick Walker (1898–1968). *Proceedings of the Geological Society of London*, no. 1651, pp. 232–233.

Dupree, A. H., 1957, *Science in the Federal Government: A History of Policies and Activities to 1940*. Cambridge, MA, Harvard University Press.

Durocher, J., 1845, Sur l'origine des roches granitiques. *Comptes Rendus Hebdomadaires des Séances de l'Académie des Sciences*, v. 20, pp. 1275–1284.

———, 1847, Recherches sur la cristallisation des roches granitiques. *Bulletin de la Societé Géologique de France*, v. 4 (2nd series), pp. 1018–1043.

———, 1857, Essai de pétrologie comparée, ou recherches sur la composition chimique et minéralogique des roches ignées, sur les phénomènes de leur émission et sur leur classification. *Annales des Mines*, v. 11, pp. 217–259.

Dutton, C. E., 1880, *Report on the Geology of the High Plateaus of Utah with Atlas*. Washington, DC, Government Printing Office.

Eckel, E. B., 1982, *The Geological Society of America: Life History of a Learned Society*. Geological Society of America Memoir 155.

Edmonds, J. M., 1974, Phillips, John, in Gillispie, C. C., ed., *Dictionary of Scientific Biography*, v. 10: New York, Charles Scribner's Sons, pp. 583–584.

Edwards, A. B., 1942, Differentiation of the dolerites of Tasmania. *Journal of Geology*, v. 50, pp. 451–480.
Élie de Beaumont, L., 1847, Note sur les émanations volcaniques et métalliferes. *Bulletin de la Societé Géologique de France*, v. 4 (2nd series), pp. 1249–1333.
Ellenberger, F., 1996, *History of Geology*, v. 1: *From Ancient Times to the First Half of the XVII Century*. Brookfield, VT, A. A. Balkema.
———, 1999, *History of Geology*, v. 2: *The Great Awakening and Its First Fruits: 1660–1810*. Brookfield, VT, A. A. Balkema.
Emmons, E., 1860, *Manual of Geology*. New York, A. S. Barnes & Burr.
Emmons, W. H., 1933, The basal regions of granitic batholiths. *Journal of Geology*, v. 41, pp. 1–11.
Engel, A.E.J., and Engel, C. G., 1964, Igneous rocks of the East Pacific Rise. *Science*, v. 146, pp. 477–485.
Engel, A.E.J., Engel, C. G., and Havens, R. G., 1965, Chemical characteristics of oceanic basalts and the upper mantle. *Geological Society of America Bulletin*, v. 76, pp. 719–734.
Epstein, S., 1997, The role of stable isotopes in geochemistries of all kinds, *in* Jeanloz, R., Albee, A. L., and Burke, K.C., eds. *Annual Review of Earth and Planetary Sciences*, v. 25, pp. 1–21.
Erdmannsdörffer, O, 1924, *Grundlagen der Petrographie*. Stuttgart, F. Enke.
Eskola, P., 1932, On the origin of granitic magmas. *Mineralogische und Petrographische Mitteilungen*, v. 42, pp. 455–481.
———, 1933, On the differential anatexis of rocks. *Comptes Rendus de la Société Géologique de Finlande*, no. 103, pp. 12–25.
———, 1950, The nature of metasomatism in the processes of granitization. *International Geological Congress Report of the 18th Session, Great Britain 1948*, Part III, pp. 5–13.
Eugster, H. P., 1957, Heterogeneous reactions involving oxidation and reduction at high temperatures and pressures. *Journal of Chemistry and Physics*, v. 26, pp. 1760–1761.
Eugster, H. P., and Wones, D. R., 1962, Stability relations of the ferruginous biotite, annite. *Journal of Petrology*, v. 3, pp. 82–125.
Eyles, J. M., 1973a, Jameson, Robert, *in* Gillispie, C. C., ed., *Dictionary of Scientific Biography*, v. 7. New York, Charles Scribner's Sons, pp. 69–71.
———, 1973b, Jukes, Joseph Beete, *in* Gillispie, C. C., ed., *Dictionary of Scientific Biography*, v. 7. New York, Charles Scribner's Sons, pp. 185–186.
Eyles, V. A., 1950, The first national geological survey. *Geological Magazine*, v. 87, pp. 373–382.
———, 1972, Hall, Sir James, *in* Gillispie, C. C., ed., *Dictionary of Scientific Biography*, v. 6. New York, Charles Scribner's Sons, pp. 53–56.
———, 1973a, Joly, John, *in* Gillispie, C. C., ed., *Dictionary of Scientific Biography*, v. 7. New York, Charles Scribner's Sons, pp. 160–161.
———, 1973b, Macculloch, John, *in* Gillispie, C. C., ed., *Dictionary of Scientific Biography*, v. 8. New York, Charles Scribner's Sons, pp. 593–595.
Fairbairn, H. W., 1953, Precision and accuracy of chemical analysis of silicate rocks. *Geochimica et Cosmochimica Acta*, v. 4, pp. 143–156.

Fairbairn, H. W., et al., 1951, *A Cooperative Investigation of Precision and Accuracy in Chemical, Spectrochemical, and Modal Analysis of Silicate Rocks*. U. S. Geological Survey Bulletin 980.

Fairclough, H. R., 1994, *Eclogues, Georgics, Aeneid I–VI*. Cambridge, MA, Harvard University Press.

Faujas de St. Fond, B., 1778, *Recherches sur les Volcans Éteints du Vivarais et du Velay*. Grenoble, J. Cuchet.

———, 1799, *A Journey Through England and Scotland to the Hebrides in 1784*, 2 v. Glasgow, Hugh Hopkins.

Faul, H., and Faul, C., 1983, *It Began with a Stone*. New York, John Wiley and Sons.

Faure, G., 1977, *Principles of Isotope Geology*. New York, John Wiley and Sons.

———, 1986, *Principles of Isotope Geology*, 2nd ed. New York, John Wiley and Sons.

———, 2001, *Origin of Igneous Rocks: The Isotopic Evidence*. Berlin, Springer-Verlag.

Faure, G., and Hurley, P. M., 1963, The isotopic composition of strontium in oceanic and continental basalts: Application to the origin of igneous rocks. *Journal of Petrology*, v. 4, pp. 31–50.

Faure, G., and Powell, J. L., 1972, *Strontium Isotope Geology*. New York, Springer-Verlag.

Fenner, C. N., 1913, The stability relations of the silica minerals. *American Journal of Science*, v. 36 (4th series), pp. 331–384.

———, 1926, The Katmai magmatic province. *Journal of Geology*, v. 34, pp. 673–772.

———, 1929, The crystallization of basalts. *American Journal of Science*, v. 18 (5th series), pp. 225–253.

Ferber, J. J., 1776, *Travels Through Italy in the Years 1771 and 1772 Described in a Series of Letters to Baron Born on the Natural History, Particularly the Mountains and Volcanos of That Country*. London, L. Davis.

Fischer, W., 1961, *Gesteins- und Lagerstättenbildung im Wandel der wissenschaftlichen Anschauung*. Stuttgart, E. Schweizerbart'sche.

———, 1971, Doelter (Cisterich y de la Torre), Cornelio August Severinus, *in* Gillispie, C. C., ed., *Dictionary of Scientific Biography*, v. 4. New York, Charles Scribner's Sons, pp. 140–142.

Fisher, D. J., 1963, *The Seventy Years of the Department of Geology, University of Chicago, 1892–1961*. Chicago, University of Chicago Press.

Fitton, J. G., and Upton, B.G.J., eds., 1987, *Alkaline Igneous Rocks*. Oxford, UK, Blackwell.

Floyd, P. A., ed., 1991, *Oceanic Basalts*. Glasgow, Blackie.

Foley, S., 1694, An account of the Giants Causway in the North of Ireland. *Philosophical Transactions of the Royal Society of London*, v. 18, pp. 170–175.

Foote, G. A., 1970, Joseph Banks, *in* Gillispie, C. C., ed., *Dictionary of Scientific Biography*, v. 1. New York, Charles Scribner's Sons, pp. 433–437.

Forbes, D., 1866, On the igneous rocks of Staffordshire. *Geological Magazine*, v. 3, pp. 23–27

———, 1867a, On the alleged hydrothermal origin of certain granites and metamorphic rocks. *Geological Magazine*, v. 4, pp. 49–58.

———, 1867b, On the chemistry of the primeval earth. *Geological Magazine*, v. 4, pp. 433–444.

———, 1867c, The microscope in geology. *Popular Science Review*, v. 6, pp. 355–368.

Förster, H.-J., Tischendorf, G., and Trumbull, R. B., 1997, An evaluation of the Rb vs. (Y + Nb) discrimination diagram to infer tectonic setting of silicic igneous rocks. *Lithos*, v. 40, pp. 261–293.

Fouqué, F., and Michel-Lévy, A., 1879, *Minéralogie Micrographique: Roches Éruptives Françaises*. Paris, A. Quantin.

———, 1882, *Synthèse des Minéraux et des Roches*. Paris, G. Masson.

Fournet, J., 1838, Sur quelques circonstances de la cristallisation dans les felons. *Annales de Chimie et Physique*, v. 58, pp. 387–415.

———, 1844, Sur l'état de surfusion du quartz dans les roches éruptives et dans les filons métalliferes. *Comptes Rendus Hebdomadaires des Séances de l'Académie des Sciences*, v. 18, pp. 1050–1057.

Francani, V., 1970, Scipione Breislak, in Gillispie, C. C., ed., *Dictionary of Scientific Biography*, v. 2. New York, Charles Scribner's Sons, pp. 439–440.

Frankel, E., 1974, Nicol, William, in Gillispie, C. C., ed., *Dictionary of Scientific Biography*, v. 10. New York, Charles Scribner's Sons, pp. 109–110.

Fritscher, B., 1991, *Vulkanismusstreit und Geochemie. Die Bedeutung der Chemie und des Experiments in der Neptunismus-Vulkanismus Kontroverse*. Stuttgart, F. Steiner.

Fritscher, B., and Henderson, F., eds., 1998, *Toward a History of Mineralogy, Petrology, and Geochemistry, Proceedings of the International Symposium on the History of Mineralogy, Petrology, and Geochemistry, Munich, March 8–9, 1996*. Munich, Institut für Geschichte der Naturwissenschaften.

G.S.B., 1896, William Richardson, in Lee, S., ed., *Dictionary of National Biography*, v. 48. London, Smith, Elder, p. 253.

Gast, P. W., 1968, Trace element fractionation and the origin of tholeiitic and alkaline magma types. *Geochimica et Cosmochimica Acta*, v. 32, pp. 1057–1086.

Gay, P., 1974, Memorial of Sir (William) Lawrence Bragg March 31, 1890–July 1, 1971. *American Mineralogist*, v. 59, pp. 408–411.

Geikie, A., ed., 1899, *Theory of the Earth with Proofs and Illustrations*, v. 3. London, The Geological Society.

———, 1903, *Text-book of Geology*, 4th ed., 2 v. New York, Macmillan.

———, 1905, *The Founders of Geology*. London, Macmillan.

Geikie, A., and Teall, J.J.H., 1894, On the banded structure of some Tertiary gabbros in the Isle of Skye. *Quarterly Journal of the Geological Society of London*, v. 50, pp. 645–659.

Geikie, J., 1866, On the metamorphic origin of certain granitoid rocks and granite in the southern uplands of Scotland. *Geological Magazine*, v. 3, pp. 529–534.

Gendler, V. E., 1956, O klassifikacii granitoidov. *Sovetsk. Geol.*, v. 51, pp. 265–279.

Geschwind, C.-H., 1994, The beginnings of microscopic petrography in the United States, 1870–1885. *Earth Sciences History*, v. 13, pp. 35–46.

———, 1995, Becoming interested in experiments: American igneous petrologists and the Geophysical Laboratory, 1905–1965. *Earth Sciences History*, v. 14, pp. 47–61.

Gibbon, D. L., and Wyllie, P. J., 1969, Experimental studies of igneous rock series: the Farrington Complex, North Carolina, and the Star Mountain Rhyolite, Texas. *Journal of Geology*, v. 77, pp. 221–239.

Gibbs, J. W., 1875, On the equilibrium of heterogeneous substances. *Transactions of the Connecticut Academy of Arts and Sciences*, v. 3, pp. 108–248.

Gilbert, G. K., 1877, *Report on the Geology of the Henry Mountains*. Washington, DC, Government Printing Office.

Gilbert, G. K., 1906, Gravitational assemblage in granite. *Geological Society of America Bulletin*, v. 17, pp. 321–328.

Gill, J. B., 1981, *Orogenic Andesites and Plate Tectonics*. Berlin, Springer-Verlag.

Gilluly, J., ed., 1948, *Origin of Granite*. Geological Society of America Memoir 28.

Gilman, D. C., 1899, *James Dwight Dana: Scientific Explorer, Mineralogist, Geologist, Zoologist, Professor in Yale University*. New York, Harper and Brothers.

Ginsburg, I. V., Efremova, S. V., Volovikova, I. M., and Eliseeva, O. P., 1962, Kolicestvenno-mineralnyi sostav granitoidov i ego snacenie dlja voprosov petrologii i nomenklaturyi (na primerach Kolskogo poluostrova, Sredneyj Azii i Kazachstana). *Sovetsk. Geol. 1962*, pp. 67–82.

Glazner, A. F., 1991, Plutonism, oblique subduction, and continental growth: An example from the Mesozoic of California. *Geology*, v. 19, pp. 784–786.

Glen, W., 1982, *The Road to Jaramillo: Critical Years of the Revolution in Earth Science*. Stanford, CA, Stanford University Press.

———, 1994, *The Mass-Extinction Debates: How Science Works in a Crisis*. Stanford, CA, Stanford University Press.

Goetzmann, W. H., 1993, *Exploration and Empire: The Explorer and the Scientist in the Winning of the American West*. Austin, TX, Texas State Historical Association.

Gohau, G., 1990, *A History of Geology*. New Brunswick, NJ, Rutgers University Press.

Goldschmidt, V. M., 1911, Die Kontaktmetamorphose im Kristianiagebiet. *Vidensk. Skrifter I. Math.-naturv. Klasse*, No. 1.

———, 1916, Geologisch-petrographische Studien im Hochgebirge des südlichen Norwegens, IV: Übersicht der Eruptivgesteine im kaledonischen Gerbirge zwischen Stavanger und Trondhjem. *Vidensk. Skrifter. I. Math.-naturv. Klasse*, No. 2.

———, 1921, Die Injektionsmetamorphose im Stavanger Gebiete (1920). *Videnskapsselskapets I Kristiania Skrifter. I. Math. naturv. Klasse*, No. 10.

———, 1922a, Stammestypen der Eruptivgesteine. *Vidensk. Skrifter. I. Math.naturv. Klasse*, No. 10.

———, 1922b, On the metasomatic processes in silicate rocks. *Economic Geology*, v. 7, pp. 105–123.

———, 1926, Geochemische Verteilungsgesetze der Elemente, 6: Über die Kristaltrukturen vom Rutiltypus, mit Bemerkungen zur Geochemie zweiwertiger and vierwertiger Elemente. *Skrifter utg. av det Norske Vidensksaps-akademi i Oslo 1926. I. Math. naturv. Klasse*, No. 1.

———, 1934, Drei Vorträge über die Geochemie. *Förh. Geol. För. Stockholm*, v. 56, pp. 385–427.

———, 1937, The principles of distribution of chemical elements in minerals and rocks. *Journal of the Chemical Society*. London, pp. 655–673.

———, 1938, Geochemische Verteilungsgesetze der Elemente, 9: Die Mengenverhältnisse der Elemente und der Atom-Arten. *Skrifter utg. av det Norske Videnskaps-akademi i Oslo 1937. I. Math naturv. Klasse*, No. 4.

———, 1954, *Geochemistry*. Oxford, UK, Clarendon Press.

Goldschmidt, V. M., Ulrich, F., and Barth, T.F.W., 1925a, Geochemische Verteilungsgesetze der Elemente, 4: Zur Krystalstruktur der Oxyde der seltenen Erdmetalle. *Skrifter utg. av det Norske Videnskaps-akademi i Oslo 1925. I. Math.-naturv. Klasse*, No. 5.

Goldschmidt, V. M., Barth, T.F.W., and Lunde, G., 1925b, Geochemische Verteilungsgesetze der Elemente, 5: Isomorphie und Polymorphie der Sesquioxyde. Die Lanthaniden-Kontraktion und ihre Konsequenzen. *Skrifter utg. av det Norsk Videnskapsakademi i Oslo 1925, I. Math.-naturv. Klasse*, No.7.

Goldsmith, J. R., 1991, Some Chicago georecollections, *in* Wetherill, G. W., Albee, A. L., and Burke, K.C., eds. *Annual Review of Earth and Planetary Sciences*, v. 19, pp. 1–16.

Good, G. A., ed., 1994, *The Earth, the Heavens and the Carnegie Institution of Washington*. Washington, DC, American Geophysical Union.

Goodspeed, G. E., 1940, Dilation and replacement dikes. *Journal of Geology*, v. 48, pp. 175–195.

———, 1948, Xenoliths and skialiths. *American Journal of Science*, v. 246, pp. 515–525.

———, 1952, Replacement and rheomorphic dikes. *Journal of Geology*, v. 60, pp. 356–363.

———, 1953, Rheomorphic breccias. *American Journal of Science*, v. 251, pp. 453–469.

Goranson, R. W., 1931, The solubility of water in granite magmas. *American Journal of science*, v. 22 (5th series), pp. 481–502.

———, 1932, Some notes on the melting of granite. *American Journal of Science*, v. 23 (5th series), pp. 227–236.

———, 1938, Phase equilibria in the $NaAlSi_3O_8–H_2O$ and $KAlSi_3O_8–H_2O$ systems. *American Journal of Science*, v. 35A, pp. 71–91.

Gouy, L. G., and Chaperon, G., 1887, Sur la concentration des dissolutions par la pesanteur. *Annales de Chimie et de Physique*, v. 12, pp. 384–393.

Green, D. H., 1976, Experimental petrology in Australia—a review. *Earth Science Reviews*, v. 12, pp. 99–138.

Green, D. H., and Ringwood, A. E., 1967, The genesis of basaltic magmas. *Contributions to Mineralogy and Petrology*, v. 15, pp. 103–190.

Green, T. H., and Ringwood, A. E., 1968, Genesis of the calc–alkaline igneous rock suite. *Contributions to Mineralogy and Petrology*, v. 18, pp. 105–162.

Greene, M., 1982, *Geology in the Nineteenth Century*. Ithaca, NY, Cornell University Press.

Greenland, L. P., 1970, An equation for trace element distribution during magmatic crystallization. *American Mineralogist*, v. 55, pp. 455–465.

Greig, J. W., 1927, Immiscibility in silicate melts. *American Journal of Science*, v. 13 (5th series), pp. 1–44, 133–154.

Greig, J. W., Shepherd, E. S., and Merwin, H. E., 1931, Melting temperatures of granite and basalt. *Carnegie Institution of Washington Year Book*, v. 30, pp. 75–78.

Gromet, L. P., and Silver, L. T., 1983, Rare earth element distributions among minerals in a granodiorite and their petrogenetic implications. *Geochimica et Cosmochimica Acta*, v. 47, pp. 925–939.

Grout, F. F., 1918a, Two phase convection currents in magmas. *Journal of Geology*, v. 26, pp. 481–499.

———, 1918b, A type of igneous differentiation. *Journal of Geology*, v. 26, pp. 626–658.

———, 1918c, The lopolith: an igneous form exemplified by the Duluth gabbro. *American Journal of Science*, v. 40 (4th series), pp. 516–522.

Grout, F. F., 1926, The use of calculations in petrology: A study for students. *Journal of Geology*, v. 34, pp. 512–558.

———, 1930, Probable extent of abyssal assimilation. *Geological Society of America Bulletin*, v. 41, pp. 675–693.

———, 1932, *Petrography and Petrology*. New York, McGraw-Hill.

———, 1941, Formation of igneous-looking rocks by metasomatism: A critical review and suggested research. *Geological Society of America Bulletin*, v. 52, pp. 1525–1576.

———, 1945, Scale models of structures related to batholiths. *American Journal of Science*, v. 243A, Daly volume, pp. 260–284.

Guettard, J.-E., 1756, Mémoire sur quelques montagnes de la France qui ont été des volcans. *Mémoires de l'Académie Royale des Sciences pour l'Année 1752*, pp. 27–59.

———, 1779, *Mémoire sur la Minéralogie au Dauphiné*, 2 v. Paris, n.p.

Gunter, W. D., 1990, Hans Peter Eugster Chemical Geologist, *in* Spencer, R. J., and Chou, I-M., eds., *Fluid–Mineral Interactions: A Tribute to H. P. Eugster*, Special Publication No. 2. San Antonio, TX, The Geochemical Society, pp. xiii–xvii.

Guthrie, F., 1884, On eutexia. *Philosophical Magazine*, v. 17, pp. 462–482.

Hague, A., and Iddings, J. P., 1883, Notes on the volcanoes of northern California, Oregon and Washington territory. *American Journal of Science*, v. 26 (3rd series), pp. 222–235.

———, 1885, *On the Development of Crystallization in the Igneous Rocks of Washoe Nevada with Notes on the Geology of the District*. U. S. Geological Survey Bulletin 17.

Hall, A. L., 1932, *The Bushveld Igneous Complex of the Central Transvaal*. Geological Survey of South Africa Memoir 28.

Hall, A. R., 1970, Bartholin, Erasmus, *in* Gillispie, C. C., ed., *Dictionary of Scientific Biography*, v. 1. New York, Charles Scribner's Sons, pp. 481–482.

Hall, B., 1815, Account of the structure of the Table Mountain, and other parts of the peninsula of the cape. *Transactions of the Royal Society of Edinburgh*, v. 7, pp. 269–278.

Hall, J., 1794, Observations on the formation of granite. *Transactions of the Royal Society of Edinburgh*, v. 3, pp. 8–12.

———, 1800, Experiments on whinstone and lava. *Journal of Natural Philosophy, Chemistry, and the Arts*, v. 4, pp. 8–18, 56–65.

Hallam, A., 1973, *A Revolution in the Earth Sciences: From Continental Drift to Plate Tectonics*. Oxford, UK, Clarendon Press.

———, 1983, *Great Geological Controversies*. Oxford, UK Oxford University Press.

Hamilton, B., 1982, The influence of the polarising microscope on late nineteenth century geology. *Janus*, v. 69, pp. 51–68.

Hamilton, E. I., 1963, The isotopic composition of strontium in the Skaergaard Intrusion, East Greenland. *Journal of Petrology*, v. 4, pp. 383–391.

Hamilton, W., 1772, *Observations of Mount Vesuvius, Mount Etna, and Other Volcanoes*. London, T. Cadell.

Hannaway, O., 1973, Nicholas Lemery, *in* Gillispie, C. C., ed., *Dictionary of Scientific Biography*, v. 8. New York, Charles Scribner's Sons, pp. 172–175.

Hansen, B., 1970, Bakewell, Robert, *in* Gillispie, C. C., ed., *Dictionary of Scientific Biography*, v. 1: New York, Charles Scribner's Sons, p. 413.

Harcourt, G. A., 1934, The minor chemical constituents of some igneous rocks. *Journal of Geology*, v. 42, pp. 585–601.

Hargreaves, J., 1991, *L. R. Wager: A Life 1904–1965.* Oxford, UK, Joshua Associates.
Harker, A., 1893, Berthelot's principle applied to magmatic concentration. *Geological Magazine,* v. 10, pp. 546–547.
———, 1894a, The evolution of igneous rocks. *Science Progress,* v. 1, pp. 152–165.
———, 1894b, Carrock Fell: A study in the variation of igneous rock-masses, part I: the gabbro. *Quarterly Journal of the Geological Society of London,* v. 50, pp. 311–337.
———, 1895a, Carrock Fell: A study in the variation of igneous rock-masses, Part II: The Carrock Fell granophyre, part III: the Grainsgill greisen. *Quarterly Journal of the Geological Society of London,* v. 51, pp. 125–148.
———, 1895b, *Petrology for Students.* Cambridge, UK, Cambridge University Press.
———, 1896a, Petrology in America. *Science Progress,* v. 5, pp. 459–483.
———, 1896b, The natural history of igneous rocks, I: their geographical and chronological distribution. *Science Progress,* v. 6, pp. 12–33.
———, 1900, Igneous rock series and mixed igneous rocks. *Journal of Geology,* v. 8, pp. 389–399.
———, 1903, A new rock-classification. *Geological Magazine,* v. 10 (4th series), pp. 173–178.
———, 1904, *The Tertiary Igneous Rocks of Skye.* Geological Survey of Scotland Memoir.
———, 1908, *The Geology of the Small Isles of Inverness-shire.* Glasgow, J. Hedderwick and Sons.
———, 1909, *The Natural History of Igneous Rocks.* New York, Macmillan.
———, 1916, Differentiation in intercrustal magma basins. *Journal of Geology,* v. 24, p 554–558.
Harmon, R. S.,, and Hoefs, J., 1984, O–isotope relationships in Cenozoic volcanic rocks: evidence for a heterogeneous mantle source and open-system magma genesis, *in* Dungan, M. A., Grove, T. L., and Hildreth, W., eds., *Proceedings of the ISEM Field Conference on Open Magmatic Systems*: pp. 66–68.
Harrell, J. A., 1995, Ancient Egyptian origins of some common rock names. *Journal of Geological Education,* v. 43, pp. 30–34.
Hatch, F. H., 1891, *An Introduction to the Study of Petrology: The Igneous Rocks.* London, Swan Sonnenschein.
———, 1908, The classification of the plutonic rocks. *Science Progress,* v. 3, pp. 244–264.
Haughton, S. H., 1956, Arthur Lewis Hall. *Biographical Memoirs of Fellows of the Royal Society,* v. 2, pp. 139–148.
Haüy, R., 1801, *Traité de Minéralogie,* 4 v. Paris, Chez Louis.
———, 1822, *Traité de Minéralogie*: 4 v. Paris, n.p.
Hedge, C. E., 1966, Variations in radiogenic strontium found in volcanic rocks. *Journal of Geophysical Research,* v. 71, pp. 6119–6126.
Heezen, B., Tharp, M., and Ewing, M., 1959, *The Floors of the Oceans, I: The North Atlantic.* Geological Society of America Special Paper 65.
Heinrich, E. W., 1966, *The Geology of Carbonatites.* Chicago, Rand McNally.
Herbert, S., 1991, Charles Darwin as a prospective geological author. *British Journal for the History of Science,* v. 24, pp. 159–192.
Hertogen, J., and Gijbels, R., 1976, Calculation of trace element fractionation during partial melting. *Geochimica et Cosmochimica Acta,* v. 40, pp. 313–322.

Hess, H. H., 1938, Plagioclase, pyroxene and olivine variation in the Stillwater complex (abstract). *American Mineralogist*, v. 21, pp. 198–199.

———, 1939, Extreme fractional crystallization of a basaltic magma, the Stillwater igneous complex. *American Geophysical Union Transactions*, part 3, pp. 430–432.

———, 1960, *Stillwater Igneous Complex, Montana: A Quantitative Mineralogical Study*. Geological Society of America Memoir 80.

———, 1962, History of ocean basins, in Engel, A.E.J., James, H. L., and Leonard, B. F., *Petrologic Studies: A Volume to Honor A. F. Buddington*. New York, Geological Society of America, pp. 599–620.

Hess, H. H., and Poldervaart, A., eds., 1967–1968, *Basalts: The Poldervaart Treatise on Rocks of Basaltic Composition*, 2 v. New York, Interscience.

Higham, N., 1963, *A Very Scientific Gentleman: The Major Achievements of Henry Clifton Sorby*. Oxford, UK, Pergamon Press.

Hildreth, W., 1979, The Bishop Tuff: evidence for the origin of compositional zonation in silicic magma chambers, in Chapin, C. E., and Elston, W. E., eds., *Ash-flow Tuffs*. Geological Society of America Special Paper 180, pp. 43–74.

Hillebrand, W. F., 1910, *The Analysis of Silicate and Carbonate Rocks*. U. S. Geological Survey Bulletin 422.

Hobbs, W. H., 1900, Suggestions regarding the classification of the igneous rocks. *Journal of Geology*, v. 8, pp. 1–31.

Hoefs, J., 1987, *Stable Isotope Geochemistry*, 3rd. ed. New York, Springer-Verlag.

Holborn, L., and Day, A. L., 1899, On the gas thermometer at high temperatures. *American Journal of Science*, v. 8 (4th series), pp. 165–193.

———, 1900, On the gas thermometer at high temperatures. *American Journal of Science*, v. 10 (4th series), pp. 171–206.

———, 1901, On the melting point of gold. *American Journal of Science*, v. 11 (4th series), pp. 145–148.

Holland, H. D., and Kulp, J. L., 1949, The distribution of accessory elements in pegmatites, I: theory. *American Mineralogist*, v. 34, pp. 35–60.

Holmes, A., 1916, The origin of igneous rocks. *Science Progress*, v. 11, pp. 67–73.

———, 1917, A mineralogical classification of igneous rocks. *Geological Magazine*, v. 4 (6th series), pp. 115–130.

———, 1920, *The Nomenclature of Petrology*. London, Thomas Murby.

———, 1921, *Petrographic Methods and Calculations: With Some Examples of Results Achieved*. London, Thomas Murby.

———, 1928, *The Nomenclature of Petrology*, 2nd ed. London, Thomas Murby.

———, 1931, The problem of the association of acid and basic rocks in central complexes. *Geological Magazine*, v. 68, pp. 241–255.

———, 1932, The origin of igneous rocks. *Geological Magazine*, v. 69, pp. 543–558.

———, 1936, The idea of contrasted differentiation. *Geological Magazine*, v. 73, pp. 228–238.

———, 1937, The idea of contrasted differentiation. *Geological Magazine*, v. 74, pp. 189–191.

Holser, W. T., 1976, Weiss, Christian Samuel, in Gillispie, C. C., ed., *Dictionary of Scientific Biography*, v. 14. New York, Charles Scribner's Sons, pp. 239–242.

Hooke, R., 1665, *Micrographia: or some Physiological Descriptions of Minute Bodies made by Magnifying Glasses with Observations and Inquiries Thereupon.* London, Martyn and Allestry.

Hooker, M., 1973, Lacroix, Alfred, *in* Gillispie, C. C., ed., *Dictionary of Scientific Biography*, v. 7. New York, Charles Scribner's Sons, pp. 548–549.

Hooykaas, R., 1972, Haüy, René-Just, *in* Gillispie, C. C., ed., *Dictionary of Scientific Biography*, v. 6. New York, Charles Scribner's Sons, pp. 178–183.

Hort, M., 1997, Cooling and crystallization in sheet-like magma bodies revisited. *Journal of Volcanology and Geothermal Research*, v. 76, pp. 297–317.

Hort, M., Marsh, B. D., Resmini, R. G., and Smith, M. K., 1999, Convection and crystallization in a liquid cooled from above: An experimental and theoretical study. *Journal of Petrology*, v. 40, pp. 1271–1300.

Hunt, C. B., Averitt, P., and Miller, R. L., 1953, *Geology and Geography of the Henry Mountains Region.* U. S. Geological Survey Professional Paper 228.

Hunt, T. S., 1859, On some points in chemical geology. *Quarterly Journal of the Geological Society of London*, v. 15, pp. 488–496.

Hunt, W. F., 1924, An improved Wentworth recording micrometer. *American Mineralogist*, v. 9, pp. 190–193.

Huppert, H. E., and Sparks, R.S.J., 1980, The fluid dynamics of a basaltic magma chamber replenished by influx of hot, dense ultrabasic magma. *Contributions to Mineralogy and Petrology*, v. 75, pp. 279–289.

———, 1985, Komatiites, I: eruption and flow. *Journal of Petrology*, v. 26, pp. 694–725.

———, 1988a, Melting the roof of a chamber containing a hot, turbulently convecting fluid. *Journal of Fluid Mechanics*, v. 188, pp. 107–131.

———, 1988b, The generation of granitic magmas by intrusion of basalt into continental crust. *Journal of Petrology*, v. 29, pp. 599–624.

Huppert, H. E., Sparks, R.S.J., and Turner, J. S., 1983, Laboratory investigations of viscous effects in replenished magma chambers. *Earth and Planetary Science Letters*, v. 65, pp. 377–381.

Huppert, H. E., and Turner, J. S., 1981, A laboratory model of a replenished magma chamber. *Earth and Planetary Science Letters*, v. 54, pp. 144–152.

———, 1991, Comments on 'On convective style and vigor in sheet-like magma chambers' by Bruce D. Marsh. *Journal of Petrology*, v. 32, pp. 851–854.

Huppert, H. E., Turner, J. S., and Sparks, R.S.J., 1982, Replenished magma chambers: effects of compositional zonation and input rates. *Earth and Planetary Science Letters*, v. 57, pp. 345–357.

Huppert, H. E., Wilson, J. R., and Hallworth, M. A., 1986, Cooling and crystallization at an inclined plane. *Earth and Planetary Science Letters*, v. 79, pp. 319–328.

Hurlbut, C. S., Jr., 1939, Igneous rocks of the Highwood Mountains, Montana, part 1: the laccoliths. *Geological Society of America Bulletin*, v. 50, pp. 1032–1112.

Hurlbut, C. S., Jr., and Thompson, J. B., Jr., 1996, Memorial of Harold Williams Fairbairn 1906–1994. *American Mineralogist*, v. 81, pp. 1018–1019.

Hutton, D.H.W., 1982, A tectonic model for the emplacement of the main Donegal granite. *Journal of the Geological Society of London*, v. 193, pp. 615–631.

Hutton, D.H.W., 1988, Granite emplacement mechanisms and tectonic controls: inferences from deformation studies. *Transactions of the Royal Society of Edinburgh: Earth Sciences*, v. 79, pp. 245–255.

Hutton, J., 1785, Abstract of a Dissertation read in the Royal Society of Edinburgh, upon the Seventh of March, and Fourth of April, MDCCLXXXV, concerning the System of the Earth, its Duration, and Stability. Edinburgh, n. p.

———, 1788, Theory of the Earth; or an Investigation of the Laws Observable in the Composition, Dissolution, and Restoration of the Land upon the Globe. *Transactions of the Royal Society of Edinburgh*, v. 1, pp. 209–304.

———, 1794, Observations on granite. *Transactions of the Royal Society of Edinburgh*, v. 3, pp. 77–85.

———, 1795, *Theory of the Earth, with Proofs and Illustrations*, 2 v. Edinburgh, William Creech.

Hyndman, D. W., 1972, *Petrology of Igneous and Metamorphic Rocks*. New York, McGraw-Hill.

———, 1985, *Petrology of Igneous and Metamorphic Rocks*, 2nd ed. New York, McGraw-Hill.

Iddings, J. P., 1888a, Obsidian Cliff, Yellowstone National Park. *U. S. Geological Survey Annual Report 7*, pp. 249–295.

———, 1888b, *Microscopical Petrography of the Rock-making Minerals: An Aid to the Microscopical Study of Rocks*. New York, John Wiley and Sons.

———, 1889, On the crystallization of igneous rocks. *Bulletin of the Philosophical Society of Washington*, v. 11, pp. 65–113.

———, 1891, The eruptive rocks of Electric Peak and Sepulchre Mountain, Yellowstone National Park. *U. S. Geological Survey Annual Report 12*, pp. 569–664.

———, 1892a, The mineral composition and geological occurrence of certain igneous rocks in the Yellowstone National Park. *Bulletin of the Philosophical Society of Washington*, v. 11, pp. 191–220.

———, 1892b, The origin of igneous rocks. *Bulletin of the Philosophical Society of Washington*, v. 12, pp. 89–213.

———, 1894, George Huntington Williams. *Journal of Geology*, v. 2, pp. 759–767.

———, 1898a, Bysmaliths. *Journal of Geology*, v. 6, pp. 704–710.

———, 1898b, On rock classification. *Journal of Geology*, v. 6, pp. 92–111.

———, 1903, *Chemical Composition of Igneous Rocks Expressed by Means of Diagrams with Reference to Rock Classification on a Quantitative Chemico-Mineralogical Basis*. U. S. Geological Survey Professional Paper 18.

———, 1904, Quartz–feldspar porphyry (granophyre liparose–alaskose) from Llano, Tex. *Journal of Geology*, v. 12, pp. 225–231.

———, 1909, *Igneous Rocks: Composition, Texture and Classification*, v. 1. New York, John Wiley and Sons.

———, 1913, *Igneous Rocks: Description and Occurrence*, v. 2. New York, John Wiley and Sons.

———, 1920, Arnold Hague. *Biographical Memoirs of the National Academy of Sciences*, v. 9, pp. 21–38.

Iddings, J. P., and Hague, A., 1899, *The Geology of the Yellowstone National Park*. U. S. Geological Survey Monograph 32.

Ingerson, E., 1935, Layered peridotitic laccoliths of the Trout River area, Newfoundland. *American Journal of Science*, v. 29 (5th series), pp. 422–440.

Irvine, T. N., 1963, Origin of the ultramafic complex at Duke Island, southeastern Alaska. *Mineralogical Society of America Special Paper 1*, pp. 36–45.

———, 1970a, Heat transfer during solidification of layered intrusions, I: sheets and sills. *Canadian Journal of Earth Sciences*, v. 7, pp. 1031–1061.

———, 1970b, Crystallization sequences in the Muskox Intrusion and other layered intrusions, I: olivine–pyroxene plagioclase relations. *Geological Society of South Africa, Special Publication 1*, pp. 441–476.

———, 1974, *Petrology of the Duke Island Ultramafic Complex, Southeastern Alaska*. Geological Society of America Memoir 138.

———, 1980a, Magmatic density currents and cumulus processes. *American Journal of Science*, v. 280-A (Jackson memorial volume), pp. 1–58.

———, 1980b, Infiltration metasomatism, adcumulate growth, and double-diffusive fractional crystallization in the Muskox Intrusion and other layered intrusions, *in* Hargraves, R. B., ed., *Physics of Magmatic Processes*. Princeton, NJ, Princeton University Press, pp. 325–383.

———, 1982, Terminology for layered intrusions. *Journal of Petrology*, v. 23, pp. 127–162.

Irvine, T. N., and Baragar, W.R.A., 1971, A guide to the chemical classification of the common volcanic rocks. *Canadian Journal of Earth Sciences*, v. 8, pp. 523–548.

Irvine, T. N., Keith, D. W., and Todd, S. G., 1983, The J-M platinum–palladium reef of the Stillwater Complex, Montana, II: origin by double-diffusive convective magma mixing and implications for the Bushveld Complex. *Economic Geology*, v. 78, pp. 1287–1334.

Irvine, T. N., Andersen, J.C.Ø, and Brooks, C. K., 1998, Included blocks (and blocks within blocks) in the Skaergaard Intrusion: geologic relations and the origins of rhythmic modally graded layers. *Geological Society of America Bulletin*, v. 110, pp. 1398–1447.

Irving, A. J., 1978, A review of experimental studies of crystal/liquid trace element partitioning. *Geochimica et Comsochimica Acta*, v. 42, pp. 743–770.

Isachsen, Y.W.,1968, *Origin of Anorthosite and Related Rocks*. New York State Museum and Science Service Memoir 18.

J. M., 1877, David Forbes, F.R.S., Sec. G. S., F.C.S., etc. *Geological Magazine* (new series, decade II), v. 4, pp. 45–48.

Jackson, E. D., 1961, *Primary Textures and Mineral Associations in the Ultramafic Zone of the Stillwater Complex, Montana*, U.S. Geological Survey Professional Paper 358.

Jaeger, J. C., and Joplin, G.,1955, Rock magnetism and the differentiation of dolerite sills. *Journal of the Geological Society of Australia*, v. 2, pp. 1–19.

James, H. L., 1971, Harry Hammond Hess May 24, 1906–August 25, 1969. *Biographical Memoirs of the National Academy of Sciences*, v. 43, pp. 109–128.

James, R. S., and Hamilton, D. L., 1969, Phase relations in the system $NaAlSi_3O_8$–$KAlSi_3O_8$–$CaAl_2Si_2O_8$–SiO_2 at 1 kilobar water vapour pressure. *Contributions to Mineralogy and Petrology*, v. 21, pp. 111–141.

Jameson, R., 1800, *Mineralogy of the Scottish Isles*, 2 v. Edinburgh, W. Creech.

———, 1802a, On granite. *Journal of Natural Philosophy, Chemistry, and the Arts*, v. 2, pp. 225–233.

Jameson, R., 1802b, On the supposed existence of mechanical deposits and petrifactions in the Primitive mountains, and an account of petrifactions which have been discovered in the newest Flötz trapp formation. *Journal of Natural Philosophy, Chemistry, and the Arts*, v. 3, pp. 13–20.

———, 1808, *System of Mineralogy*, 3 v. Edinburgh, William Blackwood.

———, 1813, *Mineralogical Travels through the Hebrides, Orkney, and Shetland Island, and Mainland of Scotland, with Dissertations upon Peat and Kelp*, 2 v. Edinburgh, Archibald Constable & Co.

———, 1833, Chemical analyses of stratified rocks altered by Plutonian agency. *Edinburgh New Philosophical Journal*, v. 15, pp. 386–389.

Jaupart, C., and Brandeis, G., 1986, The stagnant bottom layer of convecting magma chambers. *Earth and Planetary Science Letters*, v. 80, pp. 183–199.

Jaupart, C., and Tait, S., 1995, Dynamics of differentiation in magma reservoirs. *Journal of Geophysical Research*, v. 100, pp. 17615–17636.

Jenkins, R. V., 1976, Talbot, William Henry Fox, in Gillispie, C. C., ed., *Dictionary of Scientific Biography*, v. 13. New York, Charles Scribner's Sons, pp. 237–239.

Jenzsch, G., 1855, Mikroskopische und chemisch-analytische Untersuchung des bisher für Melaphyr gehaltenen Gesteines vom Hockenberge bei Neurode in Schlesien. *Annalen der Physik und Chemie*, v. 95, pp. 418–425.

———, 1856, Beiträge zur Kenntniss einiger Phonolithe des böhmischen Mittelgebirgen mit besonderer Berücksichtigung des Baues dieser Gebirges. *Zeitschrift der deutschen geologischen Gesellschaft*, v. 8, pp. 167–210.

Jevons, H. S., 1901, A systematic nomenclature for igneous rocks. *Geological Magazine*, v. 8 (4th series), pp. 304–316.

Johannes, W., and Holtz, F., 1996, *Petrogenesis and Experimental Petrology of Granitic Rocks*. Berlin, Springer-Verlag.

Johannsen, A., 1914, *Manual of Petrographic Methods*. New York, McGraw-Hill.

———, 1917, Suggestions for a quantitative mineralogical classification of igneous rocks. *Journal of Geology*, v. 25, pp. 63–97.

———, 1919, A planimeter method for the determination of the percentage compositions of rocks. *Journal of Geology*, v. 27, pp. 276–285.

———, 1920, A quantitative mineralogical classification of igneous rocks—revised. *Journal of Geology*, v. 28, pp. 38–60, 158–177, 210–232.

———, 1931–1937, *A Descriptive Petrography of the Igneous Rocks*, 4 v. Chicago, University of Chicago Press.

Johannsen, A., and Stephenson, E. A., 1919, On the accuracy of the Rosiwal method for the determination of the minerals in a rock. *Journal of Geology*, v. 27, pp. 212–220.

Johnston-Lavis, H. J., 1894, The basic eruptive rocks of Gran (Norway) and their interpretation: a criticism. *Geological Magazine*, v. 31, pp. 252–254.

Joly, J., 1891, On the determination of the melting points of minerals, part I: the uses of the meldometer. *Proceedings of the Royal Irish Academy*, v. 2 (3rd series), pp. 38–64.

———, 1903, The petrological examination of paving-sets. *Proceedings of the Royal Dublin Society*, v. 10, pp. 62–92.

Jones, B. F., 1988, Memorial of Hans P. Eugster November 19, 1925–December 17, 1987. *American Mineralogist*, v. 73, pp. 1489–1491.

Jones, H. L., 1931, *The Geography of Strabo*. London, Wm. Heinemann.
Joubin, L., 1900, *Histoire de la Faculté des Sciences*. Rennes, Simon.
Judd, J. W., 1874, On the ancient volcanoes of the Highlands and the relations of their products to the Mesozoic strata. *Quarterly Journal of the Geological Society of London*, v. 30, pp. 220–302.

———, 1876, On the ancient volcano of the district of Schemnitz, Hungary. *Quarterly Journal of the Geological Society of London*, v. 32, pp. 292–325.

———, 1886, On the gabbros, dolerites, and basalts, of Tertiary age, in Scotland and Ireland. *Quarterly Journal of the Geological Society of London*, v. 42, pp. 49–97.

———, 1893, On composite dykes in Arran. *Quarterly Journal of the Geological Society of London*, v. 49, pp. 536–564.

———, 1904, Frank Rutley (1842–1904). *Mineralogical Magazine*, v. 14, pp. 59–61.

———, 1908, Henry Clifton Sorby, and the birth of microscopical petrology. *Geological Magazine*, v. 5 (5th decade), pp. 193–204.

———, 1914, Geheimrath Prof. Karl Harry Ferdinand Rosenbusch. *Geological Magazine*, v. 51, pp. 140–141.

Jukes, J. B., 1857, *The Student's Manual of Geology*. Edinburgh, A., and C. Black.
Jung, J., and Brousse, R., 1959, *Classification Modale des Roches Eruptives*. Paris, Masson.
Kalkowsky, E., 1886, *Elemente der Lithologie*. Heidelberg, C. Winter.
Keir, J., 1776, On the crystallizations observed on glass. *Philosophical Transactions of the Royal Society of London*, v. 66, pp. 530–542.
Kemp, J. F., 1911, *A Handbook of Rocks for Use Without the Microscope*. New York, Van Nostrand.
Kendall, M. B., 1970, Beudant, François-Sulpice, *in* Gillispie, C. C., ed., *Dictionary of Scientific Biography*, v. 2. New York, Charles Scribner's Sons, p. 106.
Kennedy, R., 1805, A chemical analysis of three species of whinstone, and two of lava. *Transactions of the Royal Society of Edinburgh*, v. 5, pp. 76–98.
Kennedy, W. Q., 1933, Trends of differentiation in basaltic magmas. *American Journal of Science*, v. 25 (5th series), pp. 239–256.
Kerr, R. C., Woods, A. W., Worster, M. G., and Huppert, H. E., 1990, Solidification of an alloy cooled from above, 1: equilibrium growth. *Journal of Fluid Mechanics*, v. 216, pp. 323–342.
Kerrick, D. M., 1987, An historical perspective on high P/T research in geosciences. *Earth and Mineral Sciences*, v. 56, pp. 34–39.
Kevles, D. J., 1977, The National Science Foundation and the debate over postwar research policy 1942–1945. *Isis*, v. 68, pp. 5–26.
King, C., 1878, *Systematic Geology*. Washington, DC, Government Printing Office.
Kircher, A., 1664, *Mundus Subterraneus*, 2 v. Amsterdam, Jansson and Waesbergh.

———, 1669, *The Vulcano's or Burning and Fire-vomiting Mountains, Famous in the World: with their Remarkables*. London, John Allen.

Kirwan, R., 1784, *Elements of Mineralogy*. London, P. Elmsley.

———, 1793, Examination of the supposed igneous origin of stony substances. *Transactions of the Royal Irish Academy*, v. 5, pp. 51–87.

———, 1794, *Elements of Mineralogy*, 2nd ed., 2 v. London, J. Nichols.

———, 1799, *Geological Essays*. London, T. Bensley.

Kistler, R. W., and Peterman, Z. E., 1973, Variations in Sr, Rb, K, Na, and initial Sr^{87}/Sr^{86} in Mesozoic granitic rocks and intruded wallrocks in central California. *Geological Society of America Bulletin*, v. 84, pp. 3489–3512.

Knopf, A., 1941, Petrology, in Berkey, C. P., ed., *Geology, 1888–1938: Fiftieth Anniversary Volume*. Geological Society of America, pp. 335–363.

———, 1960, Louis Valentine Pirsson. *Biographical Memoirs of the National Academy of Sciences*, v. 34, pp. 228–248.

Koch, R. A., 1966, Die aktualistische Bedeutung der Vulkan experimente des Albertus Magnus. *Abhandlungen des Staatlichen Museums für Mineralogie und Geologie zu Dresden*, v. 11, pp. 307–314.

Kogarko, L. N., Kononova, V. A., Orlova, M. P., and Woolley, A. R., 1995, *Alkaline Rocks and Carbonatites of the World, Part Two: Former USSR*. London, Chapman & Hall.

Kohler, R. E., 1990, The Ph.D. machine: building on the collegiate base. *Isis*, v. 81, pp. 638–662.

Kornprobst, J., ed., 1984a, *Kimberlites, I: Kimberlites and Related Rocks*. Amsterdam, Elsevier.

———, ed., 1984b, *Kimberlites, II: The Mantle and Crust Mantle Relationships*. Amsterdam, Elsevier.

Koster van Groos, A. F., and Wyllie, P. J., 1963, Experimental data bearing on the role of liquid immiscibility in the genesis of carbonatites. *Nature*, v. 199, pp. 801–802.

———, 1968, Melting relationships in the system $NaAlSi_3O_8$–NaF–H_2O to 4 kilobars pressure. *Journal of Geology*, v. 76, pp. 50–70.

———, 1969, Melting relationships in the system $NaAlSi_3O_8$–$NaCl$–H_2O at 1 kilobar pressure, with petrological applications. *Journal of Geology*, v. 77, pp. 581–605.

Koyaguchi, T., and Takada, A., 1994, An experimental study on the formation of composite intrusions from zoned magma chambers. *Journal of Volcanology and Geothermal Research*, v. 59, pp. 261–267.

Kôzu, S., Watanabe, M., and Akaoka, J., 1919, Study on melting experiments of igneous rocks. *Journal of the Geological Society of Tokyo*, v. 26, pp. 57–83.

Kuno, H., Yamasaki, K., Iida, C., and Nagashima, K., 1957, Differentiation of Hawaiian magmas. *Japanese Geology and Geography Journal*, v. 28, pp. 179–218.

Kunz, G. F., 1918, The life and work of Haüy. *American Mineralogist*, v. 3, pp. 61–89.

Kupletsky, B. M., 1953, Kolicestvenno-mineralogiceskyi sostav granitoidov. *Vopr. miner. i petr.*, t. 1, Izd. AN SSSR, Moskva.

Kushiro, I., 2000, Acceptance of the Roebling medal of the Mineralogical Society of America for 1999. *American Mineralogist*, v. 85, pp. 1094–1096.

Kuslan, L. I., 1970, Berthier, Pierre, in Gillispie, C. C., ed., *Dictionary of Scientific Biography*, v. 2: New York, Charles Scribner's Sons, p. 72.

Lacroix, A., 1898, Le granite des Pyrénées et ses phénomènes de contact. *Bulletin de la Service de Carte Géologique de France*, x, No. 64., pp. 1–68.

———, 1900, Le granite des Pyrénées et ses phénomènes de contact. *Bulletin de la Service de Carte Géologique de France*, xi, No. 71, pp. 1–63.

———, 1933, Contributions à la connaissance de la composition chimique et minéralogique des roches éruptives de l'Indochine: classification des roches eruptives. *Bulletin de Service Géologique du Indochina*, v. 20, pp. 15–36, 183–206.

Lagorio, A., 1887, Über die Natur der Glasbasis, sowie der Krystallisationsvorgänge im eruptiven Magma. *Mineralogische und petrographische Mitteilungen*, v. 8, pp. 421–529.

Lambert, I. B., and Wyllie, P. J., 1972, Melting of gabbro (quartz eclogite) with excess water to 35 kilobars, with geological applications. *Journal of Geology*, v. 80, pp. 693–708.

———, 1974, Melting of tonalite and crystallization of andesite liquid with excess water to 30 kilobars. *Journal of Geology*, v. 82, pp. 88–97.

Lane, A. C., 1904, The role of possible eutectics in rock magmas. *Journal of Geology*, v. 12, pp. 83–93.

———, 1931, Size of batholiths. *Geological Society of America Bulletin*, v. 42, pp. 813–829.

La Roche, H. de, 1986, Classification et nomenclature des roches ignées: un essai de restauration de la convergence entre systématique quantitative, typologie d'usage et modélisation génétique. *Bulletin de la Societé Géologique de France*, v. 2 (8th series), pp. 337–353.

La Roche, H. de, Leterrier, J., Grandclaude, P., and Marchal, M., 1980, A classification of volcanic and plutonic rocks using R_1–R_2 diagram and major-element analyses: its relationships with current nomenclature. *Chemical Geology*, v. 29, pp. 183–210.

Larsen, E. S., Jr., 1938, Some new variation diagrams for groups of igneous rocks. *Journal of Geology*, v. 46, pp. 505–520.

———, 1958, Charles Whitman Cross September 1, 1854- April 20, 1949. *Biographical Memoirs of the National Academy of Sciences*, v. 32, pp. 100–112.

Laudan, R., 1987, *From Mineralogy to Geology: The Foundations of a Science, 1650–1830*. Chicago, University of Chicago Press.

Le Bas, M. J., 1977, *Carbonatite–Nephelinite Volcanism: An African Case History*. New York, John Wiley and Sons

———, 1989, Nephelinitic and basanitic rocks. *Journal of Petrology*, v. 30, pp. 1299–1312.

———, 2000, IUGS reclassification of the high-Mg and picritic volcanic rocks. *Journal of Petrology*, v. 41, pp. 1467–1470.

Le Bas, M. J., and Sabine, P. J., 1980, Progress in 1979 on the nomenclature of pyroclastic materials. *Geological Magazine*, v. 117, pp. 389–391.

Le Bas, M. J., and Streckeisen, A., 1991, The IUGS systematics of igneous rocks. *Journal of the Geological Society of London*, v. 148, pp. 825–833.

Le Bas, M. J., LeMaitre, R. W., Streckeisen, A., and Zanettin, B., 1986, A chemical classification of volcanic rocks based on the total alkali–silica diagram. *Journal of Petrology*, v. 27, pp. 745–750.

Lee, H.D.P., 1952, *Meteorologica*. Cambridge, MA, Harvard University Press.

Leelanandam, C., ed., 1989, *Alkaline Rocks*. Bangalore, Geological Society of India Memoir 15.

Leeman, W. P., and Dasch, E. J., 1978, Strontium, lead and oxygen isotopic investigation of the Skaergaard intrusion, East Greenland. *Earth and Planetary Science Letters*, v. 41, pp. 47–59.

Leicester, H. M., 1971, Deville, Henri Étienne Sainte-Claire, *in* Gillispie, C. C., ed., *Dictionary of Scientific Biography*, v. 4. New York, Charles Scribner's Sons, pp. 77–78.

Leicester, H. M., 1973, Le Châtelier, Henry Louis, in Gillispie, C. C., ed., *Dictionary of Scientific Biography*, v. 8. New York, Charles Scribner's Sons, pp. 116–120.

Le Maitre, R. W., 1973, Experiences with CLAIR: A computerized library of analysed igneous rocks. *Chemical Geology*, v. 12, pp. 301–308.

———, 1984, A proposal by the IUGS Subcommission on the Systematics of Igneous Rocks for a chemical classification of volcanic rocks based on the total alkali silica (TAS) diagram. *Australian Journal of Earth Sciences*, v. 31, pp. 243–255.

———, ed., 1989, *A Classification of Igneous Rocks and Glossary of Terms*. Oxford, Blackwell.

Lémery, N., 1700, Explication physique & chemique des Feux souterrains, des Tremblemens de Terre, des Ouragans, des Eclairs et du Tonneur. Histoire de l'academie Royale des Sciences.

Lewis, C., 2000, *The Dating Game: One Man's Search for the Age of the Earth*. New York, Cambridge University Press.

Lewis, J. V., 1907, The origin and relations of the Newark rocks. *New Jersey Geological Survey Annual Report*, pp. 99–129.

———, 1935, Memorial of Henry Stephens Washington. *American Mineralogist*, v. 20, pp. 179–184.

Limoges, C., 1980, The development of the Muséum d'Histoire Naturelle of Paris, c. 1800–1914, in Fox, R., and Weisz, G., eds., *The Organization of Science and Technology in France 1808–1914*. Cambridge, UK, Cambridge University Press, pp. 211–240.

Lincoln, F. C., 1913, The quantitative mineralogical classification of gradational rocks. *Economic Geology*, v. 8, pp. 551–564.

Lincoln, F. C., and Rietz, H. L. 1913, The determination of the relative volumes of the components of rocks by mensuration methods. *Economic Geology*, v. 8, pp. 120–139.

Lindberg, D. C., 1992, *The Beginnings of Western Science*. Chicago, University of Chicago Press.

Lindgren, W., 1931, Memorial tribute to Pierre Termier. *Geological Society of America Bulletin*, v. 43, pp. 116–117.

Lindsay, R. B., 1943, Carl Barus. *Biographical Memoirs of the National Academy of Sciences*, v. 22, pp. 171–211.

Lister, M., 1683, Three papers of Dr. Martin Lyster. *Philosophical Transactions of the Royal Society of London*, v. 13–14, pp. 512–519.

Loewinson-Lessing, F. Y., 1893, *Petrographisches Lexicon. Repertorium der petrographischen Termini und Benennungen*. Jurjew, C. Mattiesen.

———, 1899a, Studien über die Eruptivgesteine. VII International Geological Congress, 1897. St. Petersburg, M. Stassulewitsch.

———, 1899b, Kritische Beiträge zur Systematik der Eruptivgesteine. *Tschermaks mineralogische und petrographische Mitteilungen*, v. 18, pp. 518–524.

———, 1900, Kritische Beiträge zur Systematik der Eruptivgesteine. *Tschermaks mineralogische und petrographische Mitteilungen*, v. 19, pp. 169–181, 290–307.

———, 1901, Kritische Beiträge zur Systematik der Eruptivegesteine. *Tschermaks mineralogische und petrographische Mitteilungen*, v. 20, pp. 110–128.

———, 1902, Kritische Beiträge zur Systematik der Eruptivegesteine. *Tschermaks mineralogische und petrographische Mitteilungen*, v. 21, pp. 307–322.

———, 1954, *A Historical Survey of Petrology*. Edinburgh, Oliver and Boyd.

Loiselle, M. C., and Wones, D. R., 1979, Characteristics and origin of anorogenic granites. *Geological Society of America Abstracts with Progams*, v. 11, p. 468.

Lombaard, B. V., 1934, On the differentiation and relationships of the rocks of the Bushveld Complex. *Transactions of the Geological Society of South Africa*, v. 37, pp. 5–52.

Long, A. G., 1976, Witham, Henry, *in* Gillispie, C. C., ed., *Dictionary of Scientific Biography*, v. 14. New York, Charles Scribner's Sons, pp. 462–463.

Longwell, C. R., 1958, Clarence Edward Dutton. *Biographical Memoirs of the National Academy of Sciences*, v. 32, pp. 132–145.

Ludwig, C., 1856, Diffusion zwischen ungleich erwärmten Orten gleich zusammengesetzter Lösungen. *Sitzungsberichte Akademie Wien*, v. 20, p. 539.

Lunar Sample Preliminary Examination Team, 1969, Preliminary examination of lunar samples from Apollo 11. *Science*, v. 165, pp. 1211–1227.

Luth, W. C., 1987, Memorial of Orville Frank Tuttle June 25, 1916–December 13, 1983. *American Mineralogist*, v. 72, pp. 1020–1022.

Luth, W. C., Jahns, R. H., and Tuttle, O. F., 1964, The granite system at pressures of 4 to 10 kilobars. *Journal of Geophysical Research*, v. 69, pp. 759–773.

Lyell, C., 1830–1833, *Principles of Geology*, 3 v. London, John Murray.

Maaloe, S., and Wyllie, P. J., 1975, Water content of a granite magma deduced from the sequence of crystallization determined experimentally with water-undersaturated conditions. *Contributions to Mineralogy and Petrology*, v. 52, pp. 175–191.

MacCarthy, G. R., 1925, Some facts and theories concerning laccoliths. *Journal of Geology*, v. 33, pp. 1–18.

Macculloch, J., 1819, *A Description of the Western Islands of Scotland, including the Isle of Man*, 3 v. London, Hursts, Robinson.

———, 1821, *A Geological Classification of Rocks, with Descriptive Synopses of the Species and Varieties, Comprising the Elements of Practical Geology.* London, Longmans, Hurst, Rees, Orme, Brown.

———, 1826, Review of G. P. Scrope's 'Considerations on Volcanos.' *Westminster Review*, v. 5, pp. 356–373.

MacDonald, G. A., and Katsura, T., 1964, Chemical composition of Hawaiian lavas. *Journal of Petrology*, v. 5, pp. 82–133.

Macdougall, J. D., ed., 1988, *Continental Flood Basalts*. Dordrecht, Kluwer Academic.

Mackenzie, G. S., 1811, *Travels in the Island of Iceland, during the Summer of the Year MDCCCX*. Edinburgh, n.p.

———, 1815, An account of some geological facts observed in the Faroe Island. *Transactions of the Royal Society of Edinburgh*, v. 7, pp. 213–228.

Mahood, G. A., 1981, A summary of the geology and petrology of the Sierra La Primavera, Jalisco, Mexico. *Journal of Geophysical Research*, v. 86, pp. 10137–10152.

Maniar, P. D., and Piccoli, P. M., 1989, Tectonic discrimination of granitoids. *Geological Society of America Bulletin*, v. 101, pp. 635–643.

Manning, D.A.C., 1981, The effect of fluorine on liquidus phase relationships in the system Qz-Ab-Or with excess water at 1 kbar. *Contributions to Mineralogy and Petrology*, v. 76, pp. 206–215.

Marmo, V., 1956, On the emplacement of granites. *American Journal of Science*, v. 254, pp. 479–492.

———, 1971, *Granite Petrology and the Granite Problem*. Amsterdam, Elsevier.

Marsh, B. D., 1981, On the crystallinity, probability of occurrence, and rheology of lava and magma. *Contributions to Mineralogy and Petrology*, v. 78, pp. 85–98.

———, 1982, On the mechanics of igneous diapirism, stoping, and zone melting. *American Journal of Science*, v. 282, pp. 808–855.

———, 1984, On the mechanics of caldera resurgence. *Journal of Geophysical Research*, v. 89, pp. 8245–8251.

———, 1988a, Crystal capture, sorting and retention in convecting magma. *Geological Society of America Bulletin*, v. 100, pp. 1720–1737.

———, 1988b, Crystal size distributions (CSD) in rocks and the kinetics and dynamics of crystallization, I: theory. *Contributions to Mineralogy and Petrology*, v. 99, pp. 277–291.

———, 1989, On convective style and vigor in sheet-like magma chambers. *Journal of Petrology*, v. 30, pp. 479–530.

———, 1991, Reply. *Journal of Petrology*, v. 32, pp. 855–860.

———, 1998, On the interpretation of crystal size distributions in magmatic systems. *Journal of Petrology*, v. 39, pp. 553–599.

Marsh, B. D., and Maxey, M. R., 1985, On the distribution and separation of crystals in convecting magma. *Journal of Volcanology and Geothermal Research*, v. 24, pp. 95–150.

Martin, D., and Nokes, R., 1989, A fluid-dynamical study of crystal settling in convecting magmas. *Journal of Petrology*, v. 30, p.1471–1500.

Martin, D., Griffiths, R. W., and Campbell, I. H., 1987, Compositional and thermal convection in magma chambers. *Contributions to Mineralogy and Petrology*, v. 96, pp. 465–475.

Mason, B., 1968, Presentation of the Mineralogical Society of America award for 1967 to Alfred E. Ringwood. *American Mineralogist*, v. 53, pp. 531–532.

———, 1992, *Victor Moritz Goldschmidt: Father of Modern Geochemistry*. Special Publication #4. The Geochemical Society, San Antonio, TX.

———, 1994, Introduction of S. Ross Taylor for the 1993 V. M. Goldschmidt Award. *Geochimica et Cosmochimica Acta*, v. 58, p. 3757.

Mason, B., and Melson, W. G., 1970, *The Lunar Rocks*. New York, Wiley Interscience.

Matsuhisa, Y., 1979, Oxygen isotopic compositions of volcanic rocks from East Japan island arc and their bearing on petrogenesis. *Journal of Volcanology and Geothermal Research*, v. 5, pp. 271–296.

McBirney, A. R., 1985, Further considerations of double-diffusive stratification and layering in the Skaergaard Intrusion. *Journal of Petrology*, v. 26, pp. 993–1001.

McBirney, A. R., and Hunter, R. H., 1995, The cumulate paradigm reconsidered. *Journal of Geology*, v. 103, pp. 114–122.

McBirney, A. R., and Nicolas, A., 1997, The Skaergaard Layered Series, part II: dynamic layering. *Journal of Petrology*, v. 38, pp. 569–580.

McBirney, A. R., and Noyes, R. M., 1979, Crystallization and layering of the Skaergaard Intrusion. *Journal of Petrology*, v. 20, pp. 487–554.

McCaffrey, K.J.W., and Petford, N., 1997, Are granitic intrusions scale invariant? *Journal of the Geological Society of London*, v. 154, pp. 1–4.

McDowell, S. D., and Wyllie, P. J., 1971, Experimental studies of igneous rock series: the Kungnat Syenite Complex of southwest Greenland. *Journal of Geology*, v. 79, pp. 173–194.

McIntyre, D. B., and McKirdy, A., 1997, *James Hutton: The Founder of Modern Geology*. Edinburgh, Stationery Office.

McIntyre, D. B., and Stephenson, D., 1997, *Field Excursion Guide to Hutton Country: Scottish Highlands and Glen Tilt*. Edinburgh, Royal Society of Edinburgh.

Menard, H. W., 1986, *The Ocean of Truth: A Personal History of Global Tectonics*. Princeton, NJ, Princeton University Press.

Meniailov, A. A., 1973, Levinson-Lessing, Franz Yulevich, in Gillispie, C. C., ed, *Dictionary of Scientific Biography*, v. 8. New York, Charles Scribner's Sons, pp. 285–287.

Merrill, G. P., 1926, George Ferdinand Becker. *Biographical Memoirs of the National Academy of Sciences*, v. 21, pp. 1–19.

Meyer, H.O.A., and Leonardos, O. H., 1994a, *Kimberlites, Related Rocks, and Mantle Xenoliths*. Rio de Janeiro, CPRM.

———, 1994b, *Diamonds: Characterization, Genesis and Exploration*. Rio de Janeiro, CPRM.

Michel-Lévy, A., 1889, *Structures et Classification des Roches Éruptives*. Paris, Baudry.

———, 1893, Granite de Flamanville. *Bulletin de la Service de la Carte Géologique de France*, 5, no. 36, pp. 317–357.

———, 1897, Note sur la classification des magmas des roches eruptives. *Bulletin de la Societé Géologique de France*, v. 25, pp. 326–376.

Michel-Lévy, A., and A. Lacroix, 1888, *Les Minéraux des Roches*. Paris, Baudry.

Middlemost, E.A.K., 1971, Classification and origin of the igneous processes. *Lithos*, v. 4, pp. 105–130.

———, 1980, A contribution to the nomenclature and classification of volcanic rocks. *Geological Magazine*, v. 117, pp. 51–57.

Mikkola, T., 1968, Memorial of Pentti Eskola. *American Mineralogist*, v. 53, pp. 544–548.

Miller, F. J., 1977, *Metamorphoses*, 3rd ed., 2 v. Cambridge, MA, Harvard University Press.

Minster, J. F., and Allègre, C. J., 1978, Systematic use of trace elements in igneous processes, part III: inverse problem of batch partial melting in volcanic suites. *Contributions to Mineralogy and Petrology*, v. 68, pp. 37–52.

Minster, J. F., Minster, J. B., Allègre, C. J., and Treuil, M., 1977, Systematic use of trace elements in igneous processes, II: inverse problem of the fractional crystallization process in volcanic suites. *Contributions to Mineralogy and Petrology*, v. 61, pp. 49–77.

Misch, P., 1949a, Metasomatic granitization of batholithic dimensions. *American Journal of Science*, v. 247, pp. 209–245.

———, 1949b, Static granitization in Sheku area, NW Yunnan (China). *American Journal of Science*, v. 247, pp. 372–406.

———, 1949c, Relationships of synkinematic and static granitization. *American Journal of Science*, v. 247, pp. 673–705.

Mitchell, R. H., 1986, *Kimberlites: Mineralogy, Geochemistry, and Petrology*. New York, Plenum.

———, 1995, *Kimberlites, Orangeites, and Related Rocks*. New York, Plenum.

———, ed., 1996, *Undersaturated Alkaline Rocks: Mineralogy, Petrogenesis, and Economic Potential*. Winnipeg, Manitoba, Mineralogical Association of Canada Short Course 24.

Mitchell, R. H., and Bergman, S. C., 1991, *Petrology of Lamproites*. New York, Plenum.

Mittlefehldt, D. W., and Miller, C. F., 1983, Geochemistry of the Sweetwater Wash pluton, California: implications for "anomalous" trace element behavior during differentiation of felsic magmas. *Geochimica et Cosmochimica Acta*, v. 47, pp. 109–124.

Miyashiro, A., 1961, The evolution of metamorphic belts. *Journal of Petrology*, v. 2, pp. 277–311.

Modell, D., 1936, Ring–dike complex of the Belknap Mountains, New Hampshire. *Geological Society of America Bulletin*, v. 47, pp. 1885–1932.

Molyneux, T., 1694, Some Notes upon the foregoing Account of the Giants Causway, serving to further Illustrate the same. *Philosophical Transactions of the Royal Society of London*, v. 18, pp. 175–182.

———, 1698, A Letter from Dr. Thomas Molyneux, to Dr. Martin Lister, Fellow of the Colledge of Physicians, and of the Royal Society, in, London: Containing some additional Observations on the Giants Causway in Ireland. *Philosophical Transactions of the Royal Society of London*, v. 20, pp. 209–223.

Montel, J.-M., and Vielzeuf, D., 1997, Partial melting of metagraywackes, part II: compositions of minerals and melts. *Contributions to Mineralogy and Petrology*, v. 128, pp. 176–196.

Montlosier, F.D.R., 1802, *Essai sur laThéorie des Volcans d'Auvergne*, 2nd ed. Riom, Landriot et Rousset.

Moore, J. G., 1959, The quartz diorite boundary line in the western United States. *Journal of Geology*, v. 67, pp. 198–210.

Moorhouse, W. W., 1959, *The Study of Rocks in Thin Sections*. New York, Harper.

Morgan, M. H., 1960, *Vitruvius: The Ten Books on Architecture*. New York, Dover Books.

Moro, Anton-Lazarro, 1740, *De' Crostacei e degli altra Corpi Marini che si Truovano su' Monti*, Libri Due. Venice, S. Monti.

Morozewicz, J., 1898, Experimentelle Untersuchungen über die Bildung der Minerale im Magma. *Tschermaks mineralogische und petrographische Mitteilungen*, v. 18, pp. 1–90, 105–240.

Morse, E. W., 1970, Brewster, David, in Gillispie, C. C., ed., *Dictionary of Scientific Biography*, v. 2. New York, Charles Scribner's Sons, pp. 451–454.

Morse, S. A., 1969, *The Kiglapait Layered Intrusion, Labrador*. Geological Society of America Memoir 112.

———, 1979, Kiglapait geochemistry, I: systematics, sampling, and density. *Journal of Petrology*, v. 20, pp. 555–590.

———, 1980, *Basalts and Phase Diagrams: An Introduction to the Quantitative Use of Phase Diagrams in Igneous Petrology*. New York, Springer-Verlag.

———, 1986, Convection in aid of adcumulus growth. *Journal of Petrology*, v. 27, pp. 1183–1215.

———, 1996, Kiglapait mineralogy, III: olivine compositions and Rayleigh fractionation models. *Journal of Petrology*, v. 37, pp. 1037–1061.

Mrazec, L., 1927, Les plis diapirs et la diapirisme en general. *Comptes Rendus Seances Inst. Geol. Roumaine*, v. 6, pp. 226–270.

Muan, A., and Osborn, E. F., 1956, Phase equilibria at liquidus temperatures in the system $MgO–FeO–Fe_2O_3–SiO_2$. *Journal of the American Ceramic Society*, v. 39, pp. 121–140.

Muehlenbachs, K., and Byerly, G., 1982, ^{18}O-enrichment of silicic magma caused by crystal fractionation at the Galapagos spreading center. *Contributions to Mineralogy and Petrology*, v. 79, pp. 76–79.

Muir, I. D., 1994, Stephen Robert Nockolds. *Biographical Memoirs of Fellows of the Royal Society*, v. 40, pp. 308–317.

Muir, I. D., and Tilley, C. E., 1964, Basalts from the northern part of the Rift zone of the Mid-Atlantic Ridge. *Journal of Petrology*, v. 5, pp. 409–434.

Murase, T., and McBirney, A. R., 1973, Properties of some common igneous rocks and their melts at high temperatures. *Geological Society of America Bulletin*, v. 84, pp. 3563–3592.

Murchison, R., 1839, *The Silurian System*, 2 v. London, J. Murray.

Murray, J., 1802, *A Comparative View of the Huttonian and Neptunian Systems of Geology*. Edinburgh, Ross and Blackwood.

Naney, M. T., 1983, Phase equilibria of rock-forming ferromagnesian silicates in granitic systems. *American Journal of Science*, v. 283, pp. 993–1033.

Naumann, C. F., 1850, 1854, *Lehrbuch der Geognosie*, 2 v. Leipzig, Wilhelm Engelmann.

Necker de Saussure, L. A., 1825, Discourse on the history and progress of geology. *Edinburgh Philosophical Journal*, v. 13, pp. 292–300.

Neev, D., and Emery, K. O., 1995, *The Destruction of Sodom, Gomorrah, and Jericho: Geological, Climatological, and Archaeological Background*. New York, Oxford University Press.

Neumann, H., Mead, J., and Vitaliano, C. J., 1954, Trace element variation during fractional crystallization as calculated from the distribution law. *Geochimica et Cosmochimica Acta*, v. 6, pp. 90–99.

Nicol, W., 1829, On a method of so far increasing the divergency of the two rays in calcareous spar, that only one image may be seen at a time. *Edinburgh Philosophical Journal*, v. 6, pp. 83–84.

Nier, A.O.C., 1985, Acceptance speech for the V. M. Goldschmidt Award. *Geochimica et Cosmochimica Acta*, v. 49, pp. 1666s–1668.

Nieuwenkamp, W., 1970, Buch, [Christian] Leopold von, in Gillispie, C. C., ed., *Dictionary of Scientific Biography*, v. 2. New York, Charles Scribner's Sons, pp. 552–557.

Niggli, P., 1920, *Lehrbuch der Mineralogie*, 2 v. Berlin, Gebrüder Bornträger.

———, 1923, *Gesteine– und Mineralprovinzen*, v. 1. Berlin, Gebrüder Bornträger.

———, 1931, Die quantitative mineralogische Klassifikation der Eruptivgesteine. *Schweizerische mineralogische und petrographische Mitteilungen*, v. 11, pp. 296–364.

———, 1936, Über Molekularnormen zur Gesteinsberechnung. *Schweizerische mineralogische und petrographische Mitteilungen*, v. 16, pp. 295–317.

———, 1939, *Tabellen zur Petrographie und zum Gesteinsbestimmen*. Zürich, Mineralogisch-petrographisch Institut, E.T.H. Zürich.

———, 1942, Das Problem der Granitbildung. *Schweizerische mineralogische und petrographische Mitteilungen*, v. 22, pp. 1–84.

———, 1946, Die leukogranitischen, trondhjemitischen und leukosyenit-granitischen Magmen und die Anatexis. *Schweizerische mineralogische und petrographische Mitteilungen*, v. 26, pp. 44–78.

Noble, J. A., 1952, Evaluation of criteria for the forcible intrusion of magma. *Journal of Geology*, v. 60, pp. 34–57.

Nockolds, S. R., 1933, Some theoretical aspects of contamination in acid magmas. *Journal of Geology*, v. 41, pp. 561–589.

———, 1934, The production of normal rock types by contamination and their bearing on petrogenesis. *Geological Magazine*, v. 71, pp. 31–39.

———, 1936, The idea of contrasted differentiation: A reply. *Geological Magazine*, v. 73, pp. 529–535.

———, 1954, Average chemical composition of some igneous rocks. *Geological Society of America Bulletin*, v. 65, pp. 1007–1032.

———, 1966, The behaviour of some elements during fractional crystallization of magma. *Geochimica et Cosmochimica Acta*, v. 30, pp. 267–278.

Nockolds, S. R., and Allen, R., 1953, The geochemistry of some igneous rock series. *Geochimica et Cosmochimica Acta*, v. 4, pp. 105–142.

———, 1954, The geochemistry of some igneous rock series, part II. *Geochimica et Cosmochimica Acta*, v. 5, pp. 245–285.

———, 1956, The geochemistry of some igneous rock series, part III. *Geochimica et Cosmochimica Acta*, v. 9, pp. 34–77.

Nockolds, S. R., and Mitchell, R. L., 1948, Geochemistry of some Caledonian plutonic rocks. *Transactions of the Royal Society of Edinburgh*, v. 61, pp. 533–575.

Norton, D., and Taylor, H. P., Jr., 1979, Quantitative simulation of the hydrothermal systems of crystallizing magmas on the basis of transport theory and oxygen isotope data: An analysis of the Skaergaard Intrusion. *Journal of Petrology*, v. 20, pp. 421–486.

Nye, M. J., 1972, Hautefeuille, Paul Gabriel, in Gillispie, C. C., ed., *Dictionary of Scientific Biography*, v. 6. New York, Charles Scribner's Sons, pp. 177–178.

Oftedahl, C., 1976, Vogt, Johan Hermann Lie, in Gillispie, C. C., ed., *Dictionary of Scientific Biography*, v. 14. New York, Charles Scribner's Sons, pp. 58–59.

Ogilvie, I. H., 1945, Florence Bascom 1862–1945. *Science*, v. 102, pp. 320–321.

O'Hara, M. J., 1965, Primary magmas and the origin of basalts. *Scottish Journal of Geology*, v. 1, pp. 19–40.

———, 1968, The bearing of phase equilibria studies in synthetic and natural systems on the origin and evolution of basic and ultrabasic rocks. *Earth Science Reviews*, v. 4, pp. 69–133.

———, 1977, Geochemical evolution during fractional crystallization of a periodically refilled magma chamber. *Nature*, v. 266, pp. 503–507.

Oldfather, C. H., 1933, *Diodorus of Sicily*, 12 v. London, Wm. Heinemann.

Oldroyd, D. R., 1990, *The Highlands Controversy: Constructing Geological Knowledge Through Fieldwork in Nineteenth-Century Britain*. Chicago, University of Chicago Press.

———, 1996, *Thinking About the Earth: A History of Ideas in Geology*. Cambridge, MA, Harvard University Press.

O'Nions, R. K., Hamilton, P. J., and Evensen, N. M., 1977, Variations in $^{143}Nd/^{144}Nd$ and $^{87}Sr/^{86}Sr$ in oceanic basalts. *Earth and Planetary Science Letters*, v. 34, pp. 13–22.

Osann, C. A., 1900, Versuch einer chemischen Klassification der Eruptivgesteine. *Tschermaks mineralogische und petrographische Mitteilungen*, v. 19, pp. 351–469.

———, 1901, Versuch einer chemischen Klassification der Eruptivegesteine. *Tschermaks mineralogische und petrographische Mitteilungen*, v. 20, pp. 399–558.

———, 1902, Versuch einer chemischen Klassification der Eruptivegesteine. *Tschermaks mineralogische und petrographische Mitteilungen*, v. 21, pp. 365–448.

———, 1903, Versuch einer chemischen Klassification der Eruptivegesteine. *Tschermaks mineralogische und petrographische Mitteilungen*, v. 22, pp. 322–356, 405–436.

Osann, C. A., and Rosenbusch, H., 1922, *Elemente der Gesteinslehre*, 4th ed. Stuttgart, E. Schweizerbartsche Verlagshandlung.

Osborn, E. F., 1957, Role of oxygen pressure in the crystallization and differentiation of basaltic magma. *American Journal of Science*, v. 259, pp. 609–647.

Ospovat, A., 1971, *Short Classification and Description of the Various Rocks*. New York, Hafner.

———, 1976, The distortion of Werner in Lyell's *Principles of Geology*. *British Journal for the History of Science*, v. 9, pp. 190–198.

Ötling, C.F.W.A., 1897, Vergleichende Experimente über Verfestigung geschmolzener Gesteinsmassen unter erhöhtem und normalem Druck. *Tschermaks mineralogische und petrographische Mitteilungen*, v. 17, pp. 331–373.

Outram, D., 1984, *Georges Cuvier: Vocation, Science, and Authority in Post-Revolutionary France*. Manchester, UK, Manchester University Press.

Pabst, A., 1975, Rammelsberg, Karl (or Carl) Friedrich, in Gillispie, C. C., ed, *Dictionary of Scientific Biography*, v. 11. New York, Charles Scribner's Sons, pp. 270–271.

Page, L. E., 1975, Scrope, George Julius Poulett, in Gillispie, C. C., ed., *Dictionary of Scientific Biography*, v. 12. New York, Charles Scribner's Sons, pp. 261–264.

Paige, S., 1913, The bearing of progressive increase of viscosity during intrusion on the form of laccoliths. *Journal of Geology*, v. 21, pp. 541–549.

Paragallo, G., 1705, *Istoria naturale del monte Vesuvio*. Napoli, G. Raillard.

Parker, R. L., 1954, Memorial of Paul Niggli. *American Mineralogist*, v. 39, pp. 280–284.

Parson, L. M., Murton, B. J., and Browning, P., 1992, *Ophiolites and Their Modern Oceanic Analogues*. London, The Geological Society.

Parsons, I., ed., 1987, *Origins of Igneous Layering*. Dordrecht, Reidel.

Partington, J. R., 1964, *A History of Chemistry*, 4 v. London, Macmillan.

Patiño Douce, A. E., 1996, Effects of pressure and H_2O content on the compositions of primary crustal melts. *Transactions of the Royal Society of Edinburgh*, v. 87, pp. 11–21.

Patiño Douce, A. E., and Beard, J. S., 1995, Dehydration-melting of biotite gneiss and quartz amphibolite from 3 to 15 kbar. *Journal of Petrology*, v. 36, pp. 707–738.

———, 1996, Effects of P, $f(O_2)$, and Mg/Fe ratio on dehydration melting of model metagraywackes. *Journal of Petrology*, v. 37 pp. 999–1024.

Patiño Douce, A. E., and Johnston, A. D., 1991, Phase equilibria and melt productivity in the pelitic system: implications for the origin of peraluminous granitoids and aluminous granulites. *Contributions to Mineralogy and Petrology*, v. 107, pp. 202–218.

Peacock, M. A., 1931, Classification of igneous rock series. *Journal of Geology*, v. 39, pp. 54–67.

Pearce, J. A., 1976, Statistical analysis of major element patterns in basalt. *Journal of Petrology*, v. 17, pp. 15–43.

Pearce, J. A., and Cann, J. R., 1971, Ophiolite origin investigation by discriminant analysis using Ti, Zr, and Y. *Earth and Planetary Science Letters*, v. 12, pp. 339–349.

———, 1973, Tectonic setting of basic volcanic rocks determined using trace element analysis. *Earth and Planetary Science Letters*, v. 19, pp. 290–300.

Pearce, J. A., Harris, N.B.W., and Tindle, A. G., 1984, Trace element discrimination diagrams for the tectonic interpretation of granitic rocks. *Journal of Petrology*, v. 25, pp. 956–983.

Pearson, P., 1996, Charles Darwin on the origin and diversity of igneous rocks. *Earth Sciences History*, v. 15, pp. 49–67.

Pennant, T., 1771, *A Tour in Scotland, 1769*. Chester, UK, John Monk.

Perrin, R., 1954, Granitization, metamorphism, and volcanism. *American Journal of Science*, v. 252, pp. 449–465.

———, 1956, Granite again. *American Journal of Science*, v. 254, pp. 1–18.

———, 1957, Granites eutectiques ou métamorphiques? Discussion d'études récentes. *Bulletin de la Société Géologique de France*, v. 7 (6th series), pp. 91–113.

Peters, T., Nicolas, A., and Coleman, R. G., eds., 1991, *Ophiolite Genesis and Evolution of the Oceanic Lithosphere*. Dordrecht, Kluwer Academic.

Petersen, J. S., 1980, Rare-earth element fractionation and petrogenetic modelling in charnockitic rocks, southwest Norway. *Contributions to Mineralogy and Petrology*, v. 73, pp. 161–172.

Petford, N., 1996, Dykes or diapirs? *Transactions of the Royal Society of Edinburgh*, v. 87, pp. 105–114.

Petford, N., Kerr, R. C., and Lister, J. R., 1993, Dike transport of granitoid magmas. *Geology*, v. 21, pp. 845–848.

Petford, N., Lister, J. R., and Kerr, R. C., 1994, The ascent of felsic magmas in dykes. *Lithos*, v. 32, pp. 161–168.

Pettijohn, F. J., 1963, Memorial of Albert Johannsen. *American Mineralogist*, v. 48, pp. 454–459.

———, 1984, *Memoirs of an Unrepentant Field Geologist*. Chicago, University of Chicago Press.

———, 1988, *A Century of Geology 1885–1985 at the Johns Hopkins University*. Baltimore, MD, Gateway Press.

Philpotts, A. R., 1990, *Principles of Igneous and Metamorphic Petrology*. Englewood Cliffs, NJ, Prentice Hall.

Pichavant, M., 1981, An experimental study of the effect of boron on a water-saturated haplogranite at 1 kbar pressure: geological applications. *Contributions to Mineralogy and Petrology*, v. 76, pp. 430–439.

Pinkerton, J., 1811, *Petralogy: A Treatise on Rocks*, 2 v. London, White, Cochrane, & Co.

Pirsson, L. V., 1905a, The petrographic province of central Montana. *American Journal of Science*, v. 20 (4th series), pp. 35–49.

———, 1905b, *Petrography and Geology of the Igneous Rocks of the Highwood Mountains, Montana*. U. S. Geological Survey Bulletin 237.

———, 1918, The rise of petrology as a science. *American Journal of Science*, v.46 (4th series), pp. 222–239.

Pitcher, W. S., 1978, The anatomy of a batholith. *Journal of the Geological Society of London*, v. 135, pp. 157–182.

———, 1993, *The Nature and Origin of Granite*. London, Blackie.

———, 1997, *The Nature and Origin of Granite*, 2nd ed. London, Chapman & Hall.

Pitcher, W. S., Atherton, M. P., Cobbing, E. J., and Beckinsale, R. D., 1985, *Magmatism at a Plate Edge: The Peruvian Andes*. Glasgow, Blackie.

Piwinskii, A. J., and Wyllie, P. J., 1968a, Experimental studies of igneous rock series: A zoned pluton in the Wallowa Batholith, Oregon. *Journal of Geology*, v. 76, pp. 205–234.

———, 1968b, Experimental studies of igneous rock series: central Sierra Nevada Batholith, California. *Journal of Geology*, v. 76, pp. 548–570.

———, 1970, Experimental studies of igneous rock series: felsic body suite from the Needle Point pluton, Wallowa Batholith, Oregon. *Journal of Geology*, v. 78, pp. 52–76.

Playfair, J., 1802, *Illustrations of the Huttonian Theory of the Earth*. Edinburgh, William Creech.

Poldervaart, A., 1944, The petrology of the Elephant's Head dike and the New Amalfi sheet (Matatiele). *Transactions of the Royal Society of South Africa*, v. 30, pp. 85–119.

Poldervaart, A., and Parker, A. B., 1964, The crystallization index as a parameter of igneous differentiation in binary variation diagrams. *American Journal of Science*, v. 262, pp. 281–289.

Porter, R., 1977, *The Making of Geology: Earth Science in Britain, 1660–1815*. Cambridge, UK, Cambridge University Press.

Powell, R., 1984, Inversion of the assimilation and fractional crystallization (AFC) equations; characterization of contaminants from isotope and trace element relationships in volcanic suites. *Journal of the Geological Society of London*, v. 141, pp. 447–452.

Prendergast, M. L., 1978, *James Dwight Dana: The Life and Thought of an American Scientist*, 2 v. Ph.D. Dissertation, University of California, Los Angeles.

Prescher, H., 1971, Cotta, Carl Bernhard von, in Gillispie, C. C., ed., *Dictionary of Scientific Biography*, v. 3. New York, Charles Scribner's Sons, pp. 433–435.

Press, F., 1995, Growing up in the golden age of science. *Annual Reviews of Earth and Planetary Science*, v. 23, pp. 1–9.

Pumpelly, R., 1893, Memorial of Thomas Sterry Hunt. *Geological Society of America Bulletin*, v. 4, pp. 379–392.

Pyne, S. J., 1980, *Grove Karl Gilbert: A Great Engine of Research*. Austin, University of Texas.

Quirke, T. T., 1940, Granitization near Killarney, Ontario. *Geological Society of America Bulletin*, v. 51, pp. 237–253.

R. D., 1891, Hervey, Frederick Augustus, in Stephen, L., and Lee, S., eds., *Dictionary of National Biography*, v. 26. London, Smith, Elder, pp. 279–282.

R. G., 1885, Beddoes, Thomas, in Stephen, L., ed., *Dictionary of National Biography*, v. 4. London, Smith, Elder, pp. 94–95.

Radice, B., 1969, *Letters, and Panegyricus by Pliny*, 2 v. London, Wm. Heinemann.

Ragland, P. C., and Rogers, J.J.W., 1984, *Basalts*. New York, Van Nostrand Reinhold.

Raguin, E., 1946, *Géologie du Granite*. Paris, Masson and Cie.

Raleigh, C. B., 1980, Foreword. *American Journal of Science*, v. 280-A (Jackson memorial volume), pp. ix–xiv.

Ramberg, H., 1944, The thermodynamics of the earth's crust, I. *Norsk Geologisk Tidsskrift*, v. 24, pp. 98–111.

———, 1952, *The Origin of Metamorphic and Metasomatic Rocks*. Chicago, University of Chicago Press.

Ramdohr, P., 1975, Rosenbusch, Harry (Karl Heinrich Ferdinand), in Gillispie, C. C., ed., *Dictionary of Scientific Biography*, v. 11. New York, Charles Scribner's Sons, pp. 547–548.

Rapp, R. P., and Watson, E. B., 1995, Dehydration melting of metabasalt at 8–32 kbar: implications for continental growth and crust-mantle recycling. *Journal of Petrology*, v. 36, pp. 891–931.

Rapp, R. P., Watson, E. B., and Miller, C. F., 1991, Partial melting of amphibolite/eclogite and the origin of Archean trondhjemites and tonalities. *Precambrian Research*, v. 51, pp. 1–25.

Rappaport, R., 1972, Guettard, Jean-Étienne, in Gillispie, C. C., ed., *Dictionary of Scientific Biography*, v. 5: New York, Charles Scribner's Sons, pp. 577–579.

———, 1974, Malesherbes, Chrétien Guillaume de Lamoignon de, in Gillispie, C. C., ed., *Dictionary of Scientific Biography*, v. 9. New York, Charles Scribner's Sons, pp. 53–55.

———, 1997, *When Geologists Were Historians, 1665–1750*. Ithaca, NY, Cornell University Press.

Raspe, R. E., 1771, Nachricht von einigen niederhessischen Basalten, besonders aber einen Säulenbasaltstein Gebürge bei Felsberg und der Spuren eines verlöschten brennenden Berges am Habichtswalde über Weissenstein nahe bei Cassel. *Deutsche Schrifter der Königlicher. Societät der Wissenschaften in Göttingen*, v. 1, pp. 72–83.

———, 1776, *An Account of Some German Volcanoes*. London, Davis.

Rastall, R. H., 1945, The granite problem. *Geological Magazine*, v. 82, pp. 19–30.

Raymond, L. A., 1995, *Petrology*. Dubuque, IA, Wm. C. Brown

Read, H. H., 1920, Crystallization-differentiation in igneous magmas. By N. L. Bowen. *Geological Magazine*, v. 7 (6th series), pp. 86–87.

———, 1940, Metamorphism and igneous action. *Advancement of Science*, v. 1, pp. 223–257.

———, 1943, Meditations on granite: part one. *Proceedings of the Geologists' Association*, v. 54, pp. 64–85.

———, 1944, Meditations on granite, part two. *Proceedings of the Geologists' Association*, v. 55, pp. 45–93.

———, 1948, Granites and granites, in Gilluly, J., ed., *Origin of Granite*. Geological Society of America Memoir 28, pp. 1–19.

———, 1957, *The Granite Controversy*. London, Thomas Murby.

Recupito, G. C., 1635, *Avviso dell'incendio del Vesuvio*. Napoli, Edigio Longo.

Reilly, P. C., 1974, *Athanasius Kircher, S.J.: Master of a Hundred Arts, 1602–1680*. Wiesbaden, Edizioni del Mondo.

Reyer, E., 1877, *Beitrag zur Physik der Eruptionen und der Eruptivgesteine*. Wien, A. Hölder.

———, 1888, *Theoretische Geologie*. Stuttgart, E. Schweizerbart'sche Verlagshandlung.

Reynolds, D., 1945, Dr. Catherine Alice Raisin. *Nature*, v. 156, pp. 327–328.

———, 1946, The sequence of geochemical changes leading to granitization. *Quarterly Journal of the Geological Society of London*, v. 102, pp. 389–446.

———, 1947, The granite controversy. *Geological Magazine*, v. 84, pp. 209–223.

———, 1958, Granite: some tectonic, petrological, and physco-chemical aspects. *Geological Magazine*, v. 95, pp. 378–396.

———, 1968, Memorial of Arthur Holmes. *American Mineralogist*, v. 53, pp. 560–566.

Rhodes, F.H.T., 1991, Darwin's search for a theory of the Earth: symmetry, simplicity and speculation. *British Journal for the History of Science*, v. 24, pp. 193–229.

Richardson, W., 1805, Observations on basalts of the coast of Antrim. *Transactions of the Royal Society of Edinburgh*, v. 5, pp. 15–20.

———, 1806, On the volcanic theory. *Transactions of the Royal Irish Academy*, v. 10, pp. 35–107.

Richardson, W. A., 1923, The problem of batholithic intrusion. *Geological Magazine*, v. 60, pp. 121–128.

Richey, J. E., 1932, Tertiary ring-structures in Britain: Presidential address. *Geological Society of Glasgow Transactions*, v. 19, pp. 42–140.

Richey, J. E., and Thomas, H. H., 1930, *The Geology of Ardnamurchan, North-west Mull and Coll*. Geological Survey of Scotland Memoir.

———, 1932, The Tertiary ring complex of Slieve Gullion (Ireland). *Quarterly Journal of the Geological Society of London*, v. 88, pp. 776–847.

Ringwood, A. E., 1955a, The principles governing trace element distribution during magmatic crystallization, part I: the influence of electronegativity. *Geochimica et Cosmochimica Acta*, v. 7, pp. 189–202.

———, 1955b, The principles governing trace element distribution during magmatic crystallization, part II: the role of complex formation. *Geochimica et Cosmochimica Acta*, v. 7, pp. 242–254.

———, 1979, *Origin of the Earth and Moon*. New York, Springer-Verlag.

Rittmann, A., 1952, Nomenclature of volcanic rocks. *Bulletin Volcanologique*, v. 12 (2nd series), pp. 75–102.

Roberts, M. P., and Clemens, J. D., 1995, Feasibility of AFC models for the petrogenesis of calc-alkaline magma series. *Contributions to Mineralogy and Petrology*, v. 121, pp. 137–147.

Robertson, J. K., and Wyllie, P. J., 1971a, Experimental studies on rocks from the Deboullie Stock, northern Maine: including relations in the water-deficient environment. *Journal of Geology*, v. 79, pp. 549–571.

———, 1971b, Rock–water systems, with special reference to the water-deficient region. *American Journal of Science*, v. 271, pp. 252–277.

Rock, N.M.S., 1991, *Lamprophyres*. Glasgow, Blackie.

Rodolico, F., 1970, Arduino, Giovanni, *in* Gillispie, C. C., ed., *Dictionary of Scientific Biography*, v. 1. New York, Charles Scribner's Sons, pp. 233–234.

Rose, G., 1864, Zur Erinnerung an E. Mitscherlich. *Zeitschrift der deutschen geologischen Gesellschaft*, v. 16, pp. 21–72.

Rose, H., 1859, Über die verschiedenen Zustände der Kieselsäure. *Annalen der Physik und Chemie*, v. 108, pp. 1–40.

Rosenbusch, H., 1873, *Mikroskopische Physiographie der petrographisch wichtigen Mineralien*. Stuttgart, E. Schweizerbart'sche Verlagshandlung.

———, 1877a, *Mikroskopische Physiographie der massige Gesteine*. Stuttgart, E. Schweizerbart'sche Verlagshandlung.

———, 1877b, Die Steiger Schliefer und ihre Contactzone an den Granititen von Barr-Andlau und Hohwald. *Abhandlung Zeitschrift Geologische Specialkarte von Elsass-Lothringen*, v. 1, pp. 79–274.

———, 1887, *Mikroskopische Physiographie der massige Gesteine*, 2nd ed. Stuttgart, E. Schweizerbart'sche Verlagshandlung.

Rosenbusch, H., 1889, Über die chemischen Beziehungen der Eruptivgesteine. *Tschermaks mineralogische und petrographische Mitteilungen*, v. 11, pp. 144–178.

———, 1896, *Mikroskopische Physiographie der massige Gesteine*, 3rd ed. Stuttgart, E. Schweizerbart'sche Verlagshandlung.

———, 1898, *Elemente der Gesteinslehre*. Stuttgart, E. Schweizerbart'sche Verlagshandlung.

———, 1901, *Elemente der Gesteinslehre*, 2nd ed. Stuttgart, E. Schweizerbart'sche Verlagshandlung.

———, 1907–1908, *Mikroskopische Physiographie der massigen Gesteine*, 4th ed. Stuttgart, E. Schweizerbart'sche Verlagshandlung.

———, 1910, *Elemente der Gesteinslehre*, 3rd ed. Stuttgart, E. Schweizerbart'sche Verlagshandlung.

Rosenbusch, H., and Wulfing, E., 1904, *Mikroskopische Physiographie der petrographisch wichtigen Mineralien*. Stuttgart, E. Schweizerbart'sche Verlagshandlung.

Rosiwal, A., 1898, Uber geometrische Gesteinsanalysen. *Verhandlungen der k. k. Geolog. Reichsanstalt Wien*, No. 5, pp. 143–175.

Ross, J., ed., 1989a, *Kimberlites and Related Rocks*, vol. 1: *Their Composition, Occurrence, Origin and Emplacement*. Oxford, UK, Blackwell.

———, ed., 1989b, *Kimberlites and Related Rocks*, vol. 2: *Their Mantle/Crust Setting, Diamonds, and Diamond Exploration*. Oxford, UK, Blackwell.

Roth, J., 1861, *Die Gesteins-Analysen in tabellarischer Übersicht und mit kritischen Erläuterungen*. Berlin, W. Hertz.

———, 1869, *Die Gesteins-Analysen in tabellarischer Übersicht und mit kritischen Erläuterungen*, 2nd ed. Berlin, W. Hertz.

———, 1873, *Die Gesteins-Analysen in tabellarischer Übersicht und mit kritischen Erläuterungen*, 3rd ed. Berlin, W. Hertz.

———, 1879, *Die Gesteins-Analysen in tabellarischer Übersicht und mit kritischen Erläuterungen*, 4th ed. Berlin, W. Hertz.

———, 1883, *Allgemeine und chemische Geologie*, 3 v. Berlin, W. Hertz.

———, 1884, *Die Gesteins-Analysen in tabellarischer Übersicht und mit kritischen Erläuterungen*, 5th ed. Berlin, W. Hertz.

Rudel, A., 1962, *Les Volcans d'Auvergne*. Clermont-Ferrand, France, Volcans Editions.

Rudwick, M.J.S., 1970, Brongniart, Alexandre, in Gillispie, C. C., ed., *Dictionary of Scientific Biography*, v. 2. New York, Charles Scribner's Sons, pp. 493–497.

———, 1971, Conybeare, William Daniel, in Gillispie, C. C., ed., *Dictionary of Scientific Biography*, v. 3. New York, Charles Scribner's Sons, pp. 395–396.

———, 1972, *The Meaning of Fossils: Episodes in the History of Palaeontology*. London, Macdonald.

———, 1974, Poulett Scrope on the volcanoes of Auvergne: Lyellian time and political economy. *British Journal for the History of Science*, v. 7, pp. 205–242.

———, 1985, *The Great Devonian Controversy: The Shaping of Scientific Knowledge Among Gentlemanly Specialists*. Chicago, University of Chicago Press.

Rupke, N. A., 1983, *The Great Chain of History: William Buckland and the English School of Geology (1814–1849)*. Oxford, UK, Clarendon Press.

———, 1994, *Richard Owen: Victorian Naturalist*. New Haven, CT, Yale University Press.

Rushmer, T., 1991, Partial melting of two amphibolites: contrasting experimental results under fluid-absent conditions. *Contributions to Mineralogy and Petrology*, v. 107, pp. 41–59.

Russell, I. C., 1896a, Igneous intrusions in the neighborhood of the Black Hills of Dakota. *Journal of Geology*, v. 4, pp. 23–43.

———, 1896b, On the nature of igneous intrusion. *Journal of Geology*, v. 4, pp. 177–194.

Russell, J. K., and Stanley, C. R., 1989, *Theory and Application of Pearce Element Ratios to Geochemical Data Analysis*. Vancouver, BC, Geological Association of Canada.

Rutley, F., 1879, *The Study of Rocks: An Elementary Textbook of Petrology*. London, Longmans, Green.

Rutter, M. J., and Wyllie, P. J., 1988, Melting of vapour-absent tonalite at 10 kbar to simulate dehydration-melting in the deep crust. *Nature*, v. 331, pp. 159–160.

Salomon, W., 1903, Über die Lagerungsform und das Alter des Adamellotonalites. *K. preuss Acad. Wissenschaft, Physikalisch mathematische Klasse Sitzungsberichte*, v. 14, p. 310.

Santorelli, A., 1632, *Discorsi della natura, accidenti, e pronostici dell'incendio del monte de Somma dell'anno 1631*. Napoli, Appresso Egidio Longo.

Sarjeant, W., 1980, *Geologists and the History of Geology: An International Bibliography from the Origins to 1978*, 5 v. Melbourne, FL, Krieger.

———, 1987, *Geologists and the History of Geology: An International Bibliography from the Origins to 1978. Supplement, 1979–1984 and Additions*, 2 v. Malabar, FL, Krieger.

———, 1996, *Geologists and the History of Geology: An International Bibliography from the Origins to 1978. Supplement 2, 1985–1993 and Additions*, 3 v. Malabar, FL, Krieger.

Sartorius von Waltershausen, W., 1853, *Über die vulkanischen Gesteine in Sicilien und Island und ihre submarine Umbildung*. Göttingen, Dieterich Buchhandlung.

Scaillet, B., Pichavant, M., and Roux, J., 1995, Experimental crystallization of leucogranite magma. *Journal of Petrology*, v. 36, pp. 663–705.

Schacher, S., 1970, Bunsen, Robert Wilhelm Eberhard, *in* Gillispie, C. C., ed., *Dictionary of Scientific Biography*, v. 2. New York, Charles Scribner's Sons, pp. 586–590.

Schairer, J. F., 1942, The system $CaO-FeO-Al_2O_3-SiO_2$, I: results of quenching experiments on five joins. *Journal of the American Ceramic Society*, v. 25, pp. 241–274.

Schairer, J. F., and Bowen, N. L., 1935, Preliminary report on equilibrium relations between feldspathoids, alkali-feldspars, and silica. *Transactions of the American Geophysical Union* 16th Annual meeting, pp. 325–328.

Schairer, J. F., and Yoder, H. S., Jr., 1964, Crystal and liquid trends in simplified alkali basalts. *Carnegie Institution of Washington Year Book*, v. 63, pp. 65–74.

Scheerer, T., 1847, Discussion sur la nature plutonique du granite et des silicates cristallins qui s'y rallient. *Bulletin de la Société Géologique de France*, v. 4 (2nd series), pp. 468–495.

———, 1862, Die Gneuse des sächsischen Erzgebriges und verwandte Gesteine, nach ihrer chemischen Constitution und geologischen Bedeutung. *Zeitschrift der deutschen geologischen Gesellschaft*, v. 14, pp. 23–147.

———, 1864, Vorläufiger Bericht über krystallinische Silikatgesteine des Fassathales und benachbarter Gegenden Südtyrols. *Neues Jahrbuch*, pp. 385–411.

Schmid, R., 1981, Descriptive nomenclature and classification of pyroclastic deposits and fragments: recommendations of the IUGS subcommission on the systematics of igneous rocks. *Geology*, v. 9, pp. 41–43.

Schneer, C. J., 1979, *Two Hundred Years of Geology in America: Proceedings of the New Hampshire Bicentennial Conference on the History of Geology, October 15–19, 1976*. Hanover, NH, University Press of New England.

Schreiner, G.D.L., 1958, Comparison of the Rb-87/Sr-87 age of the red granite of the Bushveld complex from measurements on the total rock and separated mineral fractions. *Proceedings of the Royal Society of London*, Series A, v. 245, pp. 112–117.

Schwartz, G. M., 1959, Memorial of Frank Fitch Grout. *American Mineralogist*, v. 44, pp. 373–376.

Schweig, M., 1903, Untersuchungen über die Differentiation der Magmen. *Neues Jahrbuch für Mineralogie Geologie und Palaeontologie*, v. 17, pp. 516–564.

Scoates, J. S., 2000, The plagioclase–magma density paradox re-examined and the crystallization of Proterozoic anorthosites. *Journal of Petrology*, v. 41, pp. 627–649.

Scott, A., 1916, The later stages of the evolution of the igneous rocks. By N. L. Bowen. *Geological Magazine*, v. 3 (6th series), pp. 469–472.

Scott, E. L., 1973a, Keir, James, in Gillispie, C. C., ed., *Dictionary of Scientific Biography*, v. 7. New York, Charles Scribner's Sons, pp. 277–278.

———, 1973b, Kirwan, Richard, in Gillispie, C. C., ed., *Dictionary of Scientific Biography*, v. 7. New York, Charles Scribner's Sons, pp. 387–390.

Scrope, G. P., 1825, *Considerations on Volcanos*. London, W. Phillips.

———, 1827, *Memoir on the Geology of Central France*. London, Longman.

———, 1856, On the formation of craters, and the nature of the liquidity of lavas. *Quarterly Journal of the Geological Society of London*, v. 12, pp. 326–350.

———, 1858, *The Geology and Extinct Volcanos of Central France*. London, John Murray.

———, 1859, On volcanic cones and craters, *Quarterly Journal of the Geological Society of London*, v. 15, pp. 505–549.

———, 1862, *Volcanos: The Character of Their Phenomena, Their Share in the Structure and Composition of the Globe, and Their Relation to Its Internal Forces*. London: Longman, Green, Longman, and Roberts.

———, 1869, Review of Richthofen's 'The Natural System of Volcanic Rocks.' *Geological Magazine*, v. 6, pp. 510–516.

Secord, J. A., 1986, *Controversy in Victorian Geology: The Cambrian–Silurian Dispute*. Princeton, NJ, Princeton University Press.

———, 1991, The discovery of a vocation: Darwin's early geology. *British Journal for the History of Science*, v. 24, pp. 133–158.

Sederholm, J. J., 1907, Om granit och gneiss. *Bulletin de la Commission Géologique de Finlande*, no. 23.

———, 1923, On migmatites and associated Pre-Cambrian rocks of southwestern Finland, part I: the Pellinge region. *Bulletin de la Commission Géologique de Finlande*, no. 58.

Sen, C., and Dunn, T., 1994, Dehydration melting of a basaltic composition amphibolite at 1.5 and 2 Gpa: implications for the origin of adakites. *Contributions to Mineralogy and Petrology*, v. 117, pp. 394–409.

Senft, F., 1857, *Classification und Beschreibung der Felsarten*. Breslau, n.p.

Serao, F., 1738, *Istoria dell'incendio del Vesuvio accaduto nel mese di maggio del'anno MDCCXXXVII*. Napoli, Accademia della scienze.

Servos, J. W., 1983, To explore the borderland: the foundation of the Geophysical Laboratory of the Carnegie Institution of Washington. *Historical Studies in the Physical Sciences*, v. 14, pp. 147–185.

———, 1990, *Physical Chemistry from Ostwald to Pauling: The Making of a Science in America*. Princeton, NJ, Princeton University Press.

Seward, A. C., and Tilley, C. E., 1940, Alfred Harker. *Obituary Notices of Fellows of the Royal Society of London*, v. 3, pp. 197–216.

Seymour, Lord W., 1815, An account of observations made by Lord Webb Seymour and Professor Playfair upon some geological appearances in Glen Tilt and the adjacent country. *Transactions of the Royal Society of Edinburgh*, v. 7, pp. 303–375.

Shackleton, R. M., 1970, William Quarrier Kennedy—an appreciation, *in* Clifford, T. N., and Gass, I. G., *African Magmatism and Tectonics*. Darien, CT, Hafner, pp. xi–xv.

Shand, S. J., 1913, On saturated and unsaturated igneous rocks. *Geological Magazine*, v. 10 (5th series), pp. 508–514.

———, 1914, The principle of saturation in petrography. *Geological Magazine*, v. 1 (6th series), pp. 485–493.

———, 1915, The principle of saturation in petrography: A reply. *Geological Magazine*, v. 2 (6th series), pp. 575–576.

———, 1916, A recording micrometer for geometrical rock analysis. *Journal of Geology*, v. 24, pp. 394–404.

———, 1917, A system of petrography. *Geological Magazine*, v. 4 (6th series), pp. 463–469.

———, 1923, Petrographic nomenclature. *Geological Magazine*, v. 60, pp. 287–288.

———, 1949a, *Eruptive Rocks*. London, Thomas Murby.

———, 1949b, Rocks of the Mid-Atlantic Ridge. *Journal of Geology*, v. 57, pp. 89–92.

Shapiro, L., 1967, Rapid analysis of rocks and minerals by a single solution method, *in* Geological Survey Research 1967 Chapter B. *U. S. Geological Survey Professional Paper 575-B*, p. B187-B191.

Shapiro, L., and Brannock, W. W., 1952, *Rapid Analysis of Silicate Rocks*. U. S. Geological Survey Circular 165.

———, 1956, *Rapid Analysis of Silicate Rocks*. U. S. Geological Survey Bulletin 1936-C.

———, 1962, *Rapid Analysis of Silicate, Carbonate, and Phosphate Rocks*. U. S. Geological Survey Bulletin 1144-A.

Shaw, D. M., 1953, The camouflage principle and trace-element distribution in magmatic minerals. *Journal of Geology*, v. 61, pp. 142–151.

———, 1970, Trace element fractionation during anatexis. *Geochimica et Cosmochimica Acta*, v. 34, pp. 237–243.

———, 1978, Trace element behaviour during anatexis in the presence of a fluid phase. *Geochimica et Cosmochimica Acta*, v. 42, pp. 933–943.

Shaw, H. R., 1963, Obsidian–H_2O viscosities at 1000 and 2000 bars in the temperature range 700–900°C. *Journal of Geophysical Research*, v. 68, pp. 6337–6343.

———, 1964, Theoretical solubility of H_2O in silicate melts: quasi-crystalline models. *Journal of Geology*, v. 72, pp. 601–617.

Shaw, H. R., 1965, Comments on viscosity, crystal settling and convection in granitic magmas. *American Journal of Science*, v. 263, pp. 120–152.

———, 1969, Rheology of basalt in the melting range. *Journal of Petrology*, v. 10, pp. 510–535.

———, 1974, Diffusion of H_2O in granitic liquids, part 1: experimental data; part II: mass transfer in magma chambers, in Hofmann, A. W., Giletti, B. J., Yoder, H. S., Jr., and Yund, R. A., eds., *Geochemical Transport and Kinetics*. Washington, DC, Carnegie Institution of Washington, pp. 139–170.

———, 1980, Fracture mechanisms of magma transport from the mantle to the surface, in Hargraves, R. B., ed., *Physics of Magmatic Processes*. Princeton, NJ, Princeton University Press, pp. 201–246.

Shaw, H. R., Wright, T. L., Peck, D. L., and Okamura, R., 1968, The viscosity of basaltic magma: An analysis of field measurements in Makaopuhi lava lake, Hawaii. *American Journal of Science*, v. 266, pp. 225–264.

Sheets-Pyenson, S., 1996, *John William Dawson: Faith, Hope, and Science*. Montreal, McGill-Queen's University Press.

Shepherd, E. S., Rankin, G. A., and Wright, F. E., 1909, The binary systems of alumina with silica, lime and magnesia. *American Journal of Science*, v. 28 (4th series), pp. 293–333.

Sheppard, S.M.F., and Harris, C., 1985, Hydrogen and oxygen isotope geochemistry of Ascension Island lavas and granites: variation with crystal fractionation and interaction with seawater. *Contributions to Mineralogy and Petrology*, v. 91, pp. 74–81.

Shils, E., 1979, The order of learning in the United States: the ascendancy of the university, in Oleson, A., and Voss, J., eds., *The Organization of Knowledge in Modern America: 1860–1920*. Baltimore, MD, Johns Hopkins University Press, pp. 19–47.

Shimer, J. A., 1943, Spectrographic analysis of New England granites and pegmatites. *Geological Society of America Bulletin*, v. 54, pp. 1049–1066.

Shirley, D. N., 1987, Differentiation and compaction in the Palisades Sill, New Jersey. *Journal of Petrology*, v. 28, pp. 835–865.

Shrock, R. R., 1977, *Geology at M.I.T. 1865–1965: A History of the First Hundred Years of Geology at Massachusetts Institute of Technology, I: The Faculty and Supporting Staff*. Cambridge, MA, MIT Press.

———, 1982, *Geology at M.I.T. 1865–1965: A History of the First Hundred Years of Geology at Massachusetts Institute of Technology, II: Departmental Operations and Products*. Cambridge, MA, MIT Press.

Sigsby, R. J., 1966, A brief history of the petrographic microscope. *Compass*, v. 43, pp. 94–103.

Sigurdsson, H., 1999, *Melting the Earth*. New York, Oxford University Press.

Silverman, S. R., 1951, The isotope geology of oxygen. *Geochimica et Cosmochimica Acta*, v. 2, pp. 26–47.

Skjerlie, K. P., and Johnston, A. D., 1993, Fluid-absent melting behavior of an F-rich tonalitic gneiss at mid-crustal pressures: implications for the generation of anorogenic granites. *Journal of Petrology*, v. 34, pp. 785–815.

———, 1996, Vapour-absent melting from 10 to 20 kbar of crustal rocks that contain multiple hydrous phases: implications for anatexis in the deep to very deep continental crust and active continental margins. *Journal of Petrology*, v. 37, pp. 661–691.

Skjerlie, K. P., and Patiño Douce, A. E., 1995, Anatexis of interlayered amphibolite and pelite at 10 kbar: effect of diffusion of major components on phase relations and melt fraction. *Contributions to Mineralogy and Petrology*, v. 122, pp. 62–78.

Sleep, M.C.W., 1969, Sir William Hamilton (1730–1803): his work and influence in geology. *Annals of Science*, v. 25, pp. 319–338.

Smeaton, W. A., 1970, Bergman, Torbern Olof, in Gillispie, C. C., ed., *Dictionary of Scientific Biography*, v. 2. New York, Charles Scribner's Sons, pp. 4–8.

Smith, C. H., and Kapp, H. E., 1963, The Muskox intrusion, a recently discovered layered intrusion in the Coppermine River area, Northwest Territories, Canada. *Mineralogical Society of America Special Paper 1*, pp. 30–35.

Smith, I. F., 1981, *The Stone Lady: A Memoir of Florence Bascom*. Bryn Mawr, PA, Bryn Mawr College.

Smith, R. L., and Bailey, R. A., 1966, The Bandelier tuff: A study of ash-flow eruption cycles from zoned magma chambers. *Bulletin Volcanologique*, v. 29, pp. 83–103.

Snyder, D., and Tait, S., 1995, Replenishment of magma chambers: comparison of fluid-mechanic experiments with field relations. *Contributions to Mineralogy and Petrology*, v. 122, pp. 230–240.

Snyder, D., Aambes, C., Tait, S., and Wiebe, R. A., 1997, Magma mingling in dikes and sills. *Journal of Geology*, v. 105, pp. 75–86.

Sollas, W. J., 1892, Contributions to a knowledge of the granites of Leinster. *Transactions of the Royal Irish Academy*, v. 29, pp. 471–473.

Sorby, H. C., 1851, On the microscopical structure of the calcareous grit of the Yorkshire coast. *Quarterly Journal of the Geological Society of London*, v. 7, pp. 1–6.

———, 1853, On the origin of slaty cleavage. *Edinburgh New Philosophical Journal*, v. 55, pp. 137–148.

———, 1858, On the microscopical structure of crystals, indicating the origin of minerals and rocks. *Quarterly Journal of the Geological Society of London*, v. 14, pp. 453–500.

———, 1863, On the microscopical structure of Mount Sorrel [Grooby] Syenite, artificially fused and slowly cooled. *Proceedings of the Geological and Polytechnic Society of West Yorkshire*, v. 4, pp. 301–304.

———, 1876, Obituary. *Mineralogical Magazine*, v. 1, pp. 95–96.

Sørenson, H., ed., 1974, *The Alkaline Rocks*. New York, John Wiley and Sons.

Soret, C., 1879, Sur l'état d'équilibre que prend au point de vue de sa concentration une dissolution saline primitivement homogène dont deux parties sont portées a des températures différentes. *Archives des Sciences Physiques et Naturelles*, v. 27, pp. 48–61

———, 1881, Sur l'état d'équilibre que prend au point de vue de sa concentration une dissolution saline primitivement homogène dont deux parties sont portées a des températures différentes. *Annales de Chimie et de Physique*, v. 22, pp. 293–297.

Sorrentino, I., 1734, *Istoria del Monte Vesuvio*. Napoli, G. Severini.

Sosman, R. B., 1952, Temperature scales and silicate research. *American Journal of Science*, Bowen volume, pp. 517–528.

Soulavie, J.L.G., 1780–1784, *Histoire Naturelle de la France Méridionale*, 7 v. Nimes, C. Belle.

Spallanzani, L., 1798, *Travels in the Two Sicilies and some Parts of the Apennines*, 4 v. London, G. G. & J. Robinson.

Sparks, R.S.J., 1990, Discussion of "crystal capture, sorting and retention in convecting magma" by B. D. Marsh. *Geological Society of America Bulletin*, v. 102, pp. 847–850.

Sparks, R.S.J., Meyer, P., and Sigurdsson, H., 1980, Density variations amongst mid-ocean ridge basalts: implications for magma mixing and the scarcity of primitive basalts. *Earth and Planetary Science Letters*, v. 45, pp. 419–430.

Sparks, R.S.J., Huppert, H. E., and Turner, J. S., 1984, The fluid dynamics of evolving magma chambers. *Philosophical Transactions of the Royal Society of London*, v. A310, pp. 511–534.

Sparks, R.S.J., Huppert, H. E., Kerr, R. C., McKenzie, D. P., and Tait, S. R., 1985, Postcumulus processes in layered intrusions. *Geological Magazine*, v. 122, pp. 555–568.

Spencer, L., 1924, Biographical notices of mineralogists recently deceased (second series). *Mineralogical Magazine*, v. 20, pp. 252–275.

———, 1927, Biographical notices of mineralogists recently deceased (third series). *Mineralogical Magazine*, v. 21, pp. 229–257.

———, 1930, Biographical notices of mineralogists recently deceased (fourth series). *Mineralogical Magazine*, v. 22, pp. 387–412.

———, 1933, Biographical notices of mineralogists recently deceased (fifth series). *Mineralogical Magazine*, v. 23, pp. 337–366.

———, 1947, Biographical notices of mineralogists recently deceased (eighth series). *Mineralogical Magazine*, v. 28, pp. 175–229.

Spera, F. J., and Bohrson, W. A., 2001, Energy-constrained open-system magmatic processes, I: general model and energy-constrained assimilation and fractional crystallization (EC-AFC) formulation. *Journal of Petrology*, v. 42, pp. 999–1018.

Spera, F. J., Yuen, D. A., and Kirschvink, S. J., 1982, Thermal boundary layer convection in silicic magma chambers: effects of temperature-dependent rheology and implications for thermogravitational chemical fractionation. *Journal of Geophysical Research*, v. 87, pp. 8755–8767.

Spera, F. J., Yuen, D. A., Greer, J. C., and Sewell, G., 1986, Dynamics of magma withdrawal from stratified magma chambers. *Geology*, v. 14, pp. 723–726.

Spera, F. J., Oldenburg, C. M., and Yuen, D. A., 1989, Magma zonation: effects of chemical buoyancy and diffusion. *Geophysical Research Letters*, v. 16, pp. 1104–1108.

Spera, F. J., Oldenburg, C. M., Christensen, C., and Todesco, M., 1995, Simulations of convection with crystallization in the system $KAlSi_2O_6$–$CaMgSi_2O_6$: implications for compositionally zoned magma bodies. *American Mineralogist*, v. 80, pp. 1188–1207.

Stadnichenko, T. M., 1947, Memorial of Vladimir Ivanovich Vernadsky. *American Mineralogist*, v. 32, pp. 181–188.

Stafford, R. A., 1989, *Scientist of Empire: Sir Roderick Murchison, Scientific Exploration and Victorian Imperialism*. Cambridge, UK, Cambridge University Press.

Stanton, W., 1971, Dana, James Dwight, in Gillispie, C. C., ed., *Dictionary of Scientific Biography*, v. 3. New York, Charles Scribner's Sons, pp. 549–554.

Stegner, W., 1936, *Clarence Edward Dutton: An Appraisal*. Salt Lake City, n.p.

———, 1971, Dutton, Clarence Edward, in Gillispie, C. C., ed., *Dictionary of Scientific Biography*, v. 4. New York, Charles Scribner's Sons, pp. 265–266.

Stephens, L. D., 1982, *Joseph Le Conte, Gentle Prophet of Evolution*. Baton Rouge, LA, Louisiana State University Press.

Stern, C. R., and Wyllie, P. J., 1978, Phase compositions through crystallization intervals in basalt-andesite–H_2O at 30 kbar, with implications for subduction zone magmas. *American Mineralogist*, v. 63, pp. 641–663.

Stevens, G., Clemens, J. D., and Droop, G.T.R., 1997, Melt production during granulite-facies anatexis: experimental data from 'primitive' metasedimentary protoliths. *Contributions to Mineralogy and Petrology*, v. 128, pp. 352–370.

Stevens, R. E., et al., 1960, *Second Report on a Cooperative Investigation of the Composition of Two Silicate Rocks*. U. S. Gelogical Survey Bulletin 1113.

Stewart, B. W., and DePaolo, D. J., 1990, Isotopic studies of processes in mafic magma chambers, II: the Skaergaard Intrusion, East Greenland. *Contributions to Mineralogy and Petrology*, v. 104, pp. 125–141.

Stewart, F. H., 1971, The place of geology in government scientific research. *Journal of the Geological Society of London*, v. 127 pp. 437–446.

Strange, J., 1775, An account of two giants causeways, or groups of prismatic basaltine columns, and other curious vulcanic concretions, in the Venetian state in Italy; with some remarks on the characters of these and other similar bodies, and on the physical geography of the countries in which they are found. *Philosophical Transactions of the Royal Society of London*, v. 65, pp. 5–47.

Streckeisen, A., 1964, Zur Klassifikation der Eruptivgesteine. *Neues Jahrbuch für Mineralogie, Monatshefte* 1964, pp. 195–222.

———, 1965, Die Klassiffikation der Eruptivgesteine. *Geologische Rundschau*, v. 55, pp. 478–491.

———, 1967, Classification and nomenclature of igneous rocks. *Neues Jahrbuch für Mineralogie, Abhandlungen* 107, pp. 144–214, 215–240.

———, 1974, Classification and nomenclature of plutonic rocks. *Geologische Rundschau*, v. 63, pp. 773–786.

———, 1976a, To each plutonic rock its proper name. *Earth Science Reviews*, v. 12, pp. 1–33.

———, 1976b, Classification of the common igneous rocks by means of their chemical composition. *Neues Jahrbuch für Mineralogie, Monatshefte* 1976 pp. 1–15.

———, 1978, Classification and nomenclature of volcanic rocks, lamprophyres, carbonatites and melilitic rocks. *Neues Jahrbuch für Mineralogie, Abhandlungen* 134, pp. 1–14.

Streckeisen, A., and Le Maitre, R. W., 1979, A chemical approximation to the modal QAPF classification of the igneous rocks. *Neues Jahrbuch für Mineralogie, Abhandlungen* 136, pp. 169–206.

Stubblefield, J., 1965, Edward Battersby Bailey. *Biographical Memoirs of the Royal Society of London*, v. 11, pp. 1–21.

Suess, E., 1885, *Das Antlitz der Erde*, 3 v. Vienna, n.p.

———, 1895, *Sitzungsberichte der Wiener Akademie*, v. 104, p. 52.

———, 1904–1924, *The Face of the Earth*, 5 v. Oxford, UK, Clarendon Press.

Sun, S.-S., 1980, Lead isotopic study of young volcanic rocks from mid-ocean ridges, ocean islands and island arcs. *Philosophical Transactions of the Royal Society of London*, v. A297, pp. 409–445.

Sutton, J., 1970, Herbert Harold Read. *Biographical Memoirs of the Fellows of the Royal Society of London*, v. 16, pp. 479–497.

Swanson, S. E., 1979, The effect of CO_2 on phase equilibria and crystal growth in the system $KAlSi_3O_8$–$NaAlSi_3O_8$–$CaAl_2Si_2O_8$–SiO_2–CO_2 to 8000 bars. *American Journal of Science*, v. 279, pp. 703–720.

Szabadváry, F., 1974, Mitscherlich, Eilhard, in Gillispie, C. C., ed., *Dictionary of Scientific Biography*, v. 9 New York, Charles Scribner's Sons, pp. 423–426.

T.F.H., 1896, John Pinkerton, in Lee, S., ed., *Dictionary of National Biography*, v. 45. London, Smith, Elder, pp. 316–318.

T. S., 1898, Strange, John, in Lee, S., ed., *Dictionary of National Biography*, v. 55. London, Smith, Elder, p. 23.

Tait, S., and Jaupart, C., 1992, Compositional convection in a reactive crystalline mush and melt differentiation. *Journal of Geophysical Research*, v. 97, pp. 6735–6756.

Takeuchi, H., Uyeda, S., and Kanamori, H., 1967, *Debate About the Earth: Approach to Geophysics Through Analysis of Continental Drift*. San Francisco, Freeman, Cooper.

Talbot, W.H.F., 1834, Experiments on light. *London and Edinburgh Philosophical Magazine and Journal of Science*, v. 5 (3rd series), pp. 321–334.

Tatsumoto, M., Hedge, C. E., and Engel, A.E.J., 1965, Potassium, rubidium, strontium, thorium, uranium, and the ratio of strontium-87 to strontium-86 in oceanic tholeiitic basalt. *Science*, v. 150, pp. 886–888.

Taylor, H. P., Jr., 1968, The oxygen isotope geochemistry of igneous rocks. *Contributions to Mineralogy and Petrology*, v. 19, pp. 1–71.

———, 1978, Oxygen and hydrogen isotope studies of plutonic granitic rocks. *Earth and Planetary Science Letters*, v. 38, pp. 177–210.

———, 1980, The effects of assimilation of country rocks by magmas on $^{18}O/^{16}O$ and $^{87}Sr/^{86}Sr$ systematics in igneous rocks. *Earth and Planetary Science Letters*, v. 47, pp. 243–254.

Taylor, H. P., Jr., and Epstein, S., 1962, Relationship between O^{18}/O^{16} ratios in coexisting minerals of igneous and metamorphic rocks. *Geological Society of America Bulletin*, v. 73, pp. 461–480, 675–694.

———, 1963, O^{18}/O^{16} ratios in rocks and coexisting minerals of the Skaergaard intrusion, East Greenland. *Journal of Petrology*, v. 4, pp. 51–74.

Taylor, H. P., Jr., and Forester, R. W., 1979, An oxygen and hydrogen isotope study of the Skaeraard Intrusion and its country rocks: A description of a 55-m.y. old fossil hydrothermal system. *Journal of Petrology*, v. 20, pp. 355–419.

Taylor, K. L., 1969, Nicolas Desmarest and geology in the eighteenth century, in C. J. Schneer, ed., *Toward a History of Geology*. Cambridge, MA, MIT Press, pp. 339–356.

———, 1971a, Des Cloizeaux, Alfred-Louis-Olivier Legrand, in Gillispie, C. C., ed., *Dictionary of Scientific Biography*, v. 4. New York, Charles Scribner's Sons, pp. 65–67.

———, 1971b, Desmarest, Nicholas, in Gillispie, C. C., ed., *Dictionary of Scientific Biography*, v. 4. New York, Charles Scribner's Sons, pp. 70–73.

———, 1971c, Dolomieu, Dieudonné (called Déodat) de Gratet de, in Gillispie, C. C., ed., *Dictionary of Scientific Biography*, v. 4. New York, Charles Scribner's Sons, pp. 149–153.

———, 1973, Lamétherie, Jean-Claude de, in Gillispie, C. C., ed., *Dictionary of Scientific Biography*, v. 7. New York, Charles Scribner's Sons, pp. 602–604.

Taylor, S. R., 1975, *Lunar Science: A Post-Apollo View*. New York, Pergamon Press.

———, 1982, *Planetary Science: A Lunar Perspective*. Houston, TX, Lunar and Planetary Institute.
———, 1995, Memorial of Louis H. Ahrens 1918–1990. *American Mineralogist*, v. 80, pp. 410–411.
Teall, J.J.H., 1888, *British Petrography*. London, Dulau.
———, 1901, The evolution of petrological ideas. *Proceedings of the Geological Society*, v. 57, pp. lxii–lxxxvi.
Termier, P., 1904, Les schistes cristallins des Alpes occidentals. *Comptes Rendus IX International Geological Congress* (Vienna, 1903), pp. 571–586.
Thackray, A., 1971, Daubeny, Charles Giles Bridle, in Gillispie, C. C., ed., *Dictionary of Scientific Biography*, v. 3. New York, Charles Scribner's Sons, pp. 585–586.
Thompson, R. N., Morrison, M. A., Hendry, G. L., and Parry, S. J., 1984, An assessment of the relative roles of a crust and mantle in magma genesis: An elemental approach. *Philosophical Transactions of the Royal Society of London*, Series A, v. 310, pp. 549–590.
Thornton, C. P., and Tuttle, O. F., 1960, Chemistry of igneous rocks, I: differentiation index. *American Journal of Science*, v. 258, pp. 664–684.
Thorpe, R. S., ed., 1982, *Andesites: Orogenic Andesites and Related Rocks*. New York, John Wiley and Sons.
Tikhomirov, V. V., 1970, Abich, Otto Hermann Wilhelm, in Gillispie, C. C., ed., *Dictionary of Scientific Biography*, v. 1. New York, Charles Scribner's Sons, pp. 19–21.
Tikoff, B., and Teyssier, C., 1992, Crustal-scale, en-echelon 'P shear' tensional bridges: A possible solution to the batholithic room problem. *Geology*, v. 20, pp. 927–930.
Tilley, C. E., 1941, Waldemar Christofer Brögger. *Obituary Notices of Fellows of the Royal Society*, v. 3, pp. 503–517.
Tomkeieff, S. I., 1962, George Walter Tyrrell. *Proceedings of the Geological Society of London*, no. 1062, pp. 162–165.
Towell, D. G., Winchester, J. W., and Spirn, R. V., 1965, Rare-earth distributions in some rocks and associated minerals of the batholith of Southern California. *Journal of Geophysical Research*, v. 70, pp. 3485–3496.
Trager, E. A., 1973, Memorial to Joseph Volney Lewis, 1869–1969. *Geological Society of America Memorials*, v. 1, pp. 44–54.
Trial, A. F., and Spera, F. J., 1990, Mechanisms for the generation of compositional heterogeneities in magma chambers. *Geological Society of America Bulletin*, v. 102, pp. 353–367.
Troalen, M., 1943, *J. Durocher, sa vie, son oeuvre*. Unpublished manuscript, Diplôme d'Études Supérieures, Université de Rennes.
Tröger, E., 1935a, Der Gehalt an selteneren Elementen bei Eruptivgesteinen. *Chemie der Erde*, v. 11, pp. 286–310.
———, 1935b, *Spezielle Petrographie der Eruptivgesteine*. Berlin, Verlag der deutschen mineralogischen Gesellschaft.
———, 1938, Eruptivegesteinsnamen. *Fortschritte der Mineralogie*, v. 23, pp. 41–53.
Tschermak, G., 1869, Mikroskopische Unterscheidung der Mineralien aus der Augit-, Amphibol- und Biotit-Gruppe. *Sitzungsberichte Akademie Wissenschaften, Mathematisch-Naturwissenschaftliche Klasse, Wien*, v. 59, pp. 1–12.
Turner, F. J., and Verhoogen, J., 1960, *Igneous and Metamorphic Petrology*, 2nd ed. New York, McGraw-Hill.

Turner, J. S., 1980, A fluid-dynamic model of differentiation and layering in magma chambers. *Nature*, v. 285, pp. 213–215.

Turner, J. S., and Campbell, I. H., 1986, Convection and mixing in magma chambers. *Earth Science Reviews*, v. 23, pp. 255–352.

Turner, J. S., and Chen, C. F., 1974, Two-dimensional effects and double-diffusive convection. *Journal of Fluid Mechanics*, v. 63, pp. 577–592.

Turner, J. S., and Gustafson, L. B., 1978, The flow of hot saline solutions from vents in the sea floor: some implications for exhalative sulfide and other ore deposits. *Economic Geology*, v. 73, pp. 1082–1100.

Turner, J. S., Huppert, H. E., and Sparks, R.S.J., 1986, Komatiites II: experimental and theoretical investigations of post-emplacement cooling and crystallization. *Journal of Petrology*, v. 27, pp. 397–437.

Tuttle, O. F., 1952, Origin of the contrasting mineralogy of extrusive and plutonic salic rocks. *Journal of Geology*, v. 60, pp. 107–124.

Tuttle, O. F., and Bowen, N. L., 1958, *Origin of Granite in the Light of Experimental Studies in the System* $NaAlSi_3O_8$–$KAlSi_3O_8$–SiO_2–H_2O. Geological Society of America Memoir 74.

Tuttle, O. F., and Gittins, J., 1966, *Carbonatites*. New York, John Wiley and Sons.

Tyrrell, G. W., 1914, A review of igneous rock classification. *Science Progress*, v. 33, pp. 60–84.

———, 1916, The picrite-teschenite sill of Lugar (Ayrshire). *Quarterly Journal of the Geological Society of London*, v. 72, pp. 84–131.

———, 1921, Some points in petrographic nomenclature. *Geological Magazine*, v. 58, pp. 494–502.

———, 1926, *Principles of Petrology*. London, Methuen.

———, 1929, *Principles of Petrology*, 2nd ed. New York, E. P. Dutton.

Ussing, N. V., 1912, Geology of the country around Julianehaab, Greenland. *Meddelelser om Grønland*, v. 38, pp. 1–426.

van't Hoff, J. H., 1887, Die Rolle des osmotischen Druckes in der Analogie zwischen Lösungen und Gasen. *Zeitschrift für physikalische Chemie*, v. 1, pp. 481–508.

Vaughan, F. E., 1970, *Andrew C. Lawson: Scientist, Teacher, Philosopher*. Glendale, CA, A. H. Clark.

Vernadsky, V. I., 1924, *La Géochimie*. Paris, n.p.

Veysey, L. R., 1965, *The Emergence of the American University*. Chicago, University of Chicago Press.

Vielzeuf, D., and Montel, J.-M., 1994, Partial melting of metagraywackes, part I: fluid-absent experiments and phase relationships. *Contributions to Mineralogy and Petrology*, v. 117, pp. 375–393.

Vincent, E. A., 1960, Analysis by gravimetric and volumetric methods, flame photometry, colorimetry and related techniques, *in* Smales, A A., and Wager, L. R., eds., *Methods in Geochemistry*. New York, Interscience.

———, 1994, *Geology and Mineralogy at Oxford 1860–1986: History and Reminiscence*. Oxford, UK, E. A. Vincent.

Virlet d'Aoust, T., 1847, Observations sur le métamorphisme normal et la probabilité de la non-existence de véritables roches primitives à la surface du globe. *Bulletin de la Societé Géologique de France*, v. 4 (2nd series), pp. 498–505.

Vogelsang, H., 1867, *Philosophie der Geologie und mikroskopische Gesteinsstudien*. Bonn, Max Cohen & Sohn.

Vogt, J.H.L., 1891, Om Dannelsen af de vitigste in Norge og Sverige representerede grupper af jemmalmforekomster. *Geologiska Föreningens i Stockholm Förhandlingar*, v. 13, pp. 476–536, 683–735.

———, 1903–1904, *Die Silikatschmelzlösungen*, 2 v. Christiania, Norway, Jacob Dybwad.

vom Rath, G., 1860, Skizzen aus dem vulkanischen Gebiete des Niederrheins. *Zeitschrift der deutschen geologischen Gesellschaft*, v. 12, pp. 29–47.

von Buch, L., 1802–1809, *Geognostische Beobachtungen auf Reisen durch Deutschland und Italien*, 2 v. Berlin, Hande and Spener.

———, 1824, Über geognostische Erscheinungen in Fassathale. *Leonhard's Taschenbuch für 1824*, pp. 345–347.

———, 1825, *Physikalische Beschreibung der Canarischen Inseln*. Berlin, K. Akademie der Wissenschaft.

———, 1836, Über Erhebungskrater und Vulkane. *Annalen der Physik*, v. 37, pp. 169–190.

von Cotta, C. B., 1855, *Die Gesteinslehre*. Freiberg, J. G. Engelhardt.

———, 1866, *Rocks Classified and Described: A Treatise on Lithology*. London, Longmans, Green.

von Humboldt, A., 1790, *Mineralogische Beobachtungen über einige Basalte am Rhein*. Brunswick, n.p.

von Lasaulx, A., 1875, *Précis de Pétrographie: Introduction à l'Étude des Roches*. Paris, J. Rothschild.

———, 1886, *Einführung in die Gesteinslehre*. Bonn, E. Strauss.

von Leonhard, K. C., 1805–1809, *Handbuch einer allgemeinen topographischen Mineralogie*, 3 v. Frankfurt, n.p.

———, 1818, *Zur Naturgeschichte der Vulkane*. Frankfurt, n.p.

———, 1823, *Charakteristik der Felsarten*, 2 v. Heidelberg, n.p.

———, 1832, *Die Basaltgebilde*. Stuttgart, n.p.

von Leonhard, K.C.,1835, *Lehrbuch der Geognosie und Geologie*. Stuttgart, E. Schweizerbart'sche Verlagshandlung.

von Platen, H., 1965, Kristallisation granitischer Schmelzen. *Beiträge zur Mineralogie und Petrographie*, v. 11, pp. 334–381.

von Richthofen, F., 1868, The natural system of volcanic rocks. *California Academy of Sciences Memoirs*, v. 1, pp. 1–94.

von Troil, U., 1780, *Letters on Iceland: containing Observations on the Civil, Literary, Ecclesiastical, and Natural History; Antiquities, Volcanos, Basaltes, Hot Springs; Customs, Dress, Manners of the Inhabitants, &c, &c*. London, J. Robson.

von Wolff, L. F., 1922, Die Prinzipien einer quantitativen Klassifikation der Eruptivgesteine, insbesondere der jungen Ergussgesteine. *Geologische Rundschau*, v. 13, pp. 9–17.

von Zittel, K., 1899, *Geschichte der Geologie und Palaeontologie bis zum Ende des 19 Jahrhundert*. München, n. p.

W.A.S.H., 1893, Mackenzie, George Steuart, in Lee, S., ed., *Dictionary of National Biography*, v. 35. London, Smith, Elder, pp. 149–150.

W. F., 1877, Sartorius von Waltershausen. *Geological Magazine*, v. 14, pp. 141–142.

W. R., 1908, Pierre Eugène Marcellin Berthelot, 1827–1907. *Proceedings of the Royal Society of London*, Series A, v. 80, pp. iii–x.

Wager, L. R., 1959, Differing powers of crystal nucleation as a factor producing diversity in layered igneous intrusions. *Geological Magazine*, v. 96, pp. 75–80.

———, 1963, The mechanism of adcumulus growth in the layered series of the Skaergaard intrusion, in Fisher, D., Frueh, A. J., Jr., Hurlbut, C. S., Jr., and Tilley, C. E., eds., *International Mineralogical Association Papers and Proceedings of the Third General Meeting*. Mineralogical Society of America Special Paper #1, pp. 1–9

———, 1968, Rhythmic and cryptic layering in mafic and ultramafic plutons, in Hess, H. H., and Poldervaart, A., eds., *Basalts: The Poldervaart Treatise on Rocks of Basaltic Composition*, v. 2. New York, Wiley Interscience, pp. 573–622.

Wager, L. R., and Bailey, E. B., 1953, Basic magma chilled against acid magma. *Nature*, v. 172, pp. 68–70.

Wager, L. R., and Brown, G. M., 1951, A note on rhythmic layering in the ultrabasic rocks of Rhum. *Geological Magazine*, v. 88, pp. 166–168.

———, 1967, *Layered Igneous Rocks*. San Francisco, W. H. Freeman.

Wager, L. R., and Deer, W. A., 1939, Geological investigations in East Greenland, part III. the petrology of the Skaergaard Intrusion, Kangerdluqssuaq, East Greenland. *Meddelelser om Grønland*, v. 105, pp. 1–352.

Wager, L. R., and Mitchell, R. L., 1943, Preliminary observations on the distribution of trace elements in the rocks of the Skaergaard Intrusion, Greenland. *Mineralogical Magazine*, v. 26, pp. 283–296.

———, 1951, The distribution of trace elements during strong fractionation of basic magma—a further study of the Skaergaard intrusion, East Greenland. *Geochimica et Cosmochimica Acta*, v. 1, pp. 129–208.

———, 1953, Trace elements in a suite of Hawaiian lavas. *Geochimica et Cosmochimica Acta*, v. 3, pp. 217–223.

Wager, L. R., Brown, G. M., and Wadsworth, W. J., 1960, Types of igneous cumulates. *Journal of Petrology*, v. 1, pp. 73–85.

Wahlstrom, E. E., 1950, *Introduction to Theoretical Igneous Petrology*. New York, John Wiley and Sons.

Walker, F., 1930, The geology of the Shiant Isles. *Quarterly Journal of the Geological Society of London*, v. 86, pp. 355–398.

———, 1940, Differentiation of the Palisade diabase, New Jersey. *Geological Society of America Bulletin*, v. 51, pp. 1059–1106.

———, 1953, The pegmatitic differentiates of basic sheets. *American Journal of Science*, v. 251, pp. 41–60.

———, 1956, The magnetic properties and differentiation of dolerite sills—a critical discussion. *American Journal of Science*, v. 254, pp. 433–443.

Walker, F., and Poldervaart, A., 1941, The Hangnest dolerite sill, S. A. *Geological Magazine*, v. 78, pp. 429–450.

———, 1949, Karroo dolerites of the Union of South Africa. *Geological Society of America Bulletin*, v. 60, pp. 591–706.

Walker, G.P.L., and Skelhorn, R. R., 1966, Some associations of acid and basic igneous rocks. *Earth Science Reviews*, v. 2, pp. 93–109.

Walker, K. R., 1969, *The Palisades Sill, New Jersey: A Reinvestigation*. Geological Society of America Special Paper 111.

Wallace, P. J., Anderson, A. T., and Davis, A. M., 1999, Gradients in H_2O, CO_2, and exsolved gas in a large-volume silicic magma system: interpreting the record preserved in melt inclusions from the Bishop Tuff. *Journal of Geophysical Research*, v. 104, pp. 20097–20122.

Walton, M., 1955, The emplacement of "granite." *American Journal of Science*, v. 253, pp. 1–18.

Wang, P., and Glover, L. III, 1992, A tectonics test of the most commonly used geochemical discriminant diagrams and patterns. *Earth Science Reviews*, v. 33, pp. 111–131.

Ward, J. C., 1875, Notes on the comparative microscopic rock structure of some ancient and modern volcanic rocks. *Quarterly Journal of the Geological Society of London*, v. 31, pp. 388–422.

Washington, H. S., 1900, Igneous complex of Magnet Cove, Arkansas. *Geological Society of America Bulletin*, v. 11, pp. 389–416.

———, 1903, *Collected Analyses of Igneous Rocks Published from 1884–1900*. U. S. Geological Survey Professional Paper 14.

———, 1904, *The Superior Analyses of Igneous Rocks from Roth's Tabellen 1869–1884*. U. S. Geological Survey Professional Paper 28.

———, 1906, The Plauenal monzonose (syenite) of the Plauenscher Grund. *American Journal of Science*, v. 22 (3rd series), pp. 129–135.

———, 1917, *Chemical Analyses of Igneous Rocks Published from 1884–1913 Inclusive*. U. S. Geological Survey Professional Paper 99.

Watt, G., 1804, Observations on basalt, and on the transition from the vitreous to the stony texture, which occurs in the gradual refrigeration of melted basalt; with some geological remarks. *Philosophical Transactions of the Royal Society of London*, v. 94, pp. 279–314.

Weed, W. H., and Pirsson, L. V., 1895, Highwood Mountains of Montana. *Geological Society of America Bulletin*, v. 6, pp. 389–422.

Wegmann, C. E., 1930, Über Diapirismus. *Bulletin de la Commision Géologique de Finlande*, v. 92, pp. 58–76.

———, 1935, Zur Deutung der Migmatite. *Geologische Rundschau*, v. 26, pp. 305–350.

Wegmann, E., 1976, Suess, Eduard, *in* Gillispie, C. C., ed., *Dictionary of Scientific Biography*, v. 13. New York, Charles Scribner's Sons, pp. 143–149.

Weinschenk, E., 1912, *Petrographic Methods*. New York, McGraw-Hill.

Wells, A. K., 1924, A further study of the nomenclature of rocks. *Geological Magazine*, v. 61, pp. 324–327.

Wentworth, C. K., 1923, An improved recording micrometer for rock analysis. *Journal of Geology*, v. 31, pp. 228–232.

Werner, A. G., 1788, Bekanntmachung einer am Scheibenberger Hügel über die Entstehung des Basaltes gemachte Entdeckung. *Intelligenzblatt der allgemeine Litteraturzeitung*, 1788, v. 57, pp. 484–485.

Whewell, W., 1837, *History of the Inductive Sciences*, 3 v. London, John W. Parker.

White, A.J.R., 1979, Sources of granite magmas. *Geological Society of America Abstracts with Programs*, v. 11, p. 539.

White, G. W., ed., 1973, *James Hutton's System of the Earth, 1785; Theory of the Earth, 1788; Observations on Granite, 1794*. New York, Hafner.

Whitehurst, J., 1778, *An Inquiry into the Original State and Formation of the Earth: Deduced from Facts and the Laws of Nature*. London, W. Bent.

———, 1786, *An Inquiry into the Original State and Formation of the Earth; Deduced from Facts and the Laws of Nature*, 2nd ed. London, W. Bent.

Whitney, J. A., 1975, The effects of pressure, temperature, and X_{H_2O} on phase assemblage in four synthetic rock compositions. *Journal of Geology*, v. 83, pp. 1–31.

Wiebe, R. A., 1973, Relations between coexisting basaltic and granitic magmas in a composite dike. *American Journal of Science*, v. 273, pp. 130–151.

———, 1993, The Pleasant Bay layered gabbro–diorite, coastal Maine: ponding and crystallization of basaltic injections into a silicic magma chamber. *Journal of Petrology*, v. 34, pp. 461–489.

———, 1996, Mafic–silicic layered intrusions: the role of basaltic injections on magmatic processes and the evolution of silic magma chambers. *Geological Society of America Special Paper 315*, pp. 233–242.

Wilhelms, D. E., 1993, *To a Rocky Moon: A Geologist's History of Lunar Exploration*. Tucson, The University of Arizona Press.

Wilkins, T., 1988, *Clarence King: A Biography*. Albuquerque, University of New Mexico Press.

Williams, G. H., 1886, The peridotites of the 'Cortlandt Series' on the Hudson River near Peekskill, N. Y. *American Journal of Science*, v. 31 (3rd series), pp. 26–41.

———, 1888, The gabbros and diorites of the 'Cortlandt Series' on the Hudson River near Peekskill, N. Y. *American Journal of Science*, v. 35 (3rd series), pp. 438–448.

Williams, H., 1942, *The geology of Crater Lake National Park, Oregon, with a Reconnaissance of the Cascade Range Southward to Mount Shasta*. Carnegie Institution of Washington Publication 540.

Williams, H., Turner, F. J., and Gilbert, C. M., 1954, *Petrography: An Introduction to the Study of Rocks in Thin Sections*. San Francisco, W. H. Freeman.

———, 1982, *Petrography: An Introduction to the Study of Rocks in Thin Sections*, 2nd ed. San Francisco, W. H. Freeman.

Willis, B., 1907, Memorial of Israel C. Russell. *Geological Society of America Bulletin*, v. 18, pp. 582–591.

Wilsdorf, H. M., 1970, Agricola, Georgius, in Gillispie, C. C., ed., *Dictionary of Scientific Biography*, v. 1. New York, Charles Scribner's Sons, pp. 77–79.

Wilson, C.J.N., and Hildreth, W., 1997, The Bishop Tuff: new insights from eruptive stratigraphy. *Journal of Geology*, v. 105, pp. 407–439.

Wilson, L. G., 1972, *Charles Lyell, the Years to 1841: The Revolution in Geology*. New Haven, CT, Yale University Press.

———, 1998, *Lyell in America: Transatlantic Geology, 1841–1853*. Baltimore, MD, Johns Hopkins University Press.

Wilson, M., 1989, *Igneous Petrogenesis*. London, Unwin Hyman.

Winchell, A. N., 1912, Memoir of Auguste Michel-Lévy. *Geological Society of America Bulletin*, v. 23, pp. 32–34.

———, 1913, Rock classification on three coordinates. *Journal of Geology*, v. 21, pp. 208–223.

Winchilsea, H. F., 1669, *A True and Exact Relation of the Late Prodigious Earthquake & Eruption of Mount Aetna, or Monte-Gibello; as it Came in a Letter Written to His Majesty from Naples by the Right Honorable the Earle of Winchilsea...who...was an Eyewitness of that Dreadfull Spectacle. Together with a More Particular Narrative of the Same, as it is Collected out of Severall Relations sent from Catania*. London, T. Newcomb.

Winkler, H.G.F., and von Platen, H., 1961, Experimentelle Gesteinsmetamorphose IV—Bildung von anatektischer Schmelzen aus ultrametamorphisierten Grauwacken. *Geochimica et Cosmochimica Acta*, v. 24, pp. 48–69.

Winter, J. D., 2001, *An Introduction to Igneous and Metamorphic Petrology*. Upper Saddle River, NJ, Prentice Hall.

Winther, K. T., and Newton, R. C., 1991, Experimental melting of hydrous low-K tholeiite: evidence on the origin of Archean cratons. *Bulletin of the Geological Society of Denmark*, v. 39, pp. 213–228.

Witham, H., 1831, *Observation on Fossil Vegetables accompanied by Representations of their Internal Structure as seen through the Microscope*. Edinburgh, W. Blackwood.

Wolf, M. B., and Wyllie, P. J., 1991, Dehydration-melting of solid amphibolite at 10 kbar: textural development, liquid interconnectivity and applications to the segregation of magma. *Mineralogy and Petrology*, v. 44, pp. 151–179.

Wones, D. R., and Eugster, H. P., 1965, Stability of biotite: experiment, theory and application. *American Mineralogist*, v. 50, pp. 1228–1272.

Wood, D. A., Joron, J.-L., and Treuil, M., 1979, A re-appraisal of the use of trace elements to classify and discriminate between magma series erupted in different tectonic settings. *Earth and Planetary Science Letters*, v. 45, pp. 326–336.

Wood, R. M., 1985, *The Dark Side of the Earth: The Battle for the Earth Sciences, 1800–1980*. Boston, Allen and Unwin.

Woodward, A. S., 1916, The anniversary address of the president. *Proceedings of the Geological Society of London (1915–1916)*, pp. xlviii–lxxvi.

Woodward, H. B., 1910, John Roche Dakyns M. A. Born January 31, 1836. Died September 27, 1910. *Geological Magazine*, v. 47, pp. 575–576.

Woolley, A. R., 1987, *Alkaline Rocks and Carbonatites of the World, Part 1: North and South America*. London, British Museum.

Woolley, A. R., Bergman, S. C., Edgar, A. D., Le Bas, M. J., Mitchell, R. H., Rock, N.M.S., and Scott Smith, B. H., 1996, Classification of lamprophyres, lamproites, kimberlites and the kalsilitic, melilitic, and leucitic rocks: recommendations of the IUGS Subcommission on the Systematics of Igneous Rocks. *Canadian Mineralogist* (Alkaline Rocks Special Issue), v. 34, pp. 175–186.

Worster, M. G., Huppert, H. E., and Sparks, R.S.J., 1990, Convection and crystallization in magma cooled from above. *Earth and Planetary Science Letters*, v. 101, pp. 78–89.

———, 1993, The crystallization of lava lakes. *Journal of Geophysical Research*, v. 98, pp. 15891–15901.

Wright, F. E., 1951, Memorial of Clarence Norman Fenner. *American Mineralogist*, v. 36, pp. 297–303.

Würm, A., 1950, Wilhelm Salomon-Calvi. *Zeitschrift der deutschen geologischen Gesellschaft*, v. 102, pp. 141–146.

Wyllie, P. J., ed., 1967, *Ultramafic and Related Rocks*. New York, John Wiley and Sons.

———, 1974, Limestone assimilation, *in* Sørensen, H., ed., *Alkaline Rocks*. New York, John Wiley and Sons, pp. 459–474.

———, 1976, *The Way the Earth Works: An Introduction to the New Global Geology and Its Revolutionary Development*. New York, John Wiley and Sons.

Wyllie, P. J., and Huang, W. L., 1975, Peridotite, kimberlite, and carbonatite explained in the system $CaO-MgO-SiO_2-CO_2$. *Geology*, v. 3, pp. 621–624.

Wyllie, P. J., and Lee, W., 1999, Kimberlites, carbonatites, peridotites and silicate–carbonate liquid immiscibility explained in parts of the system $CaO-(MgO + FeO^*)-(Na_2O + K_2O)-(SiO_2 + Al_2O_3 + TiO_2)-CO_2$, in Gurney, J. J., Gurney, J. L., Pascoe, M. D., and Richardson, S. H., eds., *The P. H. Nixon Volume 2, Proceedings of the 7th International Kimberlite Conference*. Cape Town, pp. 923–932.

Wyllie, P. J., and Tuttle, O. F., 1959, Effect of carbon dioxide on the melting of granite and feldspars. *American Journal of Science*, v. 257, pp. 648–655.

———, 1960, The system $CaO-CO_2-H_2O$ and the origin of carbonatites. *Journal of Petrology*, v. 1, pp. 1–46.

———, 1961a, Experimental investigation of silicate systems containing two volatile components, II: the effects of NH_3 and HF, in addition to H_2O, on the melting temperatures of albite and granite. *American Journal of Science*, v. 259, pp. 128–143.

———, 1961b, Hydrothermal melting of shales. *Geological Magazine*, v. 98, pp. 56–66.

———, 1964, Experimental investigation of silicate systems containing two volatile components, III: the effect of SO_3, P_2O_5, HCl and Li_2O in addition to water, on the melting temperatures of albite and granite. *American Journal of Science*, v. 262, pp. 930–939.

Yagi, K., 1956, Memorial of Shukusuké Kôzu. *American Mineralogist*, v. 41, pp. 592–597.

Yochelson, E. L., 1998, *Charles Doolittle Walcott, Paleontologist*. Kent, OH, Kent State University Press.

Yochelson, E. L., and Yoder, H. S., Jr., 1994, Founding the Geophysical Laboratory, 1901–1905: A scientific bonanza from perception and persistence. *Geological Society of America Bulletin*, v. 106, pp. 338–350.

Yoder, H. S., Jr., 1972, Memorial to John Frank Schairer. *American Mineralogist*, v. 57, pp. 657–665.

———, 1973, Contemporaneous basaltic and rhyolitic magmas. *American Mineralogist*, v. 58, pp. 153–171.

———, 1976, *Generation of Basaltic Magma*. Washington, DC, National Academy of Sciences.

———, 1993, Timetable of petrology. *Journal of Geological Education*, v. 41, pp. 447–489.

———, 1994, Development and promotion of the initial scientific program for the Geophysical Laboratory, in Good, G. A., ed., *The Earth, the Heavens, and the Carnegie Institution of Washington*. Washington, DC, American Geophysical Union, pp. 21–28.

———, 1995, John Frank Schairer. *Biographical Memoirs of the National Academy of Sciences*, v. 66, pp. 289–320.

———, 1996, Joseph Paxson Iddings 1857–1920. *Biographical Memoirs of the National Academy of Sciences*, v. 69, pp. 3–34.

Yoder, H. S., Jr., and Tilley, C. E., 1962, Origin of basaltic magmas: An experimental study of natural and synthetic rock systems. *Journal of Petrology*, v. 3, pp. 342–532.

Young, D. A., 1998, *N. L. Bowen and Crystallization-Differentiation: The Evolution of a Theory*. Washington, DC, Mineralogical Society of America.

———, 1999a, The emergence of igneous rock diversity as a geological problem, part I: early speculations. *Earth Sciences History*, v. 18, pp. 51–77.

———, 1999b, The rise of the theory of differentiation in igneous petrology (part 2 of 2). *Earth Sciences History*, v. 18, pp. 295–320.

———, 2002, Norman Levi Bowen (1887–1956) and igneous rock diversity, in Oldroyd, D. R., ed., *The Earth Inside and Out: Some Major Contributions to Geology in the Twentieth Century*. Geological Society of London Special Publication 192, pp. 99–111.

Zavaritsky, A. N., 1964, Outline of the analysis of igneous rocks and the determination of their chemical types (in part). *International Geology Review*, v. 6, pp. 1–17.

Zirkel, F., 1859, Die trachytischen Gesteine der Eifel. *Zeitschrift der deutschen geologischen Gesellschaft*, v. 11, pp. 507–540.

———, 1863, Mikroskopische Gesteinsstudien. *Sitzungsberichte Akademie Wissenschaften Wien, Mathematisch-Naturwissenschaftliche Klasse*, v. 47, pp. 226–270.

———, 1866, *Lehrbuch der Petrographie*, 2 v. Bonn, Adolph Marcus.

———, 1870, *Untersuchungen über die mikroskopische Zusammensetzung und Struktur der Basaltgesteine*. Bonn, Adolph Marcus.

———, 1873, *Die mikroskopische Beschaffenheit der Mineralien und Gesteine*. Leipzig, Wilhelm Engelmann.

———, 1876, *Microscopical Petrography*. Washington, DC, Government Printing Office.

———, 1893, *Lehrbuch der Petrographie*, 2nd ed., 3 v. Leipzig, Wilhelm Engelmann.

INDEX OF NAMES

Abich, O.H.W., 111–12, 118, 189, 411
Agricola, 9, 20, 105–6
Ahrens, L. H.,421–22, 424–25, 464, 502
Allan, T., 55–56
Allen, E. T., 276–78
Allport, S., 161, 173–76, 181, 200
Andersen, O., 289, 309, 510
Anderson, A. T., Jr., 552
Anderson, E. M., 341–42
Anderson, G. H., 359, 363
Aristotle, 7, 10, 14
Backlund, H. G., 356–59, 362, 365, 368, 374, 378, 380, 420

Bäckström, H., 215–18, 221, 227–28, 354
Baertschi, P., 451
Bailey, E. B., 310, 337–38, 341
Balk, R., 343–45
Banks, J., 13, 23–24, 37
Baragar, W.R.A., 528–29
Barbarin, B., 543–44
Barrell, J., 336, 346–47
Barrois, C. E., 160, 352–53
Barrow, G., 223
Barth, T.F.W., 329, 339, 356, 381–82, 400, 418–19, 436, 510, 611
Barus, C., 225, 269–71, 274
Bascom, F., 164, 166, 253, 363, 434
Becker, G. F., 179–81, 215, 217–18, 221–25, 228–29, 250, 269, 274, 276, 297
Becquerel, A. C., 266, 271, 284
Beddoes, T., 70
Bergman, T., 23, 27, 43
Berthelot, P.E.M., 216, 220, 229
Berthier, P., 266
Beudant, F. S., 108, 122, 168–69
Billings, M. P., 339, 345

Bischof, G., 82–83, 85, 87, 89, 103, 112, 118, 151, 411
Blake, S., 573
Boettcher, A. L., 510, 517
Boltwood, B., 448–49
Bonney, T. G., 161, 165
Bottinga, Y., 591
Boué, A., 54, 60–61, 77, 151
Bowen, N. L., 264, 283, 285, 287–302, 304–17, 321–23, 330–32, 362, 365, 367–68, 370–77, 382–86, 417–18, 425, 437, 442, 452, 466, 483, 498, 500, 502, 505–6, 513, 515–17, 525, 559, 566–67, 577, 597, 611–12
Boyd, F. R., 499, 502
Bragg, W. H., 415
Bragg, W. L., 414–15
Brandeis, G., 556
Brannock, W. W., 473–74
Breislak, S., 12–13, 82, 84
Brögger, W. C., 163, 197, 208–9, 212–13, 215–19, 225, 227–29, 233, 254, 283, 287, 289, 347, 351, 354–55, 367, 414, 417–19, 538, 611
Brongniart, A., 105, 121–23, 161, 271
Brown, G. M., 520, 550, 561, 581–83, 586, 588
Brun, A., 272, 304
Buckland, W., 56–57, 107
Buddington, A. F., 319, 368, 373, 437, 503–5, 608, 611
Bunsen, R., 97, 100–102, 116, 118, 125, 129–33, 136–39, 144, 154, 161, 169, 179–80, 206–7, 227, 229, 239, 272, 285, 350, 374, 411–12, 420, 572, 612
Burckhardt, C., 334–35
Burnham, C. W., 507, 509, 511

INDEX OF NAMES

Campbell, I. H., 556, 563–65, 574, 588–90, 592
Cann, J. R., 490–93
Carmichael, I.S.E., 442, 484, 550
Cashman, K. V., 556–57
Chamberlin, T. C., 274, 333, 346–48
Chaperon, G., 219, 229
Chapman, C., 344
Chapman, R., 344
Chappell, B. W., 542
Chayes, F., 239, 261–62, 396–98, 489
Chen, C. F., 559
Clarke, F. W., 412–13
Clayton, R. N., 453, 458, 460
Clemens, J. D., 605–6
Clerk of Eldin, J., 65–67
Cloos, H., 336, 338, 343, 345–47, 368, 428
Conybeare, W., 56–57, 61, 77, 107
Cordier, P.-L.-A., 59, 105, 123–24
Cross, C. W., 163, 170–71, 231, 233, 240–42, 250–52, 334, 389
Cusack, R., 272–73

Dakyns, J. R., 208, 210, 213, 227, 229, 233
Daly, R. A., 191–94, 196–98, 252–53, 258, 285–86, 288, 291, 293–96, 298–99, 310, 313, 317–18, 323, 329, 333–36, 340–41, 344–47, 368, 370–71, 377–78, 386, 438, 549, 611
Dana, J. D., 81, 125, 128–31, 137, 190, 199, 227, 229, 438
Darwin, C., 54, 125, 127–31, 139, 202, 204, 206–7, 223, 227, 229, 438
Daubeny, C.G.B., 54, 57–58, 178, 189
Daubrée, G.-A., 86–87, 103, 264, 268, 304, 352, 612
d'Aubuisson de Voisins, J. F., 26, 28, 35–37, 40–41, 45, 47, 50, 52, 59, 122
Day, A. L., 270, 275–78, 286, 288, 297, 302, 309, 515
Debon, F., 541–42
Debray, H. J., 266–67, 271
Deer, W. A., 320–27, 329, 331, 414, 602, 608
Delesse, A., 83, 87, 103, 118, 120, 238–39, 352, 411
DePaolo, D., 437, 468–70, 483
de Saussure, H. B., 26, 37, 45–46, 63–64, 73, 150, 265
Des Cloizeaux, A.-L.-O. L., 129, 161, 611
Desmarest, N., 17–19, 21–23, 36, 44, 50, 59
Deville, H. E. Ste.-C.,267, 270–71, 284

Doe, B., 467–68
Doelter, C.A.S., 158, 238, 272–73, 284, 286–87, 304, 351
Dolomieu, D., 13, 19, 25–26, 37–38, 42, 44, 50, 65, 265
Durocher, J., 32, 82–83, 89, 92–98, 101, 103, 118, 120, 125, 133–37, 169, 203, 205–7, 221, 229, 266, 285, 352, 374, 611
Dutton, C. E., 203–6, 229

Ebelmen, J. J., 267
Edwards, A. B., 326–27
Élie de Beaumont, L.,83, 87, 89–90, 99, 103, 120, 189, 352, 411
Engel, A.E.J., 438–39, 465–66, 500, 512
Engel, C. G., 438–39, 465, 500
Epstein, S., 451, 453–54
Erdmannsdörffer, O., 336, 338, 368, 378, 435
Eskola, P., 297, 362, 369–70, 381, 386–87, 467, 611
Eugster, H. P., 499, 502–6

Fairbairn, H. W., 421, 428–30, 462
Faujas de St. Fond, B., 19, 24, 30, 36–38, 59
Faure, G., 462–64, 466–67
Fenner, C. N., 299–301, 305, 307, 330, 332, 567
Ferber, J. J., 12, 15, 19–20, 22–23, 37
Fersman, A. E., 416
Forbes, D., 99–100, 103, 161, 173–74, 181, 611–12
Förster, H.-J., 496–97
Fouqué, F., 159–60, 172, 208, 228, 231, 233, 238, 242, 268, 284, 287, 416
Fournet, J., 32, 82–83, 91–92, 97, 103, 120, 352
Fuchs, J., 82, 84, 411

Gast, P. W., 463, 480–82
Geikie, A., 65, 67, 88, 161, 176, 316, 333
Geikie, J., 88–89, 100, 103
Ghiorso, M., 437, 609
Gilbert, G. K., 188–90, 192, 194–95, 197, 316, 333–34, 549
Glagolev, A. A., 396–97
Glazner, A., 604
Glover, L., 494, 496
Goldschmidt, V. M., 358–59, 362, 368–70, 416–20, 423–24, 428, 477–79, 611
Goldsmith, J. R., 506, 510, 612
Goodspeed, G. E., 361–62, 373, 378

INDEX OF NAMES

Gouy, L. G., 219, 229
Green, D. H., 513–14
Green, T. H., 514–15
Greig, J. W., 297, 299, 509, 572
Griggs, D. T., 338, 502
Gromet, L. P., 488
Grout, F. F., 295–96, 310–12, 316, 321, 323, 336, 338, 345–48, 363, 368–69, 372–73, 381, 550, 604, 611
Guettard, J.-E., 16–17, 19, 22
Gustafson, L. B., 558–59
Guthrie, F., 222, 229

Hague, A., 180–81, 193, 199, 242
Hall, A. L., 317–18, 608
Hall, J., 30, 38, 42, 46–49, 54, 68–69, 73, 76, 81, 85, 91, 99, 101, 107, 150, 264–65, 350, 362, 386
Hamilton, D. L., 509, 511
Hamilton, E., 464
Hamilton, W., 11–12, 41
Harker, A., 162, 185–87, 206, 208, 211–12, 215–16, 218, 220–21, 223, 225–27, 229, 232–33, 236, 248–52, 254, 258–59, 285, 287, 294–95, 310–11, 316, 320, 335, 337–38, 351, 362, 367, 389, 414, 425, 499, 586–87, 608, 611
Hatch, F. H., 161–62, 232, 254, 256, 263, 317, 389
Hautefeuille, P. G., 267–68
Haüy, R., 52, 62, 104, 110, 120–24, 612
Hedge, C., 465–66
Hess, H. H., 319, 439, 440, 442, 550, 578–80, 583, 588, 611
Hildreth, E. W., 550–52
Hillebrand, W. F., 412–13
Hobbs, W. H., 231, 233, 236
Holborn, L., 270, 275–77, 284, 286
Holland, H. D., 479
Holmes, A., 253, 257–59, 262–63, 284, 293–94, 310–12, 321, 323, 329, 357, 362, 365, 389, 437, 462–63, 511, 538, 611
Holtz, F., 522–23
Hunt, C. B., 338
Hunt, T. S., 86, 88, 90–91, 100, 103, 164, 200, 612
Huppert, H. E., 556, 560–65, 569–71, 575
Hurley, P., 462–64, 466–67
Hutton, D.H.W., 604
Hutton, J., 30–33, 36, 41, 43, 47, 54, 56, 63–73, 75, 77, 81–82, 91, 107, 168, 218, 264, 337, 350, 362, 364, 386

Iddings, J. P., 163, 165, 180–81, 183–86, 192–93, 206, 208, 210–14, 220, 225, 227–29, 231, 233, 236, 240–42, 247, 256, 259, 273, 275, 283, 285–87, 309, 313, 340, 346–47, 351, 367, 389, 414, 608, 611–12
Ingerson, E., 319
Irvine, T. N., 528–29, 556, 559, 594–96, 598–601
Irving, A. J., 485

Jackson, E. D., 587–91, 593
Jaeger, J. C., 331
Jahns, R. H., 509–10, 521
Jameson, R., 28–30, 44–45, 51, 54–55, 57–58, 60–61, 73–77
Jaupart, C., 556, 571
Jenzsch, G., 151
Jevons, H. S., 237
Johannes, W., 437, 522
Johannsen, A., 106, 239, 253, 259–62, 284, 302, 309, 389–92, 398, 538, 608, 611–12
Johnston, A. D., 524
Johnston-Lavis, H. J., 215–16, 227–28
Joly, J., 238–39, 272–73, 284
Judd, J. W., 161, 173, 176–79, 181–83, 200, 211–12, 214, 219, 362, 572, 611
Jukes, J. B., 96, 125, 136–37, 173, 227, 229

Kalkowsky, E., 158, 231–32
Keilhau, B. F., 87
Keir, J., 47
Keith, M. L., 499, 506, 509
Kennedy, G. C., 502, 510
Kennedy, R., 46
Kennedy, W. Q., 301–2, 307, 326
Kerr, R. C., 556, 569–70, 606
King, C. R., 138–39, 153–54, 163, 179–80, 199–205, 207, 223, 227, 229, 274
Kircher, A., 9–10, 14
Kirwan, R., 15, 20, 26, 33, 35, 44, 54, 70–71, 82, 168
Kjerulf, T., 118, 209, 411, 611
Knopf, A., 368, 579
Koster van Groos, A. F., 509–10
Koto, B., 515
Koyaguchi, T., 574–75
Kôzu, S., 515
Kulp, J. L., 479
Kuno, H., 406–7, 500, 516
Kushiro, I., 505, 516
Kynaston, H., 255, 317–18

INDEX OF NAMES

Lacroix, A., 160, 196, 242, 247, 353, 370, 372, 391–92, 416, 483, 538, 611–12
Lagorio, A., 207–8, 212–13, 227, 229, 273, 414
Lane, A. C., 248, 348
Lang, O., 231, 286
La Roche, H. de, 541
Larsen, E. S., Jr., 345, 349, 368, 405–7, 437
Le Bas, M., 536–38
Le Châtelier, H., 271–72, 284, 416
Le Fort, P., 541–42
LeMaitre, R. W., 535–38
Lémery, N., 14, 264–65
Lewis, J. V., 248, 291, 314, 597
Lincoln, F. C., 239, 253–54, 261
Lindsley, D. H., 502–5
Loewinson-Lessing, F. Y., 163, 198, 215, 218, 220–21, 225, 228–29, 231–32, 259, 285–86, 291, 351, 414, 538
Lombaard, B. V., 318–19, 580
Lucretius (T. Lucretius Caro), 7
Ludwig, C., 212
Lugmair, G. W., 468
Lunde, G., 418–19
Luth, W. C., 509–10, 521
Lyell, C., 34, 54, 81, 107, 176–77, 189, 204, 337, 362

MacCarthy, G. R., 340, 348
Macculloch, J., 57, 76–77, 107, 168, 176
Mackenzie, G. S., 55–56, 62, 73
MacKenzie, W. S., 510–11
Mahood, G., 551–52
Malesherbes, C.-G., 16–17
Manning, D.A.C., 522–23
Marmo, V., 381
Marsh, B. D., 556–57, 567, 569–72, 607
Martin, D., 556, 566, 569
Mason, B., 424, 444–45
Mawer, C. K., 605–6
McBirney, A. R., 457, 556, 559, 590–95, 597–99
Mead, J., 424, 480–81
Mehnert, K. R., 531
Meunier, E. S., 268
Michel-Lévy, A., 157, 159–60, 172, 208, 215, 217, 225, 227–35, 268, 284, 287, 351–53, 370, 483, 611
Middlemost, E.A.K., 540
Misch, P., 359–61, 365, 378
Mitchell, R. L., 422–25, 578
Mitscherlich, E., 117, 266, 411

Modell, D., 341
Molengraaff, G.A.F., 255, 316–18
Montel, J.-M., 523–24
Moore, J., 441, 466
Moorhouse, W. W., 394, 528
Morozewicz, J., 273, 284, 286
Morse, S. A., 584–85, 594, 598, 601
Muan, A., 502, 506
Muehlenbachs, K., 460
Murase, T., 591–92
Murray, J., 34, 43–44, 46–48, 54, 74–75, 82

Naney, M. T., 521–22
Naumann, C. F., 96–97, 103, 107, 112–14, 153, 169, 611
Neumann, H., 424, 479–81
Newhouse, W. H., 343, 421
Nicol, W., 145–47
Nicolaysen, L. O., 464
Nier, A. O., 450–51
Niggli, P., 286, 301, 362, 368–70, 372, 374, 386–87, 389–93, 398–402, 467, 502, 510, 527, 529, 611
Noble, J., 347
Nockolds, S. R., 311–13, 392–93, 406–7, 422–23, 425–27, 477–79, 521
Noyes, R. M., 590, 592–95

O'Hara, M., 511–14, 560
Osann, C. A., 158, 231, 239–40, 253, 256, 286
Osborn, E. F., 456, 498–99, 502, 506, 509, 511
Oschatz, A., 148, 150–51, 161

Paige, S., 340
Pallas, P. S., 63, 73, 132
Patiño Douce, A. E., 524–25
Peacock, M. A., 404–5
Pearce, J. A., 490–92, 494, 497, 543
Pearce, T. H., 494
Peltier, J.C.A., 271
Perrin, R., 357–58, 365, 374, 377–81
Petersen, J. S., 486
Petford, N., 606
Phillips, W., 56, 61, 77
Pichavant, M., 523
Pinkerton, J., 40, 51, 56, 107
Pirsson, L. V., 165, 185, 217, 240–42, 263, 286, 323, 334
Pitcher, W. S., 366, 437, 603–4
Piwinskii, A., 518

INDEX OF NAMES

Playfair, J., 30, 36, 38–40, 42–48, 54, 57, 71–74, 76–77, 81–82, 91–92, 107, 188
Pliny the Elder, 6, 105–6
Poldervaart, A., 329–30, 407, 422, 597
Powell, J. L., 467
Powell, J. W., 188–89, 203

Quirke, T. T., 358–59

Raisin, C. A., 165–66, 363
Ramberg, H., 376–77, 611–12
Rammelsberg, K. F., 111, 118, 151, 411
Raspe, R. E., 19–20, 22
Rastall, R. H., 351–53, 362
Read, H. H., 50, 83, 296–97, 316, 362, 365–68, 371–75, 385–86, 388, 567, 603, 611–12
Réaumur, R.-A. F., 16, 47, 265, 268
Reynolds, D., 257, 357, 362–65, 368, 374, 381–82, 402, 434
Richardson, W., 35–39, 41, 44–45, 50, 54, 57
Richardson, W. A., 345–47
Richey, J. E., 337, 342, 344
Ringwood, A. E., 445, 478–79, 513–14
Rittmann, A., 392, 530
Robertson, J. K., 518–20
Rose, G., 86, 109, 117, 150–51, 266, 304
Rose, H., 86, 103, 150, 264
Rosenbusch, K.H.F., 154–60, 163–64, 169, 171–72, 179–81, 196, 207–8, 213, 221, 225–27, 230–35, 237, 239–40, 242, 250–54, 259, 284, 286–87, 317, 351–52, 354–55, 365–66, 368, 390, 392, 414, 416, 419, 527, 538, 608, 611–12
Rosiwal, A., 238–39
Roth, J.L.A., 117–19, 169, 231–32, 239, 283, 411, 413
Roubault, M., 357–58, 374, 377
Roy, R., 506–7
Russell, I. C., 188, 192–93, 195–96, 334
Rutherford, E., 415, 448–49
Rutley, F., 161–62, 233, 365

Salomon, W., 188, 193, 335
Sartorius von Waltershausen, W., 118, 125, 129, 131–33, 135–39, 169, 182, 201, 203, 205, 212, 229, 411
Schafhäutl, C., 85, 264
Schairer, J. F., 302–9, 312, 383, 498–99, 502, 505, 511, 515

Scheerer, T., 83–85, 87–91, 95–98, 118, 120, 169, 611
Schweig, M., 215, 219–20, 224, 227, 229
Scoates, J., 600
Scott, A., 294
Scrope, G. P., 53–55, 58, 61, 84, 90, 107, 125–31, 139, 173, 189, 207, 227, 229
Sederholm, J. J., 163, 353–55, 368, 370, 373, 375, 611
Seebeck, T. J., 270
Seger, H., 271–72
Senarmont, H. de, 161, 266
Seneca, 7, 14
Senft, K.F.F., 114–16
Seymour, W., 73, 77
Shand, S. J., 253, 255–59, 261–63, 284, 298, 317, 373, 389, 396, 439
Shapiro, L., 473–74
Shaw, D., 478, 482
Shaw, H. R., 555–56, 592
Shepherd, E. S., 278
Shirley, D. N., 597
Silver, L., 488
Silverman, S., 451–52, 454
Skjerlie, K., 524
Smith, R. L., 550, 552
Snyder, D., 575–76
Sorby, H. C., 98–100, 103, 148–51, 153, 161, 173, 176, 350, 362
Soret, C., 212, 229
Sosman, R., 277
Spallanzani, L., 12–13, 36, 38–39, 41–42, 44, 265
Sparks, R.S.J., 437, 556, 560–65, 569–71, 574–75
Spera, F., 556–57, 572
Steininger, J., 301
Strabo, 5–7, 10, 14
Strange, J., 19, 23, 37, 104
Streckeisen, A. L., 395, 529–32, 534–36, 538, 540
Streng, A., 118, 151, 411
Suess, E., 188, 191–92, 333–34, 341, 346–47
Sun, S.-S., 493
Swanson, S. E., 521–22, 524

Tait, S., 556, 572, 575
Takada, A., 574–75
Talbot, W.H.F., 145
Tatsumoto, M., 465
Taylor, H. P., Jr., 437, 453–61, 483, 590
Taylor, S. R., 424–25, 445, 597

INDEX OF NAMES

Teall, J.J.H., 161–62, 185, 207–8, 210, 212–13, 215, 218, 220–23, 225–27, 229, 263, 287, 290, 316, 351, 362, 425, 611
Termier, P., 353
Teyssier, C., 605
Thompson, R. N., 493
Thornton, C., 406–7
Tikoff, B., 605
Tilley, C. E., 320, 363–64, 368, 425, 478, 499–500, 502, 510–11, 513, 525, 541
Tröger, W. E., 391–93, 395, 423
Tschermak, G., 158–59, 277
Turner, J. S., 556, 558–66, 570–71, 575, 595
Tuttle, O. F., 380–85, 406–7, 467, 498, 500, 506–10, 513, 516, 521, 525
Tyrrell, G. W., 253, 262–63, 315–16, 338, 346, 417

Urey, H., 450–51

van't Hoff, J. H., 212, 216, 417
Vernadsky, V. I., 416
Vielzeuf, D., 523–24
Virlet d'Aoust, T., 83, 87–88, 352
Vitaliano, C. J., 424, 479, 481
Vitruvius, 6–7, 10, 14
Vogelsang, H., 154–182
Vogt, J.H.L., 208–9, 212–13, 215–16, 218–25, 227–28, 250–51, 285, 287, 290, 351, 354, 367, 572, 611
von Buch, C. L., 28, 36–37, 51–53, 107, 109, 189
von Cotta, C. B., 96–97, 103, 107, 116–17, 126, 137, 172–73, 180, 611
von Groth, P., 416–17
von Humboldt, A., 28, 51–52, 188
von Kobell, F., 268, 272
von Lasaulx, A., 158, 172, 178, 231–33, 287
von Laue, M.T.F., 414, 417
von Leonhard, K. C., 53, 61, 97, 108, 110–11, 121, 612
von Platen, H., 516
von Richthofen, F., 117–18, 125, 137–39, 169, 179–80, 199–201, 205, 229, 402, 411
von Troil, U., 13, 23, 25, 38, 43
von Wolff, L. F., 402–4

Wadsworth, W. J., 581–82, 586
Wagner, P. A., 255, 317

Wager, L. R., 319–27, 329, 331, 414, 422–25, 436, 445, 457, 464–65, 550, 563, 572–73, 578, 581–84, 586–88, 590–91, 596, 602, 608, 611
Walker, F., 326, 328–31, 414, 422, 597
Walker, K. R., 597
Wallerius, J. G., 12, 23, 106
Walton, M., 378–80
Wang, P., 494, 496
Ward, J. C., 161, 173, 176, 181
Washington, H. S., 163, 223, 233, 240–43, 247, 253, 283, 374, 400, 407, 413, 511, 515
Watt, G., 49, 265
Wegmann, C. E., 347, 363, 374, 604
Weill, D. F., 591
Weinschenk, E., 158, 231–32
Weiss, C. S., 52–53, 101
Wentworth, C. K., 261, 396–97
Werner, A. G., 26–30, 35–37, 39, 43, 45–46, 50–54, 56, 58–59, 61, 64, 73–74, 96, 104–5, 107–10, 114, 116, 122, 129, 132, 137, 167–68, 350, 611
White, A.J.R., 542–43
Whitehurst, J., 24–25
Whitney, J. A., 521–22, 524
Wiebe, R. A., 574–76
Williams, G. H., 164, 240, 259
Williams, H., 344, 393–94, 539, 590
Williamson, W. C., 98, 148
Winchell, A. N., 192, 254, 259
Winkler, H.G.F., 516, 525
Witham, H., 145–47
Wones, D. R., 503, 543
Wood, D. A., 493
Worster, M. G., 556, 570–71
Wright, F. E., 278, 317
Wyllie, P. J., 507–10, 513, 516–21, 524–25

Yagi, K., 499, 515
Yoder, H. S., Jr., 373, 499–500, 502–3, 511, 513, 525, 541, 572–73
Yuen, D. A., 556

Zachariasen, F.W.H., 415, 418–19
Zavaritsky, A. N., 401–3
Zirkel, F., 151–54, 156, 158–60, 163, 169–72, 176, 178–79, 182, 200–201, 226, 231–32, 242, 253, 259, 284, 286–87, 402, 527, 611

INDEX OF SUBJECTS

age criterion in classification, 168–81, 232
assimilation (contamination), 135, 137, 197–98, 225, 285, 291, 293–95, 297–98, 311, 323, 327, 330, 332, 347, 353, 425, 453, 457–60, 462, 464–65, 470, 479, 551, 565, 609
assimilation-fractional crystallization (AFC), 483–84, 565–66

basalt, origin of, 20–61, 172–79, 438–39, 499–502, 511–14, 525
Berthelot's principle, 216, 219–20
Bunsen's theory of two magma sources, 129–31

cauldron subsidence, 341, 344–45
chemical analysis, 46–47, 411–14, 428–30, 473–77
chemical classification, 398–407, 536–38, 541–42
CIPW (American quantitative) classification, 240–58, 284, 389, 392, 413–14, 529
classification of igneous rocks, 104–24, 231–63, 389–407, 527–45
commingled magmas, 572–77
computer technology, 609–10
consanguinity, 183–85, 206
contrasted differentiation, 311–13
convection (convection currents), 296, 321, 324–26, 331–32, 553–54, 557, 561–63, 565, 567–72, 577
craters of elevation, 188–89, 191
crystal settling, 127, 202, 204, 207, 223, 290–92, 294–95, 300, 309, 314–16, 319, 321, 324–25, 327–28, 331–32, 453, 551, 555, 559, 566–67, 578, 580–81, 583–92, 595–97, 600

crystal structure and crystal chemistry, 414–20
cumulate paradigm, 578–601

diapirism, 347, 379, 602–7
differentiated sills, 314–16, 326–32
differentiation, theory of, 205–26, 283–312
double-diffusive convection, 554, 559, 564–66, 594–98, 600

educational institutions
    Amherst College, 164, 240
    Arizona State University, 510, 556
    Australian National University, 331, 425, 444, 478, 513–14, 517, 556, 558, 561–61, 571, 597, 606
    Bergakademie of Freiberg, 22, 26, 29, 36, 51–52, 83, 96–97, 101, 107, 180, 317, 412, 611
    Brown University, 269, 488
    Bryn Mawr College, 164, 166
    California Institute of Technology (Caltech), 359, 419, 453, 468, 488, 506–7, 509, 521, 524, 590, 594
    Collège de France, 25, 159, 220, 268
    Columbia University (Columbia School of Mines), 163, 180, 192, 255, 269, 299, 328, 378, 407, 440, 479–80, 597
    École des Mines, 59, 86–87, 91–92, 266–67, 353
    École Polytechnique, 87, 92, 266, 271
    Franklin and Marshall College, 574, 584
    Harvard University, 163, 179, 196–97, 226, 279, 287, 317, 328, 404–5, 412, 428, 450, 453, 502
    Imperial College of London, 365, 435, 563, 603
    Indiana University, 424, 479

educational institutions (cont.)
- Johns Hopkins University, 163–64, 226, 240, 259, 344, 433, 503–4, 556–51, 571
- Massachusetts Institute of Technology (MIT), 91, 197, 287–88, 293, 397, 421–22, 28, 432–33, 464, 499, 502–3, 506, 556
- McGill University, 91, 584
- McMaster University, 478, 594
- Pennsylvania State University (Penn State), 382, 406, 433, 453, 456, 502, 505–11, 513, 517–18
- Polytechnicum, Delft, 154, 317
- Princeton University, 299, 319, 444, 503, 556, 579
- Queen's University (Ontario), 287, 296, 417, 428, 453, 506, 510
- Rutgers University, 314, 559, 579
- St. Andrews University, 255, 328, 507
- Stanford University, 226, 382, 510, 517, 521, 551, 574
- State University of New York at Stony Brook, 488, 505
- University of Aarhus, 486, 599
- University of Berlin, 52, 86, 108, 111–12, 117, 138, 266, 402
- University of Besançon, 82, 124, 267
- University of Blaise Pascal, 522–23
- University of Bonn, 82, 138, 157–58
- University of British Columbia, 404, 467
- University of California at Berkeley, 164, 393, 468, 551, 555–54, 590
- University of California at Los Angeles (UCLA), 433, 468, 502, 510, 517, 587, 597
- University of Cambridge, 53, 136, 165, 186, 225, 237, 287, 318, 320, 328, 351, 406, 414–15, 421–22, 425–26, 435, 444, 448, 475, 478, 490, 499, 511, 556, 558, 560–62, 565, 567, 570–71
- University of Cape Town, 328–29, 421, 425, 444, 580
- University of Chicago, 180, 226–27, 240, 259–60, 286, 309, 343, 377–78, 383, 419, 421, 432–33, 444, 450–51, 453, 460, 478, 498–99, 506, 510, 517–18, 552, 611
- University of Christiania (Oslo), 87, 209, 226, 289, 376, 417, 436, 611
- University of Dorpat, 111, 218
- University of Durham, 257, 320–21, 363, 444, 494
- University of East Anglia, 490, 492
- University of Edinburgh, 29, 34, 36, 47, 54, 57–58, 60, 88, 145, 257, 363, 444, 511, 513, 517, 560
- University of Glasgow, 301, 315, 337
- University of Göttingen, 53, 101, 164, 242, 266, 359, 419, 428, 435, 516
- University of Graz, 158, 272
- University of Hannover, 517, 522–23
- University of Heidelberg, 53, 97, 101–2, 108, 157–58, 164, 180, 226, 239, 242, 266, 272, 287, 317, 354–55, 412, 417, 612
- University of Illinois, 164, 259, 344, 358, 421
- University of Jena, 96, 117, 227
- University of Leeds, 301, 508
- University of Leipzig, 52, 96–97, 108, 138, 153, 158, 226, 242, 369, 402
- University of Liverpool, 365, 603, 606
- University of Manchester, 311, 320, 415, 419, 425, 435, 444, 510–11, 513, 517, 520, 522–23, 605
- University of Marburg, 53, 101, 129, 222
- University of Melbourne, 478, 535
- University of Michigan, 192, 233, 419
- University of Minnesota, 295, 348, 358, 450, 463, 480, 556
- University of Munich, 158, 416
- University of Nancy, 507, 522–23, 541
- University of New Mexico, 444, 605
- University of Oregon, 444, 457, 524, 556–57, 590
- University of Oxford, 56–58, 70, 148, 165, 238, 321, 421–22, 425, 435, 464, 465, 581
- University of Paris, 86, 287, 482, 556, 572, 575
- University of Stockholm, 226, 354, 417, 420
- University of Strassburg, 86, 154, 239
- University of Sydney, 499, 550, 558, 560
- University of Tokyo, 406, 444, 515–16, 574
- University of Toronto, 344, 404, 510, 564
- University of Tübingen, 117, 369
- University of Uppsala, 23, 356, 420
- University of Vienna, 151, 158, 191, 417
- University of Washington, 359, 361, 574, 609
- University of Western Australia, 563–64
- University of Wisconsin, 166, 428

INDEX OF SUBJECTS 683

University of Zürich, 301, 369, 502
Yale University, 128, 163, 199, 240, 242, 295, 302, 319, 378, 502, 579
electrical currents in differentiation, 225, 285
element distribution, 477–79
emission spectroscopy, 420–28
emplacement of plutons, mechanisms of, 194–98, 339–49
eutectic crystallization, 208, 222–23, 250–51, 285, 290–91, 297, 550
experimental petrology, 47–49, 264–79, 287–309, 498–526

filter pressing, squeezing of liquid, 127, 223, 309, 323, 598
fluid dynamics, 549–577
forceful injection, 339–40, 343–44, 349, 604
forms of igneous rock bodies, classification of, 188–94, 333–39
forms of igneous rock bodies, origins of names of
 akmolith, 336
 batholith, batholite, 191
 boss, 188
 bysmalith, 193
 cactolith, 338
 chonolith, 194
 cone sheet, 337–38
 ductolith, 338
 ethmolith, 193
 harpolith, 336
 laccolith, laccolite, 190
 lopolith, 295, 336
 phacolith, 335
 plug, 192
 ring dike, 337
 sphenolith, 334
 stock, 192
fractional crystallization (crystallization-differentiation), 215, 222–24, 283–312, 321–22, 331, 377, 417, 425, 452–53, 458, 460, 466, 477, 479–83, 552, 559, 566, 578, 609
funding in petrology, 432–34

gas transfer, 295
Gouy-Chaperon theory, 219–20, 285, 291, 551
granite, origin of, 81–103, 350–89, 507–9, 516–25
granite classification, 542–45

granite controversy, 350–98
granite emplacement, 62–77, 602–7
granitization, 358–67

institutions and organizations
 Academy of Sciences of Paris, 16–18, 108
 American Museum of Natural History, 180, 200
 Deutsche Geologische Gesellschaft, 97, 117, 150
 Geological Society of America, 373, 382
 Geological Society of London, 56, 65, 76, 99, 148–50, 165, 176, 218, 363
 Geological Survey of Canada, 90, 197, 240, 288, 358, 444, 528, 585, 594
 Geological Survey of Great Britain, 65, 88, 136, 162, 176, 210, 223, 227, 301, 337, 343, 365, 434
 Geophysical Laboratory, 242, 264, 274, 276–77, 279, 284, 286–88, 291, 296–97, 299, 302, 304, 309, 312, 317, 331–32, 368, 372, 382–83, 388, 396, 400, 415, 435, 444, 489, 498–99, 502–7, 509–11, 513, 515–17, 525–27, 552, 556, 577, 594, 611
 International Geological Congress, 90, 218, 317, 354, 531, 534, 536, 538
 Lamont-Doherty Geological Observatory, 440, 480
 Musée d'Histoire Naturelle, 24, 59, 86, 110, 120, 122–23, 160, 268, 271, 391, 612
 National Environmental Research Council (NERC), 434, 465, 511, 603
 National Science Foundation (NSF), 433
 Office of Naval Research (ONR), 421, 433
 Physikalisch-Technische Reichsanstalt, 275–77, 284, 287
 Royal Society of Edinburgh, 45–46, 64, 67–68, 76
 Royal Society of London, 11, 56–57, 108
 Service de la Carte Géologique de France, 159, 227, 352–53
 Societé Géologique de France, 60, 83, 87, 95
 United States Geographical and Geological Survey of the Rocky Mountain Region, 203, 227
 United States Geological Exploration of the Fortieth Parallel, 199–200, 227
 United States Geological Survey, 163, 179–81, 189, 192, 203, 227–28, 242, 247–

684 INDEX OF SUBJECTS

institutions and organizations (cont.)
United States Geological Survey, 48, 259, 269, 276, 286, 288, 319, 363, 378, 412–13, 421, 433, 439, 444, 458, 465, 467, 473, 550, 55, 592, 612
Woods Hole Oceanographic Institution, 438, 440, 558
internationalization of petrology, 434–35
iron enrichment (Fenner) trend, 301, 305, 332, 407, 456, 591
isotopes, 447–70
IUGS classification, 529–41

journals
American Journal of Science, 90, 128, 286, 505
Beiträge zur Mineralogie und Petrographie, 435–36
Bulletin de la Societé Géologique de France, 83, 108, 120, 286, 435
Contributions to Mineralogy and Petrology, 435–37
Edinburgh Philosophical Journal, 60–61, 77
Geochimica et Cosmochimica Acta, 423
Geological Magazine, 90, 162, 371
Geological Society of America Bulletin, 286
Geologische Rundschau, 530
Journal of Geology, 286, 435, 506
Journal of Petrology, 434–36, 508, 590, 594
Journal of the Geological Society of South Africa, 435
Lithos, 436
Mineralogical Magazine, 435
Neues Jahrbuch für Mineralogie, Geologie, und Paleontologie, 53, 151
Philosophical Transactions of the Royal Society of London, 12
Quarterly Journal of the Geological Society of London, 90, 108, 162, 165, 286
Schweizerische mineralogische und petrographische Mitteilungen, 435
Taschenbuch für die gesammte Mineralogie, 53, 108
Tschermaks mineralogische und petrographische Mitteilungen, 159, 285, 435
Zeitschrift der deutschen geologischen Gesellschaft, 108, 150–51, 158, 286

Kern theory of Rosenbusch, 207–8, 213, 221

layered intrusions, 316–26, 578–601
liquid immiscibility, liquation, 135, 207, 217, 219, 221–23, 229, 285, 291, 295–96, 299, 315–16, 551, 558, 572
lunar petrology, 443–47, 513, 613

magma, origin of term, 30–32, 91–92
magma mixing, 130–31
magmatic stoping, 197–98, 294, 339–41, 343–46, 349
magnetic attraction in differentiation, 224, 285
mineral abundances, measurement of, 237–40, 261–63, 396–98
mineral synthesis, 265–68
multiple injection, 315, 332, 586, 597

natural vs. artificial classification, 231, 248–50
neptunist-vulcanist-plutonist controversies, 16–61

osmotic exchange, 217
oxygen buffer method, 502–6

partial melting (partial fusion, anatexis), 82–99, 135, 139–40, 201–5, 225, 285, 294, 332, 350, 352, 354–55, 384, 386–87, 457–59, 462–64, 467, 481–83, 516, 523–25, 551, 609
petrographic province, 182–87, 206, 250, 295, 405
petrographic terms, origins of
accessory minerals, 111
allotriomorphic-granular texture, 158, 263
anhedral, 247, 263
CIPW norm, 243
essential minerals, 111
euhedral, 247, 263
felsic, 258
felsophyric texture, 154
femic minerals, 243, 258
granophyric texture, 154
groundmass, 111
holocrystalline, 247
holohyaline, 247
hyalopilitic texture, 158
hypidiomorphic-granular texture, 158, 263
hypocrystalline, 158, 247
idiomorphic, 263
leucocratic, 254, 258

INDEX OF SUBJECTS 685

mafic, 258
melanocratic, 254, 258
miarolitic, 158
microlite, 154
mode, 244
ocellar, 158
ophitic texture, 154
orthophyric texture, 158
panidiomorphic-granular texture, 158
phenocryst, 184
pilotaxitic texture, 158
poikilitic texture, 164
salic minerals, 243, 258
subhedral, 247, 263
vesicular, 114
vitrophyric texture, 154
petrological localities
  Ardnamurchan, 177, 182, 310, 337–38, 454–56
  Auvergne, 16–19, 21, 25, 37, 48, 51, 53–54, 58, 60–62, 95, 109, 158, 182–83, 185, 189
  Bishop Tuff, 550–53
  Bushveld Complex, 255, 311, 316–18, 325, 329, 336, 454, 559, 564, 578, 580–81, 583, 587, 597, 608
  Canary Islands, 13, 51, 109, 187, 189
  Carrock Fell, 211, 216, 220, 226, 236, 313
  Coastal batholith, Peru, 603–4
  Columbia River basalts, 442, 452, 463, 469, 496, 607
  Deccan traps, 302, 442, 607
  Duke Island Complex, 589, 594–95
  Duluth gabbro, 295, 311, 316, 321, 323, 325, 332, 336, 346, 454
  Eifel district, 53, 151–52, 185
  Euganean Hills, 18, 41, 53, 104
  Faeroe Islands, 55, 62, 92, 182–83
  Garabal Hill, 213, 233, 313, 425
  Giants Causeway, 20–22, 24–25, 35, 43, 47
  Glen Tilt, 65–68, 76, 168
  Great Dyke, 316, 326, 564, 588
  Hawaii (Kilauea), 128–29, 202, 330, 424, 426, 438, 442, 463, 557, 567, 571, 587, 591–92
  Henry Mountains, 188–90, 195, 223, 338
  Highwood Mountains, 185, 242, 338, 406
  Iceland, 11, 127, 130–31, 144, 182, 216, 438, 463
  Isle of Arran, 66, 68, 73, 168, 175, 214, 219
  Isle of Mull, 57, 177, 182, 310, 323, 337–38, 341, 343, 454, 456, 573
  Isle of Skye, 57, 177, 182, 186, 295, 316, 335, 338, 454–56, 573
  Isle of Staffa, 22–24, 37–38, 43, 47–48
  Jimberlana Intrusion, 563–64, 588–89
  Karoo basalts, 302, 327, 329–30, 427, 442, 496, 607
  Katmai volcano, 299–300
  Kiglapait Intrusion, 454, 470, 584–85, 590, 594, 601
  Lipari Islands, 4–5, 7, 12, 41, 47, 53–54
  Lugar sill, 315–16, 332
  Mount Ascutney, 196–97, 344, 455
  Mount Etna, 5–7, 9–11, 13–14, 22, 41, 46–48, 53–54, 127, 138, 189
  Mount Vesuvius, 5–7, 9–14, 22, 37–38, 40–41, 48, 51–54, 126–28, 138, 184, 216, 228, 463
  Muskox Intrusion, 454–55, 559, 584–85, 595–97
  Oslo district, 185, 417, 441
  Palisades sill, 291, 313–15, 327–29, 331, 597
  Rum (Rhum) Complex, 550, 561, 581, 586, 588, 597
  Salisbury Crags, 32, 46, 48, 60, 175
  Saxony, 20, 37, 50–53, 64
  Shonkin Sag laccolith, 190, 242, 323, 340
  Sierra Nevada batholith, 200, 316, 322, 345, 426, 454, 470, 517–18, 604–5
  Skaergaard Intrusion, 183, 319–27, 329, 331–32, 422–24, 427, 445, 453–54, 456, 461, 465, 470–71, 559, 564, 567, 569, 578, 581, 583, 589–95, 597–600, 602, 608
  Stillwater Complex, 319, 325–26, 578–81, 583–84, 587–88, 590–91, 593, 596
  Sudbury Irruptive, 302, 311, 336, 454
  Yellowstone National Park, 180, 183–84, 210–11, 220, 233, 313, 452, 461, 463
plate tectonics, 437–43, 489–97, 613
point-counting method, 396–97
polarizing microscopy, 98–100, 143–66
professionalization of petrology, 226–30
pyrometry, 268–77

radiogenic isotopes, 462–71
rare earth elements (REE), 485–89

INDEX OF SUBJECTS

reaction principle, 297–98, 417, 452
rock names, origins of
  alnöite, 158
  andesite, 109
  basalt, 105–6
  basanite, 105
  camptonite, 158
  diabase, 122
  diorite, 121
  dolerite, 122
  granite, 106
  harzburgite, 158
  liparite, 117
  melaphyre, 122
  monzosyenite, 109
  nepheline syenite, 158
  nephelinite, 118
  obsidian, 105
  oceanite, 160
  opdalite, 417
  pegmatite, 122
  perlite, 108
  phonolite, 109
  picrite, 159
  pitchstone, 108
  porphyry, 106
  pumice, 105
  propylite, 138–39
  pyroxenite, 116
  quartz trachyte, 154
  rhyolite, 117, 139
  spilite, 122
  syenite, 106
  tephra, 105
  theralite, 158
  tholeiite, 301–2
  tinguaite, 158
  trachyte, 104, 122
  troctolite, 158
  trondhjemite, 417
  tuff, 105
  vogesite, 158
  websterite, 164
room problem, 191, 347, 358, 371–72, 374, 379, 602–7
Rosiwal analysis, 239, 261, 397

saturation principle in classification, 256–59
scale of fusibility, 268–69, 272
silica enrichment (Bowen) trend, 301, 305, 332, 407, 456
Soret effect, 212–19, 223, 285, 291, 314, 551–52, 557
stable isotopes, 450–62
superheat, 298–99, 346, 567, 570, 573
surfusion theory, 91–92

trace element modeling, 479–85
trace elements and tectonic discrimination, 489–97
two magma source theory, 131–40

women in petrology, 165–66, 362–63, 434